Illustrated Handbook of Succulent Plants

Series Editors: U. Eggli, H. E. K. Hartmann

Springer

Berlin
Heidelberg
New York
Barcelona
Hong Kong
London
Milan
Paris
Tokyo

Illustrated Handbook of Succulent Plants
(Eggli/Hartmann Eds.)

Illustrated Handbook of Succulent Plants: Monocotyledons
Ed. by Urs Eggli (2001)

Illustrated Handbook of Succulent Plants: Dicotyledons
Ed. by Urs Eggli (2002)

Illustrated Handbook of Succulent Plants: Aizoaceae A–E
Ed. by H. E. K. Hartmann (2001)

Illustrated Handbook of Succulent Plants: Aizoaceae F–Z
Ed. by H. E. K. Hartmann (2001)

Illustrated Handbook of Succulent Plants: Asclepiadaceae
Ed. by F. Albers and U. Meve (2002)

Illustrated Handbook of Succulent Plants: Crassulaceae
Ed. by Urs Eggli (2003)

Focke Albers · Ulrich Meve (Eds.)

Illustrated Handbook of Succulent Plants: Asclepiadaceae

With 332 Colour Photos, Printed in 48 Colour Plates

Springer

Prof. Dr. Focke Albers
Leiter des Botanischen Gartens
Institut für Botanik und Botanischer Garten
Westfälische Wilhelms-Universität
Schlossgarten 3
48149 Münster
Germany

e-mail: albersf@uni-muenster.de

Dr. Ulrich Meve
Lehrstuhl für Pflanzensystematik
Universität Bayreuth
95440 Bayreuth
Germany

e-mail: ulrich.meve@uni-bayreuth.de

Compiled in cooperation with the International Organization for Succulent Plant Study (IOS)

With contributions by:
F. Albers, R. van Donkelaar, U. Eggli, Ch. Hoffmann, J. Kiel, S. Liede, U. Meve, B. Müller, K. Seidler, P. Stegemann, B. Willke

Coordinating editor for this volume: Dr. Urs Eggli

ISBN 3-540-41964-0 Springer-Verlag Berlin Heidelberg New York

Library of Congress Cataloging-in-Publication Data

Illustrated handbook of succulent plants. Asclepiadaceae / Focke Albers, Ulrich Meve, (eds.).
   p. cm.
   Includes bibliographical references (p.).
   ISBN 3540419640 (alk. paper)
   1. Asclepiadaceae--Classification. I. Albers, Focke. II. Meve, Ulrich, 1958-

QK495.A815 I58 2002
583'.93--cd21

2002070485

This work is subject to copyright. All rights are reserved, whether the whole or part of the material is concerned, specifically the rights of translation, reprinting, reuse of illustrations, recitation, broadcasting, reproduction on microfilms or in any other way, and storage in data banks. Duplication of this publication or parts thereof is permitted only under the provisions of the German Copyright Law of September 9, 1965, in its current version, and permission for use must always be obtained from Springer-Verlag. Violations are liable for prosecution under the German Copyright Law.

Springer-Verlag Berlin Heidelberg New York
a member of BertelsmannSpringer Science + Business Media GmbH
http://www.springer.de
© Springer-Verlag Berlin Heidelberg 2002
Printed in Germany

The use of general descriptive names, registered names, trademarks, etc. in this publication does not imply, even in the absence of a specific statement, that such names are exempt from the relevant protective laws and regulations and therefore free for general use.

Production: Colour plates by Büro Stasch, Bayreuth, Germany
Cover design: *design & production* GmbH, 69121 Heidelberg, Germany
Typesetting: Camera-ready by Urs Eggli
SPIN: 10789892   31/3130 – 5 4 3 2 1 0 – Printed on acid-free paper

Preface

With the constant flow of plant species that reached Europe in the times of the great expeditions in the 18th and 19th century, the demand for any form of cataloguing the steadily growing diversity became unevitable. Succulent plants always fascinated plant-loving people above average, and the first handbooks devoted to succulents were published in the 19th century. Initially, however, interest was focussed on the *Cactaceae*. It was only in 1954/55 when Herman Jacobsen (1898 - 1978), the former curator of the Botanic Garden of Kiel, made available a first handbook devoted to succulents (excl. *Cactaceae*), the "Handbuch der sukkulenten Pflanzen". An abridged version of this handbook was published by Jacobsen as "Sukkulentenlexikon" in 1970. It included descriptions, synonymies and numerous illustrations. The "Sukkulentenlexikon" represented the long-awaited reference work supporting both Botanical Gardens as well as amateur or professional growers in the navigation through succulent plant diversity. An English edition followed in 1975 as "Succulent Lexicon", and a revised German edition was published in 1981. This second edition had been finished by Jacobsen's successor in Kiel, Klaus Hesselbarth, who, with regard to the *Asclepiadaceae*, was supported by the senior editor of the present publication.

Over twenty years have passed since then, and the knowlegde of succulent plants has increased permanently. We saw monographs of formerly little-known groups and many revisions and re-revisions, which often change the taxonomy within single groups considerably. A large amount of new taxa has been published on one hand, and on the other hand, many taxa have been sunk into synonymy. Therefore, a new lexicon-like contribution covering the whole succulent plant world, including the *Cactaceae*, was taken into consideration by various members of the International Organization for Succulent Plant Study (IOS) more than 10 years ago. Gustav Fischer-Verlag Jena, the publisher of both Backeberg's "Kakteenlexikon" and Jacobsen's "Sukkulentenlexikon", encouraged the project, which was informally termed the "New IOS Succulent Plant Lexicon", and later "Synopsis Plantarum Succulentarum". This new work was primarily planned to consist of three volumes (*Cactaceae, Aizoaceae*, other Succulents), and after a consensus regarding style and format had been found, the compilation of the *Asclepiadaceae* taxa started in 1994. With the disappearance of Gustav Fischer-Verlag, the project was adopted by Springer-Verlag for the English edition, and by Verlag Eugen Ulmer for the German edition, and finally changed its title to "Illustrated Handbook of Succulent Plants". Also with regard to the enormous array of taxa of other succulents finally found worth to be presented, the early concept was adapted. Accordingly, the series in its English version now appears in six volumes altogether: Two volumes cover the *Aizoaceae* (Hartmann 2001), one volume each treats the Monocotyledons (Eggli 2001), the *Asclepiadaceae* (respectively *Apocynaceae* – *Asclepiadoideae* and – *Periplocoideae*; Albers & Meve 2002, this volume), the *Crassulaceae* (Eggli, in preparation for 2003), and the remaining Dicotyledons (Eggli 2002). The *Cactaceae* will not become available in time to be included in the present handbook series.

Apart from doubling the number of *Asclepiadaceae* genera covered, and their presentation in one volume solely dedicated to this plant group, this new handbook has many additional features such as the vastly expanded descriptions including typification data for all accepted taxa, full synonymy and literature references that lead to published illustrations. Keys to main groups and genera are provided. Although desirable, we do not include keys to the species. This would have made an excessive demand, especially in species-rich genera such as *Ceropegia*, which could not be performed within the frame of this handbook.

Preface

It took us tremendous efforts over the years of compilation to keep the taxonomy as current as possible. Taxonomy means change – this is especially true with regard to the last years where asclepiad systematic research saw a powerful renaissance. In 2000, we even lost "our family", the *Asclepiadaceae*, because of the formal transposition of the subfamilies *Asclepiadaceae – Asclepiadoideae* and *Asclepiadaceae – Periplocoideae* to the *Apocynaceae*. For pragmatic reasons (e.g. avoiding the clumsy new taxon names) we nevertheless stick to the *Asclepiadaceae* in the title of the present work.

Sometimes, publications like revisions or monographs that could serve as base for our compilations do not exist. This is especially true for the two largest genera of the *Asclepiadaceae – Ceropegieae*, *Ceropegia* (± 180 species) and *Brachystelma* (122 species), as well as for *Hoya* (± 200 species), the largest genus of the tribe *Marsdenieae*. While for *Ceropegia* the revison of H. Huber (1957) is outdated, a complete generic treatment has never been produced for *Brachystelma*. Own research and literature studies were unavoidable here. Although not all problems could be solved, the *Brachystelma* treatment published in this volume is the first complete "critical" presentation of this interesting genus at all. The last and only full generic treatments of *Hoya* originate from the 19th century (Decaisne 1844, Hooker 1885). Since then, due to the enormously increased number of taxa described, further revisions of *Hoya* were restricted to a regional scale, their sum being far away from representing a complete generic revision. The presentation of *Hoya* in this handbook is therefore restricted to the better known, mostly cultivated and predominantly succulent taxa.

The core and most popular asclepiadaceous succulents are represented by the stem-succulent stapeliads. Here, recent revisions often compete against each other, and made it difficult to decide which should be followed. Usually, only one revison is followed, though sometimes a pragmatic concept considering the different revisons available plus own assessments have been applied. For the genus *Caralluma* s.l., a conservative circumscription is presented despite the editors' knowledge of the necessary recircumscription of *Caralluma* and its division into a few smaller genera. However, the new taxonomy was not available within the deadline of the present work. In parts, this is also true for the recircumscription of *Cynanchum*.

Apart from the classical succulents of the tribe *Ceropegieae*, several less well-known groups or taxa are presented in this book. Inclusion has been mainly influenced by horticultural value.

Several persons helped during the production process, whether as compiler of texts (Anke Brennecke, Christiane Hoffmann, Janine Kiel, Rainer Kranz, Dr. Sigrid Liede, Birgit Müller, Petra Stegemann, Beate Willke and Ruurd van Donkelaar) or as authors of additional illustrations (Josef Bogner, Dr. Wiebe Bosma, Dr. Urs Eggli, Dr. David Goyder, Dr. Sigrid Liede, Ernst Specks, Dr. Joachim Thiede and Ruurd van Donkelaar). The support of various journals and their editors in publishing necessary new combinations or names is gratefully acknowledged. Many of the above named and further unnamed persons contributed with literature / photocopies, provided us with additional specimens or living plants for study (namely the Sukkulenten-Sammlung Zürich under its former director Dieter Supthut), or with a wide array of any kind of information on taxa, types, distribution etc. We gratefully acknowledge their long-standing interest in the project.

We would like to thank Dr. Johanna Schlüter, Gustav Fischer-Verlag Jena, for fruitful cooperation during the first years of the project. Finally, the endless patience and painstaking accuracy of the series editor responsible for this volume, Dr. Urs Eggli, in transposing our texts into the standard format, tracing the many nomenclatural problems, and composing the final layout of text and colour plates is gratefully acknowledged.

Münster and Bayreuth, April 2002   F. ALBERS, U. MEVE

Contents

Note: ♦ = family or genus completely covered; ◊ = family or genus only partially covered; ~ = family or genus only mentioned; numbers in brackets = numbers of species / infraspecific taxa covered

Introduction 1
    What is a Succulent ? 1
    How to Use This Handbook 1
    Scope of Information Presented 1

Asclepiadaceae ◊ 5
    Absolmsia ♦ (1 / 0) 9
    Asclepias ◊ (1 / 0) 9
    Aspidoglossum ♦ (36 / 0) 10
    Aspidonepsis ♦ (5 / 0) 18
    Baynesia ♦ (1 / 0) 20
    Brachystelma ♦ (122 / 3) 20
    Caralluma ♦ (53 / 9) 46
    Ceropegia ◊ (160 / 18) 63
    Cibirhiza ♦ (2 / 0) 107
    Cynanchum ◊ (57 / 4) 108
    ×Dernia ♦ 118
    Dischidia ◊ (23 / 1) 118
    Dischidiopsis ◊ (1 / 0) 123
    Duvalia ♦ (18 / 3) 124
    Duvaliandra ♦ (1 / 0) 129
    ×Duvaliaranthus ♦ (1 / 0) 129
    Echidnopsis ♦ (32 / 4) 129
    Edithcolea ♦ (1 / 0) 136
    Fanninia ♦ (1 / 0) 137
    Fockea ♦ (6 / 0) 137
    Glossostelma ♦ (12 / 0) 139
    Hoodia ♦ (14 / 4) 142
    ×Hoodialluma ♦ 146
    Hoya ◊ (55 / 6) 146
    ×Huernelia ♦ 158
    Huernia ♦ (67 / 16) 159
    ×Huernianthus ♦ 174
    Huerniopsis ♦ (2 / 0) 174
    Ischnolepis ♦ (1 / 0) 175
    Larryleachia ♦ (5 / 0) 176
    Lavrania ♦ (1 / 0) 178
    Madangia ♦ (1 / 0) 178
    Marsdenia ◊ (10 / 0) 178
    Matelea ◊ (9 / 0) 181
    Micholitzia ♦ (1 / 0) 183

Miraglossum ♦ (7 / 0)	183
Notechidnopsis ♦ (2 / 0)	185
Odontostelma ♦ (2 / 0)	186
Ophionella ♦ (2 / 1)	186
Orbea ♦ (54 / 14)	187
Orbeanthus ♦ (2 / 0)	206
×Orbelia ♦ (1 / 0)	207
Pachycarpus ~	207
Pectinaria ♦ (3 / 3)	207
Petopentia ♦ (1 / 0)	208
Piaranthus ♦ (7 / 2)	209
Pseudolithos ♦ (6 / 0)	212
Quaqua ♦ (29 / 0)	213
Raphionacme ♦ (35 / 0)	220
Rhytidocaulon ♦ (10 / 1)	229
Riocreuxia ~	232
Sarcorrhiza ♦ (1 / 0)	232
Sarcostemma ♦ (15 / 7)	233
Schizoglossum ♦ (15 / 11)	236
Schlechterella ♦ (2 / 0)	241
×Staparesia ♦ (1 / 0)	242
Stapelia ♦ (47 / 7)	242
Stapelianthus ♦ (8 / 0)	255
Stapeliopsis ♦ (6 / 1)	258
Stathmostelma ♦ (14 / 4)	260
Stenostelma ♦ (4 / 0)	263
Stomatostemma (1 /0)	264
Tavaresia ♦ (3 / 0)	265
Trachycalymma ~	265
Tridentea ♦ (8 / 2)	266
×Tromostapelia ♦	269
Tromotriche ♦ (11 / 0)	269
White-sloanea ♦ (1 / 0)	273
×Whitesloaniopsis ♦	274
Xysmalobium ~	274

References 275

Taxonomic Cross-Reference Index 283

List of Abbreviations and Symbols

∅	diameter	**Gl**	gland
±	more or less, circa	Gr.	Greek
>	greater than	**Gy**	gynostegium
≥	greater or equal than	**Ha**	hair(s)
<	less than	Herb. / herb.	herbarium
≤	less or equal than	holo	holotype
×	hybrid indicator	Hort. / hort.	of gardens
		I	illustration
		ICBN	International Code of Botanical Nomenclature, Tokyo Edition
Anth	anthers		
Ar	areole		
Art.	article (of ICBN)	illeg.	illegitimate
Ax	axil	in sched.	"in schedis", i.e. in the herbarium
BG	Botanical Garden		
Bo	body	incl.	including
BPH	Botanico-Periodicorum-Huntianum (1968)	**Inf**	inflorescence
		ING	Index Nominorum Genericorum
BPH/S	idem., Supplementum (1991)		
Br	branch	**Int**	internode
Bra	bract	inval.	invalid
Bri	bristles	**ITep**	inner tepals
C	central	KG, KGW	Karoo Garden (Worcester, RSA)
Ca	carpels		
Cal	calyx	l.c.	"loco citato", i.e. at the place cited
Cap	capitulum		
Ci	interstaminal corona	**L**	leaf, leaves
Cl	corolla	lecto	lectotype
Cn	corona	**Lit**	literature
Cs	staminal corona	ms.	manuscript
cv.	cultivar	Mt. / Mts.	Mount / Mountains
Cy	cyathium	N	North, northern
D	distribution	NBG	National Botanical Garden (RSA)
E	East, eastern		
e.g.	for example	NE	North-East, north-eastern
epi	epitype	**Nec**	nectary
esp.	especially	neo	neotype
Etym	Etymology	**NGl**	nectar gland
excl.	excluding	nom.	"nomen", i.e. botanical name
f.	figure	**NSc**	nectar scale
fa.	forma	NW	North-West, north-western
fig.	figure	**OTep**	outer tepals
Fil	filaments	**Ov**	ovary
Fl	flower	p.	page
fl.	"floruit", i.e. living	p.a.	"per annum", i.e. per year
Fr	fruit	p.p.	"pro parte", i.e. partly

List of Abbreviations

Pc	pericarpel
Ped	pedicel
Per	perianth
pers. comm.	personal comment
Pet	petals
Phy	phyllaries
pl.	plate
Poll	pollinia
pp.	pages
R	roots
Rec	receptacle
Ri	ribs
Ros	rosette
RSA	Republic of South Africa
S	South, southern
s.a.	"sine anno", i.e. without year
s.l.	"sensu lato", i.e. in a wide sense
s.n.	"sine numero", i.e. without (collection) number
s.str.	"sensu stricto", i.e. in a strict sense
Sc	scale
SE	South-East, south-eastern
Se	seeds
Sect.	section
SEM	Scanning Electron Micrograph
Sep	sepals
Ser.	series
Sp	spines
SpS	spine shield
ssp.	subspecies
St	stamens
Sti	stigma
Sty	style
Subgen.	subgenus
Subsect.	subsection
Subser.	subseries
SUG	Stellenbosch University Gardens (RSA)
SW	South-West, south-western
syn	syntype
T	(nomenclatural) type
t.	plate
Tep	tepals
TL2	Taxonomic Literature, ed. 2 (1976-2000)
tt.	plates
Tu	tubercles
unpubl.	unpublished
USA	United States of America
var.	variety
W	West, western

Abbreviations for frequently cited publications

BJS	Botanische Jahrbücher für Systematik
BMI	Bulletin of Miscellaneous Information [Kew]
BT	Bothalia
CBM	Curtis's Botanical Magazine
CSJA	Cactus and Succulent Journal (US)
CUSNH	Contributions of the US National Herbarium
EJ	Euphorbia Journal
FC	Flora Capensis (ed. Harvey & al., 1860 - 1925)
FPA	Flowering Plants of Africa
FPSA	Flowering Plants of South Africa
FTA	Flora of Tropical Africa (ed. Oliver & al., 1868 - 1937)
HIP	Hooker's Icones Plantarum
JLSB	Journal of the Linnean Society, Botany
JSAB	Journal of South African Botany
KB	Kew Bulletin
KuaS	Kakteen und andere Sukkulenten
NPF2	Die natürlichen Pflanzenfamilien Ed. 2 (ed. Engler & Prantl, 1924 - 1959)
PSRV	Prodromus Systematis Naturalis Regni Vegetabilis (De Candolle, 1824 - 1873)
RSN	Repertorium Specierum Novarum Regni Vegetabilis
SAJB	South African Journal of Botany
WDSE	Succulent Euphorbieae (South Africa) (White & al., 1941)

List of Contributors

Albers, Prof. Dr. Focke; Institut für Botanik der Westfälischen Wilhelms-Universität, Schlossgarten 3, 48149 Münster, Germany

van Donkelaar, Ruurd van; Laantje 1, 4251 EL Werkendam, The Netherlands

Eggli, Dr. Urs; Sukkulenten-Sammlung Zürich, Mythenquai 88, 8002 Zürich, Switzerland

Hoffmann, Dipl.-Biol. Christiane: See address for F. Albers

Kiel, Dr. Janine: See address for F. Albers

Liede, Prof. Dr. Sigrid; Universität Bayreuth, Lehrstuhl für Pflanzensystematik, 95440 Bayreuth, Germany

Meve, Dr. Ulrich; Universität Bayreuth, Lehrstuhl für Pflanzensystematik, 95440 Bayreuth, Germany

Müller, Dipl.-Biol. Birgit: See address for F. Albers

Seidler, Dipl.-Biol. Klaus: See address for F. Albers

Stegemann, Petra: See address for F. Albers

Willke, Beate: See address for F. Albers

Translation Credits

The following texts have been translated from the German originals by:

F. Ditsch: Duvalia, Hoodia, Larryleachia.

U. Eggli: Asclepias, Aspidoglossum, Aspidonepsis, Brachystelma, Cibirhiza, Cynanchum, Dernia, Dischidiopsis, Duvaliandra, Duvaliaranthus, Edithcolea, Fanninia, Fockea, Hoodialluma, Huernelia, Huernianthus, Orbea p.p., Petopentia, Pseudolithos, Quaqua, Raphionacme, Rhytidocaulon, Riocreuxia, Sarcorrhiza, Sarcostemma, Schizoglossum, Schlechterella, Staparesia, Stapelia, Stapelianthus, Stapeliopsis, Stathmostelma, Stenostelma, Stomatostemma, Tavaresia, Trachycalymma, Tridentea, Tromostapelia, Tromotriche, Whitesloanea, Whitesloaniopsis.

U. Meve: Baynesia, Lavrania, Piaranthus.

M. Struck: Asclepiadaceae, Dischidia, Echidnopsis, Hoya, Huernia, Ischnolepis, Marsdenia, Matelea, Micholitzia, Miraglossum, Notechidnopsis, Odontostelma, Ophionella, Orbea p.p., Orbeanthus, Orbelia, Pachycarpus, Pectinaria.

Illustration Credits

F. Albers: VII.f, XIII.g, XIII.h, XXII.d, XXIII.a, XXIX.d, XXXII.d, XXXIV.b, XXXVI.g, XLIII.g, XLIV.d, XLIV.g, XLIV.h

J. Bogner: XXII.f

W. Bosma: X.e, XII.e, XXXIX.a, XXXIX.d, XL.a, XLII.a

R. van Donkelaar: I.a, XVII.a, XVII.d, XVIII.a, XVIII.d , XVIII.e, XVIII.f, XIX.a, XIX.b, XIX.d, XIX.e, XXIV.b, XXIV.c, XXIV.d, XXIV.e, XXIV.f, XXV.a, XXV.b, XXV.c, XXV.d, XXV.e, XXVI.a, XXVI.c, XXVI.d, XXVI.e, XXVI.f, XXVII.a, XXVII.b, XXVII.c, XXVII.d, XXVII.e, XXVII.f, XXVIII.a, XXVIII.b, XXVIII.c, XXVIII.d, XXVIII.e, XXVIII.f, XXVIII.g, XXXIII.b

U. Eggli: II.b, VII.c, X.g, XIII.d, XVII.c, XIX.c, XXXIII.d, XXXIII.e, XXXV.a, XXXV.c, XXXVI.a, XXXVI.c, XXXVII.a, XXXVIII.f, XXXVIII.g, XLVIII.h

D. Goyder: I.b

S. Liede: II.a

U. Meve: I.c, I.d, I.e, I.f, I.g, II.c, II.d, II.e, II.f, II.g, III.a, III.b, III.c, III.d, III.e, III.f, III.g, III.h, IV.a, IV.b, IV.c, IV.d, IV.e, IV.f, IV.g, V.a, V.b, V.c, V.d, V.e, V.f, V.g, VI.a, VI.b, VI.c, VI.d, VI.e, VI.f, VI.g, VI.h, VII.a, VII.b, VII.d, VII.e, VIII.a, VIII.b, VIII.c, VIII.d, VIII.e, VIII.f, VIII.g, VIII.h, IX.a, IX.b, IX.c, IX.d, IX.e, IX.f, IX.g, IX.h, X.a, X.b, X.c, X.d, X.f, X.h, XI.a, XI.b, XI.c, XI.d, XI.e, XI.f, XI.g, XI.h, XII.a, XII.b, XII.c, XII.d, XIII.a, XIII.b, XIII.c, XIII.e, XIII.f, XIV.a, XIV.b, XIV.c, XIV.d, XIV.e, XV.a, XV.b, XV.c, XV.d, XV.e, XV.f, XVI.a, XVI.b, XIV.c, XVI.d, XVI.e, XVII.b, XVII.e, XVII.f, XVIII.b, XVIII.c, XX.a, XX.b, XX.c, XX.d, XX.e, XX.f, XX.g, XXI.a, XXI.b, XXI.c, XXI.d, XXI.e, XXI.f, XXII.b, XXIII.b, XXIII.c, XXIII.d, XXIII.e, XXIII.f, XXIV.a, XXV.f, XXVI.b, XXIX.a, XXIX.b, XXIX.c, XXIX.e, XXIX.f, XXIX.g, XXX.a, XXX.b, XXX.c, XXX.d, XXX.e, XXX.f, XXX.g, XXXI.a, XXXI.b, XXXI.c, XXXI.d, XXXI.e, XXXI.f, XXXI.g, XXXI.h, XXXII.a, XXXII.b, XXXII.c, XXXII.e, XXXII.f, XXXII.g, XXXIII.a, XXXIII.c, XXXIII.f, XXXIV.a, XXXIV.c, XXXIV.d, XXXIV.e, XXXIV.f, XXXIV.g, XXXV.b, XXXV.d, XXXV.e, XXXV.f, XXXV.g, XXXVI.b, XXXVI.d, XXXVI.e, XXXVI.f, XXXVII.b, XXXVII.c, XXXVII.d, XXXVII.e, XXXVII.f, XXXVII.g, XXXVIII.a, XXXVIII.b, XXXVIII.c, XXXVIII.d, XXXVIII.e, XXXIX.b, XXXIX.c, XXXIX.e, XXXIX.f, XL.b, XL.c, XL.d, XL.e, XL.f, XL.g, XL.h, XLI.a, XLI.b, XLI.c, XLI.d, XLI.f, XLI.g, XLI.h, XLII.b, XLII.c, XLII.d, XLII.e, XLII.f, XLII.g, XLII.h, XLIII.a, XLIII.b, XLIII.c, XLIII.d, XLIII.e, XLIII.f, XLIII.h, XLIV.a, XLIV.b, XLIV.c, XLIV.e, XLIV.f, XLV.a, XLV.b, XLV.c, XLV.d, XLV.e, XLV.f, XLV.g, XLV.h, XLVI.a, XLVI.b, XLVI.c, XLVI.d, XLVI.e, XLVI.f, XLVI.g, XLVII.a, XLVII.b, XLVII.d, XLVII.e, XLVII.f, XLVII.g, XLVII.h, XLVIII.a, XLVIII.b, XLVIII.c, XLVIII.d, XLVIII.e, XLVIII.f, XLVIII.g

E. Specks: XXII.a, XXII.c, XXII.e, XLVI.h

J. Thiede: XLI.e, XLVII.c

Introduction

What is a succulent ?

It is probably impossible to define what constitutes a *succulent plant* – at least in view of the several competing definitions. For the purpose of this handbook, a pragmatic approach has been selected, and apart from the multitude of unambiguous succulents, many borderline cases are included as well, especially if the species in question are encountered in cultivation together with other succulents, and if they are native to more or less semi-arid regions and consequently show some degree of xerophytic adaptation. This, then, includes most of the caudex and pachycaul plants now popular in cultivation.

Other borderline cases included are a number of bulbous and rhizomatous monocotyledons, where examples from several genera are covered, as well as several weakly developed leaf succulents from the *Gesneriaceae* (e.g. *Columnea*).

On the other hand, purely halophytic succulents (such as *Salicornia*) are omitted from these pages since they are as a whole neither adapted to climatically dry conditions nor encountered in collections devoted to succulent plants.

Finally, some families with undoubted claim to (xerophytic) succulence have been excluded from this set of volumes. This notably is the case for the *Cactaceae*, which will be treated elsewhere. In addition, the families *Bromeliaceae* and *Orchidaceae* are also excluded. Both count with a considerable number of mostly leaf succulents, but for both, vast specialist literature and numerous specialist societies are in existence, and this effort does not need to be duplicated here. For all these excluded families, however, a family description is included in the present volumes for the sake of completeness.

How to use this handbook

Since all information is presented in strictly alphabetical sequence of families, genera and species (except that monocotyledons and dicotyledons are treated separately, and that the families *Aizoaceae*, *Asclepiadaceae* and *Crassulaceae* occupy their own volumes), it is easy to find the entry for a given species as long as its family placement is known.

An alternative way is to use the taxonomic cross-reference index supplied at the end of the volume. This index contains all the names treated in the volume and for accepted names indicates the page where a treatment can be found, or in the case of synonyms gives the name of the accepted taxon and a page reference as above. For names merely mentioned in the text, the index gives the page reference and the name under which information can be found.

If a completely unknown plant is to be identified, the handbook supplies keys to the genera with succulent representatives for each family. Please note that these keys are designed to work for the succulent taxa treated, and do not necessarily include the total variation encountered in a genus. If the family is not known, the reader is referred to general botanical books that include keys to plant families. Rowley (1980) and Eggli (1994) provided keys for flowering and non-flowering succulents, and Geesink & al. (1981) produced a well-known book of keys to all flowering plants worldwide.

Scope of information presented

Families

The family names adopted are always conforming to the standard form (ending in *-aceae*); alternative names (such as *Compositae* for *Asteraceae*) are not used.

Within each family, the genera are treated in alphabetical sequence, and the same applies to the sequence of species within genera. Some genera of minimal importance or with borderline succulence are only mentioned or at the most described, but no individual species are treated.

The following families are included as a whole, i.e. with all their component species: *Agavaceae*, *Aloaceae* and *Doryanthaceae* in the present volume covering the Monocotyledons, and the *Didiereaceae*, *Fouquieriaceae* and *Nolanaceae* in the Dicotyledons volume. The *Aizoaceae* and *Crassulaceae* are covered in their entirety in separate volumes within this series, and the succulent taxa of the *Asclepiadaceae* likewise occupy a separate volume of the Handbook.

The family description characterizes the family as a whole, which often includes much more variation than that observed amongst its succulent representatives.

This is followed by notes on the distribution, classification and economic importance of the family, and the occurrence of succulence if this is not a general feature of the family as a whole. Also, a key to genera with succulents is included, and special terminology used for genera and species descriptions is discussed.

The family concept adopted more or less follows Mabberley (1987), except for the monocotyledons, where Dahlgren & al. (1985) is used as a base, with a number of small modifications.

Genera and species

The entries for genera and species follow the same layout. Names of authors are given in full, with initials added where necessary according to Brummitt & Powell (1992). The literature reference of the original description or combination is followed by information on typification (where available, see below). In the case of genera, important literature is then cited. This is followed by information on geographical distribution (including notes on ecology where available) and an explanation of the etymology for generic names.

The main part of the entry is made up by the diagnostic description of the taxon, followed by a discussion of its variability, circumscription and/or application where necessary. It should be noted that these descriptions reflect major variability only, but do not include all the reported minor variations.

For larger genera, an outline of the accepted formal or informal classification is also given, with individual taxa or groups numbered in sequence. These sequence numbers are then given at the start of each taxon description to indicate its placement within the genus.

If recent conflicting classifications are available for a given group, this is shortly discussed and the classification adopted is indicated.

Minor spelling variants of epithets are not indicated; instead, the 'corrected' spelling is used throughout for accepted names and synonyms.

Infraspecific taxa

Infraspecific taxa are given in strict alphabetic order of rank and name (i.e. ranks in the sequence cv., fa., ssp., var.). This is due to the strict alphabetical sorting used when the output for the handbook was generated from a computerized database. It also means that the typical infraspecific taxon (i.e. the one repeating the species name) is not treated first as in many handbooks, but in its appropriate alphabetical sequence.

Cultivars, hybrids

Cultivars (rank abbreviated as cv.) are not included on an exhaustive base. Cultivars not associated with a species are enumerated first, i.e. between the generic entry and the first species. Cultivars associated with a species name are included under that species, either as an entry of their own (and in the same form as subspecies etc.), or, in the case of cultivars of minor importance, in the form of a short mention in the species discussion. Cultivar nomenclature follows the guidelines of the ICBN.

Formally named hybrid genera are either included as 'genera' of their own, or dealt with in the discussion of their parent genera. The same applies to formally named hybrid species (incl. those named as cultivars). Hybrids only known with their hybrid formula are either discussed in the generic entry, or mentioned under one or the other of their parent species. No attempt has been made, however, to include all the numerous formally named hybrids.

Descriptions

The descriptions are as compact, concise and diagnostic as possible. Characters that do not vary for the group concerned are not repeated from the family or genus descriptions. In the case of genera further subdivided, information already presented in the group definitions is also not normally repeated in the descriptions of individual taxa.

Measurements

All measurements are given in metric units. Measurements without further qualifications *always refer to the long axis* of the organ described (i.e. length, height etc.); two measurements united with the ×-sign stand for length × width.

Terminology

Special terms used in descriptions are explained when first used; other botanical terminology is not further explained, and the readers are referred to the numerous botanical glossaries, of which Stearn (1992) is cited by way of a most important and useful example.

Typification

This information is included for convenience when readily available, but is lacking in numerous cases. The type citations include the country and major administrative unit where the type was collected, the collector and collection number, and the herbaria where material is said or known to be deposited. The herbarium acronyms conform to *Index Herbariorum*, Ed. 8 (Holmgren & al. 1990). Where more than one herbarium acronym is given, the first relates to the holotype, the others to isotypes. Additional information on typification is sometimes added, especially in the case of lecto- or neotypes.

Nomenclatural status of names

For all taxa treated, every attempt has been made to use only valid and legitimate names, but this was not achievable in a small number of cases. In the synonym lists, the nomenclatural status (invalid, illegitimate, rejected) is indicated by citing the ICBN articles violated (following the numbering in the "Tokyo" Code). Spelling variants are considered as invalidly published according to ICBN Art. 61.

Synonymies

The synonymies given for genera and especially species are as exhaustive as possible and include all names recognized as synonyms. The first synonym(s) – if applicable – is/are the basionym and/or later combination(s) for the accepted name of the entry. All combinations of the same basionym are given in sequence of publication and are united with the ≡-sign to indicate that these are homotypic (nomenclatural) synonyms. Please note that the ≡-sign is only used for *combinations* based on the same basionym and does not indicate other homotypic synonyms (e.g. *nomina nova*). All other synonyms are headed with 'incl.' to indicate that they are, with the exception of *nomina nova* based on the same type, taxonomic synonyms (= heterotypic synonyms). Again, groups of combinations based on the same basionym are united with ≡-signs. Basionyms are given first and in chronological order.

Geographical names

Country names are listed roughly in a North to South and West to East sequence. Every attempt has been made to standardize geographical names (of countries, administrative units, regions, etc.) as far as possible, but there is a surprising amount of change relating to such names. This is specifically the case for the names of the RSA provinces, which have changed considerably during 1995, especially affecting the former Cape Province, which has been split up into 4 units (North-West Province, Northern Cape, Western Cape, Eastern Cape). We have tried as best as we could to provide the modern names in the distribution information included in the handbook, but it has been impossible to adjust all the data for type localities, where the name "Cape Prov." is still used in some cases. This results in some inconsistencies, but it is hoped that these are tolerable under the present circumstances.

Some difficulties were also encountered in a few cases where countries have been amalgamated (as in the case of the former North Yemen and South Yemen) or divided (e.g. Eritrea, formerly part of Ethiopia). Full consistency in all these cases cannot be guaranteed.

In order to save space, geographical directions such as North, South, etc., are *always* abbreviated (N, S, etc.). Please note that *SW Africa* indicates 'southwestern Africa' and *not* the former Southwest-Africa (now Namibia). Similarly, *S Africa* indicates 'southern Africa' and *not* the Republic of South Africa, for which the abbreviation RSA is always used.

Literature references

Literature references are given for all accepted names. Normally, the publication is cited with a full abbreviation according to the standards defined in Eggli (1985) and Eggli (1998) for specialist succulent plant periodicals, or BPH (Lawrence & al. 1968) and BPH/S (Bridson & Smith 1991) for other periodicals, while TL2 of Stafleu & Cowan (1976-1988) and supplements (Stafleu & Mennega 1992-2000) are followed for book abbreviations (in both causes with some minor exceptions to conserve uniformity).

A number of frequently used titles of journals and books are further abbreviated to a short acronym, and a list of these acronyms follows the list of Abbreviations and symbols (page XI).

In the running text, literature is cited in the usual way (author and year, sometimes supplemented by a page reference), and full details can be found in the list of references at the end of the volume.

Illustrations

An attempt has been made to cite one readily accessible illustration for each species or infraspecific taxon when no illustrations are included in the literature reference for the accepted name. If the name used in the cited publication differs from the accepted name in the handbook, it is indicated (genus name abbreviated to first letter if identical, specific or infraspecific epithet omitted if identical to the accepted name).

Illustrations in the illustration section of this volume are given in bold print to distinguish them from other material cited.

Indication of authorships

For families, authorship is indicated at the end of the entry. For genera, authorship is given as a subheading after the genus heading. If more than one author has contributed species entries for a genus, each entry has its own indication of authorship as far as its authorship differs from the authorship given for the genus as a whole. It is thus possible to identify the author(s) of any entry in the handbook.

Asclepiadaceae

Shrubs, woody or herbaceous left-twining climbers, perennials with deciduous herbaceous above-ground parts or stem succulents, with watery or milky sap; **L** decussate, rarely whorled, simple and entire (rarely lobed or divided), sometimes much-reduced, absent or spinescent (in stem succulents); **Inf** cymose (terminal thyrses), rarely racemose; **Fl** actinomorphic, bisexual; **Sep** 5, basally connate; **Pet** 5, usually connate; **Cl** occasionally forming outgrowths on the upper face (→ petaloid corona, Fig. 1A); **St** 5, alternating with the **Pet**; **Fil** free, basally fused with the **Cl**, **Anth** coherent (→ *Periplocoideae*) or all **St** postgenitally fused with the **Sty** head into a column (= gynostegium, **Gy**) (→ *Secamonoideae, Asclepiadoideae*); **Gy** mostly with a simple (staminal) or double (staminal plus interstaminal) corona (**Cn**, Fig. 2A); **Anth** 4-locular (→ *Periplocoideae, Secamonoideae*) or 2-locular (→ *Asclepiadoideae*), pollen in tetrads and freely presented on pollen-carriers (translators, Fig. 1B) (→ *Periplocoideae*) or pollen grains of each pollen sac coherent into a pollinium (**Poll**) and adjacent **Poll** united into a pollinarium by means of a translator (→ *Secamonoideae, Asclepiadoideae*) (Fig. 2B); gynoeceum of 2 apically connate **Ca**, which are united by means of the 5-angular **Sty** head; **Fr** paired or single follicles, with few to many **Se**, not fleshy, slender to inflated; **Se** usually flattened, with or without wing, with a terminal silky tuft of **Ha** (coma).

Distribution: Worldwide, esp. subtropics and tropics.

Literature: Liede & Albers (1994); Sennblad & Bremer (1996); Swarupanandan & al. (1996); Liede (1997); Endress & Bruyns (2000).

The family consists of some 240 genera with 3400 species. Of these, 61 genera can be referred to as succulents in the widest sense.

In the present account, the *Asclepiadaceae* are treated as a family, which according to Schumann (1895), Bruyns & Forster (1991) and Liede & Albers (1994) is subdivided into the 3 subfamilies *Periplocoideae, Secamonoideae* and *Asclepiadoideae*. This conservative taxonomic view has been taken, since this is the system still widely accepted among scientists and amateurs alike. However, as has already been implied by earlier morphological studies, progress in molecular research suggests to include all 3 subfamilies into the *Apocynaceae* and to abandon the *Asclepiadaceae*, see Olmstead & al. (1993) or Sennblad & Bremer (1996). This concept has been formally transformed into taxonomy by Endress & Bruyns (2000).

The widely accepted suprageneric system of the *Asclepiadaceae* by Bruyns & Forster (1991), which largely relies on morphological data, is based on the classification of Brown (1810) and Schumann (1895), to which the newly described tribe *Fockeeae* by Kunze & al. (1994) has to be added. The inclusion of the tribes *Marsdenieae* into the *Stapelieae* by Swarupanandan & al. (1996) and *Gonolobeae* into the *Asclepiadeae* by Liede (1997) still needs to be corroborated and are not adopted here. The technical terms used in the keys and descriptions that follow are explained in the longitudinal flower sections and detail illustrations presented in Fig. 1 and Fig. 2.

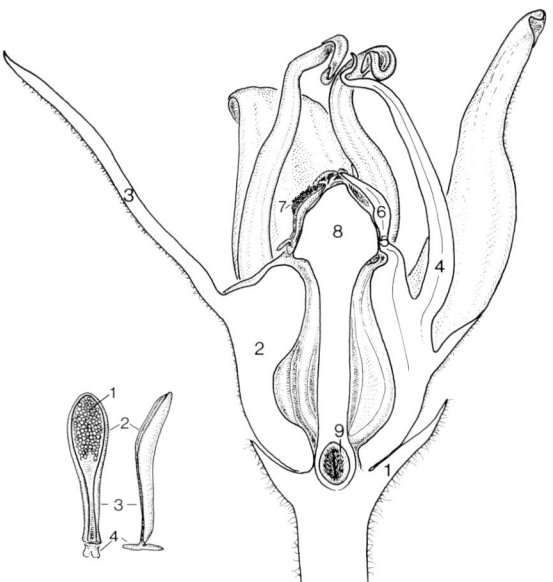

Fig. 1: *Raphionacme madiense:* – **A.** Longitudinal section through a flower (**1** sepal (**Sep**); **2** corolla (**Cl**); **3** corolla tube; **4** corolla lobe; **5** (petaloid) corona (**Cn**); **6** filament (**Fil**); **7** anther (**Anth**); **8** translator; **9** style head; **10** ovary). – **B:** Translator, left in top-view, right in side-view (K, adhesive disc = viscidium; St = stipes (stalk); Sc, scutellum; Pt, heaps of pollen tetrads).

The *Asclepiadaceae* constitute a derived family showing complex floral structures, which are dedicated to the service of a highly specialized pollination biology (→ Figs. 1 and 2). The presence of fascinating floral structures and colours in connection with facilities fostering deceit, trapping and attachment as well as the possession of pollen-masses (pollinia) warrant the *Asclepiadaceae* to be seen as the 'orchids' among the dicotyledons. This, combined with numerous forms of succulence, has made members of this family attractive objects for the plant lover – in particular ceropegias and stapeliads with their pitfall and carrion flowers. Notwithstanding, they are often fairly difficult to cultivate in comparison to cacti, which may occasionally diminish the pleasure taken in them. The question of suc-

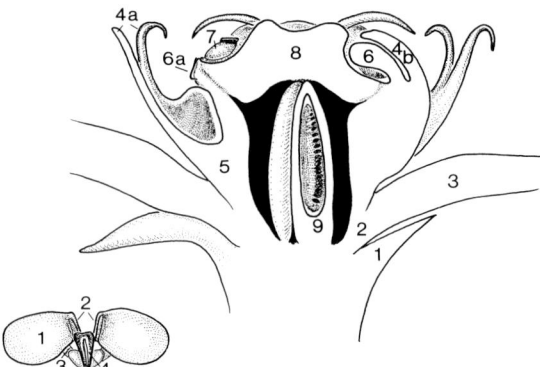

Fig. 2: *Caralluma adscendens*: – **A**: Longitudinal section through a flower (**1** sepal (**Sep**); **2** corolla tube; **3** corolla lobe; **4** staminal (gynostegial) corona (**Cs**); **5** interstaminal corona (**Ci**); **6** filament tube; **7** anther (**Anth**); **8** guide rail; **9** pollinium (**Poll**); **10** style head; **11** ovary). – **B**. Pollinarium (**1** pollinium (**Poll**); **2** germination crest; **3** caudicle; **4** corpuscle).

culence is one of definition on the one hand (e.g. whether root succulence constitutes genuine succulence or not). On the other hand, in particular cases it may prove difficult to delimitate succulents against other xerophytic life forms. The term succulence is used here in a very broad sense, not least to allow the inclusion of all taxa being worthy of cultivation or actually encountered in cultivation. Nevertheless, as a rule all taxa treated have at least one fleshy-thickened organ, be it root, stem or leaves. Thus, some 1000 succulent species, subspecies etc. of 61 genera from all tribes of the *Asclepiadaceae* (excl. *Secamoneae*) are included. Except for a few species of the genera *Asclepias*, *Marsdenia* and *Matelea*, succulence in this family is confined to the Old World. In the subfamily *Periplocoideae*, succulence occurs almost exclusively in Africa and usually in the shape of huge storage tubers (e.g. *Raphionacme*). Succulence is likewise made up by usually subterranean storage organs in the tribes *Fockeeae*, *Asclepiadeae* and *Gonolobeae* of the subfamily *Asclepiadoideae*. In contrast, the tribe *Marsdenieae* is characterized by the large number of leaf succulent members of the generally epiphytic genera *Dischidia* and *Hoya* from Australasia. Finally, most succulents, esp. stem succulents, belong to the tribe *Ceropegieae* with 34 genera of almost exclusively African origin.

Apart from their use as ornamentals and of the seed-hairs as poor-quality floss, the family has no economically important taxa. A few species are utilized as natural remedies owing to their content of alkaloids, cardenolides, pregnane esterglycosides (Hegnauer 1989).

Key to the subfamilies and tribes with succulents

1   **Anth** 4-locular, pollen tetrads loosely on a shovel-shaped translator, translator basally with an adhesive disc (viscidium): **2** (*Periplocoideae*)
–   **Anth** 2-locular, pollen coherent in masses (pollinia, **Poll**), translator with a corpuscle: **4** (*Asclepiadoideae*)
2   **Cl** rotate with very short **Cl** tube, **St** arising from the base of the **Cl**: **Periplocoideae – Periploceae**
–   **Cl** with conspicuous cylindrical to campanulate **Cl** tube, **St** arising from within the **Cl** tube: **3**
3   **St** and **Cn** originating from the upper margin of the **Cl** tube: **Periplocoideae – Gymnanthereae**
–   **St** originating from the base to the middle of the **Cl** tube, **Cn** arising from the base to the margin of the **Cl** tube: **Periplocoideae – Cryptolepideae (Stomatostemma)**
4   **Poll** with a ± asymmetrical marginal zone that is not serving as germination zone for the pollen: **Asclepiadoideae – Gonolobeae (Matelea)**
–   **Poll** without conspicuous marginal zone or with ± symmetrical marginal zone serving for the germination of pollen (germination crest): **5**
5   **Poll** pendent: **Asclepiadoideae – Asclepiadeae**
–   **Poll** erect: **6**
6   **Anth** without sterile (connective) appendage: **Asclepiadoideae – Ceropegieae**
–   **Anth** with sterile (connective) appendage: **7**
7   **Poll** attached to the corpuscle by means of a translator arm: **Asclepiadoideae – Marsdenieae**
–   **Poll** directly attached upon the corpuscle: **Asclepiadoideae – Fockeeae**

Key to the succulent genera of the *Periplocoideae*

Tribe *Periploceae*
1   Robust lianas with huge caudex, **Cl** lobes widely spreading, 12 - 15 mm long: **Petopentia**
–   Shrubs, partly epiphytic, with several ovoid tubers, **Cl** lobes ascending, overlapping, ± 5 mm long: **Sarcorrhiza**

Tribe *Gymnanthereae*
1   Shrubs with many small tubers, **L** (seemingly) 3-partite, linear, **Cn** basally ± fused to the **Fil**: **Ischnolepis**
–   Herbaceous perennials or lianas, (usually) with 1 tuber only, **Cn** only fused with the **Cl**: **2**
2   **Fl** robust, pollen loosely on the translator: **Raphionacme**
–   **Fl** delicate, short-lived, pollen in 2 pollen-masses (**Poll**) on each translator / **Anth**: **Schlechterella**

Key to the succulent genera of the *Asclepiadoideae*

Tribe *Fockeeae*
1 Cl rotate, Cn red-brown: **Cibirhiza**
− Cl with a cylindrical to campanulate tube, Cn white: **Fockea**

Tribe *Marsdenieae*
1 Inf pedunculate for 15 - 20 cm, with 2 - 6 Fl: **Absolmsia**
− Inf sessile or with a peduncle to 13 cm long, with 1 to many Fl: **2**
2 Cl globose: **3**
− Cl not globose: **4**
3 Fl hanging from long thin Ped: **Madangia**
− Fl shortly stalked, forming a globose inflorescence: **Hoya heuschkeliana**
4 Cl outside hairy: **Micholitzia**
− Cl outside glabrous: **5**
5 Cn lobes (on the lower face) with a longitudinal groove: **Hoya**
− Cn lobes without a longitudinal groove: **6**
6 Non-epiphytic shrubs, lianas or twiners: **Marsdenia**
− Epiphytic or epilithic subshrubs or herbs: **7**
7 Cn lobes leaf-like, apex ± divided: **Dischidia**
− Cn lobes shortly shovel-shaped, obtuse: **Dischidiopsis**

Tribe *Ceropegieae*
1 Stems herbaceous to fleshy, terete, erect to decumbent, scrambling or twining, usually with well-developed L (occasionally stems succulent, ± without L and 4-angled: then with delicate pitfall Fl): **2**
− Stems succulent, (3-) 4- to many-angled, L none or reduced (except for *Caralluma frerei*), caducous or forming Sp: **4 (Stapeliads)**
2 Plants with succulent or fibrous R, Cl tube 2× or many times as long as broad (if shorter then plants with fibrous R), above an inflated base usually narrowed into a tube (if open and campanulate then plants with fibrous R): **3**
− Plants with succulent R, Cl tube < 2× as long as broad, not or only little narrowed: **Brachystelma**
3 Ped fibrous-tough, Anth and Poll vertical: **Riocreuxia**
− Ped soft-fleshy, Anth and Poll ± horizontal (rarely ± vertical, but then plants succulent): **Ceropegia**
4 Stems leafy: **Caralluma frerei**
− Stems with rudimentary L transformed into Sc or Sp: **5**
5 Interstaminal Cn with free large erect lobes: **6**
− Interstaminal Cn different (if lobes large and erect then forming basal pouches): **7**
6 Staminal Cn subulate and erect (Namibia and RSA): **Tromotriche** p.p.
− Staminal Cn triangular, short (Madagascar): **Stapelianthus**
7 L rudiments tough and becoming spiny, persistent: **8**
− L rudiments inconspicuous, caducous or absent: **11**
8 Stems with parallel ribs, Sp prickly: **9**
− Stems 4- to 5-angled, with short rigid Sp (if prickly then stems with irregular Ri): **10**
9 Sp 3-partite: **Tavaresia**
− Sp simple, plants erect and cactus-like, stems 10- to many-angled: **Hoodia**
10 Plants decumbent to erect (orthotropous), Fl small, lateral, sessile: **Quaqua**
− Plants prostrate-creeping (plagiotropous), Fl very large, single: **Edithcolea**
11 Cl with small projecting lobes in the sinuses between the Cl lobes (these occasionally also present in *Tavaresia* and *Stapelianthus*); Gy with humps in front of entrance to the guiding rails: **Huernia**
− Cl without intermittent lobes in the sinuses; Gy without humps in front of entrance to the guiding rails: **12**
12 Stems distinctly tessellated (with transverse ribs between the Tu): **13**
− Stems not tessellated, 4- to 5- (to 6-) angled: **17**
13 Tu irregularly arranged, polygonal, staminal Cn linear-spatulate: **Pseudolithos**
− Tu regularly arranged, stems 4- to 6-angled, staminal Cn triangular to linear: **14**
14 L rudiments triangular to subulate, stems slender (to 1 cm ∅), prostrate-creeping (E Africa to Arabia): **Echidnopsis**
− L rudiments ± absent, stems > 1 cm ∅, usually erect or ascending (S Africa): **15**
15 Pet hairy on the upper face with simple Ha: **Notechidnopsis**
− Pet glabrous or papillose on the upper face: **16**
16 Stems 10- to 12-angled, L rudiments short, ± acute, on flat Tu: **Lavrania**
− Stems 12- to 20-angled, L rudiments thickish and sunken in a groove of the conical Tu: **Larryleachia**
17 Cn ± uniseriate (i.e. interstaminal Cn reduced to a fringe): **18**
− Cn biseriate (interstaminal Cn present): **20**
18 Stems 4-angled, L-less, pale brown-green, Cl inside glossy: **Duvaliandra**
− Stems 4- to 5- (to 6-) angled, green or green-brown, Cl inside dull or hairy: **19**
19 Cn yellow, staminal Cn with dorsal crest: **Piaranthus**
− Cn white to beige, staminal Cn without dorsal crest: **Huerniopsis**
20 Stems ± rugulose and papillose: **21**
− Stems neither rugulose nor papillose: **22**
21 L rudiments erect; Inf sunken: **Rhytidocaulon**

- L rudiments spreading; **Inf** not sunken: **Baynesia**
22 **Cl** ± urceolate, **Cl** lobes at most ⅓ as long as the **Cl**: 23
- **Cl** usually not urceolate, **Cl** lobes > ⅓ as long as the **Cl**: 24
23 Stems 4-angled, translator not winged: **Stapeliopsis**
- Stems 5- to many-angled, translator winged: **Huernia** p.p.
24 Stems 4 - 9 mm ∅, indistinctly 4-angled, creeping: 25
- Stems > 8 mm ∅ (if less, then not creeping), 4- to 6-angled: 26
25 **Cn** glabrous, staminal **Cn** dorsally enlarged into a globose or crest-like structure: **Ophionella**
- **Cn** hairy, staminal **Cn** simple: **Orbeanthus**
26 Stems 4-angled, **Tu** weakly bulging: 27
- Stems 4- to 6-angled, **Tu** usually bulging: 30
29 Stems short and clumpy, subsquare to ovoid, unbranched, **Inf** arising from the base, few-flowered: **White-sloanea**
- Stems less clumpy, branched: 28
28 **Inf** mostly basal, stalked; **Fl** large, stellate to campanulate; **Poll** large, translator winged: 29
- **Inf** sessile, in the apical regions of the stems, lateral or terminal, often as many-flowered pseudo-umbels; **Fl** rather small, of various shapes; **Poll** small, translator ± unwinged: **Caralluma**
29 **L** rudiments well developed, ovate-triangular, caducous: **Stapelia**
- **L** rudiments only slightly developed, deltoid (or ± absent): **Tromotriche**
30 Stems 6-angled, ovate to shortly cylindrical; **L** rudiments slightly developed, without stipular **Gl**: **Pectinaria**
- Stems 4- to 5-angled: 31
31 Stems conspicuously 4-angled, cylindrical; **L** rudiments subulate, stipules in the form of a few multicellular **Ha**: **Tridentea**
- Stems 4- to 5-angled, cylindrical or clavate; **L** rudiments triangular to subulate, usually with stipular **Gl**: 32
32 **Ci** disc-like, guide rails oblique, embedded in gynostegial tissue: **Duvalia**
- **Ci** lobes at least apically free, guide rails vertical and not embedded in gynostegial tissue: **Orbea**

Tribe *Asclepiadeae*
1 Erect geophytic herbaceous perennials with usually tough foliage, **Gy** incl. **Cn** ± stipitate, **Cn** usually uniseriate: **Subtribe Asclepiadinae**
- Erect or twining stem succulent herbaceous perennials (some *Cynanchum* leafy geophytes), **Gy** sessile, rarely stipitate, **Cn** sessile and uni- or biseriate, often fused to the **Gy**: **Subtribe Metastelminae**

Subtribe *Asclepiadinae*
1 **Cl** ± urceolate, **Cn** lobes subulate, erect, ± as long as the **Cl** lobes: **Stenostelma**
- **Cl** rotate to campanulate, **Cn** lobes shorter than the **Cl** lobes: 2
2 **Cn** lobes fleshy, rarely dorsally flattened: 3
- **Cn** lobes not or weakly fleshy, flattened: 4
3 **Cn** lobes almost as broad as long, appendages or keels absent, caudicles not geniculate: **Xysmalobium**
- **Cn** lobes longer than broad, often drawn out into a hooked tip, ± erect, caudicles geniculate, **Cl** ± campanulate: **Glossostelma**
4 **Cn** lobes dorsally flattened, caudicles united: 5
- **Cn** lobes laterally flattened and ± involute, caudicles united or 2-partite: 10
5 **Cn** lobes at the base fused into a ring: **Odontostelma**
- **Cn** lobes not fused into a ring: 6
6 **Cn** lobes spreading horizontally to ascending, mostly keeled on the inner face, plants robust: **Pachycarpus**
- **Cn** lobes ± erect (at least at the base): 7
7 **Cl** lobes white, strongly hairy, **Cn** lobes elongate-linear: **Fanninia**
- **Cl** lobes rarely white, **Cn** lobes ± spatulate to lanceolate, usually keeled or inside with appendage: 8
8 **Inf** pedunculate: **Schizoglossum**
- **Inf** sessile: 9
9 **Poll** with germination zone: **Aspidoglossum**
- **Poll** without germination zone: **Miraglossum**
10 **Cn** lobes papillose or shortly hairy: **Trachycalymma**
- **Cn** lobes glabrous: 11
11 Caudicles united, not winged: **Asclepias**
- Caudicles bipartite, ± winged: 12
12 Caudicles broadly winged and ± contorted: **Stathmostelma**
- Caudicles ± winged, not contorted: **Aspidonepsis**

Subtribe *Metastelminae*
1 Geophytes or ± stem succulents, with **L**: **Cynanchum** p.p.
- Stem succulent twiners, climbers or shrubs with rod-like stems, **L** ± absent: 2
2 Stems smooth, **Cn** biseriate (divided into staminal lobes and a basal outer ring composed of the fused staminal and interstaminal **Cn**): **Sarcostemma**
- Stems smooth, striate or verrucose, **Cn** uniseriate: **Cynanchum**

[F. Albers & U. Meve]

ABSOLMSIA

U. Meve

Absolmsia Kuntze (Rev. Gen. Pl. 2: 417, 1891). **T:** *Astrostemma spartioides* Bentham. – **Lit:** Green (1994: with ill.). **D:** Malaysia (Borneo: Sabah). **Etym:** For Prof. Hermann M. C. L. F. zu Graf Solms-Laubach (1842 - 1915), German botanist.
Incl. *Astrostemma* Bentham (1880) (*nom. illeg.*, Art. 53.1). **T:** *Astrostemma spartioides* Bentham.

Epiphytic subshrubs with white latex; **R** fibrous, rooting below the tree bark; **Br** horizontal to erect, ± 3 - 10 × 0.3 - 0.5 cm, glabrous, not twining, poorly branched, perennial; **L** few, shortly petiolate, elliptic to ovate, glabrous, stiff, ± succulent, 1 - 4 × 0.5 - 2 cm; **Inf** lateral, extra-axillary, 2- to 6-flowered, capitate at the tips of needle-like, green, perennial, 15 - 20 cm long peduncles; **Ped** short; **Sep** triangular-ovate; **Fl** 9 - 12 mm ∅, stiffly fleshy; **Cl** campanulate-urceolate with apically widening **Cl** tube, yellow-orange, glabrous; **Cl** tube ± as long as the **Cl** lobes, these triangular, ± 3.5 mm, horizontal or ascending; **Cn** uniseriate; **Cs** lobes stellate, fusiform, both ends acute, ± 2 mm, horizontal, resting on the **Anth**, with median longitudinal groove; **Poll** erect, ovoid-rectangular, ± 0.18 × 0.1 mm, germination mouth along the outside margins, corpuscles very voluminous; **Gy** sessile; **Fr** solitary, slender, 5 - 8 × 0.3 cm, ± terete.

This monotypic genus belongs to tribe *Marsdenieae*. It is closely related to *Hoya* and hardly distinguishable on flower morphological grounds. The reduced stems with few leaves and the persistent and obviously assimilating peduncles resembling needle leaves are characteristic. The attractive plants are difficult in cultivation and flower only rarely.

A. spartioides (Bentham) Kuntze (Rev. Gen. Pl. 2: 418, 1891). **T:** K. – **D:** Malaysia (Borneo: Sabah). **Fig. I.a**

≡ *Astrostemma spartioides* Bentham (1880) (*incorrect name*, Art. 11.4) ≡ *Hoya spartioides* (Bentham) Kloppenburg (2001).
Description as for the genus.

ASCLEPIAS

U. Meve

Asclepias Linné (Spec. Pl. [ed. 1], 214, 1753). **T:** *Asclepias syriaca* Linné. – **Lit:** Woodson (1954). **D:** America, now worldwide as neophytes. **Etym:** For Asklepios, the ancient Greek deity of medicine.
Incl. *Acerates* Elliott (1817). **T:** *Asclepias longifolia* Michaux.
Incl. *Anthonotis* Rafinesque (1817). **T:** *Anthonotis procumbens* Rafinesque.
Incl. *Podostigma* Elliott (1817). **T:** *Podostigma pubescens* Elliott.
Incl. *Anantherix* Nuttall (1818). **T:** *Acerates connivens* Decaisne.
Incl. *Stylandra* Nuttall (1818). **T:** *Stylandra pumila* Nuttall.
Incl. *Otaria* Kunth (1819) (*nom. inval.*, Art. 34.1b). **T:** *Asclepias auriculata* Kunth.
Incl. *Oligoron* Rafinesque (1836) (*nom. illeg.*, Art. 52.1). **T:** *Asclepias longifolia* Michaux.
Incl. *Otanema* Rafinesque (1836). **T:** *Otanema latifolia* Rafinesque.
Incl. *Polyotus* Nuttall (1837). **T:** *Polyotus angustifolius* Nuttall.
Incl. *Schizonotus* A. Gray (1840) (*nom. illeg.*, Art. 53.1). **T:** *Gomphocarpus purpurascens* A. Gray.
Incl. *Asclepiodora* A. Gray (1877). **T:** *Anthonotis viridis* Rafinesque.
Incl. *Solanoa* Greene (1890). **T:** *Gomphocarpus purpurascens* A. Gray.
Incl. *Solanoana* Kuntze (1891) (*nom. inval.*, Art. 61.1). **T:** *Gomphocarpus purpurascens* A. Gray.
Incl. *Oxypteryx* Greene (1897). **T:** *Asclepias arenicola* Nash.
Incl. *Podostemma* Greene (1897). **T:** *Asclepias longicornu* Bentham.
Incl. *Asclepiodella* Small (1933). **T:** *Acerates feayi* Chapman.
Incl. *Biventraria* Small (1933). **T:** *Asclepias variegata* Linné.

Sometimes geophytic perennials with deciduous herbaceous above-ground parts, shrubs or subshrubs with white partly sticky latex; **R** fibrous, irregularly thickened or in the form of fusiform to turnip-shaped storage **R**; stems erect, sometimes decumbent, normally unbranched, mostly annual, partly with underground rhizomes; **L** opposite, spiral or verticillate, sessile or petiolate, linear, elliptic, lanceolate or ovate; **Inf** 1 or several terminal, subterminal or lateral extra-axillary pseudo-umbels; **Cl** rotate with valvate buds; **Cl** lobes ± ovate, ± free, erect to reflexed; **Cn** uniseriate; **Cs** ± double with ± pouch-like basally saccate lobes, which develop cristate or horn-like secondary structures inside, with nectar; **Gy** sessile or stipitate, **Anth** appendages present, guide rails raised; **Poll** ± flat, pear-shaped, pendent, large; **Fr** as a rule solitary, erect, narrowly fusiform to thick, sometimes inflated fusiform, sometimes rostrate, smooth, hairy or softly spiny.

Most taxa are shrubs or deciduous herbaceous perennials, and can possess various forms of root succulence (often imperfectly known and here not further treated). Several species are cultivated world-wide or have escaped as weeds, e.g. *A. syriaca* ("Silk Plant", seed hairs used as a substitute for silk) or *A. curassavica* (planted as ornamental). The genus counts more than 100 species, esp. in subtropical North and South America. A modern and complete treatment is lacking. Woodson (1954) treats only the North American species and suggested a complicated infrageneric classification with 9 subgenera (incl. 9 series). While Woodson

(1954) includes the closely related genus *Gomphocarpus*, modern authors such as Nicholas & Goyder (1992) or Bruyns (1995c) prefer to separate the 2 genera, with *Asclepias* in the New World and *Gomphocarpus* in the Old World.

The following names are of unresolved application but are referred to this genus: *Asclepias nuttii* N. E. Brown (1898) ≡ *Stathmostelma nuttii* (N. E. Brown) Bullock (1953).

A. subulata Decaisne (PSRV 8: 571, 1844). **T:** (*Herb. Pavón* s.n. [FI]). – **D:** S USA (Nevada, Arizona, California), Mexico (Baja California). **Fig. II.a**

Perennial subsucculent twiggy subshrubs; **R** fibrous; stems erect, 50 - 200 cm, 3 - 5 mm ∅, smooth, glabrous, blue-green, with laxly placed opposite **L**; **L** ± sessile, linear-subulate, 2 - 8 cm × 1 - 2 mm, glabrous, blue-green, soon deciduous; **Inf** lateral near stem tips, shortly pedunculate, 5- to 12-flowered; **Ped** 1 - 1.5 cm, laxly tomentose; **Sep** lanceolate, 3 mm; **Cl** rotate, yellowish-white; **Cl** lobes almost completely reflexed, lanceolate, 9 - 11 mm; **Cn** shortly stipitate, fleshy, cream; **Cs** lobes ± erect, obovately fan-shaped, drawn together pouch-like, 9 - 10 mm, including and ± obscuring the short horn-like secondary process; **Fr** ± pendent, fusiform, ± 1 cm ∅; **Se** ovate, 6 mm, tuft of **Ha** 3 - 4 cm.

This is the only ± stem-succulent species of the genus. The dry season is survived as leafless twiggy shrub. Vernacular name (USA): "Desert Milkweed".

ASPIDOGLOSSUM

B. Müller, P. Stegemann & F. Albers

Aspidoglossum E. Meyer (Comm. Pl. Afr. Austr., 200, 1838). **T:** *Aspidoglossum biflorum* Bullock [Type according to Bullock, Kew Bull. 7: 417, 1952.]. – **Lit:** Kupicha (1984). **D:** SE Africa. **Etym:** Gr. 'aspis, aspidos', shield; and Gr. 'glossa', tongue; for the segments of the corona.

Incl. *Rhinolobium* Arnott (1838). **T:** *Rhinolobium tenue* Arnott.

Perennials with deciduous herbaceous aboveground parts with latex and **R** tubers; stems erect, rarely ascending or apically drooping, branched or not, hairy, **Ha** either transparent or brownish, short, apically globose, glandular, or then **Ha** brownish, long, delicate and appressed (only in the **Inf** region); stems evenly hairy or **Ha** in longitudinal rows; **L** decussate, rarely verticillate or alternate, shortly petiolate or sessile, lamina entire, very narrowly linear, rarely elliptic, circular, hastate or obovate, margins revolute, hairy or glabrous; **Inf** sessile in upper stem parts; **Bra** inconspicuous; **Ped** softly hairy; **Sep** triangular, outside softly hairy, inside glabrous, margins ciliate; **Cl** ± campanulate to shallow, lobes elliptic, vertical, ascending or horizontally spreading, slightly reflexed, tips very rarely remaining united, outside mostly softly hairy, rarely glabrous, inside softly pubescent, margins hairy or glabrous; **Cs** lobes delicate or paper-like, never fleshy, dorsiventrally compressed, appearing 2-parted, outer part of the lobes basally ± square, erect, inside mostly with 2 shallow longitudinal ribs, *either* of type A: apically with an oblong appendage, which is erect or falcate, and laterally often with horn- to wing-like appendages, *or* of type B: apically drawn out into a short obtuse tip, inner part of the lobes as a slender ± cylindrical appendage, much overtopping the **Sty** head and apically meeting or falcately bent outwards, sometimes with an additional ring of **Cn** elements, these lobes minute and below the guide rails (only in Sect. 2); **Anth** ± rectangular, connective appendage membranous, incumbent on the **Sty** head; **Poll** reniform, rarely pear-shaped, dorsiventrally slightly compressed, germination zone (when present) thin, mostly transparent, inserted subapically on the translators, translator arms short, thin or sometimes flattened and strap-shaped, corpuscle linear; **Sty** head concave; **Fr** mostly solitary, ± fusiform, sometimes softly prickly, pubescent; **Se** numerous, dark brown, compressed-ovate, apically with a cluster of white **Ha**.

Aspidoglossum is closely related to *Schizoglossum*, but the pollinia are inserted apically on the translator arms, instead of laterally as in *Schizoglossum*. In addition, the corona lobes of *Aspidoglossum* possess an apical and an adaxial appendage, while in *Schizoglossum*, they have only an adaxial appendage. The genus is divided as follows according to Kupicha (1984):

[1] Sect. *Aspidoglossum*: Plants hairy, **Ha** transparent (glandular in *A. breve*); **L** opposite, lamina narrowly linear, rarely elliptic, ovate or hastate; **Cs** of type A; **Poll** narrow, usually with a defined germination zone, translator arms thin; **Fr** smooth.
[2] Sect. *Latibrachium* Kupicha 1984: Plants hairy, **Ha** brown, glandular; **L** opposite, narrowly linear; **Cl** inside mostly glabrous; **Cs** of type B, variable in shape; **Poll** variable in shape and size, germination zone sometimes present, translator arms weak or flattened; **Fr** sometimes with a few **Bri**.
[3] Sect. *Verticillus* Kupicha 1984: **L** verticillate or arranged irregularly, lamina narrowly linear or much broader (oblong, elliptic, (ob-) ovate); **Cl** hairy, **Ha** brown, thin; **Cs** of type A (type B in *A. dissimile*), hairy, **Ha** brown, thin; translator arms thin, **Poll** large, without defined germination zone; **Fr** with **Bri**.
[4] Sect. *Virga* Kupicha 1984: Plants hairy, **Ha** transparent; **L** opposite, lamina narrowly linear; **Cs** mostly of type B; **Poll** narrow, with defined germination zone, translator arms thin; **Fr** glabrous.

The following names are of unresolved application but are referred to this genus: *Schizoglossum alpestre* K. Schumann (1901); *Schizoglossum aschersonianum* Schlechter (1893); *Schizoglossum aschersonianum* var. *longipes* N. E. Brown (1907); *Schizoglossum aschersonianum* var. *radiatum* N. E. Brown (1907); *Schizoglossum barbatum* Britten & Rendle (1894); *Schizoglossum filiforme* (Linné *fil.*) Druce (1917); *Schizoglossum garcianum* Schlechter (1905); *Schizoglossum linifolium* Schlechter (1894); *Schizoglossum linifolium* var. *centrirostratum* N. E. Brown (1907); *Schizoglossum montanum* R. A. Dyer (1961); *Schizoglossum peglerae* N. E. Brown (1907); *Schizoglossum pygmaeum* Schlechter (1894) ≡ *Schizoglossum aschersonianum* var. *pygmaeum* (Schlechter) N. E. Brown (1907); *Schizoglossum tenellum* (Turczaninov) Druce (1917).

A. angustissimum (K. Schumann) Bullock (KB 7: 418, 1952). **T:** Zaïre (*Schweinfurth 3879* [B]). – **D:** Congo, Central African Republic, Zaïre, Uganda, Sudan, Rwanda, Kenya, Tanzania, Malawi, Zambia, Zimbabwe; dry savanna, rarely in swamps.
≡ *Schizoglossum angustissimum* K. Schumann (1893); **incl.** *Schizoglossum elatum* K. Schumann (1893); **incl.** *Schizoglossum whytei* N. E. Brown (1907) ≡ *Aspidoglossum whytei* (N. E. Brown) Bullock (1952); **incl.** *Schizoglossum ledermannii* Schlechter (1913); **incl.** *Schizoglossum zernyi* Markgraf (1938).

[4] Stems (45-) 80 - 135 cm, normally branched; **L** mostly longer than the **Int**, lamina 4 - 13 × 0.2 - 0.6 cm, upper face appressed-pubescent, lower face glabrous, midvein pubescent; **Inf** with (2-) 4 **Fl**; **Ped** 1 - 7 mm; **Sep** 0.5 - 3 × 0.75 - 1 mm, bent inwards, softly hairy; **Cl** cup-shaped, cream to greenish-brown, lobes 3 - 4.5 × 1 - 2 mm, margins revolute, tips bent inwards, outside porrectly soft-hairy, inside pubescent, margins hairy, **Ha** white, ≤ 2 mm; **Cs** white, pale yellow or greenish, often dotted with purple, or violet to brown, 1 - 3 mm, of type B, outer part of lobes 1 - 2 mm, ± square, shoulders rounded, rarely acute, inner appendage mostly oblong-subulate, erectly bent; **Poll** 0.5 - 0.8 mm, corpuscle 0.2 - 0.33 mm; **Fr** 5 - 7 cm, narrowly fusiform, strongly hairy, **Ha** long, acute, soft.

A. araneiferum (Schlechter) Kupicha (KB 38(4): 661, 1984). **T:** RSA, KwaZulu-Natal (*Schlechter 3428* [Z]). – **D:** C Moçambique, NW Zimbabwe, RSA (KwaZulu-Natal, Free State, Mpumalanga, Northern Prov.), Lesotho; scattered in grassland, 300 - 1770 m.
≡ *Schizoglossum araneiferum* Schlechter (1895); **incl.** *Schizoglossum polynema* Schlechter (1905).

[2] Stems 30 - 60 cm, simple or little-branched; **L** 15 - 35 × 0.5 - 1.5 mm, vertical, both faces glabrous or roughly glandular-hairy; **Inf** with 3 - 7 **Fl**; **Ped** 4 - 7 mm; **Sep** 1.5 - 2 × 0.75 - 1 mm, weakly glandular-hairy, **Ha** brown, soft; **Cl** dull green or yellowish-brown, flat to inversely bowl-shaped, lobes 2.5 - 3.5 × 1 - 2 mm, spreading, margins revolute, tips bent inwards, outside weakly glandular-hairy, **Ha** brown, soft; outer part of **Cs** lobes basally ± 1 mm, erect, apically bifid, appendages long, linear, falcately bent inwards, inner appendage similar but falcately bent outwards between the outer appendages; **Poll** 0.3 - 0.45 mm, falcate, tips bent, germination zone prominent, translators flattened, corpuscle 0.1 - 0.28 mm; young **Fr** pubescent.

A. biflorum E. Meyer (Comm. Pl. Afr. Austr., 201, 1838). **T:** RSA, Cape Prov. (*Drège s.n.* [K]). – **D:** Moçambique, Zimbabwe, Namibia, Lesotho, Malawi, RSA (Northern Prov., Free State, KwaZulu-Natal); grassland.
≡ *Schizoglossum biflorum* (E. Meyer) Schlechter (1895); **incl.** *Schizoglossum excisum* Schlechter (1894); **incl.** *Schizoglossum shirense* N. E. Brown (1895); **incl.** *Schizoglossum strictum* Schlechter (1895); **incl.** *Schizoglossum tubulosum* Schlechter (1895); **incl.** *Schizoglossum venustum* Schlechter (1895); **incl.** *Schizoglossum venustum* var. *concinnum* Schlechter (1895) ≡ *Schizoglossum biflorum* var. *concinnum* (Schlechter) N. E. Brown (1907); **incl.** *Schizoglossum gwelense* N. E. Brown (1904) ≡ *Schizoglossum venustum* var. *gwelense* (N. E. Brown) N. E. Brown (1907); **incl.** *Schizoglossum conrathii* Schlechter (1905); **incl.** *Schizoglossum biflorum* var. *integrum* N. E. Brown (1907).

[1] Stems 20 - 70 cm, simple or hardly branched; **L** 25 - 75 × 1 - 4 mm, upper face weakly to densely hispid to pubescent, lower face glabrous, midvein pubescent; **Inf** with 1 - 7 **Fl**; **Ped** ≤ 3 mm or none; **Sep** 2 - 3 × 1 - 1.5 mm, softly hairy; **Cl** green or purple, ± flat, central depression ≤ 1.25 mm, lobes 3 - 5 × 1.5 - 2 mm, outside glabrous, at least basal ½ of the lobes densely pubescent, inside weakly soft-hairy; **Cs** white, yellowish or purple, 3 - 5.5 mm, outer part of the lobes basally 1 - 1.5 mm, ± square, apical appendage thin, erect, inner appendage similar but only ½ as long; **Poll** 0.5 - 0.7 mm, germination zone inconspicuous, corpuscle 0.3 - 0.5 mm; **Fr** 6 - 7 mm, densely pubescent.

A. breve Kupicha (KB 38(4): 643, 1984). **T:** Malawi (*Hilliard & Burtt 4399* [E, K]). – **D:** SW Tanzania, N Malawi; grassland, 2400 - 2780 m.

[1] Stems 11 - 32 cm, unbranched; **L** longer than the **Int**, lamina 20 - 55 × 1 - 2 mm, both faces ± strongly pubescent, **Ha** ± long; **Inf** with 2 - 7 **Fl**; **Ped** 5 - 8 mm; **Sep** 2 - 2.5 × 0.5 - 0.75 mm, weakly to densely glandular-hairy, **Ha** transparent or brownish, soft; **Cl** dull greenish-brown, campanulate, lobes 3 - 4 × 1.5 - 2.5 mm, inside glabrous, outside glandular-hairy, **Ha** transparent; **Cs** pale green or yellow, 1.5 - 2 mm, outer part of the lobes basally 1 mm, ± square, shoulders ± acute, apical appendage triangular, erect, inner appendage somewhat shor-

ter, triangular, bent slightly inwards; **Poll** 0.5 - 0.6 mm, corpuscle 0.2 - 0.3 mm.

Very similar to *A. biflorum.*

A. carinatum (Schlechter) Kupicha (KB 38(4): 640, 1984). **T:** RSA, Cape Prov. (*Flanagan* 1043 [BOL, GRA, NBG, NH, SAM]). – **D:** Namibia, Zimbabwe, RSA (Northern Cape, Eastern Cape, KwaZulu-Natal); grassland and sandy flats near coast, to 1020 m.

≡ *Schizoglossum carinatum* Schlechter (1894); **incl.** *Schizoglossum tricuspidatum* Schlechter (1895); **incl.** *Schizoglossum pentheri* Schlechter (1903).

[1] Stems (15-) 30 - 60 cm; **R** tuber turnip-shaped; stems simple, rarely branched; **L** 1 - 3.5 (-6) cm × 0.5 - 4 mm, margins revolute, upper face glabrous or (softly) pubescent, lower face glabrous, midvein softly hairy, **Ha** of midrib and **L** margin thicker; **Inf** with 1 - 7 **Fl**; **Ped** 1 - 3 mm; **Sep** 1.5 - 2.5 × 0.5 - 1.25 mm, softly hairy; **Cl** cup-shaped, dull green, lobes 2 - 3.5 × 1 - 1.5 mm, spreading, outside weakly soft-hairy, inside densely pubescent, **Ha** basally often longer, margins of lobes glabrous; **Cs** 1 - 1.7 mm, outer part of lobes 0.5 - 1 mm, basally ± square, shoulders ± rounded, apical appendage broadly triangular, erect, sometimes absent, inner appendage ± 1.5 mm, basally broad, apically ± linear, overtopping the **Sty** head and meeting each other; **Poll** 0.4 - 0.6 mm, reniform, germination zone inconspicuous, corpuscle 0.2 - 0.4 mm; **Fr** 4 - 6 cm, pubescent.

A. connatum (N. E. Brown) Bullock (KB 7: 419, 1952). **T:** Zambia (*Carson* 17 [K]). – **D:** Sudan, Uganda, Kenya, Tanzania, Zambia, Zaïre, Malawi; swampy grassland.

≡ *Schizoglossum connatum* N. E. Brown (1895); **incl.** *Schizoglossum vulcanorum* Lebrun & Taton (1943).

[4] Stems 43 - 90 cm, simple or hardly branched; **L** 17 - 90 × 0.25 - 2 (-8) mm, upper face glabrous or softly hairy, lower face glabrous, midvein pubescent; **Inf** with 3 - 7 **Fl**; **Ped** 1 - 6 mm; **Sep** 1.5 - 2.5 × 0.75 - 1 mm, softly hairy; **Cl** white, cream, yellow or pale to dark purple, lantern-shaped from the apically united **Cl** lobes or lobes sometimes flatly spreading, 5.5 - 9 × 1 - 2 mm, margins usually strongly revolute, outside weakly soft-hairy, inside densely pubescent, margins hairy, **Ha** long; **Cs** greenish, cream or purple, 1 - 1.5 mm, outer part of the lobes square, inner appendage short, irregular, sometimes rudimentary; **Poll** 0.4 - 0.8 mm, corpuscle 0.2 - 0.4 mm; **Fr** 6.5 cm, pubescent.

A. crebrum Kupicha (KB 38(4): 642, 1984). **T:** Zimbabwe (*Chase* 4228 [LISC, BM, BR, COI, K, SRGH]). – **D:** E Zimbabwe; grassland, 1050 - 1110 m.

[1] Stems 30 - 70 cm, unbranched; **L** 25 - 85 × 1 - 2 mm, upper face pubescent or glabrous, lower face glabrous, midvein pubescent; **Inf** with 4 - 10 **Fl**, very dense; **Ped** ≤ 2 mm; **Sep** 2 - 2.5 × 1 mm, softly hairy; **Cl** dark purple or brown, flat, lobes 3.5 - 4 × 1.5 - 2 mm, ascending, outside glabrous, inside densely and finely soft-hairy; **Cs** ± 2 mm, outer part of the lobes basally 1 mm, shoulders acute, apical appendage triangular to attenuate-acute, reclining, inner appendage longer, contorted; **Poll** 0.6 mm, elongate-triangular, germination zone inconspicuous, corpuscle 0.2 mm; unripe **Fr** densely pubescent.

A. delagoense (Schlechter) Kupicha (KB 38(4): 663, 1984). **T:** Moçambique (*Junod* 484 [BR, Z]). – **D:** S Moçambique, RSA (NE KwaZulu-Natal); open grassland on sandy soils.

≡ *Schizoglossum delagoense* Schlechter (1896); **incl.** *Schizoglossum biauriculatum* Schlechter (1905).

[2] Stems 20 - 45 cm, simple or branched; **L** sometimes clustered, lamina 25 - 70 × 0.25 - 4 mm, both faces glabrous or occasionally pubescent; **Inf** with 2 - 7 **Fl** (rarely 9 - 13 **Fl** in terminal **Inf**); **Ped** 5 - 8 mm; **Sep** 1.5 - 2.5 × 0.5 - 0.75 mm, softly hairy; **Cl** brown, flat, lobes 4 - 5 × 1.5 - 2 mm, outside softly hairy, inside glabrous or pubescent; **Cn** white, with 2 alternating circles, outer part of **Cs** lobes 1.5 - 2 mm, rounded-triangular, erect, margins bent inwards, inner appendage narrow, hood-shaped; alternating circle closer to the centre, lobes minute, transversally rectangular, bulging horizontally inwards; **Poll** ± 0.9 mm, linear, without germination zone, translators short, flattened, corpuscle 0.2 - 0.25 mm; **Fr** 4.5 - 5 cm, glabrous.

Closely related to *A. rhodesicum.*

A. demissum Kupicha (KB 38(4): 644, 1984). **T:** RSA, KwaZulu-Natal (*Devenish* 781 [PRE]). – **D:** RSA (KwaZulu-Natal); only known from the type locality, mountain top on rocky ground.

[1] Stems 5 - 6 cm, strongly branched, basally spreading; **L** somewhat longer than the **Int**, lamina 6 - 11 × 1 - 3 mm, linear or narrowly ovate, glabrous; **Inf** with 1 - 2 **Fl**; **Ped** 6 - 10 mm; **Sep** 2 × 0.75 mm, outside crisped-pubescent; **Cl** green or brown, flat, lobes 4 - 5 × 2 mm, margins revolute, glabrous; **Cs** white, outside purple, 3.5 - 4 mm, outer part of lobes basally 2 mm, shoulders oblique, apical appendage erect, inner appendage ½ as long, erect or slightly reclining; **Anth** appendages 1 mm, elliptic, erect; **Poll** 0.4 mm, translators bent forwards, corpuscle ± 0.25 mm.

The placement in *Aspidoglossum* is uncertain due to the atypical anther appendages.

A. difficile Hilliard (Notes Roy. Bot. Gard. Edinburgh 45(2): 179, 1989). **T:** RSA, KwaZulu-Natal (*Pooley* 1921 [E]). – **D:** RSA (KwaZulu-Natal).

[2] Stems 45 cm, weakly branched, 1 mm ∅, uniseriately hairy, **Ha** short, bent; **L** few, sessile, lami-

na 30 - 40 × 1 mm, acute, glabrous, young **L** margins weakly hairy; **Inf** with 4 - 5 pendent **Fl** from each node; **Ped** 5 - 7 mm, uniseriately pubescent; **Sep** brown, 2 × 1 mm, lanceolate; **Cl** deep purple, ± flat, incised almost to the base, lobes 9 × 2 mm, oblong, flat, obtuse, inside weakly hairy, **Ha** ≤ 1 mm; **Cn** of 2 alternating circles; **Ci** lobes ± 0.3 mm, almost circular; **Cs** red-brown, lobes 3.5 × 1.25 mm, widely overtopping the **Gy**, almost paper-like, without appendage, inside with 2 narrow fleshy keels; **Anth** appendages ± 0.4 mm, erect, just reaching the margin of the **Gy**; **Gy** 1.5 mm; **Sty** head ± 0.8 mm ∅, 5-lobed, truncate; **Poll** ± 0.6 mm, narrowly pear-shaped, strongly flattened, suspended subapically, corpuscle ± 0.3 mm.

Closely related to *A. delagoense*, but the corona circle alternating with the Cs is placed outside like the Ci. The placement in *Aspidoglossum* is questionable (Hilliard & Burtt 1989).

A. dissimile (N. E. Brown) Kupicha (KB 38(4): 646, 1984). **T**: RSA, Eastern Cape (*Sim* s.n. [BOL, K]). − **D**: RSA (Eastern Cape, E KwaZulu-Natal, Mpumalanga).

≡ *Schizoglossum dissimile* N. E. Brown (1907); **incl.** *Schizoglossum dissimile* var. *pubiflorum* N. E. Brown (1907).

[3] Stems 6 - 20 cm, simple or hardly branched; **L** basally irregularly and densely arranged, verticillate within the **Inf**, lamina 18 - 30 × 2 - 8 mm, linear (-lanceolate), both faces softly hairy; **Inf** with 3 - 5 **Fl**; **Ped** 6 - 12 mm; **Sep** 2 - 3.5 × 1 - 1.5 mm, softly hairy, **Ha** transparent or brownish; **Cl** flat, lobes 4.5 - 5 × 2.5 mm, revolute, margins revolute, outside softly hairy, **Ha** transparent, stiff, or brown, delicate, appressed; **Cl** lobes inside glabrous or densely pubescent; **Cs** 3.5 - 4 mm, outer parts of lobes basally 1 - 2 mm, ± cordate, shoulders undulate, apical appendage (when present) very short, inner appendage filiform, erect, hooked, bent outwards; **Poll** 0.5 - 0.9 mm, germination zone ± conspicuous, corpuscle 0.25 - 0.4 mm.

A. elliotii (Schlechter) Kupicha (KB 38(4): 659, 1984). **T**: (*Scott-Elliot* s.n. [BM]). − **D**: Ethiopia, Uganda, Rwanda, Burundi, Zaïre, Tanzania, Zambia, Zimbabwe; dry grassland.

≡ *Schizoglossum elliotii* Schlechter (1895); **incl.** *Schizoglossum debile* Schlechter (1895).

[4] Stems (30-) 60 - 120 cm, simple or branched; **L** shorter than the **Int**, lamina 25 - 57 × 0.5 - 2 mm, both faces weakly pubescent; **Inf** with 2 - 5 **Fl**; **Ped** 4 - 7 mm; **Sep** 1.5 - 2 × 0.5 - 0.75 mm, softly hairy; **Cl** greenish, purple or brownish, ± flat to bowl-shaped, lobes 3.5 - 4 × 1.5 - 2 mm, margins revolute, outside weakly hairy, inside delicately soft-hairy, **Ha** longest at the lobe tips; **Cs** lobes penguin-shaped, outer parts basally 1.5 mm, erect, ± square, margins bulging hump-like, apical appendage broadly triangular, incumbent over and slightly overtopping the **Sty** head; **Poll** 0.35 - 0.45 mm, corpuscle 0.25 - 0.4 mm, linear, laterally bow-like.

A. erubescens (Schlechter) Bullock (KB 9: 589, 1954). **T**: Malawi (*Scott-Elliot* 8671 [BM]). − **D**: Zimbabwe, S Malawi, N Moçambique; open grass- or woodland, 750 - 1470 m.

≡ *Schizoglossum erubescens* Schlechter (1895); **incl.** *Schizoglossum strictissimum* S. Moore (1902).

[1] Stems 37 - 110 cm, simple or somewhat branched; **L** 14 - 65 × 1 - 2.5 mm, both faces densely crisped-pubescent, sometimes glabrescent; **Inf** with 2 - 6 ± sessile **Fl**; **Sep** 1.5 - 2 × 0.75 - 1 mm, softly hairy; **Cl** dull mauve to greenish, lobes 2.5 - 3.5 × 1.5 - 2.5 mm, spreading to strongly revolute, outside softly hairy, inside densely pubescent or glabrous; **Cs** dark purple, 1 - 1.25 mm, outer part of the lobes ± square, apical and inner appendages very short, bent outwards; **Poll** 0.35 - 0.4 mm, corpuscle 0.3 - 0.33 mm, slender; **Fr** 6 - 8 cm, pubescent; **Se** 4 mm, **Ha** ± 25 mm.

A. eylesii (S. Moore) Kupicha (KB 38(4): 654, 1984). **T**: Zimbabwe (*Eyles* 500 [BM, NBG, SAM, SRGH]). − **D**: Zimbabwe, S Malawi; forests on rocks, 960 - 1800 m.

≡ *Schizoglossum eylesii* S. Moore (1914).

[4] Stems 60 - 130 cm, simple or branched; **L** longer than the **Int**, lamina (2.5-) 6 - 14 cm × 1 - 6 mm, upper face appressed-pubescent, lower face glabrous, midvein pubescent; **Inf** with 3 - 7 (-13) **Fl**; **Ped** 2 - 7 mm; **Sep** 1.5 - 2.5 × 0.75 - 1 mm, softly hairy; **Cl** greenish-red, outside sometimes brown, ± flat, lobes 3.5 - 5 × 1.5 - 2.5 mm, margins often revolute, tips bent inwards, outside softly hairy, inside normally glabrous, lobes sometimes basally or apically pubescent; **Cs** green or white, 2 - 3 mm, outer parts of lobes basally 1 - 1.5 mm, shoulders acute, inner appendage triangular, erect, or inclined, divided from the basal portion by a transversal crest; **Poll** 0.6 - 0.8 mm, germination zone inconspicuous, corpuscle 0.25 - 0.3 mm; **Fr** 4.5 - 6 cm, apically attenuate into a long straight beak, densely soft-pubescent.

A. fasciculare E. Meyer (Comm. Pl. Afr. Austr., 200, 1838). **T** [lecto]: RSA, Eastern Cape (*Drège* s.n. [not located]). − **D**: RSA (Eastern Cape, KwaZulu-Natal); grassland.

≡ *Schizoglossum fasciculare* (E. Meyer) Schlechter (1894); **incl.** *Schizoglossum ciliatum* Schlechter (1895).

[1] Shrubs with unbranched stem; stem fleshy, hispid; **L** very variable in length and width, ovate-lanceolate, pubescent; **Inf** porrectly spreading, pedunculate, nutant, to 7-flowered; **Cl** ± flat, lobes 5 - 6 × 1.5 - 2 mm, green with mauve venation, slightly reflexed; **Cs** lobes basally broad, apically attenuate, bifid, appendages short, tooth-like; **Sty** head flat; **Anth** appendages membranous; **Poll** pear-shaped.

A. flanaganii (Schlechter) Kupicha (KB 38(4): 654, 1984). **T** [lecto]: RSA, Eastern Cape (*Flanagan* 1044 [BOL]). – **D:** RSA (Eastern Cape); grassland near coast, to 30 m.
≡ *Schizoglossum flanaganii* Schlechter (1894).

[4] Stems 18 - 30 cm, branched; **L** longer than the **Int**, lamina 2 - 8 cm × 0.5 - 6 mm, both faces glabrous but midvein pubescent; **Inf** with 3 - 8 **Fl**; **Ped** 2 - 4 mm; **Sep** 2 - 2.5 × 1 mm, ± weakly softhairy; **Cl** inversely campanulate, lobes 5 × 2 mm, margins strongly revolute, glabrous; **Cs** 3 mm, outer part of lobes basally 1.5 - 2 mm, shoulders acute, inner appendage erectly inclined; **Poll** 0.6 - 0.7 mm, corpuscle 0.25 - 0.3 mm; **Fr** 6.5 cm, pubescent or glabrous; **Se** 4 mm, **Ha** 25 mm.

A. glabellum Kupicha (KB 38(4): 643, 1984). **T:** Zimbabwe (*Drummond* 4952 [PRE, BR, K, SRGH]). – **D:** E Zimbabwe; mountainous grassland, 900 - 1950 m.

[1] Stems 35 - 60 cm, usually unbranched; **L** normally shorter than the **Int**, lamina 16 - 50 × 1 - 3 mm, both faces glabrous to softly hairy; **Inf** with 1 - 4 **Fl**; **Ped** 2 - 8 mm; **Sep** 1.5 - 2 × 0.75 mm, glabrous or softly hairy; **Cl** whitish or yellowish-green, flat, lobes 4 × 2 mm, glabrous; **Cs** 3 - 4 mm, outer part of lobes 1 mm, square-shouldered, apical and inner appendage of equal length, acute, erect; **Poll** 0.3 - 0.4 mm, without visible germination zone, corpuscle 0.2 - 0.25 mm; **Fr** ± 4 cm, ± pubescent.

A. glabrescens (Schlechter) Kupicha (KB 38(4): 659, 1984). **T:** RSA, Mpumalanga (*Schlechter* 4051 [not located]). – **D:** RSA (Mpumalanga, Gauteng, KwaZulu-Natal: Drakensberg); rocky grassland, 1290 - 2100 m.
≡ *Schizoglossum glabrescens* Schlechter (1895); **incl.** *Schizoglossum longirostre* Schlechter (1895) ≡ *Schizoglossum glabrescens* var. *longirostre* (Schlechter) N. E. Brown (1907); **incl.** *Schizoglossum tenuissimum* Schlechter (1895); **incl.** *Schizoglossum barbatum* Schlechter (1895) (*nom. illeg.*, Art. 53.1); **incl.** *Schizoglossum loreum* S. Moore (1903); **incl.** *Schizoglossum schlechteri* N. E. Brown (1907); **incl.** *Schizoglossum tridens* N. E. Brown (1907); **incl.** *Schizoglossum unicum* N. E. Brown (1907); **incl.** *Schizoglossum hirtiflorum* N. E. Brown (1928).

[2] Stems 35 - 60 cm, rarely branched; **L** 15 - 35 × 0.5 - 1 mm, both faces glabrous, midvein sometimes pubescent; **Inf** with 2 - 7 **Fl**; **Ped** 4 - 8 mm; **Sep** 1.5 - 3 × 1 mm, outside glandular-hairy, **Ha** brown or transparent, soft; **Cl** dull yellow to brownish, campanulate or ± flat, lobes 2 - 4.5 × 1 - 2 mm, outside weakly glandular-hairy, **Ha** brown, rarely transparent, soft, inside densely downy or longhairy (sometimes only the lobe margins), rarely glabrous; **Cs** white or dull yellow, outer part of lobes basally 1 - 3 mm, rounded, apically bifid or lanceolate, erect, inner appendage ± long, inclined or erect; **Poll** 0.4 - 0.5 mm and without visible germination zone, corpuscle 0.3 - 0.4 mm, translators flattened, ± broadly strap-like; **Fr** 6 - 7.5 cm, slender, pubescent.

A. glabrescens shows pronounced variability in characters of the floral region, and this is correlated with the geographical distribution.

A. glanduliferum (Schlechter) Kupicha (KB 38(4): 662, 1984). **T:** RSA, KwaZulu-Natal (*Medley Wood* 4804 [Z, BOL, K]). – **D:** RSA (C KwaZulu-Natal, Mpumalanga); grassland, (semi-) dry soils.
≡ *Schizoglossum glanduliferum* Schlechter (1894); **incl.** *Schizoglossum commixtum* N. E. Brown (1907); **incl.** *Schizoglossum parile* N. E. Brown (1907); **incl.** *Schizoglossum auriculatum* N. E. Brown (1909).

[2] Stems 26 - 50 cm, erect, unbranched; **L** mostly shorter than the **Int**, lamina 25 - 33 × 1 - 2 mm, vertically arranged, both faces glabrous or roughly glandular-hairy; **Inf** with 1 - 9 **Fl**; **Ped** 5 - 8 mm; **Sep** 1.5 - 2 × 0.75 - 1 mm, outside glandular-hairy, **Ha** transparent and/or brown, soft; **Cl** pale green, slightly lineate with chestnut-brown, flat or slightly hood-shaped, lobes 2.5 - 4 × 1.5 - 2.5 mm, outside ± densely glandular-hairy (rarely glabrous), **Ha** transparent and/or brown, soft; **Cs** pale green, outer part of lobes 1 - 1.5 mm, rounded-triangular, margins bent inwards, inner appendage horizontally incumbent over the **Sty** head; guide rails conspicuously raised; **Poll** 0.25 - 0.45 mm, asymmetrically grainshaped, translators flattened in the middle, corpuscle 0.2 - 0.3 mm.

Closely related to *A. rhodesicum*.

A. gracile (E. Meyer) Kupicha (KB 38(4): 654, 1984). **T:** RSA, KwaZulu-Natal (*Drège* 4978 [K]). – **D:** RSA (Western Cape, Eastern Cape, S KwaZulu-Natal); grassland near coast, to 1710 m.
≡ *Lagarinthus gracilis* E. Meyer (1838); **incl.** *Rhinolobium tenue* Arnott (1838) ≡ *Schizoglossum tenue* (Arnott) Druce (1917); **incl.** *Lagarinthus exilis* Decaisne (1844) (*nom. illeg.*, Art. 52.1) ≡ *Schizoglossum exile* (Decaisne) Schlechter (1896); **incl.** *Rhinolobium lineare* Decaisne (1844) (*nom. illeg.*, Art. 52.1); **incl.** *Schizoglossum bolusii* Schlechter (1893); **incl.** *Schizoglossum guthriei* Schlechter (1893); **incl.** *Schizoglossum lunatum* Schlechter (1893); **incl.** *Schizoglossum filifolium* Schlechter (1894); **incl.** *Schizoglossum parvulum* Schlechter (1894); **incl.** *Schizoglossum filipes* Schlechter (1895); **incl.** *Schizoglossum monticola* Schlechter (1905); **incl.** *Schizoglossum addoense* N. E. Brown (1907); **incl.** *Schizoglossum bowkerae* N. E. Brown (1907); **incl.** *Schizoglossum buchananii* N. E. Brown (1907); **incl.** *Schizoglossum burchellii* N. E. Brown (1907); **incl.** *Schizoglossum dregei* N. E. Brown (1907); **incl.** *Schizoglossum parcum* N. E. Brown (1907); **incl.** *Schizoglossum parvulum* var. *sessile* N. E. Brown (1907).

[4] Stems 12 - 100 cm, simple or branched; **L** 10 - 80 × 0.5 - 3 mm, upper face glabrous or pubescent, lower face glabrous, midvein softly hairy; **Inf** with 2 - 13 **Fl**; **Ped** 2 - 6 mm; **Sep** 1 - 2 × 0.5 - 1 mm, softly hairy, **Ha** sometimes brown, glandular; **Cl** brownish-green, campanulate to flat, lobes 2 - 4 × 1 - 2 mm, inside glabrous, outside ± densely soft-hairy, **Ha** sometimes brown, glandular; **Cs** whitish, sometimes with purple, 1 - 3 mm, outer part of the lobes basally 0.5 - 1.5 mm, shoulders square or rounded, inner appendage erect or inclined, lobe sometimes 3-dentate, middle tooth longest; **Poll** 0.3 - 0.6 mm, corpuscle 0.1 - 0.5 mm; **Fr** 4 - 5 cm, slender, densely appressed-pubescent.

The position of the synonymized *Schizoglossum filipes* is problematical (type locality unknown).

A. grandiflorum (Schlechter) Kupicha (KB 38(4): 647, 1984). **T**: RSA, Eastern Cape (*Schlechter* 2747 [K]). – **D**: RSA (C KwaZulu-Natal, Eastern Cape); grassland.

≡ *Schizoglossum grandiflorum* Schlechter (1894); **incl.** *Schizoglossum macowanii* N. E. Brown (1907); **incl.** *Schizoglossum macowanii* var. *tugelense* N. E. Brown (1907).

[3] Stems 6 - 14 cm, 1 or occasionally 2; **L** in pairs or 3 - 5 in verticils, longer than the **Int**, lamina 18 - 45 × 4 - 14 mm, elliptic to obovate or linear, upper face roughly downy, sometimes only along the margins, lower face softly pubescent, esp. the venation; **Inf** with 4 - 7 **Fl**; **Ped** 8 - 15 mm; **Sep** 2.5 - 5 × 0.75 - 1.5 mm, softly hairy, **Ha** transparent, porrect; **Cl** jade-green, flat, lobes 7.5 - 9 × 2 - 4 mm, margins revolute, outside appressed soft-hairy, inside glabrous, basally sometimes weakly pubescent; **Cs** outside reddish with chestnut-brown dots, inside and inner appendage white, 5.5 - 8 mm, outer part of the lobes basally 2.5 - 4 mm, shoulders acute, apical appendage erect, inner appendage apically rounded, divided wrench-like, erect or slightly inclined, ± as long as the apical appendage; **Poll** 0.6 - 0.8 mm, reniform, translators stout, sometimes flattened, corpuscle 0.4 - 0.7 mm.

A. heterophyllum E. Meyer (Comm. Pl. Afr. Austr., 200, 1838). **T** [lecto]: RSA, Cape Prov. (*Drège* s.n. [not located]). – **D**: RSA (Western Cape, Eastern Cape); stony or sandy grassland near the coast, 60 - 690 m.

≡ *Schizoglossum heterophyllum* (E. Meyer) Schlechter (1893); **incl.** *Schizoglossum schinzianum* Schlechter (1893) ≡ *Schizoglossum heterophyllum* var. *schinzianum* (Schlechter) N. E. Brown (1907); **incl.** *Schizoglossum villosum* Schlechter (1894); **incl.** *Schizoglossum consimile* N. E. Brown (1907); **incl.** *Schizoglossum harveyi* N. E. Brown (1907); **incl.** *Schizoglossum heterophyllum* var. *majus* N. E. Brown (1907).

[3] Shrubs with procumbent-erect branched stems; **L** lanceolate in the basal ½ of the stems, oblong in the apical ½, pubescent; **Inf** with < 6 **Fl**, pedunculate (!); **Cl** flat, lobes 2.5 - 4.5 × 1 - 2 mm, green; **Cs** lobes basally broad, apically attenuate, untoothed; **Anth** appendage membranous; **Poll** pear-shaped.

A. hirundo Kupicha (KB 38(4): 642, 1984). **T**: Moçambique (*Torre & Correia* 16350 [LISC]). – **D**: Moçambique; *Brachystegia* woodland and inundated savanna, 30 - 500 m.

[1] Stems 40 - 82 cm, simple or branched; **L** mostly only few, lamina 45 - 85 × 1 - 1.5 mm, both faces glabrous or weakly pubescent-rough; **Inf** with 1 - 5 **Fl**; **Ped** 3 - 9 mm; **Sep** 2 × 0.75 mm; **Cl** green, flat, lobes 4 - 5 × 2 mm, outside glabrous, inside pubescent; **Cs** yellowish-green, 3.5 - 4 mm, outer part of the lobes basally ± 1 mm, without shoulders attenuating into the erect apical appendage, inner appendage ± ½ as long, slightly inclined; **Poll** 0.55 - 0.6 mm, corpuscle 0.2 mm, broader than long.

A. interruptum (E. Meyer) Bullock (KB 7: 419, 1952). **T**: RSA, Cape Prov. (*Drège* s.n. [K, BM, CGE, K, MO]). – **D**: Tropical Africa from Ivory Coast to Cameroon and S-wards to Zimbabwe (excl. Angola, Kenya), frequent in RSA.

≡ *Lagarinthus interruptus* E. Meyer (1838) ≡ *Schizoglossum interruptum* (E. Meyer) Schlechter (1896); **incl.** *Lagarinthus abyssinicus* Hochstetter ex Bentham (1876) (*nom. illeg.*, Art. 52.1); **incl.** *Schizoglossum barberae* Schlechter (1894); **incl.** *Schizoglossum altissimum* Schlechter (1895); **incl.** *Schizoglossum lasiopetalum* Schlechter (1905); **incl.** *Schizoglossum morumbenense* Schlechter (1905); **incl.** *Schizoglossum togoense* Schlechter (1905); **incl.** *Schizoglossum garuanum* Schlechter (1913); **incl.** *Schizoglossum kamerunense* Schlechter (1913); **incl.** *Schizoglossum gracile* Weimarck (1935).

[4] Stems 50 - 140 cm, always branched within the **Inf**; **L** 2 - 75 × 0.5 - 2 mm, upper face pubescent or glabrous, lower face glabrous, midvein sometimes pubescent; **Inf** with (1-) 6 - 13 **Fl**; **Ped** 2 - 10 mm; **Sep** 1 - 2 × 0.5 - 1 mm, softly hairy; **Cl** outside dull purplish-red, inside cream or yellowish, inversely campanulate, lobes 2 - 3 × 1 - 1.5 mm, margins revolute, tips bent inwards, outside softly hairy, inside glabrous or in the middle pubescent, margins and sometimes the tips softly hairy, **Ha** ≤ 1 mm; **Cs** whitish, sometimes with purple streaks, 1 - 2 mm, outer part of the lobes basally 0.75 - 1 mm, shoulders square or rounded, inner appendage ± long, triangular, beak-like, inclined, outside with or without a faint transversal crest, lobes 3-dentate, erect, middle tooth longest; **Poll** 0.3 - 0.5 mm, corpuscle 0.1 - 0.25 mm; **Fr** 5 - 6 cm, densely pubescent.

Closely related to *A. angustissimum*.

A. lamellatum (Schlechter) Kupicha (KB 38(4):

664, 1984). **T** [lecto]: RSA, Cape Prov. (*Schlechter* 605 [BOL]). – **D:** RSA (North-West Prov., Gauteng, Mpumalanga, Free State, KwaZulu-Natal, Eastern Cape), Lesotho; open grassland.

≡ *Schizoglossum lamellatum* Schlechter (1893); **incl.** *Schizoglossum bilamellatum* Schlechter (1895); **incl.** *Schizoglossum bilamellatum* var. *cordylogynoides* Schlechter (1895); **incl.** *Schizoglossum propinquum* S. Moore (1903).

[2] Stems 15 - 55 cm, unbranched; **L** mostly shorter than the **Int**, lamina 1.5 - 4.5 × 0.5 - 0.8 mm, upper face glandular-hairy, sometimes glabrescent, lower face glabrous, midvein softly hairy; **Inf** with 3 - 6 **Fl**, sometimes terminal and many-flowered on long **L**-less peduncles; **Ped** 2 - 4 mm; **Sep** 1.5 - 3 × 0.75 - 1.5 mm, glandular-hairy, **Ha** transparent and/or brown, soft; **Cl** outside pale yellow, inside green with purple, calyx-shaped, lobes 3 - 4 × 1 - 1.5 mm, erect, margins sometimes slightly revolute, inside basally sometimes delicately pubescent, outside weakly soft-hairy, **Ha** transparent and/or brownish; **Cs** white, 2 - 2.5 mm, *either* lobes simple, **Sc**-like, apically acute or emarginate, rarely trifid, inside with 2 conspicuous longitudinally arranged lobules folded in the direction of the **Fl** centre, *or* lobes ovate-acute, inner appendage triangular, inclined; **Poll** 0.35 - 0.5 mm, germination zone inconspicuous, translators not flattened, corpuscle 0.3 - 0.4 mm; **Fr** 4.5 - 7 cm, downy, without **Bri**.

A. lanatum (Weimarck) Kupicha (KB 38(4): 652, 1984). **T:** Zimbabwe, Inyanga (*Norlindh & Weimarck* 4183 [LD, BM]). – **D:** SE Tanzania, E Zimbabwe; grass- and open woodland, 200 - 1700 m.

≡ *Schizoglossum lanatum* Weimarck (1935).

[4] Stems 48 - 75 cm, simple or branched; **L** 25 - 40 × 1.5 mm, both faces glabrous, midvein appressed-pubescent; **Inf** with 3 - 11 **Fl**; **Ped** 2 - 3 mm; **Sep** ± 2.5 × 1 mm, softly hairy; **Cl** greenish, grey-brown or reddish-brown, flat, centre yellow, lobes ± 3.5 × 1.5 mm, margins revolute, both sides softly pubescent; **Cs** ± 2 mm, outer part of the lobes basally ± 1.5 mm, inner appendage erect or inclined, softly hairy like the shoulders; **Poll** ± 0.5 mm, corpuscle ± 0.25 mm.

Closely related to *A. eylesii* and *A. angustissimum*.

A. masaicum (N. E. Brown) Kupicha (KB 38(4): 656, 1984). **T:** Kenya (*Johnston* s.n. [K, BM]). – **D:** Ethiopia, Kenya, Tanzania, Uganda, Namibia, Angola, Zambia, Malawi; grassland in seasonally flooded regions.

≡ *Schizoglossum masaicum* N. E. Brown (1895); **incl.** *Schizoglossum fusco-purpureum* Schlechter & Rendle (1896); **incl.** *Schizoglossum baumii* Schlechter *ex* N. E. Brown (1902); **incl.** *Schizoglossum altum* N. E. Brown (1906); **incl.** *Schizoglossum semlikense* S. Moore (1912); **incl.** *Aspidoglossum kulsii* Cufodontis (1962).

[4] Stems (12-) 35 - 70 (-100) cm, normally branched; **L** mostly shorter than the **Int**, lamina 1 - 6 cm × 0.25 - 3 mm, both faces glabrous or weakly pubescent; **Inf** with 2 - 8 **Fl**; **Ped** 1 - 5 mm; **Sep** 1.5 - 2 × 0.5 - 1 mm, glabrous to softly hairy; **Cl** yellowish-green, dotted with rose or pale brownish-purple, or white, ± flat, lobes 2.5 - 4.5 × 1 - 2 mm, margins ± strongly revolute, tip bent inwards, outside softly hairy, rarely glabrous, inside glabrous or **Fl** centre weakly downy and/or apically with several stout **Ha**; **Cs** white, 1 - 2 mm, outer part of the lobes basally 1 mm, shoulders square or rounded, apical appendage mostly ± triangular, erect or inclined, divided from the basal part by a transversally oriented crest; **Sty** head dark purple; **Poll** (0.25-) 0.4 - 0.65 mm, corpuscle 0.2 - 0.3 mm; **Fr** 3 - 6 cm, very slender, ± densely pubescent.

A. nyasae (Britten & Rendle) Kupicha (KB 38(4): 650, 1984). **T:** Malawi (*Whyte* s.n. [BM, K]). – **D:** Zimbabwe, S Malawi, C and N Moçambique; mountainous grassland, often between rocks in scrub. **Fig. I.b**

≡ *Schizoglossum nyasae* Britten & Rendle (1894); **incl.** *Schizoglossum multifolium* N. E. Brown (1895); **incl.** *Schizoglossum leptoglossum* Weimarck (1935).

[3] Stems (36-) 45 - 110 cm, simple or branched; **L** arranged densely and ± irregularly, basally and within the **Inf** ± verticillate, lamina 40 - 85 × 4 - 13 mm, narrowly linear, linear, rarely broadly elliptic, upper face glabrous or roughly downy, lower face glabrous, midvein pubescent; **Inf** with 3 - 16 **Fl**; **Ped** 7 - 16 mm; **Sep** 4 - 6 × 1 - 1.5 mm, appressed softly-hairy, **Ha** brownish; **Cl** green, flat or inversely bowl-shaped, lobes 6.5 - 9.5 × 2.5 - 4 mm, margins often revolute, outside glabrous or appressed-hairy; **Cs** white, dotted with rose, 4 - 9 mm, outer part of the lobes basally 2 - 4 mm, shoulders acute, apical appendage erect, apically tortuous, inner appendage sometimes divided in 2 or 4, erect, shorter than or as long as the apical appendage; **Poll** 0.85 - 1 mm, corpuscle 0.45 - 0.55 mm; **Fr** 7.5 - 9 cm, fusiform, **Bri** ± 4 mm, appressed, in addition weakly pubescent.

A. ovalifolium (Schlechter) Kupicha (KB 38(4): 649, 1984). **T:** RSA, Eastern Cape (*Flanagan* 1307 [BOL, GRA, K, NBG, NH, PRE, SAM]). – **D:** C Moçambique, RSA (Eastern Cape, KwaZulu-Natal, Gauteng, Mpumalanga), Swaziland; grassland, at times near water.

≡ *Schizoglossum ovalifolium* Schlechter (1894); **incl.** *Schizoglossum striatum* Schlechter (1894); **incl.** *Schizoglossum pumilum* Schlechter (1895); **incl.** *Schizoglossum robustum* Schlechter (1895); **incl.** *Schizoglossum contracurvum* N. E. Brown (1907); **incl.** *Schizoglossum robustum* var. *inandense* N. E. Brown (1907); **incl.** *Schizoglossum robustum* var. *pubiflorum* N. E. Brown (1907).

[3] Stems 7 - 33 cm, unbranched; **L** basally irregularly arranged, verticillate within the **Inf**, mostly congested, sometimes shorter than the **Int**, lamina 14 - 35 × 0.5 - 12 mm, needle-like to oblong, broadly elliptic or ± circular, upper face glabrous, rarely roughly or softly hairy, lower face glabrous, midvein sometimes softly hairy, or both faces softly hairy; **Inf** with 2 - 3 **Fl**; **Ped** 5 - 10 mm; **Sep** 2 - 6 × 0.5 - 2 mm, softly hairy, **Ha** transparent or brownish; **Cl** green, lineate with brown or purple, flat, lobes 4 - 7 × 2 - 4 mm, margins sometimes revolute, outside softly hairy, **Ha** transparent or brownish, appressed, rarely stiff and porrect, inside glabrous or pubescent, sometimes only the margins; **Cs** white with rose, (2-) 3 - 7 mm, outer part of the lobes basally 1 - 3 mm, shoulders acute, rarely rounded or sloping, apical appendage erect, inner appendage rarely bifid, much shorter than the apical appendage, sometimes inclined; **Poll** 0.45 - 1 mm, corpuscle 0.2 - 0.5 mm; **Fr** ± 4 cm, **Bri** ± 4 mm, appressed, otherwise crisply pubescent.

A. restioides (Schlechter) Kupicha (KB 38(4): 665, 1984). **T:** RSA, Western Cape (*Schlechter* 740 [BOL [icono]]). – **D:** RSA (North-West Prov., Gauteng, Mpumalanga, Western Cape, Eastern Cape); mountainous grassland.

≡ *Schizoglossum restioides* Schlechter (1893); **incl.** *Schizoglossum pallidum* Schlechter (1895); **incl.** *Schizoglossum randii* S. Moore (1903).

[2] Stems 50 - 70 cm, simple, rarely branched; **L** shorter than the **Int**, lamina 10 - 30 × 0.5 - 1 mm, both faces glabrous or glandular-hairy; **Inf** with 2 - 5 **Fl**; **Cl** ± campanulate, lobes 2.5 - 3.5 × 1 - 1.5 mm, coriaceous, margins strongly revolute, outside weakly glandular-hairy, **Ha** brown, soft; **Cs** 2 - 4 mm, outer part of the lobes basally ovate or oblong, apically mostly bifid, inner appendage linear, inclined or reclined; **Poll** 0.45 - 0.6 mm, reniform, without visible germination zone, translators not flattened, corpuscle 0.5 - 1 mm, glistening, conspicuous, broader than the **Poll**; unripe **Fr** pubescent, **Ha** dark brown.

A. rhodesicum (Weimarck) Kupicha (KB 38(4): 663, 1984). **T:** Zimbabwe (*Fries & al.* 3441 [LD, BM, BR, PRE, SRGH]). – **D:** E Zimbabwe, frontier to Moçambique; open mountainous grassland.

≡ *Schizoglossum rhodesicum* Weimarck (1935).

[2] Stems 17 - 50 cm, unbranched; **L** mostly shorter than the **Int**, lamina 25 - 55 × 1 - 2 mm, upper face roughly or softly hairy, lower face glabrous, midvein softly hairy; **Inf** with 2 - 12 **Fl**; **Ped** 2 - 6 mm; **Sep** 1.5 - 2.5 × 1 mm, softly hairy, **Ha** transparent; **Cl** yellowish- or brownish-green, ± flat, lobes 3 × 2 mm, margins often strongly revolute, outside weakly soft-hairy, **Ha** transparent; **Cn** consisting of 2 alternating series; **Cs** lobes ± 1 mm, 3-lobed, inside continuing into a hood-like appendage, alternating series of lobes more central, lobes narrow, subtending the **Cs**; **Poll** ± 0.3 mm, broadly triangular or square, germination zone conspicuous, translators flattened, corpuscle 0.2 - 0.25 mm, **Fr** ± 2.5 cm, densely woolly-hairy, bristly.

Closely related to *A. delagoense*.

A. uncinatum (N. E. Brown) Kupicha (KB 38(4): 643, 1984). **T:** RSA, Eastern Cape (*Hutton* s.n. [K, BOL]). – **D:** RSA (Eastern Cape); grassland.

≡ *Schizoglossum uncinatum* N. E. Brown (1907).

[1] Stems 9 - 18 cm, simple or weakly branched; **L** mostly shorter than the **Int**, lamina 11 - 34 × 4 - 18 mm, linear, broadly lanceolate or hastate, upper face downy, lower face softly hairy, esp. the venation; **Inf** with 1 - 7 **Fl**; **Ped** 2 - 8 mm; **Sep** 2 - 3.5 × 1 mm, porrectly hairy; **Cl** greenish, flat, lobes 3 - 4 × 1.5 mm, margins revolute, outside glabrous; **Fl** centre porrectly hairy, **Ha** stiff, inside glabrous; **Cs** 2 mm, lobes erect, apically triangular, inner appendage somewhat longer than the outer part of the lobe, inclined, tortuous, apically hooked; **Poll** 0.45 - 0.55 mm, without visible germination zone, corpuscle 0.35 - 0.4 mm.

A. validum Kupicha (KB 38(4): 650, 1984). **T:** RSA, Mpumalanga (*Buitendag* 1032 [NBG]). – **D:** RSA (Mpumalanga: Drakensberge); grassland.

[3] Stems 20 - 100 cm, branched or not; **L** basally dense and irregularly arranged, verticillate within the **Inf**, lamina 18 - 40 × 1.5 - 9 mm, narrowly linear to oblong, broader **L** with undulate margins, upper face rough, lower face glabrous, midvein sometimes rough; **Inf** with 4 - 6 **Fl**; **Ped** 4 - 10 mm; **Sep** 3.5 - 5 × 1.5 mm, softly hairy, **Ha** delicate, appressed; **Cl** dark brown or dark purple, flat, lobes 8 - 9 × 3 - 4.5 mm, outside glabrous or softly hairy, inside hairy, **Ha** of the inside transparent, porrect; **Cs** white, 6.5 - 7.5 mm, outer part of the lobes basally 3 mm, shoulders acute, apical appendage erect, apically tortuous, inner appendage ± ½ as long, bifid, erect; **Poll** 1 - 1.25 mm, corpuscle 0.4 - 0.85 mm.

A. virgatum (E. Meyer) Kupicha (KB 38(4): 641, 1984). – **D:** RSA (Eastern Cape, KwaZulu-Natal: Loteni Reservation); grassland, often near water, 300 - 1650 m.

≡ *Lagarinthus virgatus* E. Meyer (1838) ≡ *Schizoglossum virgatum* (E. Meyer) Schlechter (1894).

[1] Stems 35 - 78 cm, simple or branched; **L** longer than the **Int**, often clustered, lamina 25 - 38 × 0.5 - 3 mm, upper face glabrous, downy or softly hairy, lower face glabrous, midvein pubescent; **Inf** with 4 - 9 **Fl**; **Ped** 1 - 3 mm; **Sep** 1.5 - 2 × 0.5 mm, softly hairy; **Cl** dark brown, lobes 2 - 2.5 × 1 mm, outside softly hairy, inside pubescent; **Cs** chestnut-brown, ± 1 mm, apical appendage of the lobe triangular, inner appendage somewhat broader, apically sometimes bifid, inclined; **Poll** 0.4 mm, corpuscle 0.2 - 0.25 mm, narrow.

A. woodii (Schlechter) Kupicha (KB 38(4): 651, 1984). **T:** RSA, KwaZulu-Natal (*Tyson* 2166 [BOL, BM, K, NBG, SAM]). — **D:** RSA (C KwaZulu-Natal); grassland.

≡ *Schizoglossum woodii* Schlechter (1895).

[4] Stems 56 - 140 cm, usually weakly branched; **L** 4 - 10 cm × 1 - 2 mm, both faces glabrous; **Inf** with 1 - 7 **Fl**; **Ped** 1 - 5 mm; **Sep** 3 - 4 × 1.5 - 2 mm, with long soft **Ha**; **Cl** brown with red and/or green, cup-shaped, lobes 3 - 4 × 1.5 - 2 mm, margins revolute, tips bent inwards, outside with long soft **Ha**, inside glabrous, margins of lobes softly hairy, **Ha** conspicuously longer towards the tips; **Cs** yellowish, 1.5 - 3 mm, outer part of the lobes basally 1 - 1.5 mm, shoulders sloping, inner appendage mostly inclined, laterally lobe-like and folded wing-like; **Poll** 0.5 - 0.63 mm, germination zone inconspicuous, corpuscle 0.2 - 0.25 mm; **Fr** 4 - 4.5 cm, narrowly fusiform, apically elongated into a long straight beak; **Se** dark brown, 3 mm, compressed-ovoid, wrinkled, **Ha** 2 cm.

Closely related to *A. interruptum*.

A. xanthosphaerum Hilliard (Notes Roy. Bot. Gard. Edinburgh 45(2): 181, 1989). **T:** RSA, KwaZulu-Natal (*Hilliard & Burtt* 18528 [E, NU]). — **D:** RSA (KwaZulu-Natal); only known from the type locality.

[2] **R** tuber 2 × 1 cm, pear-shaped; stems 5.5 - 12.5 cm, simple, **Ha** coarse, ± bent; **L** in 4 - 5 pairs, ± sessile, clearly ascending, lamina ≤ 25 × 6 mm, longer than the **Int**, lanceolate to linear, acute, basally truncate; **Inf** dense terminal pseudo-umbels; **Ped** 2 - 4 mm; **Sep** 2.5 × 1.5 mm, lanceolate; **Cl** canary-yellow, ± bowl-shaped, tube ± 0.5 mm, lobes ± 2.75 × 1.75 mm, ovate, erect, upper ½ of the lobes outside softly hairy, otherwise glabrous; **Cn** canary-yellow; **Cs** lobes ± 2 × 1.5 mm, ± as tall as the **Gy**, apical appendage of the lobes rhomboid-triangular, apical margin entire or emarginate, slightly overtopping the lobe, falcately bent and touching the lobe margin; **Anth** appendages ± 0.5 × 0.75 mm, almost circular, bent over the margin of the **Sty** head; **Gy** 1.5 mm, **Sty** head 1 mm ⌀, compressed, truncate; **Poll** 0.5 mm, bean-shaped, attached subterminally, corpuscle ± 0.2 mm.

Closely related to *A. lamellatum*.

ASPIDONEPSIS

B. Müller, P. Stegemann & F. Albers

Aspidonepsis Nicholas & Goyder (BT 22(1): 24, 1992). **T:** *Gomphocarpus diploglossus* Turczaninov. — **Lit:** Nicholas & Goyder (1992); Nicholas & Goyder (1993). **D:** SE RSA. **Etym:** Gr. 'aspis, aspidos', shield; and Gr. 'anepsia', cousin; for the close relationship with the genus *Aspidoglossum* (Asclepiadaceae).

Erect perennials with deciduous herbaceous above-ground parts, **R** tuber globose, fusiform or turnip-shaped; stems solitary (*A. flava* rarely up to 3), delicate, ≤ 65 cm, hairy or glabrous, with latex; **L** opposite, petiolate or sessile, petiole ≤ 5 mm, lamina erect, linear, lanceolate or narrowly elliptic, older **L** shorter and broader, acute or slowly attenuate; **Inf** extra-axillary or terminal, with **Bra**, lateral **Inf** pedunculate; **Ped** 5 - 15 (-21) mm; **Sep** lanceolate, triangular or narrowly ovate, acute, hairy, rarely glabrous; **Cl** yellow, green, brown or purple, ± flat or (inversely) bowl-shaped, lobes elliptic to ovate, outside glabrous or hairy, inside glabrous; **Cs** lobes basally 0.5 - 1.8 mm above the **Cl**, forming a cup- or hood-shaped pouch, with or without an inner appendage, glabrous; **Anth** winged, membranous, reniform, rounded or rhomboid, apically shallowly to deeply divided, placed against or incumbent on the **Sty** head; **Sty** head swollen, apically obtuse, concave, margins undulate; **Poll** clavate, flattened, corpuscle mostly ovate; **Fr** furred when young, fusiform, with attenuate tip; **Se** unknown.

Closely related to *Aspidoglossum*, but the inflorescences appear near the stems tips and are ± pedunculate, and the **Cs** lobes form conspicuous central pouches with a central appendage within the hollow part (*Aspidoglossum* with sessile inflorescences over the whole stem length, and **Cs** lobes different). The genus is divided into 2 subgenera according to Nicholas & Goyder (1992):

[1] Subgen. *Aspidonepsis*: **Inf** with (1-) 2 - 17 **Fl**; **Cl** yellow or green, ± flat or bowl-shaped; **Cs** lobes cup- or bowl-shaped, mostly with an inner appendage, which is raised above the pouch.

[2] Subgen. *Unguilobium* Nicholas & Goyder 1993: **Inf** with 4 - 11 **Fl**; **Cl** purple, brown or yellow, inversely bowl-shaped; **Cs** claw-like, without inner appendage.

A. cognata (N. E. Brown) Nicholas & Goyder (BT 22(1): 30, ills. (p. 31), 1992). **T:** Transkei (*Schlechter* 6469 [K, BOL, NH, PRE]). — **D:** RSA (Eastern Cape, KwaZulu-Natal: Drakensberge); mountainous grassland.

≡ *Asclepias cognata* N. E. Brown (1908).

[1] **R** tuber ± 7 mm ⌀, round; stem 18 - 55 cm; **L** (7-) 11 - 68 × (0.3-) 0.7 - 4 (-6) mm, apically acute, basally often shortly petiolate, sometimes keeled; **Inf** 1 - 2, with 1 - 7 (-9) **Fl**; peduncle 0.3 - 7.6 (-9.2) cm; **Ped** 6 - 12 mm; **Sep** 3 - 5 × 1 - 1.8 mm; **Cl** 5 - 12 × 7 - 17 mm, bowl-shaped, lobes (narrowly) elliptic, occasionally ovate, (6-) 7.5 - 10.5 × 2.5 - 6 mm, glabrous; **Cn** stipitate for 1.5 - 2 mm; **Cs** lobes 3 - 5 (-5.5) × 1.3 - 2.5 mm, ± rectangular, apically obtuse, overtopping the **Sty** head for 0.6 - 1 mm, sides bent inwards, each drawn out into a short lateral appendage (0.6-) 0.8 - 1.2 mm long, subulate or arm-like, ± incumbent on the **Sty** head, pouch 0.8 - 1.3 mm, inner appendage yellow, lingulate, overtopping the margin of the lobe for 0.8 - 1.3 mm;

Anth 0.3 - 0.6 × 0.8 - 1.3 mm, reniform, apically shallowly emarginate, placed against the side of the **Sty** head, wings 0.8 - 1.4 × 0.4 - 0.6 mm; **Sty** head 1.5 - 2.8 mm ∅; **Poll** 0.7 - 1 × 0.25 - 0.3 mm, translator 0.3 - 0.6 mm, with hook-like wings, corpuscle 0.2 - 0.3 × 0.1 mm.

A. diploglossa (Turczaninov) Nicholas & Goyder (BT 22(1): 26, ills. (p. 28), 1992). **T:** RSA, Eastern Cape (*Ecklon* 23 [KW, PRE]). – **D:** RSA (Eastern Cape, KwaZulu-Natal: Drakensberge); mountainous grassland in moist regions, 1500 - 2400 m.

≡ *Gomphocarpus diploglossus* Turczaninov (1848) ≡ *Asclepias diploglossa* (Turczaninov) Druce (1917); **incl.** *Asclepias schizoglossoides* Schlechter (1894).

[1] **R** tubers 1 or several in a chain, each 9 - 35 × 6 - 12 mm, turnip-shaped; stems 17 - 40 (-50) cm, biseriately pubescent; **L** sessile or petiolate for up to 4 mm, lamina narrowly lanceolate, sometimes falcate, rarely linear or narrowly elliptic, 0.5 - 8.4 (-13) cm × (0.3-) 0.5 - 7 mm, apically long-attenuate, basally sometimes cuneate; **Inf** 1 - 3, with 4 - 16 **Fl**; peduncle ≤ 9.5 mm; **Ped** 6 - 16 mm; **Sep** 2.5 - 4.6 × 1 - 1.5 mm, lanceolate, sometimes triangular or narrowly ovate, apically acute, ± densely pubescent; **Cl** yellow or yellow-brown, 4 - 9 × 6 - 13 mm, shallowly bowl-shaped, incised to the base, lobes 4 - 6 (-7) × 2.4 - 4.1 mm, ovate, sometimes elliptic, outside weakly silky-hairy; **Cn** 0.5 mm stipitate; **Cs** (pale) yellow, **Cs** lobes 4 - 6 (-7) × 2.4 - 4.1 mm, rectangular, apically obtuse or rounded, ± as tall as the **Sty** head, sides bent inwards, apically rounded, sometimes drawn out into short acute appendages incumbent on or slightly overtopping the **Sty** head, pouch bag-like, inner appendage ± 3.5 × 0.2 - 0.8 mm, lingulate or oblong-triangular; **Sty** head pale green to white, 1.1 - 2.1 mm ∅; **Anth** 0.3 - 0.6 × 0.6 - 0.9 mm, rhomboid, apically cleft; **Poll** 0.7 - 0.8 (-0.85) × 0.25 - 0.35 mm, translators 0.2 - 0.3 (-0.35) mm, narrow, wings transparent, narrow, hook-like, corpuscle (0.2-) 0.3 × 0.1 - 0.15 mm.

A. flava (N. E. Brown) Nicholas & Goyder (BT 22(1): 27, ills. (p. 28), 1992). **T:** Transkei (*Tyson* 1086 [K, BOL, SAM]). – **D:** Transkei, RSA (Eastern Cape near Grahamstown, KwaZulu-Natal); mountainous grassland, (450-) 600 - 2000 m.

≡ *Asclepias flava* N. E. Brown (1908).

[1] **R** tuber globose, 6 - 10 × 4 - 9 mm, sometimes several in a chain; stems solitary (rarely up to 3), slender, 18 - 47.5 cm; **L** lanceolate to narrowly elliptic, 7 - 83 × 0.5 - 6 (-7) mm, apically acute, basally petiolate to cuneate; **Inf** 1 - 3 (-6), with 4 - 18 (-24) **Fl**; peduncle (0.4-) 1 - 17.5 cm; **Bra** 2.6 - 5.3 (-7.5) × 0.2 - 0.5 mm; **Ped** 5 - 11 mm; **Sep** 2 - 3.6 (-4) × 0.7 - 1.2 mm; **Cl** 3 - 5 (-6) × 5 - 8 mm, bowl-shaped, incised to the base, lobes 3.5 - 5 × 2 - 3.2 mm, ovate, sometimes elliptic, margins sometimes slightly revolute; **Cs** lobes 0.5 - 0.8 mm above the **Cl**, 1 - 1.6 mm, broadened, bowl-shaped, marginally like a boxing glove, sides bent inwards, each elongating into a short appendage, this 0.25 - 0.7 mm, subulate or arm-like, bent inwards and crossing in the centre, basally placed against the sides of the **Sty** head, pouch funnel-shaped, 0.4 - 0.7 mm, inner appendage sausage-like, overtopping the pouch margins for 0.4 - 0.7 mm; **Sty** head 1.1 - 1.6 mm ∅; **Anth** 0.2 - 0.5 × 0.5 - 0.8 mm, reniform, apically shallowly emarginate; **Poll** 0.5 - 0.7 × 0.15 - 0.25 mm, translators 0.2 - 0.3 mm, thickened at the point of insertion, corpuscle 0.15 - 0.25 × 0.06 - 0.1 mm, with hooks.

A. reenensis (N. E. Brown) Nicholas & Goyder (BT 23(2): 237, 1993). **T:** RSA, KwaZulu-Natal (*Wood* 8635 [K, GRA, NH, PRE, SAM]). – **D:** RSA (KwaZulu-Natal: Drakensberge); dry montane grassland, sandy soils.

≡ *Asclepias reenensis* N. E. Brown (1908).

[2] **R** tuber 17 - 25 (-41) × 7 - 14 mm, turnip-shaped; stem solitary, 24 - 52 (-62.5) cm, glabrous; **L** sessile or petiolate for ≤ 1 mm, lamina linear, 10 - 56 × 0.7 - 2.5 (-4) mm, apically acute, basally cuneate; **Inf** 1 - 3 (-4), with (1-) 4 - 8 **Fl**; peduncle 0.9 - 7.5 cm; **Bra** 2.5 - 5.9 × 0.25 - 0.5 mm; **Ped** 9 - 15 (-21) mm; **Sep** 2.7 - 4.5 × 1 - 1.7 (-2.5) mm, outside hispid; **Cl** 4 - 7 × 7 - 11 mm, inversely bowl-shaped, lobes narrowly elliptic to ovate, 5.5 - 6.5 × 2.5 - 3.8 mm, outside pubescent to hispid (esp. basally); **Cn** 0.8 - 1 mm stipitate; **Cs** lobes (1.6-) 2.2 - 2.6 × 1.3 - 1.8 mm, ± rectangular, apically with an appendage ± 0.5 mm long, arm-like, bent into the pouch, sides bent inwards, tips drawn out into short appendages, these 0.4 - 1.3 × 0.4 - 1 mm, falcate, arm-like, incumbent on the **Sty** head; **Sty** head 1.8 - 2.4 mm ∅; **Anth** white, 0.8 - 1.1 × 0.3 - 0.5 mm, rounded, wrinkled or apically deeply cleft; **Anth** wings shaped like an ear lobule; **Poll** 0.85 - 1 × 0.2 - 0.3 mm, translators 0.45 - 0.65 mm, thin, transparent, corpuscle 0.3 - 0.35 (-0.4) × 0.1 - 0.2 mm.

A. shebae Nicholas & Goyder (BT 22(1): 33-34, ills., 1992). **T:** RSA, Mpumalanga (*Forrester & Gooyer* 216 [PRE]). – **D:** RSA (Mpumalanga); mountainous grassland.

[2] **R** tuber ± 15 × 7 mm, ± turnip-shaped; stems solitary, 19 - 34 cm; **L** sessile or to 5 mm petiolate, lamina linear, occasionally lanceolate, 7 - 44 × 1 - 4 mm, apically acute; **Inf** 1 - 2, with (2-) 4 - 11 **Fl**; peduncle (0.5-) 1.9 - 9 cm; **Ped** 1 - 1.5 cm, porrectly short-hairy; **Sep** 3.4 - 3.6 × 1.1 - 1.3 mm; **Cl** inversely campanulate, 4 - 6.5 × 6 - 8 mm, lobes 5.1 - 5.8 × 3 - 3.6 mm, ovate, rarely elliptic, outside pubescent; **Cs** lobes 1.8 - 3 × 2 - 2.1 mm, ± rectangular, apically obtuse, sides bent inwards, drawn out into short appendages, these falcate or subulate, acute, incumbent on the **Sty** head, pouch shallow, groove-like, ± 0.9 mm; **Sty** head 1.6 - 1.8 mm ∅; **Anth** ± 0.5 × 0.7 mm, rounded, apically deeply

cleft, wings ± 0.7 × 0.4 - 0.45 mm, ear-like; **Poll** 0.7 - 0.75 × 0.3 mm, translators 0.3 - 0.4 mm, narrow, corpuscle 0.2 - 0.25 × 0.1 - 0.12 mm.

BAYNESIA

U. Meve

Baynesia Bruyns (Novon 10: 354-356, ills., 2000). **T:** *Baynesia lophophora* Bruyns. − **D:** Namibia. **Etym:** For its occurrence in the Baynes Mts., NW Namibia, which in turn are named for Maudsley Baynes, English explorer who first investigated the area in 1911.

Clump-forming stem succulents with erect stems; stems 3 - 8 × 0.6 - 1.2 cm, 4-angled, green, soft, slightly papillate; **Tu** conical, laterally flattened; **L** rudiments caducous, 1 - 1.5 mm, cordate, flattened-conical, acute; **Inf** near stem tips, sessile, bracteolate, 1- to 3- (to 5-) flowered; **Ped** ± 2 mm; **Sep** 1.5 mm, lanceolate; **Cl** outside greenish, inside deep maroon, 3 - 4 × 6 - 8 mm, campanulate; **Cl** tube bowl-shaped, ± 1 × 3 mm; **Cl** lobes deltoid, ± 2 × 2 mm, erect, coarsely papillate along the sharply edged midrib; **Cn** biseriate, transparent to maroon, ± 1 × 2 mm; **Ci** basally fused to form a shallow bowl, **Ci** lobes deltoid, acute, ascending, fusing with the wing-like extended base of the **Cs**; free **Cs** lobes ± 0.5 mm, oblong, obtuse, decumbent on the rectangular **Anth**; **Poll** yellow, ellipsoid, ± 0.17 × 0.3 mm; **Fr** narrowly fusiform, 25 - 35 × 2.5 - 3.5 mm; **Se** pale brown, 5 × 1.5 - 2 mm.

The nearest relatives of this most-recently described and monotypic stapeliad genus from NW-most Namibia are probably *Caralluma peschii* and *Pectinaria*.

B. lophophora Bruyns (Novon 10: 354-356, ills., 2000). **T:** Namibia (*Bruyns* 8000 [BOL, K, MO, PRE, WIND]). − **D:** NW Namibia.

Description as for the genus.

BRACHYSTELMA

U. Meve

Brachystelma Sims (CBM 49: t. 2343 + text, 1822). **T:** *Stapelia tuberosa* Meerburg. − **Lit:** Dyer (1980); Walker (1982); Dyer (1983); Forster (1985); Craib (1994); Meve & Liede (2001a). **D:** Africa, India, Sri Lanka, Myanmar, Malaysia, Thailand, Philippines, Papua New Guinea, N Australia. **Etym:** Gr. 'brachys', short; and Gr. 'stelma', crown, garland, wreath; for the nature of the corona.
- **Incl.** *Microstemma* R. Brown (1810) (*nomen rejiciendum*, Art. 56.1). **T:** *Microstemma tuberosum* R. Brown.
- **Incl.** *Eriopetalum* Wight (1834). **T:** not typified.
- **Incl.** *Tenaris* E. Meyer (1837). **T:** *Tenaris rubella* E. Meyer.
- **Incl.** *Macropetalum* Burchell *ex* Decaisne (1844). **T:** *Macropetalum burchellii* Decaisne.
- **Incl.** *Decaceras* Harvey (1863). **T:** *Decaceras huttonii* Harvey.
- **Incl.** *Dichaelia* Harvey (1868). **T:** *Dichaelia gerrardii* Harvey.
- **Incl.** *Micraster* Harvey (1868). **T:** *Micraster pulchellus* Harvey.
- **Incl.** *Lasiostelma* Bentham (1876). **T:** *Lasiostelma sandersonii* Oliver.
- **Incl.** *Craterostemma* K. Schumann (1893). **T:** *Craterostemma schinzii* Schumann.
- **Incl.** *Tapeinostelma* Schlechter (1893). **T:** *Tapeinostelma caffrum* Schlechter.
- **Incl.** *Brachystelmaria* Schlechter (1895). **T:** *Dichaelia natalense* Schlechter [Typification according to Liede & Albers, Taxon 43: 225, 1994.].
- **Incl.** *Aulostephanus* Schlechter (1896). **T:** *Aulostephanus natalensis* Schlechter.
- **Incl.** *Kinepetalum* Schlechter (1913). **T:** *Kinepetalum schultzei* Schlechter.
- **Incl.** *Blepharanthera* Schlechter (1914). **T:** *Blepharanthera dinteri* Schlechter [Typification according to Liede & Albers, Taxon 43: 225, 1994.].
- **Incl.** *Siphonostelma* Schlechter (1914). **T:** *Siphonostelma stenophyllum* Schlechter.

Perennial hairy summer-green geophytic herbs with hypocotyl- or **R**-tubers or thickened fleshy **R**, tubers often edible, with white or yellow tissue; stems 1 to several, procumbent to erect, very rarely twining, branched or not, mostly annual, 2 - 100 cm; **L** petiolate or sessile, linear to ovate, entire, minute to large, stipular rudiments mostly present, flat or glandular; **Inf** extra-axillary, rarely terminal, sessile, rarely shortly pedunculate, 1- to many-flowered, ± pseudo-umbellate, often in pairs or with solitary successively appearing **Fl**; **Ped** 3 - 50 mm; **Sep** ovate-lanceolate to subulate; **Cl** ± 5 - 60 mm ⌀, with flat, campanulate or cup-shaped tube; **Cl** lobes mostly flat and rotately spreading, rarely reflexed or erect and coherent at the tips to form a cage-like structure; **Cn** biseriate but sometimes uniseriate, sessile or stipitate; **Ci** developed into 5 lobes from a bowl- to cup-like basally fused structure, lobes often bifid and hairy, rarely almost reduced; **Cs** lobes mostly rectangular, incumbent on the **Anth**, rarely long attenuate and erect; **Anth** rectangular, placed on the **Sty** head (rarely erect); **Poll** subquadrate, ovoid or pear-shaped; **Fr** narrowly or thickly fusiform; **Se** brown to black, conspicuously winged.

Apart from *Ceropegia*, this is the largest genus of the *Ceropegieae* and counts some 120 species (all succulent and here covered). *Brachystelma* predominantly inhabits savannas, open forests or sand-filled pockets of azonal places such as rock outcrops, inselbergs etc. They are in general dwarf with short, erect or decumbent growth, from seasonally very wet to very dry places. All species are

perennial geophytes with annual stems. (Sub-) succulent leaves are sometimes present. *Brachystelma* has a disjunct distribution with a centre in S Africa. The genus is absent from Madagascar, N Africa and Arabia, but occurs with 12 species in India and Sri Lanka. Only few species are found in E Asia and a single one occurs in N Australia and Papua New Guinea. A generic revision was never produced and no infrageneric taxa have been proposed. Local endemism is frequent. Recent molecular studies (Meve & Liede 2001a) support the inclusion (here accepted) of *Macropetalum* and *Tenaris* in *Brachystelma*, as earlier advocated by Peckover (1996). An informal classification can be based on the 2 different fruit forms, narrowly and thickly fusiform, respectively. Often, indications of fruit shape are lacking, and it is thus more practical here to use a superficial classification dependent on the form of root succulence:
[1] **R** fleshy, cylindrical to fusiform.
[2] **R** tuberous.

B. albipilosum A. Percy-Lancaster *ex* Peckover (CSJA 69(3): 154-156, ills., 1997). **T:** Zimbabwe (*Peckover* 256 [PRE]). – **D:** E Zimbabwe. **I:** Percy-Lancaster (1989: 74, as *B. sp.* No. 3).

[2] **R** tuber depressed-globose, 2.5 - 6 × 2 cm; stems erect or ascending, grass-like, 1 or several, to 10 cm, laxly tomentose; **L** ± sessile, linear-lanceolate, 6 - 10 cm × 2 - 5 mm, glabrous; **Inf** ± sessile, 1-flowered; **Ped** 2 - 5 cm, ± horizontal, ± pubescent; **Sep** triangular, 1.5 mm; **Cl** ± 7 mm ⌀, inside black-green to purple, outside grey-green, tomentose; **Cl** tube short, ± 1.5 × 4 mm; **Cl** lobes triangular-ovate, 3 - 3.5 mm, ascending, margins revolute, inside shaggy whitish tomentose and papillose; **Cn** ± sessile, shallowly bowl-shaped, ± 3 mm ⌀, purple-red; **Ci** basally pouch-like, apically linear, erect, deeply bifid into subulate, laterally divergent appendages, 0.7 mm, ± papillose-pubescent; **Cs** lobes short, ± pillow-shaped; **Fr** erect, linear-subulate, 4.5 - 7.5 cm × 3 mm; **Se** 10 × 7 mm.

Small grass-like plants, very closely related to *B. arnotii*.

B. alpinum R. A. Dyer (JSAB 43(1): 18, 1977). **T:** Lesotho (*Bayliss* 819 [PRE]). – **D:** Lesotho. **I:** Dyer (1983).

[2] **R** tuber to 3 cm ⌀; stems 1 - 2, decumbent-ascending, scattered-pubescent; **L** 5 - 8 mm petiolate, lamina broadly ovate to narrowly lanceolate, 10 - 20 × 5 - 8 mm, margin mostly papillate; **Inf** sessile, 1- (to 2-) flowered; **Ped** 1 cm, slender; **Sep** lanceolate, 3 mm; **Cl** 12 - 15 mm ⌀; **Cl** tube broadly campanulate, 2 mm deep, inside yellowish with purple stripes, outside slightly wrinkled; **Cl** lobes 4 - 4.5 × 3 mm, inside ± purple and long-hairy, apical margins revolute; **Cn** ± sessile, ± 3 mm ⌀, fused to form a shallow cup; **Ci** lobes pouch-like, broadly emarginate in U-shaped fashion; **Cs** lobes linear-spatulate, placed on the **Anth**, dorsally much thickened; **Poll** ± globose, ± 0.25 mm.

Consistently distinguished from *B. petraeum* (see there) only in the different habit and the deeper U-shaped emarginate interstaminal corona.

B. arenarium S. Moore (J. Bot. 50: 365-366, 1912). **T:** Angola (*Gossweiler* 2312 [BM]). – **D:** Angola.

[2] **R** tuber ± globose; stems erect, to 20 cm; **L** shortly petiolate, ± linear, 30 - 60 × 3 - 8 mm, slightly pubescent; **Inf** sessile, 3- to 5-flowered; **Ped** 7 - 10 mm; **Cl** ± 7 mm ⌀, outside pubescent, inside black-brown; **Cl** tube campanulate, ± 6 × 6 mm; **Cl** lobes triangular, ± 3 mm; **Cn** shortly stipitate; **Ci** pouch-like, erect, 0.8 mm, bifid into triangular appendages; **Cs** lobes spatulate-lingulate, ± 3 mm, apically inflexed; **Poll** broadly ovoid, ± 0.4 × 0.2 mm

Only known from the type and perhaps belonging to the *B. decipiens*-complex. The long spatulate staminal corona is unusual.

B. arnotii Baker (Refug. Bot. 1869: t. 9 + text, 1869). **T:** RSA (*Arnot* s.n. [K]). – **D:** Botswana, Namibia, RSA (North-West Prov.). **I:** Dyer (1983); Bruyns (1984).

≡ *Decaceras arnotii* (Baker) Schlechter (1894); **incl.** *Brachystelma grossartii* Dinter (1914).

[2] **R** tuber ± globose, 5 - 6 cm ⌀; stems erect, 1 or several, 7 - 20 cm, pubescent; **L** shortly petiolate, lamina linear, ovate to lanceolate, to 45 mm, lower face pubescent; **Inf** sessile or pedunculate, 2- to 6-flowered; **Ped** 1 - 4 cm, often bent downwards, hairy; **Sep** lanceolate, 1.5 mm, pubescent; **Cl** 6 - 10 mm ⌀, outside pubescent, inside glabrous, black to red-brown; **Cl** tube shallow, ± 3.5 mm ⌀; **Cl** lobes triangular, 2 - 3 × 1.2 mm, strongly reflexed; **Cn** sessile, bowl-shaped, greenish; **Ci** basally pouch-like, apically subquadrangular, erect, incised to deeply bifid into subulate laterally divergent appendages, ± hairy; **Cs** lobes short, pillow-like, laterally fused with the **Ci**; **Anth** somewhat ascending, ± freely visible.

Very closely related to *B. minimum* and *B. huttonii*, but equally to *B. dinteri* and other species from S Africa.

B. attenuatum (Wight) Hooker *fil.* (Fl. Brit. India 4: 65, 1885). **T:** India (*Royle* s.n. [K]). – **D:** NW India.

≡ *Eriopetalum attenuatum* Wight (1834).

[2?] **R** unknown; stems erect; **Inf** 1(?)-flowered; **Cl** ± 17 mm ⌀; **Cl** lobes linear-filiform, 4 - 5× as long as the **Cl** tube, long hairy; **Cn** lobes with short free teeth.

An insufficiently known taxon.

B. australe R. A. Dyer (JSAB 43(1): 12, 1977). **T:** RSA, Eastern Cape (*Leistner* 3504 [PRE]). – **D:** RSA (Eastern Cape, KwaZulu-Natal). **I:** Dyer (1983).

[2] **R** tuber irregularly globose to depressed-globose, 2 - 5 cm ⌀; stems 1 to several, ascending-decumbent, 7 - 15 (-20) cm, scatteredly pubescent; **L** shortly petiolate, elliptic to narrowly lanceolate, 10 - 30 (-40) × 3 - 8 mm, partly somewhat hairy; **Inf** sessile, 1-flowered; **Ped** 5 - 10 mm, slender; **Sep** lanceolate-subulate, 4 mm; **Cl** 15 - 22 mm ⌀, somewhat fleshy, inside glabrous or hairy; **Cl** tube broadly campanulate to bowl-shaped, 2 - 3 × 5 - 7 mm, inside yellowish, with broken circular purple stripes; **Cl** lobes 7 - 11 mm, ± horizontally spreading, lanceolate from a triangular base, apically purple-brown, margins revolute; **Cn** ± sessile, purple, ± 2 × 3 mm, fused to form a shallow cup; **Ci** lobes pouch-like, apically emarginate in U- to V-shaped fashion; **Cs** lobes linear-spatulate to lanceolate, incumbent on the **Anth** or ascending; **Poll** ovoid, ± 0.3 mm, germination mouth broad, subapical.

A very variable complex in respect to size and hairiness. Intermediate forms to other microspecies such as *B. remotum*, *B. modestum*, *B. alpinum*, *B. petraeum*, *B. pulchellum* etc. occur. *B. australe* is a characteristic species on sandstone.

B. barberae Harvey *ex* Hooker *fil.* (CBM 1866: t. 5607 + text, 1866). **T:** RSA, Transkei (*Barber* s.n. [K [icono]]). – **D:** Botswana, Zimbabwe, RSA (Eastern Cape, KwaZulu-Natal, Mpumalanga, Northern Prov., North-West Prov.). **I:** Dyer (1983).

[2] **R** tuber depressed-globose, to 7 - 20 cm ⌀; stem 1 to few, dwarf, erect to decumbent, clustered, to 10 cm, roughly hairy; **L** oblong, obovate to lanceolate, basally tapering into a short petiole, to 10 × 2.5 cm, densely hairy; **Inf** subterminal in 2 opposite head-like umbels, with up to 25 simultaneously opening intensely stinking **Fl**; **Ped** 1 - 2 cm, pubescent; **Sep** subulate, to 1 cm, pubescent; **Cl** 2 - 4.5 cm long, outside greenish, pubescent, inside red-brown or yellowish with red-brown stripes, ± densely hairy with short to long purple **Ha**; **Cl** tube ± campanulate, 4 - 5 × 5 - 7 mm, inside pale, glabrous; **Cl** lobes narrowly linear from a triangular base, apically united and forming a broadly ovoid widely opening cage; **Cn** sessile, red-brown, ± 1.5 × 4 - 5 mm, shallowly cup-shaped to campanulate, glabrous; **Ci** lobes pouch-like, somewhat incised; **Cs** lobes linear-lingulate, ± 1 mm, incumbent on the **Anth**; **Poll** broadly ovoid, ± 0.4 × 0.3 mm; **Fr** thickly fusiform, thick-walled, ± 5 × 1 cm.

The stinking flower ball with the purple cage-like flowers makes *B. barberae* one of the best-known and showiest species of the genus (see also *B. chlorozonum*).

B. bikitaense Peckover (Aloe 32(3-4): 78-79, ills., 1995). **T:** Zimbabwe (*Peckover* 242 [PRE]). – **D:** Zimbabwe. **I:** Aloe 32: 78-79, 1995.

≡ *Tenaris bikitaensis* (Peckover) J. Victor & Nicholas (1998).

[2] **R** tuber depressed-globose, 3 - 5 cm ⌀; stem solitary, erect, to 50 cm, delicate, glabrous; **L** sessile, linear-lanceolate, 70 - 90 × 2 - 3 mm, pale green; **Inf** 2- to 10-flowered, sessile, numerous near the stem tip; **Ped** 2 - 3 cm, filiform; **Cl** 3 cm ⌀, pale green or brownish; **Cl** tube campanulate, ± 1 × 3 mm, pale green with whitish papillae; **Cl** lobes linear-lanceolate, ± 13 × 1 mm, ascending, folded back along the midrib, basally with brownish papillae; **Cn** sessile, bowl-shaped, ± 1 × 3 mm, green; **Ci** lobes pouch-like, with a median V-shaped incision and divided into 2 triangular appendages, ± erect; **Cs** lobes 0.5 mm, linear, ± incumbent on the **Anth**; **Fr** 40 - 75 × 2 mm; **Se** 6 - 7 × 1.5 mm, black.

B. blepharanthera H. Huber (Mitt. Bot. Staatssamml. München 4: 33, 1961). **T** [syn]: Namibia (*Dinter* 410 [PRE]). – **D:** C Namibia. **I:** Bruyns (1995c).

Incl. *Blepharanthera dinteri* Schlechter (1913) ≡ *Brachystelma dinteri* (Schlechter) E. Phillips (1941) (*nom. illeg.*, Art. 53.1); **incl.** *Blepharanthera edulis* Schlechter (1913).

[2] Compact dwarf shrublets, 10 - 20 cm tall, with elongate **R** tuber 3 - 6 × 1.5 - 4 cm; stems erect, basal part mostly underground, above ground basally few-branched, delicately pubescent; **L** basally tapering into short petiole, lamina oblong-spatulate, 20 - 30 × 5 mm, ± hairy, margins slightly undulate; **Inf** sessile, 1- to few-flowered; **Ped** 1 cm, hairy; **Sep** narrowly triangular, 2 mm, hairy; **Cl** yellow-green or green; **Cl** tube shallow, bowl-shaped or urceolate, 2.5 - 3.5 mm ⌀; **Cl** lobes 3 - 4 × 1.5 mm, triangular, semi-erect, margins revolute, basally papillose or hairy; **Cn** greenish, sessile, basally fused to form a shallow cup, ± 4 × 3 mm; **Ci** lobes shortly pouch-like and united with the **Cs** with broadened lateral margins; **Cs** lobes ± 1.5 mm, linear-cylindrical, often clavate, erect-connivent or erect and apical ½ recurved, ± horizontal; **Anth** apically mostly long ciliate; **Poll** ovoid, ± 0.25 mm long.

One of the very few species which grow almost exclusively on stony ground and because of this, the tubers are irregularly shaped. The ciliate anthers, otherwise typical e.g. for *Sisyranthus* (*Ceropegieae*), are uncommon.

B. bracteolatum Meve (KB 52(3): 711-714, ills., 1997). **T:** Nigeria (*Specks* 549 [K, MSUN]). – **D:** Nigeria.

[2] **R** tuber ± discoid, 5 - 12 cm ⌀; stems decumbent-ascending, basally several times branched, to 25 cm, pale green-reddish, pubescent; **L** ± sessile, linear, acute, 30 - 60 × 1 mm, slightly fleshy, ± glabrous; **Inf** first terminal, umbellate, then pushed into lateral position from the shoot continuation, 8-flowered, **Fl** opening simultaneously; **Bra** to 2.5 mm; **Sep** triangular, 2 - 3 mm; **Ped** 4 - 6 mm, pubescent; **Fl** 20 - 25 mm ⌀, strongly foetid, inside ± wrinkled, glabrous; **Cl** tube bowl-shaped, mouth

somewhat narrowed, ± 3 × 4 - 5 mm, yellowish, banded with purple-brown; **Cl** lobes linear from a triangular base, ± 14 × 2 mm, acute, ascending, lamina folded back along the midrib, purple-brown; **Cn** ± sessile, purple, ± 3 × 2.2 mm, basally fused to form a short cup, glabrous; **Ci** lobes basally pouch-like, free part very short, triangular, ± 0.4 mm ∅, ascending; **Cs** lobes linear, 2 mm, straight, ascending, connivent above the **Sty** head, basally-dorsally bulging; **Poll** subquadrangular, ± 0.4 mm ∅; **Fr** thickly fusiform, ± 3 × 1 cm.

The leaves and the corolla are typical for the *B. lineare*-complex, but the extremely short interstaminal and the very long ascending staminal corona are uncommon and notable. *B. pellacibellum*, *B. johnstonii* and *B. floribundum* are also closely related.

B. brevipedicellatum Turrill (BMI 1922: 29, 1922). **T:** RSA, Gauteng (*Anonymus* s.n. [K]). – **D:** RSA (Gauteng, Northern Prov., North-West Prov.). **I:** Dyer (1983); Asklepios 65: 26, 1995.

Incl. *Brachystelma ringens* E. A. Bruce (1951).

[2] **R** tuber flatttened, ± 3 × 6 cm; stems to 4, erect, unbranched, 3 - 10 cm, delicately pubescent; **L** shortly petiolate, semi-erect, lamina ovate to lanceolate, to 15 - 50 × 10 - 20 mm, longitudinally folded, dark green, upper face glabrous, lower face and along margins densely covered with crinkled white **Ha**; **Inf** sessile, 2- to 7-flowered; **Ped** 3 - 4 mm; **Sep** triangular-lanceolate, 1.5 mm, crisped-hairy; **Cl** stellate, 8 - 9 mm ∅, outside pubescent, inside pale to dark red-brown; **Cl** tube campanulate, ± 2 × 2 mm, basally greenish; **Cl** lobes horizontal, triangular, ± 3 × 2 mm, ± flat, smooth or slightly wrinkled; **Cn** sessile, completely filling out the **Cl** tube, cup-shaped, emarginate, ± red-brown; **Ci** lobes completely integrated into the margin of the cup, inside somewhat hairy; **Cs** dorsally thickened and fused with the inner base of the cup, lobes subquadrangular, incumbent on the **Anth**, yellowish.

The crinkled hairy lower faces of the leaves are diagnostic. The structure of the corona of *B. cupulatum* is almost identical, and similarities exist also in habit. This is equally true for *B. gymnopodum* and *B. stenophyllum*.

B. brevitubulatum (Beddome) Gamble (Fl. Madras, 599, 1921). **T:** India, Deccan (*Beddome* s.n. [K]). – **D:** S India (Deccan). **I:** Beddome (1874).

≡ *Ceropegia brevitubulata* Beddome (1874); **incl.** *Brachystelma beddomei* Hooker *fil.* (1885).

[2] Stem perennial, twining, ± 50 - 100 cm, solitary, glabrous; **L** ± sessile, lamina linear-lanceolate, 7 - 10 cm × 3 - 5 mm; **Inf** pedunculate, 3-flowered; peduncle ± 5 mm, densely pubescent; **Ped** 6 mm, densely pubescent; **Sep** subulate, acute, 5 mm, pubescent; **Cl** outside yellowish-green, inside purple, ± glabrous; **Cl** tube short, shallow; **Cl** lobes linear-lanceolate, ± 25 mm, flat, apically united to form a cage; **Cn** ± sessile, cup-shaped; **Ci** lobes pouch-like, erect, poorly differentiated; **Cs** lobes lingulate, incumbent on the **Anth**; **Poll** ovoid; **Fr** 10 - 13 × 0.5 cm.

Apart from *B. volubile* (see there), *B. brevitubulatum* is the only twining species of the genus. The cage-like flowers of *B. brevitubulatum* are comparable in shape with those of *B. circinatum*.

B. brownianum (S. Moore) Meve (Pl. Syst. Evol. 228: 103, 2001). **T:** Angola (*Gossweiler* 3062 [BM]). – **D:** Angola.

≡ *Tenaris browniana* S. Moore (1912).

[1] **R** fleshy, fusiform; stems 60 - 70 cm, single, rarely branched, glabrous; **L** ± sessile, linear, ± 70 × 5 mm, glabrous; **Inf** ± sessile, umbellate, with 6 - 9 pendent **Fl**; **Ped** 3 - 4 cm, filiform; **Sep** 1 mm; **Cl** black-purple, 3 - 4 mm ∅, rotate, stout; **Cl** tube 1 mm; **Cl** lobes triangular-lanceolate, ± 1.2 × 0.8 mm, marginally with scattered **Ha**; **Cn** sessile, bowl-shaped; **Ci** lobes ± 1 mm, lanceolate, ascending, apically deeply bifid, upper face with few **Ha**; **Cs** lobes 0.5 mm, subquadrate; **Poll** ± 0.2 mm.

The minute flowers in predominantly terminal position and pendulous with long filiform pedicels, are notable. This species is very similar to *B. elegantulum* (also from Angola) in habit and flower morphology (despite the corolla being 4× as large). The latter has to be regarded as sister species of *B. browniana*.

B. bruceae R. A. Dyer (JSAB 43(1): 13, 1977). **T:** RSA, Mpumalanga (*Codd* 5756 [PRE]). – **D:** RSA (Northern Prov., Mpumalanga). **I:** Dyer (1983).

Incl. *Brachystelma bruceae* ssp. *hirsutum* R. A. Dyer (1977).

[2] **R** tuber 5 - 7 cm ∅; stems several, prostrate, to 20 cm long, porrectly hispid, purple; **L** shortly petiolate, ovate, somewhat fleshy, 10 - 15 × 5 - 10 mm, upper face and midrib hispid; **Inf** sessile, 1-flowered; **Ped** 7 - 15 mm, hispid; **Sep** lanceolate-subulate, 3 mm, spreading, roughly hairy; **Cl** rotate, 10 - 16 mm ∅, purple to pale brown-red, smooth; **Cl** tube ± flat, short; **Cl** lobes 4 - 6 mm, triangular to ovate-lanceolate, margins somewhat revolute; **Cn** sessile, overtopping the **Cl**, purple to pale brown-red, ± 3 mm ∅, basally united and cup-shaped; **Ci** lobes pouch-like, outer margin emarginate; **Cs** lobes shortly linear, incumbent on the **Anth**.

Habit and flower morphology are very similar as in *B. pulchellum*.

B. buchananii N. E. Brown (BMI 1895: 263, 1895). **T:** Malawi, Nyassaland (*Buchanan* 116 [K]). – **D:** Kongo, Malawi, Zambia, Tanzania, Zimbabwe. **I:** Asklepios 65: 26, pl. 4, 1995. **Fig. I.c**

Incl. *Brachystelma magicum* N. E. Brown (1895); **incl.** *Brachystelma shirense* Schlechter

(1895); **incl.** *Brachystelma nauseosum* De Wildeman (1904).

[2] **R** tuber depressed-globose, 5 - 10 cm ∅; stems 1 to few, erect to decumbent, clustered, to 10 cm, hispid; **L** elongate, obovate to lanceolate, 3 - 12 × 2 - 4 cm, tomentose, ciliate, basally tapering into a short petiole; **Inf** ± terminal, sessile, 5 - to 30-flowered umbels, most **Fl** opening simultaneously, cow-dung-scented; **Ped** 3 - 15 mm, pubescent; **Sep** lanceolate, 3 - 7 mm, pubescent; **Cl** bowl-shaped, 15 - 25 mm ∅, outside greenish, pubescent, inside wrinkled, cream with dark purple margin and stripes, ± densely white-hairy; **Cl** tube ± bowl-shaped, 12 - 16 mm ∅; **Cl** lobes broadly triangular, 4 - 8 × 5 - 8 mm, flat; **Cn** sessile, red-brown, ± 2 × 4 mm, only basally shortly united and cup-like; **Ci** lobes pouch-like but with deeply U-shaped emarginate outer margin, laterally elongating into subulate densely white-hairy appendages, united dorsally with the **Cs**; **Cs** lobes broadly lingulate, glabrous or hairy, apically emarginate to serrulate, incumbent on the **Anth**; **Poll** D-shaped, ± 0.5 × 0.4 mm; **Fr** thickly fusiform.

One of the commonest and most widespread of all species, and certainly closely related to *B. chlorozonum*. In some regions, the sap of the plants is used to treat wounds ("*B. magicum*").

B. burchellii (Decaisne) Peckover (Aloe 33(2-3): 43, 1996). **T:** RSA, North-West Prov. (*Burchell 2498* [K]). – **Lit:** Peckover (1993b: as *Macropetalum*); Peckover (1996); Meve & Liede (2001a). **D:** Botswana, RSA. **Fig. I.d**

≡ *Macropetalum burchellii* Burchell *ex* Decaisne (1844).

[2] Perennial herbaceous summer-green geophytes, glabrous or pubescent; with ± globose edible **R** tubers of 3 - 5 cm ∅; stems erect, 30 - 100 cm, flowering stems unbranched, delicate; **L** linearly needle-shaped, ascending or pendent, 2 - 10 cm × 1 - 3 mm, midrib prominent; **Inf** extra-axillary, nearly terminal to lateral, sessile or shortly pedunculate, 2- to 7-flowered; **Ped** filiform; **Sep** lanceolate, 1 mm; **Cl** tube very short; **Cl** lobes linear-filiform; **Cn** ± 1-seriate (**Ci** reduced to a fringe-like structure), sessile or shortly stalked; **Cs** lobes erect, linear-subulate from a broadened fleshy base, ± 2 mm; falcately erect-connivent, far overtopping the **Gy**; **Anth** ± ovate, erect or ascending, with short triangular connective appendages, ± 0.5 mm, laterally pressed against the huge **Sty** head; **Poll** ± erect and slightly impressed into the sides of the **Sty** head, ± ovoid, hardly 0.12 mm; **Sty** head huge, as long as or longer than the **Anth**, ± globose, fleshy; **Fr** in pairs, narrowly fusiform, glabrous, to 80 × 4 mm, erect; **Se** brown, winged.

The grass-like growth of these plants and the tiny flowers reminiscent of beaks are notable. The apparently uniseriate corona is significant and was the reason for more than 150 years of acceptance of the monotypic genus *Macropetalum* for this species. Molecular investigations, however, demonstrated that this taxon belongs to *Brachystelma*, and that its closest relatives are *B. filifolium* and *B. rubellum* (formerly classified as *Tenaris*) (Meve & Liede 2001a).

B. burchellii var. **burchellii** – **D:** RSA (Free State, KwaZulu-Natal, Mpumalanga).

Incl. *Macropetalum burchellii* var. *burchellii*.

[2] **Ped** glabrous, 5 - 8 mm; **Cl** lobes 2 - 3 cm, strictly reflexed shortly above the base, whitish or greenish, ± glabrous, partly finely white-ciliate; **Cs** lobes linear.

B. burchellii var. **grandiflorum** (N. E. Brown) Meve (Pl. Syst. Evol. 228: 103, 2001). **T:** RSA, North-West Prov. (*Burke s.n.* [K]). – **D:** Botswana, RSA (Free State, Gauteng, North-West Prov.).

≡ *Macropetalum burchellii* var. *grandiflorum* N. E. Brown (1908).

[2] Differs from var. *burchellii*: **Ped** pubescent; **Cl** lobes 3 - 5 cm, not reflexed, outside sparsely pubescent, inside orange; **Cs** lobes apically emarginate.

Found on rocky as well as sandy substrates. The data about the geographical and morphological distinctions is insufficient.

B. caffrum (Schlechter) N. E. Brown (Gard. Chron., ser. 3, 16: 62, 1894). **T:** RSA, Eastern Cape (*Sim 315* [K, BOL, NU, PRE]). – **D:** RSA (Eastern Cape). **I:** FPA 47: t. 1843, 1982; Dyer (1983). **Fig. I.e**

≡ *Tapeinostelma caffrum* Schlechter (1893).

[2] **R** tuber depressed-globose, 5 - 7 cm ∅; stems several, decumbent-ascending, 5 - 12 cm, scatteredly hispid; **L** shortly petiolate, lamina variable, mostly ovate-lanceolate, to 15 mm, somewhat hispid; **Inf** sessile, 1- (to 2-) flowered; **Ped** 15 - 30 mm, ascending, hispid; **Sep** lanceolate, 2 mm; **Cl** stellate, 7 - 13 mm ∅, yellow, smooth, outside papillate, with intense cow-dung-scent; **Cl** tube shallow, short; **Cl** lobes 3 - 5 mm, triangular-ovate, margins sometimes ciliate; **Cn** sessile, greenish-yellow, ± 2 × 3 mm, campanulate, united and almost urceolate; **Ci** lobes pouch-like, in the middle with a narrowly V-shaped incision, laterally fused with the broadened flattened back of the **Cs**; **Cs** lobes linear-lingulate, appressed to the ascending **Anth**; **Poll** D-shaped; **Fr** ± 4 cm; **Se** ± 6 × 2 mm.

Habit and the morphology of the corona point to *B. decipiens* as a relatively closely related species.

B. campanulatum N. E. Brown (FC 4(1): 838, 1908). **T:** RSA, Eastern Cape (*Bowie s.n.* [BM]). – **D:** RSA (Eastern Cape). **I:** Dyer (1983).

[2] **R** tuber depressed-globose, 5 - 7.5 cm ∅; stem mostly only 1, erect, ± 5 cm, pubescent; **L** ± 10 mm petiolate, lamina elliptic to ovate, 15 - 35 × 10 - 15

mm, pubescent; **Inf** sessile, 1-flowered; **Ped** ± 7 mm; **Sep** lanceolate, 4 mm, hispid; **Cl** ± 20 mm ⌀, tube campanulate, ± 10 × 10 mm, green-yellow with purple concentric stripes, papillose to shortly hairy; **Cl** lobes greenish with purple longitudinal stripes, ± 4 - 5 mm, broadly triangular, ± horizontally spreading, ± ciliate; **Cn** sessile, spotted orange-brown, basally united and shallowly bowl-shaped; **Ci** lobes pouch-like, apical ½ divided into 2 obtusely triangular erect appendages; **Cs** lobes lingulate, incumbent on the **Anth** and overtopping these; **Fr** ± 8 cm.

Hardly known locally endemic taxon from the coastal Grassveld near Bathurst; sister species of *B. kerzneri*.

B. canum R. A. Dyer (BT 12(2): 254, 1977). **T**: RSA, North-West Prov. (*Acocks* 18774 [PRE]). – **D**: RSA (North-West Prov.). **I**: Dyer (1983).

[?] **R** tuber unknown; stems solitary, erect, ± 30 cm, tomentose and hispid; **L** ± sessile, narrowly elliptic, 20 - 40 × 5 - 10 mm, tomentose; **Inf** sessile, 2-flowered; **Cl** tube short; **Cl** lobes linear-lanceolate, ± 10 mm, inside glabrous, margins revolute; **Cn** sessile, basally fused and campanulate, 2 mm ⌀; **Ci** lobes pouch-like, margin entire; **Cs** lobes short, pillow-like along the inner **Cn** margin; **Poll** ± 0.25 mm.

Little-known species, probably related to *B. brevipedicellatum*.

B. cathcartense R. A. Dyer (BT 10: 431, 1971). **T**: RSA, Eastern Cape (*Du Toit* s.n. in PRE 31309 [PRE, GRA]). – **D**: RSA (Eastern Cape: near Cathcart); Grassveld. **I**: FPA 42: t. 1667, 1973; Dyer (1983).

[2] **R** tuber depressed-globose, 4 - 5 cm ⌀; stems few, erect, 5 - 8 (-15) cm, hispid; **L** shortly petiolate, lamina linear-lanceolate, to 3 cm, ± hispid; **Inf** sessile, 1- to 2-flowered; **Ped** ± 10 mm; **Sep** lanceolate, 5 mm, hispid; **Cl** ± 32 mm ⌀, with intense carrion-scent, tube shallowly bowl-shaped, ± 18 mm ⌀, ± red-brown, densely covered with purple **Ha**; **Cl** lobes greenish, ± 12 × 8 mm, broadly triangular, ± horizontally spreading, somewhat grooved, margins somewhat revolute, apically hairy; **Cn** shortly stipitate, basally fused and shallowly bowl-shaped; **Ci** lobes rectangularly pouch-like, 1 - 1.5 mm, erect, apical ½ divided into 2 triangular appendages; **Cs** lobes linear-spatulate, erect-connivent; **Poll** D-shaped, flattened, ± 0.25 mm; **Fr** 10 × 1 cm.

Rare locally endemic taxon from the Grassveld near Cathcart. Fairly similar in habit to *B. decipiens*.

B. chloranthum (Schlechter) Peckover (Aloe 33(2-3): 43, 1996). **T** [syn]: RSA, Gauteng (*Schlechter* 4152 [K]). – **D**: RSA (Gauteng).

≡ *Tenaris chlorantha* Schlechter (1895).

[2] **R** tuber 3 - 5 cm ⌀, pale brown; stem mostly strictly erect, to 20 - 55 cm, glabrous; **L** sessile, subulate, 5 - 110 × 1 mm, ascending; **Inf** 1- to 2- (to 3-) flowered, sessile or shortly pedunculate, often clustered in terminal racemes; **Ped** 1 - 1.5 cm, horizontal to ascending; **Cl** greenish or yellowish with brown spots, often intensely speckled with purple, ± shortly papillose; **Cl** tube short, shallowly bowl-shaped to campanulate, ± 1.5 mm; **Cl** lobes narrowly triangular to linear-subulate, 5 - 8 mm, ascending-spreading, ± folded back along the midrib; **Cn** ± sessile, bowl-shaped, ± 2 mm ⌀, yellow speckled with brown; **Ci** lobes pouch-like, subquadrangular-concave, 0.3 mm, horizontal; **Cs** lobes 0.5 mm, linear-subulate, incumbent on the **Anth**; **Poll** ovoid, ± 0.2 mm; **Fr** 25 - 35 × 2 mm, grey-green.

Closely related to the more delicate *B. christianeae*, which has a very similar corona.

B. chlorozonum E. A. Bruce (HIP 1938: t. 3370 + text, 1938). **T**: RSA, Mpumalanga (*Thorncroft* s.n. [K]). – **D**: RSA (KwaZulu-Natal, Mpumalanga, Northern Prov.). **I**: Dyer (1983).

[2] **R** tuber depressed-globose, to 7 cm ⌀; stem mostly solitary, little-branched, halfway decumbent to ascending, 10 - 20 cm, densely pubescent, somewhat bristly; **L** shortly petiolate, lamina ovate-elliptic, 20 - 45 × 15 - 25 mm, densely pubescent to long hairy; **Inf** ± sessile to shortly pedunculate, 1-flowered; **Ped** 5 mm; **Cl** 10 - 30 mm ⌀, outside yellowish-green, shortly hairy, tube campanulate to shallowly bowl-shaped, 3 - 6 × 7 - 13 mm, with greenish to purple-brown concentric stripes on yellowish background; **Cl** lobes 3.5 - 8 mm, broadly triangular, often strongly reflexed, greenish-yellow, hairy; **Cn** sessile, red-brown, 1.5 × 4 mm, bowl-shaped; **Ci** lobes pouch-like, ± horizontally ovate-subquadrangular, apically obtuse or somewhat indented, margin sometimes pubescent; **Cs** lobes linear-lingulate, placed against the **Anth**; **Poll** subquadrangular, ± 0.4 × 0.25 mm; **Fr** thickly fusiform, thick-walled, ± 40 × 12 mm; **Se** broadly ovate, broadly winged.

B. chlorozonum is typical for a group of species with creeping to decumbent-ascending habit, broad, almost sessile, strongly hairy leaves, robust broadly ovoid to D-shaped pollinia and thick or thick-walled follicles. Here belong e.g. *B. barberae, B. bracteolatum, B. buchananii, B. floribundum, B. megasepalum* and *B. oianthum*.

B. christianeae Peckover (Aloe 29(3/4): 56-57, ills., 30(1): 13 [erratum], 1993). **T**: RSA, KwaZulu-Natal (*Peckover* 141 [PRE]). – **D**: RSA (KwaZulu-Natal). **I**: Aloe 29: 56-57, 1992 [publ. 1993]. **Fig. I.f**

≡ *Tenaris christianeae* (Peckover) J. Victor & Nicholas (1998).

[2] **R** tuber ± globose, 3 - 4 cm ⌀, pale brown; stems solitary to numerous, erect, apically often drooping, to 15 - 25 cm, glabrous; **L** sessile, subulate, 20 - 80 × 1 - 2 mm, ascending to erect; **Inf**

1- to 2-flowered, sessile, at the upper nodes; **Ped** 0.5 - 1 cm, ± horizontal, reddish; **Sep** to 4 mm; **Cl** 3 - 4.5 cm ⌀, glabrous; **Cl** tube slightly urceolate, ± 2.5 × 4 mm ⌀, outside whitish, inside reddish; **Cl** lobes linear-subulate, ± 16 - 22 × 0.6 mm, spreading, folded back along the midrib, yellow-green; **Cn** ± sessile, bowl-shaped, ± 1 × 3 mm, orange-yellow speckled with red-brown; **Ci** lobes pouch-like, subquadrangular-concave, horizontally ascending; **Cs** lobes 0.6 mm, triangular, incumbent on the **Anth**; **Fr** 90 - 100 × 1.5 mm, reddish.

Closely related to *B. chloranthum.*

B. circinatum E. Meyer (Comm. Pl. Afr. Austr., 196, 1837). **T**: RSA, Northern Cape (*Drège* 3440 [K]). – **D**: Namibia, Botswana, RSA (all provinces). **I**: Dyer (1983).

≡ *Dichaelia circinata* (E. Meyer) Schlechter (s.a.); **incl.** *Brachystelma filiforme* Harvey (1859) ≡ *Dichaelia filiformis* (Harvey) Schlechter (1894); **incl.** *Brachystelma ovatum* Oliver (1870) ≡ *Dichaelia ovata* (Oliver) Schlechter (1894); **incl.** *Dichaelia galpinii* Schlechter (1894) ≡ *Brachystelma galpinii* (Schlechter) N. E. Brown (1908); **incl.** *Dichaelia undulata* Schlechter (1894) ≡ *Brachystelma undulatum* (Schlechter) N. E. Brown (1908); **incl.** *Dichaelia pallida* Schlechter (1895) ≡ *Brachystelma pallidum* (Schlechter) N. E. Brown (1908); **incl.** *Dichaelia microphylla* S. Moore (1903); **incl.** *Dichaelia macra* Schlechter (1905); **incl.** *Dichaelia brachylepis* Schlechter (1907); **incl.** *Dichaelia cinerea* Schlechter (1907) ≡ *Brachystelma cinereum* (Schlechter) N. E. Brown (1908); **incl.** *Dichaelia zeyheri* Schlechter (1907) ≡ *Brachystelma zeyheri* (Schlechter) N. E. Brown (1908); **incl.** *Brachystelma bolusii* N. E. Brown (1908); **incl.** *Brachystelma commixtum* N. E. Brown (1908); **incl.** *Dichaelia forcipata* Schlechter (1913).

[2] **R** tuber mostly thickly discoid, to 10 cm ⌀, rarely oblong; stem mostly 1 richly ± porrectly branched main stem, erect, 7 - 30 cm ± pubescent; **L** sessile to shortly petiolate, linear to ovate, 5 - 20 mm, often folded together and undulate, upper face pubescent; **Inf** sessile, 1- to 2-flowered; **Ped** 2 - 5 mm, hispid; **Sep** lanceolate, 1.5 mm; **Cl** 5 - 25 mm long, cage-like, white to red-brown, inside glabrous (rarely scatteredly pubescent), outside ± pubescent; **Cl** tube short, shallow or bowl-shaped; **Cl** lobes linear-subulate, apically united; **Cn** sessile, basally fused and ± cup-shaped, ± 2.5 mm ⌀; **Ci** lobes erect, deeply divided into 2 triangular to subulate appendages, mostly hairy; **Cs** lobes oblong-lingulate, ± incumbent on the **Anth**, sometimes ascending, centrally apically overlapping; **Poll** obliquely ovoid, ± 0.35 - × 0.28 mm; **Fr** 7 - 10 mm.

Probably one of the most frequent and polymorphic taxa in S Africa. *B. circinatum* is the sister species of *B. dimorphum.* It is distinguished primarily by the corolla with glabrous inside and the much longer bifid interstaminal corona.

B. coddii R. A. Dyer (FPA 30: t. 1181 + text, 1955). **T**: Swaziland (*Codd* 7826 [PRE]). – **D**: Tanzania, Swaziland, RSA (Northern Prov.). **I**: Dyer (1983). Fig. I.g

Incl. *Brachystelma kituloense* Goyder (1990).

[2] **R** tuber 3 - 20 cm ⌀; stems several, creeping, 5 - 30 cm (- 1 m), ± pubescent; **L** shortly petiolate, ± ovate, to 6 - 20 × 5 - 15 mm, esp. veins and margins hairy; **Inf** sessile, 1- (to 2-) flowered; **Ped** 5 - 15 mm, ± glabrous; **Sep** subulate, 3 - 4 mm, porrectly hispid; **Cl** ± 20 mm ⌀, tube bowl-shaped, ± 7 mm ⌀, inside yellowish, banded or spotted with purple; **Cl** lobes 6 - 8 mm, lanceolate from a broadly triangular base, mostly ± purple, glabrous or simply hairy, margins of upper ⅔ strongly revolute; **Cn** sessile, brown, ± 2 × 3 mm ⌀, basally fused and cup-shaped; **Ci** lobes pouch-like, divided ± 1 mm deep in V-shaped manner, tips inside bearded or glabrous; **Cs** lobes shortly linear, incumbent on the **Anth**, dark brown; **Poll** D-shaped; **Fr** 35 - 65 mm.

The Tanzanian taxon *B. kituloense*, here treated as synonym, allegedly differs from the glabrous *B. coddii* primarily in the hairiness of corolla and interstaminal corona.

B. codonanthum Bruyns (BT 25(2): 157-158, ills., 1995). **T**: Namibia (*Bruyns* 5518 [BOL]). – **D**: C Namibia.

[2] Dwarf perennials with deciduous herbaceous above-ground parts, with depressed-globose **R** tubers 2 - 3 cm ⌀; stems erect, 2 - 15 cm, hairy; **L** lamina basally tapering into a short petiole, broadly elliptic, 11 - 22 × 9 - 15 mm, ± hairy; **Inf** sessile, few-flowered; **Ped** 2 - 3 mm, horizontal, hairy; **Sep** lanceolate, 2 mm, hairy; **Cl** outside pale green to brownish, inside green to brown with greenish tips, wrinkled, glabrous; **Cl** tube short, bowl-shaped, 2.5 × 4 mm; **Cl** lobes 3 - 4 × 3 mm, triangular, ascending, margins revolute, marginally and outside apically hairy; **Cn** sessile, dark brown, cup-shaped, 1.5 × 3 mm; **Ci** lobes subquadrangular, arising from the cup margins, short, ascending, obtusely triangular, apically incised; **Cs** lobes rectangular, incumbent on the **Anth**, basally-centrifugally thickened, pillow-like and united with the cup; **Poll** broadly ovoid, ± 0.25 mm.

Minute plants close to *B. dinteri*. They form a group of related species, together with *B. brevipedicellatum, B. cupulatum* (?) and *B. simplex.*

B. comptum N. E. Brown (FC 4(1): 854, 1908). **T**: RSA, Eastern Cape (*Zeyher* 9 [K]). – **D**: RSA (Eastern Cape). **I**: FPA 47: t. 1844, 1982; Dyer (1983).

[1] **R** fleshy, fusiform; stem solitary, erect, 7 - 15 cm, reddish, pubescent; **L** in only 4 - 6 pairs, shortly petiolate, lamina ovate-elliptic to almost circular, 10 - 16 × 6 - 12 mm, densely long- and white-hairy, upper face sometimes almost glabrous; **Inf** sessile, 1-flowered, near stem tip; **Ped** 7 mm; **Sep** lanceolate, 3 mm; **Cl** divided almost to the base; **Cl** lobes

± 10 × 4 mm, erect, somewhat narrowed in the middle, bulging and densely covered with purple vibratile **Ha**, upper ½ horizontally recurved, **Cn** shortly stipitate, stellate when seen from above, united only basally; **Ci** lobes triangular to linear-lingulate, obtuse, apically hairy, ascending, concave, pale purple; **Cs** lobes linear, incumbent on the **Anth**, dark purple, often hairy; **Poll** ± 0.25 mm.

Most closely related to *B. gerrardii*.

B. cupulatum R. A. Dyer (BT 10(2): 375, 1971). **T:** Namibia (*Story* 6400 [PRE]). – **D:** Namibia, Botswana, RSA (Northern Cape, Northern Prov.). **I:** Dyer (1983); Bruyns (1984).

[2] **R** tuber depressed, ± 3 × 6 cm; stem solitary or little-branched near the base, erect, 5 - 15 cm, delicately pubescent; **L** shortly petiolate, lamina linear to ovate, 30 - 80 × 4 - 15 mm, pubescent; **Inf** sessile, 1- to 2-flowered; **Ped** 2 - 3 mm; **Sep** lanceolate, 1.5 mm, scattered hairy; **Cl** 10 - 14 mm ⌀, outside pubescent, inside glabrous, green, rarely purple; **Cl** tube campanulate, ± 2 × 2.5 mm; **Cl** lobes ± horizontal, triangular, 4 - 6 mm, margins somewhat recurved; **Cn** sessile, fitting into the **Cl** tube, conspicuously cup-shaped and emarginate, yellow, glabrous; **Ci** lobes completely integrated into the margin of the cup, inside sometimes hairy; **Cs** dorsally thickened and united with the inner base of the cup; **Cs** lobe very short or absent, basally appressed to the **Anth**; **Fr** 6 - 8 cm.

The corona of *B. cupulatum* is conspicuously taller than that of *B. vahrmeijeri* from KwaZulu-Natal, but otherwise, there are hardly any significant differences between the 2 species (see also *B. brevipedicellatum* and *B. stenophyllum*).

B. decipiens N. E. Brown (FC 4(1): 842, 1908). **T:** RSA, Eastern Cape (*Bolton* s.n. [K]). – **D:** RSA (Eastern Cape), Swaziland. **I:** Dyer (1983: as *B. tuberosum*).

Incl. *Brachystelma tuberosum* R. A. Dyer (1980) (*nom. illeg.*, Art. 53.1).

[2] **R** tuber depressed-globose, ± 5 cm ⌀; stems few, erect, 8 - 15 cm tall, pubescent; **L** shortly petiolate, lamina linear-lanceolate or elliptic, to 3.5 × 1.2 cm, ciliate, lower face pubescent; **Inf** sessile, 2- to 4-flowered; **Ped** 4 - 6 mm; **Sep** lanceolate, 2 mm; **Cl** 2 cm ⌀, tube campanulate, 4 - 5 mm, yellowish with purple-brown patterning; **Cl** lobes 6 - 8 mm, linear-lanceolate, horizontally spreading, dark purple-brown, basally mostly hairy, margins ± revolute, with long vibratile white **Ha** or **Cl** glabrous; **Cn** ± sessile, 2 × 3 mm, campanulate, almost urceolate, united; **Ci** lobes pouch-like, ± triangular and with median incision, inside white-bearded; **Cs** lobes linear-spatulate, ascending, placed against the thickish ovoid **Anth**; **Poll** 0.25 - 0.4 mm; **Sty** head bulging; **Fr** fusiform, ± 6 cm × 4 mm.

Somewhat atypical for the genus in respect to the almost ovoid and somewhat ascending anthers. *B. decipiens* is closely related to *B. meyerianum* and *B. keniense*. The species was still covered under the name of *B. tuberosum* by Dyer (1980, 1983) and was regarded as type of the genus (Forster 1986).

B. delicatum R. A. Dyer (BT 12(1): 53-54, ill., 1976). **T:** RSA, Eastern Cape (*Hardy* 2198 [PRE]). – **D:** RSA (Eastern Cape: surroundings of Grahamstown only). **I:** Bruyns (1982a); Dyer (1983).

[2] **R** tuber ± globose, 2 - 4 cm ⌀; stems 1 - 2, erect, to 6 cm; **L** ± sessile, lamina linear-elliptic, to 25 × 7 mm, somewhat fleshy, keeled, margin and keel papillose; **Inf** sessile, 4- to 7-flowered; **Sep** 1.5 mm; **Ped** to 5 mm; **Cl** 7 - 8 mm ⌀, white, tube campanulate; **Cl** lobes 2 - 3 mm, oblong-ovate, tips brown or green, often with stiff white **Ha**, somewhat folded backwards along the midrib; **Cn** sessile, green, with cylindrical base, ± 1 mm, above broadened cup-like; **Ci** lobes shallowly pouch-like, marginally divided into 2 obtusely triangular appendages; **Cs** lobes triangular-ovate, ± 0.5 mm, incumbent on the **Anth**.

The contrasting coloration of the corolla and the cylindrical elongate filament tube are conspicuous and uncommon.

B. dimorphum R. A. Dyer (BT 12(4): 627, 1979). **T:** RSA, Northern Cape (*Leistner* 2935 [PRE]). – **D:** RSA (Free State, Northern Cape). **I:** Dyer (1983).

Incl. *Brachystelma dimorphum* ssp. *gratum* R. A. Dyer (1979).

[2] **R** tuber cylindrical to globose, 5 - 7 cm ⌀; mostly with 1 main stem, porrectly branched above, erect, 10 - 20 cm, pubescent; **L** shortly petiolate, ovate-elliptic, 5 - 20 × 4 - 15 mm, often folded together and undulate, lower face hispid; **Inf** sessile, 1- to few-flowered; **Ped** short, hispid; **Sep** lanceolate, 1 mm; **Cl** 6 - 7 mm long, cage-like or star-shaped, dirty cream-coloured to pale greenish-yellow, inside hairy, outside glabrous; **Cl** tube very short, shallow or bowl-shaped; **Cl** lobes linear-lanceolate, 4 - 5 mm, lamina erect and apically united or free and horizontally spreading; **Cn** sessile, basally hardly united, 1.5 - 2 × 0.75 - 1 mm; **Ci** lobes shortly pouch-like, erect, apically incised; **Cs** oblong, subquadrangular, obtuse, ± incumbent on the **Anth**; **Poll** ovoid-pear-shaped; **Fr** 7 - 10 cm.

Related to *B. pygmaeum* and *B. circinatum* (see there for notes).

B. dinteri Schlechter (BJS 51: 144, 1913). **T** [syn]: Namibia (*Dinter* 1515 [SAM]). – **D:** Namibia, Botswana, Zimbabwe, RSA (North-West Prov.); grassland. **I:** Dyer (1983); Bruyns (1984).

[2] **R** tuber flattened, 3 - 5 cm ⌀; stem solitary, unbranched, erect, 10 - 40 (-60) cm, delicately pubescent; **L** shortly petiolate, lamina ovate-elliptic, to 15 - 45 × 5 - 15 mm, ± shortly pubescent; **Inf** sessile, 2- to 12-flowered; **Ped** 2 - 4 mm; **Sep** lanceo-

late, 2.5 mm, pubescent; **Cl** stellate, 7 - 10 mm ⌀, outside pubescent, inside glabrous, green, often with black-brown spots; **Cl** tube bowl-shaped, ± 1 × 2 mm, marginally bulging somewhat ring-like, fleshy; **Cl** lobes horizontal, triangular, ± 2.5 × 2 mm, ± flat; **Cn** sessile, fitted into the **Cl** tube, shallowly bowl-shaped; **Ci** subquadrangular, erect, apically incised, inflexed; **Cs** lobes linear, incumbent on the **Anth**.

The flat star-shaped flowers are characteristic. *B. dinteri* is closely related to the tropical *B. simplex*.

B. discoideum R. A. Dyer (FPA 42: t. 1668 + text, 1973). **T:** RSA, Gauteng (*Hardy* 2440 [PRE]). – **D:** Zimbabwe, RSA (Gauteng). **I:** Dyer (1983); Boele (1989).

[2] **R** tuber strongly flattened, to 7 cm ⌀; stems few, hardly branched, erect, to 10 cm, pubescent; **L** 2 - 8 mm petiolate, lamina broadly ovate to almost circular, to 5 - 25 × 5 - 15 (-20) mm, upper face glabrous, lower face hairy; **Inf** sessile, 1- to 2-flowered; **Ped** 6 - 15 mm, bent downwards; **Sep** lanceolate, 2 mm; **Cl** 18 - 22 mm ⌀, tube shallow, ± 5 - 6 mm ⌀, often whitish; **Cl** lobes slightly ascending, lanceolate, 8 - 10 mm, margins revolute, apical ½ yellowish, basal ½ pale yellow to purple; **Cn** sessile, black-purple, fused to form a thickly discoid compact structure, deeply emarginate at **Ci** positions; **Cs** lobes short, pillow-like at the inner margin of the **Cn**; the whitish **Anth** and the **Sty** head mostly free; **Poll** 0.25 mm.

A similarly compact corona is otherwise only found in *B. schinzii*.

B. duplicatum R. A. Dyer (BT 12(4): 629, 1979). **T:** RSA, Free State (*Van Zyl* 1025 [PRE]). – **D:** RSA (Free State). **I:** Dyer (1983).

Incl. *Brachystelma floribundum* R. A. Dyer (1956) (*nom. illeg.*, Art. 53.1).

[2] **R** tuber globose to cylindrical, 8 - 9 × 8 cm; stem solitary, erect, to 15 cm, pubescent, partly underground; **L** shortly petiolate, lamina oblong-lanceolate, to 10 × 3 mm, ± delicately papillose to pubescent, somewhat undulate, blue-green; **Inf** sessile, 2- to many-flowered; **Ped** short; **Sep** lanceolate, 2.5 mm; **Cl** 20 - 26 mm ⌀, pale green, tube campanulate, 2 - 3 × 4 mm; **Cl** lobes 10 - 12 mm, ± recurved, apical ¾ ± linear, almost completely folded back along the midrib, margins of the basal ¼ revolute; **Cn** ± sessile, fused to form a broadly campanulate structure; **Ci** lobes pouch-like, inside with long white **Ha**, margin denticulate; **Cs** lobes linear-spatulate, incumbent on the **Anth**; **Fr** subcylindrical, ± 10 cm.

A local endemic of uncertain affiliation but reminiscent of *B. circinatum* in habit and foliage.

B. edule Collett & Hemsley (JLSB 28: 89, t. 14, 1890). **T:** Myanmar (*Collett & Hemsley* s.n. [K]). – **D:** India (Deccan, Karnataka, Kolhapur), Myanmar, China (Guangxi, Yunnan); forest and grasslands.

Incl. *Brachystelma elenaduense* M. B. Char (1978).

[2] **R** tuber ± globose, 1.5 - 2.5 cm ⌀; stem 1, erect, 5 - 15 cm, rarely branched, somewhat hairy or glabrous; **L** ± sessile, opposite, sometimes alternate, linear-elliptic, 10 - 40 × 2 - 4 mm, ± glabrous; **Inf** ± sessile, lateral near the stem tips, 1- to 5-flowered; **Ped** 3 - 5 mm, pubescent; **Sep** 1 - 3 mm, lanceolate, pubescent; **Cl** inside dark purple, ± hairy, tube short, shallow; **Cl** lobes triangular-lanceolate, 4 - 7 mm, ± spreading, lamina mostly folded back along the midrib, margins ± revolute, **Cn** ± sessile, ± 2 mm ⌀, purple, fused to form a cup-like structure; **Ci** lobes subquadrangularly pouch-like, ± erect, crenate or incised; **Cs** lobes triangular, short, basally placed against the **Anth**; **Poll** narrowly ovoid, 0.2 × 0.1 mm; **Fr** 8 - 11 cm × 3 - 4 mm; **Se** ± 8 × 4 mm.

A species with a wide but scattered distribution and a characteristic habit with rotate flowers. Easily confused with the sister-species *B. kerrii*.

B. elegantulum S. Moore (J. Bot. 50: 366, 1912). **T:** Angola (*Gossweiler* 2944 [BM]). – **D:** Angola.

[?] **R** unknown; stems erect, to 60 cm, glabrous; **L** sessile, narrowly linear, 6 - 10 (-12) cm × 1 - 2 mm; **Inf** terminal, 10-flowered; **Ped** 1 cm, filiform; **Cl** inside chocolate-brown; **Cl** tube campanulate, ± 2 mm ⌀; **Cl** lobes linear from a triangular base, 12 - 15 mm; **Cn** cup-like; **Ci** pouch-like, 0.6 mm; **Cs** lobes linear, 0.2 mm, incumbent on the **Anth**; **Poll** pear-shaped, ± 0.3 mm.

Only known from the type and probably forming a complex together with *B. brownianum*.

B. elongatum (Schlechter) N. E. Brown (FC 4(1): 862, 1908). **T:** RSA, Eastern Cape (*Schlechter* 2699 [K]). – **D:** RSA (Eastern Cape, Northern Prov.). **I:** Dyer (1983).

≡ *Dichaelia elongata* Schlechter (1894); **incl.** *Brachystelma distinctum* N. E. Brown (1908); **incl.** *Brachystelma gemmeum* R. A. Dyer (1979).

[2] **R** tuber 5 - 9 cm ⌀; stems 1 - 3, erect, 3 - 15 cm, pubescent, densely leafy; **L** ± sessile, elliptic-linear, 7 - 30 × 3 - 8 mm, pubescent; **Inf** sessile, 2- to 5-flowered; **Ped** 2 - 5 mm; **Sep** narrowly lanceolate, 2 - 4 mm; **Cl** 15 - 25 mm long, purple, cage-like; **Cl** tube ± cylindrical, ± 5 × 4 mm, inside glabrous, somewhat ciliate or with long purple vibratile **Ha**; **Cl** lobes linear from an ovate base, 12 - 20 mm, lamina apically united, margins somewhat revolute; **Cn** ± sessile, basally cup-like fused; **Ci** lobes bifid into linear appendages for ≤ ½ of their length, ± divergent, glabrous or apically hairy; **Cs** lobes oblong, lingulate, incumbent on the **Anth**.

B. elongatum forms a group of closely related species together with *B. gracillimum*, *B. gracile*, *B.*

tenellum, and perhaps also with *B. tenue* as well as with *B. villosum* and *B. hirtellum*.

B. exile Bullock (KB 17: 191-193, ills., 1963). **T:** Nigeria, Zaria Prov. (*Keay* s.n. in *FHI 25848* [K]). – **D:** Cameroon, Nigeria; grasslands.

[2] **R** tuber discoid, 3 - 5 cm ⌀; stem solitary, delicate, erect, 30 - 40 cm; **L** sessile, base ± amplexicaul, lanceolate, to 70 × 9 - 12 mm, scatteredly hispid and tomentose; **Inf** lateral, 1-flowered; **Inf**-stems with reduced minute **L**; **Ped** 2 - 3 cm, filiform, glabrous; **Sep** subulate, 2 mm; **Fl** 70 - 80 mm ⌀, outside green, inside purple; **Cl** tube hardly 2 mm ⌀, shallowly cup-shaped; **Cl** lobes linear-subulate, 35 - 40 × 1 - 2 mm, ciliate with purple vibratile clavate **Ha** to 4 mm; **Cn** sessile, ± 1 × 2 mm, greenish-yellow; **Ci** lobes ovate, concave pouch-like, hardly 0.5 mm ⌀; **Cs** lobes linear-clavate, obtuse, incumbent on the **Anth**, dorsally thickened, ending in the same plane as the **Ci**; **Poll** ovoid, ± 0.22 × 0.14 mm.

The delicate, long-tailed flowers swing in the wind.

B. festucifolium E. A. Bruce (HIP 34(3): t. 3369 + text, 1938). **T:** Tanzania, Tabora Distr. (*Lloyd 68* [K]). – **D:** Tanzania.

[?] Tuber not described; stems ascending, hardly branched, ± 20 × 0.3 cm, pale green-brownish, glabrous; **L** sessile, linear-subulate, acute, 40 - 80 × 1 - 1.5 mm, very slightly succulent, glabrous; **Inf** terminal, 8-flowered; **Ped** 20 - 25 mm, scatteredly hairy; **Sep** lanceolate, hairy; **Cl** black-purple, ± 3 cm ⌀, outside glabrous, tube bowl-shaped with bulging margin, ± 5 mm ⌀, inside pale yellow spotted with purple, glabrous; **Cl** lobes triangular, spreading, 8 - 12 × 8 mm, densely hairy, ± bristly, margins with purple long vibratile clavate **Ha**; **Cn** ± sessile, purple, fused to form a cup-like structure, glabrous; **Ci** segments pouch-like, bifid into narrowly triangular appendages, laterally fused with the **Cs**, erect; **Cs** segments ± linear, obtuse, ± 1.5 mm, incumbent on the **Anth**; **Poll** ovoid, ± 0.45 × 0.3 mm.

Belonging to the *B. lineare*-complex. The grass-like leaves, the terminal inflorescences and the triangular corolla segments are characteristic. The interstaminal corona is glabrous and not bearded as in *B. lineare*.

B. filifolium (Schlechter) Peckover (Aloe 33(2-3): 43, 1996). **T:** RSA, Mpumalanga (*Schlechter 11733* [B †, G, HBG, K]). – **D:** RSA (KwaZulu-Natal, Mpumalanga), Swaziland.
≡ *Macropetalum filifolium* Schlechter (1905) ≡ *Tenaris filifolia* (Schlechter) N. E. Brown (1908).

[2] **R** tuber depressed-globose, 3 - 4 cm ⌀; stems filiform, erect but apically mostly drooping, to 75 cm, glabrous; **L** sessile, subulate, 30 - 80 × 1 mm, ascending; **Inf** 1- to 2-flowered, sessile or shortly pedunculate, often clustered in branched terminal racemes, drooping; **Ped** 6 - 8 mm, filiform; **Cl** greenish-purple, glabrous, sweetish-scented; **Cl** tube campanulate, ± 1.5 mm; **Cl** lobes subulate from an oval base, 7 - 8 mm, erect-divergent, folded back along the midrib; **Cn** ± sessile, shortly bowl-shaped, ± 1 × 1.3 mm, outside ± shortly hairy; **Ci** lobes pouch-like, oval, ascending; **Cs** lobes 0.5 mm, linear, ascending; **Poll** ovoid, ± 0.18 mm.

Probably the most delicate of all species of *Brachystelma*. It has inconspicuous but intensely fragrant flowers.

B. floribundum Turrill (BMI 1922: 197, 1922). **T:** Zimbabwe (*Hislop* s.n. [K]). – **D:** Zimbabwe, Moçambique. **I:** Percy-Lancaster (1989).

[2] **R** tuber depressed-globose, to 15 cm ⌀; stem solitary, often basally several times branched and somewhat divaricately spreading, ± erect, to 10 cm tall, scatteredly hispid; **L** very shortly petiolate, narrowly elliptic, to 75 × 9 mm, densely pubescent; **Inf** sessile, 4- to 12-flowered; **Sep** triangular, pubescent; **Ped** 2 - 4 mm, hairy; **Cl** purple, to 22 mm long, outside scatteredly pubescent, inside shortly white-hairy; **Cl** tube campanulate, ± 6 × 9 mm; **Cl** lobes linear-lanceolate from a triangular base, to 25 mm, ± spreading, bent backwards; **Cn** sessile, yellowish, spotted with purple, ± 3.5 mm ⌀, cup-shaped; **Ci** lobes pouch-like, margins emarginate U-like, laterally elongating into 2 densely white-hairy appendages; **Cs** lobes linear-lingulate, incumbent on the **Anth**.

Hairiness, corolla structure and the corona construction are typical for the *B. lineare*-complex. The flowers of *B. floribundum* are very large and deep, however, and the interstaminal corona is incised U-like instead of V-like.

B. foetidum Schlechter (BJS 20(Beiblatt 51): 52, 1895). **T** [lecto]: RSA, Northern Prov. (*Schlechter 3547* [BOL]). – **D:** Botswana, RSA (Northern Prov., Northern Cape, Free State, KwaZulu-Natal, Mpumalanga). **I:** Dyer (1983).
Incl. *Brachystelma rehmannii* Schlechter (1896).

[2] **R** tuber 4 - 6 cm ⌀, upper face flattened; stems 1 to several, erect-decumbent, 7 - 15 cm, ± hispid; **L** lamina basally acutely narrowed into the petiole, linear-lanceolate to ovate or elliptic, 5 - 30 × 3 - 10 mm, lower face and margins ± hairy; **Inf** sessile, 2- to 6-flowered; **Ped** ± 1 cm, tomentose; **Sep** lanceolate, ± 5 mm, pubescent; **Cl** 25 - 50 mm ⌀, with intense scent of manure; **Cl** tube broadly campanulate, 6 - 10 mm deep, inside whitish to yellowish, spotted with purple; **Cl** lobes 10 - 25 mm, outside ± delicately pubescent, inside purple, ± glabrous, ovate-lanceolate, margins basally revolute, apical ⅔ almost completely folded back along the midrib, often greenish-purple; **Cn** sessile, dark purple, basally fused to form a cup-like structure; **Ci** lobes broadly triangular-ovate, concave, apically

incised, inside bearded or not; **Cs** lobes triangular-lingulate, ± 1 mm, incumbent on the **Anth**; **Poll** somewhat pear-shaped; **Fr** slender fusiform, 10 - 12 cm × 5 - 6 mm, tomentose.

B. foetidum is esp. variable in flower size. The intense foetid scent is characteristic. It is related to *B. megasepalum* and *B. petraeum*.

B. franksiae N. E. Brown (in J. M. Wood, Natal Pl. 6: t. 588, 1912). **T:** RSA, KwaZulu-Natal (*Franks* s.n. in *Wood* 11721 [NH, NU, PRE]). – **D:** RSA (KwaZulu-Natal). **I:** Dyer (1983).

[1] **R** fleshy, long-fusiform; stem solitary, somewhat branched, erect, to 30 cm, glabrous or nodes pubescent; **L** ± sessile, linear to narrowly elliptic, to 5 cm, margins and midrib of lower face hispid; **Inf** sessile, 1- to 3-flowered; **Ped** 6 - 10 mm; **Sep** lanceolate, 2 mm; **Cl** 6 - 8 mm long, united only basally, white flushed with pink; **Cl** tube shallowly bowl-shaped; **Cl** lobes oblong, 6 - 7 mm, erect, somewhat folded back along the midrib, margins revolute and somewhat papillose-pubescent; **Cn** ± sessile, ± 3.5 × 2.25 mm; **Ci** lobes erect, deeply divided into 2 linear-fusiform diverging appendages, papillose; **Cs** lobes linear, ± 2.25 mm, erect, somewhat connivent; **Fr** 1 or 2, narrowly fusiform, to 15 cm.

Probably related to *B. macropetalum* which has a small but overall similar corona.

B. furcatum Boele (Excelsa 16: 29, ill. (p. 32), 1994). **T:** Zimbabwe, Matobo Distr. (*Derbyshire* MRSH 3070 [SRGH]). – **D:** Zimbabwe.

[2] **R** tuber depressed-globose, ± 4.5 cm ∅, stems ± erect, divaricately branched, 10 - 12 cm, shortly pubescent; **L** sessile, lanceolate to linear, 9 - 17 × 2 - 3 mm, with few **Ha**; **Inf** sessile, 1-flowered; buds obtuse; **Ped** 2 - 3 mm, pubescent; **Cl** tube campanulate, ± 3 × 3 mm; **Cl** lobes linear from a triangular base, ± 9 × 1 mm, basally pubescent and yellowish, apically purple; **Cn** ± sessile, basally hardly united; **Ci** lobes subquadrangular, erect, ± 0.8 mm, apically incised; **Cs** lobes linear, obtuse, 0.6 mm, incumbent on the **Anth**.

Only insufficiently separated from *B. tabularium* from RSA by the short corolla tube and the shorter corolla lobes; both are probably conspecific. *B. lancasteri* and *B. tavalla* appear to be relatively closely related.

B. gerrardii Harvey (Thes. Cap. 2: 61, t. 196, 1863). **T:** RSA, KwaZulu-Natal (*Gerrard* 1318 [TCD]). – **D:** RSA (Eastern Cape, KwaZulu-Natal, Northern Prov.), Swaziland. **I:** FPA 43: t. 1686, 1976; Onderstall (1996).

Incl. *Brachystelma nigrum* R. A. Dyer (1937).

[1] **R** fleshy, long-fusiform; stem solitary, erect, 7 - 35 cm, densely long-hairy; **L** shortly petiolate, lamina broad, to 4 cm, lower face densely long-hairy, upper face glabrous; **Inf** sessile, 1-flowered; **Ped** 15 - 35 mm; **Sep** subulate, to 8 mm; **Cl** divided almost to the base, black-purple; **Cl** lobes broadly lanceolate, ± 20 × 4 - 6 mm, lower ⅓ erect, then narrowed, bulging and densely beset with purple vibratile **Ha**, apical ⅔ horizontally spreading, margins ± ciliate; **Cn** sessile, united only basally; **Ci** lobes basally pouch-like, apically divided into 2 subulate parallel ± horizontal appendages, 2 - 5 mm, pubescent; **Cs** lobes linear, to 6 mm, apically obtuse, ± emarginate, erect-connivent, dark purple; **Fr** slender fusiform, to 10 cm.

Closely related to *B. comptum* and *B. macropetalum*.

B. glabriflorum (F. Mueller) Schlechter (BJS 50: 161, 1914). **T:** Australia, Queensland (*Mueller* s.n. [K]). – **D:** Australia (Northern Territory, Western Territory), Indonesia, Myanmar, Papua New Guinea, Philippines, Thailand. **I:** Forster (1990).

≡ *Microstemma glabriflorum* F. Mueller (1858); **incl.** *Microstemma tuberosum* R. Brown (1810); **incl.** *Brachystelma kerrii* Craib (1911); **incl.** *Brachystelma papuanum* Schlechter (1914); **incl.** *Brachystelma microstemma* Schlechter (1914) (*nom. inval.*, Art. 52.1); **incl.** *Brachystelma merrillii* Schlechter (1915).

[2] **R** tuber ovoid to depressed-globose, 1 - 8 cm ∅, edible; stem solitary, rarely branched, erect, 20 - 85 cm, glabrous; **L** ± sessile, often strongly reduced, linear-lanceolate, 5 - 100 × 2 - 10 mm, glabrous; **Inf** sessile, laterally near stem tips, 1- to 5-flowered; **Ped** 7 - 25 mm, filiform, ± scatteredly hairy; **Sep** lanceolate, 2 mm, with greenish-yellow **Ha**; **Cl** 16 - 22 mm ∅, tube short, greenish; **Cl** lobes oblong-ovate, 7 - 14 × 2 - 2.5 mm, ascending to horizontal, lamina folded back along the midrib, glabrous or wrinkled, green and purple-spotted or purple, glabrous or sparsely hairy, **Ha** slender, purple; **Cn** ± sessile, ± 1 × 2.3 mm, fused to form a bowl-shaped structure, yellowish with purple margins; **Ci** lobes pouch-like, short, ascending to erect, with a median notch, inside with few white **Ha** or glabrous; **Cs** lobes triangularly pillow-shaped, basally placed against the **Anth**; **Poll** ovoid-ellipsoid, 0.27 × 0.2 mm; **Fr** slender-fusiform, 6 - 12 cm × 2 - 3 mm; **Se** ± 8 × 2 mm.

Forster (1990) compares *B. glabriflorum* with the *B. decipiens* complex. There are, however, several similarities with *B. chloranthum* in habit and the morphology of the glabrous corona, as well as with the complex around *B. edule*.

B. glabrum Hooker *fil.* (Fl. Brit. India 4: 65, 1885). **T:** India, Deccan (*Beddome* s.n. [K]). – **D:** India (Ghats, Deccan).

[2?] **R** unknown; stem solitary, erect, short, ± glabrous; **L** ± sessile, linear-lanceolate, 25 - 70 × 3 - 6 mm; **Inf** terminal, 1- to few-flowered; **Cl** ± 15 mm ∅, pubescent; **Cl** lobes 3× as long as the short **Cl** tube, obtusely linear; **Cn** sessile, very short; **Ci** lobes very short.

An insufficiently known taxon, and probably a synonym of *B. edule*.

B. glenense R. A. Dyer (BT 12(1): 54, 1976). **T:** RSA, Free State (*Mostert* 219 [PRE]). – **D:** RSA (Free State, Northern Prov.). **I:** Dyer (1983); Alp (1998).

Species rank for *B. glenense* is doubtful. The taxon is probably conspecific with *B. tuberosum*, from which it can be distinguished only by the almost glabrous stems and the ± hairy corona.

B. gracile E. A. Bruce (FPA 27: t. 1077 + 2 pp. text, 1949). **T:** Zimbabwe (*Porter* s.n. [PRE 27227]). – **D:** Botswana, Zimbabwe, RSA (KwaZulu-Natal, Mpumalanga, Northern Prov.). **I:** Dyer (1983); Meve (1993). **Fig. III.a**

[2] **R** tuber depressed-globose, 4 - 12 cm ∅; stem mostly solitary, erect, partly drooping apically, very delicate, 15 - 75 cm, pubescent; **L** sessile or shortly petiolate, lanceolate to elliptic-ovate, 20 - 120 × 1.5 - 7 mm, lower face ± hairy; **Inf** sessile, 1- to 4-flowered; **Ped** 2 - 10 mm, filiform, pubescent; **Sep** lanceolate, 1 - 2 mm; **Cl** 7 - 20 mm long, cage-like, green, brown-green or purple-green, inside glabrous, outside pubescent; **Cl** tube very short, ± flat; **Cl** lobes linear-lanceolate, 1 - 2 mm ∅, lamina apically united, somewhat folded back along the midrib, sometimes ciliate; **Cn** sessile, basally hardly fused, 3 - 5 × 3 - 4 mm, black-red, rarely greenish or spotted; **Ci** lobes bifid into narrowly triangular subulate diverging appendages, with few thickish white **Ha**; **Cs** lobes linear-spatulate, 2.5 - 3.5 mm, erect-connivent, apically partly overlapping; **Poll** ± ovoid, ± 0.35 × 0.22 mm; **Fr** slender fusiform, 8 - 13 cm × 3 - 4 mm; **Se** ± 8 - 10 × 4 mm.

B. gracile has a wide distribution and is very variable; the delicate slender plantlets are difficult to locate in grassland.

B. gracillimum R. A. Dyer (JSAB 11: 112, 1945). **T:** RSA, North-West Prov. (*Louw* 811 [PRE]). – **D:** RSA (North-West Prov.). **I:** Dyer (1983).

≡ *Dichaelia gracillima* (R. A. Dyer) Bullock (1953).

[2] **R** tuber 3 - 5 cm ∅; stems divaricately branched, erect, to 45 cm; **L** very shortly petiolate, ovate to ovate-cordate, ± 15 × 7.5 mm, lower face ± hairy; **Inf** sessile, 1-flowered; **Ped** 2 mm; **Sep** lanceolate, 3 mm; **Cl** ± 45 mm long, cage-like, attenuate into a beak, probably purple; **Cl** tube very short; **Cl** lobes linear-filiform, ± 0.5 mm ∅, apically united; **Cn** shortly stipitate, basally hardly fused; **Ci** lobes bifid into ± linear appendages with thickened tip, erect, divergent, hairy; **Cs** lobes oblong elliptic, 2.25 mm, erect-connivent, overlapping; **Fr** slender fusiform, to 9 cm.

B. gracillimum is related to *B. gracile* but unfortunately only known from the type.

B. gymnopodum (Schlechter) Bruyns (BT 25(2): 161, 1995). **T:** Namibia (*Rautanen* 82 [Z]). – **D:** Angola, Botswana, N Namibia, Zimbabwe, NE RSA; Kalahari Sandveld.

≡ *Ceropegia gymnopoda* Schlechter (1896); **incl.** *Ceropegia pygmaea* Schinz (1888); **incl.** *Ceropegia pumila* N. E. Brown (1898) ≡ *Ceropegia pygmaea* var. *pumila* (N. E. Brown) H. Huber (1957).

[2] **R** tuber globose, 2.5 - 5 cm ∅; stems 1 - 3, compactly dwarf-shrubby, only 10 - 15 cm tall, erect, annual, hardly branched, ± pubescent; **L** ascending, ± shortly petiolate, lamina linear to spatulate-obovate, 2.5 - 10 cm, ± pubescent; **Inf** sessile, 1- to 3-flowered, pedunculate to 8 mm, slender; **Ped** 3 - 15 mm, pubescent; **Sep** narrowly triangular, 2 mm, hairy; **Cl** outside greenish, pubescent, inside purple, glabrous or hairy; **Cl** tube 30 - 50 mm, ± cylindrical, basal ⅓ abruptly bent at right angle and merging into the slightly flaring **Cl** mouth near ground-level, 7 - 10 mm ∅; **Cl** lobes broadly triangular, 2 - 3 mm, erect, somewhat folded back along the midrib; **Cn** sessile, urceolate with the margins pulled up, ± 2.5 - 3.5 × 3 mm; free **Ci** lobes shortly triangular, erect or reflexed, inside somewhat hairy; **Cs** lobes united with the base of the urn, ± obtuse, placed on / against the **Anth**; **Poll** D-shaped, ± 0.3 × 0.22 mm.

The long bent corolla tube is very uncommon and the species was therefore placed in *Ceropegia* for over 100 years. It is typically associated with the "Kalahari Sandveld" (see also notes for *B. brevipedicellatum*).

B. hirtellum Weimarck (Bot. Not. 22: 406-408, ills., 1935). **T:** Zimbabwe, Inyanga (*Norlindh & Weimarck* 4399 [LD]). – **D:** Zimbabwe, RSA (Northern Prov.).

Incl. *Brachystelma pilosum* R. A. Dyer (1956).

[2] **R** tuber depressed-globose; stems solitary or several, basally 3 - 5 mm ∅, erect, robust, 20 - 30 cm, hispid; **L** ± sessile, obovate to oblong-elliptic, 10 - 35 × 2 - 20 mm, ± hispid, margins undulate; **Inf** sessile, 2- to 3-flowered; **Ped** 2 - 6 mm, ± villous; **Sep** lanceolate-subulate, 3 - 6 mm, often reflexed, hairy, ciliate; **Cl** 15 - 25 mm long, cage-like; **Cl** tube shallowly bowl-shaped, basally urceolate; **Cl** lobes filiform from a lanceolate base, apically united, inside glabrous, outside brownish-hairy; **Cn** sessile, basally fused to form a bowl-shaped structure, purple, glabrous; **Ci** lobes rectangular, apically bifid into narrowly triangular subulate appendages, 0.6 - 1.5 mm, ± erect; **Cs** lobes rectangular-lingulate from a transversely ovate thickened back, incumbent on the **Anth**; **Poll** flatly ovoid-globose, ± 0.3 × 0.2 mm.

B. hirtellum forms a complex with *B. villosum* and *B. tenue* and takes an intermediate position between these 2 taxa (see also under *B. elongatum*).

B. huttonii (Harvey) N. E. Brown (FC 4(1): 845,

1908). **T:** RSA, Eastern Cape (*Hutton* 3 [TCD, K]). – **D:** RSA (Eastern Cape). **I:** Dyer (1983). **Fig. II.b**
≡ *Decaceras huttonii* Harvey (1863); **incl.** *Brachystelma luteum* Peckover (1993).

[2] **R** tuber depressed-globose, ± 5 cm ∅; stem solitary, richly branched, erect, 5 - 12 cm, reddish, pubescent; **L** ± sessile, linear, 12 - 30 × 1 - 5 mm, greenish-red, margins ± revolute; **Inf** sessile, mostly 2 at the nodes, 1- to 10-flowered; **Ped** 5 - 10 mm, pubescent; **Sep** lanceolate, 1 - 2 mm, glabrous; **Cl** ± 8 mm ∅, greenish, yellow or brownish, tube ± bowl-shaped, 0.5 - 1 × 2 - 3 mm; **Cl** lobes lanceolate, 3 - 5 mm, margins revolute; **Cn** sessile, basally fused to form a cup-like structure, yellow, sunken into the **Cl** tube or exserted from it; **Ci** basally pouch-like, apically bifid into narrowly triangular erect appendages; **Cs** lobes short, subulate, placed on the **Anth**; **Poll** ovoid.

Related most closely to *B. arnotii* and *B. minimum*. The colour of the flowers and the divaricately branched habit is notable.

B. incanum R. A. Dyer (BT 12(1): 54-55, 1976). **T:** RSA, North-West Prov. (*Acocks* 12476 [PRE]). – **D:** Zimbabwe, RSA (North-West Prov.). **I:** Dyer (1983); FPA 53(1): t. 2090, 1994.

[2] **R** tuber depressed-globose, to 5 cm ∅; stem mostly solitary, rarely branched, semi-decumbent to ascending, 10 - 20 cm, densely pubescent, somewhat bristly; **L** petiolate to 5 mm, lamina obovate to ovate-elliptic, 15 - 20 × 10 - 15 mm, densely pubescent to long-hairy; **Inf** ± sessile, 1- to 4-flowered; **Ped** 10 - 15 mm, pubescent; **Sep** narrowly triangular, 2 - 4 mm; **Cl** 9 - 11 mm ∅, outside hairy, tube short, shallow, ± 4 mm ∅, black-purple, basally whitish, with a broad ring of pale purplish long **Ha**, partly extending to the basal **Cl** lobes, these ± 3.5 mm, triangular, horizontal, wrinkled and black-purple, apical ¼ pale red-brown or greenish-yellow and smooth; **Cn** shortly stipitate, yellowish with purple margins, basally fused to form a campanulate structure, ± 1 × 2.5 mm; **Ci** lobes shortly pouch-like, ovate-subquadrangular, apically blunt or somewhat indented; **Cs** lobes short, pillow-shaped, basally placed against the **Anth**; **Poll** pear-shaped, ± 0.25 mm; **Fr** ± 6 cm; **Se** ± 10 × 5 mm, dull olive-green, coma 2 cm.

An attractive species distinctive by the contrasting coloration of the corolla, the indumentum and the seed colour. See also comments under *B. chlorozonum*.

B. johnstonii N. E. Brown (HIP 28: t. 2754 + text, 1901). **T:** Uganda (*Johnston* s.n. [K]). – **D:** Kenya, Mali, Senegal, Tanzania, Uganda. **I:** Agnew & Agnew (1994).

Incl. *Brachystelma bagshawii* S. Moore (1907); **incl.** *Brachystelma lanceolatum* Turrill (1922); **incl.** *Brachystelma medusanthemum* J.-P. Lebrun & al. (1985).

[2] **R** tuber ± depressed-globose, 4 - 6 cm ∅; stems basally branched, erect or decumbent-ascending, 10 - 20 cm, pubescent; **L** ± sessile to shortly petiolate, linear to elliptic, 30 - 50 × 1.5 - 13 mm, upper face glabrous, lower face ± pubescent; **Inf** terminal, 4- to 10-flowered; **Ped** 2 - 5 mm, pubescent; **Sep** subulate, 5 - 8 mm, tomentose; **Fl** purple-brown; **Cl** tube ± campanulate, ± 4 × 6 mm, inside with concentric whitish spots, pubescent; **Cl** lobes linear-subulate from a triangular base, 25 - 75 mm, lamina folded back along the midrib, ± long white-hairy to pubescent; **Cn** ± sessile, purple, fused cup-shaped; **Ci** lobes pouch-like, bifid into narrowly triangular appendages, erect, inside white-bearded, laterally fused with the **Cs**; **Cs** lobes linear, ± 1 mm, incumbent on the **Anth**; **Poll** ovoid-subquadrangular, ± 0.45 × 0.3 mm; **Fr** thickly fusiform, ± 30 × 8 mm.

As in *B. lineare*, there are forms with linear (Mali, Kenya, Uganda) and elliptic (Senegal, Tanzania) leaves. Diagnostic characters for *B. johnstonii* are the terminal inflorescences and the long-tailed corolla segments.

B. keniense Schweinfurth (in Höhnel, Rudolph-Stephanie-See, 860, 1892). **T:** Kenya (*Schweinfurth* s.n. [B?]). – **D:** Kenya.

[2] **R** tuber ± globose; stems basally sparsely branched, delicate, decumbent-ascending, hardly 10 cm tall, pubescent; **L** ± sessile or shortly petiolate, elliptic-lanceolate, to 25 × 7 mm, somewhat hairy; **Inf** 1- to 3-flowered, lateral; **Ped** 4 - 15 mm; **Sep** lanceolate, pubescent; **Fl** outside greenish, inside ± dark purple, glabrous, tube campanulate; **Cl** lobes triangular, ± 6 mm, acute, spreading, ± flat; **Cn** ± sessile, purple, ± 2 × 2 mm, funnel-shaped; **Ci** lobes pouch-like, erect, truncate and incised in the middle into triangular appendages, inside glabrous or white-barbate; **Cs** lobes linear to triangular, ± 0.5 mm, incumbent on the **Anth**; **Poll** broadly ovoid, ± 0.3 × 0.24 mm.

Best distinguished from *B. lineare* by the funnel-shaped corona and the very much shorter triangular corolla lobes, but also similar to *B. decipiens*.

B. kerzneri Peckover (Aloe 31(3-4): 84-85, ills., 1995). **T:** RSA, Eastern Cape (*Peckover* 1722 [PRE]). – **D:** RSA (Eastern Cape).

[2] **R** tuber depressed-globose, 5 - 7 cm ∅; stem solitary, decumbent-ascending, to 15 cm × 4 mm, pubescent; **L** 5 - 7 mm petiolate, lamina elliptic to obovate, 2 - 6 × 1 - 3 cm, pubescent, pale green; **Inf** sessile, 1-flowered; **Ped** 5 - 8 mm; **Sep** lanceolate, 4 - 5 mm, hispid; **Cl** ± 20 mm ∅, tube campanulate, ± 6 × 9 mm, green-yellow with purple concentrically arranged spots and stripes, sparsely hairy; **Cl** lobes greenish with purple longitudinal stripes, ± 7 × 7 mm, broadly triangular, ± horizontally spreading, inside and marginally with reddish **Ha**; **Cn** sessile, basally united with the **Cl** tube, ± 2 × 5 mm, purple,

appearing uniseriate; **Ci** lobes shortly pouch-like, sunken into the **Cl** tube, deeply U-shaped, emarginate, margins laterally fused with the dorsally elongate **Cs** and overtopping the **Cs** with obtusely wing-shaped appendages, inside each with a tuft of whitish **Ha**; **Cs** lobes shortly lingulate, incumbent on the **Anth**, dorsally winged, obtuse; **Poll** D-shaped, 0.5 × 0.4 mm; **Fr** 12 cm × 7 mm; **Se** 8 × 3.5 mm.

Indistinguishable from *B. campanulatum* in habit, but the corona is unique in the genus by the conspicuously dorsally winged staminal corona being fused with the margins of the interstaminal corona. Moreover, both species appear to be close to *B. decipiens*.

B. kolarensis Arekal & Ramakrishna (Proc. Indian Acad. Sci., Pl. Sci. 90(3): 203-205, ills., 1981). **T:** Thailand (*Kerr* 1273 [CAL, K]). – **D:** India (Maharashtra, Karnataka); often in rock fissures. **I:** Yadav & al. (1993).

Incl. *Brachystelma malwanense* Yadav & N. P. Singh (1993); **incl.** *Brachystelma naorojii* P. Tetali & al. (1998).

[2] **R** tuber ± globose, 2 - 5 cm ∅; stems 1, rarely branched, erect, 8 - 50 cm, somewhat hairy or glabrous; **L** ± sessile, opposite, sometimes alternate, narrowly elliptic to ovate-lanceolate, 20 - 90 × 5 - 25 mm, ± glabrous; **Inf** pedunculate, peduncle 4 - 30 mm, pubescent, terminal or lateral near the stem tips, ± umbellate, few-flowered; **Ped** 3 - 8 mm, pubescent; **Sep** 1 - 2 mm, lanceolate, pubescent; **Cl** outside glabrous, inside dark purple or spotted with green-purple, purple-hairy to pubescent; **Cl** tube short, shallow; **Cl** lobes triangular-lanceolate, 4 - 10 mm, ± spreading to erect, sometimes apically fused and forming a cage, lamina folded back along the midrib and margins revolute; **Cn** ± sessile, ± 2 mm ∅, purple, fused to form a cup-shaped structure; **Ci** lobes subquadrangularly pouch-like, ± erect, crenate or incised, often somewhat reclining, hairy; **Cs** lobes triangular, short, basally placed against the **Anth**; **Poll** ovoid, ± 0.25 × 0.15 mm; **Fr** 10 - 15 cm.

B. laevigatum (Buchanan-Hamilton *ex* Wight) Hooker *fil.* (Fl. Brit. India 4: 65, 1885). **T:** India (*Hamilton* s.n. [K]). – **D:** India (Deccan, Karnataka). **I:** Arekal & Ramakrishna (1981: as *B. ciliatum*).

≡ *Eriopetalum laevigatum* Buchanan-Hamilton *ex* Wight (s.a.); **incl.** *Gomphocarpus laevigatus* Buchanan-Hamilton ms. (s.a.) (*nom. inval.*, Art. 32.1c); **incl.** *Brachystelma ciliatum* Arekal & Ramakrishna (1981).

[2] **R** tuber ± globose, 2 - 4 cm ∅, edible; stem solitary, rarely branched, erect, 5 - 40 cm, somewhat pubescent; **L** ± sessile, linear-subulate, 15 - 60 × 1 - 2 mm; **Inf** sessile nutant umbels, 2- to 5-flowered; **Ped** 5 - 20 mm, pubescent; **Sep** 1 mm; **Cl** pendent, ± 6 - 8 mm ∅, outside pubescent; **Cl** tube short, shallow; **Cl** lobes triangular-lanceolate, ± 2 × 1 mm, spreading or bent backwards, greenish to purple, apically with a tuft of clavate **Ha**, pale green to purple; **Cn** sessile, fused to form a cup-shaped structure, ± 1 × 2 mm; **Ci** lobes pouch-like, erect, apically incised to 2-toothed; **Cs** lobes triangular-oblong, covering the **Anth** halfway; **Poll** ovoid, ± 0.2 mm; **Fr** 5 - 9 cm; **Se** 7 × 2 mm.

The nutant umbellate inflorescences are characteristic for this species from the complex around *B. edule*. Further studies are needed to ascertain the synonymy with *B. ciliatum*.

B. lancasteri Boele (Excelsa 16: 30, ill., 1994). **T:** Zimbabwe (*Rogers* 5444 [BOL 42694]). – **D:** Zimbabwe.

[2] **R** tuber depressed-globose or globose, 3 - 7 cm ∅; stems ± erect, sparsely branched, 5 - 15 cm tall, hispid; **L** 1 - 5 mm petiolate, lamina ± elliptic, 5 - 35 × 3 - 20 mm, densely pubescent; **Inf** sessile, 1- to 6-flowered; **Ped** 1 - 4 mm, pubescent; **Sep** triangular, 2 - 3 mm; **Cl** outside green, ± papillose-hispid, inside dark purple; **Cl** tube campanulate, 2.5 - 3 × 3 - 4 mm; **Cl** lobes narrowly triangular to lanceolate, ± 8 × 2 mm, ascending, apical ⅔ bent backwards, strongly folded back along the midrib, inside glabrous to pubescent; **Cn** sessile, ± bowl-shaped, yellowish tinged ± purple or with purple margins; **Ci** lobes ascending, concave, ± 1.5 mm, apically with divergent triangular appendages; **Cs** lobes linear, ± 1 mm, incumbent on the **Anth**.

Rare endemic from Zimbabwe. See also *B. richardsii*.

B. lankanum Dassanayake & Jayasuriya (Ceylon J. Sci. 11(1): 39-41, ills., 1974). **T:** Sri Lanka, Matale Distr. (*Jayasuriya & Dassanayake* 988 [PDA, E, K, L, US]). – **D:** Sri Lanka.

[2] **R** tuber ± globose, 1 - 5 cm ∅; stem 1, rarely branched, erect, 2 - 15 cm, glabrous; **L** ± sessile, linear-elliptic, 7 - 65 × 1 - 7 mm, subsucculent, pale green, glabrous; **Inf** pedunculate to 18 mm, 1- to 3-flowered; **Ped** to 1 cm, filiform; **Sep** 1 - 2 mm; **Cl** campanulate, purple, somewhat leathery, tube campanulate, ± 3 mm ∅, glabrous; **Cl** lobes triangular, 2 mm, ascending, inside with long purple **Ha**; **Cn** ± sessile, ± 1.5 × 2 mm, cream-coloured, ± red-dotted, basally cup-like, fused, **Ci** lobes deeply divided into 2 erect divergent subulate appendages, 0.3 mm, purple; **Cs** lobes bluntly triangular, incumbent on the **Anth**; **Poll** ovoid, 0.2 × 0.15 mm; **Fr** 8 - 11 cm × 3 mm; **Se** ± 6 × 2 mm.

The only species of the genus in Sri Lanka and there endemic. The morphology of the corona is uncommon and is reminiscent of *B. swazicum* as well as several species of *Caralluma*.

B. letestui Pellegrin (Bull. Mus. Nation. Hist. Nat. 32: 393, 1926). **T:** Gabon (*Le Testu* 2352 [P]). – **D:** Gabon. **I:** Lebrun & al. (1994).

[2] Hardly distinguishable from *B. johnstonii* except by the 20- to 30-flowered but equally terminal **Inf** and the triangular, only 10 - 15 mm long **Cl** segments.

B. lineare A. Richard (Tent. Fl. Abyss. 2: 49, t. 72, 1851). **T**: Ethiopia (*Quartin Dillon* s.n. [P]). – **D**: Ethiopia, Ghana, Nigeria, Sudan. **Fig. III.b**

Incl. *Brachystelma ellipticum* A. Richard (1851); **incl.** *Brachystelma phyteumoides* K. Schumann (1893); **incl.** *Brachystelma asmarense* Chiovenda (1912); **incl.** *Brachystelma constrictum* J. B. Hall (1966).

[2] **R** tuber ± depressed-globose, 4 - 12 cm ∅; stems basally several times branched, decumbent-ascending, to 15 cm, glabrous or pubescent; **L** ± sessile to shortly petiolate, linear to elliptic, 15 - 60 × 1 - 10 mm, ± glabrous; **Inf** 1- to 2-flowered, lateral; **Ped** 4 - 6 mm, glabrous or somewhat pubescent; **Sep** triangular, 2 - 4 mm, pubescent or glabrous; **Fl** inside ± wrinkled, tube slightly urceolate, ± 4.5 × 5 mm, yellowish, banded with purple, pubescent or glabrous; **Cl** lobes linear from a triangular base, ± 9 - 12 × 1 - 2 mm, acute, spreading, lamina folded back along the midrib, ± long white-hairy to pubescent or entirely glabrous, purple; **Cn** ± sessile, purple, ± 2.5 × 3 mm, basally cup-like fused; **Ci** lobes pouch-like, erect, truncate, incised in the middle into triangular appendages, inside white-barbate; **Cs** lobes linear, ± 1 mm, ascending, acute or obtuse, placed first against the **Anth** and then ascending freely, or placed entirely on the ascending **Anth**; **Poll** ovoid-subquadrangular, ± 0.4 × 0.25 mm; **Fr** thickly fusiform, ± 30 × 8 mm.

B. constrictum matches the type of *B. lineare* except in the hairiness of the corolla and cannot be upheld as separate species. Within the complex, *B. pellacibellum*, *B. johnstonii* and *B. letestui* as well as *B. keniense* and *B. festucifolium* are closely related, and the same is the case for the complex around *B. decipiens* in RSA.

B. longifolium (Schlechter) N. E. Brown (FC 4(1): 853, 1908). **T** [lecto]: RSA, Mpumalanga (*Schlechter 3873* [BOL]). – **D**: RSA (Mpumalanga). **I**: Dyer (1983).

≡ *Brachystelmaria longifolia* Schlechter (1895) ≡ *Lasiostelma longifolium* (Schlechter) Schlechter (1899).

A very poorly known species with conspicuous leaves (30 - 70 × 4 mm). Schlechter compares it with *B. natalense* and *B. ramosissimum*.

B. macropetalum (Schlechter) N. E. Brown (FC 4(1): 852, 1908). **T** [lecto]: RSA, Mpumalanga (*Schlechter 3869* [BOL]). – **D**: RSA (Mpumalanga, Northern Prov.), Swaziland. **I**: Dyer (1983). **Fig. III.c**

≡ *Brachystelmaria macropetala* Schlechter (1895) ≡ *Lasiostelma macropetalum* (Schlechter) Schlechter (1899).

[1] **R** fleshy, long-fusiform; stem solitary, erect, 25 - 60 cm, tomentose to hispid; **L** shortly petiolate, ovate to somewhat cordate, 20 - 35 × 10 - 20 mm, pubescent; **Inf** sessile, 1-flowered; **Ped** to 25 mm, pubescent; **Sep** subulate, to 13 mm, reflexed; **Cl** united only basally, pale yellow to greenish-yellow, ± spotted green- or red-brown, with dull citrus-scent; **Cl** tube short, shallow; **Cl** lobes ± linear-lanceolate from a triangular base, 20 - 30 mm, outside long-hairy, inside shortly pubescent, lower ⅓ horizontal to ascending, apical ⅔ strongly recurved; **Cn** sessile, ± 5 × 3 mm, densely pubescent; **Ci** lobes erect, ± 2 mm, deeply divided into 2 subulate divergent appendages, dark green; **Cs** lobes spatulate-obovate, ± 5 mm, erect, straight, touching only apically, dark purple to black-green; **Poll** D-shaped to ovoid, ± 0.3 × 0.2 mm; **Fr** slender-fusiform, to 17 cm; **Se** to 10 mm, broadly winged.

Closely related to *B. comptum* and *B. gerrardii* and with similarly attractive flowers.

B. maculatum Hooker *fil.* (Fl. Brit. India 4: 65, 1885). **T**: India, Deccan ? (*Law* s.n. [K]). – **D**: S India (Ghats, Karnataka, Madurai).

Incl. *Brachystelma bourneae* Gamble (1922); **incl.** *Brachystelma rangacharii* Gamble (1922).

[2] **R** tuber fusiform; stem solitary, erect, 25 - 50 (-100) cm, glabrous; **L** sessile, linear-filiform, delicate, 70 - 100 × 2 mm; **Inf** sessile, umbellate, 2- to 3-flowered; **Ped** 5 - 15 mm, slender; **Sep** lanceolate, ± glabrous; **Cl** 16 - 20 mm ∅, tube whitish, spotted with green; **Cl** lobes triangular-lanceolate, ± 10 mm, margins with scattered purple **Ha**; **Cn** short, rounded; **Cs** lobes subulate, 1 mm, considerably overtopping the **Anth**; **Poll** ± globose.

The fusiform tuber and the tall 1-stemmed habit are good characters of *B. maculatum*.

B. mafekingense N.E.Brown (FC 4(1): 854, 1908). **T**: RSA, North-West Prov. (*Green 1683* [K, GRA]). – **D**: Namibia, N RSA. **I**: Dyer (1983: as *Ceropegia*).

≡ *Ceropegia mafekingensis* (N. E. Brown) R. A. Dyer (1977); **incl.** *Ceropegia patriciae* Rauh & Buchloh (1964).

[2] **R** tuber flattened, to 8 cm ∅; stems 1 or more, compact, of dwarf growth, only 1.5 - 6 cm, annual, branched from the base, slightly hairy; **L** ± sessile, narrowly elliptic to lanceolate, 7 - 30 × 5 - 8 mm, margins involute, upper face hairy, lower face glabrous; **Inf** sessile 10- to 20-flowered pseudo-umbels, lateral, later also terminal; **Ped** 3 - 8 mm, hairy; **Sep** narrowly lanceolate, 2 - 4 mm; **Cl** 1 - 1.4 cm, straight, erect, glabrous; **Cl** tube urceolate, 4 - 5 × 3 - 4 mm, slightly 5-angled, outside greenish with purple blotches, ± papillose, inside yellowish, stippled with purple, glabrous; **Cl** lobes 6 - 7 × 2 mm, narrowly obovate, semi-erect, margins and lamina folded back along the midrib, tip involute, inside black-brown, verrucose; **Cn** purple, sessile,

basally fused into a bowl, ± 3 × 2 mm; **Ci** lobes shallowly pouch-shaped with raised lateral margins which join into short erect appendages in front of the **Cs** base; **Cs** ± 2 mm, linear, erect-connivent, apically occasionally papillose; **Poll** globosely pear-shaped, ± 0.25 mm.

The systematic position of this distinct species (*Ceropegia* or *Brachstelma*) remains unsettled. Characters of the stems and the corolla point with considerable certainty to *Brachystelma*.

B. maritae Peckover (CSJA 68(1): 3-5, ills., 1996). **T**: Tanzania, Ruvuma Prov. (*Specks & Specks* 419 [PRE]). – **D**: Tanzania.

[2] **R** tuber depressed-globose, to 10 cm ⌀; stem solitary, ± erect, to 5 cm, laxly hispid; **L** shortly petiolate, elliptic to obovate, 20 - 45 × 15 - 20 mm, densely pubescent; **Inf** sessile, lateral, 2- to 10-flowered; **Sep** lanceolate, 5 mm, hairy; **Ped** 3 - 4 mm, hairy; **Cl** outside ± pubescent, inside glabrous, tube campanulate, 5 - 6 × 5 - 7 mm, inside yellowish with purple spots; **Cl** lobes linear-lanceolate from a triangular base, 25 - 30 mm, ± erect, yellow; **Cn** sessile, yellowish, spotted with purple, ± 2 × 3.5 mm ⌀, cup-shaped; **Ci** lobes pouch-like, margin incised in U-shaped fashion, laterally elongate into 2 white-barbate appendages and fused with the dorsally wing-like broadened **Cs**; **Cs** lobes linear-lingulate, incumbent on the **Anth**.

B. maritae is distinguished from the very closely related *B. floribundum* by the broader leaves, the predominantly yellow corolla with glabrous inside and the slightly different structure of the corona.

B. megasepalum Peckover (KuaS 47(12): 249-252, ills., 1996). **T**: Tanzania, Ruvuma Prov. (*Specks* 385 [PRE]). – **D**: Zambia, Tanzania. **Fig. III.d**

[2] **R** tuber 5 - 6 cm ⌀, with yellow tissue; stems 1 to several, decumbent-creeping to ascending, ± hispid, to 20 cm; **L** 10 - 15 mm petiolate, lamina obovate, 20 - 30 × 12 - 18 mm, shortly hairy; **Inf** sessile, from every other node, 4-flowered; **Ped** 20 - 25 mm, laxly hairy; **Sep** triangular, 8 - 10 × 2 - 3 mm, flat and horizontally spreading, pubescent; **Cl** cage-like, rarely stellate, intensely dung-scented, inside glabrous, outside pubescent; **Cl** tube shallow or somewhat convex, 10 - 12 mm ⌀, inside brown-violet, spotted with yellow-green; **Cl** lobes triangular-linear, 25 - 30 mm, apically united and forming a globose cage, or free, inside ± green-yellow, margins revolute; **Cn** sessile, 3.5 × 2 mm, yellowish, mottled with red-brown, basally united bowl-shaped, glabrous; **Ci** lobes pouch-like, emarginate in deeply U-shaped fashion, each laterally elongate into erectly divergent triangular-subulate appendages; **Cs** lobes rectangular-lingulate, ± 1 mm, incumbent on the **Anth**; **Poll** almost globose, 0.4 × 0.3 mm; **Fr** thickly fusiform, ± 50 × 15 mm; **Se** ± 10 × 7 mm.

Based on the morphology of the corona, the closest relative of *B. megasepalum* is *B. buchananii*, and *B. foetidum* also appears to be close.

B. meyerianum Schlechter (BJS 21(Beiblatt 54): 14, 1895). **T** [lecto]: RSA, Eastern Cape (*Sims* 1508 [K, BOL, GRA, NU, PRE]). – **D**: RSA (Eastern Cape). **I**: Dyer (1983). **Fig. II.c**

Incl. *Brachystelma caffrum* Schlechter (1894) (*nom. illeg.*, Art. 53.1).

[2] **R** tuber depressed-globose, 4 - 6 cm ⌀; stems few, erect, 10 - 15 cm, pubescent; **L** shortly petiolate, lamina linear-lanceolate to elliptic, to 30 × 10 mm, ± hispid; **Inf** sessile, 1- to 4-flowered; **Ped** ± 5 mm; **Sep** lanceolate, 3 mm; **Cl** ± 15 - 18 mm ⌀, outside papillose, inside whitish, pink or greenish, with cheese-like scent; **Cl** tube campanulate to urceolate, ± 6 × 3 mm, inside ± white-hairy; **Cl** lobes ± 7 × 2.5 mm, lanceolate, horizontally spreading, margins somewhat revolute, with long white cilia; **Cn** ± sessile, 2 × 3 mm, fused to form a campanulate to nearly urceolate structure; **Ci** lobes pouch-like, ± triangular and incised in the middle, apical margins on the in- and outside broadly hairy; **Cs** lobes linear-spatulate, ascending, placed against the **Anth**; **Poll** flattened-ovoid, ± 0.25 mm ⌀; **Sty** head bulging; **Fr** fusiform, ± 60 × 4 mm.

The ranges of all characters show great similarities with *B. decipiens*, from which this species can be easily distinguished, however, by flower colour, degree of hairiness and details of corona structure.

B. micranthum E. Meyer (Comm. Pl. Afr. Austr., 196, 1837). **T** [syn]: RSA, Eastern Cape (*Drège* s.n. [not located]). – **D**: RSA (Eastern Cape).

A taxon known only from the protologue, which is based on 2 syntypes which have not been located so far. It is probably synonymous with or closely related to *B. nanum*.

B. minimum R. A. Dyer (BT 12(3): 447, 1978). **T**: RSA, Eastern Cape (*Bayer* 341 [NBG]). – **D**: RSA (Eastern Cape). **I**: Dyer (1983).

[2] **R** tuber depressed-globose, 3.5 cm ⌀; stems 1 - 3, very delicate, erect, 4 - 7 cm, pubescent; **L** ± sessile, linear, 10 - 20 × 1 - 2 mm, with few **Ha**; **Inf** sessile, 1-flowered; **Ped** ± 12 mm, filiform, horizontal or slightly descending, glabrous; **Sep** triangular, 1 mm, glabrous; **Cl** 4 - 6 mm ⌀, black-brown, tube bowl-shaped, 1 × 2 mm; **Cl** lobes broadly triangular, 1.5 - 2 × 1.5 mm, completely reflexed, ciliate; **Cn** sessile, cup-shaped, 1.5 mm, pale yellow-transparent, spotted with purple; **Ci** pouch-like, integrated in the coronal cup, inside hairy, apically emarginate; **Cs** lobes short, pillow-like, laterally fused with the **Ci**; **Anth** ± freely visible; **Poll** 0.2 mm, corpuscle minute.

Very closely related to *B. arnotii* and *B. huttonii*. The last-named is also only known from the region around Grahamstown.

B. minor E. A. Bruce (FPA 28: t. 1096A + text, 1951). **T:** RSA, Northern Prov. (*Murray* 28365 [PRE]). – **D:** RSA (Northern Prov.). **I:** Dyer (1983).

[2] **R** tuber depressed, ± 5 cm ∅; stem solitary, unbranched, erect, 3 - 5 cm, delicately tomentose; **L** 8 - 16 mm petiolate, petiole semi-erect, grooved above, lamina ± spatulate, to 20 mm, marginally somewhat ciliate; **Inf** shortly pedunculate, 2- to 3-flowered; **Ped** 4 - 5 mm; **Sep** triangular-lanceolate, 1.5 mm; **Cl** 8 - 9 mm ∅, outside greenish-cream-coloured with reddish venation, inside purple, with white bristly **Ha**; **Cl** tube distinctly campanulate, 4 - 5 mm ∅; **Cl** lobes triangular, ± 2.5 mm, lamina recurved; **Cn** ± sessile, ± 1 × 1.8 mm, united only basally, ± red-brown; **Ci** lobes ovate, pouch-like, outside shaggy-hairy, apically bifid into 2 ± erect triangular-subulate appendages; **Cs** lobes subquadrangular, 0.5 mm, incumbent on the **Anth**; **Fr** ± 35 × 4 mm.

Closely related to *B. swazicum* and *B. parvulum*, but easily recognizable because of the campanulate corolla with the remarkable bristly hairs.

B. modestum R. A. Dyer (FPA 30(1): t. 1165A + text, 1954). **T:** RSA, KwaZulu-Natal (*Codd* 6984 [PRE]). – **D:** RSA (KwaZulu-Natal). **I:** Dyer (1983).

[2] **R** tuber depressed-globose, 2 - 3 cm ∅; stems 1 or 2, erect, 2 - 4 cm, glabrous or tomentose; **L** in 3 - 5 pairs, shortly petiolate, lamina oblong-elliptic, 10 - 30 × 5 - 10 mm, sometimes somewhat hairy; **Inf** sessile, 1- to 2-flowered; **Sep** lanceolate, 3 mm; **Cl** ± 15 mm ∅, tube broadly campanulate, ± 2.5 × 10 mm, inside pale brown, intermittently banded concentrically with purple; **Cl** lobes ± 4 mm, ± horizontally bent backwards, broadly triangular, apically purple-brown; **Cn** ± sessile, purple, ± 2 × 3.5 mm, united cup-like; **Ci** lobes pouch-like, apically emarginate or 2-toothed; **Cs** lobes lingulate, incumbent on the **Anth**; **Poll** ± 0.25 mm, caudicles very short.

A further micro-species from the complex around *B. australe*.

B. molaventi Peckover & A. E. van Wyk (Aloe 36(2-3): 46-48, ills., 2000). **T:** RSA, KwaZulu-Natal (*van Wyk* 12483 [PRU, NH]). – **D:** RSA (KwaZulu-Natal: Ngele Nature Reserve).

[2] **R** tuberous, flattened, to 2 × 4 cm; stems numerous, ascending-procumbent, to 15 cm × 2 mm, ± hispid; **L** shortly petiolate, lamina lanceolate to cordate, 3 - 8 × 2 - 5 mm, sparingly hairy; **Inf** sessile, 1-flowered; **Ped** 35 × 0.5 mm, horizontal, sparingly hairy; **Sep** red-brown, 1 - 2 mm, acute; **Cl** radiate, ± 25 mm ∅; **Cl** tube brown-green, ± 4 mm ∅, flat, short-hairy; **Cl** lobes brownish-green, lanceolate, to 12 × 1 mm, slightly recurved, bent back along the midrib, upper face with purple **Ha** 0.1 mm long; **Cn** yellowish to purple-brownish, glabrous, sessile, dish-shaped, ± 1 × 3 mm; **Ci** pouch-like, apically broadly U-shaped; **Cs** lobes linear-spatulate, decumbent on the back of the **Anth**, dorsally thickened; **Fr** erect, cylindrical, 20 - 30 × 3 mm; **Se** dark brown, 5 × 3 mm, tuft of **Ha** 10 - 12 mm.

Closely related to *B. petraeum*, but differentiated by its solitary root tubers and a widely reduced corolla tube.

B. montanum R. A. Dyer (BT 6(3): 540-541, 1956). **T:** RSA, Eastern Cape (*Galpin* 6278 [PRE, K]). – **D:** RSA (Eastern Cape). **I:** Dyer (1983).

[2] **R** tuber depressed-globose, 2 - 5 cm ∅; stems hardly branched, erect, ± 5 - 10 cm, ± hispid, partly underground; **L** linear-elliptic, basally narrowing and ± sessile, 15 - 25 × 2 - 4 mm, ± hispid; **Inf** sessile, 2-flowered; **Ped** 10 - 35 mm, tomentose; **Sep** lanceolate, 3 mm; **Cl** tube campanulate, ± 4.5 mm ∅; **Cl** lobes to 10 mm, ± linear from a triangular base, margins revolute, hairy below the middle; **Cn** ± sessile, 2 mm long, cylindrical-campanulate; **Ci** lobes pouch-like, apically emarginate in broadly V-shaped fashion; **Cs** lobes linear, placed on the **Anth**.

A hardly known taxon.

B. mortonii C. C. Walker (Asklepios 24: 66-67, 1981). **T:** Ghana, Northern Region (*Morton* 7381 [K]). – **D:** Ghana, Ivory Coast. **I:** Meve & Porembski (1993). **Fig. II.d**

Incl. *Brachystelma parviflorum* J. K. Morton (1978) (*nom. illeg.*, Art. 53.1).

[2] **R** tuber depressed-globose, 3 - 6 cm ∅; stems basally several times branched, decumbent-ascending, 4 - 20 cm, sparsely hairy; **L** ± sessile, pale green, narrowly elliptic to linear, 20 - 80 × 2 - 5 mm, stiff, erectly divergent, midrib hairy below; **Inf** sessile, 3- to 4-flowered; **Sep** triangular-lanceolate, 3 - 5 mm; **Cl** 7 - 16 mm ∅, opening only halfway, very coarse and fleshy, outside greenish, papillate, inside red-brown, wrinkled; **Cl** tube shallow, bowl-shaped; **Cl** lobes 3 - 9 × 3 - 5 mm, triangular-ovate, apical ½ thickened nose-like; **Cn** sessile, pale purple, 2 - 3 × 6 - 8 mm, united cup-like; **Ci** lobes basally pouch-like, inside hairy, bifid into 2 triangular-subulate broadly diverging appendages ± 2.5 mm long, ascending, several times bent; **Cs** lobes broadly triangular from a pillow-like thickened back, obtuse, 0.5 mm, incumbent on the orange **Anth**, dorsally expanded wing-like and fused with the **Ci** margins; **Poll** almost square, ± 0.3 mm, **Fr** 30 - 50 × 10 - 15 mm; **Se** ± 10 × 6 mm.

The fleshy bulges on the corolla tips are significant for *B. mortonii*. The species is very closely related to *B. plocamoides*, and when the forms are better known, the 2 will probably have to be interpreted as representing the same variable species.

B. nanum (Schlechter) N. E. Brown (FC 4(1): 848, 1908). **T:** RSA, Free State (*Zeyher* 509 [K]). – **D:** RSA (Northern Prov., North-West Prov., Eastern

Cape, Mpumalanga); sand-filled pockets on rocks. **I:** Craib (1994); Peckover (1994: as *B. angustum*). **Fig. II.f, II.g**

≡ *Lasiostelma nanum* Schlechter (1905); **incl.** *Brachystelma dyeri* K. & M.-J. Balkwill & Cadman (1988); **incl.** *Brachystelma angustum* Peckover (1995).

[2] **R** tuber ± globose, 2 - 5 cm ∅; stems 1 to several, erect, glabrous or very sparsely hairy, 3 - 12 cm; **L** shortly petiolate, lamina linear-elliptic to ovate, 10 - 25 × 3 - 7 mm, often somewhat ciliate; **Inf** sessile, 1- to 3-flowered; **Ped** 3 - 14 mm; **Sep** 1 - 3 mm; **Cl** 9 - 14 mm ∅ or 5 - 7 mm long, stellate or cage-like, whitish to rose-coloured, partly with green dots; **Cl** tube bowl-shaped; **Cl** lobes 3.5 - 6.5 mm, spreading or tips united, tips greenish, bent inwards, margins ± revolute; **Cn** sessile, ± 2 × 2 mm, white with red or green spots, basally united cup-like; **Ci** lobes deeply divided into 2 erect triangular to subulate appendages, inside usually papillose, rarely hairy; **Cs** lobes linear to slightly spatulate, incumbent on the **Anth** or erect and with reflexed tips; **Poll** ± globose, 0.2 - 0.3 mm ∅, corpuscle slender, ± 0.3 mm long; **Fr** 30 - 50 × 3 - 4 mm; **Se** ± 7 × 3 mm.

The plants are generally glabrous with elegant, partly stellate, partly cage-like flowers. *B. dyeri* is only separable on the base of the corona with a hairy inside and the longer **Cs** segments. In general, the overlap in characters is such that *B. dyeri* can at most be recognized as variety.

B. natalense (Schlechter) N. E. Brown (FC 4(1): 850, 1908). **T** [lecto]: RSA, KwaZulu-Natal (*Wood* 176 [NH]). – **D:** RSA (KwaZulu-Natal). **I:** Ngwenya & al. (1995).

≡ *Aulostephanus natalensis* Schlechter (1896); **incl.** *Brachystelma inandense* E. Phillips (1941) (*nom. illeg.*, Art. 52.1).

[1] **R** fleshy, fusiform; stem solitary, rarely branched, erect, 20 - 60 cm, with rough erect **Ha**; **L** ± sessile or petiolate, lamina ovate-elliptic to almost circular, 20 - 45 × 10 - 35 mm, densely long- and white-hairy; **Inf** ± sessile, clustered near stem tips, 2- to 5-flowered; **Ped** 15 - 30 mm, slender; **Sep** lanceolate, 2 mm; **Cl** stellate, 3.5 - 5 mm ∅, outside green, inside yellow, tube shortly cup-shaped; **Cl** lobes triangular-ovate, ± 3 × 1.25 mm; **Cn** sessile, yellow, 1.5 × 4 mm, united only basally; **Ci** lobes pouch-like, outer margin deeply incised, 2-pointed; **Cs** lobes shortly strap-shaped, ascending; **Poll** ± 0.15 × 0.1 mm; **Fr** slender-fusiform, 10 - 15 cm, often solitary.

Within the small species group with fusiform roots, *B. natalense* is probably closest to *B. comptum*. The mostly solitary fruits are unusual.

B. nepalense (Radcliffe-Smith) Meve (KB 52(4): 1012, 1997). **T:** Nepal (*Stainton & al.* 3327 [BM]). – **D:** W Nepal. **I:** Radcliffe-Smith (1967).

≡ *Riocreuxia nepalensis* Radcliffe-Smith (1967).

[1,2] **R** tubers small, 1 - 2.5 cm ∅, with numerous fleshy **R**; stem solitary, little-branched, erect, to 30 cm, hispid; **L** in only 4 - 6 pairs, sessile or very shortly petiolate, lamina broadly ovate to lanceolate, 10 - 30 × 5 - 20 mm, upper face entirely hairy, lower face hairy only on the veins; **Inf** lateral, pedunculate to 18 mm, 1- to 4-flowered; **Ped** to 17 mm, tomentose; **Sep** lanceolate, 2 - 3 mm; **Cl** campanulate, 7 - 10 mm long, horizontal to erect; **Cl** tube 3 - 5 mm deep, white; **Cl** lobes triangular, ± 5 mm, purple-brown; **Cn** 2 mm stipitate; **Ci** lobes basally united to form a short bowl-like structure, linear-lingulate, 2 mm, erect, placed against the inside of the **Cl** tube, ciliate, apically incised; **Cs** lobes subquadrangular-triangular, 0.3 mm, incumbent on the **Anth**; **Fr** glabrous.

An isolated species from the Nepalese Himalaya. Similarities in vegetative and coronal characters indicate a closer relationship with the South African *B. comptum*.

B. ngomense R. A. Dyer (BT 12(2): 255, 1977). **T:** RSA, KwaZulu-Natal (*Hilliard & Burtt* 8441 [PRE, P, S]). – **D:** RSA (KwaZulu-Natal: near Ngome). **I:** Dyer (1983).

[2] **R** tuber ± globose, 2 - 3 cm ∅; stems 1 - 3, delicate, hardly branched, decumbent, ± creeping, 10 - 20 cm, glabrous or papillate to shortly hairy; **L** shortly petiolate, lamina ovate-elliptic, 10 - 15 mm, glabrous or margins scatteredly hairy; **Inf** sessile, 1-flowered; **Ped** 5 - 10 mm; **Cl** purple or white and spotted with purple, ± 12 mm ∅; **Cl** tube shallow; **Cl** lobes triangular, ± 5 mm, widely reflexed, margins revolute; **Cn** ± 2.5 × 4 mm, united bowl-shaped; **Ci** lobes concavely pouch-like, outside toothed and directed inwards; **Cs** lobes shortly pillow-like to lingulate, placed against or on the **Anth**; **Poll** ± 0.25 mm; **Fr** ± 3 cm.

The species rank for this taxon is uncertain. It is close to *B. coddii* and *B. pulchellum* in both vegetative and flower characters and perhaps only a synonym of the last-named.

B. occidentale Schlechter (Verh. Bot. Vereins Prov. Brandenburg 35: 53, 1893). **T:** RSA, Western Cape (*Schlechter* 666 [B†]). – **D:** RSA (Western Cape). **I:** Dyer (1983). **Fig. III.e**

≡ *Brachystelmaria occidentalis* (Schlechter) Schlechter (1897) ≡ *Lasiostelma occidentale* (Schlechter) Schlechter (1905).

[2] **R** tuber oblong-globose, small; stems to 8 cm, erect, hispid; **L** shortly petiolate, lamina linear, to 15 mm; **Inf** sessile, 1- to 2-flowered; **Cl** 6 mm ∅, white with wine-red, tube campanulate; **Cl** lobes 2 mm; **Cn** sessile, basally united to form a cup-like structure; **Ci** lobes divided into 2 long linear appendages directed inwards; **Cs** lobes linear, incumbent on the **Anth**.

Poorly known only from the type.

B. oianthum Schlechter (BJS 20 (Beiblatt 51): 53, 1895). **T** [syn]: RSA, North-West Prov. (*Schlechter* 3557 [B†]). – **D**: RSA (North-West Prov., Mpumalanga, Free State). **I**: Dyer (1983).

Incl. *Brachystelma erianthum* Schlechter *ex* K. Schumann (1895) (*nom. inval.*, Art. 32.1c).

[2] **R** tuber depressed-globose, ± 5 cm ⌀; stems 1 - 2, erect to decumbent, 20 - 80 cm tall, tomentose; **L** shortly petiolate, lamina ovate-lanceolate, 10 - 60 × 5 - 15 mm, lower face paler, hairy all over; **Inf** sessile, 1-flowered; **Ped** ± 5 mm, pubescent; **Sep** narrowly triangular, 5 mm, hairy; **Cl** outside yellowish, slightly spotted with purple, inside intensely spotted with purple, papillose and long-hairy; **Cl** tube 15 - 25 mm long, ovoid-campanulate with short erect triangular **Cl** lobes, ± 3 mm long and broad, marginally with vibratile clavate **Ha**, purple or rose-coloured; **Cn** sessile, basally united to form a bowl-like structure, ± 2.5 × 3 mm; **Ci** lobes pouch-like, subquadrangular, apically somewhat incised, inside with a brush-like fringe of **Ha**, fused without suture with the base of the **Cs**; **Cs** lobes lingulate, hardly 1 mm long, incumbent on the **Anth**; **Poll** somewhat pear-shaped.

Very attractive because of the large campanulate flowers. Vegetative characters and the construction of the corona indicate a relationship to *B. decipiens* etc.

B. omissum Bullock (KB 17: 193-195, ills., 1963). **T**: Cameroon (*Maitland* 1712 [K]). – **D**: Cameroon.

[2?] **R** tuber unknown; stem solitary, little-branched, erect, to 30 cm, rather stout, hispid; **L** ± sessile, elliptic, ± 50 × 15 mm, tomentose; **Inf** terminal sessile 8- to 12-flowered umbels, most **Fl** opening simultaneously; **Ped** to 4 cm, delicate, tomentose; **Sep** lanceolate, 4 mm, tomentose; **Cl** ± to 25 mm ⌀, outside tomentose, inside velvet-purple; **Cl** tube broadly funnel-shaped; **Cl** lobes broadly triangular, flatly spreading; **Cn** sessile, red-brown, ± 2 × 3.5 mm, united shortly cup-like; **Ci** lobes pouch-like but outer margin with deeply U-shaped incision, each with lateral erect connivent appendages, subulate-acute, apically tomentose; **Cs** lobes broadly lingulate, obtuse, placed on the **Anth**.

The long pedicels and the purple flower colour represent the major differences from *B. togoense* and species rank for *B. omissum* is questionable. The taxon is also related to *B. johnstonii* and *B. letestui*.

B. pachypodium R. A. Dyer (FPA 32: t. 1269 + text, 1958). **T**: RSA, Northern Prov. (*Codd* 9456 [PRE, K]). – **D**: RSA (Northern Prov.). **I**: Dyer (1983).

[2] **R** tuber irregularly depressed-globose, ± 10 × 13 cm; stems several, semi-decumbent to ascending, 7 - 20 cm, shortly tomentose; **L** shortly petiolate, lamina broadly ovate to elliptic-ovate, 25 - 50 × 20 - 40 mm, upper face glabrous, margins and lower face on the veins with few **Ha**; **Inf** sessile, 1- to 2-flowered; **Ped** 5 - 15 mm, shortly tomentose; **Sep** lanceolate, 4 mm; **Cl** 20 - 25 mm long, cage-like, pale green-yellow, inside ± glabrous, outside with few **Ha**; **Cl** tube campanulate, 5 × 8 mm; **Cl** lobes linear-lanceolate, ± 17 mm, lamina apically united, somewhat folded back along the midrib, margins undulate, basally auriculate and bent inwards; **Cn** shortly stipitate, basally united to form a bowl-like structure, outside hairy; **Ci** lobes bifid into narrowly triangular-subulate erect appendages; **Cs** lobes linear-lingulate, erect-connivent; **Poll** ± globose, ± 0.35 mm.

Characters of flowers and corona of *B. pachypodium* are reminiscent of *Ceropegia*.

B. parviflorum (Wight) Hooker *fil.* (Fl. Brit. India 4: 65, 1885). **T**: India (*Royle* s.n. [K]). – **D**: NW India.

≡ *Eriopetalum parviflorum* Wight (1834).

[2?] **R** unknown; stem solitary, erect, ± glabrous; **L** ± sessile, linear-filiform, acute, ± 100 × 2.5 mm; **Inf** 1-flowered; **Ped** 3 - 5 mm; **Sep** subulate; **Cl** ± 17 mm ⌀, white with few purple spots; **Cl** lobes 4× as long as the **Cl** tube, long-hairy; **Cs** lobes overtopping the **Gy**; **Fr** narrowly fusiform, ± 60 × 3 mm.

Insufficiently known taxon, and probably a synonym of *B. attenuatum*.

B. parvulum R. A. Dyer (BT 12(3): 447-448, 1978). **T**: RSA, Northern Prov. (*Hardy* 4090 [PRE]). – **D**: RSA (Northern Prov.). **I**: Dyer (1983).

[2] **R** tuber ± globose and 2 cm ⌀; stem solitary, unbranched, erect, to 6 cm, delicately tomentose; **L** to 5 mm petiolate, lamina oblong-lanceolate, 25 × 4 mm, upper face somewhat tomentose; **Inf** sessile, 1- to 2-flowered; **Ped** to 12 mm, tomentose; **Sep** ovate-lanceolate, 1.5 mm; **Cl** 6 - 9 mm ⌀, tube campanulate, ± 2 × 2 mm, whitish; **Cl** lobes purple-brown, 2.3 × 2 mm, lamina and margin somewhat revolute; **Cn** sessile, united only basally; **Ci** lobes whitish, purple-margined, basal ½ shallowly pouch-like, apical ½ divided into 2 erect laterally diverging subulate appendages, apically white-hairy; **Cs** lobes brown-red, ± 1.5 mm, incumbent on the **Anth**; **Poll** globose, ± 0.2 mm.

Closely related to *B. swazicum* and *B. minor*.

B. pauciflorum Duthie (Fl. Gangetic Plain 2: 64, 1911). **T**: India (*Duthie* s.n. [K]). – **D**: India (Uttar Pradesch).

[2] **R** tuber globose, small; stem solitary, erect, 25 - 35 cm, glabrous; **L** sessile, linear-lanceolate, 5 - 10 cm, sometimes falcate, lower face with bulging midrib; **Inf** sessile, lateral, 1-flowered; **Ped** 2 - 3 mm, slender; **Cl** dark purple-brown; **Cl** lobes lanceolate, inside tomentose, margins revolute; **Cn** with 5 conspicuous appendages, subulate, overtopping the **Anth** considerably.

A hardly known taxon and closely related to or conspecific with the insufficiently known *B. parviflorum*.

B. pellacibellum L. E. Newton (Bradleya 14: 97, ills. (p. 95), 1996). **T:** Kenya, Rift Valley Prov. (*Powys* s.n. in *Newton* 4230 [K, EA]). — **D:** Kenya (highlands).

[2] **R** tuber depressed-globose, to 10 cm ∅; stems basally branched, ± erect, somewhat divaricately divergent, to 45 cm, tomentose; **L** ± sessile, linear, to 75 × 6 mm, upper face glabrous, lower face ± tomentose, esp. on midrib and along margins; **Inf** terminal, umbellate, to 8-flowered; **Sep** triangular, 2 mm, somewhat hairy; **Ped** 3 mm, tomentose; **Fl** strongly foetid; **Cl** tube bowl-shaped, ± 3 × 5 - 7 mm, inside whitish, yellowish or greenish, red-spotted, tomentose; **Cl** lobes linear from a triangular base, 13 - 18 × 2 mm, acute, ± spreading to ascending, lamina almost completely folded back along the midrib, purple or green-purple, glabrous or tomentose; **Cn** sessile, yellow, basally united bowl-shaped; **Ci** lobes erect, ± 1 mm, bifid into triangular ± divergent appendages, glabrous or white-barbate; **Cs** lobes linear, ± 1.5 mm, placed on the **Anth**; **Poll** pear-shaped, ± 0.4 × 0.25 mm.

Foliage and corolla are very similar to the *B. lineare*-complex. However, the corolla tube and the corona are broader and more open, the corolla is dotted or blotched instead of banded and the corona is yellow instead of purple. *B. johnstonii* and *B. floribundum* are also closely related.

B. perditum R. A. Dyer (JSAB 43(1): 10, 1977). **T:** RSA, KwaZulu-Natal (*Wylie* s.n. in *Medley Wood* 11229 [NH]). — **D:** RSA (KwaZulu-Natal), Lesotho. **I:** Dyer (1983).

[2] **R** tuber depressed-globose, 4 - 6 cm ∅; stems 1 to several, erect-decumbent, to 50 cm, delicately tomentose; **L** shortly petiolate, lamina ovate-elliptic, 15 - 20 × 5 - 8 mm, lower face and margins delicately hairy; **Inf** sessile, 1- to 2-flowered; **Ped** ± 1 mm, tomentose; **Sep** lanceolate, ± 3 mm, hairy; **Cl** 40 - 60 mm ∅, inside yellowish, densely purple-spotted, carrion-scented; **Cl** tube broadly campanulate, ± 10 × 4 mm; **Cl** lobes 15 - 22 mm, ovate-lanceolate, wrinkled, finely papillose, basal margins revolute and ± hairy, apical ⅔ almost completely folded back along the midrib; **Cn** shortly stipitate, ± 4 × 4.5 mm, purple-brown; **Ci** lobes pouch-like, ovate, apically somewhat incised, laterally fused with the back of the **Cs**; **Cs** lobes subquadrangular, ± 1 mm, basally with median groove, apically somewhat swollen, placed on the **Anth** and covering them completely; **Poll** obtusely broadly elliptic, ± 0.4 mm.

A very large-flowered species, related to *B. coddii* and other species from high altitudes.

B. petraeum R. A. Dyer (JSAB 43(1): 17, 1977). **T:** RSA, KwaZulu-Natal (*Hilliard* 5601 [PRE, NU]). — **D:** RSA (KwaZulu-Natal). **I:** Dyer (1983).

[1,2] **R** tubers numerous, irregularly finger-shaped, 4 - 6 cm ∅; stems numerous, erect-decumbent, ± hispid, partly rhizomatous; **L** shortly petiolate, lamina elliptic-ovate, 5 - 15 × 3 - 8 mm, normally somewhat hairy marginally; **Inf** sessile, 1-flowered; **Ped** 1 - 2 cm, tomentose; **Sep** lanceolate, 3 - 4 mm; **Cl** 12 - 20 mm ∅, tube broadly campanulate, 3 - 5 mm ∅, inside mustard-coloured, red-spotted; **Cl** lobes 5 - 9 mm, basally margins revolute, apical ⅔ almost completely folded back along the midrib, unspotted, red-brown, margins with long soft **Ha**; **Cn** ± sessile, ± 3 mm ∅, shallowly cup-shaped; **Ci** lobes pouch-like, broadly U-shaped emarginate, apically 2-toothed; **Cs** lobes linear-spatulate, incumbent on the **Anth**, dorsally conspicuously thickened; **Poll** somewhat pear-shaped, ± 0.3 mm; **Fr** cylindrical, ± 25 mm.

The numerous tuberous roots and the torulose rhizomes are notable for *B. petraeum*. Dyer (1980) places it in the vicinity of *B. alpinum* and *B. pulchellum*, but corona and corolla are much more similar esp. to *B. foetidum*.

B. plocamoides Oliver (Trans. Linn. Soc. London 29: 112, t. 77, 1875). **T:** Tanzania (*Speke & Grant* s.n. [K]). — **D:** Malawi, Tanzania, Zambia, Zimbabwe. **I:** Meve & Porembski (1993).

Incl. *Brachystelma linearifolium* Turrill (1914).

[2] **R** tuber strongly depressed-globose, 5 - 12 cm ∅, edible; stem basally several times branched, erect or decumbent-ascending, 7 - 40 cm, sparsely hairy; **L** shortly petiolate, lamina elliptic to linear, 40 - 80 × 4 - 10 mm, ± erectly divergent, midrib thickened below, ± hairy; **Inf** sessile, 1- (to 3-) flowered; **Sep** triangular-lanceolate, 3 - 4 mm; **Ped** 10 - 15 mm; **Cl** 15 - 30 (-40) mm ∅, outside pale green, glabrous or papillate, wrinkled to pustulate; **Cl** tube very short; **Cl** lobes 7 - 15 (-20) × 2 - 5 mm, triangular-ovate, slightly ascending, ± folded back along the midrib, inside red-brown, basally pale; **Cn** sessile, pale purple, 2 × 5 - 7 mm, united bowl-like; **Ci** lobes basally pouch-like, inside hairy, then divided into 2 narrowly triangular broadly divergent appendages, ± 1.5 mm, ± horizontal; **Cs** lobes ± pillow-shaped, placed against the orange **Anth**, dorsally extended wing-like and fused with the margins of the **Cs**; **Poll** almost square, ± 0.3 mm.

The rather thin-textured corolla lobes being folded longitudinally are of diagnostic value for *B. plocamoides* (see also *B. mortonii*).

B. praelongum S. Moore (J. Bot. 40: 384, 1902). **T:** RSA, Free State (*Burrett-Hamilton* s.n. [BM]). — **D:** Lesotho, RSA.

Although the 2 subspecies have a disjunct distribution, overlapping forms seem to exist (Craib 1994) and the division into 2 taxa is perhaps artificial.

B. praelongum ssp. **praelongum** − **D:** Lesotho, RSA (Western Cape, Free State, KwaZulu-Natal). **I:** Dyer (1983); Boele (1989).

[2] **R** tuber ± globose, 3 - 4 cm ∅; stems 1 - 2, erect or decumbent-ascending, 5 - 15 cm, ± tomentose; **L** shortly petiolate, lamina linear, to 30 mm, ± tomentose; **Inf** sessile, 1- to 2- or many-flowered; **Ped** 15 - 45 mm; **Sep** lanceolate, 3 mm; **Cl** ± 14 mm ∅, greenish-yellow, tube campanulate, ± 4 mm ∅; **Cl** lobes ± 5 mm, margins revolute, apically darker; **Cn** ± sessile, united cup-like, **Ci** lobes pouch-like, deeply divided into 2 triangular-falcate appendages, finely pubescent; **Cs** lobes linear-spatulate, incumbent on the **Anth**; **Fr** subcylindrical, 70 × 5 mm.

B. praelongum ssp. **thunbergii** (N. E. Brown) Boele (Asklepios No. 47: 3, ills., 1989). **T:** RSA, Western Cape (*Thunberg* s.n. [UPS [Herb. Thunberg 6295]]). − **D:** RSA (Eastern Cape, Western Cape). **I:** Dyer (1983: as *B. thunbergii*).
   ≡ *Brachystelma thunbergii* N. E. Brown (1908); **incl.** *Cynanchum crispum* Thunberg (1794).

[2] Differs from ssp. *praelongum*: **L** linear to elliptic, to 40 × 12 mm; **Ped** only 2 - 8 mm; **Fl** somewhat larger and **Cn** shorter, glabrous.

B. prostratum E. A. Bruce (KB 1948: 462-463, ills., 1948). **T:** Zambia (*Milne-Redhead* 3242 [K?]). − **D:** Zambia.

[2] **R** tuber depressed-globose, ± 3 cm ∅; stems basally branched, ± decumbent, sparsely tomentose; **L** sessile to shortly petiolate, lamina narrowly elliptic to linear, 8 - 13 × 1 - 3 mm, strictly erect, acute, margins somewhat undulate, ± glabrous; **Inf** sessile, 1- (to 2-) flowered; **Sep** lanceolate, 4 - 5 mm; **Ped** 4 - 5 mm; **Cl** ± 9 mm ∅, outside glabrous, inside shortly white-hairy; **Cl** tube campanulate, ± 2 mm ∅; **Cl** lobes triangular-ovate, ± 3.5 mm, spreading, ± folded back along the midrib, inside red-brown, basally pale; **Cn** sessile, purple-red, 2 × 5 - 7 mm, united to form a globose-urceolate structure; **Ci** lobes inside hairy, apically elongate into 2 triangular lateral appendages; **Cs** lobes ± pillow-shaped, incumbent on the orange **Anth**, dorsally extended and united with the margins of the **Cs**.

Only known from the type, this species could be confused with *B. plocamoides* from its habit. It has smaller flowers, however, and the corona is not flat and open, but almost urceolate.

B. pulchellum (Harvey) Schlechter (BJS 20 (Beiblatt 51): 53, 1895). **T:** TCD. − **D:** RSA (KwaZulu-Natal, Northern Prov.). **I:** Dyer (1983). **Fig. II.e**
   ≡ *Micraster pulchellus* Harvey (1868).

[2] **R** tuber ± 5 cm ∅; stems several, creeping, to 25 cm, porrectly hispid; **L** to 5 mm petiolate, lamina ovate to lanceolate, to 15 - 20 × 2 - 6 mm, diminishing in size towards stem tips, upper face hispid; **Inf** sessile, 1- to 2-flowered; **Ped** 3 - 7 mm, hispid; **Sep** subulate, 4 mm, porrectly hispid; **Cl** 8 - 15 mm ∅, tube flat to bowl-shaped, 5 - 6 mm ∅, inside yellowish, banded with purple; **Cl** lobes 2.5 - 4.5 mm, ± purple, glabrous; **Cn** sessile, purple, ± 3 mm ∅, basally united and shallowly bowl-shaped; **Ci** lobes flatly pouch-like, outer margin U-shaped or bidentate; free **Cs** lobes shortly linear, incumbent on the **Anth**, dorsally strongly thickened; **Poll** D-shaped, ± 0.25 mm; **Sty** head bulging; **Fr** 3 - 5 cm, green.

This species is easily recognized from its creeping growth and the leaves diminishing in size towards the stem tips, as well as from the rather small flowers (see also *B. petraeum*).

B. punctatum Boele (Excelsa 16: 31-32, ills., 1994). **T:** Zimbabwe, Chegutu Distr. (*Hornby* 3334 [SRGH]). − **D:** Zimbabwe.

[2] **R** tuber depressed-globose, to 12 cm ∅; stems irregularly branched, ± erect, to 30 cm, somewhat tomentose; **L** 4 - 6 mm petiolate, lamina narrowly obovate, 20 - 40 × 10 - 16 mm, acute, tomentose; **Inf** sessile, 2-flowered; **Sep** lanceolate, 3 mm; **Ped** ± 2 mm; **Cl** 30 - 35 mm ∅, outside tomentose, inside wrinkled to pustulate, red-brown to pale green, glabrous; **Cl** tube campanulate, ± 2 mm deep; **Cl** lobes lanceolate, 15 × 4 mm, somewhat folded back along the midrib; **Cn** sessile, purple, basally yellowish, apically purple, ± 5 mm ∅, united to form a cup-like structure; **Ci** lobes rectangular, erect, 0.8 mm, incised; **Cs** lobes ± pillow-shaped, inside united with the **Ci**.

Closely related to *B. lancasteri* and *B. tavalla*, probably also to *B. brevipedicellatum*.

B. pygmaeum (Schlechter) N. E. Brown (FC 4(1): 857, 1908). **T:** RSA, Transkei (*Barber* s.n. [B†]). − **D:** Moçambique, Zimbabwe, RSA. **I:** Dyer (1983).
   ≡ *Dichaelia pygmaea* Schlechter (1894).

B. pygmaeum ssp. **flavidum** (Schlechter) R. A. Dyer (BT 12(4): 629, 1979). **T:** RSA, KwaZulu-Natal (*Rudatis* 68 [K]). − **D:** RSA (KwaZulu-Natal, Mpumalanga). **I:** Dyer (1983).
   ≡ *Brachystelma flavidum* Schlechter (1907).

[2] Differs from ssp. *pygmaeum*: **Cl** stellate, yellow, basally incl. the tube mostly whitish with reddish **Ha**, tube shallow, 3 - 4 mm ∅; **Cl** lobes triangular-lanceolate, lamina convex but not folded along the midrib.

The differences to ssp. *pygmaeum* are not only the stellate instead of cage-like flowers, as ssp. *flavidum* moreover has an obvious corolla tube and broader corolla lobes which are not folded.

B. pygmaeum ssp. **pygmaeum** − **D:** Moçambique, Zimbabwe, RSA (Eastern Cape, KwaZulu-Natal, Mpumalanga, Northern Prov.).
   Incl. *Dichaelia breviflora* Schlechter (1895) ≡ *Brachystelma pygmaeum* var. *breviflorum* (Schlechter) N. E. Brown (1908).

[2] **R** tuber depressed-globose, 4 - 10 cm ⌀; stems few, somewhat divaricately branched, ± erect, partly somewhat divergent, ± tomentose; **L** shortly petiolate, lamina ± linear to lanceolate or oblanceolate, 5 - 20 × 3 - 7 mm, mostly somewhat hairy on the veins, margins ± hairy; **Inf** sessile, 1- to 3-flowered; **Ped** 3 - 8 mm, somewhat tomentose; **Sep** ± lanceolate; **Cl** 6 - 10 mm long, cage-like, yellow, green to purple-brown, tube absent or very short, sometimes somewhat hairy; **Cl** lobes linear-lanceolate, apically united, lamina incompletely folded back along the midrib; **Cn** considerably stipitate, ± 1 × 2 mm, basally united, ciliate; **Ci** lobes short, pouch-like, laterally united with the **Cs** base, the free margin ± elongate, subquadrangular, emarginate or incised; **Cs** lobes ± square to rectangular, obtuse or crenate, incumbent on the **Anth**; **Poll** ± pear-shaped; **Fr** 50 - 60 × 4 - 5 mm.

B. ramosissimum (Schlechter) N. E. Brown (FC 4(1): 855, 1908). **T** [lecto]: RSA, Gauteng (*Schlechter* 3554 [BOL, GRA]). – **D**: RSA (Free State, Gauteng, North-West Prov.). **I**: Dyer (1983).
≡ *Brachystelmaria ramosissima* Schlechter (1895) ≡ *Lasiostelma ramosissimum* (Schlechter) Schlechter (1899).

[1] **R** clustered, long-fusiform; main stems 1 - 2, ± grooved or angular, erect, richly branched, 10 - 15 cm, rarely sparsely tomentose; **L** ± sessile, linear-elliptic to lanceolate, 15 - 30 mm, sometimes somewhat hairy; **Inf** sessile, 2- to 4-flowered, clustered; **Ped** to 5 mm; **Sep** lanceolate, 2 mm; **Cl** 6 - 7 mm ⌀, white to pale yellow or greenish; **Cl** tube absent; **Cl** lobes oblong-ovate, reflexed, ± 3 mm, inside basally papillose to hairy; **Cn** sessile, basally united bowl-like; **Ci** lobes deeply cleft into 2 subulate erect appendages, tomentose; **Cs** lobes linear, sometimes apically incised, incumbent on the **Anth**; **Fr** very slender, ± 10 cm.

B. ramosissimum forms a species group together with *B. sandersonii* and *B. schizoglossoides*, which is characterized by fleshy roots and a similar flower morphology.

B. recurvatum Bruyns (BT 25(2): 156-157, ills., 1995). **T**: Namibia (*Bruyns* 5486 [BOL, K, WIND]). – **D**: N Namibia.

[2] Delicate tomentose dwarf shrublets; **R** tuber ± globose, to 8 cm ⌀; stems erect, 6 - 30 cm; **L** basally tapering into a short petiole, lamina narrowly lanceolate, 30 - 90 × 3 - 13 mm, tomentose; **Inf** sessile, 1- to 4-flowered; **Ped** 10 - 16 mm, horizontal; **Cl** outside green, pubescent, inside greenish, spotted with yellow, white-villous; **Cl** tube short and shallow; **Cl** lobes 7 - 9 mm, margins revolute, basal ¼ triangular, ascending, apical ¾ ± linear, bent backwards; **Cn** sessile, pale yellow with dark green spots, cup-shaped, 1.5 mm ⌀; **Ci** lobes subquadrangular, ascending, obtuse, apically incised; **Cs** lobes narrowly triangular, incumbent on the **Anth**, dorsally thickened, in one line with the **Ci**; **Poll** broadly ovoid, ± 0.25 mm.

B. recurvatum shows some similarities with *B. schultzei*, esp. because of the long corolla lobes, but it is most closely related to *B. minimum*.

B. remotum R. A. Dyer (BT 12(1): 55-56, ills., 1976). **T**: RSA, KwaZulu-Natal (*Devenish* 914 [PRE]). – **D**: RSA (KwaZulu-Natal).

[2] **R** tuber depressed-globose, 2.5 - 4 cm ⌀; stems 1 to few, ascending-erect, 3 - 6 cm, glabrous or tomentose; **L** to 15 mm petiolate, lamina lanceolate to broadly ovate, to 15 × 15 mm, partly somewhat hairy; **Inf** sessile, 1- to 2-flowered; **Sep** lanceolate, 3 mm; **Cl** ± 18 mm ⌀, violet or spotted with violet, tube campanulate, ± 3 × 6 - 7 mm; **Cl** lobes 4 - 7 mm, ± horizontally bent backwards, triangular-ovate, inside long white-barbate, apically recurved along the midrib; **Cn** ± sessile, purple, ± 2 × 3.5 mm, united to form a campanulate structure; **Ci** lobes pouch-like, apically emarginate; **Cs** lobes lingulate, incumbent on the **Anth**; **Poll** ± 0.3 mm, caudiculae very short.

A micro-species out of the *B. australe*-complex.

B. richardsii Peckover (Excelsa 17: 32-34, ills., 1997). **T**: Zimbabwe (*Peckover* 257 [PRE]). – **D**: Zimbabwe, Tanzania?.

[2] **R** tuber ± (depressed-) globose, 2.5 - 7.5 cm ⌀; stems erect, 15 - 30 cm tall, unbranched, hispid; **L** 2 mm petiolate, lamina narrowly elliptic-lanceolate, 1.5 - 10 × 0.5 - 2 cm, delicately tomentose; **Inf** ± sessile, 3- to 8-flowered; **Ped** 1 - 3 mm, tomentose; **Sep** triangular, 2 - 3 mm; **Cl** ± 15 - 30 mm ⌀, outside green, inside purple, with ± delicate white **Ha**; **Cl** tube campanulate, ± 3 × 4 mm; **Cl** lobes narrowly triangular to lanceolate, 6 - 15 mm, horizontal, often slightly contorted, lamina bent backwards along the midrib; **Cn** ± sessile, ± cup-shaped, ± 2 × 3 mm, dark purple, inside basally spotted with yellowish; **Ci** lobes rectangular, ± 1 mm, erect; **Cs** lobes narrowly triangular, ± 1 mm, incumbent on the **Anth**.

Related to *B. lancasteri*. The collection from Tanzania placed with *B. lancasteri* by Newton (1996) is probably *B. richardsii* as well, although the plants are more roughly hispid, the corolla lobes are glabrous and the corona is mostly white.

B. rubellum (E. Meyer) Peckover (Aloe 33(2-3): 43, 1996). **T**: RSA, Eastern Cape (*Drège* 2227 [K?]). – **D**: Kenya, Tanzania, Uganda, RSA (Eastern Cape, KwaZulu-Natal, Mpumalanga), Swaziland. **Fig. III.g**
≡ *Tenaris rubella* E. Meyer (1839); **incl.** *Tenaris rostrata* N. E. Brown (1885); **incl.** *Tenaris volkensii* K. Schumann (1895); **incl.** *Tenaris simulans* N. E. Brown (1908).

[2] **R** tuber 3 - 5 cm ⌀, pale brown; stems erect, to 20 - 50 cm, glabrous; **L** sessile, subulate, 20 - 60

× 1 - 2 mm, ascending; **Inf** 1- to 2-flowered, sessile or shortly pedunculate, often clustered in terminal racemes; **Ped** 5 - 12 mm, filiform; **Cl** pink, glabrous; **Cl** tube cup-shaped, ± 2 mm; **Cl** lobes linear-spatulate, 8 - 13 mm, erect-divergent, ± folded back along the midrib, with purple-red papillae; **Cn** ± sessile, bowl-shaped, 2 mm ⌀, glabrous; **Ci** lobes pouch-like, triangular-ovate, horizontal to ascending, apically ± hood-like, incised; **Cs** lobes 0.75 mm, linear to subulate, ± incumbent on the **Anth**; **Poll** ± 0.3 × 0.15 mm.

A very pretty species by virtue of its pink-coloured flowers and the spatulate corolla segments, and of easy cultivation.

B. sandersonii (Oliver) N. E. Brown (FC 4(1): 850, 1908). **T** [lecto]: RSA, KwaZulu-Natal (*Sanderson* 436 [K, NH]). – **D**: RSA (Eastern Cape, KwaZulu-Natal). **I**: Dyer (1983).

≡ *Lasiostelma sandersonii* Oliver (1883); **incl.** *Dichaelia natalensis* Schlechter (1894) ≡ *Brachystelmaria natalensis* (Schlechter) Schlechter (1895); **incl.** *Lasiostelma benthamii* K. Schumann (1895).

[1] **R** clustered, long-fusiform; stems 1 - 2 main stems, erect, branched somewhat above the middle, 15 - 45 cm, glabrous; **L** ± sessile, linear, elliptic to ovate, 20 - 40 × 3 - 12 mm, mostly somewhat ciliate; **Inf** sessile, 1- to 6-flowered, clustered; **Ped** 2 - 6 mm; **Sep** lanceolate, 2 mm; **Cl** 7 - 8 mm ⌀, inside whitish-pink, pale purple or pale greenish-yellow, tube very short; **Cl** lobes oblong to ovate, semi-erect, ± folded back along the midrib, margins revolute, ± 3.5 mm, inside papillose to hairy; **Cn** sessile, almost ovoid in outline, basally united bowl-like; **Ci** lobes deeply bifid into 2 lingulate erectly connivent appendages, tomentose; **Cs** lobes linear, incumbent on the **Anth**; **Poll** ovoid; **Fr** very slender, to 15 cm.

B. sandersonii is most closely related to *B. schizoglossoides*. See also *B. ramosissimum*.

B. schinzii (K. Schumann) N. E. Brown (FTA 4(1): 471, 1903). **T**: Namibia, Ovambo (*Schinz* s.n. [Z]). – **D**: N Namibia. **I**: Dyer (1983).

≡ *Craterostemma schinzii* K. Schumann (1893).

[2] **R** tuber conically ovoid, ± 3 cm ⌀; stems 1 to few, unbranched, erect, 8 - 15 cm, delicately tomentose; **L** shortly petiolate, lamina linear, to 50 mm; **Inf** sessile, 1- to 2-flowered; **Ped** 12 - 16 mm, thin; **Cl** stellate, to 10 mm ⌀, purple-brown, centrally white, tube very short, shallow; **Cl** lobes triangular-lanceolate, ± 4 mm, long-hairy with whitish-violet **Ha**; **Cn** yellowish, margined with purple, united to form a fleshy ring-like structure, somewhat indented at **Ci**-positions; **Cs** lobes shortly pillow-shaped, basally placed against the **Anth**.

Only known from the type collection, this species is probably related to *B. recurvatum*.

B. schizoglossoides (Schlechter) N. E. Brown (FC 4(1): 849, 1908). **T** [lecto]: RSA, Eastern Cape (*Bolus* 6694 [K, BOL]). – **D**: RSA (Eastern Cape, Free State, Northern Cape). **I**: Bruyns (1982a); Dyer (1983).

≡ *Sisyranthus schizoglossoides* Schlechter (1894).

[1] **R** clustered, slender-fusiform; stem solitary, erect, little-branched, 10 - 25 cm tall, ± glabrous, ± canaliculate; **L** sessile, linear-elliptic, 20 - 50 mm; **Inf** shortly pedunculate, 1- to 3-flowered; **Ped** 2 - 4 mm; **Sep** lanceolate, 1.5 mm; **Cl** 4 - 5 mm ⌀, yellowish-cream-coloured, glabrous, tube short, bowl-shaped to campanulate; **Cl** lobes broadly triangular, ± 1.5 mm, lamina horizontal to slightly ascending, apically bent inwards, sometimes somewhat hairy; **Cn** sessile, basally slightly united, ± 1.5 mm ⌀; **Ci** lobes deeply incised and with 2 lingulate erect appendages, hairy; **Cs** lobes linear, ± 1 mm, obtuse, ascending-connivent.

A rare and inconspicuous species of Eastern Cape grasslands. See also *B. ramosissimum*.

B. schoenlandianum Schlechter (BJS 18 (Beiblatt 45): 35, 1894). **T**: RSA, Eastern Cape (*Schlechter* 2585 [B†]). – **D**: RSA (Eastern Cape).

Insufficiently known taxon of the group of small-flowered species from the Eastern Cape.

B. schultzei (Schlechter) Bruyns (BT 25(2): 162, 1995). **T** [lecto]: Namibia (*Dinter* 2528 [SAM]). – **D**: C Namibia. **I**: Bruyns (1984: 74, as *Tenaris*).

≡ *Kinepetalum schultzei* Schlechter (1914) ≡ *Tenaris schultzei* (Schlechter) E. Phillips (1941).

[2] **R** tuber ± globose, 3 - 5 cm ⌀; stem mostly solitary, erect, 15 - 50 cm, grey-green, often reddish, densely tomentose; **L** ± sessile, linear, 20 - 60 × 1.5 - 3 mm, densely tomentose, grey-green, often reddish; **Inf** lateral, 1- to 2-flowered; **Ped** 15 - 20 mm, mostly bent downwards, pubescent; **Sep** triangular, 2 - 3 mm; **Cl** 40 - 55 mm ⌀, pubescent all over, tube bowl-shaped, margins somewhat connivent, ± 1 × 2.5 mm, cream-coloured; **Cl** lobes linear, ± 20 - 25 × 1.5 mm, basal 2 mm triangular, erect, cream-coloured with violet club-shaped **Ha**, above linear and spreading or ascending, lamina broadly folded back along the midrib; **Cn** sessile or to 0.5 mm stipitate, bowl-shaped, purple, ± 3.5 × 2 mm; **Ci** lobes rectangular, obtuse, 0.5 × 0.4 mm, or divided into 2 broadly triangular appendages, ascending, glabrous or laxly ciliate; **Cs** lobes linear-cylindrical, ± 2.5 mm, straight, long ascending, meeting apically; **Poll** ovoid, ± 0.3 × 0.3 mm, germination mouth apical; **Fr** thickly fusiform, ± 3 × 1 cm.

The grass-like growth habit has obviously evolved repeatedly in *Brachystelma*, as shown by its parallel occurrence in florally very divergent groups (*B. lineare*-complex, "*Tenaris*" group etc.). *B. schultzei* does, however, not belong to either of the groups mentioned.

B. setosum Peckover (Aloe 31(3-4): 76-78, ills., 1995). **T:** RSA, Mpumalanga (*Peckover* 186 [PRE]). – **D:** RSA (Mpumalanga, Northern Prov.).

[2] **R** tuber depressed-globose, ± 5 - 9 cm ∅; stems erect, dwarf, 5 - 7 cm, hispid and villous, to 3 mm ∅; **L** ± sessile, linear-elliptic, 50 - 70 × 4 - 6 mm, both sides densely and long white-hairy; **Inf** sessile, 2-flowered; **Ped** 5 - 6 mm, villous; **Sep** lanceolate-subulate, 4 mm, densely hairy; **Cl** 10 - 13 mm long, cage-like, outside yellowish-green, inside purple-brown, outside hairy; **Cl** tube ± flat, 5 - 6 mm ∅, inside with some long white **Ha**; **Cl** lobes linear from a triangular base, 10 - 13 mm, apically united and forming a cage, somewhat folded back along the midrib; **Cn** sessile, basally united bowl-like, ± 1.5 × 2 mm, black-purple, glabrous; **Ci** lobes bifid into subulate erect and almost horizontally spreading appendages; **Cs** lobes rectangular-lingulate from a transversely ovate thickened back, completely covering the **Anth**; **Fr** slender fusiform, 6 - 7 cm × 3 - 4 mm; **Se** grey-brown, 5 - 6 × 3 - 4 mm.

B. setosum differs from *B. tenue* only in the more compact habit, leaves that are hairy on both sides, and the broader corolla tube.

B. simplex Schlechter (BJS 38: 40-41, ills., 1905). **T:** Moçambique (*Schlechter* 12121 [B†]). – **D:** Burkina Faso, Ivory Coast, Moçambique, Nigeria, Zambia. **Fig. III.f**

Incl. *Brachystelma simplex* ssp. *banforae* J.-P. Lebrun & Stork (1989).

[2] **R** tuber 3 - 12 cm ∅; stem solitary, unbranched, erect, 10 - 40 cm, long rough-hairy; **L** shortly petiolate, lamina linear to narrowly elliptic, 10 - 80 × 2 - 10 mm, upper face fresh green, glossy, ± glabrous, lower face long and roughly hairy; **Inf** sessile, 2- to 8-flowered; **Ped** short, hispid; **Sep** lanceolate, 2 mm, hispid; **Cl** stellate, 4 - 6 mm ∅, outside long hispid or tomentose, inside densely papillose to shortly hairy, yellow to orange-yellow, often speckled with reddish; **Cl** tube bowl-shaped, ± 1 × 2 mm; **Cl** lobes broadly triangular, ± spreading, flat, 1.5 - 2 × 1.5 - 2 mm; **Cn** sessile, yellow, fitted into the **Cl** tube, bowl- to cup-shaped, ± 1 × 2 mm; **Ci** subquadrangular, erect, emarginate U-shaped, fused in leaflet-like fashion laterally and with the back of the **Cs**, inside scattered long-hairy or -pubescent, apical margin thickened, ± incised, slightly bent inwards; **Cs** lobes pillow-shaped, dorsally placed against the **Anth**; **Poll** ovoid, ± 0.3 × 0.2 mm; **Fr** slender-fusiform, ± 10 cm.

The dense, narrowly elliptic foliage and the laterally fused margins of the interstaminal corona that form separate leaflets distinguish *B. simplex* from the very similar *B. dinteri*.

B. stellatum E. A. Bruce & R. A. Dyer (FPA 30(1): t. 1165B + text, 1954). **T:** RSA, Northern Prov. (*Codd* 5648 [PRE]). – **D:** RSA (Northern Prov.). **I:** Dyer (1983). **Fig. III.h**

[2] **R** tuber ± globose, 2 - 5 cm ∅; stems few, semi-erect, divergent, basally richly branched, 2 - 8 cm, glabrous to glandular-pubescent; **L** shortly petiolate, lamina broadly ovate, to 10 × 10 mm, scatteredly glandular-pubescent; **Inf** sessile, 1- to 2-flowered; **Ped** 15 - 20 mm; **Sep** lanceolate, 2 mm; **Cl** stellate, ± 10 mm ∅, cream-coloured, densely hairy, sweet-scented; **Cl** tube campanulate, ± 2.5 mm, with purple **Ha**; **Cl** lobes ± 2 mm, triangular-ovate, densely long and white hairy; **Cn** fitted into the **Cl** tube, purple, ± 2.5 mm, united to form a campanulate-cylindrical structure; **Ci** lobes pouch-like, with V-shaped incision in the middle, laterally fused without interruption into the broadened and flattened back of the **Cs**; **Cs** lobes linear-lingulate, placed against the slightly ascending **Anth**; **Poll** D-shaped.

Most closely related to *B. caffrum*.

B. stenophyllum (Schlechter) R. A. Dyer (BT 10(2): 376, 1971). **T** [lecto]: Namibia (*Dinter* 2361 [SAM]). – **D:** Namibia, RSA (Northern Prov., North-West Prov.). **I:** Dyer (1983).

≡ *Siphonostelma stenophyllum* Schlechter (1913).

[2] **R** tuber cylindrical-ovoid, 3 - 6 cm ∅; stems densely divaricately branched, ascending to erect, 4 - 15 cm, almost glabrous to densely hispid; **L** ± sessile, ± linear, acute, 15 - 50 × 1.5 - 5 mm, hairy, lower face with bulging midrib; **Inf** sessile, 1- to 3-flowered; **Ped** 3 - 8 mm, pubescent; **Sep** triangular, ± 2 mm, pubescent; **Cl** 8 - 20 mm long, green to red-brown, inside glabrous, outside pubescent; **Cl** tube campanulate to funnel-shaped-cylindrical, 4 - 8 mm, to 7 mm ∅; **Cl** lobes linear from a broadly triangular base, 4 - 14 mm, lamina apically united and forming a cage, ± folded back along the midrib; **Cn** sessile, conspicuously cup-shaped to urceolate from the united **Ci**, ± 1.5 × 2 mm, yellow, glabrous, upper margin emarginate, **Gy** only ½ as tall as the cup; **Cs** dorsally thickened and fused with the inner base of the cup, **Cs** lobes ± obtusely triangular, densely placed against the horizontal **Anth**; **Poll** 0.25 mm; **Fr** ± 70 × 5 mm, green, striate and spotted with dark red.

Among the cage-flowered taxa, *B. stenophyllum* is the only species with a cup-like corona. This speciality of the corona could indicate a close relationship with *B. cupulatum*.

B. swarupa Kishore & Goyder (KB 56(1): 210-212, ills., 2001). **T:** India, Kerala (*Kishore Kumar* 18505 [KFRI, K, MH]). – **D:** India (Kerala, Tamil Nadu).

[2] Plants erect, 30 - 60 cm tall; tuber flattened, 3 - 8 cm, edible; stem 1, rarely branched, puberulent; **L** ± sessile, narrowly lanceolate, 80 - 120 × 2 - 7 mm; **Inf** sessile, bracteolate, umbellate, with 24 - 28 pendent **Fl**; **Ped** 3 - 4 cm, filiform, puberulent; **Sep** 1 mm; **Cl** ± 6 mm ∅, glabrous; **Cl** tube shallow, short, inner margin with 4 - 6 long **Ha**; **Cl** lobes

deltoid-lanceolate, ± 2.5 × 1 mm, recurved with the tips touching the **Ped**, striate or spotted purplish; **Cn** sessile, cup-shaped, ± 1.2 × 1 mm, bright yellow; **Ci** fused at the base to form a shallow bowl, pouch-like, erect, apically indented, ciliate with long **Ha**, **Cs** lobes subquadrate, dark green, ciliate; **Poll** broadly ovoid, ± 0.16 mm; **Fr** slender fusiform, 6 - 9 cm; **Se** 7 × 2 mm, coma 25 mm.

Closely related to *B. laevigatum*, but with glabrous and patterned corollas, hairy coronas and many-flowered inflorescences.

B. swazicum R. A. Dyer (BT 12(1): 56-57, ills., 1976). **T**: Swaziland (*Bayliss* 3733 [PRE]). – **D**: RSA (Mpumalanga), Swaziland.

[2] **R** tuber ± globose, ± 5 cm ∅; stems 2 - 3, decumbent-creeping, to 25 cm, reddish; **L** 10 - 15 mm petiolate, petioles often erect, lamina ± cordate, to 20 mm, margins and lower face somewhat hairy, dark green; **Inf** shortly pedunculate, 2- to 4-flowered; **Ped** 4 mm; **Sep** linear-lanceolate, 2 mm; **Cl** 8 - 9 mm ∅, inside dark purple, tube bowl-shaped, 1.5 - 2 × 3 mm; **Cl** lobes triangular, ± 3 mm, lamina somewhat folded back; **Cn** sessile, ± 1 × 1.8 mm, basally united to form a cup-like structure, yellowish, margined with purple; **Ci** lobes basally ovate pouch-like, apically bifid into linear-subulate falcately divergent appendages; **Cs** lobes subquadrangular, 0.5 mm, incumbent on the **Anth**, dark purple.

Very closely related to *B. minor*.

B. tabularium R. A. Dyer (BT 12(4): 629, 1979). **T**: RSA, Eastern Cape (*Paterson* s.n. [BOL]). – **D**: RSA (Eastern Cape). **I**: Dyer (1983).

[?] **R** unknown; stems basally branched, ± erect, slender, to 20 cm, shortly pubescent; **L** shortly petiolate, lamina oblong, ± 10 × 5 mm, with few **Ha**, margins ± undulate; **Inf** sessile, 1- to 2-flowered; buds cylindrical, obtuse; **Cl** 25 mm ∅, tube very short; **Cl** lobes ± 12 mm, ± linear, apically acute, inflexed; **Cn** ± sessile, basally hardly united; **Ci** lobes subquadrangular, erect, ± 1 mm, deeply bifid into triangular appendages, **Cs** lobes short, placed on the **Anth**.

This species, only known from the type, is closely related to the equally little-known *B. furcatum*.

B. tavalla K. Schumann (BJS 28: 459, 1901). **T** [neo]: Zimbabwe (*Whellan* 34949 [SRGH]). – **D**: Tanzania, NW Zimbabwe. **I**: Excelsa 13: 83, 1988.

[2] **R** tuber depressed-globose, 3 - 7 cm ∅; stems ± erect, hardly branched, 20 - 30 cm, shortly pubescent; **L** shortly petiolate, lamina lanceolate to linear-lanceolate, 10 - 50 × 5 - 12 mm, pubescent; **Inf** sessile, 1- to 5-flowered; **Ped** 4 - 7 mm, pubescent; **Cl** stellate, to 20 mm ∅, pale yellow to purple; **Cl** tube bowl-shaped, 2 - 3 × 4 - 5 mm; **Cl** lobes linear-lanceolate, ± 9 × 2 mm, inside completely or only in the apical ½ with vibratile ± 5 mm long purple **Ha**; **Cn** shortly stipitate, bowl-shaped; **Ci** lobes broadly triangular, erect, ± 0.8 mm, apically bifid into triangular appendages; **Cs** lobes linear, obtuse, incumbent on the **Anth**.

Very conspicuous through the extremely long pendent hairs in the corolla tube. See also notes for *B. furcatum* and *B. punctatum*.

B. tenellum R. A. Dyer (FPA 42: t. 1664 + text, 1973). **T**: RSA, KwaZulu-Natal (*Strey* 10050 [PRE, NH]). – **D**: RSA (Northern Prov., KwaZulu-Natal). **I**: Dyer (1983); Craib (1994).

[2] **R** tuber depressed-globose, 4 - 6 cm ∅; stems solitary or several, dwarf, semi-erect, 2 - 5 cm, almost glabrous to pubescent; **L** ± sessile, narrowly elliptic-lanceolate, 10 - 20 mm, margins and lower face scatteredly pubescent; **Inf** sessile, 1-flowered; **Ped** very slender, 10 - 15 mm; **Sep** narrowly lanceolate, 2 mm, recurved; **Cl** 3.5 - 4 mm long, purple, cage-like, outside glabrous, inside long-hairy; **Cl** tube very short; **Cl** lobes lanceolate, hardly 3 mm, lamina apically united, inside with long purple and whitish **Ha**; **Cn** shortly stipitate, basally hardly united, purple; **Ci** lobes subquadrangular, bifid for ½ into linear falcately spreading appendages, apically hairy; **Cs** lobes oblong-lingulate, incumbent on the **Anth**; **Poll** obovoid, ± 0.2 mm.

B. tenellum is one of the smallest species of the genus. It is not infrequently found in the area around the Oribi. See also *B. elongatum*.

B. tenue R. A. Dyer (BT 10(2): 376-378, ills., 1971). **T**: RSA, KwaZulu-Natal (*Vahrmeijer* 1049 [PRE]). – **D**: RSA (Northern Prov., KwaZulu-Natal). **I**: Dyer (1983); Craib (1994). **Fig. IV.a**

Incl. *Brachystelma inconspicuum* S. Venter (1989).

[2] **R** tuber depressed-globose, 2.5 - 5 cm ∅; stem mostly solitary, erect, delicate, 10 - 30 cm, hispid or villous or pubescent; **L** ± sessile or shortly petiolate, lamina elliptic to ovate, 10 - 50 × 1.5 - 15 mm, margins and lower face mostly long-hairy; **Inf** sessile, 1- to 2-flowered; **Ped** 8 - 15 (-30) mm, filiform, ± villous or pubescent; **Sep** lanceolate-subulate, 2.5 - 4 mm, concave, ± reflexed, hispid; **Cl** 10 - 25 mm long, cage-like, ± erect, yellowish-green to brownish, basally also purple, inside glabrous and smooth, outside somewhat hairy; **Cl** tube very short, ± flat; **Cl** lobes filiform from an ovate base, 9 - 23 mm, lamina apically united, somewhat folded back along the midrib; **Cn** sessile, basally united to form a bowl-like structure, ± 1.5 × 2 mm, purple, glabrous; **Ci** lobes bifid into triangular-subulate erect appendages, ± 0.8 mm, ± divergent; **Cs** lobes rectangular-lingulate from a transversely ovate back, incumbent on the **Anth**; **Poll** depressed-ovoid, ± 0.3 × 0.2 mm; **Fr** slender-fusiform, 4 - 8 cm × 3 - 5 mm.

B. tenue is very closely related to *B. setosum* and *B. villosum*, but is also close to *B. tenellum* and *B. gracile*. See *B. elongatum* as well.

B. togoense Schlechter (BJS 38: 40-41, ills., 1905). **T:** Togo (*Schlechter* 12961 [B†]). – **D:** Benin, Cameroon, Ivory Coast, Ghana, Nigeria, Zambia. **I:** Asklepios 25: 117, 1982.

Incl. *Brachystelma atacorense* A. Chevalier (1917).

[2] **R** tuber depressed, 5 - 15 cm ∅; stems solitary or several, rarely branched, sturdy, erect, 15 - 30 cm, hispid; **L** ± sessile or shortly petiolate, lamina elliptic to lanceolate, 30 - 70 × 7 - 15 mm, tomentose, ± grey-green; **Inf** mostly only terminal, sessile, 4 - to 10-flowered umbels, most **Fl** opening simultaneously; **Ped** to 8 mm, pubescent; **Sep** lanceolate, 4 mm, pubescent; **Cl** ± 15 - 20 mm ∅, inside and outside densely pubescent, outside pale green, inside yellowish to greenish-yellow or green, sometimes purple-spotted; **Cl** tube bowl- to cup-shaped, ± 8 mm ∅; **Cl** lobes broadly triangular, 6 - 8 mm, flatly spreading; **Cl** sessile, red-brown, ± 2 × 4 mm, united only basally shortly cup-like, glabrous; **Ci** lobes pouch-like but with deeply U-shaped incised margin, each with lateral subulate-pointed appendages, connivent, erect; **Cs** lobes broadly lingulate, obtuse, incumbent on the **Anth**; **Poll** D-shaped; **Fr** thickly fusiform.

Sturdy plants of wide distribution in Africa and closely related to *B. omissum* and perhaps also to *B. buchananii*.

B. tuberosum (Meerburgh) R. Brown *ex* Sims (CBM 49: t. 2343, in adnot. excl. ill., 1822). **T:** [lecto – icono]: Meerburgh, Pl. Rar. Depict., t. 54: fig. 1, 1789. – **D:** SW RSA. **I:** Dyer (1983: as *B. caudatum*). **Fig. IV.b**

≡ *Stapelia tuberosa* Meerburgh (1789); **incl.** *Stapelia caudata* Thunberg (1794) ≡ *Brachystelma caudatum* (Thunberg) N. E. Brown (1878); **incl.** *Brachystelma spathulatum* Lindley (1827); **incl.** *Brachystelma crispum* Graham (1830).

[2] **R** tuber depressed-globose, 6 - 15 cm ∅; stems 1 - 3, erect, 40 - 100 cm, ± hairy; **L** shortly petiolate, lamina ± lanceolate, 15 - 30 × 7 - 10 mm, pubescent; **Inf** ± sessile, 2- to 8-flowered; **Ped** 10 - 25 mm, pubescent; **Sep** lanceolate, 4 mm, hairy; **Cl** 45 - 95 mm ∅, outside pubescent, inside yellowish, purple-spotted; **Cl** tube cup-shaped, ± 5 × 4 mm; **Cl** lobes 20 - 40 mm, ovate-lanceolate, apically long-attenuate, margins revolute, apically greenish, ± hairy; **Cn** sessile, basally fused to form a bowl-like structure, ± 2.5 × 3 mm, purple-brown; **Ci** lobes pouch-like, subquadrangular, apically incised in triangular to U-shaped manner, laterally fused with the base of the **Cs**; **Cs** lobes lingulate, hardly 1 mm long, incumbent on the **Anth**.

An extremely large-flowered species, which is still treated under the name of *B. caudatum* in Dyer's publications (Forster 1986).

B. vahrmeijeri R. A. Dyer (BT 10(2): 378, 1971). **T:** RSA, KwaZulu-Natal (*Vahrmeijer* 1050 [PRE]).
– **D:** RSA (KwaZulu-Natal). **I:** Dyer (1983); Craib (1994).

[2] **R** tuber depressed, 5 - 10 cm ∅; stem solitary, or somewhat branched from the base, ± erect, 5 - 10 cm, shortly hairy or glabrous; **L** sessile or shortly petiolate, lamina elliptic-lanceolate, to 30 × 7 mm, ± glabrous; **Inf** sessile, 1- to 3-flowered; **Ped** 5 - 10 mm; **Sep** lanceolate, 2.5 mm, scatteredly hairy; **Cl** ± 13 mm ∅, ± glabrous, yellowish-green, lemon-yellow to brown-red; **Cl** tube funnel-shaped, ± 4 mm ∅; **Cl** lobes horizontal to ascending, ovate-triangular, 5 mm, margins somewhat revolute; **Cn** sessile, fitted into the **Cl** tube, ± 1.5 × 3 mm, cup-like, emarginate, yellow, glabrous; **Ci** lobes completely integrated into the cup; **Cs** dorsally thickened and fused with the inner base of the cup; **Cs** lobes linear, incumbent on the **Anth**; **Fr** 6 - 8 cm.

Sister-species of *B. cupulatum* or probably even conspecific (see there and under *B. brevipedicellatum*).

B. villosum (Schlechter) N. E. Brown (FC 4(1): 863, 1908). **T** [lecto]: RSA, Northern Prov. (*Galpin* 588 [PRE]). – **D:** RSA (KwaZulu-Natal, Mpumalanga, Northern Prov.), Swaziland. **I:** Dyer (1983).

≡ *Dichaelia villosa* Schlechter (1899).

[?] **R** unknown; stems solitary or several, erect, 10 - 30 cm, densely villous; **L** ± sessile, elliptic to oblong-elliptic, 10 - 25 × 5 - 7 mm, marginally and lower face villous; **Inf** sessile, 2- to 3-flowered; **Ped** 2 - 6 mm, ± villous; **Sep** lanceolate-subulate, 3 mm; **Cl** 20 - 30 mm long, cage-like, ± erect; **Cl** tube shallowly campanulate, 4 - 5 mm deep, inside shortly white-hairy, outside ± glabrous; **Cl** lobes filiform from a triangular base, 12 - 25 mm, lamina apically united, inside glabrous, outside ± pubescent; **Cn** sessile, basally fused to form a bowl-like structure, ± 1.5 × 2 mm, purple, glabrous; **Ci** lobes bifid into narrowly triangular-subulate erect appendages, ± divergent; **Cs** lobes rectangular-lingulate, incumbent on the **Anth**.

B. villosum is very closely related to *B. tenue* and *B. setosum* or even conspecific. Habit and hairiness are much more robust, however, and the corolla tube is deeper and hairy inside.

B. volubile Hooker *fil.* (Fl. Brit. India 4: 65, 1885). **T:** K. – **D:** India (Deccan, Ghats, Karnataka). **I:** Stevens (1976: 438).

[?] **R** tuber unknown; stems perennial, twining, ± 50 - 100 cm, solitary, pubescent; **L** shortly petiolate, lamina linear-lanceolate, to 10 cm; **Inf** ± 8 mm long pedunculate, 2-flowered; **Ped** 7 - 10 mm; **Sep** lanceolate, acute, 5 mm; **Cl** ± 30 mm ∅, pink, white-hairy; **Cl** tube short, shallow; **Cl** lobes ovate-lanceolate, ± 13 mm, horizontally spreading; **Cn** ± sessile, cup-like; **Cs** lobes placed against the **Anth**; **Poll** ovoid with broad membranous germination mouth.

B. volubile was incompletely described and is insufficiently known. The twining habit is a rare exception within the genus and otherwise only occurs with *B. brevitubulatum,* which is most closely related.

CARALLUMA

B. Müller & F. Albers

Caralluma R. Brown (Mem. Wern. Nat. Hist. Soc. 1: 14, 1810). **T:** *Stapelia adscendens* Roxburg. — **Lit:** Gravely & Mayuranathan (1931); Bruyns (1987a); Bruyns (1987b); Gilbert (1990); Bruyns & Jonkers (1994); Plowes (1995). **D:** Dry regions of tropical Asia, S Mediterranean, Near East and in N, C and E Africa. **Etym:** From Arabian 'qarh alluhum', wound in the flesh, abscess; for the floral odour of some taxa (Genaust 1996).

Incl. *Desmidorchis* Ehrenberg (1829) (*nom. illeg.*, Art. 53.1). **T:** *Desmidorchis retrospiciens* Ehrenberg.

Incl. *Boucerosia* Wight & Arnott (1834). **T:** *Caralluma umbellata* Haworth [Lectotype, designated by M. G. Gilbert, Bradleya 8: 15, 1990.].

Incl. *Hutchinia* Wight & Arnott (1834). **T:** *Hutchinia indica* Wight & Arnott.

Incl. *Apteranthes* Mikan (1835). **T:** *Apteranthes gussoneana* Mikan.

Incl. *Frerea* Dalzell (1865). **T:** *Frerea indica* Dalzell.

Incl. *Sarcocodon* N. E. Brown (1878). **T:** *Sarcocodon speciosa* N. E. Brown.

Incl. *Spathulopetalum* Chiovenda (1912). **T:** *Spathulopetalum dicapuae* Chiovenda.

Incl. *Australluma* Plowes (1995). **T:** *Caralluma peschii* Nel.

Incl. *Borealluma* Plowes (1995). **T:** *Boucerosia munbyana* Decaisne.

Incl. *Caudanthera* Plowes (1995). **T:** *Boucerosia sinaica* Decaisne.

Incl. *Crenulluma* Plowes (1995). **T:** *Boucerosia awdeliana* Deflers.

Incl. *Cryptolluma* Plowes (1995). **T:** *Boucerosia edulis* Edgeworth.

Incl. *Cylindrilluma* Plowes (1995). **T:** *Caralluma solenophora* Lavranos.

Incl. *Monolluma* Plowes (1995). **T:** *Stapelia quadrangula* Forskål.

Incl. *Sanguilluma* Plowes (1995). **T:** *Boucerosia socotrana* Balfour fil.

Incl. *Saurolluma* Plowes (1995). **T:** *Caralluma furta* P. R. O. Bally.

Incl. *Somalluma* Plowes (1995). **T:** *Caralluma baradii* Lavranos.

Incl. *Spiralluma* Plowes (1995). **T:** *Caralluma longidens* N. E. Brown.

Incl. *Sulcolluma* Plowes (1995). **T:** *Caralluma hexagona* Lavranos.

Perennial stem succulents; stems 4-ribbed or 4-angled, often ± strongly tapering towards the tips, glabrous; **Ri** continuous or divided into **Tu**, these normally laterally compressed, rarely rounded or conical; **L** rudiments ± absent, **Sc**-like to lanceolate, mostly short-lived; **Inf** 1-flowered to umbellate, rachis cushion-like or much elongate, **Ped** terete, glabrous, mostly bracteate; **Sep** short, deltoid to lanceolate, acute, mostly glabrous; **Cl** flat to campanulate, partly ± deeply incised; **Cl** lobes ± triangular to lanceolate; **Cn** stipitate or sessile, of the C(is)+Cs-type, ± cup-shaped; **Ci** segments mostly bifid, sometimes basally pouch-forming, rarely reduced; **Cs** segments simple, mostly ± strap-shaped, placed on the **Anth**, sometimes broadened where fused with the **Ci**; **Anth** rectangular, basally broadened, rarely with apical processes; **Poll** oblong, broadly D-shaped or ovoid; **Fr** (narrowly) fusiform, glabrous, follicles of a pair spreading at acute to right angles; **Se** brown and/or ochre, ovate, flattened, with voluminous convex-bulging margin, margin ± smooth, with a cluster of white **Ha** 2 - 3× as long as the **Se**.

According to Gilbert (1990) the genus can be divided into 4 subgenera on the base of stem and flower morphology. Plowes (1995) over-divides the genus on the base of an exaggeratingly differentiating morphological concept and recognizes 17 genera and ± 70 species. *Caralluma* embraces a large and diverse assemblage of species, but this diversity can be classified into few main groups, which could be accepted as separate genera (*Caralluma* s.str., *Boucerosia* s.lat., and *Desmidorchis* s.str.). Under the present circumstances, and awaiting a useful generic concept, the treatment by Gilbert (1990) is here used as a base, recognizing that e.g. Subgen. *Urmalcala* cannot be sufficiently distinguished from Subgen. *Boucerosia* and uniting both could be discussed because of overlapping characters. Of all the mono- or bitypic genera separated by Plowes (1995), *Australluma* (for *C. peschii*) and *Caudanthera* (for *C. sinaica* and *C. mireillae*) merit further considerations. Both are very distinct on the base of the morphology of flowers and pollinaria as well as phytogeographically.

Following Gilbert (1990) the genus can be divided as follows:

[1] Subgen. *Caralluma*: Stems mostly ± erect, frequently with paler coloured **Ri** or angles, 4-ribbed or -angled, sometimes almost terete, apically ± strongly tapering; **Br** partly conspicuously heteromorphic, then basal parts ± stoutly robust, **L** rudiments conspicuous, mostly oblong-lanceolate, apical floriferous part long, slender, mostly ± terete, whip-like, often withering after flowering, scars of spent **Ped** distant and irregularly spaced; **Inf** few-flowered (2 or more **Fl**), opening in succession; **Fl** mostly pendent; **Cl** rather small, campanulate or mostly with completely free **Cl** lobes; **Cn** mostly ± cup-like, often stipitate; **Ci** segments ± deeply bifid, processes often horn-like, ± la-

terally diverging; **Poll** small, transversely rectangular.

[2] Subgen. *Urmalcala* M. G. Gilbert 1990: Plants rather small, clustering; stems erect, green, unspotted, 4-angled; **Ri** rounded; **Tu** oblong, ± conical; **L** rudiments small, sessile, triangular to ovate, ± appressed, long-lived; **Inf** with several ± simultaneously opening **Fl** in small terminal to subapical false umbels without conspicuous contortion of the **Ri**, appearing extra-axillary from a slight lateral overtopping growth; **Cl** with conspicuous tube; **Cn** cup-shaped, shortly stipitate; **Ci** segments erect, basally pouch-forming, apically bifid, processes horn-like, ± horizontally diverging, ± spreading outwards; **Poll** ± broadly ovoid.

[3] Subgen. *Boucerosia* (Wight & Arnott) M. G. Gilbert 1990: Stems uniformly coloured, erect, 4-ribbed; **Ri** or **Tu** frequently laterally ± strongly compressed; **L** rudiments small, ovate-deltoid, horizontally spreading or bent downwards; **Inf** mostly sessile many-flowered apical false umbels, often appearing laterally between the **Ri** through overtopping growth (then **Ri** conspicuously contorted); **Fl** in a ± globose false umbel, ± simultaneously opening; **Cl** mostly flat to ± deeply campanulate, incised for ± ½; **Cl** lobes mostly triangular-ovate; **Cn** ± sessile, basally ± broadly cup-shaped; **Ci** segments mostly deeply bifid, processes mostly laterally diverging, rarely absent; **Cs** segments mostly linear-strap-shaped; **Poll** broadly D-shaped to ovoid.

[4] Subgen. *Desmidorchis* (Ehrenberg) M. G. Gilbert 1990: Plants of medium size; stems erect, 4-angled; **Tu** flat, rounded; **L** rudiments very delicate, ± erect, perennial; **Inf** terminal or subapically-axillary along the angles, 1- to 2-flowered, successively opening, leaving conspicuous scars; **Cl** of medium size with short tube; **Cn** sessile, ± as broad as tall, cup- to bowl-shaped; **Ci** absent or segments divided ± so deeply as to appear united with the **Cs** and apparently forming processes of the **Cs**; **Poll** ± D-shaped; **Fr** often winged.

An intergeneric hybrid with *Hoodia* is covered under ×*Hoodialluma*. In addition, the following intrageneric hybrids are known: *C. adscendens* var. *attenuata* × *C. adscendens* var. *gracilis*, *C. arabica* × *C. penicillata*, *C. diffusa* × *C. umbellata*.

C. chlorantha Schlechter is poorly understood and could belong either to *Piaranthus* or to *Pectinaria*.

The following names are of unresolved application but are referred to this genus: *Boucerosia aucheriana* Decaisne (1844) ≡ *Desmidorchis aucheriana* (Decaisne) Kuntze (1891) ≡ *Caralluma aucheriana* (Decaisne) N. E. Brown (1892) ≡ *Crenulluma aucheriana* (Decaisne) Plowes (1995); *Caralluma cucullata* hort. (s.a.) (*nom. inval.*, Art. 29.1); *Caralluma stapeliiformis* hort. (s.a.) (*nom. inval.*, Art. 29.1); *Lasiostelma somalense* Schlechter (1899) ≡ *Tenaris somalensis* (Schlechter) N. E. Brown (1903); *Stapelia ango* A. Richard (1851) ≡ *Caralluma ango* (A. Richard) N. E. Brown (1892).

C. acutangula (Decaisne) N. E. Brown (Gard. Chron., ser. 3, 12: 369, 1892). **T** [neo]: Mali (*de Wailly* 4872 [P]). – **D**: Africa, Sahel region from Mauritania to Ethiopia and Somalia, Kenya; Saudi Arabia, N Yemen. **I**: FPA t. 1062, 1949, as *C. retrospiciens*; Plowes (1995). **Fig. IV.c, IV.e**

≡ *Desmidorchis acutangula* Decaisne (1838) ≡ *Boucerosia acutangula* (Decaisne) Decaisne (1844) ≡ *Caralluma retrospiciens* var. *acutangula* (Decaisne) A. Chevalier (1934); **incl.** *Stapelia desmidorchis* Steudel (s.a.); **incl.** *Desmidorchis retrospiciens* Ehrenberg (1832) (*nom. inval.*); **incl.** *Boucerosia russelliana* Courbon *ex* Brongniart (1860) ≡ *Caralluma russelliana* (Courbon *ex* Brongniart) Cufodontis (1961); **incl.** *Caralluma hirtiflora* N. E. Brown (1895) ≡ *Caralluma retrospiciens* var. *hirtiflora* (N. E. Brown) A. Berger (1910); **incl.** *Boucerosia tombuctuensis* A. Chevalier (1900) ≡ *Caralluma tombuctuensis* (A. Chevalier) N. E. Brown (1904) ≡ *Caralluma retrospiciens* ssp. *tombuctuensis* (A. Chevalier) A. Chevalier (1934) ≡ *Caralluma retrospiciens* var. *tombuctuensis* (A. Chevalier) A. C. White & B. Sloane (1937); **incl.** *Caralluma retrospiciens* Ehrenberg *ex* N. E. Brown (1904); **incl.** *Caralluma retrospiciens* var. *glabra* N. E. Brown (1904); **incl.** *Caralluma retrospiciens* var. *laxiflora* Maire (1939).

[3] Plants robust, densely branched, to 75 cm ∅; stems robust, 40 - 80 (- 150) × 2 - 5 cm, pale green or whitish, 4-angled or with very concave sides; **Tu** acute, broadly triangular, bent downwards; **L** rudiments 1 × 1 mm, persistent; **Inf** extremely many-flowered (up to > 100 open **Fl** at the same time), globose umbel-like to 12 cm ∅; **Ped** 0.5 - 3.5 cm; **Sep** 2 mm, outside sometimes finely hairy; **Cl** black-purple, flat or funnel-shaped, ± 2 cm ∅; **Cl** lobes 8 × 8 mm, broadly ovate-triangular, acute, inside densely tuberculate, mostly papillate-pubescent, **Ha** long, purple, rims of the **Cl** lobes mostly ciliate, **Ha** purple, vibratile, ± clavate; **Cn** brownish-purple, ± 5 mm ∅, weakly pubescent; **Ci** lobes deeply bifid, appendages falcately curved inwards, basal-medium region broadened, sometimes with a small appendage, ± ciliate; **Cs** lobes blunt, not surpassing the **Anth**, rarely glabrous; **Poll** inversely pear-shaped; **Fr** to 14 × 1 cm; **Se** brown, 11 × 8 mm.

One of the mightiest stapeliads; closely related to *C. edithae* and *C. somalica*.

C. adenensis (Deflers) A. Berger (Stapel. & Klein., 79, 1910). **T**: P. – **D**: Yemen. **I**: Deflers (1896).

≡ *Boucerosia adenensis* Deflers (1896) ≡ *Crenulluma adenensis* (Deflers) Plowes (1995); **incl.** *Caralluma kalmbacheriana* Lavranos (1965) ≡ *Crenulluma kalmbacheriana* (Lavranos) Plowes (1995); **incl.** *Caralluma rauhii* Lavranos (1965) ≡ *Crenulluma rauhii* (Lavranos) Plowes (1995).

[3] Massive richly branched perennials, 20 - 60 × 50 cm; stems upright or ascending, (greyish-, yellowish- or brownish-) green, ≤ 3.5 cm ∅, 4-ribbed, edges bluntish; **Tu** ≤ 2 (upper ½ of stem 1.2) cm, edges between the **Tu** slightly crenate, sometimes ± entire; **L** rudiments small, pointing upwards; **Inf** 15- to 50-flowered, globose, with strong foetid odour; **Bra** 4 - 8 mm, linear or fusiform, numerous; **Ped** (yellowish-) green, 7 - 20 × ≤ 2 mm; **Sep** ≤ 8 mm, fusiform-acuminate, apically bent outwards, outside finely papillate; **Cl** outside greenish or yellowish-cream, dotted with maroon, inside dark purple or yellow dotted with orange-red, fleshy, 2 - 5 cm ∅; **Cl** tube 6 - 10 mm deep, basally 7 - 15, apically (4 -) 5 - 7 mm ∅, campanulate or ± urceolate, basally flattened; **Cl** lobes 5 - 18 × 8 - 18 mm, horizontally spreading or slightly ascending, inside finely tuberculate, tip sometimes with a bunch of **Ha**, **Ha** purple, 3 - 4 mm, clavate, vibratile; **Cn** pale pink, dark maroon or shiny orange-red, 2 × 3.5 - 6 mm, ± broadly cylindrical, cup- or bowl-shaped, finely pubescent; **Ci** lobes 2.5 - ± 3.5 mm, apically elongated, deeply divided, appendages linear, ± spreading or touching each other in one point, upright or ascending at an angle, sometimes touching the opening of the **Cl** tube; **Cs** lobes 1 - 3.5 mm, blunt or acuminate; **Poll** D-shaped, ± 0.3 × 0.28 mm; **Fr** grey-green, 10 - 20 × 0.5 - 1 cm.

According to Gilbert (1990) the species belongs to the *adenensis-speciosa*-group and is very closely related to *C. awdeliana* and *C. lavrani* (see there).

C. adscendens (Roxburgh) R. Brown (Mem. Wern. Nat. Hist. Soc. 1: 14, 1810). **T:** [lecto – icono]: Roxburgh, Pl. Coast Coromandel, t. 30, 1795. – **D:** India, Sri Lanka.

≡ *Stapelia adscendens* Roxburgh (1795).

This highly variable complex with a geographically wide distribution needs a complete revision. Var. *fimbriata* and var. *gracilis* seem to be hardly discernible from var. *adscendens* (Meve, pers. comm.).

C. adscendens var. **adscendens** – **D:** India. **Fig. IV.d**

[1] Stems basally procumbent, then upright, 30 - 65 cm, basally ≤ 2 cm ∅, shrubby, branched, apically ± strongly tapering and only ≤ 8 mm ∅, green, often striped reddish, concavely 4-angled, edges rounded or sharp; **Tu** blunt, protruding, horizontally spreading; **L** rudiments ≤ 4 mm, lanceolate, ± acute; **Inf** 1- to 2-flowered in the **L** axils, loosely scattered; **Fl** drooping; **Ped** 1 - 4 mm, thin, vertical or bent downwards; **Sep** 2 - 3 mm, acute; **Cl** (pale) green, finely dotted with purple, sometimes horizontally striped with purple or chestnut-brown, ± 2.3 cm ∅, ± flat to shortly campanulate (then tube 1.5 mm); **Cl** lobes 1 - 3 × 1.5 cm, lanceolate, bluntly acuminate, apically brownish, spreading outwards, margins slightly revolute, glabrous; **Cn** brownish to dark purple, bowl-shaped; **Ci** basally pouch-shaped, lobes ≤ 1 mm, deeply divided, appendages filiform, falcate, bent towards each other; **Cs** lobes broadly triangular, ± blunt; **Poll** ovoid.

Most closely related to the Afro-Arabian *C. subulata*, which has already been put into synonymy by Bruyns (1992).

C. adscendens var. **attenuata** (Wight) Gravely & Mayuranathan (Bull. Madras Gov. Mus. 4(1): 13, 1931). **T:** [lecto – icono]: Wight, Icon. Pl. Ind. Orient., t. 1268, 1848. – **D:** India.

≡ *Caralluma attenuata* Wight (1848).

[1] Differs from var. *adscendens*: **Tu** less distinctly set off, more strongly bent upwards; **Fl** not quite as drooping as in var. *adscendens* and var. *fimbriata*; **Ped** ± horizontally spreading; **Cl** 1.5 cm ∅, hairy.

The dominating variety on the Indian subcontinent.

C. adscendens var. **carinata** Gravely & Mayuranathan (Bull. Madras Gov. Mus. 4(1): 16, t. 2: 6, 1931). **T:** [lecto – icono]: l.c. t. 2: 6. – **D:** India.

[1] Differs from var. *adscendens*: ± unbranched; stem edges (at least basally) acute; **Tu** as in var. *attenuata*; **Fl** rarely ± erect; **Ped** horizontally spreading; **Cl** purple, **Cl** lobes light maroon banded with yellow, otherwise like var. *attenuata*; **Cn** yellowish.

Intermediate between var. *attenuata* and var. *fimbriata* in habit.

C. adscendens var. **fimbriata** (Wallich) Gravely & Mayuranathan (Bull. Madras Gov. Mus. 4(1): 13, 1931). **T** [lecto]: Burma (*Wallich* s.n. [[icono]: Wallich, Pl. Asiat. Rar., t. 8, 1830]). – **D:** India, Sri Lanka.

≡ *Caralluma fimbriata* Wallich (1830).

[1] Differs from var. *adscendens*: Stems (at least close to the base) with rounded edges; **Tu** as in var. *attenuata*; **Cl** rather small, similar to var. *attenuata*.

C. adscendens var. **geniculata** Gravely & Mayuranathan (Bull. Madras Gov. Mus. 4(1): 16, t. 2: 7, 1931). **T:** [lecto – icono]: l.c. t. 2: 7. – **D:** S India. **Fig. IV.f**

[1] Compact plants with ascending-erect stems, occasionally branched; stems green, 4-angled, slender, apically tapering, **Br** apically more strongly tapering and branched, edges (at least basally) acute; **Tu** indistinctly set off, bent upwards; **L** rudiments small, lanceolate, only on the growing tip of the stems; **Fl** in upright position; **Ped** ascending or ± horizontally spreading; **Cl** yellow striped with dark

chestnut-brown, 1.1 cm ⌀; **Cl** tube opening widely, twisted upwards, with abrupt transition to the **Ped**; **Cl** lobes lanceolate, margins in the upper ½ strongly revolute, inside ± very hairy (rim of the tube protruding); **Cn** ± 2 mm ⌀; **Ci** brownish, bifid, appendages 2 mm, horn-like; **Cs** lobes black, margin translucent-yellowish, apically undulate, not surpassing the **Anth**.

Probably better treated as a species of its own. This taxon is more closely related to *C. bhupinderiana* than to *C. adscendens*.

C. adscendens var. **gracilis** Gravely & Mayuranathan (Bull. Madras Gov. Mus. 4(1): 14, t. 2: 5, 1931). **T:** [lecto – icono]: l.c. t. 2: 5. – **D:** India.

[1] Differs from var. *attenuata*: Stems very slender and strongly branched in the upper ½, (at least close to the base) with acute edges; **Inf** with 2 erect **Fl**; **Ped** long, slender, ascending; **Cl** ≤ 1 cm ⌀, **Cl** lobes strongly revolute.

C. arabica N. E. Brown (BMI 1895: 318, 1895). **T:** Yemen (*Hirsch* 28 [K]). – **Lit:** Bruyns & Jonkers (1994: 60-62, ills.). **D:** N and S Yemen, N Oman.

≡ *Crenulluma arabica* (N. E. Brown) Plowes (1995).

[3] Shrubby, 6 - 18 × 30 - 40 (-75) cm ⌀; stems upright, green, often tinged purple, 35 × 1.5 - 2 cm, 4-ribbed (similar to *C. flava*); **Tu** slightly projecting, apically with a distinct furrow; **Inf** 20- to 40-flowered; **Fl** with foetid odour; **Ped** ± 2 - 5 × ± 1 - 2 mm; **Sep** ± 1.5 × ± 1 mm; **Cl** ± maroon, sometimes spotted with yellow (Yemen), flat or slightly campanulate, ± 0.5 - 2 cm ⌀; **Cl** lobes ± 3 - 5 × ± 3 mm, triangular, acute, ± strongly spreading outwards, margins reflexed, horizontally wrinkled on the inside, apically often with purple clavate **Ha** (Oman); **Cn** red to purple, ± 2.8 - 3.5 mm ⌀, ± circular, bowl-shaped; **Ci** lobes ± 1.5 - 2 × ± 1 mm, erect, deeply bifid, appendages short, narrow, straight or laterally diverging, sometimes strongly reduced, rarely shortly hairy on the inside; **Cs** lobes apically ± 0.5 - 0.7 × ± 0.5 mm, ± elongate-triangular, incumbent on the **Anth** but not surpassing them, much broader where adnate to the short **Cs** lobes; **Anth** ± 0.4 - 0.5 mm broad, apically crenate, distinctly separate from each other; **Poll** ± 0.6 × ± 0.25 mm, roundish; **Fr** 6.5 - 1.5 cm.

A widely distributed species. The populations now treated as *C. arabica* in part differ strongly from the type material (Bruyns & Jonkers 1994). According to Gilbert (1990) the species belongs to the *europaea-hexagona*-group.

C. arachnoidea (P. R. O. Bally) M. G. Gilbert (Nation. Cact. Succ. J. 32(2): 26-29, ills., 1977). **T:** Uganda (*Eggeling* 5692 in *Bally* 6294 [G]). – **D:** E Africa.

≡ *Caralluma gracilipes* ssp. *arachnoidea* P. R. O. Bally (1969) ≡ *Spathulopetalum arachnoideum* (P. R. O. Bally) Plowes (1995).

C. arachnoidea var. **arachnoidea** – **D:** Ethiopia, Kenya, Uganda, Tanzania. **Fig. IV.g**

[1] Stems distinctly heteromorphic, basally 4-ribbed, **Ri** acute, straight, oriented upwards, ± serrate, **Fl**-bearing apical region ± terete to 4-angled; **Cl** whitish, yellow, spotted with maroon to purple, deeply incised; **Cl** tube strongly reduced; **Cl** lobes stiff, horizontally spreading, almost completely reflexed along the midrib, upper face pubescent, ciliate; **Cn** 3 mm ⌀, low, ± bowl-shaped, short-stalked, evenly broadened towards the bases of the **Ci**; **Ci** lobes deeply bifid, appendages < 0.8 mm, horn-like or slender-fusiform, acute, erect; **Cs** lobes ribbon-like, ascending over the **Sty** head, coming together almost along their entire length, apically rounded, slightly incised, ± hiding the **Anth**.

C. arachnoidea var. **breviloba** (P. R. O. Bally) M. G. Gilbert (Nation. Cact. Succ. J. 32(2): 29, ills. (pp. 26-30), 1977). **T:** Kenya (*Bally* 12267 [K, ZSS]). – **D:** Ethiopia, Kenya.

≡ *Caralluma gracilipes* ssp. *breviloba* P. R. O. Bally (1969).

[1] Differs from var. *arachnoidea*: **Fl**-bearing apical parts of stems ± 4-angled; **Cn** 2.5 mm ⌀; **Ci** lobes bifurcate, appendages ≥ 1 mm, at first erect, then curved inwards; **Cs** lobes shorter, not rising above the **Sty** head, deltoid, acute, partly touching apically, as long as or shorter than the **Anth**.

C. awdeliana (Deflers) A. Berger (Stapel. & Klein., 81, 1910). **T:** Yemen (*Deflers* 485 [P]). – **D:** Yemen. **I:** Deflers (1896).

≡ *Boucerosia awdeliana* Deflers (1896) ≡ *Crenulluma awdeliana* (Deflers) Plowes (1995); **incl.** *Caralluma petraea* Lavranos (1983) ≡ *Crenulluma petraea* (Lavranos) Plowes (1995).

[3] Plants of medium size, growing in clumps; stems basally branching, ± 20 - 30 × 2 - 3 cm, green (older stems spotted with red), 4-ribbed or 4-angled; **Tu** blunt; **L** rudiments acute, basally with stipular rudiments; **Inf** 5- to 15-flowered; **Ped** ± 5 - 7 × ± 2.4 mm; **Sep** ± 1 - 1.5 × ± 0.8 - 1 mm, bluntly triangular; **Cl** outside reddish-green spotted with red, inside sulphur-yellow to ochre, densely spotted with brownish-purple, deeply campanulate or urceolate, ± 1.2 cm ⌀, tube ± 1 × ± 1 cm; **Cl** lobes ± 15 × ± 5 - 8 mm, erect, glabrous, margins strongly revolute; **Cn** 2.5 - 3 mm ⌀, bowl-shaped; **Ci** reddish-violet, apically violet, lobes 1.5 - 2 × ± 0.8 mm, apical ½ bifid, appendages narrowly linear; **Cs** dark violet, lobes ± 0.3 - 0.5 × ± 0.3 mm, broadly triangular, not surpassing the **Anth**; **Anth** ± 0.5 - 0.7 mm broad, apically touching each other laterally.

Belonging to the *adenensis-speciosa*-group (Gilbert 1990).

C. baradii Lavranos (CSJA 65(5): 246-247, ills., 1993). **T:** Somalia (*Lavranos & al.* 23338 [UPS]). – **D:** Somalia. **Fig. V.b**

≡ *Somalluma baradii* (Lavranos) Plowes (1995).

[1] Plants with subterranean runners; stems procumbent to erect, clustered, olive-green to brownish, 2 - 8 × ± 1 cm, tapering towards the base, apically ± blunt, 4-angled, edges rounded; **L** rudiments 1 - 2 mm, drying and persistent; **Inf** few-flowered, erect, with 1 - 2 peduncles, to 12 × 0.1 - 0.15 cm, filiform, drying after fruiting; **Bra** solitary, triangular; **Ped** ≤ 30 × 0.3 mm, wiry, horizontally spreading, apically abruptly thickened and ascending; **Sep** ± 1 mm; **Cl** basally dark chestnut-brown, then yellowish striped with chestnut-brown, upper ⅔ unicoloured yellowish, 0.8 - 1 cm ⌀, divided ± to the bottom; **Cl** lobes linear, ascending, curved inwards, apically linked with each other forming a broad ellipsoid cage, basally strongly folded inwards, apically tapering, both sides glabrous; **Ci** wine-red, appendages bifid, ± 1.3 mm, fusiform, ascending, wrinkled, tip with solitary **Ha**, dark wine-red, 3 - 4 mm, simple, vibratile; **Cs** dark maroon, lobes ± 2 × 0.25 mm, touching each other in upright position above the **Sty** head and together forming a column, apically obtuse and slightly emarginate.

The relationships of this striking species are still unclear.

C. bhupinderiana Sarkaria (Nation. Cact. Succ. J. 35(3): 68-72, ills., 1980). **T:** India (*Sarkaria* J93-78 [PAN]). – **D:** India (Tamil Nadu).

[1] Plants compact, stems ascending to upright, unbranched, light green spotted with brownish-purple, to 32 × 1.5 cm, gradually tapering towards the tip, acutely 4-angled; **L** rudiments 3 × 1.5 mm, lanceolate, mostly only on growing young stem tips; **Inf** in pairs; **Bra** 3, pale yellow, < 0.5 mm, very thin; **Ped** pale to brownish-yellow, 6 - 8 (-11) mm, ascending, sometimes horizontally spreading from the stems; **Sep** 1.5 mm; **Cl** inside pale green, 1.5 - 1.6 cm ⌀, flat, central depression < 0.5 × 3 mm ⌀, rim finely ciliate; **Cl** lobes basally light yellow to greenish, ± distinctly striped with light brown, apically pale yellowish to dark brown, curved outwards, 6 × 3 mm, lanceolate, apically ending in a tooth-like appendage, the latter translucent, ± 0.5 mm, occasionally with a ciliate midrib, margins of the upper ½ strongly revolute, apically densely, basally weakly hairy, **Ha** flat, acuminate, soft; **Cn** yellowish, 2 mm ⌀; **Ci** translucent, deeply bifid, lobes ascending, pale brown, 3 mm, forming loops; **Cs** translucent or weakly spotted with brown, lobes spatulate, blunt, undulate, slightly recurved, not surpassing the **Anth**; **Poll** pale yellow, ovoid; **Fr** dark brown with purple markings, 8.5 cm.

Forming a complex with *C. adscendens* var. *geniculata*.

C. burchardii N. E. Brown (BMI 1913: 121, 1913). **T:** Canary Islands, Fuerteventura (*Burchard* 385 [K]). – **Lit:** Meve (1995: with ill.). **D:** Canary Islands, Morocco.

≡ *Apteranthes burchardii* (N. E. Brown) Plowes (1995).

C. burchardii ssp. **burchardii** – **D:** E Canary Islands (Fuerteventura, Lanzarote, Gran Canaria). **Fig. V.c**

Incl. *Caralluma burchardii* var. *purpurascens* Gattefossé & Maire (1945); **incl.** *Caralluma burchardii* var. *sventenii* E. & B. M. Lamb (1956).

[3] Forming clumps or heaps, to 60 cm ⌀, with slender subterranean runners; stems unbranched, green, sometimes weakly spotted with reddish, 15 - 30 (-50) × 1.5 - 2.5 cm ⌀, acutely 4-angled, apically 4-ribbed, after flowering esp. increasing in diameter and thus extremely irregularly shaped; **L** rudiments 1.5 mm ⌀, curved downwards; **Inf** 3- to 9-flowered; **Ped** 1 - 3 mm; **Sep** 3 × 1 mm; **Cl** inside olive-brown to ochre, 13 - 16 mm ⌀, deeply divided, flat or weakly campanulate caused by the slightly erect lobes, tube containing the elongated **Cn**; **Cl** lobes 4 × 3.5 mm, oblong-ovate, acute, margin slightly revolute, inside densely pubescent, **Ha** whitish-translucent, thick, cylindrical, giving a shaggy appearance to the **Fl**, rarely ± glabrous; **Cn** yellow, deeply tubular (similar to *C. quadrangula* and *C. cicatricosa*); **Ci** lobes inserted at the rim of the **Cn**, 1 mm, erect, 2-parted, appendages fusiform, blunt, nectar cavities deeply sunken into the **Fil** tube; **Cs** lobes 1 - 1.25 mm, as long as the **Anth**, linear, blunt; **Anth** 0.6 - 0.8 mm broad, distinctly separate from each other; **Poll** ± 0.5 × ± 0.4 mm, translator wings 0.1 mm; **Fr** green striped with purple, 7 - 8 cm × 7 - 8 mm; **Se** pale brown, 6 × 3.5 mm, transversely oblong-ovate, flat, with a broad rim.

Belonging to the *europaea-hexagona*-group (Gilbert 1990), though rather isolated because of its unusual corona and gynostegium.

C. burchardii ssp. **maura** (Maire) Meve & F. Albers (Nordic J. Bot. 15(5): 465, 1995). **T:** Morocco (*Maire* s.n. [MPU]). – **D:** Morocco.

≡ *Caralluma burchardii* var. *maura* Maire (1923); **incl.** *Caralluma burchardii* fa. *grandiflora* Maire (1942); **incl.** *Caralluma burchardii* fa. *sordida* Maire (1942); **incl.** *Caralluma burchardii* fa. *viridis* Gattefossé & Maire (1945).

[3] Differs from ssp. *burchardii*: Plants ± 15 - 20 cm tall; stems blue-green, ± 1.5 - 2 cm ⌀, 4-ribbed, **Ri** laterally more compressed; **Tu** conical; **L** rudiments inconspicuous, < 1 mm; **Ped** 5 - 9 mm; **Sep** ± 2 - 2.5 mm; **Cl** inside dark olive-green, broadly campanulate caused by the ascending lobes, 0.7 - 1.1 (-1.3) cm ⌀; **Cl** lobes 4 - 5 × 3 - 4 mm; **Cn** golden yellow, ± 4 - 6 mm ⌀, pentagonal when viewed from above; **Ci** lobes ± 1 - 1.5 × ± 1 mm; **Cs** lobes ± 1.5 mm, not surpassing the **Anth**; **Poll** ± 0.4 × 0.3 mm.

C. cicatricosa (Deflers) N. E. Brown (Gard. Chron.,

ser. 3, 12: 369, 1892). **T:** Yemen (*Deflers* 435 [P]).
– **Lit:** Plowes (1990). **D:** Yemen. **I:** Gilbert (1990). **Fig. V.a**

≡ *Boucerosia cicatricosa* Deflers (1889) ≡ *Monolluma cicatricosa* (Deflers) Plowes (1995).

[4] Stems shrubbily branched, green, 50 - 60 × 1 - 2 cm, edges with white mostly round scars from **Ped** that have fallen off, with strong odour; **Tu** indistinct, callously toothed, ≤ 5 mm, oblong-linear; **Inf** 1-flowered, close to the stem tips, seemingly axillary, sometimes opening in close sequence, **Fl** with or without foetid odour; **Bra** 1 - 2, small, triangular, **Sc**-like; **Ped** thick, 2 - 3 × 2 mm; **Sep** ± 2 mm, ascending; **Cl** outside light green, inside intensively chestnut-brown or dark maroon to blackish-purple, rim of the tube lemon-yellow or orange-red (colour variants: inside yellowish dotted with chestnut-brown, or rim of the tube red/pink or pale brick-red, or rim of the tube and **Cl** lobes pale yellow with brownish **Tu**); **Cl** ± flat to slightly bowl-shaped, ± 1 cm ⌀, tube (when present) ≤ 2 mm deep, basally enclosing the **Cn**, rim almost annulus-like, inside smooth, margin with scattered papillae; **Cl** lobes 8 mm, oblong-lanceolate, blunt, fleshy, ± spreading upright or slightly curved inwards, margins ± reflexed, inside ± tuberculate-papillate, apically finely hairy; **Cn** (orange-) yellow, **Cs** finely spotted with ochre, long-stalked, appearing to be composed of 5 narrowly rectangular lobes pointing towards the centre of the **Fl** with a central ribbon-like appendage; **Ci** lobes basally pouch-like, ± deeply bifurcate, appendages narrowly triangular, apically ± setose, laterally spreading, connected to the backside of the **Cs** along their entire length; **Cs** ribbon-like lanceolate, surpassing the **Anth**; **Fr** 6 - 7 cm, 3-angled.

Belonging to the *C. quadrangula*-complex (Plowes 1990).

C. congestiflora P. R. O. Bally (Candollea 20: 13-15, ills., 1965). **T:** Somalia, Northern Prov. (*Bally* 11996 [K, G]). – **D:** Somalia. **I:** FPA t. 1715, 1976.

≡ *Spathulopetalum congestiflorum* (P. R. O. Bally) Plowes (1995); **incl.** *Caralluma plurifasciculata* P. R. O. Bally *ex* M. G. Gilbert (1990) (*nom. inval.*, Art. 34.1(a)).

[1] Basally strongly branched; stems 10 - 25, to 28 cm, occasionally with 2. order side-**Br**, distinctly heteromorphic, lower part to 23 × 2 - 2.2 cm, 4-angled, then tapering, edges acute, sides with ± deep longitudinal furrows; **Tu** 1 - 2 mm, ascending, 8 - 10 mm apart from each other; **L** rudiments delicate, fusiform; **Fl**-bearing upper part of the stems 7 - 11 × ± 0.4 cm; **Inf** 24- to 30-flowered, each node with 1 - 3 peduncles ≤ 3 mm, very fleshy, apically rounded; **Bra** ≤ 5 mm, narrowly triangular, mucronate, occasionally weakly serrate at the base; **Ped** 7 - 13 × 1 - 1.5 mm, ascending; **Sep** 2.5 × 1 mm; **Cl** outside green, inside yellow striped with green, ± 1.5 cm ⌀, flat, divided to the base; **Cl** lobes ± 7 × ± 3 mm, lanceolate, margins reflexed, apically hairy with 2 - 5 clavate **Ha**, **Ha** purple, 4 mm; **Cn** yellow, ± 2.5 × 3 mm, glabrous; **Ci** lobes 2 - 2.5 mm, bluntly triangular, shortly bifid, appendages erect, reaching far above the staminal column; **Cs** lobes 1.5 - 2 mm, narrowly triangular, geniculate, hardly surpassing the **Anth**.

Closely related to *C. priogonium*.

C. crenulata Wallich (Pl. Asiat. Rar. 1: 6, t. 7, 1830). **T** [lecto]: Myanmar (*Wallich* s.n. [icono: l.c. t. 7]). – **D:** India, Myanmar. **Fig. V.d**

≡ *Boucerosia crenulata* (Wallich) Wight & Arnott (1834) ≡ *Desmidorchis crenulata* (Wallich) Decaisne (1838); **incl.** *Caralluma truncatocorona* hort. (s.a.) (*nom. inval.*, Art. 29.1); **incl.** *Boucerosia truncato-coronata* Sedgwick (1921) ≡ *Caralluma truncato-coronata* (Sedgwick) Gravely & Mayuranathan (1931); **incl.** *Caralluma nilagiriana* Kumari & Subba Rao (1976) ≡ *Boucerosia nilagiriana* (Kumari & Subba Rao) Plowes (1995).

[3] Plants small, growing in clumps, with subterranean runners; stems ± erect, laxly branched, ± 12 - 15 × 0.6 cm, 4-ribbed, laterally furrowed; **Tu** oblong-conical, acutely tapering, horizontally spreading; **L** rudiments ovate; **Inf** 8- to 10- (to 15-) flowered; buds 8.5 mm ⌀, pentagonal, apically flattened; **Bra** 2.5 mm, linear; **Ped** ± 8 - 20 × 1.5 - 3 mm; **Sep** ± 2 - 4 × ± 1.5 - 2 mm; **Cl** outside green spotted with violet, inside white or yellowish with broken concentrical yellow circles, ± 1.2 - 2.2 cm ⌀, campanulate, tube 4 - 6 × 6 - 9 mm, cup-shaped, enclosing the **Cn**; **Cl** lobes ± 5 - 8 × ± 5 - 9 mm, ovate-triangular, horizontal to erect, brown, margins slightly revolute and weakly hairy, **Ha** clavate; **Cn** white or yellowish dotted with reddish, or dark purple, ± 4.5 - 5.5 mm ⌀; **Ci** lobes ± 1.5 - 2 × ± 1.3 mm, obtuse, slightly incised, appendages ± 0.4 - 0.6 mm, triangular, spreading outwards; **Cs** lobes ± 1.5 - 2 × ± 0.5 mm, ± triangular to ribbon-like, ± covering the **Anth**, apically bluntly dentate or furrowed; **Anth** yellowish, ± 0.8 × ± 0.6 mm, rectangular, blunt, often touching each other laterally; **Poll** orange, ± 0.4 × ± 0.5 mm, roundish, translator wings 1.5 mm, erect.

Because of the great similarities, *C. truncato-coronata* and *C. nilagiriana* are treated here as synonyms. According to Gilbert (1990), this species belongs to the *umbellata-indica*-group.

C. dicapuae (Chiovenda) Chiovenda (in A. C. White & B. Sloane, Stapelieae, ed. 2, 1: 187, ills., 1937). **T:** Eritrea (*Terracciano & Pappi* 498 [FT]). – **D:** Ethiopia, Eritrea, Somalia, Kenya. **Fig. V.e**

≡ *Spathulopetalum dicapuae* Chiovenda (1912); **incl.** *Caralluma dicapuae* ssp. *dicapuae*; **incl.** *Caralluma dicapuae* ssp. *seticorona* P. R. O. Bally (1969).

[1] Dwarf shrubs; stems numerous, weakly branching, to 60 cm, grey-green, heteromorphic, lo-

wer part to 30 cm, robust, ± 4-angled, edges blunt; **Tu** 5 - 8 mm, distinctly conical, ascending, 2 - 3 cm apart from each other, ± short-lived; **Fl**-bearing upper part ± 20 cm, long-extended, ± terete, **Int** 2.5 - 4 cm; **Inf** 2-flowered; **Fl** buds basally clavate-cylindrical, apically with abruptly articulate tip, 8 - 10 × 2 - 3 mm, tip 0.7 - 1 mm; **Bra** 2 - 3 × 0.5 mm; **Ped** 1 - 1.6 cm, slender, curved downwards; **Sep** linear-fusiform, very acute; **Cl** outside dark purple, inside light green finely dotted or horizontally striped with chestnut-brown, 2.2 cm ∅, campanulate to narrowly columnar, deeply divided, central depression 0.5 mm; **Cl** lobes 11 - 14 × 1 (base) - 3 (tip) mm, narrowly spoon-shaped to linear, mucronate, at first spreading, then flexible, drooping, freely swinging, lamina folded back along midrib, outside often setose, inside densely hairy, **Ha** clavate, spreading; **Cn** 2.4 × 3 mm; **Ci** lobes rectangular, upright-spreading, shortly bifid, appendages bluntly deltoid, partly laterally spreading, rarely finely hairy on the outside, **Ha** 0.5 mm, stiff; **Cs** white dotted with chestnut-brown, lobes 3 × 0.9 mm, ribbon-like, ± linear, apically bluntly triangular, far overtopping the **Sty** head, apically spreading, glabrous; **Fr** 11 - 13 × 0.4 - 0.5 cm.

C. diffusa (Wight) N. E. Brown (Gard. Chron., ser. 3, 12: 369, 1892). **T** [lecto]: India (*Anonymus* s.n. [icono: Icon. Pl. Ind. Orient., t. 1599, 1850]). – **D:** S India. **I:** Gravely & Mayuranathan (1931). **Fig. VI.a**

≡ *Boucerosia diffusa* Wight (1850) ≡ *Desmidorchis diffusa* (Wight) Kuntze (1891).

[3] Stems in dense large clumps, ± 10 - 15 × 1.5 cm, fresh green, concave acutely 4-angled; **Tu** ± wing-like; **Inf** 12- to 30-flowered; **Fl** buds apically ± acutely conical; **Cl** inside yellowish marked with dark purple stripes, 2.5 cm ∅, deeply campanulate, deeply divided; **Cl** lobes (narrowly) triangular, ± acute, purple stripes finely papillate, margins hairy, **Ha** long, caducous; **Cn** bowl- or cup-shaped; **Ci** lobes strongly reduced, bifurcate for > ½; **Cs** lobes variable in shape, apically slightly undulate.

According to Gilbert (1990), the species belongs to the *umbellata-indica*-group.

C. dolichocarpa O. Schwartz (Mitt. Inst. Allg. Bot. Hamburg 10: 194, 1939). **T:** Yemen (*von Wissmann* 1225 [HBG]). – **D:** S Yemen (Hadhramaut).

≡ *Crenulluma dolichocarpa* (O. Schwartz) Plowes (1995).

[4] Plants low, forming dense mats, strongly branching; stems greenish, to 7 × 1.2 cm, edges acute; **Tu** very small, thus edges of young shoots strongly undulate, **Ped** scars hardly visible; **Inf** 1-flowered, upright, growing from the stem tip or uppermost **Ax**; **Ped** 9 - 12 mm, strongly thickened at **Fr** time, stout; **Sep** 3 × 1.8 mm, ovate; **Cl** outside greenish, inside dark maroon, ± 2 cm ∅, ± campanulate, divided for up to ⅗; **Cl** lobes 7 × 6 mm, ovate, apically acuminate; **Cn** like in *C. quadrangula*; **Fr** 16 - 18.5 × 0.3 - 0,5 cm, both ends blunt.

Insufficiently known, perhaps closely related to *C. cicatricosa*.

C. edithae N. E. Brown (BMI 1895: 219, 1895). **T:** Somalia (*Cole* s.n. [K]). – **D:** Ethiopia, Somalia. **I:** FPA 36: t. 1430, 1964; Gilbert (1990).

≡ *Desmidorchis edithae* (N. E. Brown) Plowes (1995).

[3] Branched in a spreading-shrubby fashion; stems 10 - 30 cm, 4- (rarely 5-) ribbed; **Tu** conical-triangular, distinctly projecting, with hardened **L** scars; **Inf** 30- to 70-flowered, 5 - 7 cm ∅; **Fl** with strong odour like manure; **Ped** 15 - 20 × 1.5 - 2 mm; **Sep** 3 - 5 × 1 - 1.5 mm, sometimes apically distinctly curved outwards; **Cl** ± purple, ± campanulate, ± 1.5 cm ∅; **Cl** tube 3 - 5 × 5 - 7 mm; **Cl** lobes oblong-triangular, apically slightly curved outwards, upright, inside tuberculate, tips with a bunch of clavate **Ha**, **Ha** purple, ± 2 mm; **Cn** 3 - 4.5 mm ∅, glabrous; **Ci** lobes 1 - 1.5 × ± 1 mm, apically distinctly bifid, appendages apically rounded, strongly spreading laterally; **Cs** lobes 1 - 1.2 × ± 0.3 mm, apically irregularly slightly crenate, not or hardly surpassing the **Anth**; **Anth** ± 0.5 × 0.5 - 0.7 mm; **Poll** ± 0.3 × ± 0.35 mm, roundish to transversely ovate, translator wings very short.

Closely related to *C. penicillata*. According to Gilbert (1990), the species belongs to the *adenensis-speciosa*-group.

C. edulis (Edgeworth) Bentham & Hooker *fil.* (Gen. Pl. 2: 782, 1876). **T:** India (*Edgeworth* 6035 [K]). – **D:** India (Punjab), Pakistan, Iran, Oman, Saudi Arabia, Ethiopia, Somalia, Sudan, Mauritania, Dhalak Archipelago (Dalcos). **I:** Bruyns (1989). **Fig. VI.b**

≡ *Boucerosia edulis* Edgeworth (1862) ≡ *Cryptolluma edulis* (Edgeworth) Plowes (1995); **incl.** *Boucerosia stocksiana* Boissier (1879) ≡ *Desmidorchis stocksiana* (Boissier) Kuntze (1891); **incl.** *Caralluma longidens* N. E. Brown (1892) ≡ *Spiralluma longidens* (N. E. Brown) Plowes (1995); **incl.** *Caralluma vittata* N. E. Brown (1904); **incl.** *Caralluma mouretii* A. Chevalier (1934) ≡ *Spiralluma mouretii* (A. Chevalier) Plowes (1995); **incl.** *Caralluma edulis* A. Chevalier (1934) (nom. inval., Art. 34.1c).

[1] Stems halfway climbing, ± 60 cm, with long subterranean runners, dull greenish-brown, 4-angled to terete, **Fl**-bearing upper part of stems gradually tapering; **L** rudiments ≤ 1.2 cm, unusually large, cuneate or elliptic; **Inf** 2-flowered, appearing from the **Ax**; **Bra** 2, delicate, fusiform; **Ped** slender, ± ascending; **Sep** acute, margins translucent; **Cl** inside whitish or sand-coloured, 0.8 - 1.1 cm ∅; **Cl** tube semiglobose-campanulate, inside purple-striped, ± 2× as long as the **Cl** lobes; **Cl** lobes fusiform, very narrow, curved outwards, glabrous, distinctly veined; **Cn** ± cup- to bowl-shaped, ochre,

with nectar; **Ci** lobes basally pouch-shaped, deeply bifid, appendages falcate; **Cs** lobes oblong-lanceolate to filiform, geniculate, reaching above the **Sty** head and joining there.

Easily confused in habit with *C. sinaica*.

C. edwardsiae (M. G. Gilbert) M. G. Gilbert (Bradleya 8: 13, 1990). **T:** Ethiopia (*Gilbert* 1730 [K, ETH]). – **D:** Ethiopia. **I:** Gilbert (1977: as *C. gracilipes* ssp.).

≡ *Caralluma gracilipes* ssp. *edwardsiae* M. G. Gilbert (1977) ≡ *Spathulopetalum edwardsiae* (M. G. Gilbert) Plowes (1995).

[1] Stems heteromorphic, basally more slender than in *C. gracilipes*, bluntly 4-angled; **Tu** conical, ascending, distinctly articulate (as in *C. dicapuae*); **Fl**-bearing upper part gradually tapering, few-flowered; **Cl** maroon, finely spotted with reddish-brown or purple, 1.5 - 2 cm; **Cl** tube reduced to a rim; **Cl** lobes delicate but stiff, horizontally spreading, basally narrowing down nail-like, finely hairy and ciliate inside; **Cn** thin, stalked ± 1.5 mm, 1.2 × 2 mm, **Ci** lobes basally pouch-like, apically triangularly extended, slightly folded within the transitional region to the **Cs**; **Cs** lobes broadly spatulate, apically bluntly dentate, hardly surpassing the **Anth**.

Closely related to *C. peckii*.

C. europaea (Gussone) N. E. Brown (Gard. Chron., ser. 3, 12: 369, 1892). **T:** Italy, Lampedusa (*Gussone s.n.* [NAP]). – **Lit:** Bruyns (1987a: ills.). **D:** Italy, Spain, Morocco, Algeria, Tunesia, Libya, Egypt, Israel, Jordania. **Fig. V.f**

≡ *Stapelia europaea* Gussone (1832) ≡ *Boucerosia europaea* (Gussone) Caruel (1886) ≡ *Desmidorchis europaea* (Gussone) Kuntze (1891) ≡ *Apteranthes europaea* (Gussone) Plowes (1995); **incl.** *Stapelia italica* Tenore *ex* Salm-Dyck (s.a.); **incl.** *Stapelia lampadosa* Jacquin *ex* Salm-Dyck (s.a.); **incl.** *Stapelia quadrangula* Schousboe (s.a.) (*nom. illeg.*, Art. 53.1); **incl.** *Apteranthes gussoneana* Mikan (1835) ≡ *Stapelia gussoneana* (Mikan) Lindley (1835) ≡ *Boucerosia gussoneana* (Mikan) Hooker *fil.* (1874) ≡ *Caralluma europaea* ssp. *gussoneana* (Mikan) Maire (1924) ≡ *Apteranthes europaea* ssp. *gussoneana* (Mikan) Plowes (1995); **incl.** *Boucerosia maroccana* Hooker *fil.* (1874) ≡ *Caralluma maroccana* (Hooker *fil.*) N. E. Brown (1892) ≡ *Caralluma europaea* var. *maroccana* (Hooker *fil.*) A. Berger (1910) ≡ *Caralluma europaea* ssp. *maroccana* (Hooker *fil.*) Maire (1924) ≡ *Apteranthes europaea* ssp. *maroccana* (Hooker *fil.*) Plowes (1995); **incl.** *Caralluma affinis* De Wildeman (1904) ≡ *Caralluma europaea* var. *affinis* (De Wildeman) A. Berger (1910) ≡ *Apteranthes europaea* var. *affinis* (De Wildeman) Plowes (1995); **incl.** *Caralluma simonis* hort. *ex* A. Berger (1904) ≡ *Caralluma europaea* var. *simonis* (hort. *ex* A. Berger) A. Berger (1910) ≡ *Apteranthes europaea* var. *simonis* (hort. *ex* A. Berger) Plowes (1995); **incl.** *Caralluma europaea* var. *marmaricensis* A. Berger (1910) ≡ *Apteranthes europaea* var. *marmaricensis* (A. Berger) Plowes (1995); **incl.** *Caralluma confusa* Font Quer (1922) ≡ *Caralluma europaea* var. *confusa* (Font Quer) Font Quer (1924) ≡ *Apteranthes europaea* var. *confusa* (Font Quer) Plowes (1995); **incl.** *Caralluma europaea* var. *barrueliana* Maire (1935) ≡ *Apteranthes europaea* var. *barrueliana* (Maire) Plowes (1995); **incl.** *Caralluma europaea* var. *albotigrina* Maire (1936) ≡ *Apteranthes europaea* var. *albotigrina* (Maire) Plowes (1995); **incl.** *Caralluma europaea* var. *gattefossei* Maire (1936) ≡ *Apteranthes europaea* var. *gattefossei* (Maire) Plowes (1995); **incl.** *Caralluma europaea* var. *decipiens* Maire (1938) ≡ *Apteranthes europaea* var. *decipiens* (Maire) Plowes (1995); **incl.** *Caralluma europaea* fa. *parviflora* Maire (1939); **incl.** *Caralluma europaea* var. *micrantha* Maire (1941) ≡ *Apteranthes europaea* var. *micrantha* (Maire) Plowes (1995); **incl.** *Caralluma europaea* var. *tristis* Maire (1941) ≡ *Apteranthes europaea* var. *tristis* (Maire) Plowes (1995); **incl.** *Caralluma europaea* var. *schmuckiana* Gattefossé & Maire (1943) ≡ *Apteranthes europaea* var. *schmuckiana* (Gattefossé & Maire) Plowes (1995).

A difficult complex in which at least the 2 taxa here accepted are involved. Closely related to *C. joannis*.

C. europaea var. **europaea** – **D:** Italy (Lampedusa), Spain, Morocco, Algeria, Tunesia, Libya, Egypt. **I:** Asklepios 59: 14-21, 1993.

[3] Stems forming mats, 5 - 20 × 1 - 2 cm, green or purplish-brown, distinctly 4-ribbed, concave between the **Ri**; **Tu** oblong-conical, slightly porrect; **Inf** 5- to 15-flowered; **Ped** 2 - 3 × ± 1.5 mm; **Sep** 1.5 - 2.5 × ± 1 mm; **Cl** outside greenish, inside yellowish, horizontally streaked with dark maroon to purple, 1.5 - 2 cm ⌀, funnel-shaped, campanulate or flat; **Cl** tube 1 - 5 mm deep, bowl-shaped to campanulate, basally enclosing the **Cn**; **Cl** lobes 5 - 8 × ± 5 mm, broadly triangular, ± horizontally spreading, often unicoloured purple, tips and margins ± strongly curved outwards, inside glabrous to densely covered with long **Ha**, **Ha** purple, ± 1 mm, simple; **Cn** 3.5 - 4 mm ⌀; **Ci** dark purple, bifid, appendages 1.2 - 2 × ± 1 mm, apically retuse to incised, mostly swollen and yellow; **Cs** dark brown, apically lighter, lobes 1 - 1.3 × ± 0.6 (apically 0.7) mm, shorter than the **Anth**, ± upright, apically deeply retuse or 2-parted, blunt; **Anth** ± 0.8 × 0.6 mm, laterally well separated from each other; **Poll** ± 0.4 × ± 0.4 mm, roundish, translator wings 0.1 mm.

Belonging to the *europaea-hexagona*-group according to Gilbert (1990).

C. europaea var. **judaica** M. Zohary (Palestine J. Bot., Jerusalem Ser., 2: 173, ills. (p. 176), 1941). **T:**

HUJ. – **D:** Israel, Jordania, Egypt?. **I:** Bruyns (1987b).

≡ *Apteranthes europaea* var. *judaica* (Zohary) Plowes (1995); **incl.** *Boucerosia aaronis* Hart (1885) ≡ *Caralluma aaronis* (Hart) Bruyns (1987) (*nom. inval.*, Art. 34.1); **incl.** *Caralluma israelitica* M. Zohary & Chaouat (1976) (*nom. inval.*, Art. 37.1?); **incl.** *Caralluma negevensis* M. Zohary (1978) ≡ *Apteranthes negevensis* (Zohary) Plowes (1995).

[3] Stems procumbent, forming small mats with subterranean runners, 3 - 20 × ± 1.2 - 2.5 cm; **Tu** acute, without stipular rudiments; **Inf** 2- to 12-flowered; **Ped** 1 - 1.5 mm; **Sep** 3 - 4 × 1 - 2 mm; **Cl** markings as in var. *europaea*; **Cl** tube basally narrow, enclosing the **Cn** stalk; **Cl** lobes 4 - 6 × ± 3.5 mm, margins slightly revolute, inside almost glabrous to covered with long **Ha**; **Cn** dark purple to reddish or yellowish-purple, apically often yellowish, 3 - 3.5 mm ∅, bowl-shaped, outside sometimes hairy, stalk 0.25 - 1 mm, squat; **Ci** lobes ± 1 × ± 1 mm, ± halfway bifid, appendages apically flattened, apically rounded, slightly spreading laterally, curved outwards; **Cs** lobes ± 1 × ± 0.5 mm, apically obtuse, rounded or retuse; **Anth** yellow, < 1 mm, thick, apically obtuse; translator wings 0.05 mm; otherwise like var. *europaea*.

C. flava N. E. Brown (BMI 1894: 335, 1894). **T:** Yemen, Hadhramaut (*Lunt* s.n. [K]). – **Lit:** Bruyns & Jonkers (1994: 56, 59, ills.). **D:** S Yemen, Oman. **Fig. VI.c**

≡ *Crenulluma flava* (N. E. Brown) Plowes (1995).

[3] Plants shrubbily branched, to 30 (-100) cm ∅; stems grey-green or purple-brown, 2.5 cm ∅, 4-ribbed, smooth, **Ri** winged; **L** rudiments flat; **Inf** 20- to 100-flowered; **Fl** with musty-rancid odour; **Ped** 10 - 30 × 1 - 1.5 mm; **Sep** ± 2 mm; **Cl** yellow, sometimes white (coast of Dhofar), (10-) 15 - 30 (-35) mm ∅, ± deeply divided, tube flat (N Oman) to ± deep and narrow (coast of Dhofar); **Cl** lobes horizontal, ovoid-lanceolate, margins ± strongly curved outwards, inside wrinkled, papillate, papillae partly with **Bri**, tips of lobes rarely with a bunch of carmine clavate **Ha**; **Cn** intensely yellow, glabrous, round or pentagonal; **Ci** lobes erect, bifid ± to the base, appendages straight or falcate and curved towards each other; **Cs** linear, shorter than the **Anth**; **Poll** D-shaped, ± 0.25 mm ∅; **Fr** 9 - 10 cm.

The size of the flowers is very variable within this species. According to Gilbert (1990) it belongs to the *adenensis-speciosa*-group. There is a very close relationship to *C. arabica*.

C. flavovirens L. E. Newton (Asklepios No. 74: 24-25, ills., 1998). **T:** Kenya, Coast Prov. (*Newton* 5589 [K, EA]). – **D:** Kenya.

[2] Stems laxly and irregularly branched, 5 - 12 × ± 0.7 cm, green, 4-ribbed with acute angles, **Tu** inconspicuous; **L** rudiments subulate, 4 × 2 mm, folded along the midrib, acute, persistent; apical **Fl**-bearing stem part to 26 cm, ± terete; **Inf** usually 1-flowered; **Ped** ± 1 cm, directed downwards; **Sep** 2 × 0.5 mm, triangular; **Cl** pale yellow-green, basally spotted with red, stellate; **Cl** lobes free, ± 8 × 1 mm, basal 1 mm erect and with vibratile 1 mm long purple clavate **Ha**, then recurved horizontally, linear-lanceolate, rolled outwards, margins with simple 1 mm long **Ha**; **Cn** 1 mm stipitate, stalk pale yellow; **Ci** lobes dark red, 0.75 mm, deeply bifid, spreading almost horizontally; **Cs** lobes pale red to pink to dark red, 2 × 0.5 mm, linear, erect-connivent; **Fr** to 7.5 cm, apically obtuse; **Se** pale brown, 5 × 2 mm, tuft of **Ha** 25 mm.

Endemic to inselbergs and belonging to the group around *C. arachnoidea* and *C. gracilipes*. – [U. Meve]

C. foetida E. A. Bruce (HIP 34: t. 3371 + text, 1938). **T:** Uganda (*Eggeling* 2955 [K]). – **D:** Kenya, Uganda. **I:** FPA t. 1714, 1976.

≡ *Desmidorchis foetida* (E. A. Bruce) Plowes (1995).

[3] Stems 10 - 15 cm, growing in clumps, 2 - 3.5 cm ∅, robust, 4-angled; **Tu** oblong to acutely triangular; **L** rudiments short, acute, becoming thorny; **Inf** 30- to 40-flowered, ± 4 cm ∅, with foetid odour; **Ped** 6 - 12 × 2 - 3 mm; **Sep** ± 4 × 1.5 - 2.5 mm, apically curved outwards, inside tuberculate; **Cl** inside dark maroon, 1.3 - 1.8 cm ∅, campanulate; **Cl** tube 5 - 7 mm, 1 - 1.3 cm ∅; **Cl** lobes ± 5 × ± 4 mm, ovoid-triangular, outside with distinct venation, inside tuberculate-papillate, margins densely hairy, **Ha** ± 3 mm, clavate, vibratile; **Cn** 0.8 - 1 cm ∅; **Ci** lobes 2 - 3 mm, deeply bifid, appendages triangular, blunt, often overlapping with neighbouring lobes, curved outwards; **Cs** lobes 1.2 - 1.5 × 0.8 - 1 mm, apically bluntly rounded, not surpassing the **Anth**; **Anth** 1 - 1.2 mm broad, slightly touching each other laterally; **Poll** ± 0.5 × 0.4 mm, roundish, translator wings short; **Fr** erect, follicles parallel, ± 12 × 0.8 cm.

Belonging to the *adenensis-speciosa*-group (Gilbert 1990) and showing a very close relationship to the aforementioned species.

C. frerei G. D. Rowley (Nation. Cact. Succ. J. 13(4): 78, 1958). **T:** K. – **D:** NW India (Poona, Maharashtra). **Fig. V.g**

Incl. *Frerea indica* Dalziel (1865).

[3] Stems with small roundish **R** tuber, ascending or growing as a ground cover, shrubbily branched, ± 10 cm, green, tinged whitish, terete or indistinctly 4-angled, fleshy; **L** ± 2.5 × 1 cm, oblong-ovate, fleshy, short-stalked, crowded in the upper ½ of the stems; **Inf** 1-flowered, lateral close to the stem tips; **Ped** 2 - 4 × ± 1.5 mm; **Sep** 3 - 4 × 2 - 3 mm, triangular-ovate, acuminate; **Cl** green, roughly streaked and spotted with purple, 2 - 3.5 cm ∅, flat with a

central depression enclosing the **Cn**, slightly convex around the **Cn**; **Cl** lobes centrally with irregular markings of small yellow dots, 0.6 - 1 × 1 - 1.3 cm, broadly triangular, acute, finely hairy, margins ± densely covered with vibratile clavate **Ha**; **Ha** dark purple, 1 - 2 mm; **Cn** dark purple, 4.5 - 5 mm ∅, cup- to bowl-shaped; **Ci** lobes 1.5 - 2 × 1.3 - 1.5 m, undulate, obtuse, ± erect, appendages ± triangular, laterally (where fused with the **Cs**) with small fusiform appendages; **Cs** lobes 1.4 - 1.6 × ± 0.5 mm, ribbon-like, not surpassing the **Anth**, apically irregularly furrowed or crenate; **Anth** light ruby-red, inner margins translucent gold- or amber-coloured, ± 1 × 0.7 - 0.9 mm, inversely trapeziform, laterally distinctly separated from each other; **Poll** ± 0.4 × 0.4 mm, roundish, translator wings short.

This is the only leafy stapeliad. Gilbert (1990) continues to separate it as the monotypic genus *Frerea*, but this is not justified on account of its floral structure. It is classified as a member of Subgen. *Boucerosia* on account of its floral morphology.

C. furta P. R. O. Bally (Candollea 18: 345-347, ills., 1963). **T:** Somalia, Northern Prov. (*Bally* 7129 [K]). – **D:** Ethiopia, Somalia. **I:** FPA 37: t. 1457, 1965; Plowes (1995). **Fig. VI.d**

≡ *Saurolluma furta* (P. R. O. Bally) Plowes (1995).

[1] Stems richly branched, small, in clumps, pale greenish-grey (**Ri** matt greyish-purple), ± 7 × 1 cm, **Br** 3 - 5 × 0.6 - 0.8 cm, apically tapering, bluntly 4-ribbed; **Tu** ≤ 2 mm, conical, ascending-protruding, basally 10 - 15 mm apart, crowded towards the apex (± 1 mm apart); **L** rudiments 0.6 mm, **Sc**-like; apical **Fl**-bearing stem region ≤ 35 × 2 mm, fleshy, persistent, **Ped** scars densely and regularly spread; **Inf** with 2 - 4 ± sessile **Fl**, erect, standing close together; **Ped** < 1 mm; **Sep** 2 × 0.5 mm, margins and tips purple; **Cl** 6 × 10 mm, inside pale green finely dotted with purple, deeply divided; **Cl** tube apically pale purple, 1 × 2.5 mm, cup-shaped, only basally enclosing the **Cn**; **Cl** lobes white in the basal ⅔, 5 × 2 mm, lower ½ ovate, upper ½ linear-spatulate, apically abruptly acuminate, erectly spreading, margins folded outwards, glabrous; **Cn** whitish to light pink, 3 × 2 mm, short-stalked; **Ci** lobes ± rectangular, erect, shortly bifid; **Cs** lobes spatulate, blunt, dorsally with 2 erect horn-like appendages; **Poll** ovoid, 0.3 × 0.15 mm; **Sty** head green; **Fr** pale leather-brown sparsely striped with red, ≤ 65 × 5 mm, upright; **Se** pale yellowish-brown, 5 × 3.2 mm, tuft of **Ha** > 2 cm.

There are vegetative similarities to *C. adscendens* p.p. and *C. socotrana*. However, inflorescence and coronal morphology as well as the more slender fruits identify this species as a true *Caralluma*. The monotypic genus *Saurolluma* set up by Plowes (1995) for this species is therefore superfluous.

C. gracilipes K. Schumann (in Engler, Pfl.-welt Ost-Afr., Teil C, 328, 1895). **T:** Kenya, Kitui Distr. (*Hildebrandt* 2700 [B [†]]). – **D:** N Kenya, Tanzania. **I:** Gilbert (1977).

≡ *Spathulopetalum gracilipes* (K. Schumann) Plowes (1995).

[1] Stems 12 - 30 cm, distinctly heteromorphic, basally 8 - 15 mm ∅, ± sharply 4-angled, distantly dentate; **L** rudiments ± 5 mm, narrow, acute; apical **Fl**-bearing stem region 3 - 4× as long as the basal region, slender, ± terete; **Inf** with ± 15 laxly spread **Fl**; **Ped** 7 - 8 mm; **Cl** whitish or yellow, spotted with maroon or purple, 0.9 × ± 2.5 cm ∅, deeply lobed; **Cl** tube reduced, inflated; **Cl** lobes 1 cm, stiff, ± horizontally spreading, margins revolute, hairy, basal ½ with clavate **Ha**, apically with simple **Ha**; **Cn** inserting high up on the staminal column, ± distinctly stalked, stalk narrowly cylindrical, abruptly broadened just under the bases of the lobes, surpassing the **Cl** at anthesis; **Ci** lobes basally pouch-like, bluntly deltoid, sometimes shortly 2-parted, or entirely lacking; **Cs** lobes oblong spoon-shaped, apically rounded, erect, rising high above the **Sty** head, apically touching each other; **Fr** round in cross-section.

Closely related to *C. arachnoidea* and *C. priogonium*.

C. hexagona Lavranos (JSAB 29: 105-107, ills. (pl. 14: 1-2), 1963). **T:** Yemen (*Lavranos* 1829 [K, PRE]). – **D:** Saudi Arabia, Yemen, Oman. **I:** FPA 44: t. 1743, 1977, as *C. shadhbana*; Bruyns (1987a).

≡ *Sulcolluma hexagona* (Lavranos) Plowes (1995); **incl.** *Caralluma foulcheri-delboscii* Lavranos (1964) ≡ *Sulcolluma foulcheri-delboscii* (Lavranos) Plowes (1995); **incl.** *Caralluma foulcheri-delboscii* var. *greenbergiana* Lavranos (1967) ≡ *Sulcolluma foulcheri-delboscii* var. *greenbergiana* (Lavranos) Plowes (1995); **incl.** *Caralluma shadhbana* Lavranos (1977) ≡ *Sulcolluma shadhbana* (Lavranos) Plowes (1995); **incl.** *Caralluma hexagona* var. *septentrionalis* Lavranos & L. E. Newton (1979) ≡ *Sulcolluma hexagona* var. *septentrionalis* (Lavranos & L. E. Newton) Plowes (1995); **incl.** *Caralluma shadhbana* var. *barhana* Lavranos & L. E. Newton (1979) ≡ *Sulcolluma shadhbana* var. *barhana* (Lavranos & L. E. Newton) Plowes (1995).

[3] Stems upright or ascending, ± strongly branching, forming dense mats or clumps to 20 cm tall, with subterranean runners; stems (grey-) green, 3 - 5 (-8) × 1.5 - 2 cm, bluntly 4-ribbed; **Ri** flattened, undulate-dentate; **Tu** oblong, slightly spreading, apically with hard scar-tissue; **Inf** 5- to 15- (to 20-) flowered; peduncle short, long-tapering; **Bra** short, linear; **Ped** green, 4 - 10 × 1.5 - 3 mm; **Sep** finely dotted with red, 1 - 3 × ± 1 mm, deltoid or narrowly lanceolate or linear, acute, keeled; **Cl** outside greenish-white, densely dotted with red, dots small, convex, inside (greenish-) cream-coloured, densely dotted with purplish-red, ≤ 2.2 cm ∅, flat or campa-

nulate; **Cl** tube ≤ 4 × 6 mm, enclosing or just basally encircling the **Cn**; **Cl** lobes ± broadly triangular, ≤ 8 × 8 mm, margins slightly revolute, inside tuberculate, lobes apically hairy, **Ha** dark purple, ≤ 2 mm, simple, compressed, vibratile; **Cn** ± 6 mm ⌀; **Ci** pale red, brownish-purple to bluish, margins pink, basally ± white, lobes ± 1.5 × 1.5 mm, deeply bifid, curved outwards, appendages of neighbouring lobes sometimes overlapping, with distinct furrows median and in the area where fused with the **Cs**; **Cs** pale red or pink, apically yellowish, lobes < 1 mm, narrowly lanceolate to ribbon-like, apically tapering, obtuse, strongly broadened in the area where fused with the **Ci**; **Anth** ± 0.5 mm broad, apically ovoid.

An extremely variable species, which can possibly be split up into several taxa (Gilbert 1990). In the prologue it is placed in the *europaea-umbellata*-group, by Gilbert (l.c.) in the *europaea-hexagona*-group.

C. indica (Wight & Arnott) N. E. Brown (Gard. Chron., ser. 3, 12: 369, 1892). **T:** India (*Wight* s.n. [K]). – **D:** S India (Tamil Nadu). **I:** Gravely & Mayuranathan (1931).

≡ *Hutchinia indica* Wight & Arnott (1834) ≡ *Desmidorchis indica* (Wight & Arnott) Kuntze (1891) ≡ *Boucerosia indica* (Wight & Arnott) Plowes (1995); **incl.** *Boucerosia hutchinia* Decaisne (1844) (*nom. illeg.*, Art. 52.1).

[3] Stems ± upright, ≤ 10 cm (longer when partly procumbent), loosely branched, with subterranean runners; stems 10 - 15 × ± 0.3 cm, 4-ribbed; **Ri** rounded, **L** rudiments curved downwards; **Inf** with 3 - 12 **Fl** with fungus-like odour; **Fl** buds conical, apically with weak central depression; **Cl** inside one-coloured light green, 2 cm ⌀, flat; **Cl** lobes rather narrow, triangular, apically very acute and ± glabrous, basally hairy, **Ha** simple, dark purple; **Ci** light yellow, outermost tips occasionally purple, lobes erect, pouch-like, apically sawtooth-like bifid, appendages falcate, curved towards each other, basally widely separated, in-between sometimes with small tooth-like appendage; **Cs** apically yellow, lobes basally broadened, with transition to the **Ci**, widely separated from each other; **Poll** D-shaped, translator arms short.

Closely related to *C. pauciflora* and belonging to the *umbellata-indica*-group (Gilbert 1990).

C. joannis Maire (Bull. Soc. Hist. Nat. Afr. Nord 31: 27, 1940). **T:** Morocco (*Pitault* s.n. [not located]). – **D:** Morocco (plain of Soussa). **I:** Asklepios 59: 14-21, 1993.

≡ *Apteranthes joannis* (Maire) Plowes (1995).

[3] Stems ± erect or hanging depending on habitat, shrubbily branched, subterranean runners ≤ 70 cm; stems green tinged reddish, ± 40 (hanging ≤ 90) × 1.5 cm, 4-angled (4-ribbed when young); **Tu** conical; **L** rudiments 3 × 2 mm, cordate, curved downwards, persistent when dry; **Inf** 5- to 12-flowered; **Ped** 5 - 8 × 1.5 mm; **Cl** outside green, densely and finely spotted with purple, inside cream-coloured spotted with dark purple, spots in interrupted concentrical circles, 1.5 - 2 cm ⌀, funnel-shaped; **Cl** tube 3 - 6 mm, broadly campanulate, enclosing the **Cn**; **Cl** lobes purple, ≤ 8 × 5 mm, broadly triangular, ascending, margins slightly curved outwards, inside roughly papillate; **Cn** ± 4 mm ⌀; **Ci** dark red, lobes 1.5 × 1 mm, apically retuse or incised; **Cs** light red to brownish, lobes ≤ 1.5 × 0.5 mm, apically 0.7 mm, crenate; **Poll** yellow, roundish.

The hexaploid sister species of *C. europaea*.

C. lavrani Rauh & Wertel (KuaS 16(4): 62-64, ills., 1965). **T:** Yemen (*Rauh* 13300 [HEID]). – **D:** S Yemen. **Fig. VI.g**

≡ *Crenulluma lavrani* (Rauh & Wertel) Plowes (1995).

[3] Plants 30 - 40 × 30 cm, mat- to clump-forming, irregularly and densely branched like a coral colony; stems grey to green, waxy, 5 - 30 × 1 - 3 cm, 4-angled; **Tu** blunt, crenate (caused by hardly protruding **L** cushions); **Inf** 1- to 2- (to 3-) flowered, at the stem tips; **Ped** 5 × 2 - 3 mm; **Sep** 1.5 - 2 mm; **Cl** outside grey-green, inside yellowish-ochre, dotted and spotted with wine-red, 2 - 2.5 cm ⌀; **Cl** tube flatly bowl-shaped; **Cl** lobes basally pale green, apically yellow, with tiny red dots, ± 1 × 0.5 cm, acute, margins curved outwards, inside finely wrinkled, tips of lobes hairy, **Ha** wine-red, clavate, vibratile, in bunches; **Cn** dark wine-red, relatively large, ± 4 × 7 mm, hairy, **Ha** white, ± 0.7 mm, **Bri**-like; **Ci** lobes reduced to a short basally dented edge, glabrous; **Cs** lobes in total 5 mm long, for ½ of their length split into 2 outer, wing-like, 2 mm long appendages and 1 inner cone-shaped, 2.5 mm long appendage, these erect or slightly bent over the **Sty** head; **Poll** D-shaped, ± 0.5 × 0.3 mm, germination mouth almost as long as the **Poll**.

A compact, easily flowering species, placed by Gilbert (1990) in Subgen. *Desmidorchis*, although probably most closely related to *C. adenensis* (Subgen. *Boucerosia*).

C. longiflora M. G. Gilbert (Bradleya 8: 12-14, ills., 1990). **T:** Somalia (*Thulin & Warfa* 4604 [K]). – **D:** Somalia.

≡ *Spathulopetalum longiflorum* (M. G. Gilbert) Plowes (1995).

[1] Stems loosely branched in irregular clusters, green, distinctly heteromorphic, basally ≤ 35 × ± 0.6 - 0.7 cm, ± 4-angled, edges weakly rounded, slightly paler; **Tu** indistinct, peg-like; **L** rudiments 2 mm, narrowly triangular, lanceolate, acute, margins hairy, **Ha** ± large, unicellular; upper **Fl**-bearing stem area ≤ 40 cm, ± terete, usually curved and with horizontal tip, sometimes branched; **Inf** to 4-flowered; tips of **Fl** buds clockwisely contorted; **Ped** 0.8 - 1 cm, pointing downwards at anthesis, **Fl**

pendent; **Sep** 1.5 - 2 mm, curved outwards; **Cl** pale ochre dotted with pale purple, 5.5 - 7.5 cm ∅, very deeply divided, central depression enclosing the **Cn** only basally; **Cl** lobes 25 - 35 × 1.5 - 1.8 mm, linear-lanceolate, curved outwards at ± 45°, margins revolute, finely ciliate; **Cn** narrowly stalked; **Ci** yellow, apically dark, lobes horizontally concave, apical ½ bilobed, appendages short, ± obtuse; **Cs** lobes pale yellow, ± 5 mm, basally ribbon-like, apically broadly lanceolate, geniculate, reaching far above the **Sty** head, acuminate, placed together column-like, partly twisted around each other.

Closely related to *C. gracilipes*, *C. edwardsiae* and *C. arachnoidea*, but slightly isolated on account of its unusual corona.

C. mireillae Lavranos (Nation. Cact. Succ. J. 24(4): 78-80, ills., 1969). **T:** Djibouti (*Lavranos* 6842 [PRE]). – **D:** Djibouti; only known from the type collection.

≡ *Caudanthera mireillae* (Lavranos) Plowes (1995).

[1] Dwarfish; stems richly branched, shrubby, grey-green, size not stated, dotted brownish esp. on the **Ri**, apically gradually tapering, 4-ribbed; **Ri** blunt, rounded; **L** rudiments 2 - 3 mm, ovate-triangular, acute; **Inf** few-flowered, laterally, close to the stem tips; **Ped** 3 mm; **Sep** 2.5 × 1.5 mm; **Cl** pale yellow, ± 1 cm ∅, campanulate; **Cl** tube 2 mm; **Cl** lobes basally dotted with dark red, 3.5 × 2 mm, deltoid, acute, ascending, margins slightly curved outwards, glabrous; **Ci** reddish, lobes oblong-triangular, running down the staminal column, tips recurved upwards; **Cs** yellowish, lobes 2.5 mm, basally with protruding dorsal appendage, this horn-like, terete, curved outwards, basally linear, flattened, apically erect, filiform, reaching over the staminal column and touching each other, tips undulate, twisted around each other; **Poll** pear-shaped, obtuse, with dorsal furrows.

Closely related to *C. sinaica*, but with glabrous corolla and entire interstaminal corona.

C. moniliformis P. R. O. Bally (Candollea 20: 17-19, ills., 1965). **T:** Somalia, Northern Prov. (*Bally* 11018 [G]). – **D:** N Somalia. **I:** FPA 45: t. 1766, 1978.

≡ *Spathulopetalum moniliforme* (P. R. O. Bally) Plowes (1995).

[1] Stems erect, sparsely branched, grey-green, distinctly heteromorphic, 10 - 30 × basally 2 cm, lateral stems a little shorter, bluntly 4-ribbed; **Ri** finely dotted with dark purple; **Tu** 2 - 4 mm, apically blunt, curved upwards; **L** rudiments small, **Sc**-like, on growing stem tips; upper **Fl**-bearing stem parts ≤ 15 × 0.4 cm, ± terete, with regular constrictions at intervals of 6 - 8 mm; **Inf** few-flowered, **Fl** in groups of 2 - 3 at the constrictions within the apical **Inf** region; **Bra** 1.6 mm, subulate; **Ped** 3 - 5 mm, very thin, apically thickened, curved downwards and **Fl** pendent; **Sep** 1.2 × 0.4 mm, acute; **Cl** outside grey-green dotted with chestnut-brown, inside yellow to very dark maroon, ± 7 × 3.5 mm, campanulate; **Cl** tube 1.3 × 2 mm, cup-shaped, basally enclosing the **Cn**; **Cl** lobes 6 × ± 2.5 mm, oblong triangular-fusiform, erect, basally cordate, margins revolute, tips curved outwards, laterally touching each other but not clinging together, with weakly protruding midrib, margins of lobes densely hairy, **Ha** purple, 0.4 - 2 mm, stiff, spreading; **Cn** 3 × 2.2 mm, glabrous; **Ci** greenish-white, lobes 0.5 × 0.5 mm, broadly ovate to oblong-triangular, erect or slightly curved inwards; **Cs** lobes 2.5 × 0.5 mm, narrowly rectangular, apically blunt, reaching over the **Sty** head and touching each other; **Poll** ovoid.

C. munbyana (Decaisne) N. E. Brown (Gard. Chron., ser. 3, 12: 278, 1892). **T:** Algeria (*Munby* s.n. [G]). – **D:** S Spain, Algeria. **I:** Plowes (1995). **Fig. VI.h**

≡ *Boucerosia munbyana* Decaisne (1847) ≡ *Borealluma munbyana* (Decaisne) Plowes (1995); **incl.** *Boucerosia munbyana* var. *hispanica* de Coincy (1898) ≡ *Boucerosia hispanica* (de Coincy) de Coincy (1899).

[2] Stems 10 - 20 cm, often with subterranean runners, green, ± 4-ribbed; **Tu** slightly spreading; **L** rudiments 2 × 2 mm, cordate, fleshy, curved upwards, long-lived; **Inf** with 4 - 10 (-15) **Fl** with extremely foetid odour; **Ped** 1 - 1.5 × 1 mm; **Sep** 1.5 - 2 × 0.8 mm, oblong-triangular, apically rounded; **Cl** outside pale red, basally whitish, inside dark maroon, campanulate, 6 - 8 mm ∅; **Cl** tube ≤ 3 mm deep, basally enclosing the **Cn**, greenish spotted with dark maroon; **Cl** lobes ± 6 × ± 2 mm ∅; **Ci** black-purple, lobes 0.3 - 0.5 mm, apically bifid, appendages narrowly triangular, enclosing the sides of the area of fusion towards the **Cs**; **Cs** reddish-brown, lobes 0.7 × ± 0.4 mm, triangular-lanceolate to rectangular, reddish-brown; **Anth** ± 0.3 × 0.5 mm, apically weakly crenate; **Poll** ± 0.3 × 0.4 mm, ovoid.

C. pauciflora (Wight) N. E. Brown (Gard. Chron., ser. 3, 12: 370, 1892). **T** [lecto]: India, Tamil Nadu (*Wight* 2429 [K]). – **D:** S India (Tamil Nadu). **I:** Gravely & Mayuranathan (1931).

≡ *Boucerosia pauciflora* Wight (1837) ≡ *Desmidorchis pauciflora* (Wight) Decaisne (1838).

[3] Stems ± erect or procumbent, loosely branching, with subterranean runners, ± 6 (in cultivation often much longer) × ± 0.3 cm, 4-ribbed; **Ri** rounded; **L** rudiments 1 × 1 mm, triangular, curved downwards; **Inf** at stem tips, with 1 - 3 **Fl** with unpleasant odour; buds conical, apically slightly convex; **Ped** short; **Cl** cupular, 2.5 cm ∅, pale yellow with purple markings in concentric circles, most densely in the middle of the **Fl**; **Cl** lobes with broad horizontal stripes, triangular, ± strongly curved out-

wards, inside hairy, **Ha** purple, simple; **Cn** 7 mm ⌀; **Ci** whitish with purple tips, lobes pouch-like, apically bifid, appendages short, fused along their entire length with the **Cs**, sometimes with a basal-median tooth-like appendage; **Cs** apically ± whitish, obtuse, rounded or 2-parted, broadened within the fusion area to the **Ci**, reaching over the **Anth** or incumbent on them.

Closely related to *C. indica* and belonging to the *umbellata-indica*-group (Gilbert 1990).

C. peckii P. R. O. Bally (Candollea 18: 14-15, ill., 1962). **T:** Kenya, Northern Frontier Prov. (*Peck* s.n. in *Bally* S61 [K]). – **D:** Ethiopia, W Kenya, E Uganda. **I:** FPA 35: t. 1394, 1962. **Fig. VII.b**

≡ *Spathulopetalum peckii* (P. R. O. Bally) Plowes (1995).

[1] Stems small, ± shrubbily branched, green to grey-green, finely spotted with purple between the **Ri**, apically lighter green, distinctly heteromorphic, ± 8 × 0.8 cm, 4-ribbed, with strong odour; **Tu** conical-subulate; **L** rudiments ≤ 2 mm, lanceolate, acute, curved upwards; upper **Fl**-bearing stem region ± 3 mm ⌀, weakly 4-angled, long-tapering; **Bra** ± 3 mm, delicate, lanceolate; **Ped** often reddish, ± 10 × 0.5 mm, ± recurved and **Fl** pendent; **Sep** ± 3 × 1 mm; **Cl** yellow, with fine red dots forming horizontal ribbon-like markings, ± 1.5 cm ⌀, divided to the base, basally with swollen **Rec**; **Cl** lobes appearing stalked basally, lanceolate, flexible, drooping, middle folded outwards, appearing flatly double-layered, both sides glabrous; **Cn** cream-coloured, clavate ± 2 mm stalked; **Ci** dark purple, lobes reduced to 2 **Sc**-like appendages; **Cs** apically spotted with dark purple, lobes short, ribbon-like or broadly triangular, erect, only basally attached to the white **Anth**; **Poll** yellow, 0.2 × 0.2 mm, roundish.

C. penicillata (Deflers) N. E. Brown (Gard. Chron., ser. 3, 12: 370, 1892). **T:** Yemen (*Deflers* 597 [P]). – **D:** Sudan, Ethiopia, Somalia?, Yemen, Saudi Arabia, Oman. **I:** Collenette (1985).

≡ *Boucerosia penicillata* Deflers (1889) ≡ *Desmidorchis penicillata* (Deflers) Plowes (1995); **incl.** *Echidnopsis golathii* Schweinfurth *ex* Deflers (1896) (*nom. illeg.*, Art. 52.1); **incl.** *Caralluma robusta* N. E. Brown (1904) ≡ *Caralluma penicillata* var. *robusta* (N. E. Brown) A. C. White & B. Sloane (1937).

[3] Stems erect, curved downwards when old, forming clumps of 0.6 - 1 × ≤ 2 m; stems light green, 3 - 6 cm ⌀, 4-ribbed, flowering stems apically tapering, with distinct **Ped** scars; **Tu** oblong-triangular, wing-like; **Inf** with densely arranged **Fl** with very weak odour; **Ped** 10 × 1 - 1.5 mm; **Sep** apically blunt; **Cl** outside pale green, inside green to yellow spotted ± with dark purple, 1 - 1.5 cm ⌀, flat; **Cl** lobes ± 4 × 2.5 - 3 mm, inside papillate, papillae irregularly rounded, with small **Bri**, tips of lobes with a dense bunch of **Ha**, **Ha** dark brownish-purple; **Cn** green or (rarely) reddish-brown, covered with nectar, ± 3.5 mm ⌀, **Ci** inside and **Cs** apically pubescent, **Ha** white or light pink; **Ci** lobes ± lacking or shortly bifid or bilobed, then ± 1.5 × 0.8 mm, appendages ± triangular, erect; **Cs** lobes 0.5 - 0.8 × ± 0.5 mm, broadly triangular, apically blunt, ≤ ½ as long as the **Anth**; **Poll** roundish, flat, pollinarium ± 0.3 × 0.6 mm, translator wings 0.1 mm; **Fr** 5.5 - 7 cm.

An isolated species with relationships to Subgen. *Desmidorchis* (Gilbert 1990).

C. peschii Nel (Jahrb. Deutsche Kakt.-Ges. 1: 41-42, ill., 1935). **T:** Namibia (*Pesch* s.n. in *SUG* 7082 [BOL [holo ?]]). – **D:** Namibia. **I:** Bruyns (1982c). **Fig. VI.e**

≡ *Australluma peschii* (Nel) Plowes (1995).

[1] Stems branched and shrubby, forming loose mats to 30 cm tall with numerous subterranean runners (≤ 12 cm), edible; stems grey-green, 2 - 5 (runners ≤ 8) mm ⌀, apically rounded, rounded 4-angled; **L** rudiments ≤ 1 mm, triangular; **Inf** with 1 - 2 nodding **Fl**, spread in the upper stem regions; **Ped** 1 - 2 mm, curved downwards; **Sep** ± 4 × 2 mm; **Cl** both sides yellowish-green, 9 mm ⌀, flat; **Cl** lobes 4 - 5 × 2 - 3 mm, ovate, acute, apically curved outwards, inside densely papillate, papillae brown, globose, apically hairy, **Ha** white to purple, stiff, giving the **Fl** a shaggy appearance; **Cn** 4 mm ⌀, basally fused to form a bowl-shaped structure; **Ci** yellowish-green, apically dark brown, lobes 1 mm, bluntly triangular, tips bi- or trifid, inside with distinct longitudinal furrows; **Cs** whitish edged brown, lobes triangular to subulate, sometimes apically touching each other; **Poll** 0.34 × 0.25 mm, bean-shaped.

A morphologically and geographically isolated species. The genus *Australluma* set up for this species by Plowes (1995) may have to be accepted.

C. priogonium K. Schumann (BJS 34: 327, 1905). **T:** Tanzania (*Engler* 1521/a [B†]). – **D:** Ethiopia, Kenya, Somalia, Tanzania. **I:** Gilbert (1977).

≡ *Spathulopetalum priogonium* (K. Schumann) Plowes (1995); **incl.** *Caralluma mogadoxensis* Chiovenda (1932) ≡ *Spathulopetalum mogadoxense* (Chiovenda) Plowes (1995); **incl.** *Caralluma elata* Chiovenda (1939).

[1] Stems erect or ascending, shrubby, freely branched (similar to *C. congestiflora*); stems 30 - 40 cm, distinctly heteromorphic, basally ≤ 4 cm ⌀, acutely 4-angled; **Ri** distinctly serrate, **L** rudiments 3.5 - 4 mm, oblong-ovate, acuminate, basally rounded; upper **Fl**-bearing stem region 1 - 2 mm ⌀, ± 4-angled, with 1 - 3 **Fl** per node; **Fl** buds 1.2 - 1.6 cm, grey-green finely dotted with purple; **Ped** 4 - 6 mm, slender, or absent; **Sep** 2 mm, acute, outside weakly papillate; **Cl** outside silvery-green with small purple dots, inside dark purple, sometimes tinged olive, 1.9 - 3.4 cm ⌀, flat, deeply divided; **Cl**

lobes on the lower ⅓ distinctly marked with white stripes and dots, lanceolate, stiff, horizontally spreading, apically mucronate, margins hairy, clavate **Ha** delicate, tip of lobes setose, **Ha** and clavate **Ha** apically with vesicular thickening; **Cn** dark purple to cream-coloured, glabrous or with fine purple **Ha**; **Ci** lobes 0.5 mm, oblong-ovate, erect, bifid, appendages horn-like, subulate; **Cs** lobes spatulate, apically blunt, shortly bifid, sometimes apically touching each other, tips sometimes finely hairy; **Fr** flattened in cross-section, lozenge-shaped.

A very variable species and closely related to *C. congestiflora* and *C. gracilipes*.

C. procumbens Gravely & Mayuranathan (Bull. Madras Gov. Mus. 4(1): 26, fig. 4: 13-17, 1931). **T** [lecto]: India (*Mayuranathan* s.n. [icono: l.c. fig. 4: 13-17]). – **D**: S India (Travancore). **I**: Gravely & Mayuranathan (1931).

≡ *Boucerosia procumbens* (Gravely & Mayuranathan) Plowes (1995).

[3] Stems ascending, rampantly growing in cushions to 1.8 m ∅, weakly branched, rooting at soil contact, 7 - 12 × 1.2 - 2 cm, acutely 4-angled; **L** scars indistinct; **Inf** few-flowered; **Cl** outside green with few purple spots on the lobes, inside whitish striped purple, 2.3 cm ∅, deeply campanulate, fleshy; **Cl** tube 2 × 2 cm, sometimes weakly spotted with purple; **Cl** lobes densely dotted with purple, 1 - 1.5 × 1 cm, bluntly triangular, ascending, margins recurved, hairy, clavate **Ha** purple, ≤ 4 mm, vibratile; **Cn** whitish tinged pale purple, bowl-shaped; **Ci** lobes 5 × 2 mm, deeply bifid, lanceolate, appendages facing each other, closely adjacent to the **Cl** tube; **Cs** lobes 1.5 × 1 mm, triangular, sometimes apically touching each other, blunt; **Anth** yellowish, margin partly undulate; **Poll** yellow, 0.3 × 0.2 mm, roundish.

Closely related to *C. diffusa* and *C. umbellata* and belonging to the *umbellata-indica*-group (Gilbert 1990).

C. quadrangula (Forsskål) N. E. Brown (Gard. Chron., ser. 3, 12: 370, 1892). **T**: [lecto – icono]: Forsskål, Icones, t. 6, 1776. – **D**: Saudi Arabia, Yemen, Oman (Dhofar). **I**: Gilbert (1990).

≡ *Stapelia quadrangula* Forsskål (1775) ≡ *Boucerosia quadrangula* (Forsskål) Decaisne (1844) ≡ *Echidnopsis quadrangula* (Forsskål) Deflers (1896) ≡ *Desmidorchis quadrangula* (Forsskål) M. G. Gilbert & Raynal (1980) ≡ *Monolluma quadrangula* (Forsskål) Plowes (1995); **incl.** *Stapelia quadrangula* var. *ramosa* Forsskål (1775); **incl.** *Boucerosia forskaolii* Decaisne (1844) (*nom. illeg.*, Art. 52.1).

[4] Stems shrubbily branched in clumps ≤ 40 × 75 cm, 2 cm ∅, green to brownish-green, with strong odour; **Tu** indistinct; **Inf** sessile close to the stem tips, 1-flowered, appearing axillary; **Fl** with fruity-sweetish odour; **Bra** solitary, acute, delicate; **Ped** thick, ≤ 6 × 3 mm; **Cl** greenish or lemon-coloured (material from Taizz incl. **Cn** white, then pinkish), 1.2 - 1.5 cm ∅, flat to slightly bowl-shaped; **Cl** lobes ascending, recurved along the midrib for ± ½, glabrous, 6 - 10 mm; **Cn** darker yellow than the **Cl** (**Ci** rarely red), long-stalked, surpassing the **Cl** tube; **Ci** lobes erect, basally pouch-shaped, deeply bifid, appendages inserted close to each other, apically rounded, margins folded back within the fusion area to the **Cs**, inside sometimes shortly hairy (material from Saudi Arabia); **Cs** lobes broadly triangular, apically blunt, ± ½ as long as the **Anth**; **Fr** 5 - 8.5 (-18) cm (extremely variable in length), 3-ridged (ribbed).

Closely related to *C. cicatricosa*. Both species have 1-flowered seemingly axillary inflorescences, extremely thick and short pedicels, a characteristic odour of the stems and 3-ridged fruits.

C. sarkariae Lavranos & R. Frandsen (CSJA 50(5): 211-213, ills., 1978). **T**: India (*Sarkaria* J64-77 [E]). – **D**: India (Tamil Nadu).

[1] Stems ascending, basally richly branched, green spotted with purple, ≤ 30 × 1.5 cm, 4-angled, tapering towards the tips, edges ± acute; **L** rudiments 2 × 1 mm, lanceolate; **Inf** 2-flowered in the **Ax** of the slender upper stem region; **Ped** green weakly marked with purple, ± 3 mm, ascending; **Cl** outside green, inside pale yellow, sometimes horizontally striped with darker yellow, 9 mm ∅, shallowly campanulate; **Cl** tube 1 mm; **Cl** lobes ± 3 × 2 mm, oblong-triangular, basally ovate, ascending, margins apically revolute, rarely hairy, **Ha** yellowish, thin; **Cn** whitish; **Ci** basally shortly fused with the **Cs**, lobes ascending, deeply bifid, appendages horizontally spreading outwards; **Cs** lobes ribbon-like, apically ± retuse, not surpassing the **Anth**; **Fr** pale purple with darker markings, ± 6 cm.

Very closely related to *C. stalagmifera* and *C. adscendens*.

C. sinaica (Decaisne) Bentham & Hooker *fil.* (Gen. Pl. 2: 782, 1876). **T**: Egypt (*Aucher-Eloy* 2850 [P]). – **D**: Israel, Egypt (Sinai Region), NW Saudi Arabia, Sudan. **I**: Bruyns (1987b). **Fig. VI.f**

≡ *Boucerosia sinaica* Decaisne (1844) ≡ *Caudanthera sinaica* (Decaisne) Plowes (1995); **incl.** *Caralluma sinaica* var. *sinaica*; **incl.** *Caralluma maris-mortui* Zohary (1941); **incl.** *Caralluma sinaica* var. *baradii* Lavranos & L. E. Newton (1979).

[1] Stems ascending-erect, richly branched, shrubby, forming mats or clumps (15 - 30 × 50 cm), grey- to pale green spotted with dark maroon, distinctly heteromorphic, basally 1 cm ∅, bluntly 4-ribbed; **L** rudiments 4 × 1.5 mm, ovoid-acute or linear-subulate, stipular rudiments minute; upper **Fl**-bearing stem region 2 - 5 mm ∅, terete; **Inf** with 1 - 5 **Fl** from the **Ax**; **Ped** 3 - 4 mm, recurved; **Sep** ≤ 2.5 × 1 mm; **Cl** outside pale green, inside greenish, yellow, sand-coloured to light pink, sometimes with scattered dark brownish-red dots, (7-) 12 - 16 mm

⌀, campanulate to flat, deeply divided; **Cl** tube 1.5 - 2 mm, broadly campanulate; **Cl** lobes 6 × 2 mm, acute-ovate to lanceolate-subulate, spreading outwards, margins strongly revolute, inside strongly hairy, **Ha** pale purple or white, ≤ 1.2 mm, cylindrical, arising from a small papilla; **Cn** yellow to white, basally often tinged reddish, stalk 0.5 - 1 × 2.2 - 2.7 mm; **Ci** basally purple-red, apically whitish, lobes ± 1 mm, basally fused with the **Cs**, there with clear nectar drops, deeply bifid, appendages apically ± acute; **Cs** lobes 3 × 0.25 mm, triangular-lanceolate, flattened, only basally touching the **Anth**, apically blunt, basally with dorsal horn-like appendage; **Anth** appendage whitish to purple-red, 2 - 4.5 mm, subulate-linear, basally thickened, erect-connivent, margin crinkled and bent outwards, reaching far above the **Sty** head, apically spreading; **Poll** ± pear-shaped; **Fr** grey-green spotted with purple, ≤ 6 cm.

This species is distinguished by the possession of sterile connective appendages, a primitive feature within the family. Closely related to *C. edulis*.

C. socotrana (Balfour *fil.*) N. E. Brown (Gard. Chron., ser. 3, 12: 370, 1892). **T** [lecto]: Socotra (*Balfour* s.n. [K]). – **D:** Socotra, Ethiopia, Somalia, Kenya. **I:** FPA 41: t. 1607, 1970; Plowes (1995).

≡ *Boucerosia socotrana* Balfour *fil.* (1884) ≡ *Sanguilluma socotrana* (Balfour *fil.*) Plowes (1995); **incl.** *Caralluma rosengrenii* Vierhapper (1905); **incl.** *Caralluma corrugata* N. E. Brown (1912); **incl.** *Caralluma rivae* Chiovenda (1929).

[4] Clump-forming, apically irregularly branching; stems ± 15 × 0.8 - 1 cm (lateral stems apically 3 - 5 mm ⌀), with pungent odour, edges blunt; **Tu** oblong-conical and stems undulate in outline; **L** rudiments small, deltoid, compressed, cartilaginous, apically brown, papillate; **Inf** 1-flowered, subterminally from **Ax**, or few-flowered pseudo-umbels; **Ped** 1 - 3 mm, ≤ 7 mm at **Fr** ripening; **Sep** 2 - 2.5 mm, alternating with small **Gl**; **Cl** dark red, 2.5 - 3 cm ⌀, divided for ± ½; **Cl** tube 6 - 20 × 9 - 11 mm ⌀, campanulate, margins inside with distinct concentrical furrows; **Cl** lobes ± 5 × 5 mm, ovate-triangular, shortly acuminate, ± horizontally spreading, margins strongly recurved, basally weakly papillate, towards the tips densely papillate; **Cn** ± 4.5 mm ⌀; **Ci** lobes 1.5 - 2 × 1 mm, erect, deeply bifid (± to the base), appendages 0.5 mm, narrowly triangular, acute, falcately curved towards each other, fused with the **Cs** along their entire length, apical margins densely white-ciliate; **Cs** lobes ± 1 × 0.5 (apically ± 0.2) mm, linear-lanceolate or narrowly triangular, apically blunt, slightly overtopping the **Anth**; **Anth** ± 0.8 × 0.5 mm; **Poll** ± 0.35 × 0.3 mm, roundish-ovate, flat, translator wings 0.1 mm; **Fr** pale green-yellow spotted with purple, ± 10 × 0.5 - 0.8 cm.

Closely related to *C. cicatricosa*.

C. solenophora Lavranos (JSAB 29: 107-109, ill. (pl. 13), 1963). **T:** Yemen (*Lavranos* 1860 [K, PRE]). – **D:** S Yemen. **I:** Plowes (1995).

≡ *Cylindrilluma solenophora* (Lavranos) Plowes (1995).

[3] Stems erect or ascending, densely branched, (brownish-) green, ≤ 20 × 2 cm, lateral stems < 8 cm, ± bluntly 4-ribbed; **Tu** tips pointing downwards; **Inf** pseudo-umbels, close to the stem tips, few-flowered; peduncle ≤ 5 mm, ascending; **Bra** 1 - 2 mm, linear, acute; **Fl** pendent; **Ped** ± 6 × 1.5 mm; **Sep** 4 - 5 × 1 mm; **Cl** outside yellowish-brown with indistinct longitudinal lines or furrows, inside light green to yellowish, with narrow dark purple horizontal stripes (confluent at the lobe tips with narrow dark purplish-brown lobe margin), very long tubular, ± 2.2 cm long; **Cl** tube basally dotted with dark purplish-brown, ± 1.9 × ± 9 (basally) or 6 (median) or 8 (apically) mm ⌀, globose, upper 1.1 cm cylindrical; **Cl** lobes 5 - 6 × 7 mm, broadly triangular, acute, ± horizontally spreading, ± curved outwards, lobe tips with a bunch of **Ha**, **Ha** dark purple, 1 - 2 mm, simple, laterally compressed, vibratile; **Cn** outside greenish, inside ± purple, ± 3 × 5 mm, shortly fused to form a cup-like structure, bluntly pentagonal in top-view; **Ci** lobes ovate-triangular, erect, concave, laterally fused with the back of the **Cs** lobes, the latter greenish, with ± purple margins, 1 - 2 mm ⌀, almost square to triangular, ascending or incumbent on the **Anth**, bases of the guide-rail extremely spreading.

Belonging to the *europaea-hexagona*-group (Gilbert 1990).

C. somalica N. E. Brown (BMI 1895: 264, 1895). **T:** Somalia (*Kirk* s.n. [K]). – **D:** Somalia. **I:** FPA 26: t. 1008, 1947.

≡ *Desmidorchis somalica* (N. E. Brown) Plowes (1995).

[3] Stems erect, robust, in clumps, to 30 cm, 4-angled; **Tu** small, conical; **L** rudiments ± reduced; **Inf** globose, ± 5 cm ⌀; **Fl** with sweet odour; **Ped** 6.5 - 8 × 2 - 3 mm; **Sep** 3 - 4.5 × ± 1.2 mm, outside very weakly hairy; **Cl** yellow, campanulate, 1.5 - 2 cm ⌀; **Cl** tube ± 4 mm, enclosing the **Cn**; **Cl** lobes ± 5 × 3 - 4 mm, oblong-triangular, apically acuminate, inside velvety; **Cn** 5 - 6 mm ⌀; **Ci** lobes ± 2 × ± 1 mm, apically 3-parted for ± ½, rarely 2-parted, outer appendages laterally spreading, curved outwards, margins slightly revolute, outside very weakly hairy; **Cs** lobes 1 - 1.5 × ± 0.6 mm, oblong-triangular to ribbon-shaped, apically irregularly crenate, not surpassing the **Anth**; **Anth** ± 0.8 × ± 0.6 mm, apically slightly crenate; **Poll** ± 0.4 × ± 0.3 mm, roundish, flat, translator wings 0.1 mm.

Belonging to the *adenensis-speciosa*-group (Gilbert 1990). *C. acutangula* and *C. speciosa* are closely related.

C. speciosa (N. E. Brown) N. E. Brown (Gard.

Chron., ser. 3, 12: 370, 1892). **T:** Somalia (*Kirk* s.n. [K]). – **D:** Ethiopia, Sudan, Somalia, Uganda, Kenya, Tanzania. **I:** FPA 38: t. 1485, 1967. **Fig. VII.c**

≡ *Sarcocodon speciosa* N.E.Brown (1878) ≡ *Desmidorchis speciosa* (N. E. Brown) Plowes (1995); **incl.** *Caralluma codonoides* K. Schumann (1895); **incl.** *Caralluma oxydonta* Chiovenda (1932).

[3] Stems 10 - 30 cm, in large clumps, rather soft, 4-angled; **Tu** conical, acute; **L** rudiments spreading laterally, thorn-like; **Inf** richly flowered pseudoumbels, globose; **Bra** ± 2 mm, subulate, weakly hairy, **Ha** small, thick; **Fl** probably with foetid odour; **Ped** 4.5 - 6.5 × 2.5 - 3.5 mm, stout; **Sep** ± 3 × ± 1.5 mm, acuminate, outside very weakly hairy, **Ha** thick; **Cl** 3.5 - 4.5 cm ⌀; **Cl** tube inside yellow, campanulate, ± 2 × ± 2 cm; **Cl** lobes 1.2 - 1.5 × ± 1.3 cm, inside dark blackish-purple, broadly ovate-triangular, apically acuminate, ± horizontally spreading, margins slightly revolute, hairy, clavate **Ha** short; **Cn** 4 - 4.5 mm ⌀, flatly bowl-shaped; **Ci** lobes 4 - 5 × ± 1 mm, narrowly lanceolate-rectangular, upper ⅓ bifid, appendages parallel, pressed against the inside of the **Cl** tube; **Cs** lobes 1.5 - 2 × ± 1 mm, apically with irregular furrows, bluntly obtuse or retuse, not surpassing the **Anth**; **Anth** 1.5 - 2 × ± 1 mm, apically irregularly crenate, laterally slightly touching each other; **Fr** ± 12 cm; **Se** 12 × 9 mm, broad-rimmed.

Belonging to the *adenensis-speciosa*-group (Gilbert 1990). *C. foetida* and *C. somalensis* show a great resemblance in vegetative characters.

C. staintonii H. Hara (J. Jap. Bot. 52(12): 357, 1977). **T:** Nepal (*Stainton 5410* [BM]). – **D:** W Nepal (Bheri River Valley). **I:** Bruyns (1989).

≡ *Borealluma staintonii* (H. Hara) Plowes (1995).

[2] Stems ± 15 × 0.5 - 1 cm, green, sometimes apically tinged reddish; **L** rudiments horizontally spreading; **Inf** 2- to 10-flowered; **Fl** mostly pointing a little downwards; **Ped** ± 2 - 3 × ± 1.5 mm; **Sep** 1 - 1.5 × 1 mm, ovate-triangular, blunt; **Cl** outside pale green, inside yellowish dotted with dark maroon, dots fewer towards the centre, campanulate, ± 8 mm ⌀; **Cl** tube ± 2 × 4 - 5 mm, basally enclosing the **Cn**; **Cl** lobes 2.5 - 4 × ± 3 mm, broadly triangular, ascending, rugose-papillate, tip ± blunt, margins hairy, **Ha** 0.5 - 3 mm, ± flattened; **Cn** dark maroon, ± 3 mm ⌀; **Ci** lobes 1 - 1.3 × ± 1 mm, apically bifid, appendages ± 0.5 mm, narrow; **Cs** 1.2 - 1.5 × ± 0.6 (apically 0.3) mm, oblong-triangular, apically blunt or irregularly crenate, not surpassing the **Anth**; **Anth** yellow, 0.5 - 0.8 × ± 0.5 mm; **Poll** ± 0.3 × ± 0.3 mm, roundish, translator wings ± 0.05 mm.

Endemic to Nepal with a very isolated distribution at the southern border of the Himalayas. The closest relative is probably *C. munbyana* from SW Europe and NW Africa.

C. stalagmifera C. E. C. Fischer (BMI 1925: 430, 1925). **T:** India, Madras (*Mayuranathan* s.n. [K]). – **D:** India (Madras). **I:** Gravely & Mayuranathan (1931).

[1] Stems few-branched, ± 60 cm, (grey-) green, mostly tinged reddish esp. along the edges, strongly tapering towards the tips, basally 5 - 8 mm ⌀, 4-angled, edges rounded; **L** rudiments 2 - 3 mm, narrowly triangular, acute, ascending; upper **Fl**-bearing stem region ± terete; **Inf** mostly only 1- to 2-flowered; **Ped** ± 5 mm, slender, spreading; **Bra** delicate, triangular-subulate; **Sep** 1.5 - 2 mm; **Cl** outside greenish, inside uniformly chestnut-brown or dark purple, ± 1 cm ⌀, shortly funnel-shaped, fleshy; **Cl** tube cream-coloured; **Cl** lobes ovate-lanceolate, margins curved outwards, inside with silky gloss, margins apically hairy, **Ha** 2 mm, purple or white, fusiform, pendent like stalagmites, vibratile, margins of the **Cl** lobes basally sometimes with simple **Ha**; **Cn** dark purple; **Ci** lobes basally bowl-shaped, with nectar, apically linear-lanceolate, erect, deeply bifid, appendages subulate, pointing outwards; **Cs** lobes triangular-lanceolate, apically retuse or finely 3-toothed, sometimes overlapping, mostly pubescent; **Poll** 0.25 × 0.3 mm; **Fr** green striped with purple.

Hybridizing with *C. adscendens* var. *gracilis* and belonging to this complex.

C. subulata Forsskål *ex* Decaisne (Ann. Sci. Nat. Bot., sér. 2, 9: 267-268, ills., 1838). **T:** [lecto – icono]: Forsskål, Icones, t. 7, 1776. – **D:** Saudi Arabia, N and S Yemen, Africa N of the equator.

Incl. *Stapelia subulata* Forsskål (1775) (*nom. inval.*, Art. 32.1c); **incl.** *Caralluma dalzielii* N. E. Brown (1912).

[1] Differs from *C. adscendens*: Stems 3 - 15 cm; stipular rudiments as lateral small bunches of translucent thick minute **Ha**; **Fl** short-stalked; **Cl** inside whitish, horizontally striped with wine-red (Yemen) or dark purple (Saudi Arabia); **Cl** tube inside finely hairy, **Ha** whitish, crinkled; **Cl** lobes wine-red or dark purple in the middle, partly ± strongly hairy, margin densely purple-ciliate, **Ha** dark pink, ≤ 5 mm, thick, spreading, **Cl** tips green, glabrous; appendages of the **Ci** lobes sometimes with a small tip between them.

Closely related to or even conspecific with *C. adscendens*.

C. tuberculata N. E. Brown (Gard. Chron., ser. 3, 12: 370, 1892). **T:** Pakistan, Baluchistan (*Stocks 596* [K]). – **D:** Jordania, Saudi Arabia, S Yemen, Pakistan, Afghanistan, NW India. **I:** Ricánek & Hanácek (2001). **Fig. VII.a**

≡ *Borealluma tuberculata* (N. E. Brown) Plowes (1995); **incl.** *Caralluma plicatiloba* Lavranos (1962) ≡ *Borealluma plicatiloba* (Lavranos) Plowes (1995).

[2] Stems 6 - 15 × 0.8 - 1.2 cm, basally branching; **Ri** acutely concave; **Tu** apically slightly curved

upwards; **L** rudiments fleshy, upwards-curved towards the stem; **Inf** of 3 - 8 terminal **Fl**; **Ped** 2 - 2.5 × ± 1.5 mm; **Sep** 1.5 - 1 × ± 1 mm, acuminate; **Cl** sand-coloured yellowish ± spotted with dark maroon or blackish-purple, campanulate, ± 1.6 cm ∅; **Cl** tube 5 - 7 × 3 - 5 mm, opening pentagonal; **Cl** lobes 6 - 8 × 3 - 4 mm, horizontally spreading, basally ovate, apically linear, inside strongly structured by dark tuberculate spots and stripes; **Cn** 2.5 - 3 mm ∅, deeply sunken into the **Cl** tube; **Ci** lobes 0.7 - 1 × 0.7 - 1 mm, apically bifid, appendages 0.2 - 0.5 mm, sometimes tuberculate like the **Cl**; **Cs** lobes 0.8 - 1 × ± 0.4 mm, rectangular, apically rounded, surpassing the **Anth**; **Anth** ± 0.6 × ± 0.4 mm; **Poll** ± 0.3 × ± 0.3 mm, roundish-ovoid, translator wings short; **Fr** ≤ 10 cm.

Ricánek & Hanácek (2001) understand the Asian *C. plicatiloba* as distinct species.

C. turneri E. A. Bruce (HIP 34: t. 3339 + text, 1937). **T:** Kenya (*Allen Turner* s.n. in *CM* 3692 [K]). – **D:** Ethiopia, Kenya, Uganda.

≡ *Caralluma dicapuae* ssp. *turneri* (E. A. Bruce) P. R. O. Bally (1969) ≡ *Spathulopetalum turneri* (E. A. Bruce) Plowes (1995).

C. turneri ssp. **turneri** – **D:** Ethiopia, Kenya, E Uganda. **Fig. VII.e**

[1] Stems in clumps of ± 45 × 50 cm; stems 5 - 15 × 0.8 - 2 cm, grey-green to brown, distinctly heteromorphic, acutely 4-angled, apically tapering; **Ri** serrate; **Tu** 8 - 14 mm; **L** rudiments 2 mm, lanceolate or subulate, fleshy; upper **Fl**-bearing stem region ≤ 30 × 0.3 - 0.4 cm, ± 4-angled; **Inf** 1 - 4 cm apart from each other, with 2 - 4 **Fl**; **Ped** 3 - 6 mm, spreading; **Sep** ± 2 mm, long acuminate; **Cl** 8 - 12 mm long, narrowly campanulate, divided almost to the centre; **Cl** lobes cream-coloured striped with blackish- to brownish-purple, rarely entirely dark, 8 - 12 × 2 - 3 mm, oblong-lanceolate to spatulate, mobile, pendent, lamina ± completely folded back along the midrib, basally ciliate, apically rounded, base claw-like, inside slightly pubescent, clavate **Ha** purple, vibratile; **Cn** 0.7 mm, ± short-stalked; **Ci** lobes 0.7 mm, rectangular or ± square, deeply bifid, appendages fusiform; **Cs** lobes 1.5 mm, broadly ribbon-shaped, apically weakly irregularly dentate, geniculate, apically reaching far over the **Sty** head, touching each other in the upper ½; **Fr** grey to straw-coloured, 8 × 0.5 cm.

Closely related to *C. priogonium* and *C. gracilipes*. An exquisite and distinct species.

C. turneri ssp. **ukambensis** (P. R. O. Bally) L. E. Newton (Asklepios No. 72: 9, ills. (p. 10), 1998). **T:** Kenya, Southern Prov. (*MacArthur* s.n. in *Bally* S135 [ZSS]). – **D:** Kenya.

≡ *Caralluma dicapuae* ssp. *ukambensis* P. R. O. Bally (1969).

[1] Smaller than ssp. *turneri*; **Br** almost terete and **L** rudiments shorter; **Cl** lobes hairy; stalk of the **Cn** clavate.

C. umbellata Haworth (Synops. Pl. Succ., 47, 1812). – **D:** India (Tamil Nadu). **I:** Plowes (1995).

≡ *Stapelia umbellata* (Haworth) Roxburgh (1819) ≡ *Boucerosia umbellata* (Haworth) Wight & Arnott (1834) ≡ *Desmidorchis umbellata* (Haworth) Decaisne (1838); **incl.** *Boucerosia campanulata* Wight (1848) ≡ *Caralluma campanulata* (Wight) N. E. Brown (1892); **incl.** *Boucerosia lasiantha* Wight (1848) ≡ *Caralluma lasiantha* (Wight) N. E. Brown (1892); **incl.** *Stapelia callamulia* Buchanan-Hamilton *ex* Hooker *fil.* (s.a.).

[3] Stems weakly branched, 30 - 60 cm, forming large dense clumps; stems 10 - 20 × 1 cm, concave 4-angled to -ribbed; **Tu** oblong-conical, wing-like, sometimes apically slightly spreading horizontally; **L** rudiments horizontally spreading or ascending; **Inf** 20- to 60-flowered, 6 - 10 cm ∅; buds apically and basally flattened; **Ped** ± 7 × 2 - 4 mm; **Sep** 2 - 3 × 1 - 2 mm, apically acuminate; **Cl** outside pale green, inside yellowish, finely dotted and striped with dark purple, sometimes one-coloured dark maroon, 2.5 - 3 cm ∅, flat to slightly funnel-shaped, weakly bulging around the **Cn**; **Cl** lobes 6 - 10 × 8 mm, broadly triangular, horizontally spreading, flat, only the margins slightly revolute, inside ± strongly short-hairy, locally occasionally finely tuberculate-papillate, **Ha** > 1 mm, very fine, dark, longer **Ha** freely swinging from their bases, highly ephemeral; **Cn** ± 4 mm ∅, purple to maroon; **Ci** lobes ± 1 × ± 1.2 mm, apically ± strongly divided, appendages short; **Cs** lobes ± 1.2 × ± 0.5 mm, apically strongly furrowed, surpassing the **Anth**; **Anth** ± 0.8 × ± 0.5 mm; **Poll** ± 0.4 × ± 0.3 mm, roundish-ovoid, translator wings small; **Fr** 15 cm.

The list of synonyms follows Gravely & Mayuranathan (1931); the status of some forms still has to be clarified. The taxon belongs to the *umbellata-indica*-group (Gilbert 1990).

C. vaduliae Lavranos (CSJA 63(4): 170-172, ills., 1991). **T:** Somalia (*Lavranos & al.* 23275 [UPS]). – **D:** Somalia.

≡ *Spathulopetalum vaduliae* (Lavranos) Plowes (1995).

[1] Stems basally branched to form dense groups, (grey-) green, distinctly heteromorphic, basally ≤ 20 × 1 cm, acutely 4-angled; **Tu** 2 - 3 mm, subulate, projecting, ascending; **L** rudiments acute; upper **Fl**-bearing stem region 15 - 25 × 0.5 cm, ± 4-angled, apically round, lasting for several years; **Inf** with 1 - 3 sessile **Fl**; **Bra** 1 - 3, 2 × 0.5 mm, subulate, fleshy; buds erect; **Ped** 4 × 0.5 mm; **Sep** 2.5 × 0.8 mm; **Cl** white to pink, 1 - 1.2 cm ∅; **Cl** tube campanulate, white, 4 × ≤ 2.75 mm; **Cl** lobes apically purple, both sides glabrous, 8 × 1 - 2 mm, stiff, fleshy, linear-spatulate from a triangular base, tips tapering to subulate, ± 2 mm, curved, margins

weakly recurved; **Cn** 1.5 × 1.6 mm, glabrous; **Ci** lobes rectangular, deeply bifurcate, appendages linear, erect; **Cs** lobes ± 1 × 0.5 mm, ribbon-shaped, apically weakly 2-parted; **Poll** pear-shaped.

Closely related to *C. gracilipes, C. priogonium* and *C. dicapuae*, and hardly discernible from the latter when in bud. This beautiful and distinct species partly corresponds with *C. edulis* in coronal structure. The sharply 4-angled stems, however, clearly indicate that this species belongs to Subgen. *Caralluma*.

CEROPEGIA

U. Meve

Ceropegia Linné (Spec. Pl. [ed. 1], 1: 211, 1753). **T:** *Ceropegia candelabrum* Linné. − **Lit:** Huber (1957); Dyer (1980); Dyer (1983). **D:** Africa, Arabia, Asia, N Australia. **Etym:** Gr. 'keros', wax, wax candle; and Gr. 'pegynnai', assemble, unite; perhaps for the chandelier-like inflorescences of some species.

Incl. *Niota* Adanson (1763) (*nom. illeg.*, Art. 52.1). **T:** *Ceropegia candelabrum* Linné.
Incl. *Apegia* Necker (1790). **T:** not typified.
Incl. *Systrepha* Burchell (1822). **T:** *Systrepha filiforme* Burchell.
Incl. *Cinclia* Hoffmannsegg (1833). **T:** not typified.

Perennial herbs, geophytes, **L** and/or stem succulents with fibrous **R**, fleshy lateral **R** or tubers arising from the hypocotyl or the **R**; stems prostrate to erect, often twining, terete, compressed or 4- (to 6-) angled, 1 - 20 (-40) mm ∅, twining species to 10 m tall; **L** petiolate or sessile, triangular, linear to ovate, ± simple, minute to large, ± short- or long-lived; stipular rudiments usually present, flat or glandular; **Inf** extra-axillary, usually shortly pedunculate, 1- to many-flowered, umbel-like, racemose or with successively developing **Fl**; **Ped** 1 - 20 × 1 - 2 mm, horizontally spreading or erect; **Sep** shortly triangular to subulate; **Cl** ± 1 - 10 cm long, tube longer than the lobes, basally inflated (**Cl** inflation), narrowed above (**Cl** tube), mouth in the area transitional to the **Cl** lobes usually widening again, **Cl** lobes reflexed, pendulous or mostly erect and apically united into a cage-like structure, or with tips dilated in an umbrella-like fashion and connate, tips of lobes often conspicuously ciliate; **Cl** mostly multi-coloured, blotched or striped, outside glabrous, rarely hairy, **Cl** tube inside often with **Ha** reminiscent of a fish trap; **Cn** biseriate, sessile to shortly stipitate; **Ci** from a bowl- to cup-shaped connate basal structure extending into 5 (respectively 2× 5) triangular to subulate spreading to erect lobes, or merely forming a narrow rim, often hairy; **Cs** lobes mostly extensively elongated and overtopping the **Ci**, erect or with revolute tips; **Anth** rectangular, incumbent on the **Sty** head or slightly erect; **Poll** ovoid to pear-shaped, germination crests situated along the inner margins or near the apex; **Fr** usually long and narrowly fusiform to filiform, 4 - 40 cm; **Se** brown, distinctly winged.

Boasting ± 180 species, *Ceropegia* is the largest genus of the *Ceropegieae*. Climbing herbs of tropical forests and scrubs are not dealt with here, since those are not succulent. Succulents showing the various forms of root, stem and leaf succulence (or combinations thereof) are particularly represented in Africa and Madagascar. The subdivision by Huber (1957) into 19 sections and 23 series appears to be artificial to a large extent and a modern infrageneric concept is wanting. Therefore, the ± succulent taxa as covered here are not assigned to sections. Instead, the most closely related species or groups are given, as far as possible, and growth forms are indicated (combinations are possible):

[1] **R** fleshy, cylindrical to fusiform.
[2] **R** tuberous.
[3] Stem succulents.
[4] **L** succulents.

The following names are of unresolved application but are referred to this genus: *Brachystelma subaphyllum* K. Schumann (1898) ≡ *Tenaris subaphylla* (K. Schumann) N. E. Brown (1903); *Ceropegia aphylla* Haworth (1812); *Ceropegia boerhaaviifolia* Deflers (1896); *Ceropegia candelabrum* Loureiro (1793) (*nom. illeg.*, Art. 53.1); *Ceropegia cordata* Loureiro (1790); *Ceropegia deflersii* Schwartz (1939) (*nom. inval.*); *Ceropegia farrokhii* McCann (1945); *Ceropegia loureiroi* G. Don (1838); *Ceropegia obtusa* Loureiro (1790).

C. abyssinica Decaisne (PSRV 8: 644, 1844). **T:** Ethiopia (*Schimper* 1416 [G, BR, FT, K, S, W]). − **D:** Ethiopia, Eritrea, Kenya, Tanzania, Zaïre, Zambia, Central African Republic, Zimbabwe, Angola. **I:** Archer (1992: I).

Incl. *Ceropegia hirsuta* Hochstetter *ex* Decaisne (1844) (*nom. inval.*, Art. 34.1c); **incl.** *Ceropegia steudneri* Vatke (1876); **incl.** *Ceropegia leucotaenia* K. Schumann (1893); **incl.** *Ceropegia steudneriana* K. Schumann (1895); **incl.** *Ceropegia gilletii* De Wildeman & Durand (1899); **incl.** *Ceropegia hispidipes* S. Moore (1908); **incl.** *Ceropegia bequaertii* De Wildeman (1920); **incl.** *Ceropegia filicalyx* Bullock (1933); **incl.** *Ceropegia abyssinica* var. *songeensis* H. Huber (1957); **incl.** *Ceropegia bonafouxii* var. *linearifolia* Stopp (1971).

[2] **R** tuber ± globose, 3 cm ∅, stems erect, only rarely apically twining, ± shortly villose, ± 25 × 2.5 mm, pale green; **L** 3 - 5 mm petiolate, lamina narrowly elliptic, ovate to broadly lanceolate, usually bent-erect, 20 - 50 × 3 - 20 mm, delicate, hairy, margins occasionally undulate; **Inf** very shortly pedunculate, 1- to 4-flowered; **Ped** 0.5 - 1 cm; **Sep** subulate, 6 - 12 mm; **Cl** 1.8 - 2.5 (-3) cm, bottle-

shaped, outside rarely hairy; **Cl** inflation 5 - 12 × 3 - 5 mm, ellipsoid-barrel-shaped, outside whitish-grey, with age purple with a grey stripe on the upper part, inside whitish and purple, apically ± continuously constricted to ± 3 mm ⌀; **Cl** tube 3 - 5 mm long, apically 3 - 5 mm ⌀, outside grey, occasionally maculate with purple; **Cl** lobes 5 - 10 (-15) mm, linear, apically fused to form an ovoid to ellipsoid cage, lamina folded back along the midrib or at the margins only, also at the basal corners, outside white, occasionally hairy, inside velvety-black; **Cn** ± purple, ± sessile, fused into a bowl; **Ci** lobes ± rectangular, semi-erect, obtuse, emarginate or apically bifid to ½, ± 1 mm, with short black **Ha**; **Cs** ± 2 mm, linear-spatulate, erect-connivent, apically inflexed, basally and apically often with blackish **Ha**; **Fr** narrowly fusiform, ± 11 cm × 3 mm; **Se** 6 × 2 mm.

See note under *C. achtenii*.

C. achtenii De Wildeman (Pl. Bequaert. 4: 356, 1928). **T:** Zaïre (*Achten* 589 [BR]). — **D:** Angola, Zimbabwe, Togo, Zaïre. **I:** Malaisse & Schaijes (1993).

Incl. *Ceropegia adolfii* Schlechter *ex* Werdermann (1939) ≡ *Ceropegia achtenii* ssp. *adolfii* (Werdermann) H. Huber (1957); **incl.** *Ceropegia adolfii* var. *gracillima* Werdermann (1939); **incl.** *Ceropegia achtenii* ssp. *togoensis* H. Huber (1957).

[2] **R** tuber depressed-globose, ± 2.5 cm ⌀; stems annual, erect, rarely twining, 5 - 10 cm, pilose, unbranched; **L** shortly petiolate, lamina linear-lanceolate, 4 - 6 × 0.5 - 1.2 cm; **Inf** sessile, 1- to many-flowered; **Ped** 0.5 - 1 cm; **Sep** narrowly lanceolate, ± 4 mm; **Cl** 2 - 3.5 cm, curved, greenish, ± blotched with brown-red, outside hairy or ± glabrous; **Cl** inflation ovoid, ± 4 × 3.5 mm; **Cl** tube narrowing to ± 2 mm, apically scarcely widening; **Cl** lobes narrowly lanceolate-spatulate, 4 - 12 mm, apically connate to form a cylindrical cage, lamina folded back along the midrib and margins, inside green to brownish, ± blotched with purple-brown, margins mostly with projecting **Ha**; **Cn** sessile, reddish, only basally dish-like connate, ± 2 × 2 mm; **Ci** lobes squarely pouch-like, ascending, apically emarginate to incised; **Cs** linear to spatulate, erect-connivent; **Fr** 8 - 10 cm × 2.5 mm, erect.

Huber (1957) recognizes 3 subspecies differing only in the indumentum and the size of the corolla. *C. achtenii* is closely allied to other tuberous, erect and hairy species like *C. abyssinica* and *C. umbraticola*.

C. affinis Vatke (Linnaea 40: 218, 1876). **T:** Ethiopia (*Schimper* 301 [K, G, P, S]). — **D:** Tropical and subtropical Africa from Ethiopia to RSA and from Guinea to Kenya, Madagascar. **I:** Dyer (1983: as *C. racemosa*). **Fig. VII.d**

Incl. *Ceropegia angusta* N. E. Brown (1895); **incl.** *Ceropegia racemosa* N. E. Brown (1895); **incl.** *Ceropegia setifera* Schlechter (1895) ≡ *Ceropegia racemosa* ssp. *setifera* (Schlechter) H. Huber (1957); **incl.** *Ceropegia biddumana* K. Schumann (1897); **incl.** *Ceropegia ruspoliana* K. Schumann (1897); **incl.** *Ceropegia setifera* var. *natalensis* N. E. Brown (1908); **incl.** *Ceropegia hochstetteri* Chiovenda (1912); **incl.** *Ceropegia secamonoides* S. Moore (1912) ≡ *Ceropegia racemosa* ssp. *secamonoides* (S. Moore) H. Huber (1957); **incl.** *Ceropegia cynanchoides* Schlechter (1913); **incl.** *Ceropegia kamerunensis* Schlechter (1913); **incl.** *Ceropegia atacorensis* A. Chevalier (1917); **incl.** *Ceropegia gourmacea* A. Chevalier (1917); **incl.** *Ceropegia glabripedicellata* De Wildeman (1920); **incl.** *Ceropegia pedunculata* Turrill (1921); **incl.** *Ceropegia butaguensis* De Wildeman (1928); **incl.** *Ceropegia racemosa* ssp. *glabra* H. Huber (1957).

[1] **R** fusiform-fleshy, in dense clusters; stems persistent, slender, twining, 1 - 2 m, ± terete, 1 - 2 mm ⌀, dark green, often slightly villose; **L** 5 - 15 mm petiolate, lamina narrowly ovate to broadly lanceolate or obovate, weakly succulent, tough, ± glossy, 2.5 - 3.8 × 1 - 4.5 cm, acute or mucronate, petiole, margins and occasionally the lamina with scattered **Ha**, margins sharp, mostly weakly undulate; **Inf** 2- to many-flowered, only 1 **Fl** open at a time, with mushroom-like odour; peduncle 0.5 - 2 cm, slender; **Ped** 0.5 - 1 cm; **Sep** lanceolate, acute, 2 - 3 mm; **Cl** 2 - 3 (-4) cm, whitish-green, yellowish, orange to red-brown, blotched and striped with red-brown, outside glabrous, rarely hairy, inside ± purple, with scattered **Ha**; **Cl** inflation slender, ellipsoid, basally weakly constricted and white, ± 5 × 5 mm, scarcely broader than the curved or crooked **Cl** tube, tube gradually widening to 8 - 10 mm ⌀; **Cl** lobes yellow or green-yellow tipped with brown-red or green, 4 - 6 mm long, fused only at the tips, triangular-lanceolate, ± erect, forming a subglobose to broadly ovoid cage, lamina folded back along the midrib except for the broadly triangular base, margins and/or keel simple, with white or purple **Ha**; **Cn** white to yellowish, basally connate with the **Cl** tube, shortly cup-shaped, ± 4.5 × 3 - 4 mm; **Ci** lobes triangular to nearly square-ovate, 2 mm broad, deeply incised, terminating in erect ± connivent triangular white-bearded appendages, 1 - 1.5 mm long, forming pouches at the base, margins winged and trapezoidally fused with the base of the **Ci**; **Cs** 2 - 3 mm, subulate, erect-connivent, apically slightly spreading, glabrous; **Poll** ellipsoid, 0.3 × 0.2 mm, corpuscle apically widened into a hammer-like shape, 0.15 mm broad; **Fr** very narrowly fusiform, 9 - 15 cm × 2 - 3 mm.

The species is widely distributed and common and accordingly exhibits a corresponding variability; an infraspecific subdivision does not seem feasible at present. *C. affinis* is better known under the synonymous name *C. racemosa*. See under *C. carnosa* for further notes.

C. africana R. Brown (Bot. Reg. 1822: t. 626 + text, 1822). **T:** [icono]: l.c. t. 626. – **D:** RSA.

C. africana ssp. **africana** – **D:** RSA (Western Cape, Eastern Cape). **I:** Dyer (1980); Bruyns (1986a).

[2,4] **R** tuber flattened; stems creeping or (weakly) twining, usually annual, 1 - 2 mm ⌀, partly forming rhizomes, often with swollen nodes; **L** 5 mm petiolate, lamina linear-lanceolate, acute, 15 - 25 × 10 mm, slightly fleshy, upper face concave, lower face convex; **Inf** 2 - 15 mm pedunculate, few-flowered; **Sep** triangular to subulate, 3 - 4 mm; **Cl** 2 - 2.5 (-3) cm, greenish-white, striped with purple, ± glabrous; **Cl** inflation globose, ± 4 mm ⌀, abruptly constricted to 1 - 2 mm ⌀; upper **Cl** tube funnel-shaped, widening to 4 - 5 mm ⌀, inside with purple **Ha**; **Cl** lobes linear and recurved in longitudinal direction along the middle, 6 - 12 mm, erect, connivent in the lower ⅓, then curved outwards to form an ovoid cage, inside purple, keel and margin hairy; **Cn** cup-shaped, shortly stipitate, ± 2 × 2 mm; **Ci** lobes extended to form deep ovoid pouches, margins often crenate, shortly hairy, fused directly with the base of the **Cs**; **Cs** 1.5 - 2 mm, broadly falcate, erect-connivent, strongly revolute from the middle; **Poll** ± 0.25 mm ⌀.

C. africana ssp. **barklyi** Bruyns (Bradleya 3: 35, ill. (p. 34), 1985). **T:** RSA, Eastern Cape (*Bowker & Barkly* s.n. [K]). – **D:** RSA (Eastern Cape).

Incl. *Ceropegia barkleyi* Hooker *fil.* (1877); **incl.** *Ceropegia barkleyi* var. *tugelensis* N. E. Brown (1908).

[2,4] Differs from ssp. *africana*: **L** lanceolate, 1.5 - 4 × 0.5 - 2 cm, upper face green, lower face purple; **Sep** subulate, ± 2.5 mm, **Cal** base swollen; **Cl** 2.5 - 5 cm; **Cl** inflation globose, slightly 5-angled and constricted in the lower ½, ± 3.5 mm ⌀, inside purple, glabrous; **Cl** lobes green, basally purple, linear, 15 - 25 mm, partially twisted together, often meeting each other above the middle, basally and often also at the margins hairy; **Cn** ± sessile; **Cs** 1 - 1.5 mm, broader than the **Ci**.

Based on *C. barkleyi*, but republished as a new name in order to retain the customary spelling.

C. ahmarensis Masinde (KB 55: 225-228, ills., 2000). **T:** Somalia (*Bailes* 81 [K]). – **D:** Somalia (Ahmar Mts.).

[3] Stems succulent, scrambling, ± 3 mm ⌀, glabrous; **L** 3 - 4 mm petiolate, margins pubescent, lamina subsucculent, narrowly ovate, acute, 12 - 17 × 5 - 6 mm, shortly ciliate; **Inf** shortly pedunculate, pseudo-umbellate, up to 4-flowered; **Bra** subulate, 1 mm; **Ped** 3 mm, glabrous; **Sep** subulate, ± 2 × 0.5 mm, glabrous; **Cl** ± 24 mm, colour unknown, erect, curved; **Cl** inflation narrowly ovoid, ± 17 × 4 mm, narrowing gradually to 2 mm ⌀, mouth widening to 3 mm ⌀, glabrous; **Cl** lobes linear, 7 × 1.5 mm, apically fused, forming an ellipsoid cage-like structure, ± 6.5 mm ⌀; **Cn** shortly stipitate, basally cup-shaped, ± 3.2 × 1.5 mm, with a tuft of **Ha** at each **Ci** position, **Ci** forming concave pouches, free lobes bifid, appendages erect, ± 0.7 mm, inside with long cilia; **Cs** lobes ± 2.5 × 0.3 mm, spatulate, erect-connivent, apically shortly pubescent; **Poll** ellipsoid, 0.3 × 0.15 mm.

This taxon, only known from the type, is probably related to *C. stenantha*, although the latter has wiry stems.

C. albisepta Jumelle & H. Perrier (Ann. Inst. Bot.-Géol. Colon. Marseille, sér. 2, 16: 227-228, 1908). **T:** Madagascar (*Perrier* 1726 [P]). – **D:** Madagascar, Kenya, Uganda, Zaïre. **I:** FPA 36(2): t. 1431, 1964, as *C. succulenta*; Malaisse & Schaijes (1993). **Fig. VIII.a**

Incl. *Ceropegia decaryi* Choux (1925); **incl.** *Ceropegia helicoidea* Choux (1925); **incl.** *Ceropegia verrucosa* Choux (1925); **incl.** *Ceropegia viridis* Choux (1925) ≡ *Ceropegia albisepta* var. *viridis* (Choux) H. Huber (1957); **incl.** *Ceropegia robynsiana* Werdermann (1938) ≡ *Ceropegia albisepta* var. *robynsiana* (Werdermann) H. Huber (1957); **incl.** *Ceropegia succulenta* E. A. Bruce (1941); **incl.** *Ceropegia evelynae* E. A. Bruce & P. R. O. Bally (1950); **incl.** *Ceropegia albisepta* var. *bruceana* H. Huber (1957); **incl.** *Ceropegia albisepta* var. *truncata* H. Huber (1957) ≡ *Ceropegia viridis* var. *truncata* (H. Huber) H. Huber (1970) (*nom. inval.*, Art. 33.2).

[3,4] **R** fibrous; stems persistent, twining, to 6 m, ± terete, succulent, 4 - 7 mm ⌀, partly even thicker at the base, green to glaucous-green, glabrous or faintly verrucose-uneven; **L** 5 - 15 mm petiolate, lamina linear to ovate, 1 - 8 × 1 - 6 cm, ± succulent, obtuse to acute, partly acuminate, fresh green, shining, caducous; **Inf** with a fleshy peduncle 1 - 10 (-15) cm long, many-flowered, yet mostly with 1 **Fl** open at a time; **Ped** 0.5 - 3 cm; **Sep** narrowly lanceolate, 0.5 - 1 cm; **Cl** 3.5 - 6 cm, outside glabrous, either side whitish-green, greenish to yellowish, dotted or blotched with ± red-brown; **Cl** inflation 9 - 13 × 7 - 11 mm, globose to cylindrical, apically conical to abruptly horizontally constricted; **Cl** tube 3 - 5 mm ⌀, then gradually widening to 14 - 20 mm ⌀, inside faintly hairy; **Cl** lobes 10 - 30 mm, ± triangular, apex obtuse or with linear extensions that are ± connate, lamina folded back along the midrib, broadly keeled, inside white to green, partly with green or red-brown reticulate markings, apically usually purple, green, brown, or yellow, glabrous or hairy, margins ± with white or purple **Ha**, partly with purple clavate **Ha**; **Cn** sessile or shortly stipitate, fused into a shallow bowl, 4 - 5 mm ⌀, greenish, very variable as to hairiness; **Ci** lobes ascending, bilobed into linear appendages, ± 2 mm, apically fused; **Cs** ± 2 - 3 mm, ± linear to spatulate, erect-connivent; **Poll** ovoid, ± 0.4 × 0.28 mm; **Fr** parallel or only weakly divergent, 10 - 24 cm × 4 -

8 mm ⌀, glaucous-green, glabrous; **Se** 10 - 15 × 2 - 4 mm, **Ha** tuft 2.5 - 5.5 cm.

A species exhibiting a complex variability. Differences are found particularly in the size of the leaves, the length of the corolla lobes, and the indumentum of the corona. The inclusion of African taxa such as *C. robynsiana* / *C. succulenta* is questionable. They seem to be phylogenetically closer to the E African *C. ballyana* than the Madagascan *C. albisepta*.

C. ambovombensis Rauh & Gerold (Succulentes 20(2): 3-6, ills., 1997). **T:** Madagascar (*Gerold* s.n. [HEID 74872]). – **D:** S and W Madagascar.

[2] **R** tuber depressed-globose, to 7 cm ⌀; stems twining, delicate, ± annual, 1 - 2 mm ⌀; **L** petiole 3 - 8 mm long, upper face canaliculate, lamina lanceolate, acute, 3 - 30 × 5 - 10 mm, succulent, upper face dark green, shining; **Inf** with a peduncle 5 mm long, borne on leafy short shoots, 2- to 3-flowered; **Ped** short; **Sep** 3 - 4 mm; **Cl** 25 - 35 mm long; **Cl** inflation obtusely ovoid, 7 - 10 × 4 - 5.5 mm, whitish, inside striped with purple; **Cl** tube slightly bent, ± 8 × 2 mm, whitish, purple-striped and -tinged, apically widened to 5 - 7 mm, white; **Cl** lobes linear-elliptic, ± 15 × 5 mm, fused at the tips to form a cylindrical cage, outside whitish, inside dark purple, margins revolute in the middle, densely covered with vibratile purple **Ha** 2 - 5 mm long; **Cn** subsessile, fused into a shallow bowl, ± 2.5 mm ⌀, pale green and purple, ± hairy; **Ci** lobes ± bilobed into erect linear appendages, ± 1.5 mm, apically connate; **Cs** ± 1.5 mm, linear-cylindrical, erect-connivent.

A leaf succulent climber of the *C. albisepta* alliance; most closely related to *C. hermannii*.

C. ampliata E. Meyer (Comm. Pl. Afr. Austr., 194, 1837). **T:** W, K, P, MEL. – **D:** Tropical and subtropical Africa, RSA, Madagascar. **I:** Bruyns (1986a). **Fig. VIII.b**

Incl. *Ceropegia ampliata* ssp. *insulicola* Lavranos *in schedis* (s.a.) (*nom. inval.*, Art. 29.1); **incl.** *Ceropegia triebneri* Dinter ex Suessenguth & Merxmüller (1952) (*nom. nud.*); **incl.** *Ceropegia ampliata* var. *oxyloba* H. Huber (1957); **incl.** *Ceropegia ampliata* ssp. *madagascariensis* Lavranos (1973).

[1,3] Stem succulents to 2 m, creeping or slightly twining, sparsely branched; **R** clustered, fleshy, fusiform; stems ± terete, 4 - 5 mm ⌀, green, finely striate, rough; **L** lanceolate, ± 5 - 10 × 2 - 3 mm, rapidly caducous; **Inf** sessile, 1- to 4-flowered; **Ped** 0.5 - 2 cm; **Sep** triangular, acute, ± 3 mm; **Cl** 5 - 7 cm, white or greenish-white, striped with green; **Cl** inflation globose to ovoid, 1.5 - 2.5 × 2 - 3 cm, often weakly indented; **Cl** tube cylindrical in the middle, 12 - 15 mm ⌀, entrance to the inflation lined with a purple band, hairy; **Cl** lobes inside green to yellow-green, narrowly triangular to linear, 0.6 - 2 cm, apically connate, free margins ± revolute, glabrous; **Cn** ± sessile, white, in total ± 8 - 10 × 6 - 7 mm, shallowly bowl-shaped; **Ci** lobes deeply bifid, with triangular spreading appendages oriented towards the **Cs**, 0.7 - 1.5 mm, inside ± finely hairy; **Cs** filiform, basally swollen, 4 - 7 × 0.2 mm, erect, connivent; **Poll** narrowly ovoid, 0.5 - 0.6 × 0.3 mm, corpuscle 0.4 mm.

This species shows an unsual disjunct distribution. Its main area is situated in E South Africa whereas a few populations are represented in SE Madagascar, but are morphologically inseparable. *C. ampliata* is an attractive and easily growing greenhouse plant, often flowering twice or even more often in the year.

C. antennifera Schlechter (BJS 20(Beiblatt 51): 46-47, 1895). **T:** RSA, KwaZulu-Natal (*Schlechter 3426* [Z, BOL]). – **D:** RSA (KwaZulu-Natal). **I:** Dyer (1983).

Incl. *Ceropegia craibii* Victor (2001).

[2] **R** tuber flattened, 2 - 3.5 cm ⌀; stems erect, not twining, 15 - 20 cm, delicate, with short **Int**, ± unbranched, glabrous; **L** sessile, linear-filiform, 20 - 30 × 1 - 2 mm, margins revolute; **Inf** sessile, 1-flowered, first terminal, later lateral; **Ped** erect, 2.5 - 3 cm; **Sep** subulate, ± 6 mm; **Cl** outside whitish, striped with purple, inside hairy, 7.5 - 9.5 cm, straight; **Cl** inflation ovoid, ± 4 mm ⌀, gradually merging into the **Cl** tube, 2 - 3 mm ⌀, apically widening to 3 - 4 mm ⌀; **Cl** lobes inside whitish, 5 - 6 cm, linear-lanceolate, apically free, erect-spreading, lamina mostly folded back along the midrib, apically globose (similar to butterfly antennae), slightly pendent, furrowed, with purple clavate **Ha** along the margin; **Cn** black-purple, cup-shaped, ± sessile; **Ci** lobes forming ovoid pouches, denticulate, hairy; **Cs** linear, acute, erect-connivent, tips revolute.

The recently described *C. craibii* differs only by corolla lobes lacking a knobby tip. With regard to the often poor taxonomic value of corolla lobe morphology in *Ceropegia* because of considerable variability, differences in minor structure of the tips of the corolla lobes do not justify the acceptance of a species of its own. The short distance of ± 100 km between the type localities of the two taxa in KwaZulu-Natal, as well as the lack of further populations, supports the idea that these are two relictual populations of one and the same species.

See *C. dinteri* for further comments.

C. arabica H. Huber (Mem. Soc. Brot. 12: 138, 1957). **T:** Arabia (*Ogilvie-Grant 75a* [E]). – **Lit:** Bruyns (1988b). **D:** Arabian Peninsula. **I:** Collenette (1985).

[1,3] Stem succulents to 2 m, twining, sparsely branched; **R** clustered, fleshy, fusiform; stems ± terete, 2 - 4 mm ⌀, green, finely striate, rough; **L** lanceolate, acute, 4 - 8 mm, rapidly caducous; **Inf**

sessile, 1- to 15-flowered; **Ped** 4 - 8 mm; **Sep** subulate, acute, 2 - 4 mm; **Cl** 3 - 8 cm; **Cl** inflation ± rectangular in lateral view, ± 2 × 0.5 - 1 cm, outside striate, inside papillose, upper margin hairy; **Cl** tube cylindrical, 1.5 - 3 mm ⌀, gradually widening into a funnel-shaped mouth; **Cl** lobes very variable in length, apically fused, partially twisted together, mostly folded back along the midrib; **Cn** sessile or shortly stipitate, white with a red spot on the **Ci**, 3.5 - 4 × 3 - 3.5 mm, cup-shaped; **Ci** lobes deeply bifid terminating in triangular erect or spreading appendages, glabrous; **Cs** linear-cylindrical, erect, connivent; **Poll** ovoid, ± 0.4 - 0.5 × 0.2 mm.

C. arabica var. **abbreviata** Bruyns (Notes Roy. Bot. Gard. Edinburgh 45(2): 318-320, ills., 1988). **T:** Yemen (*Wood* 3295 [K]). – **D:** Yemen, Saudi Arabia.

[1,3] **Sep** 1.5 - 2 mm; **Cl** 3 - 4.5 cm, with faint musky odour; **Cl** inflation greenish-white, cylindrical, ± 9 - 11 × 5 - 9 mm, with conical elongated base; **Cl** tube cylindrical, 2 - 3 mm ⌀, upper rim striped with green or blotched with pale red; **Cl** lobes marked with yellow-brown, 10 - 13 mm, lamina folded back, keeled, keel and margins hairy; **Ci** lobes ± erect, appendages ± 1 mm; **Cs** lobes 1.5 - 2 mm.

C. arabica var. **arabica** – **D:** N Yemen, Saudi Arabia. **Fig. VIII.c**

[1,3] **Sep** 2.5 - 4 mm; **Cl** 3.5 - 7.5 cm; **Cl** inflation greenish-yellowish, ± 7 - 9 mm ⌀; **Cl** tube 2 - 2.5 mm ⌀ in the middle, upper margin only slightly widening, here striped with green or blotched with pale red on white background; **Cl** lobes whitish, narrowly triangular, to 35 mm, often connivent in the upper ⅔, faintly keeled, keel blackish-greenish, tips usually spirally intertwined, margins revolute, hairy; **Ci** and **Cs** lobes ± 1.5 - 2 mm.

C. arabica var. **powysii** (D. V. Field) Meve & R. Mangelsdorff (Bot. J. Linn. Soc. 137: 105, ill. (p. 103), 2001). **T:** Kenya (*Field & Powys* 115 [K]). – **D:** Kenya. **I:** Archer (1992: XVIII). **Fig. VIII.d** ('*C. barbigera*')

≡ *Ceropegia powysii* D. V. Field (1982); **incl.** *Ceropegia barbigera* Bruyns (1989).

[1,3] **Sep** 3 - 4 mm; **Cl** 28 - 47 mm, pale grey-green with red-brown dots and spots; **Cl** inflation globose to ovoid, ± 7 (-11) × 9 (-11) mm, basally constricted; **Cl** tube 1.5 - 3 mm ⌀, gradually widening towards the upper funnel-shaped margin to 6 - 10 (-12) mm ⌀, inside occasionally hairy; **Cl** lobes purple-brown inside, occasionally with a yellowish spot, deltoid or ovate, 9 - 12 mmm, folded back along the midrib, margins often undulate-serrate or revolute, keeled or not, on the keel or at the basal margins with purple vibratile clavate **Ha**, ± 3 mm; **Ci** lobes ± erect, appendages ± 1.5 mm; **Cs** lobes 0.5 - 2 (-4) mm.

This higly variable taxon represents the African form of *C. arabica*. In the S of Kenya, stout forms are found with esp. long and erect corona lobes. *C. galeata* is closely related. In the S part of the continent, *C. arabica* is replaced by the *C. fimbriata* group.

C. arabica var. **superba** (D. V. Field & Collenette) Bruyns (Notes Roy. Bot. Gard. Edinburgh 45(2): 317-318, ills., 1988). **T:** Saudi Arabia (*Collenette* 3159 [K, E]). – **D:** Yemen, Saudi Arabia.

≡ *Ceropegia superba* D. V. Field & Collenette (1984).

[1,3] **Sep** 2.5 - 4 mm; **Cl** 4.5 - 8 cm, outside grey-white spotted with red; **Cl** inflation ± 7 - 22 × 7 - 9 mm; **Cl** tube cylindrical, 2.5 - 3 mm ⌀, upper margin gradually widening in a funnel-shaped manner, inside greenish striped with red-brown; **Cl** lobes 17 - 32 mm, basally white, hairy, distinctly keeled, keel blackish, hairy, above with a white patch, then merging into brown very acute tips, margins glabrous; **Ci** and **Cs** lobes 1.5 - 2 mm.

C. arenaria R. A. Dyer (BT 12(3): 444, 1978). **T:** RSA, KwaZulu-Natal (*Strey* 5031 [PRE, NH]). – **D:** RSA (KwaZulu-Natal); dune forest. **I:** Peckover (1993a).

[1,3,4] **R** ± fleshy and thickened, clustered; stems creeping and climbing, not twining, 3 - 4 mm ⌀, terete, succulent, marked with contrasting dark and pale green, dull; **L** ± 2 mm petiolate, lamina ± ovate to lanceolate, rapidly caducous, to 4 × 2.5 cm, weakly succulent; **Inf** 5 - 15 mm pedunculate, 1- to 3-flowered, **Fl** opening in succession; **Ped** 5 - 10 mm; **Sep** lanceolate, 4 - 5 mm; **Cl** to 6 cm, whitish-yellowish, blotched with purple; **Cl** inflation basally narrowly ovoid, to 14 mm long, slightly bent and narrowed, then widening again to a ± globose section, 6 - 7 mm ⌀, narrowed above to form the **Cl** tube, gradually widening into a funnel-shaped mouth of ± 15 mm ⌀; **Cl** lobes linear-spatulate from a triangular base, 20 - 25 mm, lamina entirely folded back along the midrib, interior basal ½ pale yellow with brown margins, ± hairy, broadly keeled, keels touching each other in the middle, upper ½ divaricate, brown-red, margins with whitish-purple vibratile **Ha**, these ± 6 mm; **Cn** ± sessile, ± 5 × 4 mm, fused into a short cup; **Ci** to 3 mm, deeply 3-partite, central appendage very short, lateral appendages triangular, erect; **Cs** linear, ± 3 mm, erect-connivent.

An unusual plant from an unusual habitat (dune forest) and possibly of hybrid origin – with *C. denticulata* and *C. cimiciodora* as possible parents.

C. aridicola W. W. Smith (Notes Roy. Bot. Gard. Edinburgh 12: 197, 1920). **T:** China, Yunnan (*Forrest* s.n. [BM, K]). – **D:** China (Yunnan); grassland. **I:** Tsiang (1939).

[2] **R** tuber to 3 cm ⌀, ovoid-globose, with thi-

ckened **R**; stems almost glabrous, single or in pairs, branched; **L** to 6 mm petiolate, lamina triangular-ovate, 5 - 15 × 3 - 10 mm, acute, weakly fleshy, upper face villose, lower face with pubescent veins; **Inf** sessile, 1- to 3-flowered; **Ped** 3 - 10 mm; sparsely pubescent; **Sep** lanceolate, 2 × 1 mm; **Cl** 1 - 1.5 cm, straight, outside sparsely hairy, inside glabrous; **Cl** tube 7 - 10 mm, **Cl** inflation broadly ovoid; **Cl** lobes broadly ovate to rhomboid, 3 - 5 mm, upper ½ connate to form a flat umbrella with an acute central appendage; **Cn** sessile, fused into a cup, glabrous; **Ci** lobes triangular; **Cs** linear-spatulate; **Fr** ± 5 cm; **Se** ± 5 mm.

Allied to *C. pusilla*.

C. aristolochioides Decaisne (Ann. Sci. Nat. Bot., sér. 2, 9: 263-264, 1838). **T:** Senegal (*Heudelot* s.n. [G, P]). – **D:** Arabia, tropical Africa. **I:** Asklepios 53: 19-23, 1991.

[3,4] Stem and **L** succulent climbers to 3 m; **R** fibrous; stems ± terete, robust, 3 - 6 mm ⌀, glaucous-green, slightly verrucose; **L** petiolate, cordate to ovate-lanceolate, 1 - 4 × 0.5 - 2.5 cm, weakly succulent, acute, partly apiculate; **Inf** multiflowered, drepanoid, usually with a single **Fl** open at a time; peduncle fleshy, 1 - 10 cm; **Ped** 5 - 15 mm; **Sep** subulate, 2 - 5 mm; **Cl** 2.5 - 4.5 cm, outside glabrous, rarely pubescent, either uniformly yellowish or with red-brown dots or blotches on whitish-green background or uniformly red-brown; **Cl** lobes often green-yellow; **Cl** inflation 7 - 10 × 6 - 9 mm, globose to ovoid, basally weakly compressed, inside red-purple with red-purple **Ha** at least on the upper margin, merging ± gradually into the **Cl** tube, tube 3 - 5 mm ⌀ in the middle, then widening gradually to ≤ 1 cm ⌀, inside greenish with red-brown stripes or uniformly red-brown, hairy, **Ha** fine, red-purple; **Cl** lobes ± greenish-yellow with red-brown reticulate markings, broadly lanceolate to narrowly linear, 5 - 11 mm, tips connivent, usually obtuse, lamina folded back, with few short simple **Ha** on apex and keel; **Cn** broadly bowl-shaped, ± 4 - 5 mm, ± sessile; **Ci** lobes whitish-pellucid to green-yellow, margins and appendages with purple **Ha**, triangular-ovate, concave, apically bifid into 2 ± spreading obtuse to acute appendages; **Cs** yellowish, 2 - 3 mm, linear, erect-connivent; **Poll** D-shaped, ± 0.35 × 0.25 mm, germination crests triangular, membranous, corpuscle ± 0.35 mm, basally with triangular processes.

Forming a widespread complex from Senegal to East Africa and Arabia, and most closely related to *C. rupicola*. Characteristics are the robust, blue-green stems and the massive peduncles.

C. aristolochioides ssp. **aristolochioides** – **D:** Tropical Africa, Senegal to Kenya. **Fig. VIII.e**

Incl. *Ceropegia beccariana* Martelli (1896); **incl.** *Ceropegia perrottetii* N. E. Brown (1903); **incl.** *Ceropegia albertina* S. Moore (1907) ≡ *Ceropegia aristolochioides* ssp. *albertina* (S. Moore) H. Huber (1957); **incl.** *Ceropegia crassula* Schlechter (1913); **incl.** *Ceropegia aristolochioides* var. *wittei* Staner in Werdermann (1938); **incl.** *Ceropegia seticorona* E. A. Bruce (1941); **incl.** *Ceropegia volubilis* var. *crassicaulis* H. Huber (1957); **incl.** *Ceropegia seticorona* var. *dilatiloba* P. R. O. Bally (1965); **incl.** *Ceropegia maasaiorum* Halda & Prokes (2000).

[3,4] **Cl** 5 - 15 mm, lobes broadened in the middle with truncate apex.

C. aristolochioides ssp. **deflersiana** Bruyns (Notes Roy. Bot. Gard. Edinburgh 45(2): 296-299, ills., 1988). **T:** Yemen, Ibb District (*Wood & Hepper* 5887 [K]). – **D:** N Yemen, Saudi Arabia. **Fig. VIII.f**

[3,4] Differs from ssp. *aristolochioides*: Peduncle tending to be shorter; **Fl** often more slender and delicate; **Ca** hairy.

The hairy carpels of this subspecies are highly significant.

C. armandii Rauh (Adansonia, n.s., 4: 419-425, ills., 1964). **T:** Madagascar (*Rauh* 10564 [HEID]). – **D:** S Madagascar.

[3] Stem succulents; stems creeping-decumbent, occasionally with tuberous hypocotyl, 4-angled, flattened, 2 - 3 cm thick, thickest at the nodes, brown-grey-green, rough-verrucose, with persistent short recurved bases of petioles; **L** narrowly ovate, acute, divaricate, dark green, succulent, 5 - 10 × 2 - 4 mm, deciduous; **Inf** with fleshy peduncle 8 - 10 mm long, only on twining thin terete stems, 1- to 4-flowered; **Ped** 5 - 10 mm; **Sep** narrowly lanceolate, ± 2 mm; **Cl** 1.4 - 1.8 cm; **Cl** inflation inversely bulbiform, ± 5 × 5 mm, outside grey-green with dark venation, inside white, mottled with black-purple, apically abruptly constricted, here inside hairy, and immediately merging into **Cl** lobes; **Cl** lobes 10 - 12 mm, linear, apically fused to form a globose widely opening cage, lamina completely folded back along the midrib, inside yellowish-green, basally ± hairy; **Cn** shortly stipitate, purple-yellowish, ± 3.5 mm ⌀, basally fused into a shallow bowl; **Ci** lobes bilobed into narrow triangular appendages, ± 2 mm, erect, inside with short **Ha**, tips connivent; **Cs** ± 2 × 0.4 mm, linear-linguiform, erect-connivent, tips partly overlapping; **Poll** ellipsoid-ovoid; **Fr** narrowly fusiform, completely spreading, 10 - 15 cm, grey-green, warty.

A striking member of the 'dimorphic' Ceropegias, and closely allied to *C. dimorpha*.

C. arnottiana Wight (Contr. Bot. India, 32, 1834). **T:** Myanmar (*Wight* s.n. [K]). – **D:** Myanmar, India, Thailand. **I:** Huber (1957).

[2] Glabrous twining herbs with globose **R** tubers; **L** shortly petiolate, lamina narrowly lanceolate to linear; **Inf** sessile or shortly pedunculate,

peduncle weakly hairy, 2- to 4-flowered; **Cl** 2.7 - 5.4 cm; **Cl** tube 1.2 - 2.2 cm, basal ⅓ to ¼ ovoidly widened; **Cl** lobes narrowly linear, 1.5 - 3.2 cm, acute, apically connate to form an acutely ovoid cage, lamina ± strongly folded back along the midrib, inside glabrous, margins revolute down to the base, ciliate; **Cn** sessile; **Ci** lobes triangular, bifid into triangular processes; **Cs** linear, erect.

Allied to *C. pusilla* and *C. aridicola*. The flower is also similar in shape to that of *C. stenantha*.

C. attenuata Hooker (Icon. Pl. ser. 2, 5: t. 867 + text, 1852). **T:** India, Maharashtra (*Dalzell* s.n. [K]). – **D:** India (Maharashtra, Karnataka). **I:** Ansari (1984).

[2] Slender erect non-twining herbs with a **R** tuber; stems delicate, ± hairy; **L** ± linear, 5 - 8 cm, ± sessile; **Inf** 1-flowered, shortly pedunculate; **Ped** ± 5 mm; **Sep** subulate, 2 - 4 mm; **Cl** 5 - 7.5 cm, glabrous; **Cl** inflation ovoid, ± 4 mm ∅, inside glabrous, merging gradually into the **Cl** tube, tube ± 3 mm ∅, mouth widening to ± 6 mm ∅; **Cl** lobes to 37 mm, linear, apically connate, lamina folded back along the midrib, hairy; **Cn** bowl-shaped, sessile; **Ci** lobes broadly triangular, largely fused, ± erect, crenate or bifid into triangular processes, ciliate; **Cs** ± 2 mm, linear-subulate, 2 - 3 mm, erect-connivent, tips weakly revolute.

An attractive species allied to *C. jainii* and *C. pusilla*.

C. ballyana Bullock (KB 1955(4): 625, 1955). **T:** Kenya (*Heudelot* s.n. [G]). – **D:** Kenya. **I:** FPA 41: t. 1614, 1970; Archer (1992: XXIX). **Fig. IX.a**

Incl. *Ceropegia helicoides* E. A. Bruce & P. R. O. Bally (1950) (*nom. illeg.*, Art. 53.1).

[3,4] **R** fibrous; stems persistent, succulent, to 3 m, twining, ± terete, 5 - 10 mm ∅, glaucous-green, glabrous; **L** petiole 5 - 15 mm, upper face canaliculate, with **Ha** along margins, lamina ovate to elliptic to obovate, 3 - 9 × 2 - 6 cm, weakly succulent, obtuse to acute, partly acuminate, dark to pale green, shining, veins ± whitish; **Inf** with fleshy peduncle, 3 - 10 cm, many-flowered, but mostly only 1 **Fl** open at a time; **Ped** 1.5 - 3 cm; **Sep** narrowly lanceolate, 1 - 3 cm, spreading, ciliate; **Cl** 7 - 12 cm, outside glabrous, inside and outside whitish-green to yellowish, dotted or blotched with ± red-brown; **Cl** inflation 10 - 25 × 9 - 16 mm, ± cylindrical, often faintly 5-angled, inside apically short-hairy, abruptly horizontally constricted; **Cl** tube 2 - 4 mm ∅, then widening gradually to 12 - 25 mm, inside with scattered **Ha**; **Cl** lobes ± green-yellow with red-brown reticulation, narrowly triangular, ± filiform and elongated, 3 - 6 cm, tips fused, lamina strongly folded back along the midrib, margins with white or purple **Ha**, inner ½ of lamina white to greenish, partly dotted with red, partly hairy, apical ½ ± spiralling, green to dark purple, partly blotched, ± hairy; **Cn** shortly stipitate, bowl-shaped, ± 5 × 5 mm ∅, greenish, blotched with ± purple, glabrous; **Ci** lobes triangular, ± 2 mm, horizontally spreading to ascending, deeply bifid into acutely triangular appendages, apically fused; **Cs** ± 3 mm, linear, erect-connivent; **Poll** broadly ovoid, ± 0.4 × 0.28 mm; **Fr** parallel, 10 - 20 × 0.8 - 1 cm, often one follicle shorter than the other, glaucous-green, glabrous; **Se** 10 - 15 × 3 - 5 mm; tuft of **Ha** 2 - 6 cm.

A splendid species closely related to *C. albisepta* from which it also differs by its striking size. The lower part of the stems may attain thumb's width.

C. bonafouxii K. Schumann (BJS 33: 327, 1904). **T** [neo]: Angola (*Gossweiler* 2439 [BM]). – **D:** Angola, Botswana, Namibia, Zambia, Zimbabwe. **I:** Dyer (1983).

Incl. *Ceropegia saxatilis* S. Moore (1908) (*nom. illeg.*, Art. 53.1).

[2] **R** tuber flattened, 2 - 4 cm ∅, with fissured bark; stems twining, 1 - 2 m, 1 - 3 mm ∅, annual, with short **Ha**, some **Ha** longer; **L** 0.8 - 1 cm petiolate, lamina elliptic to ovate, 3 - 4 (-7) × 1.4 - 2.5 cm, hairy, lower face more densely so; **Inf** ± sessile, 1- to 3-flowered; **Ped** 1 - 2 cm, villose; **Sep** subulate, ± 4 mm, hairy; **Cl** 3 - 4 cm, bottle-shaped, outside hairy; **Cl** inflation 11 - 15 × 5 mm, ellipsoid, outside whitish-greenish, striped a little with red-brown, fairly abruptly constricted to 2 mm ∅; **Cl** tube inside purple, then gradually widening to 4 mm; **Cl** lobes 5 - 8 mm, linear, lamina strongly folded back along the midrib down to the basal edges, apically connate to form an ovoid cage, inside velvety green; **Cn** ± shortly stipitate, white, fused into a shallow cup; **Ci** lobes forming pouches, ± 1 mm; **Cs** ± 2 mm, linear, erect-connivent, tips occasionally curved inwards, hairy; **Fr** narrowly fusiform, ± 10 cm × 3 mm.

Closely related to *C. meyeri* and *C. abyssinica*.

C. bosseri Rauh & Buchloh (KuaS 16(12): 226-229, ills., 1965). **T:** Madagascar (*Bosser* s.n. in *BG Heidelberg* 10668 [HEID]). – **D:** S Madagascar.

Incl. *Ceropegia bosseri* var. *razafindratsirana* Rauh & Buchloh (1989) ≡ *Ceropegia razafindratsirana* (Rauh & Buchloh) Rauh (1993); incl. *Ceropegia adrienneae* Rauh & Gerold (1998).

[3] Stem succulents, creeping-decumbent; stems acutely 4-angled, compressed, 1 - 4 cm thick, thickest at the nodes, brown to grey-green, rough-verrucose, furnished with **Tu** being horizontally elongated to 10 - 15 mm; **L** sessile, dark green, succulent, ovate to lanceolate, 8 - 12 × 7 - 10 mm, mostly mucronate, margins occasionally crenulate, with hooked **Ha**, upper face canaliculate at the base, stipules present; **Inf** with a fleshy peduncle 7 - 10 mm long, only on twining thin terete stems, 1- to 6-flowered; **Ped** ± 1 cm; **Sep** subulate, 2 mm; **Cl** 3.2 - 4 cm, pale green to yellowish, partly dotted and striped with red; **Cl** inflation inversely bulbiform to cylindrical, 9 - 12 × 8 - 10 mm ∅, apically ± hori-

zontally constricted, inside blotched ± with red, most notably on the veins, often weakly hairy, changing abruptly into the **Cl** tube, tube ± 3 mm ∅, mouth gradually widening to 12 - 15 mm ∅; **Cl** lobes 12 - 22 mm, narrowly triangular, ± folded back along the midrib, basally keeled, extending a little into the **Cl** tube, apically sometimes linguiform, tips connate to form a cone-shaped to cylindrical cage, basal ½ yellow, inside glabrous or ± finely hairy, apical ½ purple-brown to greenish, margins with vibratile purple clavate **Ha**, 2 - 3 mm; **Cn** ± sessile to shortly stipitate, greenish-white, ± 3 × 3 mm, shallowly bowl-shaped; **Ci** lobes deeply bifid down to the limb of the **Cn** into appendages, ± 2 mm, ± erect; **Cs** 2 × 0.3 mm, subulate, erect-connivent; **Poll** ovoid, ± 0.4 × 0.25 mm.

Distinguished from *C. petignatii* only by the longer corolla lobes and the stipitate gynostegium. Both belong probably to the same 'bio-species'.

C. bowkeri Harvey (Thes. Cap. 1: 9, t. 14, 1859). **T:** RSA, Eastern Cape (*Bowker* s.n. [K]). – **D:** RSA (Eastern Cape). **I:** Dyer (1983).

Incl. *Ceropegia sororia* Harvey *ex* Hooker *fil.* (1866) ≡ *Ceropegia bowkeri* ssp. *sororia* (Harvey *ex* Hooker *fil.*) R. A. Dyer (1980).

[1] Erect non-twining herbs, 20 - 50 cm tall, with fleshy thickened **R**; stems annual, sparsely branching, ± glabrous; **L** ± sessile, linear to linear-lanceolate, acute, 4 - 13.5 cm × 2 - 3 mm; **Inf** sessile, 1-flowered, lateral; **Ped** to 2 cm; **Sep** subulate, 4 - 8 mm; **Cl** 3.5 - 6 cm, ascending, ± straight, outside greenish-white, inside greenish, often faintly marked with purple; **Cl** inflation globose to ovoid, 5 - 8 mm ∅, merging ± gradually into the **Cl** tube, tube ± 2 - 3 mm ∅, apically widening to 3 - 4 mm ∅; **Cl** lobes free, ± drooping, linear to narrowly elliptic from a narrowed base, 18 - 22 mm, greenish, marked with purple, in the middle with sculptured surface, with white **Ha**, ± folded back along the midrib, laterally with yellowish or purple clavate **Ha**; **Cn** ± sessile, basally fused into a cup; **Ci** lobes triangular, ± erect, ± 1.5 mm, bilobed into acute narrowly triangular appendages, hairy; **Cs** ± 1.5 mm, linear-lanceolate, erect-connivent; **Fr** ± 5 cm; **Se** narrowly ovate, ± 3 × 1.75 mm.

Hardly separable from *C. dinteri* except for the roots. In *C. bowkeri* these are numerous, thick and fleshy, whereas *C. dinteri* possesses a tuber. There is also a close relation to *C. tomentosa*.

C. bulbosa Roxburgh (Pl. Coast Coromandel 1: 11, t. 7, 1795). **T** [lecto]: India, Coromandel Coast (*Roxburgh* s.n. [icono: l.c. t. 7]). – **D:** India, Bangladesh, Pakistan, Saudi Arabia, Oman, N Yemen, Ethiopia, Somalia, Kenya, Tanzania. **I:** Ali (1983); Ansari (1984); Bruyns (1988b). **Fig. IX.b**

Incl. *Ceropegia acuminata* Roxburgh (1795); **incl.** *Ceropegia lushii* Graham (1834) ≡ *Ceropegia bulbosa* var. *lushii* (Graham) Hooker *fil.* (1883); **incl.** *Ceropegia vignaldiana* A. Richard (1851); **incl.** *Ceropegia tuberosa* Dalziel & Gibson (1861) (*nom. illeg.*, Art. 53.1); **incl.** *Ceropegia esculenta* Edgeworth (1862) ≡ *Ceropegia bulbosa* var. *esculenta* (Edgeworth) Hooker *fil.* (1883); **incl.** *Ceropegia brosima* E. A. Bruce & P. R. O. Bally (1951).

[2] Delicate geophytes, strongly twining, little-branched, 0.5 - 1 m, with ± flattened **R** tuber with rough bark; stems terete, 1 - 2 mm ∅, slightly hairy; **L** petiolate, ovate-lanceolate to elliptic to linear, 1 - 10 × 0.3 - 1.2 cm, sparsely villose; **Inf** with a peduncle 2 - 25 mm long, pseudo-umbellate, 2- to 8-flowered; **Ped** 5 - 15 mm; **Sep** subulate, 2 - 4 mm; **Cl** 1.2 - 3 cm, pale grey-green, usually striped or blotched with red-brown; **Cl** inflation depressed-globose, 3 - 4 × 4 - 5 mm, inside greenish-white, merging gradually into the **Cl** tube, tube 1.5 - 2 mm ∅, apically widening to 5 - 10 mm ∅, inside pale purple, basally hairy; **Cl** lobes inside ± uniformly purple, basally with a white blotch, apically often greenish, 4 - 10 × basally 3 - 6 mm, apically 0.5 - 1 mm broad, lamina ± entirely folded back along the midrib, erect, tips joined in a cage-like fashion, sometimes spiralled, inside and laterally hairy; **Cn** sessile to shortly stipitate, shallowly bowl-shaped, 2 - 3 × 2 - 3 mm; **Ci** lobes 0.5 - 0.8 × 0.8 - 1.1 mm, broadly 4-angled, channelled, with raised margins, horizontal; **Cs** 1.5 - 2 mm, ± cylindrical, erect, vertical and free or meeting each other above the centre; **Poll** tear-shaped to ovoid, ± 0.25 × 0.2 mm with very short germination crest near the tip, corpuscle small, elliptic; **Fr** grey-green, ± 10 × 3 mm.

This species is treated here in a broad sense to form an extraordinarily variable and widespread taxon. A supraregional taxonomic re-assessment is needed, since several valid names are currently used in Africa (*C. vignaldiana*, *C. brosima*) and Arabia and India (*C. bulbosa* and *C. lushii*). The ovate (*C. bulbosa* s.str.) vs. linear leaves (*C. vignaldiana*, *C. lushii*) have often been used for a subdivision. Due to fairly frequent intermediates exhibiting ± elliptic leaves, leaf characters are of little value and may at best be suggestive of 2 ecotypes (upland form with linear leaves).

C. campanulata G. Don (Gen. Hist. 4: 112, 1838). **T:** Ghana (*Don* s.n. [BM]). – **D:** Equatorial-Guinea, Ivory Coast, Ghana, Mali, Nigeria, Togo. **I:** Bullock (1963: 101).

Incl. *Ceropegia kerstingii* K. Schumann (1903); **incl.** *Ceropegia dalzielii* N. E. Brown (1913); **incl.** *Ceropegia abinsica* N. E. Brown (1914) ≡ *Ceropegia campanulata* var. *abinsica* (N. E. Brown) H. Huber (1957); **incl.** *Ceropegia tamalensis* W. W. Smith (1922); **incl.** *Ceropegia hepburnii* Hutchinson & Dalziel (1931).

[2] Erect non-twining herbs, 10 - 25 cm tall, with a flattened **R** tuber; stems annual, ± unbranched, with scattered **Ha**; **L** ± sessile, lanceolate to linear, acute, ± 50 - 70 × 6 mm, upper face with scattered

Ha; **Inf** sessile, terminal or lateral, 1-flowered; **Cl** ± brown-purple, 4 - 13 cm, straight, outside finely hairy or glabrous; **Cl** inflation globose, inside purple-haired, 7 - 18 mm ⌀, merging gradually into the **Cl** tube, tube 4 mm ⌀, apically widening to ± 12 mm ⌀, inside purple-haired; **Cl** lobes erect and straight, joined at the tips, linear to filiform, 1.5 - 6 cm, margins revolute, inside ± hairy; **Cn** sessile, shallowly cup-shaped; **Ci** lobes horizontal-ascending, bilobed into narrowly triangular to linear processes, finely hairy; **Cs** lobes linear, erect-connivent, tips revolute.

A variable species showing a scattered but wide distribution within the tropics. It is very closely related to *C. insignis*.

C. cancellata Reichenbach (Iconogr. Bot. Exot. 3: t. 207, 1830). **T**: W. – **D**: RSA (Eastern Cape). **I**: Dyer (1983); Bruyns (1985: 5, J-K).

Incl. *Ceropegia assimilis* N. E. Brown (1908).

[2,4] **R** tuber dark brown, flat, 5 - 10 cm ⌀; stems annual, twining; **L** shortly petiolate, broadly ovate-cordate to lanceolate, to 25 mm, succulent, occasionally pubescent, apiculate, upper face concave, lower face convex; **Inf** pedunculate, 1- (to few-) flowered; **Cl** 2.3 - 3.5 cm, whitish-yellowish, striped with brown-violet, **Cl** lobes outside brown-violet; **Cl** inflation globose to ovoid, ± 4 - 5 × 4 mm, inside purple, glabrous, above **Cl** tube abruptly narrowed to 1.3 - 2 mm ⌀, apically widening to 4 - 5 mm ⌀, inside hairy; **Cl** lobes linear, 10 - 15 mm, tips fused, lamina somewhat folded back along the midrib, forming an ovoid cage, inside pale yellow, glabrous; **Cn** white, bowl-shaped, sessile or shortly stipitate, ± 2.5 - 3 mm ⌀; **Ci** lobes forming tall oval pouches with horizontal margin, laterally fused with the base of the **Cs**; **Cs** lanceolate, falcate, ± 1.5 mm, erect-connivent but strongly reflexed above the middle.

A species closely related to *C. africana* and to the *C. linearis* alliance.

C. candelabrum Linné (Spec. Pl. [ed. 1], 211, 1753). **T**: [icono]: Rheede, Hort. Malab. 9: t. 16. – **D**: India, Sri Lanka. **I**: Matthew (1983).

Incl. *Ceropegia biflora* Linné (1753) ≡ *Ceropegia candelabrum* var. *biflora* (Linné) M. Ansari (1984); **incl.** *Ceropegia tuberosa* Roxburgh (1795) ≡ *Ceropegia candelabrum* ssp. *tuberosa* (Roxburgh) H. Huber (1983) ≡ *Ceropegia candelabrum* var. *tuberosa* (Roxburgh) N. P. Singh (1988); **incl.** *Ceropegia mucronata* Roth (1821); **incl.** *Ceropegia longiflora* Poiret in Roemer & Schultes (1838); **incl.** *Ceropegia candelabriformis* Saint-Lager (1880); **incl.** *Ceropegia elliotii* Hooker fil. (1883); **incl.** *Ceropegia discreta* N. E. Brown (1909).

[2] **R** tuber roundish; stems robust, twining, glabrous, ± 2 - 3 mm ⌀; **L** with a 3 - 20 mm long petiole, lamina linear to roundish, usually elliptic, 2 - 7 × 0.8 - 3.5 cm, acute, weakly fleshy; **Inf** with a peduncle 3 cm long, pseudo-umbellate, 5- to 12-flowered; **Ped** 3 - 10 mm; **Sep** subulate, ± 4 mm; **Cl** 2.5 - 4.5 cm, greenish-yellow, usually striped with red-brown; **Cl** inflation ± ovoid, merging gradually into the **Cl** tube, tube 3 - 4 mm ⌀, straight or bent, apically widening to ± 10 mm ⌀, inside hairy in the middle; **Cl** lobes 4 - 15 mm, triangular-ovate to linear from a triangular base, truncate or elongated and acute, lamina ± entirely folded back along the midrib, basally keeled, erect, fused to form a cage, inside purple-green, densely hairy; **Cn** stipitate; **Ci** lobes basally connate, a short rim only, ± 0.7 mm ⌀, laterally raised and connate to the **Cs** base, hairy; **Cs** 2 - 3 mm, linear-spatulate, erect-connivent; **Poll** broadly ovoid, ± 0.3 × 0.23 mm.

Very much like *C. intermedia* and often confused with it. The most striking attributes of *C. candelabrum* are the tubers and the hairy corona. Ansari (1984) separates var. *biflora* and places *C. intermedia* as a synonym, although the latter shows fibrous roots.

C. carnosa E. Meyer (Comm. Pl. Afr. Austr., 193, 1837). **T**: RSA, Eastern Cape (*Drège* 4946 [W]). – **D**: Kenya, RSA (Eastern Cape, KwaZulu-Natal), Swaziland. **I**: Dyer (1983). **Fig. VIII.g**

[1,4] Climbers to 2 m; **R** fusiform-fleshy; stems ± terete, twining, 1 - 2 mm ⌀, dark green, slightly succulent and rough; **L** 5 - 15 mm petiolate, lamina narrowly ovate to lanceolate, slightly succulent, 2 - 3 × 1 - 2 cm, acute, petiole and margins often hairy; **Inf** with slender peduncle 5 - 20 mm long, 3- to 5-flowered, always only 1 **Fl** open at a time; **Ped** 5 - 10 mm; **Sep** subulate, acute, 2 - 4 mm; **Cl** 1.5 - 2.2 cm, whitish-green blotched or striped with red-brown, inside with ± scattered long **Ha**, with mushroom-like odour; **Cl** inflation slender, inside purple, ± 4 × 4 mm ⌀, scarcely broader than the curved **Cl** tube, tube occasionally laterally compressed, basally slightly constricted, inside red-violet, mouth gradually widening to 6 - 10 mm ⌀; **Cl** lobes whitish with purple stripes or blotches, or greenish with brown tip, narrowed in the middle, 4 - 6 mm long, only tips joined to form a cage, shortly triangular and horizontally constricted or lanceolate, erect, lamina folded back along the midrib, laterally only at the base or over the entire length with simple white **Ha**; **Cn** white with purple margins, shallowly cup-shaped, 3.5 × 3.5 mm, basally connate to the **Cl** tube; **Ci** lobes deeply incised, with erect convergent triangular white-bearded appendages, basally forming pouches, margins fused in a wing-like fashion with the base of the **Cs**; **Cs** 1.8 - 2.2 mm, subulate, erect-connivent, tips usually divergent, basally hairy; **Poll** ± rectangular, 0.4 - 0.2 mm.

Not always clearly separable (by its floral structure) from *C. affinis* (syn. *C. racemosa*), which is often more robust in its foliage. Both are presumably forming a single good species.

C. cataphyllaris Bullock (KB 1955(4): 625, 1955). **T:** Zambia (*Milne-Redhead* 3241 [K]). – **D:** Zambia. **I:** Bruce (1948: as *C. filiformis*).

Incl. *Ceropegia filiformis* E. A. Bruce (1948) (*nom. illeg.*, Art. 53.1).

[2] **R** tuber 1.5 - 2.5 cm ∅; stems erect, delicate and inconspicuous, 15 - 30 cm, glabrous, 1 mm ∅, unbranched; **L** sessile, erect, narrowly lanceolate to needle-shaped, increasing in length from the base to the top of the stem, 2 - 30 × 0.3 - 0.7 mm, margins reflexed; **Inf** sessile near the stem tip, 1-flowered; **Ped** filiform, 1.5 - 2 cm; **Sep** linear-subulate, 2 mm, reddish; **Cl** 3 - 3.5 cm, ± straight, erect, whitish, glabrous; **Cl** inflation obovoid, ± 7 × 5 mm ∅; **Cl** tube cylindrical, 2 - 3 mm ∅, apically scarcely widening; **Cl** lobes brownish, linear, 20 × 1 mm, folded back along the midrib, apically connate and arching out to form an ellipsoid cage; **Cn** sessile, whitish-hyaline, fused into a shallow cup; **Ci** bilobed into linear appendages, 1 mm, each tipped with 1 - 2 whitish **Ha**; **Cs** linear-subulate, 1.5 - 2 mm, erect-connivent, apically slightly revolute.

In habit reminiscent of *C. illegitima* from Zaïre, but the latter shows fleshy roots instead of a tuber, red flowers and a reduced interstaminal corona.

C. ciliata Wight (Icon. Pl. Ind. Orient. 4: 15, t. 1262, 1848). **T:** India, Tamil Nadu (*Wight* 1927 [K, G]). – **D:** India (Tamil Nadu). **I:** Huber (1957).

Incl. *Ceropegia hirsuta* Decaisne (1844) (*nom. illeg.*, Art. 53.1).

[2] **R** tuber 3 - 4 cm; stems delicate, hairy, twining; **L** ± sessile to shortly petiolate, lamina linear-lanceolate, acute; **Inf** many-flowered, peduncle and **Ped** hairy; **Cl** 2.5 - 3 cm; **Cl** tube ± 18 mm, basally with an ovoid inflation 7 - 9 mm long, apically only a little widening; **Cl** lobes 6 - 10 mm, lamina folded back along the midrib; **Cn** sessile; **Ci** lobes basally fused, ± triangular, apically emarginate or bifid, hairy; **Cs** spatulate, erect-spreading or tips only revolute; **Fr** 8 × 0.4 cm.

See note under *C. ensifolia*.

C. cimiciodora Obermeyer (FPSA 13: t. 488 + 2 pp. of text, 1933). **T:** RSA, Northern Prov. (*Obermeyer & al.* 322 [PRE]). – **D:** RSA (Northern Prov., KwaZulu-Natal), Swaziland. **I:** Dyer (1983).

[3] Scrambling stem succulents, occasionally twining; **R** fibrous; stems ± terete, 15 - 150 × 1 - 1.5 cm, grey and often spotted with olive-green, slightly shiny; **L** triangular to cordate, ± 2.5 × 3 mm, with stipular **Gl**; **Inf** cymose; peduncle 1 - 20 mm, 1 to several **Fl** on each of the twining peduncles of the part-**Inf**, which often attain several 10 cm in length; **Ped** 8 × 2.5 mm; **Sep** lanceolate, ± 2 mm; **Cl** ± 5 cm, robust, grey-white stippled and blotched with red-brown; **Cl** inflation ovoid, ± 8 × 6 mm, with whitish basal constriction, inside hairy, merging gradually into the **Cl** tube, tube 15 × 3.5 mm, bent, apically widening into a short funnel; **Cl** lobes inside white, apically and laterally often red-violet, greenish or brownish, broadly triangular, 6 × 14 mm, apically connivent and folded back along the midrib, basal margins standing out (auriculate), thus appearing star-like in top view, inside finely hairy, **Ha** white; **Cn** shortly stipitate or sessile, whitish or brownish, ± 6 × 4 mm, fused into a cup, margin slightly undulate, inside and outside ± ciliate; **Ci** lobes very shortly triangular, abruptly projecting from the upper margin of the coronal bowl, ± erect, bifid into triangular appendages, ± 0.3 mm; **Cs** ± 4 × 0.4 mm, subulate, erect-connivent with revolute papillose tips; **Poll** ovoid, ± 0.4 × 0.27 mm; **Fr** grey-green, erect, thickly fusiform, ± 10 cm, finely tuberculate.

A magnificent plant, together with other pronounced succulents like *C. stapeliiformis* and *C. variegata* representing a small group of closely allied species.

C. claviloba Werdermann (BJS 70: 221-222, 1939). **T:** Zimbabwe (*Exell & al.* 147 [BM]). – **D:** Tanzania, Malawi, Zaïre, Zambia, Zimbabwe. **I:** Malaisse (1984: 226, as *C. floribunda*). **Fig. IX.c**

[2] **R** tuber small, ± 2 cm ∅; stems sparsely branched, delicate, twining, 0.5 - 1 (-2) m tall, 1.5 mm ∅; **L** 2 - 4 cm petiolate, lamina cordate-lanceolate, acuminate, 2.5 - 4.5 (-7) × 2 - 3 cm, hairy, margin basally ± lobed; **Inf** 5 - 30 mm pedunculate, 1- to 10-flowered; **Ped** 3 - 8 mm, usually glabrous; **Sep** subulate, 2 - 4 mm, villose, rarely glabrous; **Cl** bent, 1 - 1.8 cm, outside whitish-green; **Cl** inflation globose, basally narrowed, 4 - 5 mm ∅, merging gradually into the **Cl** tube, tube 1.5 - 2 mm ∅, mouth widening to 3 - 4 mm ∅; **Cl** lobes 4 - 7 × 3 mm, broadly spatulate from a triangular base, lamina folded back along the midrib, apically joined into a globose cage, inside velvety purple or green, outside greenish; **Cn** sessile, ± 3 × 2.5 mm, white-hyaline, basally fused into a bowl or cup; **Ci** lobes pouch-shaped, ± horizontal and ovate to rectangular and erect, ± 1 mm, apically crenate, glabrous, papillose or hairy; **Cs** 1.5 - 2 mm, cylindrical-spatulate, straight, erect, papillose; **Poll** globose, ± 0.1 mm; **Fr** narrowly fusiform, 7 - 8 cm × 2 - 3 mm.

A species with very large leaves and small flowers; very closely allied to *C. papillata* and *C. muzingana*.

C. conrathii Schlechter (BJS 38: 45, 1905). **T:** RSA, Gauteng (*Conrath* 1008 [K]). – **D:** RSA (KwaZulu-Natal, Gauteng, Northern Prov.). **I:** Dyer (1983). **Fig. IX.d**

[2] Herbaceous dwarf perennials with erect crowded stems, not twining, with a large **R** tuber, 3 - 10 × 2 - 7 cm, flattened; stems perennial, terete, 6 - 12 cm × 1 - 2 mm; **L** ± sessile, erect-spreading, faintly succulent, narrowly lanceolate to ovate, acute, to 30 × 12 mm, with undulate ± hairy margins; **Inf** sessile, lateral, 1- to 6-flowered, all **Fl** opening simul-

taneously; **Ped** 7 - 15 mm; **Sep** lanceolate, 2 mm; **Cl** ± 3.5 cm, weakly bent, cinnamon-coloured; **Cl** inflation ellipsoid, ± 8 - 10 × 3 - 4 mm, inside papillose, papillae with **Bri**; **Cl** tube cylindrical, ± 2.5 mm ⌀, inside with scattered **Ha**; **Cl** lobes to 8 mm, narrowly linear, flat, apically fused to form a bulbiform cage, suffused with purple; **Cn** shortly stipitate, white, shallowly campanulate, glabrous; **Ci** of 5 basal pouches with U-shaped upper margins, ± 0.5 mm deep, laterally fused with the bulging **Cs** base; **Cs** ± 1.5 × 0.3 mm, cylindrical, erect-connivent; **Fr** ± 10 cm × 4 mm.

C. conrathii is closely related to the twining *C. floribunda*. The treatment of *C. floribunda* by Huber (1957) as a synonym of *C. conrathii* has been dismissed by Dyer (1980) who recognized both as separate species.

C. convolvuloides A. Richard (Tent. Fl. Abyss. 2: 47, 1851). **T**: Ethiopia (*Quartin Dillon & Petit* s.n. [P]). – **D**: Ethiopia, Eritrea. **I**: Huber (1957).

Incl. *Ceropegia ringens* Vatke (1876) (*nom. illeg.*, Art. 53.1); **incl.** *Ceropegia ellenbeckii* K. Schumann (1903).

[1] **R** fleshy, fusiform; stems hairy, sparsely branching, twining, terete, sparsely hairy; petioles 1 - 3 cm, pubescent, lamina oval to lanceolate, acute, basally cordate, 4 - 10 × 2 - 5 cm, pubescent; **Inf** pseudo-umbellate, 12- to 20-flowered, pedunculate; peduncle 7 - 15 mm, hairy; **Ped** 7 - 15 mm, hairy; **Sep** subulate, ± 2 mm, hairy; **Cl** 2.5 - 3 cm, outside setulose, inside glabrous; **Cl** inflation ovoid, ± 5 × 4 mm, narrowed abruptly into the bent **Cl** tube of 2 mm ⌀, tube widening apically to 5 mm ⌀, inside ± black-purple; **Cl** lobes 9 - 14 mm, linear-lanceolate from an ovate base, basally keeled, keel somewhat tuberculate, lamina folded back along the midrib, erect, apically fused, basal margins projecting with auricles 3 - 4 mm long; **Cn** sessile, cup-shaped, glabrous or scattered-ciliate; **Ci** ± as long as the **Cs**, **Ci** lobes triangular-ovate, concave, laterally fused with one another and with the base of the **Cs** lobes, erect, bilobed into ± 1 mm long triangular-linear appendages, tips spreading sidewards; **Cs** linear, ± vertically ascending, 2.5 mm.

A little-known species, but well characterized by the auriculate corolla lobes (similar to *C. variegata*) and the leaves being reminiscent of those of the bindweed. The Arabian *C. sepium* and *C. foliosa* are probably most closely related.

C. crassifolia Schlechter (J. Bot. 33: 273, 1895). **T**: RSA, Eastern Cape (*Sim* s.n. [BOL]). – **D**: Namibia, Kenya, Botswana, Zimbabwe, RSA, Swaziland. **I**: Dyer (1983).

C. crassifolia var. **copleyae** (E. A. Bruce & P. R. O. Bally) H. Huber (Mem. Soc. Brot. 12: 102, 1957). **T**: Kenya (*Armstrong* s.n. [EA]). – **D**: Kenya. **I**: Archer (1992: XIV).

≡ *Ceropegia copleyae* E. A. Bruce & P. R. O. Bally (1950).

[1,3,4] Differs from var. *crassifolia*: **L** linear, 3 - 5 mm broad, V-shaped in cross-section; **Cl** pale green-yellow; **Cl** lobes ± 5 × 2 mm, lamina halfways folded back along the midrib, white-ciliate along the margins; **Cn** cup-shaped, as tall as the **Gy**; **Ci** apically incised or emarginate; **Cs** shortly rectangular, incumbent on the **Anth**; **Poll** ovoid, 0.28 × 0.2 mm.

C. crassifolia var. **crassifolia** – **D**: Namibia, Kenya, Botswana, Zimbabwe, RSA, Swaziland. **I**: Dyer (1983). **Fig. VII.f, IX.e**

Incl. *Ceropegia brachyceras* Schlechter (1905); **incl.** *Ceropegia crispata* N. E. Brown (1908); **incl.** *Ceropegia thorncroftii* N. E. Brown (1912); **incl.** *Ceropegia tuberculata* Dinter (1923).

[1,3,4] **R** clustered, fleshy, fusiform; stems erect, creeping or twining, to 2 m, usually annual, 2 - 5 mm ⌀, succulent, sulcate; **L** shortly petiolate, linear to broadly elliptic-lanceolate, 3 - 10 × 0.5 - 4.5 cm, ± succulent, lower face with a prominent midrib, margins often undulate or serrulate and hairy; **Inf** 3 - 30 mm pedunculate, 3- to 7-flowered, **Fl** opening successively; **Ped** 2 - 5 × 1 - 2 mm; **Sep** narrowly lanceolate, to 5 mm; **Cl** 2 - 5 cm, weakly bent, pale green-yellow, blotched with ± purple; **Cl** inflation globose to ovoid, 6 - 9 mm ⌀, usually abruptly narrowed into the **Cl** tube, tube basally 3 - 5 mm ⌀, inside hairy, apically widening to 9 mm ⌀; **Cl** lobes 4 - 15 × 2 - 4 mm, narrowly ovate, fused into an ovoid to subglobose cage, lamina ± folded back to the margins, glabrous; **Cn** white, ± sessile, basally fused into a bowl- to cup-like structure; **Ci** ± rectangular to ovate and forming pouches, almost horizontal to erect, ± 1 mm, apically obtuse or incised, usually hairy; **Cs** rectangular to linear-subulate, 0.5 - 1.5 mm, horizontally incumbent on the **Anth** to erect-connivent; **Poll** globose-ovoid, ± 0.4 × 0.25 mm; **Fr** ± 7 - 10 cm × 3 - 4 mm.

A very variable taxon, allied to *C. nilotica* and *C. stenantha*.

C. cufodontii Chiovenda (Miss. Biol. Borana, Racc. Bot. Angiosp.-Gymnosp., 167, 1939). **T**: Ethiopia (*Cufodontis* 337 [FT]). – **D**: Ethiopia, Kenya, Uganda. **I**: FPA 35: t. 1370, 1962; Archer (1992: XII). **Fig. IX.f**

[1] **R** clustered, fleshy, fusiform; stems perennial, twining, 1 - 2 m × 1 mm, hairy; **L** 1 - 3.5 cm petiolate, lamina ovate to lanceolate, 4 - 15 × 2.2 - 5 cm, delicate, hairy; **Inf** 1 - 2 mm pedunculate, 4- to 10- (to 20-) flowered; **Ped** 4 - 10 mm, hairy; **Sep** narrowly lanceolate, 2 - 5 mm, hairy; **Cl** 2.8 - 4 cm, weakly bent, greenish-white, ± longitudinally striate with red-brown and sulcate; **Cl** inflation ± ovoid, 9 - 12 × 6 - 10 mm; **Cl** tube gradually narrowing to 2 mm ⌀, apically widening to 6 - 12 mm, inside hairy; **Cl** lobes triangular-lanceolate, 8 - 13

mm, apically fused to form a slender ovoid cage, only the margins somewhat folded back, weakly keeled, inside yellow-green to brownish, hairy along the middle of the keel; **Cn** sessile, bowl-shaped, ± 3.5 × 3 mm, ± red-brown; **Ci** lobes broadly ovate, ± obtuse, scarcely overtopping the hairy **Cn** limb; **Cs** ± 3 × 0.3 mm, linear, connivent, apically joining; **Fr** 4.5 - 9 cm × 2 - 4 mm.

Allied to *C. meyeri-johannis*, which also prefers shady habitats.

C. cycniflora R. A. Dyer (BT 12(3): 444-445, ill., 1978). **T:** RSA, KwaZulu-Natal (*Cronwright* 26 [PRE]). – **D:** RSA (KwaZulu-Natal). **I:** Dyer (1983).

[1] **R** fleshy, fusiform, clustered; stems annual, climbing and twining, to 3 m tall, weakly 4-angled, slightly succulent, scattered-hairy; **L** with a hairy petiole 5 mm long, lamina ovate-lanceolate, basally cordate, ± 1 cm; **Inf** many-flowered; peduncle 1 - 2.5 cm, finely hairy; **Ped** ± 1 cm; **Sep** linear-lanceolate, ± 4 mm; **Cl** 3 - 4 cm, pale green, strongly bent; **Cl** inflation ellipsoid, 6 - 8 mm ∅; **Cl** tube narrowing gradually to 2.5 mm ∅, apically widening to ± 7 mm; **Cl** lobes linear from a flat triangular-ovate base, 12 - 15 mm, erect, apically fused, laterally only slightly divergent, margins slightly folded back along the midrib, slightly rolled inwards, basal ½ with purple vibratile clavate **Ha**; **Cn** ± sessile, only basally fused; **Ci** lobes bilobed into linear-subulate appendages, ± 2.5 mm; **Cs** linear-subulate, ± 3 mm, erect-connivent, tips slightly revolute; **Poll** ovoid; **Fr** ± 6 cm × 4 mm.

A rare species of unknown relationship. There are at best similarities to *C. zeyheri*.

C. damannii Stopp (BJS 83: 122-123, ills., 1964). **T:** Angola (*Damann* 1222 [K]). – **D:** Angola.

[1] **R** fleshily thickened; stem annual, erect, unbranched, not twining, 30 - 40 cm tall; **L** sessile, lanceolate-elliptic, acute, ± 8 × 3 cm; **Inf** sessile, 1-flowered, lateral; **Ped** ± 1 cm; **Sep** subulate, 8 - 12 mm; **Cl** ± 12 cm, slightly bent and ascending, greenish-yellow; **Cl** inflation ± globose, ± 8 mm ∅, merging gradually into the narrowed **Cl** tube, tube ± 4 mm ∅, apically widening to 5 - 6 mm ∅; **Cl** lobes ± 6 × (in the middle) 0.9 cm, narrowly lanceolate, basal angles slightly projecting, apically fused to form an ellipsoid cage, inside black-green, glabrous; **Cn** sessile, ± 5 × 3 mm, basally fused into a bowl; **Ci** lobes triangular, ± erect, ± 2 mm, bilobed into subulate appendages, hairy; **Cs** subulate, ± 3 mm, erect-connivent.

Very closely allied to *C. filipendula*, yet entirely glabrous and with corolla lobes as long as the whole corolla tube.

C. decaisneana Wight (Icon. Pl. Ind. Orient. 4: t. 1259, 1848). **T:** India, Tamil Nadu (*Wight* s.n. [BM, K]). – **D:** S India (Tamil Nadu). **I:** Ansari (1984).

Incl. *Ceropegia gracilis* Beddome (1864); **incl.** *Ceropegia brevicollis* Hooker *fil.* (1883) ≡ *Ceropegia decaisneana* var. *brevicollis* (Hooker *fil.*) H. Huber (1957).

[2] **R** tuber ± globose; stems twining, ± hairy (at least at the nodes); **L** narrowly elliptic to lanceolate or ovate, upper face slightly hairy; **Inf** 3- to 8-flowered, peduncle longer than the **Ped**, hairy; **Cl** 4 - 7.7 cm, whitish-green, marked with red-brown, outside occasionally faintly hairy; **Cl** inflation ovoid, to 20 × 15 mm, inside with a ring of **Ha** near the upper end, merging abruptly into the **Cl** tube, tube 3 - 4 mm ∅, apically widening to ± 15 mm ∅; **Cl** lobes 2 - 4 cm, narrowly linear-spatulate from a triangular base, acute, fused, lamina ± entirely folded back along the midrib, basally keeled, erect, joined to form a cage, inside dark or with dark venation on pale background; **Cn** sessile, orange-yellowish; **Ci** lobes basally fused, triangular, erect, bifid into triangular appendages, hairy; **Cs** 2 - 3 mm, linear-spatulate, erect, incompletely connivent.

This species, newly circumscribed by Ansari (1984), is next of kin to *C. metziana*, a herbaceous climber from India, the latter presumably lacking tubers. *C. decaisneana* possesses large attractive flowers reminiscent of *C. scabra* from Madagascar.

C. decidua E. A. Bruce (BT 6(1): 213, ills., 1951). **T:** RSA, Northern Prov. (*Codd* 47 [PRE]). – **D:** RSA, Swaziland.

C. decidua ssp. **decidua** – **D:** RSA (Northern Prov.), Swaziland. **I:** Bruyns (1980); Dyer (1983).

[2,4] **R** tuber flattened, ± 3 - 6 cm ∅; stems slightly twining, to 20 (-40) cm, 1 - 2 mm ∅, sparsely villose; **L** shortly petiolate, ovate to lanceolate, fleshy, to 2 × 1 cm, villose, rapidly caducous; **Inf** with short peduncle, 2- to 4-flowered; **Sep** lanceolate, ± 2 mm; **Cl** 2.2 - 2.6 cm, erect, brownish; **Cl** inflation ovoid, 5 - 7 × 3.5 mm ∅, inside greenish-purple, papillose; **Cl** tube basally narrowed to 2 mm ∅, apically hardly widening, inside hairy; **Cl** lobes linearly spoon-shaped, lamina only in the middle folded back along the midrib, 4 - 6 mm, erect, forming an ovoid cage with flattened tip, inside purple, margins hairy, **Ha** purple; **Cn** sessile, cup-shaped, subcircular in top view, ± 2.5 × 2.5 mm; **Ci** lobes drawn out into deep ovoid pouches, apically obtusely triangular, erect, overtopping the **Gy**, margins often crenate, laterally fused with the **Ci** base; **Cs** ± 2 mm, spoon-shaped, erect-connivent with tips revolute; **Poll** ± 0.25 mm ∅; **Fr** narrowly fusiform, ± 5 - 6 cm × 2 - 3 mm.

Easily confused with *C. occulta* due to the morphology of the corolla.

C. decidua ssp. **pretoriensis** R. A. Dyer (BT 12(4): 630, 1979). **T:** RSA, Gauteng (*Codd* 9565 [PRE]). – **D:** RSA (Gauteng). **I:** Bruyns (1980); Dyer (1983).

[2,4] Differs from ssp. *decidua*: Stems weakly twining, to 10 (-25) cm; **L** ovate to elliptic, acute, to 1.5 × 1 cm; **Inf** 1- to 3-flowered; **Sep** acutely triangular, ± 2 mm; **Cl** 2 - 2.4 cm; **Cl** inflation globose to ovoid, ± 3.5 × 3.5 mm, inside coarsely papillose; **Cl** tube narrowed to 1.5 mm ∅ in the middle; **Cl** lobes linear, lamina basally folded back along the midrib, apically scarcely so, forming an ovoid apically globose cage, basally purple, apically greenish, ± glabrous; **Cn** bowl-shaped, 5-angled in top view, 2.5 × 2.5 mm; **Ci** lobes drawn out into flat prominent pouches, ± as long as the **Gy**, margins horizontal, glabrous; **Cs** ± 1.5 mm.

C. deightonii Hutchinson & Dalziel (Fl. West Trop. Afr. 2: 61-62, 1931). **T:** Sierra Leone (*Deighton* 2075 [K]). – **D:** Equatorial Guinea, Ivory Coast, Ghana, Liberia, Nigeria, Sierra Leone, Central African Republic. **I:** Berhaut (1971); Newton (1974).
Incl. *Ceropegia deightonii* ssp. *conjuncta* H. Huber (1957); **incl.** *Ceropegia deightonii* ssp. *tisserantii* H. Huber (1957).

[2] **R** tuber flattened, ± 2 - 3 × 1.5 - 2 cm; stems annual, erect, not twining, 10 - 70 cm tall, little-branched, sparsely hairy; **L** sessile, linear, acute, 25 - 70 × 1 - 3 mm, ± sparsely hairy, particularly along the margins; **Inf** sessile, terminal or lateral, 1- to 2- (to 3-) flowered; **Ped** ± 5 mm; **Sep** narrowly triangular to subulate, 3 - 4 mm; **Cl** ± red-brown-purple, 3 - 4.5 cm, straight; **Cl** inflation basally greenish, ellipsoid to ovoid, 7 - 12 × 3 - 5 mm, merging gradually into the **Cl** tube, tube ± 2 mm ∅, mouth widening to 3 - 4 mm ∅, inside glabrous or with purple **Ha**; **Cl** lobes ± erect, apically free or fused, often spiralled, narrowly linear to elliptic from a triangular base, 5 - 22 × 1 - 2 mm, margins folded back, usually with purple **Ha**; **Cn** sessile, cup-shaped, ± 3 × 3 mm; **Ci** lobes pouch-like, erect, apically entire, dentate or bifid into appendages < 0.5 mm long, glabrous to long-hairy; **Cs** lobes 1.5 - 2 × 0.3 mm, linear-cylindrical, erect-connivent, glabrous; **Poll** ovoid, 0.28 × 0.17 mm; **Fr** narrowly fusiform, 10 - 12 cm × 3 mm.

Closely related to *C. campanulata* and to the species group around *C. dinteri* (see there for further notes).

C. denticulata K. Schumann (in Engler, Pfl.-welt Ost-Afr., Teil C, 327, ill., 1895). **T:** Tanzania (*Holst* 3583 [not located]). – **D:** Kenya, Tanzania, Uganda, RSA (KwaZulu-Natal). **I:** Archer (1992: XXI); Peckover (1993a); both as *C. nilotica*. **Fig. X.a**
Incl. *Ceropegia brownii* Ledger (1909) ≡ *Ceropegia denticulata* var. *brownii* (Ledger) P. R. O. Bally (1965) (*nom. inval.*, Art. 34.1); **incl.** *Ceropegia nilotica* var. *simplex* H. Huber (1957).

[1,3,4] Climbing and twining, to 3 m tall, branching, latex slightly milky; **R** partly ± fleshily thickened, cylindrical to fusiform; stems usually annual, 3 - 6 mm ∅, often thicker at the base, terete, succulent, glabrous, shiny; **L** 1.5 cm petiolate, lamina narrowly to broadly lanceolate, 2 - 8 × 1 - 4 cm, succulent, acute, margins usually undulate, serrulate or ciliate; **Inf** 5 - 25 mm pedunculate, 1- to many-flowered, **Fl** opening successively, scent unpleasant, fruity; **Ped** 4 - 8 × 2 mm, fleshy; **Sep** subulate, 2 - 5 mm, reflexed; **Cl** 2.5 - 5.5 cm, slightly bent, outside ± glabrous, shining; **Cl** inflation obovoid to cylindrical, without or at best with a faint constriction, 6 - 15 × 5 - 10 mm, outside and inside greenish, glabrous, gradually narrowed to 2.5 - 4 mm ∅ in the middle of the **Cl** tube, tube widening into a funnel-shaped mouth of 12 - 20 mm ∅, outside whitish-green, blotched with purple; **Cl** lobes 10 - 24 × 4 - 6 mm, ovate-lanceolate from a broad base, striped with green-brown-purple-white or blotched, lamina entirely folded back along the midrib, lower margins spreading ± horizontally, fused into a ± pyramid-shaped yet blunt structure, inside broadly keeled, keels joining above the middle, apically slightly divaricate, usually with pale brown, white or purple **Ha**, apical margins with vibratile **Ha** 3 - 6 mm long, purple or rarely white; **Cn** yellowish, ± sessile, ± 4 × 3 mm, fused into a shallow cup, glabrous; **Ci** broadly ovate with undulate margin, concave, basally forming pouches, ± 0.5 × 1 mm, often purple, lateral margins connate to the **Cs** base; **Cs** linear-lanceolate, basally falcate, ± 2.5 mm, erect-connivent; **Poll** ellipsoid-ovoid, ± 0.35 × 0.2 mm; **Fr** ± 10 - 20 cm × 3 mm.

This is the sister species of *C. nilotica* from which it is separable by the corolla inflation lacking a median constriction, a shining exterior of the corolla, and long vibratile hairs at the tips of the corolla lobes.

C. dichotoma Haworth (Synops. Pl. Succ., 13, 1812). **T:** [lecto – icono]: Link & Otto, Ic. Pl. Select. Hort. Berol., t. 18, 1821. – **Lit:** Bruyns (1986a). **D:** W Canary Islands.

C. dichotoma ssp. **dichotoma** – **D:** Canary Islands (Tenerife, La Palma, Hierro). **I:** Bruyns (1986a). **Fig. VIII.h**
Incl. *Ceropegia aphylla* Link (1820) (*nom. illeg.*, Art. 53.1); **incl.** *Ceropegia hians* Sventenius (1960); **incl.** *Ceropegia hians* var. *striata* Sventenius (1960).

[3,4] Stem and **L** succulents growing to 1 m tall, usually erect, basally branching, partly prostrate-decumbent; stems ± terete, 5 - 20 mm ∅, green respectively whitish-green due to a wax layer; **L** sessile, linear-elliptic, 2 - 120 × 3 - 8 mm, slightly succulent, green, veins paler, margins revolute; **Inf** ± sessile pseudo-umbels, on upper stem sections, (1- to) 2- to 10 (to 15-) flowered; **Ped** 2 - 6 mm; **Sep** triangular, 1 - 2 mm, acute; **Cl** 3 - 4 cm, yellow; **Cl** tube 10 - 16 mm, inside basally ± hairy, ± as broad as the cylindrical to ovoid **Cl** inflation, inflation 4 -

5 mm ⌀, apically scarcely widening; **Cl** lobes 8 - 12 × 1 - 3 mm, yellow, narrowly triangular, apically fused to form a conical to ovoid cage, margins slightly folded back, basally projecting; **Cn** shortly stipitate, 3 - 4 × 2 mm, pale yellow, only at the base fused into a cup; **Ci** lobes ± erect, deeply bifid into appendages spreading laterally, hardly 0.5 mm; **Cs** 2 - 3 mm, linear-cylindrical, erect-connivent; **Poll** narrowly ovoid, ± 0.3 × 0.18 mm; **Fr** fusiform, ± 11 cm × 4 mm, erect.

C. dichotoma is the most succulent *Ceropegia*. It is related to *C. fusca* and probably to *C. rupicola* as well.

C. dichotoma ssp. **krainzii** (Sventenius) Bruyns (Beitr. Biol. Pfl. 60(3): 454, ills. (pp. 454-457), 1986). **T:** Canary Islands, Gomera (*Sventenius* s.n. [not located]). – **D:** Canary Islands (Gomera, Tenerife).

≡ *Ceropegia krainzii* Sventenius (1954); **incl.** *Ceropegia krainziana* hort. (s.a.) (*nom. inval.*, Art. 61.1); **incl.** *Ceropegia ceratophora* Sventenius (1960); **incl.** *Ceropegia chrysantha* Sventenius (1960).

[3,4] Differs from ssp. *dichotoma*: More strongly growing with stems to 1.5 m × 3 cm; **L** smaller, nearly needle-shaped, 20 - 40 × 1.5 - 2 mm; **Inf** with 10 - 50 (-70) **Fl** from gall-like thickened peduncles; **Cl** slender, 3 - 4 mm ⌀, lobes apically brownish; **Cn** stipitate.

C. dimorpha Humbert (Bull. Mus. Nation. Hist. Nat., Sér. 2, 29(6): 503-507, 1957). **T:** Madagascar (*Humbert* 28763 [P]). – **D:** S Madagascar. **I:** FPA 36: t. 1437, 1964.

Incl. *Ceropegia pseudodimorpha* Rauh (1995).

[3] Erect stem succulents, 8 - 20 cm tall; **R** fibrous; stems sparsely branched, 1 - 2 cm ⌀, ± terete-fusiform, olive-green, tessellate by thickened **Tu** arranged in longitudinal rows, **Tu** 5 - 15 × 4 - 7 mm, persistent, spreading, succulent; **L** linear-lanceolate, acute, 20 - 30 × 2 - 5 mm, slightly succulent, caducous, upper face slightly canaliculate, lower face with raised midrib, margin revolute, with curved **Ha**; **Inf** a terminal, erect, branched and much thinner stem section, 1 - 2 mm ⌀, slightly succulent, ± leafy, part-**Inf** with 1 - 3 mm long peduncles, 2- (to 3-) flowered; **Ped** 4 - 10 mm; **Sep** narrowly triangular, 2 - 4 mm; **Cl** 1 - 1.6 cm; **Cl** inflation globose, 7 - 10 × 5 - 8 mm ⌀, whitish-green, veins red-brown, inside rarely blotched with purple, apically narrowed and projecting directly into the **Cl** lobes; **Cl** lobes 8 - 10 (-18) mm, linear, apically connate to form a globose or rarely ovoid broad cage, lamina entirely folded back along the midrib, inside dark brown, margins with red-brown **Ha** usually at the base only; **Cn** ± sessile, whitish-purple, ± 3 × 3 mm, shallowly bowl-shaped, fused at the base only; **Ci** lobes triangular, deeply bifid into triangular appendages, ± 0.5 mm, ciliate; **Cs** 2 × 0.3 mm, linear-acuminate, erect-connivent; **Poll** broadly ellipsoid, ± 0.35 × 0.25 mm; **Fr** narrowly fusiform, ± 5 cm.

Probably most closely related to *C. armandii*, the stems of which are, however, 4-angled and the leaves decussate as found in all dimorphic Ceropegias except *C. dimorpha*.

C. dinteri Schlechter (BJS 51: 153, 1913). **T:** Namibia (*Dinter* 2527 [SAM]). – **D:** Namibia; summer-rainfall regions. **I:** Dyer (1983).

[2] Erect non-twining herbs, 10 - 35 cm tall; **R** tuber 1 - 4 cm ⌀; stems delicate, with short **Int**, ± unbranched, glabrous; **L** sessile, narrowly linear, 12 - 20 cm × 4 mm, margins revolute, midrib prominent on the lower face; **Inf** sessile, 1-flowered, first terminal, later lateral; **Ped** 5 - 8 mm; **Sep** narrowly lanceolate, 2 - 4 mm; **Cl** outside greenish-white, inside greenish, (3-) 4 - 10 cm, straight, glabrous; **Cl** inflation ovoid, ± 5 mm ⌀, merging gradually into the **Cl** tube, tube 3 - 4 mm ⌀, apically widening to 6 - 7 mm ⌀; **Cl** lobes inside whitish, 9 - 13 mm, linear from a triangular base, apically free, erect, ± twisted together, lamina largely folded back along the midrib, laterally with purple clavate **Ha**; **Cn** shallowly cup-shaped, ± sessile; **Ci** lobes triangular, ± erect, bifid into spreading triangular appendages, hairy; **Cs** ± 2 mm, linear, erect-connivent; **Poll** ovoid, 0.35 × 0.25 mm.

C. dinteri is closely allied to, if not even conspecific with, *C. deightonii* from W Africa. However, *C. antennifera*, *C. insignis*, *C. turricula*, and *C. ledermanni* as well as the Indian *C. attenuata* are also very similar.

C. distincta N. E. Brown (BMI 1895: 262, 1895). **T:** Tanzania, "Zanzibar" (*Kirk* 28 [K]). – **D:** Kenya, Tanzania. **I:** Field (1981b). **Fig. X.b**

Incl. *Ceropegia cyrtoidea* Werdermann (1939); **incl.** *Ceropegia brevirostris* P. R. O. Bally & D. V. Field (1981).

[3] Leafy stem succulent climbers to 3 m; **R** fibrous; stems ± terete, 3 - 5 mm ⌀, glaucous-green, slightly verrucose; **L** petiolate, narrowly to broadly ovate, to 7.5 × 3 cm, acute, partly acuminate, basally ± cordate, partly auriculate, slightly fleshy; **Inf** pedunculate, ± umbellate, 3- to 8-flowered with only one **Fl** open at a time; peduncle 1 - 4 cm; **Ped** 4 - 15 mm; **Sep** subulate, acute, 3 - 8 mm, yellowish blotched with red, glabrous; **Cl** 2.5 - 4 cm, whitish-yellowish dotted or blotched with red-brown; **Cl** inflation outside often greenish, 4 - 5 mm ⌀, laterally flattended, basally slightly constricted, inside red-purple, (basally) hairy, merging abruptly into the narrowed **Cl** tube, tube widening into a funnel-shaped mouth of 12 - 17 mm ⌀, inside often purple or striped, glabrous; **Cl** lobes yellowish/greenish marked with (red-) brown reticulation on the broadened base, apically red-brown or green, narrowed in the middle, 9 - 16 mm long, entirely free and fused with the tips only or joined for up to ½ the

length to form a blunt tip, lamina folded back along the midrib, with white cilia towards the apex; **Cn** yellowish with purple, broadly cup-shaped, 4 - 4.5 × 4 mm, basally fused with the **Cl** tube; **Ci** lobes ovate, forming pouches, incised in the middle, margin with erect inwards-bent **Ha**; **Cs** 2.5 - 3.5 mm, linear, erect-connivent, apically slightly spreading, basally hairy; **Poll** ± 0.3 × 0.2 mm; **Fr** narrowly fusiform, ± 8 - 12 cm.

Forming an exceptionally variable species complex together with *C. haygarthii*, *C. lugardiae*, *C. verruculosa*, and *C. volubilis*. The taxonomic position of this group has remained ambiguous since decades.

C. dolichophylla Schlechter (Notes Roy. Bot. Gard. Edinburgh 8: 17, 1913). **T:** China, Yunnan (*Forrest* 4738 [N]). – **D:** China (Yunnan); dense forests. **I:** Tsiang (1939).

Incl. *Ceropegia dolichophylla* var. *brachyloba* Handel-Mazzetti (s.a.); **incl.** *Ceropegia dolichophylla* var. *purpureobarbata* W. W. Smith (1920); **incl.** *Ceropegia profundorum* Handel-Mazzetii (1923); **incl.** *Ceropegia longifolia* ssp. *sinensis* H. Huber (1957).

[?] Twining; very similar to *C. longifolia*, but with glabrous and pale grey stems, narrower leaves and narrowly triangular elongated **Ci** lobes.

The presence of root succulence needs to be corroborated. See also under *C. exigua*.

C. ensifolia Beddome (Madras J. Lit. Sci., ser. 3, 1(1): 52, 1864). **T:** India, Tamil Nadu (*Beddome* s.n. [K]). – **D:** S India (Tamil Nadu). **I:** Beddome (1874: t. 173); Huber (1957).

≡ *Ceropegia ciliata* ssp. *ensifolia* (Beddome) H. Huber (1957); **incl.** *Ceropegia albiflora* Hooker *fil.* (1883).

[2] **R** tuber roundish; stems delicate, twining, branched, ± glabrous; **L** ovate-lanceolate to linear, nearly glabrous; **Inf** many-flowered, peduncle and **Ped** hairy; **Cl** to 4 cm, white or whitish-green; **Cl** tube ± 2 cm, basally inflated; **Cl** lobes 1 - 2 cm, linear, erect, apically fused to form an ovoid cage, lamina slightly folded back along the midrib; **Cn** sessile; **Ci** lobes basally connate, ± triangular and hairy or scarcely differentiated; **Cs** linear-spatulate, erect, straight; **Fr** 12 - 16 cm.

Treated as a subspecies of *C. ciliata* by Huber (1957). In contrast to that taxon, *C. ensifolia* possesses tubers, which is the reason why Ansari (1984) recognized it as a separate species again.

C. evansii McCann (J. Bombay Nat. Hist. Soc. 45: 209, 1945). **T:** India, Maharashtra (*Anonymus* s.n. [not located]). – **D:** C India (Maharashtra). **I:** Ansari (1984).

[2] **R** tuber 2 - 6 × 1 - 3 cm; stems twining to 4 m, glabrous; **L** ± 2 cm petiolate, lamina ovate-lanceolate to ovate, 7 - 14 × 3 - 7 cm, scattered-hairy; **Inf** few-flowered, 4 cm pedunculate, villose; **Ped** 1 - 1.2 cm, glabrous; **Cl** 3 - 4 cm, whitish-green; **Cl** tube 2 - 2.8 cm, outside grey-white, basal ½ purple inside, with globose **Cl** inflation, ± 7 mm ⌀, gradually merging into the **Cl** tube, tube basally ± 3 mm ⌀, apically widening to ± 8 mm, outside blotched with purple; **Cl** lobes 8 - 12 × 8 mm, ovate, ± entirely folded back along the midrib, inside whitish, apically lemon-yellow, ± pubescent, margin dark; **Cn** sessile, basally fused into a cup; **Ci** lobes triangular-ovate, bifid into triangular hairy appendages; **Ci** linear, erect.

See also note under *C. media*.

C. exigua (H. Huber) M. G. Gilbert & P. T. Li (Novon 5(1): 3-4, 1995). **T:** China, Sichuan (*Wilson* 4112 [BM, K, P]). – **D:** China (Sichuan). **I:** Huber (1957).

≡ *Ceropegia longifolia* var. *exigua* H. Huber (1957).

[?] A form of *C. longifolia* with particularly short **Cl** lobes; **R** unknown.

See note under *C. longifolia*.

C. fantastica Sedgwick (J. Indian Bot. 2: 124-125, 1921). **T:** India, Karnataka (*Bell* 4252 [K]). – **D:** India (Karnataka). **I:** Ansari (1984).

[2] **R** tuber depressed-globose; stems twining, glabrous; **L** petiolate, ovate to lanceolate, ciliate; **Inf** 4- to 7-flowered, peduncles very long, ± hairy; **Ped** ± 1 cm, glabrous; **Sep** subulate, 1.5 - 5 cm, overtopping the **Cl**; **Cl** 2 - 4 cm, purple, partially blotched with yellow; **Cl** inflation asymmetrically ovoid-cylindrical, 6 - 11 × 4 - 6 mm, inside glabrous, merging gradually into the ± cylindrical **Cl** tube, tube ± 3 mm ⌀, apically scarcely widening, upper ½ of the inside hairy; **Cl** lobes 5 - 10 × 3 - 4 mm, ovate, ± folded back along the midrib, inside hairy, margin ciliate; **Cn** sessile, basally connate, glabrous; **Ci** lobes triangular-ovate, bifid into narrowly triangular appendages; **Cs** linear, erect, tips ± revolute; **Fr** slender, 9 × 0.3 cm; **Se** 8 × 3 mm.

The calyx of *C. fantastica* is unique for the whole genus: in no other species do the calyx lobes overtop the corolla.

C. filiformis (Burchell) Schlechter (BJS 21(Beiblatt 54): 13, 1896). **T:** RSA, Eastern Cape (*Burchell* 2092 [K]). – **D:** Namibia, RSA (Northern Cape, Western Cape, Eastern Cape). **I:** Bruyns (1985); Bruyns (1995c).

≡ *Systrepha filiforme* Burchell (1822); **incl.** *Ceropegia infundibuliformis* E. Meyer (1837); **incl.** *Ceropegia barbata* R. A. Dyer (1979).

[1] **R** fleshy, clustered, fusiform; stems inconspicuous, erect-twining, 15 - 80 cm, annual, delicate, not or little branched, 1 - 2 mm ⌀, glabrous; **L** shortly petiolate, narrowly linear-lanceolate, to 7 ×

1 cm; **Inf** lateral, ± 7 mm pedunculate, 1- to 3-flowered; **Ped** ± 1 cm; **Sep** narrowly triangular, 3 - 5 mm; **Cl** 5 - 6 cm, entirely glabrous; **Cl** inflation outside grey-green, striped with purple, globose to ellipsoid, with longitudinal furrows due to the prominent veins on the inside, inside white, veins green or with green papillae, to 12 mm ⌀, abruptly merging into the **Cl** tube, tube 3 - 4 mm ⌀, outside grey-green, blotched with purple, apically widening to 7 - 12 mm ⌀; **Cl** lobes lanceolate, 15 - 20 mm, erect, then abruptly bent inwards and upper 1 - 6 mm twisted together, entirely folded back along the midrib, keeled, inside red-brown, white along the middle of the keel, margins black-purple, exuding tiny droplets; **Cn** sessile, fused at the base only, ± 3 mm ⌀, yellowish-white, blotched with purple; **Ci** lobes erect, outside hairy, bilobed into parallel cylindrical appendages, ± 2 mm, apically purple-brown, swollen; **Cs** ± 2 mm, linear, erect-connivent; **Poll** ovoid, 0.33 × 0.24 mm; **Sty** head elongated, overtopping the **Anth**.

A species of unknown relationship. Striking features are the droplets exuded along the margins of the corolla lobes, which may be attractive to the pollinators, and the elongated style head. The inclusion of *C. barbata* is preliminary; Dyer separates it from *C. filiformis* in particular because of the bearded corolla lobes.

C. filipendula K. Schumann (BJS 17: 150-151, 1893). **T** [neo]: Angola (*Damann* 1228 [K]). – **D:** Angola, Malawi, Moçambique, Tanzania, Zaïre, Zambia. **I:** Malaisse & Schaijes (1993).

Incl. *Ceropegia medoensis* N. E. Brown (1895); **incl.** *Ceropegia dichroantha* K. Schumann (1901); **incl.** *Ceropegia peteri* Werdermann (1939); **incl.** *Ceropegia mirabilis* H. Huber (1957); **incl.** *Ceropegia renzii* Stopp (1964).

[1] **R** thickened and fleshy, to 15 × 0.7 cm; stems erect, hairy, not twining, 25 - 45 cm, annual, unbranched; **L** sessile or to 4 mm petiolate, lamina ovate to elliptic, ± 3 - 6 × 1 - 3 cm; **Inf** sessile, 1- (to 3-) flowered, lateral; **Ped** 7 - 10 mm; **Sep** subulate, 6 - 14 mm; **Cl** 2.5 - 15 cm, slightly bent and ascending, pale yellow, becoming brown after 1 day; **Cl** inflation ± globose-ovoid, 5 - 10 mm ⌀, inside purple, merging gradually into the narrowed **Cl** tube, tube 3 - 4 mm ⌀, gradually widening towards the mouth to 6 - 8 mm ⌀; **Cl** lobes 1 - 7 cm × 4 - 7 mm, ovate to lanceolate, basal sinus slightly projecting, apically connate, ususally folded back along the midrib in the middle part only, thus forming a square to ovoid cage with a broad roof, rarely elongated into a linear shape above the roof and fused into a stalk and diverging once more into another small ovoid cage, inside bright yellow, glabrous or ciliate; **Cn** sessile, ± 5 × 3.5 mm, basally fused into a bowl; **Ci** lobes triangular, ± erect, 2.5 mm, bilobed into subulate-triangular appendages, glabrous, papillose or hairy; **Cs** subulate, ± 4 mm, erect-connivent; **Fr** 10 - 12 cm × 4 - 5 mm, ± erect, spreading at 35°; **Se** ± 6 × 3 mm.

A complex of great variety within the *C. umbraticola* group. Huber's *C. mirabilis* deviates from typical *C. filipendula* merely by the tips of the corolla lobes forming a "lantern" – a feature that does not warrant taxonomic consideration in other species such as, e.g., *C. somalensis*.

C. fimbriata E. Meyer (Comm. Pl. Afr. Austr., 194, 1837). **T:** RSA, Eastern Cape (*Drège* 4948 [K]). – **D:** RSA.

C. fimbriata ssp. **connivens** (R. A. Dyer) Bruyns (Bradleya 3: 25-26, ills., 1985). **T:** RSA, Western Cape (*Stayner* s.n. in *KG* 59/490 [NBG, PRE]). – **D:** RSA (Western Cape). **I:** Dyer (1983); Bruyns (1985).

≡ *Ceropegia connivens* R. A. Dyer (1979); **incl.** *Ceropegia connivens* fa. *angustata* R. A. Dyer (1983).

[1,3] Differs from ssp. *fimbriata*: **Cl** 3 - 4 cm; **Cl** inflation globose, ± 7 mm ⌀; **Cl** lobes apically with less broad parachute-like projection, 10 - 12 mm ⌀, basal margin without ascending teeth, margins at the most narrow section of the **Cl** lobes with purple vibratile clavate **Ha**; **Cn** ± sessile, ± 3.5 × 3 mm, fused into a shallow cup; **Ci** lobes ± 1.5 mm; **Cs** 2 mm.

C. fimbriata ssp. **fimbriata** – **D:** RSA (Eastern Cape). **I:** Dyer (1983); Bruyns (1985).

Incl. *Ceropegia estelleana* R. A. Dyer (1964).

[1,3] Stem succulents, climbing and twining to 1 m, little-branched; **R** fleshy, fusiform, clustered; stems terete, 2 - 4 mm ⌀, grey-green, smooth, glabrous; **L** lanceolate to linear-subulate, to 5 × 1 mm, rapidly caducous; **Inf** sessile, 1- to 2-flowered; **Ped** ± 1 cm; **Sep** lanceolate-subulate, 4 - 6 mm; **Cl** 5 - 6 cm, outside glabrous; **Cl** inflation globose to ovoid, 1 - 1.4 × ± 1 cm, outside pale green, often with longitudinal stripes, inside verrucose; **Cl** tube narrowed to ± 3 mm ⌀, mouth gradually widening to 15 mm with slightly involute margins, outside marked with green-brown reticulations; **Cl** lobes broadly lanceolate, 10 - 12 × 9 mm, base narrowed to ± 3 × 2 mm, erect with involute margins, basally divided into small ascending teeth and connate, apically conspicuously widening and fused into a parachute-like roof, 15 - 18 mm ⌀, greenish, often reticulate, slightly depressed along the midrib, upper free margins with purple vibratile clavate **Ha** ± 2 mm; **Cn** shortly stipitate, purple, ± 5 × 4 mm, only basally connate and 2.5 mm ⌀; **Ci** lobes triangular, ± erect, bifid for ¾ into parallel subulate appendages, ± 3 mm; **Cs** 3 - 4 × 0.4 mm, subulate, erect-connivent, tips irregularly bent.

Forming a complex of great variety together with *C. zeyheri*. *C. galeata* from Kenya is also very similar.

C. fimbriata ssp. **geniculata** (R. A. Dyer) Bruyns (Bradleya 3: 25-29, ills., 1985). **T:** RSA, Western Cape (*Joubert* s.n. in *KG* 46/1973 [NBG]). – **D:** RSA (Western Cape). **I:** Dyer (1983).

≡ *Ceropegia geniculata* R. A. Dyer (1979).

[1,3] Differs from ssp. *fimbriata*: **Cl** 4 - 6 cm; **Cl** inflation globose, 8 - 11 × 7 - 8 mm ⌀, merging fairly gradually into the elongated **Cl** tube, tube apically widening to 5 - 6 mm; **Cl** lobes much longer, 1.5 - 2.5 (-3) cm, basally triangular to linear, flat, lower ⅓ abruptly bent and slightly keeled, apically widening to 3 - 7 mm, ± spatulate, inside purple, often finely hairy, basal margins not divided into ascending teeth, forming deep acute slits, margins of the slender middle part of the **Cl** lobes with purple vibratile clavate **Ha**; **Cn** ± sessile, ± 3 × 3 mm; **Ci** lobes 0.5 - 2 mm; **Cs** 1.5 - 2 mm.

C. fimbriifera Beddome (Madras J. Lit. Sci., n.s., 3(1): 53, 1861). **T:** India, Tamil Nadu (*Beddome* s.n. [K]). – **D:** S India (Tamil Nadu). **I:** Beddome (1874: t. 172).

[2] **R** tuber depressed-globose; stems erect, 20 - 30 cm, hairy; **L** sessile, linear, pubescent on the upper face, with hairy veins on the lower face; **Inf** 1- to 4-flowered; peduncle 1 - 2 cm, almost glabrous; **Cl** ± 4.5 cm, straight, greenish with purple longitudinal stripes; **Cl** tube ± cylindrical; **Cl** inflation scarcely broader; **Cl** lobes ± 25 mm, linear-spatulate from a triangular base, inside pale green-purple, margins purple, glabrous, at the base laterally with purple clavate **Ha** 3 - 4 mm long; **Cn** shortly stipitate, basally fused into a cup; **Ci** lobes bilobed into triangular appendages, laterally spreading and fused with the **Cs** base; **Cs** lobes linear-spatulate, erect-connivent; **Fr** narrowly spatulate, 7 - 10 cm.

C. fimbriifera belongs to the group of species showing tubers, an erect short growth and linear leaves. *C. attenuata* and *C. noorjahaniae* are most closely related.

C. floribunda N. E. Brown (FTA 4(1): 460, 1903). **T:** Botswana (*Lugard* 161 [K?]). – **D:** Botswana, Namibia. **I:** Dyer (1983).

[2] **R** tuber 2 - 3 cm ⌀; stems twining, mostly annual, **Int** 3 - 6 cm × 1.5 - 2 mm; **L** 2 - 5 mm petiolate, lamina ± elliptic, partly with a mucro, 15 - 25 × 8 - 15 mm, lower face and margins slightly hairy; **Inf** sessile dense globose pseudo-umbels, many-flowered, several **Fl** opening simultaneously; **Ped** 2 - 4 mm; **Sep** lanceolate-subulate, 2 - 3 mm; **Cl** 2.5 - 3 cm, not or slightly bent, green-yellowish tinged with ± purple; **Cl** inflation globose, ± 6 mm ⌀; **Cl** tube cylindrical, basally ± 2 mm ⌀, inside scattered-hairy; **Cl** lobes ± 5 × 0.8 mm, linear, flat, apically fused to form a bulbiform cage; **Cn** white, ± sessile, shallowly cup-shaped, glabrous; **Ci** forming 5 basal pouches with U-shaped margin, ± 0.5 mm deep, laterally fused with the bulging **Cs** base; **Cs** ± 1.5 × 0.3 mm, linear-spatulate, erect-connivent; **Fr** ± 12 - 16 cm × 6 mm.

The twining sister species of *C. conrathii*.

C. foliosa Bruyns (Notes Roy. Bot. Gard. Edinburgh 45(2): 293-294, ills., 1988). **T:** Yemen (*Wood* 1829 [K]). – **D:** Yemen.

[1] **R** clustered, fleshy, fusiform; stems climbing, terete, 1 - 2 mm ⌀, hairy, relatively short-lived; **L** petiolate, cordate-lanceolate, 3.5 - 9.5 × 1.5 - 5 cm, finely hairy; **Inf** appearing axillary, pedunculate, few-flowered; peduncle 1 - 3 cm; **Ped** 0.8 - 2 cm, finely hairy; **Sep** lanceolate, ± 4 mm, finely hairy; **Cl** 2.5 - 3.5 cm, pale green, white at the base, outside finely hairy; **Cl** inflation ovoid, ± 20 × 6 mm, mouth lined with **Ha** on the inside, merging abruptly into the **Cl** tube, tube 3 mm ⌀, apically widening to 15 mm ⌀; **Cl** lobes green at the base, then whitish marked with pink reticulation and a black patch, apically yellow-green, 10 - 12 mm, ± fully reflexed above a triangular base and 5 mm broad, partially hairy; **Cn** sessile, shallowly cup-shaped; **Ci** lobes broadly triangular, ± 1.5 mm, bifid into short erect triangular appendages, glabrous; **Cs** ± cylindrical, ± 2.8 × 0.3 mm, erect, ventrally in contact almost over the entire length.

This species seems to be close to the African *C. convolvuloides*.

C. fortuita R. A. Dyer (FPSA 24: t. 925 + text, 1944). **T:** RSA, KwaZulu-Natal (*Carnegie* s.n. [PRE 27134]). – **D:** SE RSA, Swaziland. **I:** Dyer (1983). **Fig. X.c**

≡ *Ceropegia africana* ssp. *fortuita* (R. A. Dyer) H. Huber (1957).

[2] **R** tuber ± 3 × 5 cm; stems delicate, twining, to 1 m, annual, 1.5 - 2 mm ⌀; **L** 2 - 3 mm petiolate, lamina slightly succulent, ± ovate-lanceolate, partly with a mucro, ± 25 × 12 - 15 mm, laterally and on the midrib of the lower face hairy; **Inf** 3 - 4 mm pedunculate, usually 2-flowered; **Ped** 3 mm; **Sep** lanceolate-subulate, 3 - 4 mm; **Cl** ± 3 cm, not or slightly bent, outside pale brown-green, inside ± purple; **Cl** inflation ellipsoid, ± 6 × 4 mm, merging gradually into the **Cl** tube, tube cylindrical, basally 1.5 - 2 mm ⌀, inside hairy, apically ± 3 mm ⌀; **Cl** lobes 10 - 15 × 0.5 - 1 mm, linear-lanceolate, erect, apically fused to form a fusiform cage, margins ± revolute, inside purple-brown, finely hairy; **Cn** ± sessile; **Ci** completely connate to form a cup-shaped **C(is)**, ± 1 - 1.5 mm tall, margin denticulate; **Cs** connate with the **C(is)** for ½ of its length, ± 2 × 0.3 mm, linear-spatulate, erect-connivent with revolute tips; **Poll** broadly ovoid, apically strongly acuminate, 0.2 × 0.12 mm.

In habit and corolla structure similar to *C. conrathii*, *C. africana* and *C. linearis*. The fused coronal cup of extraordinary hight opposes an unequivocal placement of *C. fortuita*.

C. fusca Bolle (Bonplandia 9: 51-52, 1861). **T** [lecto]: Canary Islands, Gran Canaria (*Bolle* s.n. [Z]). – **D:** Canary Islands (Gran Canaria, Tenerife). **I:** Bruyns (1986a). **Fig. X.d**

≡ *Ceropegia dichotoma* ssp. *fusca* (Bolle) G. D. Rowley (1998).

[3,4] Stem and **L** succulents growing to 1 m tall, usually erect, basally branched, partly decumbent; stems ± terete, 8 - 15 mm ∅, brown-green to whitish-green; **L** sessile, narrowly linear-elliptic, 25 - 50 × 2 - 4 mm, slightly succulent, reddish-green with paler midrib; **Inf** ± sessile pseudo-umbels from upper stem sections, many-flowered; **Ped** 2 - 4 mm; **Sep** narrowly triangular, 2 - 3 mm, acute; **Cl** 2 - 3 cm, red-brown, with ± dark longitudinal stripes; **Cl** tube 12 - 16 mm, basal ½ hairy inside, as broad as the cylindrical to ovoid **Cl** inflation or only slightly less broad, gradually widening to 4 - 6 mm towards the apex; **Cl** lobes 6 - 11 mm, inside black-purple, sometimes yellow in the middle, triangular, apically connate to form an ovoid to bulbiform cage, margins slightly folded back, projecting at the base; **Cn** pale yellow, ± 3 × 2.5 mm, ± sessile, fused into a cup only at the very base; **Ci** lobes ± erect, bilobed into triangular appendages spreading in lateral directions, 0.5 mm; **Cs** ± 3 - 4.5 mm, linear-cylindrical, erect-connivent; **Poll** narrowly ovate, ± 0.28 × 0.16 mm.

This is the more drought-resistant relative of the similarly Canarian *C. dichotoma*.

C. galeata H. Huber (Mem. Soc. Brot. 12: 200-201, 1957). **T:** Kenya (*Bayliss* 31 [K, G]). – **D:** SE Kenya. **I:** FPA 37: t. 1443, 1965; Archer (1992: XVI).

[1,3] Stem succulents, 0.5 - 1 (-2) m, twining, sparsely branched; **R** fleshy, fusiform, in clustered groups; stems terete, 3 - 4.5 mm ∅, succulent, greygreen, finely striate, rough; **L** narrowly subulate, 3 - 5 mm, rapidly caducous; **Inf** 1 - 2 mm pedunculate, 1- to 3-flowered; **Ped** 8 - 18 mm; **Sep** subulate, ± 6 mm; **Cl** 3.5 - 5.5 cm, straight or slightly bent; **Cl** inflation broadly ovoid, ± 18 - 24 × 14 - 16 mm, outside pale grey-green, finely striate with red-brown, slightly rough; **Cl** tube abruptly narrowed to 3 - 4 mm ∅, towards the apex gradually widening to 12 - 16 mm, margin slightly involute, whitish-green, apically with subdued green-brown-red patches; **Cl** lobes 18 - 25 mm long from a triangular base, lower ¼ narrowed to 1 - 2 mm, erect with involute margins, upper ¾ much broadened and fused into an umbrella-like roof, this 2 - 2.5 cm ∅, depressed along the midribs, outside whitish-green, inside yellowish, margins slightly revolute, lower face dark brown and white, hairy, upper face with purple vibratile clavate **Ha** 3 mm long; **Cn** sessile, white, ± 4 × 4 mm, cup-shaped; **Ci** lobes triangular, ± erect, ± 2 × 1 mm, apically incised; **Cs** ± 2 - 3.5 × 0.4 mm, subulate, erect-connivent; **Fr** 10 - 20 × 0.6 - 0.9 cm; **Se** 7 - 9 × 4 mm, tuft of **Ha** 3 - 4.5 cm.

The 'helmet' formed by the tips of the corolla lobes is significant. Otherwise, *C. galeata* is closely related to the very variable *C. arabica* and *C. fimbriata*.

C. gikyi Rauh & Gerold (Succulentes 21(3): 3-4, 6-7, ills., 1998). **T:** Madagascar, Androy (*Gerold & Robivelo* s.n. in BG Heidelberg 75007 [HEID]). – **D:** SW Madagascar.

[3] Stem succulents; stems decumbent-erect, 4-angled, flattened, 7 - 20 mm thick, thickest around the nodes because of elongate downwards-curved **Tu** (L bases), brown-green, roughly papillose, stipules minute, acute; **L** sessile to shortly petiolate, lamina 4 - 20 × 1 - 4 mm, dark green, succulent, narrowly lanceolate, acute; **Inf** on elongate cylindrical twining stems 2 - 4 mm ∅, peduncles 5 - 20 mm, succulent, with subulate **Bra**, part-**Inf** 2- to 4-flowered, cymose; **Ped** 8 - 12 mm; **Sep** subulate, ± 3 mm; **Cl** 40 - 52 mm, with intensely sweet evil scent; **Cl** inflation pale green, inflated-cylindrical, ± 10 × 8 mm, apically constricted horizontally and merging abruptly into the **Cl** tube, tube inside spotted with red, basally 2.5 mm ∅, apically widening in a funnel-shaped manner to ± 10 mm ∅, outside papillose; **Cl** lobes inside yellow, basally brown-purple (spotted), 20 - 25 mm, apically fused and forming a narrowly ellipsoid cage, slightly contorted to the right, keel to 3 mm broad, inside with white **Ha**, somewhat decurrent into the **Cl** tube, folded back along the midrib, margins with purple vibratile 4 mm long **Ha**; **Cn** whitish-green, sessile, ± 4 × 3 mm, bowl-shaped; **Ci** lobes deeply bifid to the **Cn** margin into subulate appendages, tips glued together, ± 3 × 0.2 mm, ± erect; **Cs** ± 2 × 0.3 mm, subulate, erect-connivent; **Poll** ± ovoid, ± 0.35 × 0.25 mm.

The extremely strong overpoweringly evil scent is very uncommon for this species of Sect. *Dimorpha*.

C. gilgiana Werdermann (BJS 70: 205, ill., 1939). **T** [neo]: Tanzania (*Bullock* 2363 [K]). – **D:** Tanzania, Zambia; moist savannas. **I:** Asklepios 53: 25, 1991.

[1] **R** thickened and fleshy; stems erect, grass-like, 20 - 30 cm, annual, unbranched, glabrous or shortly uniseriately hairy; **L** sessile, ascending, narrowly linear-lanceolate, acute, 40 - 60 × 3 - 9 mm, ± ciliate; **Inf** sessile, 1- (to 2-) flowered, lateral; **Ped** thin, 5 - 10 mm; **Cl** 2.5 - 5 cm, very slender, straight, green to yellow, outside glabrous or slightly hairy; **Cl** inflation globose to ovoid, 3 - 4 mm ∅, merging ± gradually into the cylindrical **Cl** tube, tube 10 - 22 × 1 - 2 mm, apically widening to 3 mm ∅; **Cl** lobes linear-filiform from a narrowly triangular base, folded back along the midrib, 2 - 3.5 cm, vertical, apically fused to form a cylindrical often spiralling structure, ± ciliate; **Cn** ± sessile, whitish-hyaline, basally fused into a cup, glabrous; **Ci** lobes

triangular-ovate, concavely pouch-like, ± erect, 1 mm, apically incised or bifid into triangular appendages spreading in lateral direction; **Cs** ± 2 mm, linear-spatulate, erect.

Related to *C. filipendula* and *C. umbraticola*.

C. haygarthii Schlechter (BJS 38: 46, 1905). **T** [neo]: RSA, Eastern Cape (*Scully* 196 [K]). – **D**: Angola, Moçambique, RSA (Eastern Cape, KwaZulu-Natal, Gauteng, Mpumalanga, Northern Prov.).

≡ *Ceropegia distincta* ssp. *haygarthii* (Schlechter) H. Huber (1957); **incl.** *Ceropegia tristis* Hutchinson (1922).

[3] Leafy stem-succulent climbers to 3 m; **R** fibrous or fleshy; stems ± terete, 3 - 4 mm ⌀, green, covered with a glaucous bloom, glabrous or slightly verrucose; **L** petiolate, narrowly to broadly ovate, 3 - 7 × 2 - 5 cm, acute, partly apiculate; **Inf** with a ± long peduncle, 1- to 3- (to few-) flowered with only 1 **Fl** open at a time; **Ped** 4 - 15 mm; **Sep** subulate, acute, 3 - 8 mm, yellowish, often blotched with red, glabrous; **Cl** 3 - 5 cm, whitish-yellowish stippled or blotched with red-brown; **Cl** inflation outside often greenish, 4 - 5 mm ⌀, basally slightly narrowed, inside red-purple, (basally) hairy, merging abruptly into the **Cl** tube and at the point of merger narrowed to 2 - 3 mm and ± rectangularly bent, tube gradually widening to 14 - 20 mm towards the mouth, inside often purple or striped, glabrous; **Cl** lobes whitish with fine (red-) brown reticulation at the broad base, apically red-brown, narrowed in the middle, 10 - 30 mm, above the broadenend base conspicuously narrowed and elongated into an erect thin stalk (5 - 20 mm), at the tip abruptly widening into a winged globose small head (1 - 5 mm ⌀), head at the tip ususally with simple white **Ha** at the margin; **Cn** yellowish with purple, broadly cup-shaped, 4 - 4.5 × 4 mm, basally connate with the **Cl** tube; **Ci** lobes ovate, forming pouches, incised in the middle, margin with erect-connivent **Ha**; **Cs** 2.5 - 3.5 mm, linear, erect-connivent, apically usually divaricate, basally hairy; **Poll** ± 0.3 × 0.2 mm; **Fr** narrowly subulate, ± 8 - 12 cm.

Probaby nothing more than a S variety of *C. distincta* (see note there). Furthermore, *C. haygarthii* is closely related to *C. somalensis* but is always separable on account of the glabrous outside of the corolla.

C. hermannii Rauh & M. Teissier (Succulentes 19(4): 22-25, ills., 1996). **T**: Madagascar, Toliara (*Petignat* 455 [HEID]). – **D**: S Madagascar.

[2,4] **R** tuber ± globose, 2 - 4 cm ⌀, stems twining 3 - 4 m tall, annual, terete, 1.5 - 2 mm ⌀ (hypocotyl and nodes to 5 mm ⌀), dark green, often slightly rough; **L** petiole 8 - 15 mm, upper face canaliculate, lamina broadly ovate, obtuse and mucronate to lanceolate, acute, 1 - 2 × 0.5 - 1 cm, ± succulent, green, margin and petiole often reddish, ± hairy; **Inf** with fleshy peduncle 1 - 2 cm long, with **Bra**, 2- to 4- (to 7-) flowered, often 2 **Fl** simultaneously open; **Ped** 4 - 8 mm; **Sep** narrowly triangular, 2 - 5 mm; **Cl** 2 - 2.6 cm, inside and outside whitish with fine red-brown blotches or stripes; **Cl** inflation 4 - 6 × 4 - 5 mm, cylindrical, apically abruptly horizontally narrowed, inside ± hairy; **Cl** tube 2 - 3 mm ⌀, then gradually widening to 7 - 8 mm ⌀, inside glabrous; **Cl** lobes 6 - 9 mm, triangular-linear from a broadly triangular base, only slightly divaricate, apically connate, lamina halfways folded back along the midrib, basally broadly keeled and white, transition to the narrow section with a purple hairy belt, then green, or apically entirely black-purple, glabrous; **Cn** sessile, fused into a short bowl, 4 × 4 mm, greenish, outside hairy; **Ci** lobes ascending, ± bilobed into linear appendages, ± 15 mm, apically fused, with long and scattered **Ha**; **Cs** ± 2 mm, linear-spatulate, erect-connivent; **Poll** ovoid, 0.13 × 0.28 mm; **Se** 5 × 3 mm.

A species well-characterized by the tubers and the slender subsucculent stems. The slightly succulent leaves, the fleshy and many-flowered inflorescences and white as the dominant colour of the corolla are suggestive of a relationship to the *C. albisepta* complex.

C. hirsuta Wight & Arnott (Contr. Bot. India, 30, 1834). **T**: India, Tamil Nadu (*Wight* s.n. [K]). – **D**: India (Tamil Nadu). **I**: Ansari (1984). **Fig. X.e**

Incl. *Ceropegia jacquemontiana* Decaisne (1844) ≡ *Ceropegia hirsuta* var. *jacquemontiana* (Decaisne) Hooker *fil.* (1883); **incl.** *Ceropegia ophiocephala* Dalziel (1850) ≡ *Ceropegia hirsuta* var. *ophiocephala* (Dalziel) Hooker *fil.* (1883); **incl.** *Ceropegia hirsuta* var. *stenophylla* Hooker *fil.* (1883); **incl.** *Ceropegia hispida* Blatter & McCann (1931).

[2] **R** tubers flattened; stems twining, hispid; **L** ± 1 cm petiolate, lamina linear-lanceolate to ovate, 4 - 7 × 2 - 3.5 cm, paper-thin, acute; **Inf** few- to many-flowered, with a peduncle ± 2 cm long; **Ped** 5 - 10 mm, hairy as the peduncle; **Sep** 6 - 8 × 2 mm; **Cl** 2.8 - 5.8 cm, yellowish to greenish with purple stripes or blotches in the upper ½; **Cl** tube 2 - 4 cm, basal ¼ - ⅓ broadly conically inflated to 5 - 8 × 6 - 9 mm, inside with a ring of **Ha** in the middle part; **Cl** tube gradually widening towards the mouth; **Cl** lobes 8 - 18 mm, triangular-ovate to obovate with acute tips, fused, lamina folded back along the midrib, inside hairy, margins ± purple, hairy; **Cn** sessile, shallowly cup-shaped; **Ci** ± rectangular, emarginate in the middle and each tapering laterally into triangular hairy appendages; **Cs** linear-subulate, ± 2.5 mm, erect-connivent, tips revolute; **Poll** ovoid, ± 0.4 mm long; **Fr** erect.

One of the more common and fairly variable species in India. Presumably related to *C. evansii* and *C. fantastica*.

C. hofstaetteri Rauh (Trop. subtrop. Pfl.-welt 85: 17-20, ills., 1993). **T**: Madagascar, Mahajanga

(*Hofstätter* s.n. in *BG Heidelberg* 73340 [HEID]). – **D:** S Madagascar.

[3] Stem-succulents, creeping-decumbent; stems 4-angled, flattened, 1 - 2 cm ∅, thickest at the nodes due to broadened **Tu**, brown- to grey-green, red-verrucose; **L** ± distinctly petiolate, dark green, succulent, narrowly ovate to spatulate, acuminate, 1 - 3 × 0.5 - 1 cm, usually mucronate, margin occasionally undulate, with curved **Ha**, petiole canaliculate on the upper face; **Inf** with fleshy peduncles 0.8 - 1 cm long, borne only on twining thin terete stems, 1- to 4-flowered; **Ped** ± 1 cm; **Sep** narrowly triangular, 3 mm; **Cl** ± 3 cm, pale yellow to grey-green, glabrous; **Cl** inflation depressed-globose, 5 - 6 × 5 - 7 mm, apically almost horizontally narrowed, merging abruptly into the **Cl** tube, tube 2 - 3 mm ∅, widening to ± 7 mm towards the mouth; **Cl** lobes 10 - 15 × 3 - 5 mm, narrowly triangular, basally keeled, ± folded back along the midrib, basal margins projecting, apically fused to form an acute-tipped ovoid cage, inside of lower ⅓ whitish-green, upper ⅔ yellow-green; **Cn** ± sessile, greenish-white, ± 2.5 × 2.5 mm, shallowly bowl-shaped; **Ci** lobes bilobed into subulate appendages down to the margin of the **Cn**, ± 1.5 mm, ± erect, apically connivent; **Cs** ± 2 × 0.3 mm, erect-connivent.

Within the group of dimorphic ceropegias (Sect. *Dimorpha*) very close to the complex of *C. bosseri* and *C. petignatii*. The completely glabrous flowers are a striking feature.

C. hookeri C. B. Clarke *ex* Hooker *fil.* (Fl. Brit. India 4: 73, 1883). **T:** India, Sikkim (*Hooker* s.n. [K, BM, FT, G, GH, W]). – **D:** India (Sikkim), Nepal, Tibet, Bhutan; highland around ± 3000 m. **I:** Hart (1988).

Incl. *Ceropegia hookeri* var. *mollis* H. Huber (1957).

[1] **R** fleshy, fusiform; stems erect, prostrate or twining, fairly stiff, with **Ha** arranged in 1 or 2 rows; **L** 1 - 15 mm petiolate, lamina ovate-acuminate, rarely reniform, 1 - 10 cm, upper face hairy; **Inf** sessile or with a short hairy peduncle, 1- to 7-flowered; **Ped** 3 - 8 mm, ± glabrous; **Sep** lanceolate, 1.5 mm; **Cl** 1.5 - 3 cm, green or also ± purple, ± straight; **Cl** tube elliptic in overall shape, 7 - 20 × 6 mm, or gradually narrowed from the basal **Cl** inflation (± 6 mm ∅) to 3 - 5 mm at the mouth, inside hairy; **Cl** lobes 3 - 15 mm, ovate to broadly lanceolate-spatulate, only slightly folded back along the midrib, apically obtuse, fused into a depressed-globose cage; **Cn** shortly stipitate, basally fused into a very shallow bowl; **Ci** lobes pouch-like, short, concave, obtuse, laterally fused wing-like with the **Cs** base, laterally hairy; **Cs** narrowly linear-cylindrical, erect-connivent; **Poll** D-shaped, germination crest triangular.

This species shows an erect or prostrate habit in nature (compare the closely related *C. meleagris*), but is always erect in cultivation.

C. humbertii H. Huber (Mitt. Bot. Staatssamml. München 12: 72, 1955). **T:** Madagascar (*Humbert* 18542 [P]). – **D:** N Madagascar; moist and shady forests, rare. **Fig. IX.h**

[2] **R** tuber globose, 1 - 3 cm ∅; stems twining, to 1 m tall, delicate, annual, 1 - 2 mm ∅, pale green often suffused with red, prostrate stems sometimes with stem tubers; **L** petiole 1 - 3 mm, upper face canaliculate, sparsely villose, lamina ovate-lanceolate to elliptic, base ± obtuse, 3 - 5 × 2 - 3.5 cm, glabrous, faintly succulent, upper face fresh green, venation impressed, lower face pale green, venation prominent; **Inf** 2 - 8 mm pedunculate, 1- to 4-flowered; **Ped** 5 - 10 mm; **Sep** subulate, ± 1.5 mm; **Cl** 2.8 - 4 cm, slightly curved; **Cl** inflation globose-ovoid, 6 - 10 × 5 - 7 mm, pale green; **Cl** tube basally narrowed to 2 - 3 mm, inside finely hairy, mouth widening to ± 5 mm; **Cl** lobes basally pale yellow blotched with red-brown, apically red-brown, 5 - 10 mm, with the tips joined, triangular-ovate, broadly keeled, erect, lamina completely folded back along the midrib, white-ciliate, lower ½ pale green, upper ½ dark purple; **Cn** sessile, white, ± 4 × 3.5 mm, basally fused into a shallow bowl; **Ci** lobes bifid into narrow triangular appendages, 2 mm, ascending, tips glued togther; **Cs** ± 3 mm, strap-shaped, basally cylindrical, erect, parallel-connivent, apically papillose; **Poll** broadly ovoid, 0.35 × 0.25 mm.

Most readily separable by its distinct leaf morphology, and presumably most closely related to *C. madagascariensis*.

C. illegitima H. Huber (Mitt. Bot. Staatssamml. München 12: 72, 1955). **T:** Zaïre (*Schmitz* 2116 [BR]). – **D:** Zaïre, Central African Republic, Malawi; grasslands. **I:** Malaisse & Schaijes (1993).

[1] **R** 6 - 9, fleshy, shortly fusiform, ± 9 × 0.6 cm; stems erect, delicate and inconspicuous, to 60 cm, 1 - 2 mm ∅, glabrous, unbranched; **L** sessile, narrowly lanceolate to needle-shaped, 3 - 5 (-20) × ± 1 mm, acute, caducous; **Inf** pseudo-umbellate, on ascending peduncles 5 - 7 cm long; **Ped** 1 - 4 mm; **Sep** narrowly triangular, 2 mm, reddish; **Cl** 1.8 - 2.5 cm, ± straight, pink to salmon; **Cl** inflation inversely bulbous, ± 5 mm ∅, glabrous; **Cl** tube ± cylindrical, 3 mm ∅, mouth a little broader; **Cl** lobes linear from a narrowly triangular base, 5 - 9 × 0.5 mm, flat, glabrous or whitish-ciliate, ± erect, apically fused and inside yellowish-white, diverging into an ellipsoid to bulbous cage; **Cn** sessile, dish-shaped; **Ci** short, edged with 2 short teeth, sometimes inside slightly hairy; **Cs** linear, apically slightly spatulate, straight, ascending, only tips meeting apically; **Fr** ± 5 cm × 3.5 mm; **Se** ± 11 × 3 mm, tuft of **Ha** 1.2 cm.

The needle-shaped leaves are a typical feature of this species, which is a native of grasslands. The pink to salmon flowers render the plants unmistakable.

C. imbricata E. A. Bruce & P. R. O. Bally (KB 1950: 372, 1951). **T:** Kenya (*Copley* s.n. in *Bally* S48 [K]). – **D:** Kenya. **I:** Bradleya 13: 35-37, 1995, as *C. intracolor*. **Fig. X.f**

Incl. *Ceropegia intracolor* L. E. Newton (1995).

[2] **R** tuber globose to flattened, to 8 cm ⌀; stems 10 - 50 cm, usually twining, little-branched, 1 - 2 mm ⌀, short-lived, glabrous; **L** linear-lanceolate, 1 - 6.5 × 0.5 - 1.2 cm, ± sessile or with petioles to 3 mm long, acute; **Inf** 1- to 4-flowered, peduncle 1 - 2 mm long, slightly pointing downwards; **Ped** 3 - 8 mm; **Sep** ± 2 mm, straight but pointing downwards; **Cl** 1.6 - 2.5 cm, pale green ± striped and blotched with purple; **Cl** inflation 5 - 6 mm ⌀, globose, inside basally yellowish, remainder purple, slightly verrucose, abruptly bent for ± 90° and merging into the ascending **Cl** tube, tube 2 - 3 mm ⌀, towards the mouth gradually widening to 6 - 7 mm, inside purple, with whitish **Ha**; **Cl** lobes outside greenish, inside purple, basally also whitish with purple reticulate markings, 5 - 9 mm, lamina ± linear and folded back along the midrib, with purple cilia; **Cn** cup-shaped, shortly stipitate; **Ci** lobes triangular, ± erect, 1 - 2 × 2 - 3 mm, emarginate or bifid into triangular appendages; **Ci** 2.3 - 3 × 0.2 mm, linear-subulate, erect-connivent.

C. imbricata is closely related to *C. bulbosa* and *C. linearis*, but readily separable by its curved corolla and raised corona.

C. inflata Hochstetter *ex* Chiovenda (Ann. Bot. (Roma) 10: 396, 1912). **T:** Ethiopia (*Schimper* 2053 [FT, P, S]). – **D:** Ethiopia; only known from the type.

Incl. *Ceropegia ventricosa* Hochstetter *ex* Chiovenda (1912) (*nom. inval.*, Art. 34.1c).

[1] **R** fleshy, fusiform; stems twining, 1 - 2 mm ⌀; **L** 5 - 12 × 1 - 2 cm, broadly lanceolate, acute, basally cuneate, hairy; **Inf** 1 - 3 cm pedunculate, 1- to 3-flowered; **Cl** 4 - 5 cm, bottle-shaped, ± straight, greenish; **Cl** lobes 5 - 10 mm, linear, purple.

Insufficiently known and of dubious status but presumably closely related to the very variable *C. meyeri-johannis*.

C. inornata P. R. O. Bally *ex* Masinde (KB 53(4): 949-953, ills., 1998). **T:** Kenya, Coast Prov. (*Masinde & al.* 838 [EA, K, MSUN]). – **D:** Ethiopia, Kenya, N Tanzania.

[2,4] **R** tuber depressed-globose, 7 - 18 (-30) × 18 - 25 (-80) mm, smooth, light brown; stems to 2 m, sparsely branched, 1 - 2 mm ⌀, wiry, glabrous or hairy, nodes often swollen to form globose stem tubers; **L** stiffly succulent, lamina 2 - 6 × 0.5 - 2.3 (-4) cm, cordate to elliptic-ovate, acuminate, apiculate, basally ± rounded, glabrous or whitish-pubescent, petiole 2 - 6 × 0.5 - 1.25 mm; **Inf** pseudo-umbellate, up to 4 **Fl** simultaneously open; peduncle 2 - 9 mm, ± glabrous, annual; **Bra** subulate, 1.5 × 0.5 mm; **Ped** 2.5 - 8 (-15) × 0.4 - 1 mm; **Sep** subulate, 0.5 - 2.5 × 0.25 - 0.5 mm; **Cl** 1.5 - 3.2 cm, greenish or greenish-yellow / purple; **Cl** inflation globular, 4 - 9 × 3 - 8 mm; **Cl** tube narrowing abruptly above the inflation, 1 - 2.5 mm ⌀, geniculate at 30° - 90°, then widening to 2 - 5 mm ⌀, outside glabrous, inside with a few white **Ha** at the base; **Cl** lobes ± linear, 10 - 13 × 1 - 2 mm, apically connate to form an ellipsoid cage-like structure; **Cn** white, cup-shaped, ± 4 × 3 mm; **Ci** suberectly radiating to form ovate concave pouches, ± 1 × 1 mm; **Cs** lobes ± 2 × 0.6 mm, strap-shaped, connivent-erect, apically spreading; **Poll** ovoid, 0.27 × 0.2 mm; **Fr** 7 - 16 cm × 2 - 3 mm, light green, streaked with maroon; **Se** 6 × 1.5 mm, brown, tuft of **Ha** 25 (-40) mm.

C. insignis R. A. Dyer (FPSA 1943: t. 902 + text, 1943). **T:** RSA, North-West Prov. (*Louw* 812 [PRE]). – **D:** Uganda, RSA (North-West Prov.). **I:** Dyer (1983).

Incl. *Ceropegia campanulata* var. *pulchella* H. Huber (1957).

[2] **R** tuber flattened, 4 - 5 cm ⌀; stems erect, not twining, herbaceous, 20 - 30 cm, annual, ± unbranched, scatteredly hairy; **L** ± sessile, linear, acute, 12 - 14 cm × 7 - 8 mm, scatteredly hairy, midrib prominent on the lower face; **Inf** sessile, terminal or lateral, 1-flowered; **Ped** ± 5 mm; **Sep** narrowly lanceolate, 6 mm; **Cl** ± red-brown, 10 - 13 cm, straight; **Cl** inflation ovoid, 9 - 11 × 7 - 9 mm, merging gradually into the **Cl** tube, tube 4 mm ⌀, widening apically to 14 - 17 mm ⌀, inside with purple **Ha**; **Cl** lobes erect, apically fused, ± twisted together, linear from a triangular base, 4 - 6 cm, margins folded back; **Cn** sessile, 5-rayed-bowl-shaped, ± 5 - 6 × 5 mm; **Ci** horizontally ascending, triangular, ± 2 mm, apically deeply incised, upper face hairy; **Cs** lobes ± 5 mm, linear-filiform, erect-connivent, apically strongly revolute.

Closely related to *C. campanulata* and the *C. dinteri* species group.

C. jainii Ansari & Kulkarni (Bull. Bot. Surv. India 22(1-4): 221-222, 1982). **T:** India, Maharashtra (*Kulkarni* 121885 [CAL, BLAT, BSI, K]). – **D:** C India (Maharashtra). **I:** Ansari (1984).

Incl. *Ceropegia prainii* Repert. Pl. Succ. (1982) (*nom. inval.*, Art. 61.1c).

[2] **R** tuber flattened; stems erect, not twining, only 6 - 10 cm tall, unbranched, scatteredly hairy, becoming glabrous with age; **L** elliptic, 2 - 4 cm, ± sessile, upper face and margins hairy, green, lower face pale green and glabrous except for the veins; **Inf** 1-flowered, with a short peduncle; **Ped** 4 - 6 mm, hairy; **Sep** 2.5 mm; **Cl** to 2 cm; **Cl** lobes nearly as long as the **Cl** tube, purple except for the green base of the **Cl** inflation, inflation ± 5 mm ⌀, glabrous, merging gradually into the **Cl** tube, tube ± 3 mm ⌀, gradually widening towards the mouth, inside with red-violet longitudinal stripes on pale green background; **Cl** lobes purple, 9 - 10 mm, ±

linear, forming a broadly ovoid cage, largely folded back, inside with white rigid **Ha** at the base; **Cn** cup-shaped, sessile; **Ci** lobes triangular, emerging from the upper rim of the coronal cup, ± erect, 1.5 mm, deeply bifid into triangular hairy appendages; **Cs** ± 2.5 mm, linear to slightly spatulate, erect-connivent.

A very delicate plant. Closely related to *C. pusilla*.

C. johnsonii N. E. Brown (FTA 4(1): 451, 1903). **T:** Ghana (*Johnson* 768 [K]). – **D:** Ghana, Congo, Liberia, Uganda, Zaïre. **I:** Huber (1957).

Incl. *Ceropegia lujai* De Wildeman (1920).

[3] **R** fibrous; stems perennial, twining, terete, robust and slightly succulent, 2.5 - 4 mm ⌀, glaucous-green, glabrous; **L** petiole 1 - 2 cm, shortly pubescent, lamina 3 - 8 × 2 - 5 cm, ± ovate, obtuse or acute, mucronate, basally cordate, ± scatteredly pubescent; **Inf** 3- to 6-flowered, pseudo-umbellate, peduncle 1 - 2.5 cm, shortly hairy; **Ped** 7 - 15 mm, shortly hairy; **Sep** subulate, 4 - 5 mm, slightly hairy, tips revolute; **Cl** 3.5 - 5 cm, bent, outside slightly pubescent; **Cl** inflation ovoid, ± 5 mm ⌀, whitish, inside ± purple and hairy at the base, **Cl** tube then narrowed to ± 2 mm ⌀, towards the mouth gradually widening to 12 mm ⌀, whitish blotched with red-brown, inside greenish, pubescent; **Cl** lobes 1 - 2 cm, ovate, lamina folded back along the midrib, basal margins projecting, apically fused to form a broadly ovoid cage, whitish with a green base, apically with purple venation; **Cn** sessile, fused into a cup, ± 4 × 3 mm; **Ci** lobes triangular-ovate, concave, erect, apically slightly incised, margins ciliate; **Cs** ± 3 mm, linear-subulate, erect-connivent, tips revolute; **Fr** 20 - 28 cm × 4 mm, beaked; **Se** 11 × 4 mm.

Allied to *C. aristolochioides*.

C. juncea Roxburgh (Pl. Coast Coromandel 1: 12, t. 10, 1795). **T:** India (*Roxburgh* s.n. [icono: l.c. t. 10]). – **D:** C India, Sri Lanka.

[3] **R** fibrous or thickened; stems persistent, succulent, climbing and twining, to 6 m, glabrous, terete, 3 - 4 mm ⌀, green, often suffused with reddish, glabrous but finely striate; **L** ± sessile or to 1 cm petiolate, lamina lanceolate-elliptic, 4 - 10 × 1 - 2 mm, acute, slightly succulent, caducous; **Inf** with a ± short peduncle, few- to many-flowered; **Ped** 4 - 10 mm; **Sep** ± 4 mm; **Cl** 4 - 5.5 cm, whitish-green spotted with red-brown or ± brown-purple; **Cl** inflation ± 5 mm ⌀, globose to ovoid, merging ± gradually into the **Cl** tube, tube ± 3 mm at the base, then gradually widening to 12 - 14 mm ⌀, inside hairy at the base; **Cl** lobes triangular, keeled, apically also linear, 15 - 25 mm, lamina entirely folded back along the midrib, apically joined to form a cage being triangular in side view, basal ½ white, upper ½ greenish or purple, inside finely hairy; **Cn** sessile, shallowly cup-shaped; **Ci** lobes ± lanceolate-ovate, concave, ascending, pouch-like, ± 1.5 mm, apically incised, slightly ciliate; **Cs** linear-cylindrical, ± 4 mm, erect, tips involute or revolute, outside ± hairy at the base; **Poll** ovoid.

On account of its nearly leafless succulent stems, *C. juncea* hardly differs from *Cynanchum gerrardii* or *C. insigne*. The wide-ranging species makes an attractive plant in cultivation.

C. keniensis Masinde (KB 54(2): 477-481, ills., 1999). **T:** Kenya, Rift Valley Prov. (*Harvey* 1374 [EA, K]). – **D:** Kenya.

[1] **R** clustered, fusiform, fleshy; stems erect, to 40 cm, branched, 1 - 2 mm ⌀, hairy; **L** sessile, lamina narrowly elliptic-linear, 3 - 3.3 cm × 1.5 - 3 mm, upper face glabrous, margins and midrib of the lower face with white **Ha**; **Inf** sessile, pseudo-umbellate, 1- to 2-flowered; **Bra** subulate, 1 mm; **Ped** 8 - 11 mm, glabrous; **Sep** subulate, 2 × 0.4 mm; **Cl** greenish-purple, erect, slightly bent, 1.2 - 1.7 cm; **Cl** inflation ovoid, 4.5 × 3.5 mm; **Cl** tube narrowing gradually to 1.5 mm ⌀, mouth widening to 2.5 mm ⌀, glabrous; **Cl** lobes greenish, linear, 6 × 4 mm, inner face hairy except the glabrous base, apically connate to form an ellipsoid cage-like structure, ± 3.5 mm ⌀; **Cn** white, shortly stipitate, cup-shaped, ± 3 × 2 mm; **Ci** erectly radiating outwards to form rectangular concave pockets, ± 1 × 0.5 mm, ciliate; **Cs** lobes 2 × 0.4 mm, spatulate, connivent-erect; **Poll** ovoid-subrectangular, 0.23 × 0.19 mm; **Fr** 8 - 9 × 0.3 cm; **Se** 6 - 7 × 3 mm, tuft of **Ha** 7 mm.

As to flower morphology, *C. keniensis* is similar to *C. bulbosa*, but the latter possesses root tubers.

C. kituloensis Masinde & F. Albers (BJS 122(2): 161-167, ills., 2000). **T:** Tanzania (*Leedal* 5293 [K]). – **D:** S Tanzania.

[2] **R** tuber flattened-globose, ± 10 × 8 mm, bark smooth; stems erect, 70 - 180 × 1.5 mm, rarely branched, biseriately hairy; **L** 1 - 2 mm petiolate, ovate-lanceolate to narrowly elliptic, acute, (8-) 12 - 33 × 3 - 9 mm, ± pubescent; **Inf** 1-flowerred; **Ped** 5 - 11 × 0.8 mm, pubescent; **Sep** narrowly lanceolate, ± 5 mm, pubescent; **Cl** greenish, 26 - 38 mm, erect, slightly bent, glabrous; **Cl** inflation ovoid, ± 7 × 6 mm; **Cl** tube narrowing gradually to 2.5 mm ⌀, mouth widening to 3.5 mm; **Cl** lobes greenish with purplish tips, linear, 10 - 22 × 1 mm, margins recurved, apically connate to form an ellipsoid cage-like structure; **Cn** shortly stipitate, cup-shaped, ± 3 × 3 mm, glabrous; **Ci** as horizontally radiating concave pouches, ± 1.4 × 1.4 mm; **Cs** lobes 2 × 0.3 mm, spatulate, erect-connivent, apically diverging; **Poll** broadly elliptic, ± 0.28 × 0.2 mm.

C. kituloensis is the Tanzanian sister taxon of *C. achtenii*, distinguished only by the biseriately hairy stems and the glabrous corolla, hence rather representing a variety.

C. konasita Masinde (CSJA 71(3): 146-149, ills.,

1999). **T:** Kenya, Coast Prov. (*Masinde* 819 [EA, K, MSUN]). – **D:** S Kenya.

[2,3] **R** tubers globose to ovoid, often clustered, bark smooth; stems dark green, erect, twining, to 2 m, sparsely branched, 6-angled, 1.5 - 2.5 mm ∅, smooth; **L** sessile, linear, acute, 10 - 25 × 1 - 2 mm, subsucculent, glabrous, deciduous; **Inf** 4 - 25 mm pedunculate, pseudo-umbellate, 1- to 6-flowered; **Bra** subulate, 1 mm; **Ped** (2-) 5 - 9 mm, glabrous; **Sep** subulate, 1 - 2 × 0.5 - 0.7 mm, glabrous; **Cl** whitish to greenish with purple dots or longitudinal streaks, 25 - 32 mm, erect, bent at an angle of 45 - 85°; **Cl** inflation invertedly funnel-shaped, ± 7 × 6 mm; **Cl** tube narrowing gradually to ± 2 mm ∅, mouth widening to 3.5 - 6 mm, glabrous; **Cl** lobes in the apical ⅔ ± dark purple, with purple vibratile 1 mm long **Ha**, lanceolate from a triangular base, lamina in the middle ± folded back along the midrib, 6.5 - 10 × 1 - 2 mm, apically connate to form a broadly ellipsoid cage; **Cn** white, glabrous, stipitate, dish-shaped, ± 3 × 3 mm; **Ci** horizontally radiating rectangular concave pouches, ± 1 × 1 mm; **Cs** lobes 2 × 0.25 mm, cylindrical, erect-connivent; **Poll** ovoid, 0.26 × 0.15 mm; **Fr** 6 - 10 cm × 1 - 3 mm; **Se** dark brown, 6 - 8 × 2.5 - 3 mm, tuft of **Ha** ± 3 cm.

A member of the *C. linearis* complex. The succulent 6-angled stems, the reduced leaves and the bulbous basal inflation of the corolla tube characterize this as one of the most easily recognizable species with great potential for cultivation.

C. kundelunguensis Malaisse (Bull. Jard. Bot. Belg. 54(1-2): 217-219, ills., 1984). **T:** Zaïre (*Malaisse* 11369 [BR]). – **D:** Zaïre.

[1] **R** fleshy, thickened, irregularly fusiform, to 8 × 0.5 cm; stems erect, hairy, not twining, 20 - 25 cm, probably annual, unbranched; **L** 3 - 5 mm petiolate, lamina ovate, ± 4 × 2 cm; **Inf** sessile, 1-flowered, lateral; **Ped** 2 cm; **Sep** subulate, 6 mm; **Cl** 6 - 7 cm, erect, ± straight, outside yellowish-green, shortly hairy; **Cl** inflation ± globose, ± 1 cm ∅, purple, merging fairly abruptly into the **Cl** tube, tube ± 3 mm ∅ at the base, towards apex widening to 5 mm ∅; **Cl** lobes narrowly triangular, ± 25 mm, apically connate into an ovoid cage, base flat, remainder folded back along the midrib, greenish, upper ⅓ purple; **Cn** shortly stipitate, 6 mm ∅, basally fused into a shallow dish; **Ci** lobes triangularly pouch-like, semi-erect, ± 3 mm, deeply bifid into subulate-triangular appendages; **Cs** linear, ± 4 mm, erect-connivent.

Separable from *C. umbraticola* merely on account of the corona lacking hairs and the long lobes of the staminal corona, and resembling the S African *C. rudatisii* in habit.

C. lawii Hooker *fil.* (Fl. Brit. India 4: 67, 1883). **T:** India, Maharashtra (*Stocks & Law* s.n. [K, BM, FT, G, GH, P, W]). – **D:** C India. **I:** Ansari (1984).

Incl. *Ceropegia panchganiensis* Blatter & McCann (1933); **incl.** *Ceropegia rollae* Hemadri (1969); **incl.** *Ceropegia sahyadrica* Ansari & Kulkarni (1971); **incl.** *Ceropegia maccannii* Ansari (1982).

[2] **R** tuber ± globose, 2 - 3 cm ∅, edible; stems erect, not twining, herbaceous, 25 - 100 cm, 1 - 3 mm ∅, furnished with short globose **Ha**; **L** 0.3 - 2 cm petiolate, lamina broadly ovate to lanceolate, ± 2.5 - 12 × 2 - 6 cm; **Inf** long-pedunculate, lateral and terminal, many-flowered; **Ped** 4 - 15 mm; **Sep** subulate, 3 - 6 mm; **Cl** 1.7 - 5.5 cm, ± straight, outside whitish to faintly pink, turning to lilac-purple at the base; **Cl** inflation narrowly to broadly ovoid, 5 - 10 × 4 - 7 mm, inside basally with white **Ha**, merging gradually into the ± narrowed **Cl** tube, tube 2 - 4 mm ∅, apically widening to 3 - 8 mm, inside greenish to purple, also striped; **Cl** lobes linear-elliptic to broadly lanceolate, 2 - 13 × 2 - 7 mm, apically fused into a broadly ovoid scarcely opened cage, lamina folded back only marginally-basally, inside white, greenish or yellowish, glabrous; **Cn** ± stipitate, 4 - 6 × 2.5 mm, basally fused into a bowl, hairy; **Ci** lobes scarcely elaborated, yellowish with purple, ± entire or incised, hairy; **Cs** linear-spatulate, 2 - 4 mm, erect-connivent, yellow tipped with white, dorsally hairy, rarely glabrous; **Poll** ovoid, ± 0.3 × 0.2 mm; **Fr** narrowly fusiform, ± 6 - 15 cm × 5 mm, divaricate; **Se** ± 6 × 2.5 mm.

Ansari (1984) recognizes all 4 taxa of this complex (here treated as synonyms) as separate species. However, all known collections are native to Maharashtra, are flowering simultaneously from July to September and show the same general floral morphology including the corona (though the quantitative differences may be of great magnitude). Furthermore, all specimens exhibit the highly significant feature of the inflation being basally hairy inside.

C. ledermannii Schlechter (BJS 51: 154, 1913). **T:** Cameroon (*Ledermann* 4237 [B?]). – **D:** Cameroon, Nigeria. **I:** Hutchinson (1921).

[2] **R** tuber flattened, 3 - 6 cm ∅; stems erect, not twining, herbaceous, 20 - 30 cm, annual, with short **Int**, ± unbranched, scatteredly hairy; **L** sessile, narrowly elliptic to linear, acute, 3 - 8.5 cm × 2 - 7 mm, scatteredly hairy, margins revolute; **Inf** sessile, lateral, 1-flowered; **Ped** 1 - 2 cm, hairy; **Sep** subulate, 4 mm, hairy; **Cl** outside whitish, mottled with purple, 8 - 15 cm, straight; **Cl** inflation ellipsoid, 6 - 8 mm ∅, merging gradually into the **Cl** tube, tube 3 - 4 mm ∅, towards the apex widening to ± 12 mm ∅; **Cl** lobes inside pale green, apically free, erect, linear-filiform from a triangular base, 6 - 8 cm, revolute, hairy, within each angle between the **Cl** lobes furnished with an erect subulate tooth; **Cn** sessile, cup-shaped-cylindrical, ± 6 × 3 mm; **Ci** lobes produced from the upper margin of the **Cn**, bifid into erect narrowly triangular appendages, 1 - 1.5 mm,

glabrous; **Cs** lobes 3 - 4 mm, linear, obtuse, erect; **Poll** obliquely ellipsoid.

See note under *C. dinteri*.

C. leroyi Rauh & Marnier-Lapostolle (KuaS 15(9): 178-181, ills., 1964). **T:** Madagascar (*Rauh* 10866 [HEID, P]). – **D:** C Madagascar.

[3] Stem succulents of creeping-decumbent habit; stems ± 4-angled, flattened, < 10 × 0.3 - 0.6 cm ∅, thickest at the nodes due to persistent triangular **Tu**, grey- to reddish-green, with purple verrucose stripes; **L** sessile, narrowly linear, acute, divaricate, dark green, slightly succulent, 15 - 25 × 2 mm, caducous, with raised midrib on the lower face, margins scatteredly hairy; **Inf** with fleshy peduncle 3 - 10 mm long, borne only on twining thin terete stems of up to 50 cm, 1- to 2- (to 4-) flowered; **Ped** ± 1 cm; **Sep** narrowly lanceolate, 3 - 5 mm; **Cl** 3 - 4.5 cm, scarcely bent, grey-green; **Cl** inflation cylindrical to narrowly ovoid, 3 - 4 mm ∅, outside grey-green, blotched with ± wine-red, inside basally hairy, almost continuously narrowing into the slender **Cl** tube, tube 1.5 - 2.5 mm ∅, apically widening to ± 3 mm; **Cl** lobes 6 - 12 mm, linear, apically mostly connate into an ovoid cage, occasionally twisted or free, lamina ± entirely folded back along the midrib, inside grey-brown to olive-green, hairy, margins basally with vibratile purple clavate **Ha**; **Cn** shortly stipitate, ± 4 × 3 mm, basally fused into a shallow bowl; **Ci** lobes bilobed into narrowly triangular appendages, ± 1 mm, linear with slightly swollen papillose tips, erect-connivent.

Very closely related to *C. armandii* and *C. dimorpha*.

C. lindenii Lavranos (CSJA 63(4): 168-171, ills., 1991). **T:** Somalia, Shebeellaha Dhexe Region (*Lavranos* 23222 [UPS, K, MO]). – **D:** Somalia.

[2,3] **R** tuberous, pear-shaped, to 2.5 × 2 cm; stems succulent, climbing and twining in scrub, little-branched, terete, 3 - 5 mm ∅, brown-green, blotched with purple; **L** minute, caducous; **Inf** ± sessile or with a peduncle up to 5 mm long, 3- to 5-flowered; **Ped** ± 5 mm; **Cl** 4 - 5 cm, pale golden-yellow; **Cl** inflation ± ovoid, ± 10 × 4 mm, inside longitudinally sulcate with roundish papillae, gradually narrowed to 2 mm ∅; **Cl** tube towards the mouth gradually widening to ± 6 mm ∅, inside hairy; **Cl** lobes 20 - 25 × 2 mm, horizontally spreading, lamina slightly folded back along the midrib, yellow with white base, glabrous; **Cn** shortly stipitate, white, ± 2.5 × 2 mm, shallowly cup-shaped, glabrous; **Ci** lobes obtusely triangular, basally ± 0.6 × 0.9 mm; **Cs** filiform, ± erect, at the base with 1 ventral and 2 lateral elongated to roundish appendages; **Poll** ± pear-shaped.

A species still insufficiently known. The thick stems furnished with purple blotches and the free corolla lobes are suggestive of a relationship with *C. stapeliiformis* and *C. variegata*.

C. linearis E. Meyer (Comm. Pl. Afr. Austr., 194, 1837). **T:** RSA, KwaZulu-Natal (*Drège* 4947 [P]). – **D:** Kenya, Zimbabwe, Malawi, Moçambique, RSA, Swaziland.

A complex "collective" species comprising numerous local forms within the subspecies presently accepted. Ssp. *debilis* appears to be a mere form or variety of the particularly variable ssp. *woodii*.

C. linearis ssp. **debilis** (N. E. Brown) H. Huber (Mem. Soc. Brot. 12: 118, 1957). **T:** Malawi (*O'Brien* s.n. [K ?]). – **Lit:** Huber (1957). **D:** Malawi, Zimbabwe. **I:** Werdermann (1939).

≡ *Ceropegia debilis* N. E. Brown (1895).

[2,4] Differs from ssp. *woodii*: **L** ± sessile, narrowly linear, succulent, semicircular in cross-section; **Cl** lobes narrowly spatulate, apically connate, lamina folded back along the midrib, inside black-purple, with violet **Ha**; **Cn** glabrous.

C. linearis ssp. **linearis** – **Lit:** Dyer (1980). **D:** Moçambique, RSA (Eastern Cape, KwaZulu-Natal). **I:** Bruyns (1985). **Fig. XI.a**

Incl. *Ceropegia caffrorum* Schlechter (1894); incl. *Ceropegia caffrorum* var. *dubia* N. E. Brown (1908).

[2,4] **R** tuber ± globose, bark rough; stems twining, creeping or pendent, terete, 1 - 2 mm ∅, nodes often with globose stem tubers; **L** petiolate, ovate-lanceolate, elliptic to linear, 1 - 5 × 0.4 - 1.2 cm, slightly fleshy; **Inf** 6 - 12 mm pedunculate, few-flowered; **Sep** subulate, 2 - 3 mm; **Cl** 2 - 2.5 cm, whitish-purple; **Cl** inflation globose, ± 4 × 4 mm, inside (striped with) purple, merging gradually into the **Cl** tube, tube 2 mm ∅, apically widening to 3 mm ∅, inside hairy; **Cl** lobes lanceolate, lamina entirely folded back along the midrib, purple, erect, 3 - 5 × 1 - 2 mm, usually glabrous, apically fused and often drawn out into a small tip; **Cn** white, sessile or shortly stipitate, ± 2.5 - 3 × 2 mm; **Ci** lobes drawn out into ovate concave pouches, lateral margins raised and fused shield-like with the bulging **Cs** base; **Cs** 1.5 - 2 mm, spatulate-lanceolate, acute, erect-connivent but above the middle strongly recurved; **Poll** ovoid, ± 0.15 × 0.12 mm, with very short germination crest near the tip, corpuscle small, elliptic; **Fr** grey-green, 4 - 7 × 2 - 3 mm; **Se** ± 5 mm, with broad wing.

C. linearis ssp. **tenuis** (N. E. Brown) Bruyns (Bradleya 3: 40, ills., 1985). **T:** RSA (*Pegler* 665 [K]). – **D:** RSA (Eastern Cape). **I:** Bruyns (1985).

≡ *Ceropegia tenuis* N. E. Brown (1908).

[2,4] **R** tuber present; stems occasionally with small stem tubers at the nodes; **L** ± triangular to ovate, 10 - 25 × 7 - 17 mm, acuminate, slightly succulent, upper face slightly canaliculate, green, petiole 3 - 6 mm; **Cl** 2 - 2.4 cm; **Cl** inflation ± globose, ± 4 × 4 mm, inside pubescent, merging gradually into

the **Cl** tube, tube 2 mm ⌀, apically scarcely widening, inside hairy; **Cl** lobes dark violet, narrowly spatulate-linear, apically fused, ± folded back along the midrib, erect and forming an ovoid cage, ± 4 mm, inside rarely with few purple **Ha** along the basal keel; **Cn** glabrous; **Cs** ± 2 mm.

C. linearis ssp. **woodii** (Schlechter) H. Huber (Mem. Soc. Brot. 12: 118, 1957). **T**: RSA, KwaZulu-Natal (*Wood* 1317 [K]). – **D**: Kenya, Zimbabwe, RSA (Western Cape, Eastern Cape, KwaZulu-Natal), Swaziland. **I**: Bruyns (1985). **Fig. IX.g, X.g**

≡ *Ceropegia woodii* Schlechter (1894); **incl.** *Ceropegia euryacme* Schlechter (1905); **incl.** *Ceropegia leptocarpa* Schlechter (1905); **incl.** *Ceropegia barbertonensis* N. E. Brown (1908); **incl.** *Ceropegia hastata* N. E. Brown (1908); **incl.** *Ceropegia schoenlandii* N. E. Brown (1913); **incl.** *Ceropegia corallicorona* Werdermann (1939).

[2,4] **R** tuber 2.5 - 5 (-10) cm ⌀, often irregular with age; stems at numerous nodes usually with globose stem tubers; **L** broadly cordate, ovate to lanceolate, 6 - 18 × 3 - 16 mm, succulent, often acuminate, upper face flat, marked with dark green or whitish-dark-green, lower face convex, pale green often suffused with red, petiole 3 - 10 mm; **Cl** 1.8 - 2.5 cm, pale green-white, whitish-rose, often striped with dark violet; **Cl** lobes dark purple, rarely greenish; **Cl** inflation globose, ± 4 × 4 mm, inside (striped with) purple, merging ± gradually into the **Cl** tube, tube 2 mm ⌀, widening to 3 mm ⌀ towards the mouth, basally (striped with) purple, apically whitish, inside hairy; **Cl** lobes narrowly spatulate (in Kenya also broadly spatulate), at least basally folded back along the midrib, erect, 6 - 7 mm, forming an elongated cage with a blunt (rarely cone-shaped) tip, inside with purple **Ha**; **Cn** white, sessile or shortly stipitate, ± 2.5 - 3 × 2 mm; **Ci** lobes with upper margin often setulose; **Cs** 2 - 2.5 mm.

Presumably the most widely distributed taxon of the genus in cultivation and very popular among beginners (very decorative in hanging baskets, in greenhouses occasionally used as ground cover). The stem tubers are used as stocks for grafting difficult stem succulent stapeliads. There are ± white-variegated forms in cultivation as well as a form with pink foliage. Vernacular name: "Chain of Hearts".

C. linophylla H. Huber (Mem. Soc. Brot. 12: 115, 1957). **T**: Central African Republic (*Tisserant* 3166 [P]). – **D**: Equatorial Guinea, Cameroon, Ghana, Nigeria, Senegal, Central African Republic. **I**: Berhaut (1971).

Incl. *Ceropegia senegalensis* H. Huber (1957).

[2] **R** tuber ovoid, 2 - 4 mm ⌀; stems delicate, slightly twining, 5 - 10 cm, ± glabrous, annual, ± 1 mm ⌀; **L** ± sessile, linear to narrowly elliptic, acute, 4 - 10 × 0.3 - 1 cm, slightly fleshy; **Inf** 1 - 2 cm pedunculate, 2- to 3-flowered; **Ped** 4 - 6 mm; **Sep** subulate, 3 mm; **Cl** ± erect and straight, 1.4 - 2.5 cm, whitish-greenish suffused with purple, inside glabrous; **Cl** inflation globose-ovoid, 3 - 4 mm ⌀, inside ± hairy, merging gradually into the **Cl** tube, tube 2 mm ⌀, apically widening to 3 mm ⌀, purple; **Cl** lobes 5 - 15 × 1 mm, linear to filiform from a triangular base, lamina strongly folded back along the midrib, apically connate to form an ellipsoid cage, inside basally with purple **Ha**, apically greenish, glabrous; **Cn** ± sessile, bowl-shaped; **Ci** lobes almost square, concave, apically incised; **Cs** ± 2 mm, narrowly lanceolate to slightly falcate, erect, tips revolute; **Fr** fusiform, ± 6 cm × 2.5 mm.

To be placed in the *C. bulbosa* alliance and with its habit suggestive of the Asian form "*C. lushii*" (a synonym of *C. bulbosa*).

C. longifolia Wallich (Pl. Asiat. Rar. 1: 56, t. 73, 1830). **T** [lecto]: Nepal (*Wallich* s.n. [K]). – **D**: Bangladesh, Bhutan, Nepal, India, Myanmar, China. **I**: Ansari (1984).

Incl. *Ceropegia angustifolia* Wight (1834); **incl.** *Ceropegia lanceolata* Wight (1834); **incl.** *Ceropegia borii* Raizada (1941).

[1] **R** fibrous, some thickened and fusiform; stems robust, to 4 m, twining, ± 2 - 3 mm ⌀, with **Ha** arranged in 1 or 2 rows; **L** ± 1.2 cm petiolate, pubescent, lamina linear-lanceolate, 5 - 15 × 1 - 3 cm, acute, upper face finely hairy, lower face hairy on the midrib; **Inf** pseudo-umbellate, 2- to 8-flowered, 0.2 - 3 cm pedunculate, hairy; **Ped** 1 - 1.5 cm; **Sep** linear-lanceolate, 4 - 7 mm; **Cl** 2.2 - 4 cm, (whitish-) greenish and striped with ± brown-purple and blotched with brown-purple; **Cl** inflation globose to cylindrical, 5 - 7 mm ⌀, basal ⅓ ± narrowed, with a circle of **Ha** around the mouth, ± gradually merging into the **Cl** tube, tube 1.5 - 4 mm ⌀, apically widening to 5 - 12 mm ⌀; **Cl** lobes 7 - 15 mm, triangular-ovate, quite obtuse, lamina folded back along the midrib except for the lower ⅓, keeled, obliquely erect, tips joined into a broadly ovoid cage, inside whitish with ± brown-purple reticulate pattern, tips green, centre line of keel and margins ciliate or glabrous; **Cn** sessile, fused into a bowl; **Ci** lobes ovate, basally drawn out into pouches, apically bifid into 2 obtusely triangular appendages, ± ciliate; **Cs** linear, 2 - 2.5 mm, erect, dorsal base usually hairy; **Poll** broadly ovoid; **Fr** slender, to 15 cm.

Belonging to the highly variable complex comprising *C. mairei*, *C. christenseniana* (not succulent), *C. dolichophylla*, *C. stenophylla* and *C. exigua*. The latter species has been described by Huber (1957) as a variety of *C. longifolia* ssp. *sinensis* (= *C. dolichophylla*) but has lately been proposed as separate species by Gilbert & al. (1995).

C. loranthiflora K. Schumann (BJS 17: 150, 1893).

T [neo]: Ethiopia (*Schimper* 723 [P]). – **D:** Ethiopia.

Incl. *Ceropegia abyssinica* Vatke (1876) (*nom. illeg.*, Art. 53.1).

[2] **R** tuber globose; stems delicate, slightly twining, ± 5 cm, hairy, annual, ± 1 mm ∅; **L** 5 - 7 mm petiolate, lamina 3 - 4 × 1 - 2 cm, V-shaped in cross-section; **Inf** sessile, 2-flowered, with yellowish **Ha**; **Ped** 5 - 7 mm; **Sep** narrowly lanceolate, 7 mm; **Cl** colour not recorded; **Cl** inflation cylindrical, 5 mm long, outside hairy; **Cl** lobes ± 18 mm, linear, apically slightly widening, fused, diverging; **Cn** 1 mm stipitate; **Ci** lobes nearly square, short, hairy; **Cs** ± 0.5 mm.

Insufficiently known and possibly merely a twining form of *C. abyssinica*.

C. lugardiae N. E. Brown (Gard. Chron., ser. 3, 30: 302, 1901). **T:** Botswana (*Lugard* 262 [K]). – **D:** Angola, Botswana, Kenya, Tanzania, Moçambique, Zimbabwe, Namibia, RSA (Northern Prov.). **I:** Dyer (1983); Bruyns (1984). **Fig. XI.b**

≡ *Ceropegia distincta* ssp. *lugardiae* (N. E. Brown) H. Huber (1957); **incl.** *Ceropegia apiculata* Schlechter (1913); **incl.** *Ceropegia distincta* fa. *pubescens* H. Huber (1957); **incl.** *Ceropegia distincta* ssp. *verruculosa* R. A. Dyer (1977) ≡ *Ceropegia verruculosa* (R. A. Dyer) D. V. Field (1981).

[1,3] **R** fusiform-fleshy; stems robust, persistent, climbing and twining, 1 - 10 m tall, glabrous, often with an acrid smell, ± terete, 3 - 5 mm ∅, ± succulent, glaucous-green, ± verrucose; **L** 0.4 - 4 cm petiolate, lamina narrowly ovate to broadly ovate-elliptic, 1.5 - 5.5 × 0.6 - 2 cm, acute, usually acuminate, basally cordate, often auriculate, ± caducous; **Inf** 0.5 - 5 cm pedunculate, almost umbellate, 2- to 9-flowered, 1 - 2 **Fl** simultaneously open; **Ped** 1 - 2 cm; **Sep** subulate, acute, 3 - 6 mm, stippled with reddish; **Cl** 4 - 6.5 cm, greenish-whitish-yellowish stippled and blotched with red-brown, outside sometimes pubescent; **Cl** inflation outside often greenish, blotched with ± purple, inside purple, ± 4 - 7 (-15) × 4 - 6 mm, globose to ovoid, basally slightly constricted, inside hairy; **Cl** tube narrowed to ± 3 mm, straight to bent, gradually widening to 12 - 16 mm ∅ towards the mouth, outside whitish blotched with purple, inside green, striped with ± purple; **Cl** lobes very variable, 8 - 35 mm, ± linear to ovate from a triangular keeled base, narrowed in the middle, lamina folded back along the midrib, apically fused, erect, straight, spreading or spiralling (and often crowned by an additional cage-like structure), basally greenish (rarely even white), partly with red-brown reticulate markings, apically red-brown or green-brown, margin with long **Ha**; **Cn** whitish-green with purple, broadly cup-shaped, 4 - 5 mm ∅, basally fused with the **Cl** tube; **Ci** lobes ovate, forming pouches, ascending, 1.5 mm, incised in the middle, ciliate; **Cs** 3 - 4 mm, linear, erect-connivent, usually with spreading tips; **Poll** ± 0.3 × 0.2 mm; **Fr** narrowly fusiform, 10 - 25 cm × 4 - 5 mm.

C. lugardii is extremely variable and widely distributed. Characteristic features include an often pungent smell of the vegetative parts and even the flowers, the verrucose-rough stems, and the fleshy roots by which it is separable from its next-of-kin *C. distincta*. This variable taxon demands a broad taxonomic approach, however. The inclusion of *C. verruculosa* is preliminary.

C. macrantha Wight (Contr. Bot. India, 31, 1834). **T:** India, Uttar Pradesh (*Royle* s.n. [K]). – **D:** India, Bhutan, Nepal, Pakistan; forests. **I:** Hart (1988).

Incl. *Ceropegia thorelii* Costantin (1912) ≡ *Ceropegia macrantha* var. *thorelii* (Costantin) H. Huber (1957); **incl.** *Ceropegia raizadiana* Babu (1977).

[1] **R** fusiform, fleshy; stems robust, twining, with **Ha** arranged in 1 - 2 rows; **L** petiolate, ovate to lanceolate, upper face hairy; **Inf** few- to many-flowered; peduncle short, with **Ha** arranged in 1 - 2 rows; **Ped** ± 1 cm, glabrous; **Sep** subulate, 4 - 5 mm, ± revolute; **Cl** 5 - 6 cm, whitish to yellowish, often with reddish longitudinal stripes; **Cl** inflation ± globose, ± 6 - 8 × 6 mm, outside finely stippled with purple, inside with a circle of **Ha** around the mouth, abruptly narrowed into the **Cl** tube to ± 2 mm ∅, tube gradually widening to 10 - 13 mm towards the mouth, outside blotched with purple; **Cl** lobes 1.5 - 3 cm, linear from a triangular-ovate base, keeled, basally flat, then folded back along the midrib, inside basal ½ pale with reticulate markings, apical ½ darker; **Cn** shortly stipitate, whitish, ± 4 mm ∅; **Ci** lobes almost free to the base and deeply bifid into triangular-subulate appendages, ± 2 mm, erect, ciliate; **Cs** subulate, 2 - 3 mm, erect-connivent, tips sometimes revolute; **Poll** narrowly ovoid.

To be placed in the vicinity of *C. longifolia*.

C. madagascariensis Decaisne (PSRV 8: 642, 1844). **T:** Madagascar (*Goudot* s.n. [G]). – **Lit:** Meve & Liede (1994: with ill.). **D:** E Madagascar; forest, rare. **Fig. XI.c**

Incl. *Ceropegia breviloba* Jumelle & H. Perrier (1908); **incl.** *Ceropegia petiolata* Jumelle & H. Perrier (1908); **incl.** *Ceropegia perrieri* Choux (1923).

[2] **R** tuber globose, some 2 cm ∅, also forming subterranean stem tubers; stems very delicate, 0.5 - 1.5 m tall, twining, glabrous, annual, 1 - 2 mm ∅, pale green suffused with reddish; **L** 7 - 20 mm petiolate, lamina ovate-lanceolate to slightly cordate, very thin, 1 - 4.5 × 0.8 - 2.5 cm, acute; **Inf** 2 - 10 mm pedunculate, delicate, 1- to 3-flowered; **Ped** 4 - 15 mm; **Sep** triangular-lanceolate, ± 1 mm; **Cl** 1.5 - 3 cm; **Cl** inflation ovoid, pale green, 6 - 10 × 4 - 7 mm; **Cl** tube bent, basally narrowed, inside finely hairy, apically widening to 4 mm, pale green, blotched with red-brown; **Cl** lobes basally pale yellow blotched with red-brown, apically red-brown, 6

- 9 mm, fused at the tips only, triangular-lanceolate with a keel 2 mm broad, erect, only the middle part of the lamina folded back along the midrib, white-ciliate; **Cn** shortly stipitate, whitish, ± 3 × 2.5 mm, basally fused into a shallow cup; **Ci** lobes triangular-ovate, 1 mm, ascending, bilobed into erect ± connivent triangular-subulate appendages, with long and scattered **Ha**; **Cs** 2 - 3 mm, linear-clavate, 2 - 2.5 mm, erect-connivent, apically papillose; **Poll** broadly ovoid, 0.3 × 0.25 mm.

C. mahabalei Hemadri & Ansari (Indian Forester 97(2): 105-108, ills., 1971). **T:** India, Maharashtra (*Hemadri* 107266A [CAL, BSI, K, L, MO]). – **D:** S India (Maharashtra).

[2] **R** tuber ± 3.5 cm ⌀, edible; stems persistent, erect, not twining, herbaceous, unbranched, hairy; **L** ± sessile, linear-lanceolate, 3 - 15 × 0.3 - 1 cm; **Inf** 1-flowered, shortly pedunculate; **Ped** 0.5 - 1 cm; **Sep** subulate, 1 - 1.5 cm; **Cl** 5.5 - 11 cm, only weakly bent or straight, basally green, remainder purple-brown, glabrous; **Cl** inflation ovoid-ellipsoid, ± 20 × 14 mm, inside with purple longitudinal stripes, glabrous, merging gradually into the **Cl** tube, tube 3 - 4 mm ⌀ in the middle, apically widening to ± 12 mm; **Cl** lobes 17 - 35 mm, narrowly triangular-linear from a broadly triangular base, apically joined into a beak-shaped cage, lamina folded back along the midrib, inside dark green and brown-yellow, finely hairy; **Cn** bowl-shaped, sessile, ± 4 mm ⌀, glabrous; **Ci** lobes broadly triangular, ± erect, bifid into triangular appendages; **Cs** 3 - 4 mm, linear, erect-connivent; **Fr** ± 4 cm; **Se** 5 × 2.5 mm, tuft of **Ha** 6 mm.

A very attractive and large-flowered species. Closely related to *C. fimbriifera* and the other erect species with linear leaves.

C. mairei (Leveillé) H. Huber (Mem. Soc. Brot. 12: 43, 1957). **T:** China, Yunnan (*Maire* 3536 [E]). – **D:** China (Yunnan).

≡ *Aristolochia mairei* Leveillé (1912); **incl.** *Aristolochia blinii* Leveillé (1913); **incl.** *Aristolochia viridiflora* Leveillé (1913); **incl.** *Aristolochia viridiflora* var. *occlusa* Leveillé (1913); **incl.** *Ceropegia balfouriana* Schlechter (1913); **incl.** *Ceropegia mairei* var. *tenella* H. Huber (1957).

[1] **R** fusiform, fleshy, clustered; stems erect to decumbent, partly twining, ± 35 cm, with scattered fine **Ha**; **L** 3 - 10 mm petiolate, petiole acutely wing-like, lamina elliptic to lanceolate, 1 - 5 × 0.4 - 2 cm, margins slightly undulate, lower face hairy on the midrib only, upper face ± scatteredly hairy; **Inf** with a short peduncle, 1- to 2- (to 5-) flowered; **Ped** 4 - 17 mm; **Sep** triangular-linear, 7 - 9 mm, with few **Ha**; **Cl** (2.3-) 4 - 5 cm, yellow to green, stippled with purple; **Cl** inflation 5 - 7 mm ⌀, merging gradually into the **Cl** tube, tube 2 - 3.5 mm ⌀, apically widening to (6-) 9 - 12 mm; **Cl** lobes (7-) 14 - 25 mm, lamina folded back entirely along the midrib, erect with tips inflexed and joined, inside basally light with dark reticulating markings, apically dark, densely hairy; **Cn** sessile, fused into a bowl; **Ci** lobes bifid into triangular appendages, hairy; **Cs** linear, apically obtusely spatulate.

To be placed within the alliance around *C. dolichophylla* and *C. longifolia*, but most closely related to *C. sinoerecta*.

C. maiuscula H. Huber (Mem. Soc. Brot. 12: 116, ill., 1957). **T:** Tanzania (*Vaughan* 3123 [BM]). – **D:** Tanzania.

[2?] Leafy climbers, presumably tuberous; stems weakly succulent; **L** ± sessile, linear, 5 - 10 cm × 1 - 5 mm, weakly succulent; **Inf** 2 - 2.5 cm pedunculate; **Sep** 2 - 3 mm; **Cl** 3 - 4 cm, greenish; **Cl** inflation ovoid, ± 8 × 6 mm; **Cl** tube ± straight, narrowed to 2 - 3 mm in the middle, inside finely hairy, widening to ± 5 mm towards the mouth; **Cl** lobes linear, 12 - 15 mm, apically fused, inside densely hairy; **Cn** ± sessile, bowl-shaped; **Ci** lobes almost square, concave, apically ± incised; **Cs** ± 2 mm, narrowly lanceolate to slightly falcate, erect, tips revolute.

Seemingly only represented by the type specimen. Apparently allied to the complex comprising *C. linophylla* and *C. bulbosa* and possibly merely a large-flowered form of the latter.

C. media (H. Huber) Ansari (Bull. Bot. Surv. India 11(1/2): 199, 1971). **T:** India, Maharashtra (*Irani* 1194 [BLAT]). – **D:** India (Maharashtra). **I:** Ansari (1984).

≡ *Ceropegia evansii* var. *media* H. Huber (1957).

[2] **R** tuber present; stems delicate, twining, glabrous; **L** linear-lanceolate, upper face pubescent; **Inf** few-flowered, peduncle and **Ped** hairy; **Cl** 2.5 - 3 cm, ± cream-coloured; **Cl** inflation globose, ± 5 mm ⌀, merging gradually into the 2 cm long **Cl** tube, tube ± 3 mm ⌀ in the middle, apically widening to 7 mm ⌀; **Cl** lobes 8 × 2.5 mm, ovate, ± completely folded back along the midrib, inside whitish to apically purple-brown, glabrous; **Cn** stipitate, fused into a cup; **Ci** lobes completely integrated into the cup, pouch-shaped, crenate in the middle, ciliate; **Cs** basally integrated into the margin of the cup, linear, L-shaped (almost horizontally bent inwards, upper ½ changing abruptly into erect position), apically parallel.

In contrast to *C. evansii*, *C. media* is more hairy, has narrower leaves, smaller flowers, and a distinct cup-shaped corona with the staminal lobes bent into an almost rectangular position.

C. meleagris H. Huber (Mem. Soc. Brot. 12: 48, 1957). **T:** Nepal (*Stainton & al.* 1548 [BM, UPS]). – **D:** Nepal (Bheri River). **I:** Asklepios 43: 71-74, 1988. **Fig. X.h**

[1] **R** forming an irregular cluster, fleshy, fusi-

form; stems decumbent-creeping, to ± 50 cm, terete, 1 - 2 mm ⌀, hairy, annual; **L** petiole 1 - 1.5 cm, upper face distinctly canaliculate, hairy, lamina broadly ovate (-cordate), 3.5 - 5 × 3.5 - 4 cm, slightly fleshy, upper face fresh green, sparsely villose, lower face pale green, glabrous except for the veins, margins villose; **Inf** sessile or with a peduncle a few mm long, 2- to 4- (to 6-) flowered; **Ped** 3 - 6 mm, sparsely villose; **Sep** subulate, ± 3 mm, villose; **Cl** 1.2 - 1.7 cm, urceolate, pale green, outside scatteredly villose at base and apex, scent sweet-sour; **Cl** tube globose without constrictions, ± 8 - 10 × 9 mm, inside with a circle of **Ha** near the mouth, abruptly narrowed to 5 mm ⌀ and directly drawn out into the erect **Cl** lobes, these fused into a cone, **Cl** tube inside whitish stippled with purple, scattered with long **Ha**; **Cl** lobes inside purple-brown, velvety, tipped with few **Ha**, lanceolate, 5 - 7 × 2.5 mm, margins joined with each other except for the lower 2 mm, there slightly folded back to form an ovate opening; **Cn** whitish, dotted purple, 1 mm stipitate, ± 3.5 × 3.5 mm, shallowly bowl-shaped drawn out into widely spreading **Ci** lobes and **Cs** bases, **Ci** lobes ± 1.2 × 1.5 mm, incised to ½, margins joined with the base of the **Cs** in a wing-like fashion and ascending at an angle of 90°, with long white **Ha**; **Cs** basally vertically linear, apically subulate, ± 2 mm, ± erect-connivent; **Poll** broadly ovoid, 0.3 × 0.22 mm; **Fr** 4 - 5 cm, fusiform, weakly moniliform.

With its decumbent stems as well as its urceolate flowers, which are colourful inside only and which present but minute openings, *C. meleagris* makes a pretty plant for hanging baskets. Within the genus, it is isolated from the other species also in terms of geographical distribution.

C. mendesii Stopp (BJS 90(4): 471-472, ills., 1971). **T:** Angola, Benguela (*Mendes* 573 [LISC]). – **D:** Angola

Merely known from the type specimen. The taxon appears to represent an aberrant form of *C. filipendula* or another closely related species.

C. meyeri Decaisne (PSRV 8: 645, 1844). **T:** RSA, Eastern Cape (*Drège* 4945 [P]). – **D:** Malawi, Moçambique, Zambia, Zimbabwe, Namibia, RSA. **I:** Dyer (1983); Asklepios 55: 4, figs. 6-7, 1992.

Incl. *Ceropegia pubescens* E. Meyer (1837) (*nom. illeg.*, Art. 53.1).

[2] **R** tuber flattened, 4 - 7 cm ⌀, with fissured bark; stems twining, 1 - 2 m, annual, 1 - 3 mm ⌀, shortly hairy; **L** 0.8 - 3 cm petiolate, lamina elliptic, ovate to broadly lanceolate, margins often emarginate, crenulate or lobed, 2 - 5 × 0.8 - 3 cm, delicate, hairy; **Inf** sessile or with a short thickened peduncle, 2- to 4- (to 10-) flowered; **Ped** 0.5 - 2 cm, villose; **Sep** subulate, 0.7 - 1 cm, villose; **Cl** 4 - 6 cm, bottle-shaped; **Cl** inflation outside whitish-greenish, upper ½ with longitudinal red-brown stripes and blotches, 25 - 40 × 7 - 9 mm, elliptic, apically gradually narrowed to 2 - 3 mm ⌀ and abruptly widening to 6 - 8 mm; **Cl** lobes ± 10 × 3 mm, linear, folded back along the midrib including the basal margins, apically fused to form a broadly ovoid cage, inside velvety-black with greenish longitudinal stripes, hairy; **Cn** ± shortly stipitate, white, basally fused into a cup; **Ci** lobes triangular, acute, ± 1 mm, ± erect; **Cs** ± 2 mm, linear-spatulate, erect-connivent, in the middle at times inwardly bulging, basally black-purple; **Fr** narrowly fusiform, 8 - 12 cm × 3 mm.

Closely allied to *C. bonafouxii*, less so to *C. paricyma* and *C. stenoloba*.

C. meyeri-johannis Engler (Hochgebirgsfl. Afrika, 343, 1892). **T** [neo]: Tanzania (*Johnston* s.n. [BM]). – **D:** Kenya, Tanzania, Uganda, Zaïre, Malawi, Moçambique, Zambia, RSA (Eastern Cape). **I:** FPA 36: t. 1410, 1963; Archer (1992: XI). **Fig. XI.d**

Incl. *Ceropegia calcarata* N. E. Brown (1903); **incl.** *Ceropegia verdickii* De Wildeman (1903) ≡ *Ceropegia meyeri-johannis* var. *verdickii* (De Wildeman) H. Huber (1957); **incl.** *Ceropegia angiensis* De Wildeman (1928) ≡ *Ceropegia meyeri-johannis* var. *angiensis* (De Wildeman) H. Huber (1957); **incl.** *Ceropegia criniticaulis* Werdermann (1938); **incl.** *Ceropegia dubia* R. A. Dyer (1980).

[1] **R** clustered, fleshy, fusiform or cylindrical, to 30 cm long; stems persistent, twining, to 3 m × 2 mm, hairy; **L** 1.5 - 3.5 cm petiolate, basally obtuse to cordate, lamina ovate to broadly lanceolate, 3 - 10 × 2.5 - 7 cm, delicate, hairy; **Inf** with a peduncle 0.5 - 8 cm long, hairy, with several membranous **Bra**, richly flowering with 1 - 5 **Fl** concurrently open; **Ped** 8 - 15 mm, hairy; **Sep** narrowly lanceolate, 2 - 6 mm, hairy; **Cl** 2 - 3 cm, strongly bent, whitish, stippled or striped with ± red-brown; **Cl** inflation globose to bulbous, weakly narrowed at the base, 5 - 7 × 4 - 7 mm; **Cl** tube narrowed to 2 mm ⌀, towards the mouth gradually widening to 8 - 10 mm, inside white or purple; **Cl** lobes narrowly lanceolate from a triangular base, 4 - 10 mm, acute, apically fused, widely divergent to form a globose to broadly cylindrical cage, lamina folded back along the midrib, weakly keeled, keels widely projecting down into the **Cl** tube, inside and on the margins green to brownish, basally even whitish, inside ± hairy, ciliate, lower margins folded back; **Cn** ± sessile to shortly stipitate, bowl-shaped, ± 3 × 3.5 mm, white; **Ci** lobes broadly ovate to rectangular, ascending to erect, concave, 1 - 1.5 mm, apically emarginate to deeply sinuate, appendages triangular, papillose or hairy; **Cs** ± 2 mm, linear-spatulate, erect-parallel, connivent, tips papillose; **Poll** ovoid, ± 0.2 × 0.12 mm; **Fr** 9 - 11 cm × 3 - 4 mm; **Se** ± 10 × 2 mm, tuft of **Ha** 1 - 2.5 cm.

The variability of this species is amazing. However, the corolla lobes being deeply inserted in the corolla tube are significant. The 3 varieties as still recognized by Huber (1957) can be found in Kenya alone. They would not nearly be sufficient to name all known forms.

C. muliensis W. W. Smith (Notes Roy. Bot. Gard. Edinburgh 12: 199, 1920). **T:** China, Yunnan (*Forrest* 13117 [E]). – **D:** China (Sichuan, Yunnan). **I:** Tsiang (1939).

[1] **R** clustered, fusiform, fleshy; stems to 1 m, twining, delicate, densely curly-pubescent; **L** petiole ± 1 cm, pubescent, lamina lanceolate, 10 - 13 × 2 - 3 cm, acuminate, upper face woolly-pubescent, lower face on the midrib only; **Inf** with pendent peduncle, 1.5 - 4 cm, many-flowered; **Ped** to 12 mm; **Sep** linear-lanceolate, 3 - 4 mm; **Cl** 2.5 - 3 cm, green or blotched with green-purple, glabrous; **Cl** inflation ellipsoid, ± 6 × 2.5 - 3.5 mm; **Cl** tube gradually narrowed towards the mouth to ± 2 mm ∅; **Cl** lobes 7 - 9 mm, linear-elliptic, lamina only slightly folded back along the midrib, tips fused into an ellipsoid cage; **Cn** basally connate; **Ci** lobes bifid into triangular appendages, ± hairy; **Cs** ± linear; **Fr** ± 9 cm.

Endemic to China, yet presumably most closely related to *C. ciliata* from India.

C. multiflora Baker (Refug. Bot. 1: t. 10 + text, 1869). **T:** RSA, Northern Cape (*Arnott* s.n. [K]). – **D:** Angola, Namibia, Zimbabwe, Botswana, RSA.

Incl. *Systrepha multiflorum* Burchell *ex* Baker (1869) (*nom. inval.*, Art. 34.1c).

C. multiflora ssp. **multiflora** – **D:** RSA (Northern Prov., North-West Prov., Free State, Gauteng). **I:** Dyer (1980); Dyer (1983). **Fig. XI.g**

Incl. *Ceropegia multiflora* var. *latifolia* N. E. Brown (1908); **incl.** *Ceropegia multiflora* fa. *pubescens* H. Huber (1957).

[2] **R** tuber slightly flattened, 1 - 2.5 cm ∅; stems twining, 50 - 200 cm, terete, 1 - 2 mm ∅, glabrous or hairy, base perennial and each forming 1 - 2 annual stems; **L** to 8 mm petiolate, lamina broadly ovate, obtuse to sublinear, partly mucronate, 1 - 7 × 0.3 - 1.2 cm, slightly fleshy, glabrous or scatteredly hairy, margin often undulate; **Inf** sessile, 1- to many-flowered; **Ped** 5 - 10 mm; **Sep** lanceolate, 2 - 4 mm; **Cl** 1.5 - 2.8 cm, greenish; **Cl** inflation globose, ± 3.5 - 5 × 3.5 - 5 mm, inside with purple papillae, merging gradually into the **Cl** tube, tube basally 2 mm ∅, widening towards the mouth to 4 - 5 mm ∅, inside hairy; **Cl** lobes 5 - 7 mm, basally lanceolate, ± horizontal, then abruptly bent inwards, filiform, tips connate, upper face white, yellow or green, finely hairy or papillose; **Cn** sessile, shallowly cup-shaped, ± 2 × 2 mm, white, glabrous; **Ci** lobes drawn out into deep ovoid pouches, laterally fused to the **Cs**, margins denticulate, ± 1 mm; **Cs** lobes ± 1.5 mm, linear to spatulate, erect-connivent; **Poll** attenuate-ovoid, ± 0.25 × 0.2 mm; **Fr** narrowly fusiform, to 16 cm.

See note under *C. rendallii*.

C. multiflora ssp. **tentaculata** (N. E. Brown) H. Huber (Mem. Soc. Brot. 12: 122, 1957). **T:** Angola (*Welwitsch* 4277 [BM, K, P]). – **D:** Angola, Namibia, Botswana, Zimbabwe, RSA (Northern Prov.). **I:** Dyer (1980); Dyer (1983); Bruyns (1984).

≡ *Ceropegia tentaculata* N. E. Brown (1895); **incl.** *Ceropegia tentaculata* var. *puberula* Hiern (1898) ≡ *Ceropegia multiflora* fa. *puberula* (Hiern) H. Huber (1957).

[2] Differs from ssp. *multiflora*: **Cl** lobes free and spreading, 7 - 12 mm; **Ci** lobes drawn out into short ± horizontal and open pouches, lateral margins robust, raised, joined with the bulging **Cs** base.

C. muzingana Malaisse (Bull. Jard. Bot. Belg. 54 (1-2): 227-228, ills., 1984). **T:** Zaïre (*Malaisse* 8974 [BR]). – **D:** Zaïre.

[2] **R** tuber small, ± 1 × 2 cm; stems sparsely branched, glabrous, twining, to 1 m, 2 mm ∅; **L** 1 - 4 cm petiolate, lamina cordate, acuminate, 4 - 5.5 × 2.5 - 4.5 cm; **Inf** ± sessile, 1-flowered; **Ped** 5 - 15 mm; **Sep** subulate, 2 - 3 mm, villose; **Cl** straight or slightly bent, 1.2 - 1.6 cm, whitish-green; **Cl** inflation 4 - 5 × 3 - 4 mm, gradually narrowed into the cylindrical **Cl** tube, tube 2 mm ∅ at the mouth, outside suffused with purple; **Cl** lobes 5 - 8 mm, linear to narrowly lanceolate, lamina folded back along the midrib (including the basal angles), apically fused into an ovoid cage, inside velvety-purple, finely hairy, outside greenish; **Cn** sessile, basally fused into a disc-like structure, ± stippled purple, remainder whitish-hyaline; **Ci** lobes merely a narrow entire fringe; **Cs** ± 2 mm, spatulate, erect, tips inflexed, dorsal base hairy; **Fr** pendent, narrowly fusiform, 7 - 8 cm × 2 - 3 mm.

Closely related to *C. claviloba* and *C. bonafouxii*. An unusual feature is the shallow almost reduced outer corona.

C. nana Collett & Hemsley (JLSB 28: 89, t. 13, 1890). **T:** Myanmar (*Collett & Hemsley* 809 [K]). – **D:** Myanmar; only known from the type.

[2] **R** tuber irregularly flattened; stems dwarf-erect, slightly pubescent, herbaceous; **L** lanceolate; **Inf** sessile, 1- (to 2-) flowered; **Cl** 4 - 5 cm, erect-straight, occasionally outside slightly hairy; **Cl** inflation 10 - 15 mm, narrowly ovoid, merging gradually into the **Cl** tube and narrowed to ± 2 mm, widening towards the mouth to 3 - 4 mm; **Cl** lobes linear-spatulate, 15 - 20 mm, lamina folded back along the midrib, acute, tips fused into an ovoid cage, inside glabrous; **Cn** sessile, together with the **Ci** lobes fused into a bowl-shaped structure, inside shortly hairy.

Presumably closely allied to *C. arnottiana* and *C. pusilla*.

C. nigra N. E. Brown (BMI 1895: 261, 1895). **T:** Nigeria (*Baikie* s.n. [K]). – **D:** Senegal, Ghana, Equatorial Guinea, Nigeria, Cameroon, Central African Republic, Sudan, Ethiopia. **I:** Berhaut (1971).
Incl. *Ceropegia kroboensis* N. E. Brown (1903); **incl.** *Ceropegia melanops* H. Huber (1957).

[2] **R** tuber globose, ± 2 - 3 cm ∅; stems annual, twining, ± 5 - 20 cm × 2 mm, shortly hairy; **L** shortly hairy, 1 - 4 cm petiolate, lamina elliptic to broadly triangular-ovate, basally ± cordate, margins occasionally sinuate or crenulate, 2 - 8 × 1 - 4.5 cm, delicate; **Inf** shortly pedunculate, many-flowered; **Ped** 5 - 25 mm, ± villose; **Sep** subulate, 2 - 6 mm, ± villose; **Cl** 8 - 20 mm long, outside whitish-green, glabrous or slightly hairy; **Cl** inflation globose to ellipsoid, 5 - 12 × 4 - 6 mm, inside glabrous, merging ± gradually into the obliquely bent **Cl** tube, upper ½ of the tube inside blackish (-green) with longitudinal stripes; **Cl** lobes triangular-ovate, 2 - 6 mm, ± horizontally (stellately) spreading, margins folded back, inside black-red to black-green, at the base (and continuing down the tube) with whitish central double stripes, ± pubescent or glabrous or only margins or angles scatteredly hairy to bearded; **Cn** sessile to shortly stipitate, base fused into a ± bowl-shaped structure; **Ci** lobes short and roundish-entire to rectangular and ± 2-lobed, 0.5 - 2.5 mm, ± erect, margins mostly ciliate; **Cs** ± 2 mm, linear, slightly S-shaped, erect-connivent; **Poll** ± globose-rectangular, 0.2 × 0.15 mm.

Closely related to *C. abyssinica* and *C. meyeri*. The stellate free corolla with the basal double stripes is unusual. This is another species that is variable in leaf shape, size and indumentum of the flowers and length of the outer corona lobes. The latter character was used by Huber (1957) to segregate *C. melanops*, which could be ranked as a variety of *C. nigra*.

C. nilotica Kotschy (Sitzungsber. Kaiserl. Akad. Wiss., Math.-Naturwiss. Cl., Abt. 1, 51: 356, 1865). **T:** Sudan (*Knoblecher* 35 [W, G]). – **D:** Tropical and subtropical Africa (from Senegal to RSA). **I:** Dyer (1983); Archer (1992: XX). **Fig. XI.e**

Incl. *Ceropegia constricta* N. E. Brown (1895); **incl.** *Ceropegia mozambicensis* Schlechter (1895); **incl.** *Ceropegia gemmifera* K. Schumann (1903); **incl.** *Ceropegia boussingaultifolia* Dinter (1914); **incl.** *Ceropegia gossweileri* S. Moore (1929); **incl.** *Ceropegia plicata* E. A. Bruce (1937) ≡ *Ceropegia nilotica* var. *plicata* (E. A. Bruce) H. Huber (1957); **incl.** *Ceropegia mozambicensis* var. *ulugurensis* Werdermann (1939); **incl.** *Ceropegia grandis* E. A. Bruce (1952); **incl.** *Ceropegia decumbens* P. R. O. Bally (1957).

[1,3,4] **R** partially ± fleshy and thickened, cylindrical to fusiform; stems climbing and twining to 2 m tall, with slightly milky latex, usually annual, 2 - 5 mm ∅, terete, angular to acutely 4-angled, succulent, ± glabrous or striate, dull, base of stems often thickened; **L** to 1.5 cm petiolate, lamina narrowly to broadly lanceolate, 1.5 - 10 × 0.6 - 3 cm, ± succulent, acute, margins often undulate or sometimes serrulate or hairy; **Inf** 2 - 60 mm pedunculate, 1- to many-flowered, **Fl** opening consecutively, with unpleasant fruity scent; **Ped** 3 - 25 × 2 mm, fleshy; **Sep** narrowly lanceolate, 1 - 4 mm; **Cl** 2.5 - 6 cm, slightly bent, outside slightly striate; **Cl** inflation 10 - 14 × 5 - 9 mm, base globose to ovoid, outside whitish, inside ± purple, papillose, glabrous or scatteredly hairy, inflation constricted above the middle to form a ring-like, globose or cylindrical inflation, outside whitish, striped or blotched with purple, inside ridged, striped with purple, hairy at both ends, only slightly narrowed at the transition to the **Cl** tube, 3 - 4.5 mm ∅, tube inside hairy at the base and the mouth, gradually widening towards the mouth, outside whitish, blotched with purple; **Cl** lobes 7 - 25 × 5 - 8 mm, broadly ovate, triangular to lanceolate, lamina entirely folded back along the midrib, lower margins projecting ± horizontally, tips fused into a ± pyramid-shaped, rarely apically concave structure, inside broadly keeled, keels ± coming into contact in the middle, striped with green-brown-purple-white, rarely blotched, with white or purple **Ha**; **Cn** white, ± sessile, ± 5 × 3 mm, fused into a shallow cup, glabrous; **Ci** broadly ovate with undulate margin, concave, basally widening into a pouch-like structure, ± 0.5 × 1 mm, lateral margins connate with the **Cs** base; **Cs** linear-lanceolate, 2 - 3.5 mm, incumbent on the **Anth**, then erect in a parallel manner, obtuse, often with a small hump at the base; **Poll** ellipsoid, ± 0.4 × 0.25 mm; **Fr** ± 12 - 22 × 2 - 3 mm, spreading.

A nearly pan-African, very variable species, at present impossible to divide into infraspecific taxa. Closely related to *C. denticulata* (see there for distinctions).

C. noorjahaniae Ansari (J. Bombay Nat. Hist. Soc. 69(1): 249-253, ills., 1972). **T:** India, Maharashtra (*Ansari* 104880 [CAL, BSI, K, L, MO]). – **D:** C India (Maharashtra). **I:** Ansari (1984).

[2] **R** tuber present; stems erect, not twining, herbaceous, only 5 - 10 cm tall, delicate, with short **Int**, ± unbranched, ± hairy; **L** ± linear, 5 - 8 cm, ± sessile, upper face hairy; **Inf** 3-flowered, shortly pedunculate; **Ped** 6 - 7 mm; **Cl** 2 - 2.8 cm, glabrous; **Cl** inflation ovoid, ± 5 mm ∅, inside glabrous, merging gradually into the **Cl** tube, tube ± 3 mm ∅, mouth widening to ± 6 mm ∅; **Cl** lobes 9 - 13 mm, linear from a triangular base, tips fused, lamina largely folded back along the midrib, glabrous; **Cn** cup-shaped, sessile; **Ci** lobes narrowly triangular, borne on the upper margin of the coronal cup, ± erect, 1.5 mm, incised or bifid into triangular appendages; **Cs** ± 2 mm, linear, slightly clavate, erect-connivent; **Fr** 9 cm × 4 mm; **Se** 4 × 2.5 mm, tuft of **Ha** 2 cm.

Forming a group of closely related species together with *C. attenuata*, *C. jainii*, *C. mahabalei*, *C. pusilla* and *C. spiralis*. All are tuberous and of dwarf erect growth and possess linear leaves.

C. occidentalis R. A. Dyer (BT 12(3): 445, ills., 1978). **T**: RSA, Western Cape (*Hall* 3679 [PRE, NBG]). – **D**: RSA (Western Cape). **I**: Bruyns (1985). **Fig. XI.f**

[2,4] **R** tuber ± globose; stems creeping, not twining, olive-green, terete, 1 - 2 mm ⌀, occasionally forming globose stem tubers at the nodes; **L** ± sessile, fleshy, narrowly ovate-lanceolate, lower face rounded, upper face furrowed along the midrib, 1 - 2 × 0.2 - 0.5 cm; **Inf** shortly pedunculate, 1- to 2-flowered; **Sep** subulate, 2 mm; **Cl** 2.5 - 3 cm, pale green striped and blotched with purple; **Cl** inflation ovoid, ± 5 - 7 × 4 - 5 mm, inside (striped with) purple, then narrowed to < 2 mm; **Cl** tube apically widening to 5 - 7 mm, inside scatteredly hairy; **Cl** lobes lanceolate, tips fused, lamina ± folded back along the midrib, erect, 5 - 7 mm, lower ½ concolorous with the tube, upper ½ purple and hairy inside, midrib entirely purple; **Cn** bowl-shaped, whitish, shortly stipitate, ± 3 × 3 mm; **Ci** lobes drawn out into ovate convex pouches, ciliate, laterally fused with the bulging **Cs** base; **Cs** 1.2 × 0.3 mm, linear-lanceolate, acute, erect-connivent, tips not or only slightly revolute; **Poll** D-shaped, ± 0.27 × 0.2 mm.

C. occidentalis is presumably better ranked as a western microspecies or subspecies of *C. linearis*, from which it is distinguished merely by small differences in habit and corona structure. The species is only known from 2 collections and is thus obviously very rare.

C. occulta R. A. Dyer (BT 7(1): 21-22, ills., 1958). **T**: RSA, Western Cape (*Van Breda* 85 [PRE]). – **D**: RSA (Western Cape: Little Karoo). **I**: Dyer (1983); Bruyns (1985).

[2,4] **R** tuber ± globose; stems annual, creeping and twining, subsucculent, ± 2 mm ⌀, occasionally forming globose stem tubers at the nodes; **L** ± sessile, fleshy, ovate-lanceolate, lower face rounded, upper face furrowed along the midrib, 1 - 2 × 0.3 - 1 cm; **Inf** ± sessile, 1- to 3-flowered; **Sep** lanceolate, ± 3 mm; **Cl** 2 - 2.8 cm, outside finely blotched with brown on whitish background, outside glabrous; **Cl** inflation ovoid, inside finely vertically ridged, ± 5 × 4 mm, then narrowed to 2.5 - 3 mm ⌀, widening towards the mouth to 5 - 6 mm ⌀; **Cl** lobes 5 - 8 mm, lanceolate, margins slightly hairy, scarcely folded back, forming a cage of nearly square shape and 4 mm ⌀, inside with strong keel, ± purple; **Cn** stipitate, shallowly bowl-shaped, ± 3 × 2 mm, whitish, glabrous; **Ci** lobes drawn out into short entire pouches, laterally fused to the bulging **Cs** base; **Cs** 2 - 2.5 mm, elliptic-lanceolate, erect-connivent; **Poll** subglobose, ± 0.25 mm ⌀.

Within the *C. linearis* complex, *C. occulta* is the element native to the Little Karoo.

C. oculata Hooker (CBM 1844: t. 4093 + text, 1844). **T**: [lecto – icono]: l.c., t. 4093. – **D**: C India (Maharashtra). **I**: Ansari (1984).

[2] **R** tuber depressed-globose; stems twining; **L** broadly to narrowly ovate, upper face scatteredly hairy; **Inf** few-flowered, with a short hairy peduncle; **Ped** 5 - 7 mm, glabrous; **Cl** ± 5 - 6.5 cm, glabrous; **Cl** inflation ovoid to cylindrical, 10 - 15 × 8 - 9 mm, whitish-green, even becoming striped with purple in the upper ½, inside glabrous, merging gradually or abruptly into the **Cl** tube, tube 3 - 4 mm ⌀, mouth widening to 9 mm ⌀; **Cl** lobes 15 mm, basally purple, then with a white patch, then purple again and greenish at the tip, linear-elongated, triangular-linear from an ovate base, basally keeled, lamina ± folded back along the midrib, glabrous; **Cn** cup-shaped, shortly stipitate; **Ci** lobes narrowly triangular, from the upper margin of the coronal cup, becoming conspicuously narrow, ascending, apically incised, glabrous; **Cs** linear-clavate, erect-connivent.

With magnificent, yet slender flowers.

C. odorata Nimmo *ex* Hooker *fil.* (Fl. Brit. India 4: 75, 1883). **T** [lecto]: India, Maharashtra (*Law* s.n. [K]). – **D**: C India. **I**: Ansari (1984).

Incl. *Ceropegia blatteri* McCann (1945).

[2] **R** tuber globose; stems delicate, twining, ± glabrous; **L** narrowly lanceolate to linear, hairy on the upper face; **Inf** few- to many-flowered, peduncle hairy; **Ped** glabrous; **Fl** scented; **Cl** 3 - 4 cm, pale yellow; **Cl** tube 1.8 - 2 cm, basally slightly ovoid and inflated to ± 4 mm ⌀, then narrowed to 2 mm ⌀, mouth widening to ± 4 mm; **Cl** lobes 1 - 2 cm, narrowly linear, with margins projecting at the base, tips fused into a narrowly ovoid cage, lamina slightly folded back along the midrib, glabrous; **Cn** sessile, fused with the short **Ci** lobes into a bowl-shaped structure, ± glabrous; **Cs** narrowly triangular or linear, erect-connivent, tips slightly revolute.

The flower shape is reminiscent of that of *C. attenuata* as well as the African *C. stenantha*.

C. pachystelma Schlechter (BJS 20(Beiblatt 51): 47, 1895). **T**: RSA, Northern Prov. (*Schlechter* 4511 [BOL]). – **D**: Moçambique, Namibia, Botswana, Zimbabwe, RSA (Eastern Cape, KwaZulu-Natal, Mpumalanga, Northern Prov., North-West Prov.), Swaziland. **I**: Dyer (1980); Bruyns (1984: 52). **Fig. XII.a**

Incl. *Ceropegia undulata* N. E. Brown (1908) ≡ *Ceropegia pachystelma* ssp. *undulata* (N. E. Brown) H. Huber (1957); **incl.** *Ceropegia acacietorum* Schlechter *ex* Dinter (1914); **incl.** *Ceropegia boerhaaviifolia* Schinz (1926) (*nom. illeg.*, Art. 53.1); **incl.** *Ceropegia schinziana* Bullock (1955).

[2] **R** tuber flattened, 4 - 12 cm ⌀, verrucose;

stems twining to several m tall, usually annual, 1 - 2 mm ⌀, pubescent; **L** 2 - 6 mm petiolate, lamina lanceolate, ovate to narrowly elliptic, mucronate, 22 - 50 × 8 - 15 mm, slightly fleshy, margin usually undulate, hairy; **Inf** 2- to many-flowered, with a hairy peduncle 5 - 15 mm long; **Sep** triangular-lanceolate, ± 2 mm; **Cl** 2 - 3 cm, greenish, partially with violet markings, outside hairy; **Cl** inflation globose to ovoid, ± 4 - 5 × 4 mm, inside and outside pubescent, narrowed ± gradually to 1.5 mm ⌀; **Cl** tube slender, scarcely bent, 15 - 25 mm, mouth widening to some 3 mm ⌀; **Cl** lobes linear, 4 - 12 mm, ± flat, tips fused into an ovoid cage, 5 - 7 × 5 - 9 mm; **Cn** shortly stipitate, shallowly bowl-shaped, ± 2 × 2 mm, glabrous; **Ci** lobes robust, ovate, forming ± horizontal pouches, ± 0.7 mm, entire, laterally directly fused with the **Cs** base; **Cs** 1.5 - 2 mm, lanceolate, with revolute tips; **Poll** attenuate-ovoid, 0.22 × 0.14 mm; **Fr** narrowly fusiform, 6 - 14 cm.

C. pachystelma is inconspicuous due to its greenish hairy flowers. It belongs to the complex comprising *C. africana, C. linearis, C. rendallii* etc.

C. papillata N. E. Brown (BMI 1898: 308, 1898). **T:** Malawi (*Whyte* s.n. [K]). – **D:** Tanzania, Zaïre, Zambia, Malawi. **I:** FPA 43: t. 1716, 1976, as var. *cordiloba*. **Fig. XI.h**

Incl. *Ceropegia cordiloba* Werdermann (1939) ≡ *Ceropegia papillata* var. *cordiloba* (Werdermann) H. Huber (1957).

[2] **R** tuber 2 - 3 cm ⌀, with fissured bark; stems sparsely branched, hairy, twining, 0.5 - 1 (-2) m, 1.5 mm ⌀; **L** 12 - 25 mm petiolate, lamina cordate-lanceolate, ± distinctly acuminate, 2.5 - 6.5 × 1.3 - 3 cm; **Inf** ± sessile, 5- to 12-flowered; **Ped** 8 - 12 mm; **Sep** narrowly triangular, 3 - 5 mm, villose; **Cl** slightly bent, 1.2 - 2.2 cm, outside whitish-green; **Cl** inflation globose to bulbous, narrowed at the base, inside with stout papillae along the veins, ± 4 × 5 mm, merging fairly abruptly into the nearly cylindrical **Cl** tube, tube ± 1.5 mm ⌀, suffused with purple, mouth widening to 3.5 - 4 mm ⌀, inside glabrous; **Cl** lobes 4 - 7 × 2 mm, linear-spatulate from a triangular base, lamina folded back along the midrib, basal margins revolute, tips fused into a globose cage, inside velvety, with fine white **Ha**, upper ½ and margins purple, keel white; **Cn** sessile, white-hyaline, ± 2 × 2.2 mm, ± cup-shaped; **Ci** erect, linear to ovate, 1 - 1.5 mm, apically incised, basal ½ often broadened, pouch-like, occasionally the entire **Cn** fused into a cup with emarginate margin; **Cs** 1.5 - 2 × 0.3 mm, linear, erect-connivent, tips slightly clavate, often revolute and papillose; **Poll** ovoid, ± 0.25 × 0.12 mm; **Fr** narrowly fusiform, 10 - 12 cm × 2 - 3 mm.

The coronal structure is extremely variable. *C. papillata* is closely related to *C. claviloba*.

C. paricyma N. E. Brown (BMI 1898: 309, 1898). **T:** Malawi (*Simons* s.n. [K]). – **D:** Tanzania, Zambia, Zimbabwe, Moçambique, Namibia, Malawi. **I:** Bruyns (1984).

Incl. *Ceropegia dentata* N. E. Brown (1909); **incl.** *Ceropegia mutabilis* Werdermann (1939).

[2] **R** tuber small, roundish, with fissured bark; stems twining, 1 - 2 m tall, annual, ± 2 mm ⌀; **L** 1 - 2 cm petiolate, lamina elliptic, ovate to broadly lanceolate, nearly triangular, ± 3 - 10 cm, delicate, glabrous or slightly hairy; **Inf** usually shortly pedunculate, usually in pairs, 2- to 6-flowered; **Sep** narrowly triangular, ± 2 mm; **Cl** 1.5 - 2.5, cream-coloured with purple; **Cl** inflation globose, ± 5 mm ⌀, inside and outside pubescent, ± gradually narrowed to 2 mm ⌀; **Cl** tube slender, mouth not expanding; **Cl** lobes linear-spatulate, 4 - 12 mm, tips fused, lamina entirely folded back along the midrib, basal margins revolute, inside dark green to black, hairy; **Cn** sessile, dish-shaped, ± 3 × 3 mm; **Ci** lobes triangular, ± 1 mm, bilobed into subulate spreading appendages, shortly hairy on the upper face; **Cs** 2.5 - 3 mm, subulate, erect-connivent with recurved tips; **Fr** narrowly fusiform, ± 10 cm.

Remarkable by its large leaves and the roughly fissured surface of the tubers. Most closely related to *C. sobolifera* and *C. papillata*.

C. petignatii Rauh (Trop. subtrop. Pfl.-welt 85: 8-17, ills., 1993). **T:** Madagascar, Toliara (*Rauh* 72226 [HEID]). – **D:** S Madagascar. **Fig. XII.b**

Incl. *Ceropegia armandii* var. *petignatii* Rauh ms. (s.a.) (*nom. inval.*, Art. 29.1).

[3] Stem succulents of creeping-decumbent habit, occasionally with tuberous **R**; stems acutely 4-angled, compressed and ± rectangular in cross-section, 1 - 4 cm thick, thickest at the nodes, grey-green, rough-verrucose, with deflexed elongated **L** bases, **L** petiole canaliculate on the upper face, with stipules, lamina dark green, succulent, obovate to lanceolate, 1 - 4 × 0.7 - 2 cm, shortly mucronate, margins occasionally undulate, with hooked **Ha**, caducous; **Inf** only on twining thin terete stems to several m long; peduncle fleshy, 5 - 20 mm, horizontally spreading, with subulate **Bra**; **Inf** cymose, 2- to 4-flowered; **Ped** 8 - 15 mm; **Sep** subulate, 3 - 5 mm; **Cl** 3.2 - 4 cm, pale yellow to yellow-green, partly stippled and striped with red; **Cl** inflation inflated-cylindrical, 7 - 10 × 7 - 10 mm, transversely narrowed and merging abruptly into the **Cl** tube, tube basally ± 3 mm ⌀, towards the mouth gradually widening to 9 - 12 mm ⌀, inside scatteredly hairy; **Cl** lobes 6 - 9 mm, folded back along the midrib, keel pale yellow, to 4 mm broad, extending a little down the **Cl** tube, with fine white **Ha**, mouth deep yellow with purple, with vibratile **Ha** 2 mm long, inside with fine white **Ha**; **Cn** sessile, greenish-white, ± 3 × 3 mm, shallowly bowl-shaped; **Ci** lobes bilobed down to the **Cn** margin into subulate appendages, ± 2 × 0.2 mm, ± erect; **Cs** 3 × 0.3 mm, subulate, erect-connivent with papillose tips; **Poll** broadly ovoid-rectangular, ± 0.3 mm ⌀; **Fr** brown-

green, mottled, erect, narrowly fusiform, ± 12 cm × 4 mm; **Se** dark brown, ± 7 × 3 mm.

Together with *C. bosseri* and *C. hofstaetteri* forming a complex within the natural clade of 'dimorphic' Ceropegias (= Sect. *Dimorpha*).

C. poluniniana Bruyns (KB 44: 723-726, ills., 1989). **T:** Nepal (*Bruyns* 2497 [K]). – **D:** Nepal.

[1] **R** fleshy, fusiform, in small clusters; stems unbranched, 25 (-60) cm, ± terete, ± 1 mm ⌀, finely hairy, short-lived; **L** linear, papery, 30 - 55 × 3 - 4 mm, margins revolute, midrib raised on the lower face, very finely hairy; **Inf** 8 - 11 mm pedunculate, 1-flowered; **Ped** 5 - 7 mm, finely hairy; **Sep** filiform, acute, ± 4 mm, finely hairy; **Cl** 2.7 - 3.6 cm, greenish-white; **Cl** inflation little widened, 7 - 9 × 5 - 6 mm; **Cl** tube gradually narrowed to 2.5 mm, mouth little expanding, ± 4 mm ⌀, glabrous; **Cl** lobes whitish at the base, then brown-violet, apically becoming yellow-green, 14 - 17 mm, above a triangular base lamina ± entirely folded back along the midrib, basal ½ of the margins with 2 - 3 mm long **Ha**; **Cn** very shortly stipitate, shallowly cup-shaped, white, but **Ci** appendages and **Cs** base blackish; **Ci** lobes broadly triangular, bifid into short triangular appendages, with erect **Ha**; **Cs** ± cylindrical, ± 3 × 0.3 mm, erect, ventrally coming into contact almost for their entire length, tips scarcely recurved; **Poll** ovoid, ± 0.27 × 0.21 mm.

A fairly isolated species. The flowers are, however, reminiscent of those of the W and C African *C. linophylla*.

C. porphyrotricha W. W. Smith (Notes Roy. Bot. Gard. Edinburgh 13: 307-308, 1922). **T:** Ghana (*Dalziel* 60 p.p. [K]). – **D:** Equatorial Guinea, Benin, Ghana, Nigeria. **I:** Bullock (1963).

≡ *Ceropegia campanulata* var. *porphyrotricha* (W. W. Smith) H. Huber (1957).

[2] **R** tuber flattened; stems erect, not twining, herbaceous, 20 - 30 cm, annual, ± unbranched, scatteredly hairy; **L** ± sessile, lanceolate to linear, acute, ± 5 - 7 × 0.6 cm, scatteredly hairy on the upper face, one **L** pair conspicuously larger than the others; **Inf** sessile, terminal or lateral, 1-flowered; **Cl** greenish with dark purple, 6 - 14 cm, straight, outside hairy; **Cl** inflation ovoid, 4 - 8 mm ⌀, merging very gradually into the **Cl** tube, tube 3 - 4 mm ⌀; **Cl** lobes erect and straight, tips joined, narrowly triangular to linear, 2.5 - 6 (-9) cm, margins revolute, with purple **Ha** at the base; **Cn** sessile, shallowly cup-shaped; **Ci** lobes horizontal-ascending, bifid into narrowly triangular to linear appendages, finely hairy; **Cs** lobes linear, erect-connivent, tips recurved.

Another element from the *C. dinteri*-group, which should be united into a single species together with *C. campanulata*.

C. praetermissa J. Raynal & A. Raynal (Adansonia, n.s., 7: 307-309, ills., 1967). **T:** Senegal (*Raynal* 6367 [ALF, P]). – **D:** Senegal, Tanzania.

[2] **R** tuber flattened to beetroot-shaped, 3 - 6 mm ⌀; stems annual, delicate, strongly twining, to 10 - 25 cm, 1.5 mm ⌀, little hairy; **L** 0.5 - 2 cm petiolate, lamina ovate-lanceolate with cordate base, 3 - 9 × 2 - 8 cm, slightly fleshy, shortly pointed, ± scatteredly hairy; **Inf** pseudo-umbels, 3- to 10-flowered; peduncle 1 - 2 cm, hairy; **Ped** ± 1 cm, hairy; **Sep** subulate, 2 mm, hairy; **Cl** slightly bent, 1.5 - 3 cm, whitish-greenish, weakly striped or blotched with grey-green and red-brown, outside glabrous or finely and scatteredly hairy, inside glabrous; **Cl** inflation depressed-globose, 3 - 5 mm ⌀, inside greenish-white, merging gradually into the **Cl** tube, tube 2 mm ⌀, mouth widening hardly to 5 mm ⌀; **Cl** lobes inside purple at the base, green or white at the tip, 6 - 8 mm, linear, erect, lamina strongly folded back along the midrib, tips fused into a cage, sometimes slightly spiralling, ± hairy; **Cn** sessile, shallowly bowl-shaped, ± 3 × 2 mm; **Ci** lobes broadly 4-angled, concave, with slightly drawn-out tips and subulately drawn out erect lateral margins; **Cs** 2 - 2.5 mm, narrowly lanceolate, flat, erect, ± without coming into contact; **Fr** greenish, mottled with red-brown, 8 - 12 cm × 3 mm; **Se** obovate, 5.5 × 2.2 mm.

Together with *C. linophylla* belonging to the *C. bulbosa*-group. *C. praetermissa* is most notably distinguished from *C. bulbosa* by the indumentum on the exterior of the sepals and the corolla.

C. pubescens Wallich (Pl. Asiat. Rar. 2: 81, t. 187, 1831). **T:** Nepal (*Wallich* s.n. [K, W]). – **D:** India, Nepal, Bhutan, Myanmar, China; highlands. **I:** Hart (1988).

Incl. *Ceropegia tsaiana* Tsiang (1939).

[1] **R** only very slightly thickened and fleshy; stems robust, twining, to 1.5 m, ± glabrous; **L** petiole 3 - 4.5 cm, villose or glabrous, lamina ovate, ± 3 - 16 × 0.1 - 6 cm, with drip-tip, very delicate, sometimes hairy on the upper fac; **Inf** few- or many-flowered; peduncle 15 - 25 mm, usually with **Ha** arranged in 1 row; **Ped** ± 1 - 1.5 cm; **Sep** lanceolate, 1.5 - 3 mm, ± spreading; **Cl** 4 - 6.5 cm, whitish-yellow to greenish, often with longitudinal reddish stripes; **Cl** inflation ovoid, basally narrowed, ± 10 × 6 mm; **Cl** tube gradually narrowed to ± 3 mm ⌀, mouth 3 - 5 mm ⌀, inside with long **Ha**, outside occasionally villose; **Cl** lobes 20 - 33 mm, basal ⅓ triangular-ovate, margins revolute and basally projecting, inside pale yellow, yellow, orange and/or green, upper ⅔ linear, ± entirely folded back along the midrib, inside purple-brown, tips fused into an ellipsoid cage, sometimes also spiralling; **Cn** shortly stipitate, ± 4 × 3 mm, basally fused into a very shallow cup; **Ci** lobes pouch-like, very short, concave, obtuse, ± setulose; **Cs** linear-spatulate to cylindrical, to 3.5 mm, parallel-erect-connivent, base dorsally slightly broadened, ± hairy; **Poll** D-

shaped; **Fr** very narrowly fusiform, 13 cm × 2 mm; **Se** lanceolate, 10 × 1 mm, tuft of **Ha** 35 mm.

A species of the highlands with conspicuous intensely coloured flowers. Allied to *C. hookeri* and *C. meleagris*.

C. purpurascens K. Schumann (BJS 17: 152, 1893). **T:** Angola (*Mechow* 122 [Z]). – **D:** Tanzania, Zaïre, Zimbabwe, Botswana, Angola, Namibia. **I:** Bruyns (1984). **Fig. XII.c**

Incl. *Ceropegia kwebensis* N. E. Brown (1903); **incl.** *Ceropegia kassneri* S. Moore (1910); **incl.** *Ceropegia thysanotos* Werdermann (1939) ≡ *Ceropegia purpurascens* ssp. *thysanotos* (Werdermann) H. Huber (1957).

[2] **R** tuber flattened with soft delicate bark, 2 - 6 cm ∅; stems twining, 1 - 2 m tall, annual, 2 - 3 mm ∅, partly with underground stem tubers; **L** to 1 cm petiolate, elliptic to cordate-attenuate, 2 - 4 (-7) × 1.5 - 2.5 cm, slightly fleshy, rarely weakly hairy; **Inf** few- to many-flowered; peduncle 1 - 2 cm, slightly descending, pubescent (?); **Ped** 5 - 8 mm; **Sep** subulate, ± 2 mm; **Cl** ± 2 cm, outside greenish, inside purple and scatteredly hairy; **Cl** inflation 4 - 5 mm ∅, ovoid; **Cl** tube narrowed to ± 2 mm, bent almost at a right angle, towards the mouth gradually widening to 6 - 7 mm; **Cl** lobes ± linear, 5 - 9 mm, tips fused into a widely opened cage, lamina folded back along the midrib, inside scatteredly hairy; **Cn** cup-shaped; **Ci** lobes triangular, ± erect, 1 - 2 × 2 - 3 mm, bifid into glabrous or hairy appendages; **Cs** lobes ± 2 mm, narrowly spatulate to falcate, erect-connivent, tips rounded or denticulate, ± recurved; **Fr** narrowly fusiform, ± 10 cm; **Se** ± 8 × 2 mm.

Within the *C. africana / C. linearis* group, *C. purpurascens* is very closely allied to *C. imbricata*, which can be separated from *C. purpurascens* by its linear leaves, the almost glabrous interior of the corolla and the purple marginal hairs of the corolla lobes.

C. pusilla Wight & Arnott (Contr. Bot. India, 31, 1834). **T:** India, Tamil Nadu (*Wight* 1932 [CAL, W]). – **D:** S India (Tamil Nadu). **I:** Ansari (1984).

[2] **R** tuber 2 - 3 cm ∅; stems delicate, erect, not twining, merely 5 - 10 cm tall, with short **Int**, ± unbranched, ± hairy; **L** ± linear, 5 - 8 cm, ± sessile, upper face hairy; **Inf** 1-flowered, shortly pedunculate; **Ped** 6 - 7 mm, hairy; **Cl** 2.5 - 2.8 cm, glabrous, colour unknown; **Cl** inflation ovoid, ± 5 mm ∅, inside glabrous, merging gradually into the **Cl** tube, tube ± 3 mm ∅, apically widening to ± 6 mm ∅; **Cl** lobes 9 - 13 mm, linear from a triangular base, largely reflexed, glabrous; **Cn** cup-shaped, sessile; **Ci** lobes narrowly triangular, from the upper margin of the coronal cup, ± erect, 1.5 mm, incised or bifid into triangular appendages; **Cs** ± 2 mm, linear, slightly clavate, erect-connivent; **Fr** 4 × 0.5 cm; **Se** broadly ovate.

See note under *C. noorjahaniae*.

C. radicans Schlechter (BJS 18(Beiblatt 45): 12, 1894). **T:** RSA, Eastern Cape (*Flanagan* 348 [K, BOL, NH, SAM, PRE]). – **D:** RSA (Eastern Cape).

C. radicans ssp. **radicans** – **D:** RSA (Eastern Cape); creeping in and below shrubs. **I:** Bruyns (1985).

[1,3,4] Creeping stem and **L** succulents, robust, little branched, easily rooting at the nodes, glabrous; **R** fleshy, fusiform, clustered (also at the nodes); stems terete, 3 - 5 mm ∅, dark green, slightly rough, striate; **L** 0.3 - 1 cm petiolate, lamina ovate-elliptic to nearly orbicular, to 4 × 2.5 cm, acute, fleshy, persistent; **Inf** peduncle 5 - 10 mm × 2 - 3 mm, 1- to 3-flowered; **Ped** 1 - 3 cm × 2 mm; **Sep** linear-lanceolate, 4 - 6 mm; **Cl** 5.5 - 8 cm; **Cl** inflation cylindrical, weakly 5-angled, 12 - 17 × 5 - 7 mm, whitish-green, inside hairy at the base; **Cl** tube basally narrowed to ± 3 mm ∅ and hairy inside, widening gradually to a funnel-shaped mouth of 15 mm, apically blotched with purple; **Cl** lobes obtusely lanceolate to triangular, 20 - 25 mm, broadly keeled, tips fused, folded back along the midrib, basal angles projecting, inside basally purple-brown with a white and then a black-purple transverse stripe and hairy, upper ½ pale green, margins with purple vibratile clavate **Ha**; **Cn** shortly stipitate, fused into a cup with the upper margin ± straight and slightly revolute; **Ci** lobes pouch-like; **Cs** ± 4 × 0.4 mm, linear, erect, ± connivent, tips sometimes recurved.

Most closely related to *C. sandersonii* and thus allied to *C. fimbriata* and *C. zeyheri*, all of which are characterized by succulent stems, leaves, and roots. The robust long-lived plants are easily cultivated.

C. radicans ssp. **smithii** (M. R. Henderson) R. A. Dyer (in Leistner & al. (ed.), Fl. South Afr. 27(4): 62, 1980). **T:** RSA, Eastern Cape (*Smith* 5318 [NBG]). – **D:** RSA (Eastern Cape). **I:** Bruyns (1985).

≡ *Ceropegia smithii* M. R. Henderson (1945) ≡ *Ceropegia radicans* var. *smithii* (M. R. Henderson) H. Huber (1957).

[1,3,4] Differs from ssp. *radicans*: **L** smaller with slightly sinuate-undulate margins; **Cl** lobes shorter and ± obtuse.

C. rendallii N. E. Brown (BMI 1894: 100, 1894). **T:** RSA (*Rendall* s.n. [K]). – **D:** Zimbabwe, RSA (KwaZulu-Natal, Northern Prov.), Swaziland. **I:** Dyer (1980); Dyer (1983). **Fig. XIII.a**

Incl. *Ceropegia galpinii* Schlechter (1894).

[2,4] **R** tuber flattened, 2 - 7 cm ∅; stems to 1 m, twining, terete, 1 - 2 mm ∅, green, red with age; **L** short and narrowly petiolate, broadly ovate to cordate, ± mucronate, 1 - 3 × 1 - 1.5 cm, upper face green, lower face grey-green, slightly fleshy; **Inf** shortly pedunculate, 1- to 5-flowered, long-lived; **Ped** 3 - 6 mm, **Sep** subulate, 2 - 3 mm; **Cl** 1.8 - 2.8

cm, whitish-greenish; **Cl** inflation ± 5 × 3 mm, basally cylindrical and white, then constricted, upper ½ globose to 5-angled and greenish with 5 purple stripes, inside purple, glabrous; **Cl** tube basally 1 mm ⌀, straight or slightly bent, towards the mouth gradually widening to 4 - 6 mm, inside whitish with purple stripes, with white **Ha**; **Cl** lobes ± 3 mm, basal ½ linear, erect, folded back along the midrib, inside purple, upper ½ cordate, spreading horizontally and fused into an umbrella-like structure, 6 - 12 mm ⌀, brown-green, all free margins with fine purple cilia; **Cn** sessile, cup-shaped, ± 2 × 2.5 mm, white; **Ci** lobes drawn out into deep ovoid pouches that are directly connate with the **Cs** base, margins entire or divided into short triangular appendages, usually purple, hairy on either side; **Cs** lobes ± 2 × 2 mm, broadly falcate, erect-connivent, strongly reflexed from the middle; **Poll** ovoid-attenuate, ± 0.2 × 0.14 mm.

C. rendallii belongs to the complex of *C. africana*, *C. multiflora* and *C. linearis*, and is as easily cultivated and rewarding as the latter.

C. ringens A. Richard (Tent. Fl. Abyss. 2: 47-48, 1851). **T:** Ethiopia (*Quartin Dillon & Petit* s.n. [P]). – **D:** Ethiopia, Eritrea.

Incl. *Ceropegia sinuata* Decaisne *ex* A. Richard (1851).

Except for the apically free corolla lobes, the protologue matches the later described *C. stenoloba*, which thus would become a younger synonym of *C. ringens*. However, the type specimen has not been seen to corroborate this view. The synonymy of *C. sinuata* under *C. ringens* as accepted since Brown (1903) is questionable.

C. ringoetii De Wildeman (Bull. Jard. Bot. État 4: 394, 1914). **T:** Zaïre (*Ringoet* s.n. in *Homblé* 553 [BR]). – **D:** Tanzania, Zaïre. **I:** Malaisse & Schaijes (1993).

Incl. *Ceropegia schlechteriana* Werdermann (1939).

[2] **R** tuber flat; stems with erect base, twining above, hairy, 1 - 3 mm ⌀; **L** 0.8 - 2 cm petiolate, ovate to lanceolate, to 10 × 4 cm, acute; **Inf** ± sessile, 3- to 12-flowered; **Ped** 4 - 12 mm; **Sep** subulate, villose; **Cl** 1.5 - 3 cm, straight, outside hairy; **Cl** inflation ovoid, 2 - 4 mm ⌀, merging gradually into the cylindrical **Cl** tube, tube apically 1 - 2 mm ⌀, inside black-purple; **Cl** lobes 1 - 2 cm, whip-shaped from a narrowly triangular base, tips free, ± revolute, inside cream-coloured, glabrous; **Cn** shortly stipitate, fused into a shallow cup; **Ci** lobes ± rectangularly pouch-like, ± erect, obtuse or slightly incised; **Cs** ± 1.2 mm, linear-spatulate, erect-connivent, tips involute, base and tips often hairy; **Fr** narrowly fusiform, ± 7 cm × 3 mm.

Related to *C. abyssinica*. The hairy exterior of the corolla and the flagellate free corolla lobes are significant characters.

C. robivelonae Rauh & Gerold (Succulentes 21(4): 3-9, ills., SEM-ills., 1998). **T:** Madagascar (*Robivelo* s.n. in *BG Heidelberg* 74007 [HEID]). – **D:** S Madagascar; Didiereaceae scrub.

[2] **R** tuber depressed-globose, 3 - 6 (-10) cm ⌀; stems twining, delicate, ± glabrous, ± annual, 1 - 2 mm ⌀; **L** petiole 3 - 5 mm, canaliculate above, lamina ± 2 × 1 cm, broadly lanceolate to rounded-triangular, cuspidate to acuminate, succulent; **Inf** ± 1 cm pedunculate, 2- to 4-flowered; **Ped** 15 mm; **Sep** 3 - 4 mm; **Cl** 3 - 4 cm, outside whitish, for the greater part dotted or blotched with red-brown, with large papillae; **Cl** inflation shortly cylindrical-ovoid, 6 mm ⌀; **Cl** tube somewhat geniculate, ± 15 × 2 - 3 mm, apically widening to ± 20 mm; **Cl** lobes 6 - 7 mm, tips fused and forming a cylindrical cage, inside finely papillose, basally whitish, apical ⅔ dark purple, margins recurved, with vibratile purple **Ha**; **Cn** subsessile, shortly fused into a bowl-shaped structure; **Ci** lobes erect, triangular, ± 1 mm, apically shortly hairy; **Cs** narrowly lanceolate-cylindrical, erect-connivent, ± 2.5 mm.

A leaf-succulent twining taxon with conspicuously papillose corolla; closely related to *C. ambovombensis*.

C. rudatisii Schlechter (BJS 40: 94, 1907). **T** [lecto]: RSA, KwaZulu-Natal (*Rudatis* 372 [K]). – **D:** RSA (Eastern Cape, KwaZulu-Natal). **I:** Dyer (1983).

[1] **R** fleshy and thickened, fusiform; stems erect, glabrous, not twining, herbaceous, 10 - 50 cm × 2 - 4 mm, annual, unbranched, robust; **L** ± sessile, ovate to elliptic, obtuse, often mucronate, 4 - 6 × 2 cm; **Inf** sessile, 1-flowered, lateral; **Ped** to 5 cm; **Sep** linear, acute, 1 - 1.5 cm; **Cl** 8 - 9 cm, erect, straight, outside greenish; **Cl** inflation globose to ovoid, 5 - 7 mm ⌀, merging very gradually into the **Cl** tube, tube ± 3 mm ⌀, mouth widening to 7 mm ⌀, inside veins papillose, purple; **Cl** lobes free, erect to divaricate, linear, 25 - 35 mm, folded back along the midrib, inside greenish-yellow, hairy, at the base with tufts of vibratile purple clavate **Ha**; **Cn** purple-brown, shortly stipitate, basally fused into a cup; **Ci** lobes triangularly pouch-like, erect, 1.5 - 2 mm, bilobed into subulate appendages, with white **Ha**; **Cs** ± 3 mm, erect-connivent; **Fr** narrowly fusiform, erect, ± 18 cm × 5 mm; **Se** ± 12 × 5 mm, with broad margin.

Plants of similar habit occur particularly in the *Umbraticola* group native to C Africa, e.g. *C. kundelunguensis* from Zaïre.

C. rupicola Deflers (Voy. Yemen, 2: 167-168, 1889). **T:** Yemen (*Deflers* 405 [P]). – **D:** Yemen. **I:** FPA 47: t. 1847, 1982. **Fig. XIII.b**

[3,4] Stem and **L** succulents, stems basally branched, ± terete with **Int** slightly compressed, 30 - 70 × 0.8 - 1.2 cm, grey-green; **L** shortly petiolate, ovate-lanceolate, 1.5 - 6 × 1 - 4 cm, slightly succu-

lent, caducous; **Inf** shortly pedunculate pseudo-umbels, many-flowered, 2 - 4 **Fl** simultaneously open; **Ped** ± 5 × 2 mm; **Sep** lanceolate-linear, 2 - 5 mm, acute; **Cl** 3.5 - 6 cm, with red-brown blotches on whitish-greenish background, base whitish without blotches; **Cl** inflation ovoid, 6 - 9 mm ⌀, basally slightly compressed, enclosing the **Gy**; **Cl** tube ± 25 × 3.5 - 4.5 mm ⌀, inside hairy, towards the mouth gradually widening to ± 12 mm; **Cl** lobes whitish, outside blotched with red-brown, inside marked with a fine reticulation, triangular, ± 7 mm, tips fused into a pyramidal roof, openings slit-shaped to ovate; **Cn** pale pink, ± 5 × 4.5 mm, ± sessile, fused into a cup, with long **Ha** on the margin; **Ci** lobes triangular, ± erect, bifid into triangular appendages, ± 1 × 0.5 mm, ciliate; **Cs** ± 3 - 4.5 × 0.4 mm, linear, erect-connivent with recurved tips; **Poll** ovoid, slightly tapering at the base, ± 0.35 × 0.25 mm, germination crest triangular, membranous, corpuscle ± 0.4 mm, apically expanding shield-like to dome-shaped; **Fr** pale green, cylindrical, ± 12 cm × 4 mm; **Se** 8 - 10 × 3 - 3.5 mm.

Morphologically fairly isolated, but most closely related to *C. aristolochioides*.

C. rupicola var. **stictantha** N. P. Taylor (Cact. Succ. J. Gr. Brit. 42(4): 111-112, ill., 1980). **T:** Yemen (*Radliffe-Smith & Henchie* 4437 [K]).

Presumably the natural hybrid between *C. rupicola* and *C. aristolochioides* (Bruyns 1988b).

C. salicifolia H. Huber (Mem. Soc. Brot. 12: 51-52, ill., 1957). **T:** China, Yunnan (*Henry* 9729 [K]). − **D:** China (Sichuan, Guangxi, Yunnan); mountain forests.

[1] **R** fusiform, fleshy, clustered; stems twining, to 1.5 m, glabrous; **L** petiole ± 1 cm, with soft **Ha** arranged in 2 rows, lamina lanceolate, 6 - 15 × 1 - 2 cm, acuminate, delicate, upper face scatteredly pubescent, lower face ciliate on the midrib; **Inf** 13- to 20-flowered; peduncle 3 - 10 mm, with soft **Ha** arranged in 2 rows; **Ped** to 1.5 cm, pubescent; **Sep** narrowly lanceolate, 5 mm, glabrous; **Cl** 3 - 4 cm, whitish; **Cl** inflation 4 - 6 mm ⌀; **Cl** tube narrowed to 2 - 3.5 mm ⌀, towards the mouth widening to 5 - 8.5 mm; **Cl** lobes 13 - 19 mm, broadly ovate, lamina folded back along the midrib for ± ½, broadly keeled, apically fused into a depressed-globose structure, inside with short **Ha**, ± ciliate; **Cn** fused into a cup; **Ci** lobes triangular with crenulate apex to deeply bifid into triangular appendages, ± hairy; **Cs** linear, erect.

The relationships of this taxon are obscure. It is presumably allied to *C. longifolia*, *C. macrantha* etc.

C. sandersonii Decaisne *ex* Hooker *fil.* (CBM 1869: t. 5792, 1869). **T:** RSA, KwaZulu-Natal (*Sanderson* s.n. [K]). − **D:** E RSA, Swaziland. **I:** Dyer (1983).

Incl. *Ceropegia monteiroae* Hooker *fil.* (1887).

[1,3,4] **R** fleshy, narrowly fusiform, clustered; stems climbing and twining, to 2 m tall, robust, sparsely branched, glabrous, terete, 4 - 5 mm ⌀, dark green, rough; **L** 6 mm petiolate, lamina ovate-lanceolate to cordate-ovate, 2 - 5 × 1.2 - 2.5 cm, acute, fleshy, persistent; **Inf** pedunculate, 5 - 10 × 2 - 3 mm, 1- to 4-flowered; **Ped** ± 10 × 2 mm; **Sep** linear-lanceolate, ± 6 mm; **Cl** 4 - 7 cm, straight to slightly bent; **Cl** inflation ovoid-cylindrical, ± 6 mm ⌀, whitish-green, inside hairy at the base; **Cl** tube basally narrowed to ± 3 mm ⌀ and inside hairy, towards the mouth gradually widening to 15 - 25 mm, striped with greenish or whitish; **Cl** lobes lanceolate-cordate, first erect, then horizontal and fused into a broad umbrella of 2.5 - 5 cm ⌀, umbrella outside whitish, stippled with green, centrally often depressed, outer margins slightly pointing outwards and with purple vibratile clavate **Ha**; **Cn** shortly stipitate, fused into a shallow cup with ± entire margin; **Ci** lobes pouch-like; **Cs** ± 3 × 0.3 mm, linear, erect, ± connivent, tips sometimes recurved; **Poll** ovoid, 0.4 × 0.2 mm; follicles spreading horizontally, 7.5 - 13 cm × 7 - 8 mm, verrucose, blotched with purple.

See under *C. radicans* for a note. The thick verrucose fruits are reminiscent of those of *C. stapeliiformis*.

C. santapaui Wadhwa & Ansari (Bull. Bot. Surv. India 10(1): 95-97, ills., 1968). **T:** India, Maharashtra (*Wadhwa* 109640 [CAL, BSI, K]). − **D:** C India. **I:** Ansari (1984).

Incl. *Ceropegia huberi* Ansari (1969).

[2] **R** tuber ovoid to globose, 2 - 4.5 × 1.5 - 4 cm; stems twining, herbaceous, to 1.5 m, nearly glabrous; **L** petiole to 3.5 cm, glabrous, upper face canaliculate, lamina broadly to narrowly lanceolate, 5 - 12 × 2 - 4 cm, somewhat robust, with basally thickened cilia; **Inf** 4- to 9-flowered, pseudo-umbellate, lateral or terminal; peduncle 2 - 10 cm, with long **Ha**; **Ped** 5 - 15 mm, pubescent; **Sep** narrowly triangular, 3 - 4 mm, lower face ± hairy; **Cl** white to pale pink, 1 - 1.5 cm, ± straight, outside with slightly villose veins, inside glabrous; **Cl** inflation ovoid, ± 5 mm ⌀, basal ½ with purple inside, merging very gradually into the ± narrowed **Cl** tube, tube ± 3 mm ⌀ at the mouth, scarcely or not widening; **Cl** lobes broadly ovate to cordate, 4 - 7 × 4 - 10 mm, tips fused, acuminate, forming a depressed-globose and scarcely opened cage or widening into a broadly umbellate structure, lamina flat and margins folded back only basally, glabrous; **Cn** sessile, pale yellow to cream-coloured, ± 2.5 × 2 mm; **Ci** lobes reduced to a fringe hardly visible; **Cs** narrowly triangular, 2 mm, erect-connivent, dorsally hairy; **Poll** ellipsoid to almost square; **Fr** narrowly fusiform, 6 - 7 cm × 5 mm, nearly parallel; **Se** ± 6 × 3 mm, tuft of **Ha** 10 - 15 mm.

A very exceptional species due to the indumentum, the flower size and the reduction of the interstaminal corona. *C. huberi* appears distinct on first glance but, nevertheless, is to be treated as a synonym characterized by broader corolla lobes and less reduced interstaminal corona. In addition, both taxa are based on single specimens collected from the same region. The extremely conical pollinium and the narrowly elliptic corpuscle as shown by Ansari (1971) for *C. huberi* probably belong to a *Pentasacme* species. Furthermore, the flowers are reminiscent of *C. hookeri*.

C. saxatilis Jumelle & H. Perrier (Ann. Inst. Bot.-Géol. Colon. Marseille, sér. 2, 6: 223-224, 1908). **T:** Madagascar (*Perrier* 1688 [P]). – **D:** N Madagascar. **I:** Huber (1957). **Fig. XIII.c**

Incl. *Ceropegia contorta* Jumelle & H. Perrier (1908).

[2] **R** tuber small; stems twining, ± 0.5 - 1 (-2) m, ± 2 mm ⌀, green, glabrous, **L** petiole 0.5 - 2 cm, upper face with **Ha** arranged in 2 rows, lamina ovate-cordate, soft, 3 - 6 × 1.5 - 2.2 cm, acute, upper face green, lower face pale green, veins slightly prominent; **Inf** 2- to 4-flowered, pseudo-umbellate, shortly pedunculate; **Ped** 0.5 - 1 cm; **Cl** 2 - 3.8 cm, slightly bent, green-white, coarsely blotched with purple; **Cl** inflation with globose inflation, 5 - 9 mm ⌀, basally weakly constricted; **Cl** tube in the middle narrowed to 2 - 3 mm, towards the mouth widening to 7 - 9 mm; **Cl** lobes 11 - 18 mm, ± triangular, upper ⅓ distinctly narrowed, acute or obtuse, tips fused, lamina largely folded back along the midrib, upper ⅓ purple, ± pubescent; **Cn** sessile, 3 - 4 × 2 - 3 mm, basally bowl-shaped and fused to the **Cl**, glabrous; **Ci** lobes bilobed into linear obtuse appendages, ± 1.5 mm, purple; **Cs** ± 2 - 3 mm, linear, often slightly spatulate, obtuse, erect-connivent, purple at the base, otherwise white; **Poll** ovoid; **Fr** narrowly fusiform, follicles spreading at ± 180°, ± 10 cm × 4 mm; **Se** ovate, ± 6 × 2.5 mm, dark brown, margin denticulate.

The dense and coarse purple blotches covering the entire corolla tube are particularly striking. In habit very similar to *C. humbertii* and *C. scabra*.

C. scabra Jumelle & H. Perrier (Ann. Inst. Bot.-Géol. Colon. Marseille, sér. 2, 16: 220-223, 1908). **T:** Madagascar (*Perrier* 11626 [P]). – **D:** W Madagascar.

[2] **R** tuber depressed-globose, 3 - 4 cm ⌀; stems persistent, twining, 1 - 2 mm ⌀, dark; **L** 8 - 11 mm petiolate, lamina ovate to lanceolate, 7 - 16 × 2 - 4 cm, acute, base round to ± obtuse; **Inf** shortly pedunculate, 2- to 3-flowered; **Ped** long; **Cl** 7 - 8 cm (at times perhaps also shorter), strongly bent, pale green with purple dots; **Cl** with an ovoid inflation, ± 30 × 12 mm, inside with papillose veins; **Cl** tube 10 mm long, narrowed to 3 - 4 mm in the middle, widening apically to ± 15 mm; **Cl** lobes 2.5 - 3 cm, very narrowly triangular, nearly filiform from a triangular base, erect, very acute, tips fused into a large open ellipsoid-cylindrical cage, lamina entirely folded back along the midrib, hairy; **Cn** sessile, basally fused into a bowl; **Cs** bifid into triangular appendages, ± 1 mm, hairy; **Cs** ± 4 mm, linear, obtuse, erect-connivent, hairy at the base; follicles parallel, 18 cm, slender.

C. scabra is closely related to *C. saxatilis* and is forming a group of its own with *C. humbertii*, *C. madagascariensis* and *C. striata*.

C. sepium Deflers (Voy. Yemen, 2: 167, 1889). **T:** Yemen (*Deflers* 382 [P]). – **D:** N Yemen. **I:** Bruyns (1988b).

[1] Primary **R** somewhat thickened, with a cluster of fleshy fusiform lateral **R**; stems strongly twining, few-branched, terete, 1 - 2 (-4) mm ⌀, sparsely hairy; **L** petiolate, ovate-lanceolate (to linear), 4 - 8 × 1 - 3.5 cm, sparsely villose; **Inf** pedunculate, pseudo-umbellate, many-flowered; peduncle 1 - 3 cm; **Ped** 5 - 15 mm; **Sep** subulate, ± 4 mm, finely hairy; **Cl** 2 - 2.5 cm, outside white except the grey-green base, hairy; **Cl** inflation inside purple, widening into an ovoid form, 10 - 15 × 3.5 - 4 mm, inside with a circle of **Ha** at the mouth, abruptly merging into the straight **Cl** tube, tube 2 mm ⌀, towards the mouth widening to 4 mm; **Cl** lobes yellow-green, 8 - 12 mm, tips fused, folded back along the midrib and then 1 - 1.5 mm wide, keel rounded; **Cn** sessile, cup-shaped, ± 3 × 2.5 mm; **Ci** lobes triangular-ovate, concave, bifid into short erect triangular round-tipped appendages, inside with erect **Ha**; **Cs** ± cylindrical (-clavate), erect, 2 - 3 mm, ventrally coming into contact nearly over the entire length, apically papillose; **Poll** ovoid, ± 0.2 × 0.18 mm, with short germination crest near the tip, corpuscle apically widening.

This species is somewhat isolated in Arabia and presumably more closely related to the African *C. convolvuloides*.

C. simoneae Rauh (Trop. subtrop. Pfl.-welt 85: 20-25, ills., 1993). **T:** Madagascar, Toliara (*Rauh* 73113 [HEID]). – **D:** S Madagascar. **Fig. XIII.d**

[1,3] Stem succulents with a small tuber; stems creeping-decumbent, acutely 4-angled, compressed, < 1 cm thick, greatest width at the nodes because of descending elongated **L** bases, grey-green or grey-violet, verrucose; **L** shortly petiolate, dark green, succulent, narrowly ovate, acuminate, ± 5 × 3 mm, pointing downwards, margins with hooked **Ha**; **Inf** only from twining cylindrical stems; peduncle fleshy, ± 1 (-7) cm; **Inf** cymose, few-flowered; **Ped** ± 1 cm; **Sep** narrowly triangular, 3 mm; **Cl** 6 - 7 cm, grey-green, blotched ± with purple; **Cl** inflation inversely bulbous to obovoid, 8 - 10 × 6 - 7 mm, inside dark wine-red; **Cl** tube narrowed ± abruptly to ± 3 mm ⌀ in the middle, mouth widening again to ± 8 mm ⌀; **Cl** lobes 2 - 2.5 mm, narrowly linear and folded back along the midrib, with dense purple **Ha**,

tips free, ± spiralling and pendulous, from a broadly triangular base and there broadly keeled, centre of the keel with white rigid **Ha**, inside grey-green stippled with purple; **Cn** ± sessile, greenish-white, blotched with purple, ± 3 × 3 mm, shallowly bowl-shaped, outside hairy; **Ci** lobes bilobed down to the margin of the **Cn** into subulate appendages, ± 2 mm, ± erect, tips glued together; **Cs** 2 × 0.3 mm, linear, erect-connivent; **Poll** ovoid, ± 0.3 × 0.2 mm.

A magnificient member of the group of stem succulent dimorphic Ceropegias.

C. sinoerecta M. G. Gilbert & P. T. Li (Novon 5(1): 4, 1995). **T:** China, Yunnan (*Delavay* 2625 [P]). – **D:** China (Yunnan); on limestone.

[1] **R** fleshy, fusiform, clustered; stems erect, to 20 cm, finely pubescent; **L** petiole 4 - 6 mm, weakly winged, lamina elliptic, 2 - 5 × 0.6 - 1.6 cm, acute, upper face densely pubescent, lower face pale and with **Ha** along the midrib; **Inf** 4 - 17 mm pedunculate, pseudo-umbellate, 2- to 4-flowered; **Ped** 5 - 17 mm; **Sep** linear-lanceolate, 3 mm; **Cl** 3.6 - 4.3 cm, glabrous; **Cl** inflation ovoid-ellipsoid, 4 - 6 mm ⌀, lower ½ dark, upper ½ pale; **Cl** tube towards the mouth narrowed gradually to 3.5 - 5.5 mm ⌀; **Cl** lobes 14 - 15 mm, linear, inside dark and sparsely pubescent, tips fused and abruptly bent inwards, forming a cylindrical cage; **Cn** basally fused; **Ci** lobes bilobed into linear appendages, ± ciliate; **Cs** linear, ± 2.5 mm, erect.

Presumably most closely allied to *C. mairei*.

C. sobolifera N. E. Brown (BMI 1895: 261, 1895). **T:** Ethiopia (*Schimper* 463 [K, G, P]). – **D:** Ethiopia, Kenya, Tanzania, Zimbabwe.

C. sobolifera var. **nephroloba** H. Huber (Mem. Soc. Brot. 12: 201, 1957). **T:** Tanzania (*Milne-Redhead & Taylor* 8863 [K, G]). – **D:** Tanzania, Zimbabwe.

[2] Differs from var. *sobolifera*: More delicate and smaller; **Cl** lobes reniform, almost as broad as long, 3 - 4.5 × 3.5 - 4.5 mm.

C. sobolifera var. **sobolifera** – **D:** Ethiopia, Kenya. **I:** Werdermann (1939). **Fig. XIII.e**
  Incl. *Ceropegia subtruncata* N. E. Brown (1895).

[2] **R** tuber small, roundish; stems twining, annual, ± 1 mm ⌀, with **Ha** arranged in 1 row; **L** ovate to lanceolate, 1 - 5 cm, shortly petiolate, delicate, glabrous or sparsely hairy; **Inf** sessile, rarely shortly pedunculate, 1- to 5-flowered; **Ped** 3 - 5 mm, hairy; **Sep** lanceolate, 2 mm, scatteredly hairy; **Cl** 1 - 2.5 cm, cream-coloured to greenish; **Cl** inflation ± ½ as long as the **Cl**, ovoid, 4 - 6 × 3 - 4 mm, mouth inside with white **Ha**; **Cl** tube very gradually narrowed to apically 2 mm ⌀, ± straight, abruptly widening into a cage 5 - 7 mm wide; **Cl** lobes 7 - 12 × 3.5 mm, broadly ovate to cuneate, tips fused, folded back along the midrib, margins revolute at the base, outside yellow-green, inside dark red, with ± white or purple cilia; **Cn** sessile, dish-shaped, ± 3.5 × 3 mm, glabrous; **Ci** lobes triangular, ± 1.5 mm, bilobed into subulate spreading appendages, hardly 1 mm, shortly hairy; **Cs** ± 3 × 0.3 mm, linear-subulate, acute, vertical, not or faintly connivent, tips hook-like inflexed, touching each other; **Poll** ovoid, 0.2 - 0.25 × 0.15 mm.

Closely allied to *C. paricyma* and *C. papillata*. Valuable diagnostic characters include the broad corolla lobes and the long strictly vertical staminal corona.

C. somalensis Chiovenda (Result. Sci. Miss. Stefanini-Paoli Somalia Ital. 1: 116, 1916). **T:** Somalia (*Paoli* 889 [FT]). – **D:** N Yemen, Somalia, Kenya. **I:** Bruyns (1988b). **Fig. XII.d**
  Incl. *Ceropegia somalensis* fa. *erostrata* H. Huber (1957).

[3] Stem and **L** succulents to 1.5 m, twining; **R** fibrous; stems terete, 3 - 4 mm ⌀, grey-green, to 8 mm ⌀ at the base and corky; **L** broadly lanceolate, acute, 1 - 2 × 1.5 cm; **Inf** pedunculate, few-flowered; peduncle ± 1 cm, glabrous; **Ped** 8 - 12 mm, glabrous or hairy; **Sep** subulate, 5 - 8 mm; **Cl** 3 - 4.5 cm, outside pale green, velvety-hairy; **Cl** inflation inside white at the base, remainder purple, globose to ovoid, inside hairy at the base, lower ⅓ constricted, 4 - 5 × 5 - 6 mm, horizontally oriented to descending and then merging into the ± vertical **Cl** tube, tube inside purple at the base, then pale green with fine dark green or violet blotches and scatteredly hairy, basally 3 mm ⌀, towards the mouth widening to ± 12 mm, abruptly merging into the **Cl** lobes; **Cl** lobes 2 - 3 cm, broadly triangular, then linear-lanceolate, 1 - 5 mm broad, lower 5 - 12 mm free, then fused, above again diverging into a cage-like structure and apically fused again, lamina ± folded back along the midrib, with straight **Ha** on the lower ⅓ of the keel, without (Arabia) or with nose-like purple humps (= osmophores) (Africa); **Cn** ± 4 - 5 × 4 mm, sessile, bowl-shaped; **Ci** lobes purple or yellowish with purple edges, highly connate, bulging pouch-like, broadly triangular, ± erect, ± 1.2 × 1.5 mm, bifid almost down to the middle into triangular appendages, apically connivent, with straight whitish **Ha** along the inner margins; **Cs** white, ± 0.25 mm, linear-cylindrical, basally hairy, erect-connivent, tips diverging; **Poll** ovoid (to almost rectangular), ± 0.35 × 0.25 mm, germinating crest membranous, triangular; corpuscle ± 0.35 mm with broadly triangular lateral appendages.

Belonging to the group of *C. distincta*, *C. haygarthii* and *C. lugardiae*. The pale green and velvety exterior of the flowers is remarkable, as well as the (rarely absent) double cage structure of the corolla, which, in the African collections only, also bears purple osmophores on the inner keels.

C. sootepensis Craib (BMI 1911: 420, 1911). **T:**

Thailand (*Kerr* 695 [K]). – **D:** Thailand. **I:** Huber (1957). **Fig. XII.e**

[2] **R** tuber globose; stems erect-twining, delicate, glabrous; **L** narrowly linear to lanceolate; **Inf** ± sessile, 1- (to 3-) flowered; **Cl** 3 - 4.5 cm, ± straight, outside glabrous; **Cl** inflation ovoid to globose, ± 12 × 7 - 12 mm, fairly abruptly merging into the **Cl** tube, tube narrowed to ± 3 mm, towards the mouth widening to 5 - 8 mm, yellowish with fine purple blotches; **Cl** lobes linear to lanceolate, 15 - 25 mm, tips fused into an ellipsoid cage, inside hairy, basally yellowish, apically brownish, lamina folded back along the midrib; **Cn** sessile, basally fused; **Ci** lobes bifid into glabrous or hairy appendages which are overtopping the **Gy** appendages.

C. speciosa H. Huber (Mem. Soc. Brot. 12: 144, ill., 1957). **T:** Tanzania (*Burtt* 4632 [K]). – **D:** Tanzania.

[3] Stems twining, glabrous, subsucculent; **L** 1 - 2.5 cm petiolate, lamina ovate, 5 - 10 × 2 - 5 cm, acute, delicate; **Inf** 2 - 3 cm, with a slender peduncle, few- to many-flowered; **Sep** subulate, 7 - 10 mm; **Cl** 4.5 - 6.9 cm, strongly bent, glabrous, outside whitish; **Cl** inflation ovoid-inflated, ± 2 × 1 cm; bent **Cl** tube narrowed to ± 2 mm ⌀, apically abruptly widening to ± 15 mm ⌀ and margins curved outwards; **Cl** lobes narrowly linear, 2 - 3 cm, lamina folded back along the midrib, inserted on the inner margin of the **Cl** tube, vertical, inside brown-purple, lower ¾ stalk-like and villose, upper ¼ forming a small ellipsoid cage, tips fused; **Cn** basally fused, glabrous; **Ci** lobes bilobed into 2 appendages overtopping the **Gy**; **Cs** linear-spatulate, erect-connivent.

Plants with delightful flowers, but unfortunately not in cultivation and practically unknown.

C. spiralis Wight (Icon. Pl. Ind. Orient. 4: t. 1267, 1848). **T:** India, Tamil Nadu (*Wight* s.n. [K]). – **D:** S India (Tamil Nadu).

Incl. *Ceropegia munronii* Wight (1848).

[2] **R** tuber flattened; stems delicate, erect, rarely twining, ± 10 cm, ± unbranched, ± glabrous; **L** sessile, linear, 7 - 10 × 2 - 3 mm; **Inf** 1-flowered, shortly pedunculate; **Ped** 6 - 7 mm, glabrous; **Cl** 4 - 5 cm, glabrous, greenish-yellow with purple stripes on the lower ½; **Cl** tube 2 - 2.5 cm; **Cl** inflation inside glabrous, merging gradually into the **Cl** tube, tube hairy inside; **Cl** lobes 2 - 2.5 cm, linear-filiform from a triangular base, often spiralling, lamina largely folded back along the midrib, inside hairy; **Cn** cup-shaped, sessile; **Ci** lobes tringular, ± 0.7 mm, bifid into triangular appendages; **Cs** linear, erect-connivent.

Forming a group of closely related species with *C. attenuata*, *C. jainii*, *C. mahabalei*, *C. noorjohaniae* and *C. pusilla*, sharing tubers, dwarf and erect growth and with linear leaves.

C. stapeliiformis Haworth (Philos. Mag. Ann. Chem. 1827: 121, 1827). **T:** RSA, Eastern Cape (*Bowie* s.n. [not located]). – **D:** RSA, Swaziland.

C. stapeliiformis ssp. **serpentina** (E. A. Bruce) R. A. Dyer (in Leistner & al. (ed.), Fl. South Afr. 27(4): 51, 1980). **T:** RSA, Gauteng (*Erens & Phillips* 2176 [PRE]). – **D:** RSA (Northern Prov., Gauteng, Mpumalanga), Swaziland. **I:** Dyer (1983).

≡ *Ceropegia serpentina* E. A. Bruce (1949) ≡ *Ceropegia stapeliiformis* var. *serpentina* (E. A. Bruce) H. Huber (1957).

[3] Differs from ssp. *stapeliiformis*: Stem **Tu** with 2 golden-yellow stipular **Gl** accompanied by 2 **Gl**-like structures; peduncles usually twining; **Cl** lobes not or only a little spreading, usually connivent and spiralling; **Ci** lobes ± 1mm.

C. stapeliiformis ssp. **stapeliiformis** – **D:** RSA (Eastern Cape). **I:** Dyer (1983). **Fig. XIII.g**

[3] Stem succulents; stems creeping, scrambling or occasionally twining, ± terete, with prominent **Tu**, 1.5 - 15 × 1.2 - 1.8 cm, olive-green, often striped or blotched with red-brown, slightly glossy; **L** triangular to cordate, ± 3 × 3 mm, with robust milky-cloudy stipular **Gl**; **Inf** 0.2 - 2 cm pedunculate, with 1 to many **Fl** on the peduncles of each part-**Inf** which reach several 10 cm length and scarcely twine; **Ped** 6 - 10 × 2.5 mm; **Sep** lanceolate, ± 4 mm; **Cl** ± 7 cm, robust, greenish-white stippled and blotched with red-brown; **Cl** inflation slender, merely ± 5 - 7 mm ⌀, with a narrrow basal constriction, inside hairy, merging gradually into the **Cl** tube, tube ± 3 mm ⌀, mouth abruptly widening into a funnel-like structure; **Cl** lobes inside white, apically and laterally often red-violet, greenish, or brownish, 20 - 35 mm, free, semi-erect, tips revolute, broadly triangular at the base, margins at the base somewhat folded back, apically fully folded back along the midrib, inside with fine white **Ha**; **Cn** shortly stipitate or sessile, whitish or brownish, ± 5 × 3.5 mm, fused into a cup; **Ci** lobes very shortly triangular, abruptly inserted on the upper margin of the coronal cup, ± erect, bifid into triangular appendages, ± 0.8 mm, inside ± hairy; **Cs** 3.5 - 4.5 × 0.4 mm, subulate, erect-connivent, tips papillose, recurved; **Poll** reniform, ±0.4 × 0.25 mm, corpuscle 0.25 × 0.2 mm with tips widening into a shield; **Fr** grey-green, erect, broadly fusiform, ± 10 cm, verruculose.

The 2 subspecies show a disjunct distribution, with the typical subspecies native to the Eastern Cape (only S of 31° S) and ssp. *serpentina* to the NE of RSA and Swaziland (only N of 28° S).

C. stenantha K. Schumann (BJS 17: 152-153, 1893). **T:** Sudan (*Schweinfurth* 2104 [K]). – **D:** S and E Africa (Sudan to RSA and Namibia). **I:** Dyer (1983).

Incl. *Riocreuxia longiflora* K. Schumann (1901);

incl. *Ceropegia infausta* N. E. Brown (1903); **incl.** *Ceropegia stenantha* var. *parviflora* N. E. Brown (1903); **incl.** *Ceropegia angustifolia* De Wildeman (1903) (*nom. illeg.*, Art. 53.1); **incl.** *Ceropegia mazoensis* S. Moore (1908); **incl.** *Ceropegia quarrei* De Wildeman (1927); **incl.** *Ceropegia tenuissima* S. Moore (1939).

[1,3,4] **R** fleshy, fusiform; stems robust, much-branched, twining, to several m tall, often only annual, 2 - 5 mm ∅, ± succulent, striate; **L** 4 - 8 mm petiolate, lamina linear to lanceolate, to 3 - 15 × 0.3 - 1 cm, subsucculent, rarely ciliate; **Inf** 3 - 12 mm pedunculate, pseudo-umbellate, 3- to 6-flowered; **Ped** 1 - 5 mm; **Sep** subulate, 2 - 3 mm; **Cl** 2.5 - 3 cm, straight or slightly bent, green-yellowish to pale yellow; **Cl** inflation globose-ovoid, 4 - 8 × 2 - 3 mm, inside hairy at the base; **Cl** tube ± cylindrical, 2 mm ∅, inside hairy at the base; **Cl** lobes 14 - 20 × 1 mm, linear, tips fused into an ellipsoid to obovoid cage, lamina folded back along the midrib, slightly keeled at the base, with projecting margins, even in the basal angles (auriculate); **Cn** white, ± sessile, fused into a shallow cup, glabrous; **Ci** shortly pouch-like, lateral margins fused with the **Cs** base; **Cs** linear-lanceolate, ± 1.5 × 0.3 mm, erect, connivent up to the middle, then often recurved; **Fr** ± 6 - 10 cm × 3 mm; **Se** ± 6 × 2 mm.

Allied to *C. crassifolia* and *C. nilotica*.

C. stenoloba Hochstetter *ex* Chiovenda (Ann. Bot. (Roma) 10: 395-396, 1912). **T:** Ethiopia (*Schimper* 2048 [BM, S]). – **D:** Ethiopia, Kenya, Rwanda, Tanzania, Uganda, Zambia, Namibia. **I:** Dyer (1983); Archer (1992: VIII).

Incl. *Ceropegia aberrans* Schlechter (1913); **incl.** *Ceropegia schliebenii* Markgraf (1932); **incl.** *Ceropegia chortophylla* Werdermann (1939); **incl.** *Ceropegia stenoloba* var. *australis* H. Huber (1957); **incl.** *Ceropegia stenoloba* var. *moyalensis* H. Huber (1957).

[2] **R** tuber globose, slightly flattened, ± 2.5 × 3 - 5 cm, with fissured bark; stems 0.5 - 2 m tall, twining, annual, 1 - 3 mm ∅, shortly hairy; **L** to 15 mm petiolate, elliptic, ovate to broadly lanceolate, margins often crenulate, (1-) 2 - 7.5 × (0.5-) 1 - 4 cm, delicate, ± hairy; **Inf** ± sessile, 1- to many-flowered; **Ped** 5 - 15 mm; **Sep** narrowly triangular, ± 3 mm; **Cl** 1.2 - 3 cm, pale yellow; **Cl** inflation ± globose, 4 - 6 mm ∅, outside blotched or striped with red-brown, inside red-brown, hairy, ± gradually narrowed to 3 mm ∅ and bent at 20 - 45° with the **Cl** tube slightly irregularly shaped, tube inside yellowish, blotched with red-brown, hairy, towards the mouth widening to 4 - 8 mm, with oblique margin; **Cl** lobes linear, 6 - 10 mm, folded back along the midrib (incl. the margins at the basal angles), inside green to brown, ± hairy or glabrous, tips fused into a ± globose cage; **Cn** shortly stipitate, at the base fused into a dish-like structure; **Ci** lobes forming transversely ovate pouches, < 0.5 mm ∅, apically occasionally incised, often ciliate, with purple blotches, otherwise **Cn** white; **Cs** 1.5 - 3.5 mm, linear, erect-connivent; **Poll** globose to nearly square, 0.2 mm ∅; **Fr** narrowly fusiform, 9 - 13 cm × 3 mm; **Se** ± 9 × 2.5 mm.

A variable and diverse complex to which Huber's varieties can only partially be assigned (see also *C. ringens*). *C. stenoloba* is allied to *C. paricyma* and *C. claviloba*, which also exhibit similarly large and lobed leaves and tubers with a fissured bark.

C. stenophylla C. K. Schneider (Pl. Wilson. 3: 350, 1916). **T:** China, Sichuan (*Wilson* 2313 [K, BM, GH]). – **D:** China (Sichuan).

[1] **R** fusiform, fleshy, clustered; stems robust, to 2 m, twining, ± 2 mm ∅, glabrous or scatteredly hairy; **L** petiolate, linear or linear-lanceolate, 3.5 - 9 × 0.2 - 0.7 mm, ± scatteredly hairy; **Inf** 1 - 6 mm pedunculate, 1- to few-flowered; **Ped** to 1.5 cm; **Sep** linear-lanceolate, 5 - 7 mm, with several **Gl** at the base; **Cl** 3 - 5 cm; **Cl** inflation 5.5 - 8 mm ∅, abruptly merging into the **Cl** tube, tube 3 mm ∅, apically widening to 8 - 10 mm; **Cl** lobes 9 - 20 mm, narrowly elliptic, tips fused into a cage of unknown shape, lamina folded back along the midrib, inside hairy, margins ciliate; **Cn** sessile, fused into a bowl; **Ci** lobes bilobed into triangular obtuse appendages, hairy; **Cs** linear, obtuse; **Fr** fusiform, ± 9 cm × 3 mm, striate; **Se** ± 7 × 2.5 mm, tuft of **Ha** 2 mm.

Closely related to *C. dolichophylla* and *C. longifolia*.

C. stentiae E. A. Bruce (BMI 1936: 490, 1936). **T:** RSA, Gauteng (*Stent* s.n. in *PRE* 10179 [K, PRE]). – **D:** RSA (Gauteng, North-West Prov.). **I:** Dyer (1983); Venter (1993). **Fig. XIII.h**

[2] **R** tuber 3 - 6 × 2 - 3 cm, flattened; stems delicate, 7 - 12 cm, annual, terete, 1 - 2 mm ∅, with short **Int**, hairy when young; **L** ± sessile, narrowly linear, 4 - 6 cm × 2 - 3 mm, margins revolute; **Inf** sessile, lateral, 1-flowered; **Ped** ± 1 cm; **Sep** lanceolate, 2 mm; **Cl** 4.5 - 7 cm, brown-purple, partly striped with greenish, outside glabrous, rarely pubescent, inside ± purple and scatteredly hairy; **Cl** inflation ellipsoid-cylindrical, ± 6 × 4 mm, greenish-white, with purple longitudinal stripes, merging gradually into the **Cl** tube with its upper ⅓ bent at right angle, **Cl** tube 2 - 3 mm ∅ in the middle, widening into a funnel-shaped mouth of 6 - 7 mm, inside scatteredly hairy; **Cl** lobes 2.5 - 4 cm, erect, lanceolate in the lower ⅕ and nearly flat, inside cream-coloured, margins usually with vibratile purple **Ha**, upper ⅘ narrowly linear, ± 1 mm broad, purple-brown, glabrous, tips slightly broadened and joined to form a narrow cage, lamina halfways folded back along the midrib; **Cn** white, cup-shaped, ± 3.5 × 2.5 mm; **Ci** lobes forming triangular-ovate pouches, 1.5 - 2 mm long, incised for 1 mm, with erect-parallel triangular white-bearded appendages, margins winged and trapezoid, fused to

the base of the **Cs**; **Cs** ± 2.5 mm, subulate with obtuse tips, erect-connivent, apically slightly spreading, glabrous; **Poll** ellipsoid, 0.4 × 0.23 mm; follicles erect-parallel, fusiform, ± 6 cm, pale green.

This rare species is allied to *C. dinteri* and *C. insignis*.

C. striata Meve & Masinde (Novon 8(1): 38-40, ills., 1998). **T:** Madagascar (*Grubenmann* s.n. [ZSS, MO, MSUN]). – **D:** C Madagascar; granite inselbergs. **Fig. XIII.f**

[2] **R** tuber depressed-globose, hardly > 2 - 3 cm ∅; stems annual, 50 - 100 cm tall, twining, delicate, glabrous, 1 - 2.5 mm ∅, green suffused with reddish; **L** shortly petiolate, lamina lanceolate, acute, 10 - 30 × 5 - 8 mm; **Inf** sessile to shortly pedunculate, usually 1-flowered; **Ped** 5 mm; **Sep** lanceolate, ± 1 mm; **Cl** 2.5 - 3.5 cm, whitish-green, striped with purple-red, inside glabrous except for the circle of **Ha** around the upper margin of the **Cl** inflation, inflation ovoid, 7 × 6 mm; **Cl** tube bent, 12 - 15 mm, narrowed to 1 - 2 mm at the base, apically widening to 6 - 8 mm; **Cl** lobes linear, ± 10 × 0.7 mm, tips fused into an elliptical cage with acute end, outside green, inside black-purple to dark green, margins slightly revolute, densely covered with purple **Ha**; **Cn** ± sessile, ± 3.5 mm ∅, fused into a cup, whitish, basally also purple, glabrous; **Ci** lobes bifid into linear appendages, 1.5 × 0.3 mm, ascending; **Cs** spatulate, ± 2.2 × 0.6 mm, slightly concave, erect-connivent; **Poll** broadly ellipsoid, 0.3 × 0.15 mm.

A delicate climber with exquisite flowers, allied to *C. madagascariensis*.

C. subaphylla K. Schumann (BJS 33: 329, 1903). **T:** Somalia (*Ellenbeck* 190 [K]). – **D:** Ethiopia, Kenya, Somalia, Saudi Arabia, Yemen, Oman. **I:** Bruyns (1988b: 305, as *C. botrys*). **Fig. XIV.a**

Incl. *Ceropegia botrys* K. Schumann (1903); **incl.** *Ceropegia nuda* Hutchinson & E. A. Bruce (1941); **incl.** *Ceropegia mansouriana* Chaudhary & Lavranos (1985).

[2,3] Stem succulents, to 2 m, twining; **R** a globose-ovoid or also irregularly shaped tuber; stem ± terete, 2 - 3 mm ∅, grey-green, faintly rough; **L** linear-lanceolate, 4 - 10 mm, rapidly caducous; **Inf** 2 - 5 mm pedunculate, 1- to 4-flowered; **Ped** 8 - 12 mm; **Sep** subulate, 4 - 8 mm; **Cl** 3 - 5 cm; **Cl** inflation greenish, ovoid, ± 7 × 5 mm, ± horizontally oriented, inside slightly papillose, narrowed into the **Cl** tube, tube greenish-white, blotched with red-brown, 2 mm ∅ at the base, inside with a circle of dense **Ha**, 8 - 12 mm long, at ½ of its length usually abruptly bent, above ascending ± vertically, apically widening to 4 - 5 mm ∅ and abruptly merging into the **Cl** lobes which are broadly triangular at first; **Cl** lobes 1.5 - 3 cm, linear-spatulate, free for the lower 1.5 - 1.8 cm, then fused (if **Cl** lobes long: diverging above again into a cage-like structure), lamina ± folded back along the midrib, inside yellowish-white with red-brown reticulate markings, ± hairy, often with purple median band, apically brown, glabrous; **Cn** ± 4 × 3 mm, sessile, greenish-white/purple, cup-shaped; **Ci** lobes highly connate, at the base pouch-like, triangular, ± erect, 1 - 1.5 mm, at least for ½ bifid into triangular appendages, apically connivent; **Cs** ± 2 - 3 mm, linear-cylindrical, erect-connivent; **Poll** ovoid, ± 0.35 × 0.25 mm.

An exquisite species; remarkable features include the sometimes elongate tuber and the usually geniculate corolla. Allied to *C. arabica* var. *powysii*.

Ellenbeck collected 2 plants at the same site under his no. 190, which Schumann choose as types for *C. subaphylla* and *C. botrys*. Following Huber (1957) both are here treated as conspecific, though this has been disputed by Field (1982).

C. swaziorum D. V. Field (KB 37(2): 305-308, ills., 1982). **T:** Swaziland (*Kemp* 1300 [K, PRE]). – **D:** Swaziland. **I:** Dyer (1983: 203).

[2] **R** tuber 4 - 8 × 3 cm ∅, bark rough; stem sparsely branched, hairy, twining, 0.5 - 1 m, ± 1 mm ∅; **L** 1 - 2 cm petiolate, lamina lanceolate, acute, base subcordate, 5 - 7 × 1.5 - 2.5 cm, margin often serrate or sinuate; **Inf** ± sessile, 2- to 4-flowered; **Ped** 8 - 12 mm; **Sep** subulate, 5 - 7 mm, villose; **Cl** almost straight, 2.5 - 3 cm, greenish-cream-coloured; **Cl** inflation globose, 6 - 8 × 3 - 4 mm, merging fairly gradually into the **Cl** tube, tube 1 - 1.5 mm ∅ at the base, towards the mouth widening to 2 - 3 mm ∅, inside cream-coloured, striped with purple, glabrous; **Cl** lobes 5 - 6 × 2 mm, linear-spatulate from a triangular base, lamina weakly folded back along the midrib, basal margins revolute, tips fused into an ovoid cage, inside purple, with fine purple **Ha**; **Cn** sessile, white, ± 2 × 2 mm, basally fused into a cup; **Ci** lobes ovate-triangular, ± erect, concave, ± 0.5 mm, glabrous, tips incised; **Cs** 1.5 × 0.3 mm, linear-spatulate, hairy at the base, erect-connivent, apically ± involute and papillose; **Poll** ovoid, ± 0.25 × 0.12 mm; **Fr** narrowly fusiform, ± 7 cm × 2 mm.

Very close to *C. papillata*, but with ± serrate leaves and larger flowers.

C. taprobanica H. Huber (Revised Handb. Fl. Ceylon 4: 116-117, 1983). **T:** Sri Lanka (*Trimen* s.n. [PDA]). – **D:** Sri Lanka.

[1] **R** fusiform, fleshy, clustered; stems delicate, glabrous; **L** 1 - 3 cm petiolate, lamina ovate, elliptic to lanceolate, 3 - 16 × 1 - 7 cm, acute; **Inf** 1- to 8-flowered; peduncle much longer than the **Ped** and with **Ha** arranged in 1 row; **Sep** 3 - 7 mm, glabrous; **Cl** (3-) 5 - 8 cm, greenish-white, stippled with purple; **Cl** inflation large, ovoidly inflated; **Cl** tube abruptly constricted in the middle, widening into a funnel-shaped mouth; **Cl** lobes 2 - 3.5 cm, linear from a triangular base, tips fused, inside with long and purple **Ha**.

An insufficiently known endemic from Sri Lanka, which has been separated from *C. decaisneana* by Huber.

C. tihamana Chaudhary & Lavranos (Notes Roy. Bot. Gard. Edinburgh 42(2): 316, ill., 1985). **T:** Saudi Arabia (*Chaudhary* 901A [RIY, E]). – **D:** SW Saudi Arabia. **I:** Collenette (1985).

[1,3] Twining stem succulents to 2 m; **R** tuber irregularly thickened and fleshy with finger-shaped appendages; stems ± terete, 2 - 3 mm ⌀, grey-green, slightly verrucose; **L** linear, 6 - 10 × 2 - 3 mm, margins slightly hairy, rapidly caducous; **Inf** almost axillary, 3 - 5 mm pedunculate; **Ped** 8 - 12 mm; **Sep** subulate, 3 - 4 mm; **Cl** 1.5 - 2.2 cm; **Cl** inflation greenish, inside striped with red, ovoid, ± 5 × 3 mm, at the transition to the **Cl** tube thickened, densely hairy, **Ha** on purple papillae; **Cl** tube 1.5 mm ⌀ at the base, greenish-whitish, occasionally striped with red-brown, densely hairy at the base, upper part widening to 2 mm and here slightly bent, greenish-white with red blotches, inside laxly hairy, above widening into a funnel-shaped mouth of 5 - 6 mm ⌀; **Cl** lobes 4 - 5 mm, lower 2 mm whitish with red-brown reticulation (at least at the base), nearly horizontal, folded back along the midrib, margins and in the middle hairy, upper 2 - 3 mm grey, nearly vertical, not folded back, tips fused, inside finely hairy; **Cn** 2.5 - 3 × 2 mm, ± sessile, cup-shaped; **Ci** lobes erect, bulging pouch-like, ovate, ± 1.5 mm long, apically bifid into triangular appendages; **Cs** ± 1.5 mm, linear-clavate, erect-connivent; **Poll** ovoid, ± 0.32 × 0.23 mm; **Fr** 9 cm × 6 mm.

Vegetatively, *C. tihamana* shows great resemblance to *C. subaphylla*, and even the slender corona is very similar.

C. tomentosa Schlechter (BJS 18(Beiblatt 45): 33, 1894). **T:** RSA, Eastern Cape (*Barber* 372 [K]). – **D:** RSA (Eastern Cape, KwaZulu-Natal); grassland. **I:** Dyer (1983).

Incl. *Ceropegia scabriflora* N. E. Brown (1908).

[1] **R** fleshy and thickened, to 10 cm; stems erect, not twining, 20 - 40 cm, annual, sparsely branched, glabrous; **L** ± sessile and erect, linear, acute, 25 - 100 × 1 - 3 mm; **Inf** sessile, lateral, 1- (to 2-) flowered; **Ped** 5 - 10 mm; **Sep** subulate, 5 - 8 mm; **Cl** 3.5 - 6.5 cm, ascending, ± straight, outside greenish to yellowish, partly marked with purple; **Cl** inflation globose to ovoid, outside glabrous or roughly papillose, 5 - 8 mm ⌀, merging gradually into the **Cl** tube, tube ± 3 mm ⌀, apically widening to 4 - 5 mm, ± purple; **Cl** lobes free, erect to spreading, linear from a narrowly triangular base or filiform when folded back along the midrib, 2.5 - 3.5 cm, greenish, marked with purple, inside finely hairy, with a tuft of purple clavate **Ha** just above the base; **Cn** ± sessile, white-yellowish, basally fused into a cup; **Ci** lobes triangularly pouch-like, ± erect, 1 - 2 mm, bilobed into narrowly triangular appendages; **Cs** ± 2 mm, linear-spatulate, erect-connivent.

Huber (1957) and Dyer (1980) separate *C. scabriflora* from *C. tomentosa*. Dyer, however, points out that *C. scabriflora* is merely distinguishable from *C. tomentosa* by the corolla being rough outside. Adjacent distribution areas in grasslands and, due to their rarity, an insufficiently known variability of these taxa, render the present treatment adequate. *C. bowkeri* (see note there) presumably falls under this species as well.

C. trichantha Hemsley (J. Bot. 23: 286, 1885). **T:** Hongkong (*Ford* s.n. [not located]). – **D:** Myanmar, China, Hongkong, Malaysia, Thailand.

Incl. *Ceropegia angustilimba* Merrill (1932); **incl.** *Ceropegia jucunda* Kerr (1939).

[1] **R** fusiform, fleshy, clustered; stems to 1.5 m tall, twining, straw-yellow, hairy on the nodes; **L** petiole to 2.8 cm, winged, ± pubescent, lamina narrowly elliptic to lanceolate, ± 4.5 × 2.5 cm, delicate, upper face pubescent, lower face bluish and hairy on the midrib; **Inf** 1.4 - 3.5 cm pedunculate, 1- to 2-flowered; **Ped** 1 - 1.5 cm; **Sep** linear-lanceolate, 3 - 4 (-6) mm; **Cl** 3 - 4.5 cm, white and green, glabrous; **Cl** inflation oblique-ovoid, 3.5 - 4.5 mm ⌀; **Cl** tube at the base narrowed to ± 2 mm, apically gradually widening to ± 5 mm ⌀; **Cl** lobes 1.5 - 3 cm, filiform from a triangular base, keeled at the base, apically spatulate, fused, lamina ± folded back along the midrib; **Cn** shortly stipitate, ± 4 × 3 mm, basally fused into a cup; **Ci** lobes triangular-ovate, bifid into 2 triangular densely ciliate appendages; **Cs** linear-spatulate, to 3 mm, erect, villose; **Fr** ± 15 cm, slender.

Closely allied to (the non-succulent and thus here not treated) *C. lucida* Wallich (incl. *C. driophila* C. K. Schneider *sensu* Gilbert & al. (1995)).

C. turricula E. A. Bruce (BMI 1936: 419, 1937). **T:** RSA, Northern Prov. (*Galpin* 697 [K, PRE]). – **D:** RSA (Northern Prov.). **I:** Dyer (1983).

[2] **R** tuber flattened, 3 - 5 cm ⌀; stems erect, 35 - 45 cm, not twining, delicate, annual, with short **Int**, ± unbranched, ± pubescent; **L** ± sessile, narrowly lanceolate-linear, ± 12 × 0.5 cm, midrib ± raised on the lower face; **Inf** sessile, lateral, 1-flowered; **Ped** ± 17 mm; **Sep** linear, acute, to 1 cm; **Cl** pale green, blotched with red-brown, to 6.5 cm, buds 5-angled; **Cl** inflation ovoid, ± 7 mm ⌀, inside white at the base, merging gradually into the **Cl** tube, tube 3 - 4 mm ⌀, inside purple and green, apically widening to ± 17 mm ⌀; **Cl** lobes fused into an ovoid cage, ± 17 × 6 mm, erect but slightly contorted in the middle, folded back along the midrib, strongly keeled, inside basally pale green, then with transverse white bands, apically green-purple, with long purple cilia; **Cn** cup-shaped, sessile; **Ci** lobes triangular, ± erect, laterally fused with the **Cs** base, apically incised into subulate appendages, ± 1

mm, ciliate; **Cs** ± 4 mm, linear-subulate, erect-connivent.

For a note see under *C. dinteri*.

C. umbraticola K. Schumann (BJS 17: 153, 1893). **T** [neo]: Zaïre (*Schmitz* 1055 [BR]). — **D**: Zaïre, Zambia, Tanzania, Angola, Malawi. **I**: Malaisse & Schaijes (1993).

Incl. *Ceropegia wellmanii* N. E. Brown (1908); **incl.** *Ceropegia rostrata* E. A. Bruce (1948); **incl.** *Ceropegia chipiaensis* Stopp (1964); **incl.** *Ceropegia schaijesiorum* Malaisse (1986).

[1] **R** fleshy and thickened to 20 × 0.3 cm; stems erect, 10 - 30 cm, not twining, herbaceous, hairy, annual, unbranched; **L** sessile or with a petiole to 5 mm, lamina broadly ovate to narrowly elliptic, ± 4 × 2 cm; **Inf** sessile, lateral, 1- (to 2-) flowered; **Ped** 15 mm; **Sep** subulate, 4 - 6 mm; **Cl** 5 - 8 cm, ± horizontal to ascending, straight or slightly bent, outside whitish-green; **Cl** inflation cylindrical, globose to ovoid, to 15 mm ∅, inside glabrous or hairy, outside occasionally shortly pubescent, merging gradually into the often slightly narrowed **Cl** tube, tube apically often slightly constricted, suffused with reddish; **Cl** lobes linear from a triangular base, 1 - 4 × 0.5 cm, tips fused into a tapering-ovoid cage, folded back along the midrib (incl. the basal margins), inside greenish or brownish, often suffused or striped with purple, at the base occasionally whitish or red, ± glabrous; **Cn** sessile, 5 mm ∅, basally dish-like fused; **Ci** lobes triangularly pouch-like, horizontal to nearly erect, 3 - 4 mm, bilobed into subulate-triangular appendages, occasionally with a small appendage in between, with long **Ha** to glabrous; **Cs** narrowly lanceolate, hardly 1 mm, incumbent on the **Anth** to ascending; **Fr** 20 - 40 cm × 4 - 6 mm, erect, almost parallel; **Se** ± 8 × 4 mm, tuft of **Ha** 3 cm.

A variable complex, which remains insufficiently known, due to lack of material. *C. schaijesiorum*, characterized by nearly white flowers with brown corolla lobes, is here treated as synonym, but this placement needs corroboration.

C. vanderystii De Wildeman (Bull. Jard. Bot. État 4: 393, 1914). **T**: Zaïre (*Vanderyst* 3024 [BR]). — **D**: Angola, Zaïre. **I**: Malaisse & Schaijes (1993).

Incl. *Ceropegia rara* S. Moore (1929).

[1] **R** tuber small, basal **Int** of the stems underground, rhizome-like, with thickened lateral **R**; stems robust, unbranched, twining, 1 - 3 m, hairy, 2 - 3 mm ∅; **L** 2 - 5 cm petiolate, lamina cordate, acuminate, 5 - 7 × 2 - 4 cm; **Inf** with a short thickened peduncle, 3- to many-flowered; **Ped** 1 - 1.5 cm; **Sep** subulate, 2 - 5 mm, blotched with red, villose; **Cl** 1.2 - 2.5 cm, bottle-shaped, outside hairy or glabrous; **Cl** inflation 7 - 15 × 4 - 6 mm, outside whitish, ± striped with red-brown, ovoid, occasionally asymmetrical, merging gradually into the **Cl** tube, tube apically 1 mm ∅; **Cl** lobes 5 - 8 mm, linear to narrowly lanceolate, lamina folded back along the midrib (incl. the basal angles), tips fused into a small ovoid cage, inside velvety-black, usually finely hairy, outside green; **Cn** sessile, purple at the base, otherwise whitish-hyaline, fused into a short cup; **Ci** lobes erect, 0.5 - 1 mm, nearly square, obtuse, denticulate and/or apically incised; **Cs** ± 2 mm, linear-cylindrical, erect, apically involute, hairy at the base.

Closely related to *C. meyeri* and *C. abyssinica*.

C. variegata Decaisne (Ann. Sci. Nat. Bot., sér. 2, 9: 262, ills., 1838). **T**: Yemen (*Botta* s.n. [P]). — **D**: Saudi Arabia, Yemen, Ethiopia, Somalia, Kenya, Tanzania. **I**: Collenette (1985: p.p. as *C. devecchii*).
Fig. XIV.c

Incl. *Stapelia variegata* Forsskål (1776) (*nom. illeg.*, Art. 53.1); **incl.** *Ceropegia tubulifera* Deflers (1896); **incl.** *Ceropegia devecchii* Chiovenda (1932); **incl.** *Ceropegia variegata* var. *cornigera* H. Huber (1957); **incl.** *Ceropegia devecchii* var. *adelaidae* P. R. O. Bally (1974) ≡ *Ceropegia variegata* var. *adelaidae* (P. R. O. Bally) Cufodontis (s.a.); **incl.** *Stapelia sarmentosa* Steudel (s.a.).

[3] Stem succulents, scrambling or twining in bushes to 3 m high; stems terete, with ± projecting **Tu**, 8 - 13 mm ∅, dark green to glaucous-green, striped or blotched with red-brown, shining when young; **L** triangular-attenuate, ± 2 × 2 mm, with membranous hairy stipules; **Inf** shortly pedunculate, with 1 to several **Fl** opening successively, **Ped** 3 - 5 mm; **Sep** subulate, 3 - 5 mm, tips usually recurved; **Cl** ± 7 cm, robust, outside grey-white, finely stippled and blotched with violet, inside grey-white striped with violet and green, ± unscented, short-lived; **Cl** inflation constricted in the middle, 2 - 3 mm ∅, basal part ovoid, 5 - 20 × 5 - 7 mm, inside usually wrinkled, upper part of the inflation often yellowish, striate, 4 - 5 × 5 - 7 mm, then gradually merging into the **Cl** tube, tube ± 3 mm, apically abruptly widening in a funnel-shaped manner into the appendages of the **Cl** lobes, inside with white **Ha**; **Cl** lobes 5 - 18 mm, with green, red-brown and green bands, erect, apically fused into a short tip, bases of adjacent lobes united into grooved ± horizontally spreading processes (auricles), lobes 5 - 35 mm, inside whitish and ± completely hairy; **Cn** shortly stipitate or sessile, pale yellow, 4 - 5 × 4 - 5 mm, fused into a cup; **Ci** lobes narrowly triangular to linear-subulate, ± erect, bilobed into triangular appendages, 1 - 2 mm; **Cs** 2 - 3 × 0.4 mm, linear, erect-connivent; **Poll** ovoid to reniform, ± 0.45 × 0.25 mm; **Fr** 9 cm.

An extremely variable complex, which had been reduced to a single species by Bruyns (1988b). The taxa here subsumed are uniform as to the morphology of the stems, but the basal processes of the corolla lobes in particular are very variable (few to 35 mm long) thus giving the flowers of *C. variegata* very different appearances.

C. verticillata Masinde (CSJA 72(3): 155-158, ills., 2000). **T:** Kenya, Taita-Taveta Distr. (*Bytebier* 1160 [EA, BR]). – **D:** S Kenya.

[1] **R** clustered, fusiform, fleshy; stems erect, to 2 m, sparsely branched, 1 - 1.5 mm ⌀, hairy; **L** 9 - 20 mm petiolate, in whorls of 3, lamina narrowly ovate-lanceolate, 12 - 40 × 6 - 13 mm, pubescent, margins ciliate; **Inf** shortly pedunculate, pseudo-umbellate, 1- to 2-flowered; **Ped** 2 - 6 mm, ± glabrous; **Sep** subulate, ± 2 - 2.5 mm, hairy; **Cl** whitish-cream-coloured, in the middle with reddish dots and streaks, 11 - 20 mm, erect, slightly bent; **Cl** inflation ovoid, ± 3 × 3 mm; **Cl** tube narrowing gradually to 1.5 mm ⌀, mouth widening to 4 mm ⌀, glabrous; **Cl** lobes in the apical ⅔ dark green, linear, ± 4 × 1.4 mm, lamina folded back along the midrib, inner face hairy, apically connate to form an ellipsoid cage-like structure with ± 4 mm ⌀; **Cn** shortly stipitate, cup-shaped, ± 2.3 × 3 mm; **Ci** ± 1.4 × 1.2 mm, semi-erectly radiating out to form rectangular concave pouches, bifid into deltoid teeth, ± ciliate; **Cs** lobes 2.8 × 0.4 mm, spatulate, ± erect-connivent; **Poll** ovoid-ellipsoid, 0.2 × 0.12 mm; **Fr** 8.5 - 10 cm × 3 mm; **Se** black, 12 × 2 mm, tuft of **Ha** ± 24 mm.

This is the only *Ceropegia* with verticillate leaves. It is closely related to *C. meyeri-johannis*.

C. vincifolia Hooker (CBM 1839: t. 3740 + text, 1839). **T:** [lecto – icono]: l.c. t. 3740. – **D:** India. **I:** Ansari (1984).

≡ *Ceropegia hirsuta* var. *vincifolia* (Hooker) Hooker *fil.* (1883); **incl.** *Ceropegia stocksii* Hooker *fil.* (1883); **incl.** *Ceropegia polyantha* Blatter & McCann (1931); **incl.** *Ceropegia oculata* var. *subhirsuta* H. Huber (1957).

[2] **R** tuber flattened; stem twining, glabrous; **L** ovate with cordate base and acuminate tip, upper face hairy; **Inf** many-flowered, peduncle hairy; **Ped** glabrous; **Cl** 3 - 8 cm, yellowish; **Cl** inflation occupying the lower ¼ or ⅓, 4 - 5 mm ⌀, ovoid, merging gradually into the **Cl** tube, tube with purple stripes in the upper ½, widening in a funnel-shaped manner towards the mouth; **Cl** lobes 1.5 - 3.5 cm, linear from a triangular-ovate base, apically acute, fused into a narrowly ovoid cage, lamina folded back along the midrib, inside pubescent, ciliate, basal ½ pale green, upper ½ dark green; **Cn** shortly stipitate, fused into a bowl; **Ci** lobes nearly fully connate and entire or sinuate or slightly incised, concave, hairy; **Cs** linguiform-spatulate, erect, tips ± recurved.

Most closely related to *C. hirsuta*.

C. volubilis N. E. Brown (BMI 1895: 261, 1895). **T:** Angola (*Welwitsch* 4272 [BM, COI, K, P]). – **D:** Angola. **I:** Bally (1965).

Incl. *Ceropegia scandens* N. E. Brown (1895); **incl.** *Ceropegia dewevrei* De Wildeman (1904).

[3] **R** fibrous; stems glabrous, twining, 2 - 4 mm ⌀, succulent, glaucous-green; **L** ovate to elliptic, 3 - 7 × 1.5 - 3 cm, base usually strongly cordate, apically mucronate; **Inf** pseudo-umbellate, 2- to 4-flowered; **Cl** ± 2.5 - 4.5 cm, otherwise as *C. distincta*, but **Cl** tube inside glabrous.

A dubious taxon, presumably to be relegated into the synonymy under *C. distincta*.

C. wallichii Wight (Contr. Bot. India, 32, 1834). **T:** Nepal (*Wallich* s.n. [BM]). – **D:** Nepal. **I:** Huber (1957).

Incl. *Ceropegia erecta* Wallich *ex* Wight (1834) (*nom. inval.*, Art. 34.1c).

[1] **R** fusiform-cylindrical, fleshy, in clusters; stems erect, robust, firm, faintly pubescent; **L** sessile or shortly petiolate, lamina elliptic to ovate, upper face densely pubescent, lower face pale and midrib hairy; **Inf** sessile or shortly pedunculate, 1- to many-flowered, developing together with the sprouting stem; **Sep** subulate; **Cl** 3 - 5 cm, whitish-green with purple longitudinal stripes; **Cl** inflation globose-ovoid, 6 - 10 mm long, inside at the transition to the **Cl** tube hairy; **Cl** tube basally narrowed to ± 3 mm ⌀, widening into a funnel-shaped mouth of 12 mm ⌀; **Cl** lobes 12 - 17 mm, ovate from a broadly triangular base, broadly keeled, lamina folded back along the midrib except for the base, margins with purple **Ha**, upper ½ purple, tips fused; **Cn** shortly stipitate, ± 4 × 3 mm, at the base fused into a cup; **Ci** lobes bilobed into triangular ciliate appendages; **Cs** linear-subulate, ± 3.5 mm, erect-parallel.

Endemic to Nepal and most easily distinguishable by its robust erect stems.

C. yemenensis Meve & R. Mangelsdorff (Bot. J. Linn. Soc. 137: 100, ills. (p. 101, 103), 2001). **T:** Yemen (*Mangelsdorff* Y24 [B, FR, UBT]). – **D:** N Yemen. **Fig. XIV.b**

[1] **R** fusiform-fleshy, clustered; stems erect, twining to 50 cm, sparsely branched, ≤ 1 mm ⌀; **L** shortly pedicellate, narrowly linear to lanceolate, acute, 15 - 90 × 1 - 2 mm, slightly subsucculent, ± ciliate; **Inf** 3 - 10 mm pedunculate, pseudo-umbellate, 1- to 2- (to 5-) flowered; **Bra** minute; **Ped** ± 10 mm, glabrous; **Sep** lanceolate-subulate, 1 - 2 × 0.5 mm, glabrous; **Cl** pale yellow with pale greenish base, 25 - 35 mm, erect, slightly curved, with intense scent like Lily of the Valley flowers; **Cl** inflation ovoid, 6 - 9 × 3 - 4 mm, inside dark red, with a ring of white **Ha**; **Cl** tube narrowing gradually to 1.5 mm ⌀, mouth widening to 2 - 3 mm ⌀, glabrous; **Cl** lobes narrowly linear-lanceolate, lamina ± bent back along the midrib, apically fused to form an ovoid cage-like structure, basally slightly auriculate; **Cn** yellowish, sessile, cup-shaped, ± 3 × 3.5 mm; **Ci** deltoid, ± 0.5 × 0.5 mm, bifid, marginally reddish with long white **Ha**; **Cs** lobes 2.5 × 0.4 mm, laterally compressed, erect-connivent, apically papillose; **Poll** ovoid-subrectangular, 0.3 × 0.2 mm.

The relationships of this recently described species from the Yemen remain unresolved. The flowers are superficially very similar to those of *C. stenantha*.

C. yorubana Schlechter (BJS 38: 47-48, ills., 1905). **T** [neo]: Nigeria (*Meikle* 1256 [K, P]). – **D:** Nigeria, Ghana.

[3] **R** fibrous; stems perennial, weakly succulent, 2 - 3 m, twining, ± terete, ± 3 mm ∅; **L** petiolate, broadly cordate, 2 - 3.5 × 1.5 - 2.5 cm, acute, ± glabrous; **Inf** ± 2 cm pedunculate, 1- to 3-flowered; **Ped** 5 mm; **Sep** narrowly lanceolate, 3 mm; **Cl** ± 2.5 cm, ± straight, yellow-green, blotched with ± red-brown; **Cl** inflation ± 4 mm ∅, globose, inside scatteredly hairy, merging ± gradually into the **Cl** tube, tube ± 2 mm ∅, towards the mouth widening to 4 mm ∅; **Cl** lobes linear-spatulate, 8 mm, tips fused, lamina folded back along the midrib, inside apically with black-purple **Ha**; **Cn** cup-shaped, sessile; **Ci** lobes shortly triangular, hardly 1 mm, concave, apex bifid into triangular acute hairy appendages; **Cs** ± 4 mm, linear with clavate tips curving outwards, erect, ± connivent; **Poll** ± pear-shaped.

Well-separable by its broadly cordate leaves and the very long staminal corona. The species belongs into the group of *C. distincta*.

C. zeyheri Schlechter (BJS 38: 48-49, 1905). **T:** RSA, Eastern Cape (*Zeyher* s.n. [G]). – **D:** RSA (Eastern Cape). **I:** Dyer (1983); Bruyns (1985).

Incl. *Ceropegia patersoniae* N. E. Brown (1913).

[1,3,4] Stem succulents, climbing and twining to 1 m, few-branched; **R** fleshy, fusiform, clustered; stems terete, 2 - 4 mm ∅, green, smooth, glabrous; **L** narrowly ovate, 5 - 10 mm, succulent, rapidly caducous; **Inf** sessile or with a peduncle 1 cm long, 1- to 2- (to 3-) flowered; **Ped** ± 7 mm; **Sep** lanceolate, acute, ± 4 mm; **Cl** 4.5 - 6 cm, ± erect, only slightly bent, outside pale green; **Cl** inflation globose to ovoid, 5 - 8 mm ∅, inside blotched with red; **Cl** tube gradually narrowed to ± 3 mm ∅, towards the mouth widening to 8 - 11 mm, inside whitish-green, faintly striped with red-brown, glabrous; **Cl** lobes linear, 3 - 4 cm × 1 mm, erect, forming a delicate cage, green, inside and outside ± hairy, margins ± involute; **Cl** lobes abruptly ovate in the basal ⅓, with red-brown reticulate markings, tip spatulate and widening to 2 - 3 mm, glabrous, margins with purple vibratile **Ha** in the narrowed section; **Cn** sessile, ± 4.5 × 3 mm, only basally fused; **Ci** lobes narrowly triangular, ± erect, for ¾ bifid into parallel subulate appendages, ± 2 mm; **Cs** ± 3.5 × 0.4 - 0.6 mm, linear-subulate, erect-connivent, tips slightly recurved, ± acute or truncate and denticulate; **Poll** ellipsoid; **Fr** ± 9 cm.

Very closely allied to *C. fimbriata* s.str. The two taxa occur in mixed populations.

CIBIRHIZA

U. Meve

Cibirhiza Bruyns (Notes Roy. Bot. Gard. Edinburgh 45(1): 51-54, ills., 1988). **T:** *Cibirhiza dhofarensis* Bruyns. – **Lit:** Kunze & al. (1994). **D:** Oman, Tanzania, Zambia. **Etym:** Lat. 'cibus', nourishment, food; and Gr. 'rhiza', root; for the edible root tubers.

Perennial geophytes with **R** tubers, becoming woody; latex white; stems erect, sometimes decumbent, mostly twining, perennial; **L** petiolate, elliptic to oblong-ovate, large, coarse, basally with colleters, hairy; **Inf** extra-axillary, lateral, sessile, many-flowered, pseudo-umbellately clustered; **Ped** short; **Cl** rotate with shallow to shallowly campanulate tube shorter than the **Cl** lobes; **Cl** lobes triangular, flatly spreading, inside papillose; **Cn** biseriate, sessile, red-brown, glossy from nectar; **Ci** fused fringe- or disc-like, ± indented at intrastaminal positions; **Cs** differentiated into 2 or 3 processes arranged one behind the other, each ± subulate and erect; **Gy** sessile; **Anth** appendages broadly ovate, whitish-transparent, lying on and obscuring the **Sty** head; translator without caudicles; **Poll** broadly ellipsoid, ± 0.2 × 0.12 mm, flat; **Fr** solitary.

This small genus has an uncommonly disjunct distribution with 1 species in Oman and 1 species in Zambia and Tanzania. The lianas with massive root tubers occupy a basal position within the *Asclepiadoideae*. *Cibirhiza*, together with *Fockea*, belongs to tribe *Fockeeae* (Kunze & al. 1994).

The two species of the genus are closely related. Vegetative differences are in the leaf shape, florally there are differences in characters of the corona.

C. albersiana H. Kunze & al. (Taxon 43(3): 368-374, ills., 1994). **T:** Zambia (*White* 6969 [K]). – **D:** Tanzania, Zambia; esp. Mopane woodland.

Stems climbing and twining; **R** tubers partly above-ground, globose to turnip-shaped or irregularly shaped, to 30 cm ∅; stems to 2 m, yellowish-brown, somewhat hairy; **L** 4 - 7 cm petiolate, lamina ovate to obovate, basally cuneate to somewhat cordate, apically ± acute, 6 - 14 × 4 - 9 cm, sparsely pubescent; **Inf** 5- to 10- (to 20-) flowered; **Ped** hardly 1 mm, very sparsely pubescent; **Sep** ovate, obtuse, ± 1.5 mm, pubescent; **Cl** yellowish to greenish with red-brown spots and dots, ± 1 cm ∅, outside ± glabrous, inside papillose; **Cl** lobes triangular, ± 3.5 × 3 mm; **Cn** purple-brown, ± 2.5 × 2 mm; **Ci** fringe-like, erect; **Cs** differentiated into 2 processes, outer process cuneate-subulate, ± 0.8 mm, basally embracing the inner process, inner **Cs** process bifid into an outer segment, this triangular-subulate, ± 0.7 mm, ascending, and an inner segment, this subulate-filiform, ± 1.5 mm, erect-connivent.

C. dhofarensis Bruyns (Notes Roy. Bot. Gard. Edinburgh 45(1): 51-54, ills., 1988). **T:** Oman,

Dhofar (*Miller* 7525 [E, K, KWT, ON, OXF, UPS]). – **D:** Oman (Dhofar); rocky places. **Fig. XIV.d**

Stems decumbent or twining; **R** tuber globose to turnip-shaped, > 15 cm ⌀, edible; stems to 1.5 m, 3 - 8 mm ⌀, bark uneven; **L** 1 - 7 cm petiolate, lamina ovate to oblong-ovate, basally obtuse to cordate, apically ± obtuse, 4 - 15 × 1.5 - 10 cm, ± pubescent; **Inf** 15- to 25-flowered; **Ped** 2 - 3 mm, pubescent; **Sep** linear-subulate, ± 2.5 mm, hairy; **Cl** pale green with red-brown spots and dots, 1.2 - 1.4 cm ⌀, outside pubescent, inside papillose; **Cl** lobes triangular-ovate, ± 4.5 × 4 mm; **Cn** purple, ± 3 mm ⌀; **Ci** united disc-like; **Cs** differentiated into 2 processes, outer processes shortly hooked, inner ± 1.5 mm, subulate, erect with reflexed tip; **Fr** ± 7 × 1 cm, fusiform, acute, smooth, glabrous.

CYNANCHUM

S. Liede

Cynanchum Linné (Spec. Pl. [ed. 1], 212, 1753). **T:** *Cynanchum acutum* Linné. – **Lit:** Jumelle & Perrier (1908); Jumelle & Perrier (1911); Choux (1914); Choux (1923); Descoings (1961); Liede (1993a); Liede (1993b); Liede (1996a); Liede & Meve (2001); Liede & Täuber (2002). **D:** Tropics and subtropics worldwide, few species in temperate climates. **Etym:** Gr. 'kynos', dog; and Gr. 'anchein', to choke; for the toxicity of the plants.

Incl. *Cyathella* Decaisne (1838). **T:** not typified.
Incl. *Cynoctonum* E. Meyer (1838) (*nom. illeg.*, Art. 53.1). **T:** not typified.
Incl. *Decanema* Decaisne (1838). **T:** *Decanema bojerianum* Decaisne.
Incl. *Pycnoneurum* Decaisne (1838). **T:** *Pycnoneurum junciforme* Decaisne.
Incl. *Sarcocyphula* Harvey (1863). **T:** *Sarcocyphula gerrardii* Harvey.
Incl. *Vohemaria* Buchenau (1889). **T:** *Vohemaria messeri* Buchenau.
Incl. *Flanagania* Schlechter (1894). **T:** *Flanagania orangeana* Schlechter.
Incl. *Platykeleba* N. E. Brown (1895). **T:** *Platykeleba insignis* N. E. Brown.
Incl. *Folotsia* Costantin & Bois (1908). **T:** *Folotsia sarcostemmoides* Costantin & Bois.
Incl. *Voharanga* Costantin & Bois (1908). **T:** *Voharanga madagascariensis* Costantin & Bois.
Incl. *Mahafalia* Jumelle & H. Perrier (1911). **T:** *Mahafalia nodosa* Jumelle & H. Perrier.
Incl. *Nematostemma* Choux (1921). **T:** *Nematostemma perrieri* Choux.
Incl. *Karimbolea* Descoings (1960). **T:** *Karimbolea verrucosa* Descoings.

Erect scrambling or twining **R** or stem succulents or mesophytic **L** plants without thickened **R**, with milky, white to yellow latex; stems green, 1.5 - 15 mm ⌀, smooth, delicately ribbed or tuberculate, wax-covered, glabrous or hairy; **L** reduced to **Sc** or flat, then petiolate or sessile, usually with 1 - 7 **Gl**, linear or not, margins entire, basally often cordate to reniform; **Inf** extra-axillary, few- to many-flowered, bostrychoid to pseudo-umbellate; **Fl** frequently sweet-scented, with nectar; **Cl** white to brown, sometimes rose-coloured, 3 - 20 mm ⌀, stellate to tubular, rarely broadly campanulate (*C. insigne*); **Cn** white, rarely red, lower than or as tall as or taller than the **Gy**, consisting of fused staminal (**Cs**) and interstaminal (**Ci**) parts; **Cs** occasionally with adaxial appendages (= ligule), or with additional free **Cs** (*C. insigne*); **Gy** sessile or stipitate; **Anth** with apical appendage and conspicuously differentiated lateral **Anth** wings, these wings usually consisting of a proximal and a distal rail, space between the rails usually filled with upwards-directed **Bri**, proximal rail curved; **Poll** pendent, rarely ± erect; **Sty** head forming a conspicuous bulge above the corpuscles, upper part flat, conical, drawn out or tabular; **Fr** usually only 1 per **Fl**, thin-walled, winged or not, smooth or tuberculate, glabrous, weakly longitudinally ribbed; **Se** brown, pear-shaped to ovoid, smooth, sculptured, papillose or with **Ha**, winged or not, with apical tuft of **Ha**.

The genus *Cynanchum* embraces some 300 species throughout the world, of which most are non-succulent climbers and twiners. All succulents covered below belong to Sect. *Cynanchum*.

Karimbolea, *Folotsia* and *Platykeleba* belong to the complex of succulent Madagascan *Cynanchum* as shown by molecular studies (Liede & Täuber 2002). The necessary taxonomic changes were published by Liede & Meve (2001). *Sarcostemma* his its roots here, too, and will have to be transferred to *Cynanchum* (see further comments for *Sarcostemma*).

The following name is of unresolved application but is referred to this genus: *Cynanchum tuberosum* hort. (s.a.) (*nom. inval.*, Art. 29.1).

C. aculeatum (Descoings) Liede & Meve (Novon 6(1): 59, ills. (p. 60), 1996). **T:** Madagascar, Toliara (*Descoings* 1013 [P, TAN]). – **D:** S Madagascar (Toliara).

≡ *Prosopostelma aculeatum* Descoings (1957) ≡ *Folotsia aculeata* (Descoings) Descoings (1961).

Stems 30 - 40 cm, ascending, not twining, 1.5 - 2 mm ⌀, delicately striate, inconspicuously wax-covered, glabrous, latex yellow; **L** rudiments scaly, early caducous; **Inf** with 1 - 5 sessile **Fl**, pseudo-umbellate; **Fl** buds elongate-conical; **Cl** yellow, trumpet-like; **Cl** lobes united for almost the whole length, 6 - 7 mm, free lobes short, recurved; **Cn** 2.5 - 3 mm, surpassing the stipitate **Gy** (excl. appendages of the **Sty** head), **Cs** and **Ci** fused for ± the whole **Cn** length; **Ci** planar, oblong, tip bifid; **Cs** placed against the back of the **St**, not differentiated; **Anth** wings as long as the **Anth**; **Sty** head cream-coloured to yellow, long-elongate.

Belongs to the species group of *C. folotsioides, C. luteifluens, C. mahafalense* and *C. messeri.*

C. ambovombense (Liede) Liede & Meve (Adansonia, sér. 3, 23(2): 350, 2001). **T:** Madagascar, Toliara (*Decary* 8374 [P]). – **D:** S Madagascar (Toliara). **I:** Liede (1996b).

≡ *Folotsia ambovombensis* Liede (1996).

Stems 4 - 5 mm ⌀, with conspicuous wax cover, glabrous; **L** rudiments ± 0.8 mm petiolate, lamina ± 1.6 × 1.2 mm, triangular, early deciduous; **Inf** 6- to 12-flowered, sessile; buds ovoid; **Cl** almost globose, lobes basally united for ± 2 mm, tips bent inwards; **Cn** ± 1.2 mm, shorter than the stipitate **Gy**, **Cs** and **Ci** basally united; **Ci** laminar, ovate, flat; **Cs** not differentiated; **Anth** wings as long as the **Anth**; **Sty** head flatly conical; **Fr** ± 6 cm, inversely club-shaped, apically obtuse.

C. ampanihense Jumelle & H. Perrier (Rev. Gén. Bot. 23: 258, 1911). **T:** Madagascar (*Decary* 4534 [P]). – **D:** S Madagascar (Toliara). **Fig. XIV.e**

Incl. *Cynanchum humbertii* Choux (1926).

R torulose; stems 40 - 130 cm, ascending, twining, 2 - 2.5 mm ⌀, delicately longitudinally striate, inconspicuously wax-covered, becoming glabrous, latex ivory-coloured; **L** rudiments scaly, ± 0.5 × ± 0.5 mm; **Inf** 2- to 5-flowered, pseudo-umbellate; peduncle persistent; **Ped** 1 - 4 mm; **Fl** unscented; buds cylindrical; **Cl** yellow to brown, basally united; **Cl** lobes 3 - 3.5 mm, horizontal to recurved; **Cn** 2.5 - 3 mm, as tall as the sessile **Gy**; **Cs** and **Ci** fused for ± ½ of the **Cn** length, with a conspicuous constriction in the middle; **Ci** not differentiated; **Cs** planar, pointed; **Anth** wings as long as the **Anth**; **Sty** head white to cream-coloured, flatly conical; **Fr** 5 - 8 cm, obclavate, tip strongly beaked; **Se** ± 3 mm, pear-shaped, densely hairy, without wings.

Closely related to *C. lecomtei.*

C. andringitrense Choux (Ann. Inst. Bot.-Géol. Colon. Marseille, sér. 4, 1(2): 21, 1923). **T:** Madagascar (*Perrier* 14477 [P]). – **D:** Madagascar (Antananarivo, Fianarantsoa).

R tubers 1 - 3 × 0.5 - 1 cm; stems 0.3 - 1 (-2) m, ascending, twining, 0.7 - 1.3 mm ⌀, becoming glabrous; **L** petiole 13 - 22 mm, lamina 23 - 45 × 12 - 22 mm, ovate, rarely elliptic; **Inf** 5- to 12-flowered, bostrychoid to pseudo-umbellate; **Fl** faintly honey-scented, buds oblong-conical; **Ped** 16 - 31 mm; **Cl** green, along the main veins rose-coloured, cup-shaped, united for ± ¼; **Cl** lobes 4 - 4.7 mm, curved inwards; **Cn** ± 3 mm, surpassing the **Gy** and completely covering it; **Cs** and **Ci** fused for > ¾ of the **Cn** length; **Ci** planar, triangular, keeled; **Cs** not differentiated; **Anth** wings shorter than the **Anth**; **Sty** head umbonate; **Fr** ± 6.5 cm, tip strongly beaked.

Belonging to the *C. lineare* species complex.

C. angavokeliense Choux (Ann. Inst. Bot.-Géol. Colon. Marseille, sér. 4, 5(2): 31, 1928). **T:** Madagascar, Antananarivo (*Perrier* 12937 [P]). – **D:** C Madagascar (Antananarivo). **Fig. XV.b**

R tubers 10 - 15 × 2 - 3 cm; stems 30 - 70 cm, ascending, twining, 0.75 - 1.5 mm ⌀, becoming glabrous; **L** petiole 2 - 7 mm, lamina 42 - 60 × 1 - 3.8 mm, linear; **Inf** 4- to 8-flowered; **Fl** buds (oblong-) conical; **Ped** 8 - 17 mm; **Cl** greenish-brown; **Cl** lobes basally united, ± 3.5 mm, curved inwards to spreading; **Cn** ± 1.5 mm, as tall as the **Gy**, **Cs** and **Ci** completely fused; **Ci** planar, triangular, keeled; **Cs** not differentiated; **Anth** wings longer than the **Anth**; **Sty** head conical to capitate.

Belonging to the *C. lineare* species complex.

C. ansamalense Liede (Bull. Mus. Nation. Hist. Nat., Sect. B, Adansonia 18(1-2): 112-114, ills., 1996). **T:** Madagascar, Toliara (*Rauh* 21850 [HEID]). – **D:** S Madagascar (Toliara).

Stems ascending, twining, 1 - 1.5 mm ⌀, delicately striate, glabrous; **L** scaly, ovate, 0.8 - 1 × 0.8 - 1 mm; **Inf** 1- to 4-flowered; **Fl** sessile, buds globose; **Cl** white, along the main veins purple; **Cl** lobes basally united, 1.8 - 2 mm, down-curved; **Cn** ± 1.2 mm, shorter than the stipitate **Gy**, **Cs** and **Ci** completely fused; **Ci** thicker than the **Cs**, planar; **Cs** fused with the **Fil**, placed against the back of the **St**; **Anth** wings as long as the **Anth**; **Sty** head shortly inversely funnel-shaped.

Belonging to the *C. gerrardii* species complex.

C. appendiculatopsis Liede (Bull. Mus. Nation. Hist. Nat., Sect. B, Adansonia 18(1-2): 114-116, ills., 1996). **T:** Madagascar, Fianarantsoa (*Bosser* 16816 [P]). – **D:** E Madagascar (Fianarantsoa).

Stems 40 - 50 cm, erect to decumbent, not twining, 1.5 - 3 mm ⌀, finely striate, inconspicuously wax-covered, becoming glabrous; latex ivory-coloured; **L** scaly, 1 - 1.2 × 0.6 - 0.8 mm; **Inf** with 2 - 5 almost sessile **Fl**, pseudo-umbellate; **Fl** buds conical; **Cl** cream-coloured, along the main veins purple, campanulate; **Cl** lobes basally united, 2.5 - 4.5 mm, curved inwards to spreading; **Cn** pale greenish-yellow, ± 1 mm, a little shorter than the stipitate **Gy**; **Cs** and **Ci** fused for more than ¾ of the **Cn** length; **Ci** planar, rectangular, forming a conspicuous convex fold; **Cs** fused with the **Fil**, planar, ovate; **Anth** wings as long as the **Anth**; **Sty** head furcate; **Fr** ± 5 cm, obclavate to oblong, tip strongly beaked.

Forming part of the *C. gerrardii* species complex.

C. arenarium Jumelle & H. Perrier (Ann. Inst. Bot.-Géol. Colon. Marseille, sér. 2, 6: 189-190, 1908). **T:** Madagascar (*Perrier* 1316 [P]). – **D:** Madagascar (Fianarantsoa, Mahajanga, Toliara).

Incl. *Voharanga madagascariensis* Costantin & Bois (1908); **incl.** *Mahafalia nodosa* Jumelle & H. Perrier (1911) ≡ *Cynanchum nodosum* (Jumelle &

H. Perrier) Descoings (1961); **incl.** *Cynanchum madecassum* Descoings (1961).

Stems 2 - 3 m, ascending, twining, 4 - 10 mm ∅, tuberculate, conspicuously wax-covered, glabrous; latex white, presumably poisonous; **L** scaly, 1.5 - 2 × 0.8 - 1 mm; **Inf** with 10 - 16 (-32) almost sessile **Fl**, bostrychoid; **Fl** weakly sweet-scented, buds conical; **Cl** yellowish-green; **Cl** lobes basally united, ± 3 mm, horizontal; **Cn** ± 2 mm, surpassing the sessile **Gy**, **Cs** and **Ci** fused for more than ¾ of the **Cn** length, divided into a lower and an upper part; **Ci** planar, oblong when spread out, forming a conspicuous convex fold; **Cs** planar, triangular, ligule shorter than the **Ci**; **Anth** wings as long as the **Anth**; **Sty** head white, tabular; **Fr** 5 - 9 cm, obclavate, apically obtuse; **Se** ± 7 mm, pear-shaped, with inconspicuous longitudinal ribs, winged.

Closely related to *C. hardyi*.

C. bisinuatum Jumelle & H. Perrier (Rev. Gén. Bot. 23: 258, 1911). **T:** Madagascar (*Decary* 9003 [P]). – **D:** Madagascar (Fianarantsoa, Toliara). **I:** Asklepios No. 74: 18, 1998.

Stems 2 - 3 m, creeping to ascending, rarely twining, 4 - 6 mm ∅, weakly tuberculate, delicately striate, strongly wax-covered, glabrous; **L** scaly, early caducous; **Inf** 6- to 11-flowered, bostrychoid; **Ped** slender, long, curved; **Fl** buds ovoid; **Cl** greenish-yellow; **Cl** lobes united basally, ± 7 mm, down-curved; **Cn** ± 2.5 mm, shorter than the sessile **Gy**, **Cs** and **Ci** fused for ± ½ of the **Cn** length, with a conspicuous constriction and thus divided into a lower and an upper part; **Ci** thinner than the **Cs**; **Cs** planar, bifid, ligule almost as long as the **Ci**; **Anth** wings longer than the **Anth**; **Sty** head green, flat.

Closely related to *C. perrieri*.

C. compactum Choux (Ann. Inst. Bot.-Géol. Colon. Marseille, sér. 3, 2: 310, 1914). **T:** Madagascar, Fianarantsoa (*Perrier* 13175 [P]). – **D:** Madagascar.

C. compactum var. **compactum** – **D:** E Madagascar (Fianarantsoa).

Stems 8 - 15 cm, erect to decumbent, not twining, 3 - 5 mm ∅, inconspicuously wax-covered, glabrous; latex ivory-coloured; **L** scaly, 1 - 1.3 × 0.7 - 1 mm; **Inf** with 5 - 15 (-27) almost sessile **Fl**, pseudo-umbellate; **Fl** buds oblong-conical; **Cl** yellowish-green, along the main veins purple, almost globose; **Cl** lobes basally united, ± 4 mm, curved inwards to reflexed; **Cn** 1.5 - 2 mm, somewhat shorter than the sessile **Gy**, **Cs** and **Ci** fused for ± ½ of the **Cn** length; **Ci** not differentiated; **Cs** fused with the **Fil**, planar, oblong to triangular, with a basal bulge; **Anth** wings as long as the **Anth**; **Sty** head white, flat (to conical); **Fr** 8 - 10 cm, fusiform, tip shortly beaked; **Se** ovate.

Closely related to *C. nematostemma*. Easily cultivated and floriferous, but the compact growth is lost in cultivation.

C. compactum var. **imerinense** Descoings (Adansonia, n.s., 1(2): 321-325, ills. (p. 316), 1961). **T:** Madagascar (*Descoings* 3228 [P]). – **D:** C Madagascar (Antananarivo).

Differs from ssp. *compactum*: Stems 2.5 - 3 mm ∅; **Cl** lobes ± 2 mm; **Fr** 4 - 5 cm, fusiform, tips strongly beaked; **Se** ± 5 mm, ovate, winged, side of the raphe smooth, backside with **Ha** arranged in groups.

C. crassipedicellatum Meve & Liede (Novon 4(3): 276-279, ills., 1994). **T:** Madagascar, Toliara (*Hardy* 2852 [K, MSUN]). – **D:** S Madagascar (Toliara). **Fig. XV.a**

Stems 30 - 40 cm, erect, diverging, not twining, 4 - 5 mm ∅, tuberculate, with strong wax covering; latex white; **L** scaly, early caducous, ± 1.2 × ± 0.5 mm, ovate; **Inf** with 8 - 15 almost sessile **Fl**, bostrychoid; **Fl** buds conical; **Ped** ± 10 × 3 (-5) mm, fleshy, persistent; **Cl** purple; **Cl** lobes basally united, 2.5 - 3.5 mm, twisted; **Cn** papillose, ± 2.2 mm, surpassing the sessile **Gy**, **Cs** and **Ci** only basally fused; **Ci** not differentiated; **Cs** fused with the **Fil**, planar, oblong, twisted; **Anth** wings as long as the **Anth**; **Sty** head green, capitate.

The uncommon character of the twisted **Cs** is otherwise only found in *C. descoingsii*.

C. cucullatum N. E. Brown (BMI 1897(128): 272, 1897). **T:** Madagascar, Antananarivo (*Baron* 2036 [K, P]). – **D:** C Madagascar (Antananarivo). **I:** Rauh (1995a).

R turnip-like, 10 - 15 × 1.5 - 2 cm; stems 45 - 60 cm, ascending, twining, 0.8 - 1 mm ∅; **L** petiole 2.5 - 5 mm, lamina 5 - 10 cm × 1.7 - 8 mm, linear; **Inf** 20- to 25-flowered, bostrychoid; **Fl** sweet-scented, buds depressed-globose; **Ped** 6.5 - 13.5 mm; **Cl** white, depressed-globose; **Cl** lobes basally united, 2.8 - 3.9 mm, incurved; **Cn** red, 0.7 - 1 mm, as tall as the sessile **Gy**, **Cs** and **Ci** completely fused; **Cs** fused with the **Fil**, placed against the back of the **St**, fleshy, thicker than the **Ci**; **Anth** wings shorter than the **Anth**; **Sty** head green, flat (to conical); **Fr** ± 4 cm, obclavate, tip strongly beaked; **Se** 4.5 - 5 mm, ovate, longitudinally ribbed, winged.

This is the only *Cynanchum* species known to have a fleshy staminal corona.

C. descoingsii Rauh (Trop. subtrop. Pfl.-welt 85: 37-41, ills., 1993). **T:** Madagascar, Toliara (*Rauh* 68639 [HEID]). – **D:** S Madagascar (Toliara).

Stems 30 - 80 cm, erect, not twining, tuberculate, strongly wax-covered, 3.5 - 4 mm ∅; **L** scaly, 1.5 - 2.2 × ± 0.7 mm, oblong, early caducous; **Inf** 5- to 20-flowered, bostrychoid, **Inf** axis persistent, almost sessile; **Fl** buds conical; **Cl** brown to purple, cup-shaped; **Cl** lobes basally united, ± 7 mm, recurved, twisted; **Cn** 2.5 - 4 mm, surpassing the sessile **Gy**, papillose, **Cs** and **Ci** fused for > ½ of the **Cn** length; **Cs** fused with the **Fil**, robust, twisted, with a ligule as long as the **Ci**; **Anth** wings shorter than the **Anth**; **Sty** head yellow to brown, cylindrical, with a papillose conical cap.

Closely related to *C. crassipedicellatum*.

C. fimbriatum Choux (Ann. Inst. Bot.-Géol. Colon. Marseille, sér. 4, 5(2): 32, 1928). **T:** Madagascar, Antananarivo (*Perrier* 16756 [P]). – **D:** Madagascar (Antananarivo).

R turnip-shaped, ± 10 × 1 - 1.5 cm; stems 35 - 40 (-60) cm, erect to ascending, weakly twining, 1.5 - 2 mm ⌀, becoming glabrous; **L** petiole 1 - 2 mm, lamina 30 - 42 × 2 - 4 mm, linear; **Inf** 5- to 8-flowered, pseudo-umbellate; **Fl** buds oblong-conical; **Ped** 3.5 - 4 mm; **Cl** brown, cup-shaped; **Cl** lobes basally united, ± 5 mm, spreading; **Cn** ± 3 mm, surpassing the **Gy**, **Cs** and **Ci** fused for > ¾ of the **Cn** length; **Ci** planar, triangular, forming a conspicuous convex fold; **Cs** not differentiated; **Anth** wings as long as the **Anth**; **Sty** head capitate.

Belonging to the *C. lineare* species complex.

C. floriferum Liede & Meve (Adansonia, sér. 3, 23(2): 350, 2001). **T:** Madagascar (*Perrier* 12104 [P]). – **D:** N Madagascar (Antsiranana).

Incl. *Prosopostelma grandiflorum* Choux (1914); incl. *Folotsia floribunda* Descoings (1961).

Stems 1 - 1.3 cm ⌀, with indistinct wax cover, glabrous; **L** rudiments early deciduous; **Inf** 20- to 30-flowered, almost sessile; **Fl** with scent like roses; **Cl** brown, lobes basally united for ± 8 mm, reflexed; **Cn** ± 4.5 mm, overtopping the stipitate **Gy**, **Cs** and **Ci** united for ± ½ of the **Cn** length, **Cs** not differentiated; **Ci** laminar, triangular; **Anth** wings as long as the **Anth**; **St** head rose-coloured, flatly conical.

C. folotsioides Liede & Meve (Novon 6(1): 60-63, fig. 2, 1996). **T:** Madagascar, Toliara (*Rauh* 21847 [K, MSUN]). – **D:** S Madagascar (Toliara). **Fig. XV.c**

Stems ascending, twining, delicately striate, 5 - 8 mm ⌀, becoming glabrous; latex yellow; **L** scaly, 2 - 2.4 × ± 1.2 mm; **Inf** 15- to 20-flowered, pseudo-umbellate; **Fl** scented, buds cylindrical; **Cl** yellow, along the main veins purple; **Cl** lobes basally united, ± 4 mm, horizontal to recurved; **Cn** ± 4.5 mm, surpassing the sessile **Gy**, **Cs** and **Ci** fused for > ¾ of the **Cn** length; **Ci** longer and thinner than the **Cs**, planar, ovate; **Cs** fused with the **Fil**, planar, very shortly bifid; **Anth** wings longer than the **Anth**; **Sty** head white, flatly conical.

Related to *C. aculeatum*.

C. gerrardii (Harvey) Liede (Taxon 40(1): 117, 1991). **T:** RSA, KwaZulu-Natal (*Gerrard* 1321 [TCD]). – **D:** Arabia, S and E Africa, Madagascar.

≡ *Sarcocyphula gerrardii* Harvey (1863).

C. gerrardii ssp. **bekinolense** (Choux) Liede & Meve (Adansonia, sér. 3, 23(2): 349, 2001). **T:** Madagascar (*Perrier* 11740 [P]). – **D:** Comoros, Madagascar (Mahajanga).

≡ *Cynanchum bekinolense* Choux (1914).

Differs from ssp. *gerrardii*: Stems 1 - 1.5 mm ⌀, glabrous; **L** scaly, ± 1 × 1 mm; **Inf** 3- to 10-flowered; **Fl** buds conical; **Cl** yellow, along the main veins purple, campanulate; **Cl** lobes basally united, 1.5 - 2 mm, horizontal or ± curved downwards; **Cn** ± 1.2 mm, as tall as the **Gy**, **Cs** and **Ci** fused for ± ½ of the **Cn** length; **Ci** not differentiated; **Cs** fused with the **Fil**, oblong; **Sty** head umbonate to tabular; **Fr** ± 7 cm, obclavate, keeled; **Se** ± 3 mm.

C. gerrardii ssp. **gerrardii** – **D:** Arabia, Comoros, Kenya, RSA (Eastern Cape, KwaZulu-Natal, Northern Prov.), Madagascar (Antsiranana, Fianarantsoa, Toliara). **Fig. XV.d**

Incl. *Cynanchum sarcostemmatoides* K. Schumann (1895); incl. *Cynanchum edule* Jumelle & H. Perrier (1911).

Stems 0.5 - 3 m, ascending, twining, delicately striate, 1.5 - 2.5 mm ⌀, inconspicuously wax-covered, becoming glabrous; latex slightly ivory-coloured; **L** scaly, 0.8 - 1.2 × 0.5 - 0.8 mm; **Inf** 4- to 7-flowered, bostrychoid to pseudo-umbellate; **Fl** sweet-scented, buds globose to ovoid; **Ped** ≤ 2.5 mm; **Cl** green; **Cl** lobes 2 - 3 mm, united for ± ¼ of their length, curved downwards; **Cn** ± 1.5 mm, slightly surpassing the sessile **Gy**, **Cs** and **Ci** fused for > ¼ of the **Cn** length; **Ci** thinner than the **Cs**; **Cs** fused highly with the **Fil**, placed against the back of the **St**, planar, triangular; **Anth** wings longer than the **Anth**; **Sty** head white, flat to flatly conical; **Fr** 8 - 12 cm, oblong, tip shortly beaked; **Se** 5 - 6 mm, pear-shaped, densely hairy, without wings.

The only taxon of the genus that occurs in Madagascar as well as on the African continent.

C. grandidieri Liede & Meve (Adansonia, sér. 3, 23(2): 350-351, 2001). **T:** Madagascar, Toliara (*Perrier* 1442 [P]). – **D:** S Madagascar (Toliara). **I:** Rauh (1998). **Fig. XXII.f**

Incl. *Decanema grandiflorum* Jumelle & H. Perrier (1908) ≡ *Folotsia grandiflora* (Jumelle & H. Perrier) Jumelle & H. Perrier (1911); incl. *Folotsia sarcostemmoides* Costantin & Bois (1908).

Stems 4 - 5 m, ascending, twining, 8 - 15 mm ⌀, with conspicuous wax cover, glabrous; **L** rudiments with 3 - 4 mm long petiole, lamina 6 - 8 × ± 3.5 mm, soon deciduous; **Inf** 8- to 20-flowered, sessile; buds oblong-conical; **Cl** white, lobes basally united for 6 - 7 mm, spreading; **Cn** ± 5 mm, overtopping the sessile **Gy**, **Cs** and **Ci** united for ± ½ of the **Cn** length, **Ci** 2× as long as the **Cs**, flattened, tapering filiform; **Cs** placed against the back of the **St**, flat, tapering filiform; **Anth** wings as long as the **Anth**; **Sty** head rose-coloured, capitate; **Fr** ± 20 cm, oblong, apically blunt; **Se** ± 7 mm, ovoid, indistinctly sculptured with longitudinal ribs, winged.

C. hardyi Liede & Meve (Novon 6(1): 63, fig. 3 (p. 62), 1996). **T:** Madagascar, Toliara (*Hardy & Jacobsen* 3571 [K, MSUN, PRE]). – **D:** S Madagascar; near coasts.

Scrambling to 4 m tall; stems not twining, to 12 mm ∅, tuberculate, strongly wax-covered; **L** scaly, early caducous; **Inf** 12- to 20-flowered, bostrychoid, almost sessile; **Fl** buds ovoid; **Cl** basally yellow, apically brown, campanulate; **Cl** lobes ± 4 mm, united for ± ¼ of their length, spreading, lobes yellow or basally yellow and apically brown; **Cn** ± 2.5 mm, surpassing the sessile **Gy**, **Cs** and **Ci** completely fused, divided into a lower and an upper part; **Ci** planar, bifid when spread out, keeled; **Cs** with a ligule; **Anth** wings shorter than the **Anth**; **Sty** head white, umbonate.

Closely related to *C. arenarium*.

C. humbert-capuronii Liede & Meve (Adansonia, sér. 3, 23(2): 351, 2001). **T:** Madagascar, Toliara (*Humbert & Capuron* 29382 [P]). – **D:** S Madagascar (Toliara). **I:** Liede (1996b).

Incl. *Folotsia humbertii* Liede (1996).

Stems ascending, twining, 4 - 8.5 mm ∅, with conspicuous wax cover, glabrous; **L** rudiments soon deciduous; **Inf** 6- to 10-flowered, sessile; **Cl** cup-shaped, lobes basally united for 5 mm, tips bent backwards; **Cn** ± 4 mm, somewhat overtopping the sessile **Gy**, **Cs** and **Ci** united for ± ½ of the **Cn** length; **Ci** laminar, broadly triangular; **Cs** not differentiated; **Anth** wings as long as the **Anth**; **Sty** head bulging.

C. implicatum (Jumelle & H. Perrier) Jumelle & H. Perrier (Rev. Gén. Bot. 23: 251, 1911). **T:** Madagascar (*Perrier* 8994 [P]). – **D:** Madagascar (Antsiranana, Mahajanga).

≡ *Sarcostemma implicatum* Jumelle & H. Perrier (1908) ≡ *Vohemaria implicata* (Jumelle & H. Perrier) Jumelle & H. Perrier (1909); **incl.** *Cynanchum aequilongum* Choux (1914).

Stems ascending, twining, delicately striate, 1.5 - 3 mm ∅, inconspicuously wax-covered, glabrous; **L** scaly, ± 0.9 × ± 0.6 mm; **Inf** 3- to 9-flowered, pseudo-umbellate; **Fl** buds cylindrical; **Ped** ± 2 mm; **Cl** yellowish-green, along the main veins purple; **Cl** lobes basally united, ± 12 mm, spreading to reflexed; **Cn** ± 3.5 mm, as tall as the stipitate **Gy**, **Cs** and **Ci** fused for ± ½ of the **Cn** length; **Ci** as long as the **Cs**, planar, weakly bifid; **Cs** widely fused with the **Fil**, placed against the back of the **St**, planar, triangular; **Anth** wings as long as the **Anth**; **Sty** head umbonate; **Fr** 8 - 9 cm; **Se** hairy.

C. insigne (N. E. Brown) Liede & Meve (Adansonia, sér. 3, 23(2): 351, 2001). **T:** Madagascar, Antananarivo (*Baron* 973 [P]). – **D:** C Madagascar.
Fig. XXXVIII.a

≡ *Platykeleba insignis* N. E. Brown (1895) ≡ *Sarcostemma insigne* (N. E. Brown) Descoings (1961).

Stems erect to prostrate, not twining, strongly waxy, glabrous, 30 - 35 cm, 2.5 - 3.5 mm ∅; **L** rudiments reduced to minute **Sc**, 0.6 × 0.7 mm; **Inf** 2- to 4-flowered, pseudo-umbellate, sessile; buds shortly conical; **Cl** white to green, veined with dark red, broadly campanulate, 1.5 - 2 cm ∅; **Pet** almost completely united with somewhat recurved tips; **Cn** glabrous, ivory-coloured to yellow; **Cs** ± 1.5 mm, free, peg-like, shorter than the sessile **Gy**, surrounded by a ring of fused staminal and interstaminal parts, the ring fused with the **Cs**; **Anth** wings as long as the **Anth**, with distal rail only; **Sty** head white, conical; **Fr** only 1 per **Fl**, erect, obclavate; **Se** ± 3.5 mm, ovate, longitudinally ribbed, winged.

The flowers are similar to those of *Oxystelma* R. Brown and *Philibertia* Kunth (all species herbaceous).

C. juliani-marnieri Descoings (Bull. Soc. Bot. France 118(1-2): 105-108, ills., 1971). **T:** Madagascar, Toliara (*Rauh* 21989 [MPU]). – **D:** S Madagascar (Toliara).

Stems 15 - 80 cm, ascending, twining, inconspicuously striate, 1.5 - 2 mm ∅, strongly wax-covered, glabrous; **L** scaly, 0.5 - 1.5 × ± 0.4 mm; **Inf** 3- to 5-flowered, pseudo-umbellate, almost sessile; buds globose; **Cl** cup-shaped; **Cl** lobes basally united, ± 1.5 mm, ± horizontal; **Cn** ± 1.5 mm, surpassing the sessile **Gy**, **Cs** and **Ci** fused for > ½ of the **Cn** length, with a conspicuous constriction and thus divided into a lower and an upper part; **Ci** not differentiated; **Cs** planar, bifid, apically twisted; **Anth** wings shorter than the **Anth**; **Sty** head capitate.

Belonging to the *C. ampanihense* species group.

C. jumellei Choux (Ann. Inst. Bot.-Géol. Colon. Marseille, sér. 4, 5(2): 29, 1928). **T:** Madagascar, Antananarivo (*Perrier* 11624 [P]). – **D:** Madagascar (Antananarivo).

R turnip-like, 9 - 12 × 1.5 - 3 cm; stems 20 - 60 cm, ascending, twining, 0.8 - 1 mm ∅, becoming glabrous; **L** petiole 2.5 - 7 (-12) mm, lamina 2 - 6 cm × 1.5 - 3.8 (-5) mm, linear; **Inf** 3- to 7-flowered, pseudo-umbellate; **Fl** buds ovoid to globose; **Ped** 2.5 - 6 cm; **Cl** brown to purple, cup-shaped; **Cl** lobes basally united, ± 5 mm, curved inwards; **Cn** ± 2 mm, as tall as the sessile **Gy**, **Cs** and **Ci** fused for ± ¾ of the **Cn** length; **Cs** basally planar, apically drawn out into a subulate tip; **Anth** wings as long as the **Anth**; **Sty** head flatly conical.

C. junciforme (Decaisne) Liede (Bull. Mus. Nation. Hist. Nat., Sect. B, Adansonia 14(3-4): 449, 1993). **T:** Madagascar (*Bojer* s.n. [P]). – **D:** Madagascar (Antananarivo, Fianarantsoa, Toliara).

≡ *Pycnoneurum junciforme* Decaisne (1838); **incl.** *Cynanchum rusillonii* Hochreutiner (1908).

R turnip-like, 20 - 25 × 1 - 2 cm; stems 30 - 60 (-100) cm, erect, not twining; **L** petiole 0 - 2.5 mm, lamina 7 - 11 cm × 2 - 3 mm, linear; **Inf** 14- to 18-flowered, pseudo-umbellate, sessile; **Fl** buds elongate-conical; **Cl** white to rose-coloured, cup-shaped; **Cl** lobes united for > ½ of their length, ± 3.5 mm, curved inwards, apically strongly twisted; **Cn** ± 2 mm, as tall as the **Gy**, **Cs** and **Ci** completely fused; **Anth** wings shorter than the **Anth**; **Sty** head flat; **Fr** ± 12 cm, obclavate, tip strongly beaked; **Se** ± 3.5 mm, pear-shaped, with **Ha** arranged in groups, winged.

Closely related to *C. sessiliflorum* and *C. papillatum*.

C. lecomtei Choux (Bull. Mus. Nation. Hist. Nat. 33: 196, 1927). **T:** Madagascar, Toliara (*Decary* 3220 [P]). – **D:** S Madagascar (Toliara).

R torulose; Stems 1.5 - 2 m, ascending, twining, delicately striate, 1 - 2.5 mm ∅; **L** scaly, ± 2.5 × 0.6 mm; **Inf** 2- to 4-flowered, pseudo-umbellate; **Fl** unscented, buds (oblong-) conical; **Ped** 1.5 - 2 mm; **Cl** green, lobes basally united, ± 4.5 mm; **Cn** ± 3.5 mm, surpassing the **Gy**, **Cs** and **Ci** fused for ± ½ of the **Cn** length, divided into a lower and an upper part; **Ci** not differentiated; **Cs** tips filiform, twisted; **Anth** wings as long as the **Anth**; **Sty** head white, flat to conspicuously conical; **Fr** ± 4 cm, obclavate, tip shortly beaked; **Se** ± 3.5 mm, ovate, papillose, densely hairy, wingless.

Closely related to *C. ampanihense*.

C. lenewtonii Liede (KB 49(1): 119-123, ills., 1994). **T:** Kenya, Northern Frontier Prov. (*Gillett* 14031 [K]). – **D:** Ethiopia (Bale), Kenya (Northern Frontier Prov.).

Stems 2.5 m, ascending, twining, delicately striate, 1.5 - 2.5 mm ∅, inconspicuously wax-covered, becoming glabrous; latex white; **L** scaly, 1.2 - 1.7 × ± 0.7 mm; **Inf** 2- to 6-flowered, pseudo-umbellate, sessile; **Fl** buds ovoid to cylindrical; **Cl** green to yellow; **Cl** lobes united for ± ¼ of their length, 2.5 - 3.5 mm, down-curved; **Cn** ± 2.5 mm, surpassing the sessile **Gy**, **Cs** and **Ci** fused for ± ¾ of the **Cn** length, **Ci** thinner than the **Cs**, planar, equal-sided to very oblong-triangular; **Cs** widely fused with the **Fil**, placed against the back of the **St**, apically filiform; **Anth** wings as long as the **Anth**; **Sty** head flat to flatly conical; **Fr** ± 6 cm, obclavate.

Closely related to *C. gerrardii*.

C. lineare N. E. Brown (BMI 1897: 273, 1897). **T:** Madagascar (*Baron* 109 [K, P]). – **D:** Comoros, Madagascar.

C. lineare ssp. **keraudreniae** Liede (Bull. Mus. Nation. Hist. Nat., Sect. B, Adansonia 18(1-2): 108, fig. 2 (p. 107), 1996). **T:** Madagascar, Nosy Be (*Keraudren* 1594 [P]). – **D:** Comoros, Madagascar (Antsiranana).

Differs from ssp. *lineare*: **R** unknown; **L** 5 - 9 × 2 - 5 cm, ovate instead of linear; **Ped** 4.5 - 7 cm.

C. lineare ssp. **lineare** – **D:** Madagascar (Antananarivo, Fianarantsoa, Mahajanga).

R turnip-shaped, 12 - 15 × 2 - 2.5 cm; stems 25 - 50 cm, ascending, twining, 1.5 - 2 mm ∅; **L** petiole 1 - 5 mm, lamina 4 - 8 cm × 0.7 - 15 mm, linear, rarely ovate-oblong; **Inf** 7- to 20-flowered, bostrychoid to pseudo-umbellate; **Fl** buds conical; **Ped** 13 - 25 mm; **Cl** greenish-yellow, brown along the main veins; **Cl** lobes basally united, 2 - 5 mm, recurved; **Cn** 2 - 3.5 mm, surpassing or completely covering the **Gy**, **Cs** and **Ci** fused for > ¾ of the **Cn** length; **Ci** planar, triangular, keeled; **Cs** placed against the back of the **St**, not differentiated; **Anth** wings shorter than the **Anth**; **Sty** head (flatly) conical; **Fr** ± 6 cm, obclavate, tip shortly beaked; **Se** 4 - 5 mm, ovate, longitudinally ribbed, winged.

C. luteifluens (Jumelle & H. Perrier) Descoings (Adansonia, n.s., 1(2): 314, 1961). **T:** Madagascar (*Bojer* s.n. [P]). – **D:** Madagascar, widespread in all dry parts except the uppermost N and the E coast. **I:** Rauh (1998: 250).

≡ *Decanema luteifluens* Jumelle & H. Perrier (1911); **incl.** *Cynanchum decaisneanum* Descoings (1961) (*nom. illeg.*, Art. 53.1).

C. luteifluens var. **longicoronae** Liede (Bull. Mus. Nation. Hist. Nat., Sect. B, Adansonia 18(1-2): 118-119, ills., 1996). **T:** Madagascar (*Keraudren* 306 [P]). – **D:** Madagascar (Fianarantsoa, Toliara).

Differs from var. *luteifluens*: Stems delicately striate, often attacked by a black mould, 2.5 - 4 mm ∅, at least the terminal stem parts thicker than in the typical var.; **Inf** 2- to 7-flowered; **Cl** greenish-yellow, brown along the main veins; **Cl** lobes 4.5 - 6 mm, down-curved; **Cn** 4.5 - 6.5 mm.

C. luteifluens var. **luteifluens** – **D:** Madagascar. Fig. XV.e

Incl. *Sarcostemma mauritianum* Bojer (1837) (*nom. illeg.*, Art. 53.1); **incl.** *Decanema bojerianum* Decaisne (1838) ≡ *Cynanchum bojerianum* (Decaisne) Choux (1927); **incl.** *Asclepias aphylla* Bojer *ex* Decaisne (1844) (*nom. inval.*, Art. 34.1c).

Stems 2 - 3 m, ascending, twining, 2 - 3 mm ∅, very conspicuously striate, inconspicuously wax-covered, glabrous; latex yellow; **L** scaly, 1.2 - 1.5 × 0.8 - 1 mm; **Inf** with 1 - 4 sessile **Fl**, pseudo-umbellate; **Fl** buds conical; **Cl** yellow, lobes basally united, 1.5 - 3 mm, horizontal; **Cn** 1.7 - 3 (-3.5) mm, surpassing the sessile **Gy**, **Cs** and **Ci** fused for ± ½ of the **Cn** length; **Ci** thinner than and as long as or longer than the **Cs**, apically filiform, twisted; **Cs** fused with the **Fil**, placed against the back of the **St**, apically filiform; **Anth** wings as long as the **Anth**; **Sty** head white, flatly conical or umbonate; **Fr** 8 - 18 cm, oblong, apically obtuse; **Se** 3.5 - 5 mm, pear-shaped, hairy, wingless.

The most common and most variable species of *Cynanchum* in Madagascar.

C. macranthum Jumelle & H. Perrier (Rev. Bot. 23: 260, 1911). **T** [neo]: Madagascar, Toliara (*Decary* 9062 [P]). – **Lit:** Liede & Meve (1996: with ill.). **D:** S Madagascar (Toliara: Ambovombe). **Fig. XXXI.f, XXXI.g**

≡ *Karimbolea macrantha* (Jumelle & H. Perrier) Liede & Meve (1996).

Stems ascending, twining, 4 - 7 mm ∅, tuberculate, with strong wax cover, otherwise glabrous; **L** 1 mm petiolate, lamina 3 × 1.5 mm, ovate; **Inf** 4- to 7-flowered, bostrychoid to pseudo-umbellate, almost sessile; buds conical; **Cl** brown, cup-shaped; **Pet** united for ± ¼, 8 - 10 mm, lobes bent inwards; **Cn** ± 3.5 mm, surpassing the sessile **Gy**; **Cs** and **Ci** united for > ½ of the **Cn** length; **Ci** shorter than the **Cs**, laminar, carinate; **Cs** laminar, ovate, with a ligula of equal length; **Anth** wings as long as the **Anth**; **Sty** head tabular; **Fr** ± 10 cm, tip shortly beaked; **Se** ± 5 mm, ovoid, slightly papillate, winged.

C. macrolobum Jumelle & H. Perrier (Rev. Gén. Bot. 23: 257, 1911). **T:** Madagascar (*Perrier* 11644 [P]). – **D:** Madagascar (Fianarantsoa, Toliara). **I:** Rauh (1998: 251-252).

Stems 20 - 40 cm, erect to ascending, weakly twining, 3.5 - 5 mm ∅, roughly but delicately tuberculate, rarely almost smooth, strongly wax-covered; **L** scaly, early caducous; **Inf** 5- to 40-flowered, pseudo-umbellate; **Inf** axis persistent, almost sessile; **Fl** buds globose; **Cl** yellowish-green to brown, purple along the main veins, urceolate; **Cl** lobes united for ± ¼ of their length, 4 - 5 mm, curved downwards; **Cn** 1.5 - 2 mm, as tall as the sessile **Gy**, papillose, **Cs** and **Ci** fused for > ½ of the **Cn** length; **Ci** shorter than the **Cs**, planar, triangular; **Cs** planar, rectangular, with a ligule; **Anth** wings as long as the **Anth**; **Sty** head white, flatly conical; **Fr** 4 - 7 cm, fusiform, tip strongly beaked; **Se** ± 4.5 mm, ovate, smooth, winged.

Closely related to *C. rauhianum*.

C. madagascariense (K. Schumann) K. Schumann (in Engler & Prantl (eds.), Nat. Pfl.-fam. [ed. 1], 4(2): 253, 1895). **T:** Madagascar (*Scott-Elliot* 2882 [P]). – **D:** Madagascar (Antananarivo, Fianarantsoa, Toliara). **Fig. XVI.a**

≡ *Vincetoxicum madagascariense* K. Schumann (1893); **incl.** *Cynanchum helicoideum* Choux (1914).

R tubers 2 - 2.5 × 1.5 - 2 cm; stems 30 - 60 cm, ascending, twining; **L** petiole 2 - 12 mm, lamina 1.5 - 5 (-17) × 0.7 - 2 cm, linear, rarely elliptic or obovate; **Inf** 25- to 35-flowered, bostrychoid; **Fl** buds oblong-conical; **Ped** 2 - 5.8 cm; **Cl** brown, urceolate, fused for > ½ of its length; **Cl** lobes 4 - 7.5 mm, recurved; **Cn** ± 2 mm, surpassing and completely covering the sessile **Gy**, **Cs** and **Ci** fused for > ½ of the **Cn** length, both planar and each forming a conspicuous convex fold; **Cs** placed against the back of the **St**; **Anth** wings shorter than the **Anth**; **Sty** head umbonate; **Fr** obclavate, tip strongly beaked.

Related to the *C. papillatum* species group.

C. mahafalense Jumelle & H. Perrier (Rev. Gén. Bot. 23: 260, 1911). **T:** Madagascar (*Perrier* 14877 [P]). – **D:** Madagascar (Antananarivo, Fianarantsoa, Toliara). **I:** Rauh (1998: 253).

Incl. *Cynanchum ambositrense* Choux (1914).

Stems 2 - 3 m, ascending, twining, finely striate, 3 - 6 mm ∅, strongly wax-covered, glabrous to glabrescent; latex yellow; **L** scaly, 1.2 - 1.5 × 1.2 - 1.5 mm; **Inf** 3- to 10-flowered, bostrychoid, almost sessile; **Inf** axis persistent; **Fl** buds cylindrical; **Cl** yellow, purple along the main veins, campanulate, fused for ± ½ of its length; **Cl** lobes 4 - 5 mm, densely short-hairy, horizontal; **Cn** 3 - 4 mm, as tall as the stipitate **Gy**, **Cs** and **Ci** fused for > ½ of the **Cn** length; **Ci** thinner than the **Cs**, planar, triangular to rectangular, rarely oblong; **Cs** widely fused with the **Fil**, placed against the back of the **St**, planar, triangular; **Anth** wings as long as the **Anth**; **Sty** head green, umbonate; **Fr** 5 - 8 cm, oblong, apically obtuse; **Se** 6 - 7 mm, ovate, winged, side of the raphe smooth, back with **Ha** arranged in groups.

Closely related to *C. messeri*.

C. mariense (Meve & Liede) Liede & Meve (Adansonia, sér. 3, 23(2): 351, 2001). **T:** Madagascar, Toliara (*Liede & Conrad* 2825 [K, MSUN, ULM]). – **D:** S Madagascar (Toliara). **I:** Meve & Liede (1998).

≡ *Karimbolea mariensis* Meve & Liede (1998).

R fibrous; stems erect to ascending, not twining, sparsely branched, 5 - 15 × 0.3 - 0.7 cm (in cultivation longer), blue-green, verrucose, glaucous; **L** rudiments ovate, ± 1 mm; **Inf** shortly pedunculate, 3- to 7-flowered, **Fl** opening in succession; **Ped** 3 - 4 × 2 mm, fleshy; **Sep** 2 × 1 mm, triangular; **Fl** sweet-scented; **Cl** cup-shaped, basally fused; **Cl** lobes 6 - 7 × 3 mm, stout and fleshy, outside yellowish-green, warty, inside olive-green, with brown reticulation, ascending, apically twisted, with revolute margins and strong median keel; **Cn** ± 3.5 × 4 mm, sessile, pure white, pentagonally barrel-shaped, **Ci** and **Cs** fused high up, closed; **Ci** cucullate with involute margins; **Cs** ovate, concave, apically involute, adaxially with an appendage (ligule); **Anth** appendages ± 1.25 × 1 mm, ovate, slightly apiculate; **Poll** 0.35 × 0.27 mm, ovoid; **Sty** head rose, 2 × 1.5 mm, pyramidal.

Sister species of *C. verrucosum* and with similar stem morphology, but differentiated through the green-brown flower colour, the massive pentagonally barrel-shaped corona and the rose-coloured stylar head. – [U. Meve]

C. marnierianum Rauh (CSJA 42: 104-106, ills., 1970). **T:** Madagascar, Toliara (*Rauh* 9395 [HEID]). – **D:** S Madagascar (Toliara).

Stems 20 - 40 (-50) cm, decumbent, rarely creeping, not twining, 3.5 - 5 (-7) mm ∅, tuberculate-uneven; **L** scaly, ± 1.5 × ± 1.5 mm; **Inf** 3- to 6-flowered, pseudo-umbellate, sessile; **Fl** faintly honey-scented, buds oblong-conical; **Cl** greenish-yellow; **Cl** lobes free to the base, 5 - 6 mm, curved inwards, with fused tips at the beginning of anthesis; **Cn** 1.2 - 2 mm, surpassing the sessile **Gy** (except the extremely long **Sty** head), **Cs** and **Ci** fused for > ¾ of the **Cn** length, **Ci** not differentiated; **Cs** placed against the back of the **St**, planar, triangular; **Anth** wings as long as the **Anth**; **Sty** head white, inversely funnel-shaped; **Fr** ± 3.5 cm, obclavate, tip strongly beaked, unwinged; **Se** 4 mm, pear-shaped, smooth, winged.

A very attractive easily cultivated and easily flowering species.

C. menarandrense Jumelle & H. Perrier (Rev. Gén. Bot. 23: 257, 1911). **T** [neo]: Madagascar, Toliara (*Rauh* 10634 [MPU]). — **D:** S Madagascar (Toliara). **I:** Rauh (1998: 253-254).

Incl. *Cynanchum antandroy* Descoings (1971).

Stems 35 - 50 cm, erect, not twining, 7 mm ∅, tuberculate, strongly wax-covered, glabrous; **L** scaly, early caducous; **Inf** 12- to 17-flowered, bostrychoid; **Inf** axis persistent; **Fl** buds cylindrical; **Ped** 3 - 5 mm; **Cl** basally green, apically yellow to purple, purple along the main veins, cup-shaped; **Cl** lobes ± 3.5 mm, basally united, horizontal to incurved; **Cn** 2.5 mm, as tall as the sessile **Gy**, papillose, **Cs** and **Ci** fused for ± ½ of the **Cn** length, divided into a lower and an upper part; **Ci** not differentiated; **Cs** fused with the **Fil**, planar, rectangular, with a ligule; **Anth** wings longer than the **Anth**; **Sty** head white, umbonate.

Belonging to the *C. macrolobum* species group.

C. messeri (Buchenau) Jumelle & H. Perrier (Rev. Gén. Bot. 11: 252, 1911). **T:** Madagascar (*Rutenberg* s.n. [not traced]). — **D:** Madagascar (Antananarivo, Antsiranana, Toliara). **Fig. XV.f**

≡ *Vohemaria messeri* Buchenau (1889).

Stems 2 - 3 m, ascending, twining, finely striate, 2 - 4 mm ∅, strongly wax-covered, glabrous to glabrescent; latex yellow; **L** scaly, ± 1.5 × ± 1.5 mm; **Inf** 3- to 10-flowered, bostrychoid, almost sessile; **Fl** buds cylindrical; **Cl** yellow, purple along the main veins, campanulate, united for ± ¼ of its length; **Cl** lobes 4 - 5 mm, densely short-hairy, horizontal; **Cn** 3 - 3.5 mm, as tall as the long-stipitate **Gy**, **Cs** and **Ci** fused for ± ¼ of the **Cn** length; **Ci** as long as the **Cs**, planar, bifid; **Cs** planar, oblong; **Anth** wings as long as the **Anth**; **Sty** head green, flat to umbonate; **Fr** 8 - 10 cm, oblong, apically obtuse; **Se** ± 6.5 mm, ovate, winged, side of the raphe smooth, back with **Ha** arranged in groups.

Closely related to *C. mahafalense*.

C. mevei Liede (Bull. Mus. Nation. Hist. Nat., Sect. B, Adansonia 18(1-2): 120-123, ills., 1996). **T:** Madagascar, Toliara (*Liede & Conrad* 2780 [MO, MSUN, TAN]). — **D:** S Madagascar (Toliara). **Fig. XVI.d**

Stems 0.8 - 1.5 m, ascending, twining, finely striate, 1 - 2 mm ∅, inconspicuously wax-covered, becoming glabrous; **L** scaly, 2 - 2.5 × ± 0.6 mm; **Inf** 2- to 3-flowered, pseudo-umbellate, almost sessile; **Fl** buds conical; **Cl** greenish-yellow, urceolate, fused for ± ¼ of its length; **Cl** lobes ± 5 mm, basally curved inwards, apically curved outwards; **Cn** 2 - 2.5 mm, as tall as the sessile **Gy**, **Cs** and **Ci** fused for the whole length of the **Cn** (excl. the long filiform appendages of the **Cs**), divided into a lower and an upper part; **Ci** shorter than the **Cs**, planar, strongly keeled; **Cs** fused with the **Fil**, placed against the back of the **St**, planar, conspicuously pointed, filiform recurved; **Anth** wings as long as the **Anth**; **Sty** head white, conical; **Fr** ± 7.5 cm, obclavate, apically obtuse; **Se** ± 2.5 mm, pear-shaped, strongly hairy, wingless.

Belonging to the *C. gerrardii* species group.

C. moramangense Choux (Ann. Inst. Bot.-Géol. Colon. Marseille, sér. 4, 5(2): 34, 1928). **T:** Madagascar (*Perrier* 16887 [P]). — **D:** C Madagascar (Antananarivo, Toamasina).

R turnip-shaped; stems 40 - 80 cm, ascending, twining; **L** petiole 2 - 4.5 mm, lamina 15 - 60 × 1 - 3.5 mm, linear; **Inf** 3- to 5-flowered, pseudo-umbellate; **Fl** buds globose to ovoid; **Ped** 5 - 12 mm; **Cl** green or brown, cup-shaped; **Cl** lobes basally united, ± 3 mm, curved inwards; **Cn** ± 1.5 mm, as tall as the sessile **Gy**, **Cs** and **Ci** completely fused; **Ci** planar, forming a conspicuous convex fold; **Cs** placed against the back of the **St**; **Anth** wings shorter than the **Anth**; **Sty** head capitate; **Fr** 3 - 5 cm, obclavate, tip strongly beaked; **Se** ± 3.5 mm, pear-shaped, longitudinally ribbed, winged.

Belonging to the *C. lineare* species group.

C. napiferum Choux (Ann. Inst. Bot.-Géol. Colon. Marseille, sér. 3, 2: 353, 1914). **T:** Madagascar, Antananarivo (*Perrier* 11689 [P]). — **D:** C Madagascar (Antananarivo).

R turnip-shaped, 9 - 10 × 1.2 - 1.5 cm; stems 9 - 12 cm, erect, not twining; **L** petiole ± 1.8 mm, lamina 10 - 35 × 2.5 - 9.5 mm, ovate; **Inf** 5- to 7- (to 12-) flowered, pseudo-umbellate; **Fl** buds oblong-conical; **Ped** 10 - 25 mm; **Cl** cup-shaped; **Cl** lobes basally united, ± 5 mm, tips spreading; **Cn** ± 2.5 mm, surpassing the sessile **Gy**, **Cs** and **Ci** completely fused; **Ci** planar, keeled; **Cs** placed against the back of the **St**; **Anth** wings longer than the **Anth**; **Sty** head conical.

Belonging to the *C. lineare* species group.

C. nematostemma Liede (Bull. Mus. Nation. Hist. Nat., Sect. B, Adansonia 18(1-2): 123, ills. (p. 124), 1996). **T:** Madagascar, Mahajanga (*Perrier* 13227 [P]). — **D:** E Madagascar (Mahajanga).

Incl. *Nematostemma perrieri* Choux (1921).

Stems 30 - 60 cm, erect or drooping, not twining, finely striate, 1.5 - 2 mm ⌀, glabrous; **L** scaly, ± 1 × ± 0.4 mm, triangular, early caducous; **Inf** 5- to 7-flowered, pseudo-umbellate, sessile; **Fl** buds oblong-conical; **Cl** white, purple along the main veins and in the centre, campanulate, fused for ± ¼ of its length; **Cl** lobes 5 - 7 mm, spreading; **Cn** ± 4.5 mm, surpassing the sessile **Gy**, **Cs** and **Ci** fused only basally; **Ci** not differentiated; **Cs** fused basally with the **Fil**, basally planar, apically filiform, not twisted; **Anth** wings shorter than the **Anth**; **Sty** head flatly conical; **Fr** ± 6 cm, obclavate, tip strongly beaked.

Closely related to *C. compactum*.

C. orangeanum (Schlechter) N. E. Brown (FC 4(1): 745, 1908). **T:** RSA, Free State (*Flanagan* 1502 [SAM, BOL]). – **D:** Botswana, Namibia, Zimbabwe, RSA (Northern Prov., North-West Prov., Eastern Cape, Free State).

≡ *Flanagania orangeana* Schlechter (1894).

Rhizome 3 - 5 mm ⌀; stems 10 - 20 cm, erect, not twining; **L** 3 - 5 cm × 0.7 - 1.5 mm, linear; **Inf** 1- to 5-flowered, pseudo-umbellate, almost sessile; **Fl** buds flatly conical; **Cl** brown; **Cl** lobes basally united, 4 - 6 mm, curved inwards; **Cn** 4 - 5 mm, slightly surpassing the sessile **Gy**, **Cs** and **Ci** fused for ± ½ of the **Cn** length; **Ci** as long as the **Cs**, filiform; **Cs** placed against the back of the **St**, planar, oblong; **Anth** wings as long as the **Anth**; **Sty** head capitate; **Fr** 4 - 7 cm, fusiform, tip strongly beaked; **Se** 5 - 6 mm, papillate, winged.

Closely related to *C. praecox*.

C. pachycladon Choux (Ann. Inst. Bot.-Géol. Colon. Marseille, sér. 4, 5(2): 54, 1928). **T:** Madagascar, Toliara (*Perrier* 16602 [P]). – **D:** S Madagascar (Toliara). **I:** Rauh (1998: 256).

Stems 5 - 6 m, ascending, twining, ± 1.5 cm ⌀, pale grey-brown, tuberculate, becoming glabrous; **L** petiole 2 - 4 cm, lamina 5.5 - 8 × 4 - 6.5 cm, ovate, basally lobed, with unpleasant scent; **Inf** 15- to 33-flowered, bostrychoid; **Fl** buds oblong-conical; **Ped** 4 - 7 mm; **Cl** basally greenish-yellow, apically brown; **Cl** lobes basally united, ± 4.5 mm, horizontal to down-curved; **Cn** ± 3.5 mm, surpassing the very shortly stipitate **Gy**, **Cs** and **Ci** fused for ± ¾ of the **Cn** length; **Ci** shorter than the **Cs**, planar, ovate; **Cs** planar, elongate-triangular, ligule much shorter than the **Cs**; **Anth** wings as long as the **Anth**; **Sty** head flatly conical; **Fr** ± 12 cm, obclavate, tip shortly beaked, winged; **Se** ± 9 mm, pear-shaped, papillose and with **Ha** arranged in groups, winged.

C. papillatum Choux (Ann. Inst. Bot.-Géol. Colon. Marseille, sér. 4, 1(2): 11, 1923). **T:** Madagascar (*Perrier* 17424 [P]). – **D:** Madagascar (Antananarivo, Antsiranana, Fianarantsoa). **Fig. XVI.e**

R turnip-shaped, 10 - 12 × 1 - 1.5 cm; stems 10 - 40 cm, ascending, twining; **L** petiole 2 - 5 mm, lamina 4 - 7 cm × 1 - 4 mm, linear; **Inf** 11- to 15-flowered, pseudo-umbellate; **Fl** buds ovoid; **Ped** 5 - 15 mm; **Cl** white to rose-coloured, urceolate, united for ± ½ of its length; **Cl** lobes ± 5 mm, outside papillate, curved inwards; **Cn** red, ± 3 mm, surpassing the sessile **Gy**, **Cs** and **Ci** fused for > ¾ of the **Cn** length; **Ci** planar, trifid; **Cs** not differentiated; **Anth** wings as long as the **Anth**; **Sty** head flatly conical.

Closely related to *C. junciforme* and *C. sessiliflorum*, but much more delicate and twining.

C. perrieri Choux (Ann. Inst. Bot.-Géol. Colon. Marseille, sér. 3, 3: 307, 1914). **T:** Madagascar (*Perrier* 11742 [P]). – **D:** Madagascar (Antananarivo, Fianarantsoa, Toliara).

Stems 80 - 120 cm, erect, not twining, 9 - 13 mm ⌀, blue-green, tuberculate, strongly wax-covered, glabrous; **L** scaly, ± 5.5 × 1 mm, early caducous; **Inf** 10- to 25-flowered, bostrychoid, almost sessile; **Inf** axis persistent; **Fl** buds globose; **Cl** greenish-yellow, cup-shaped, fused for ± ¼ of its length; **Cl** lobes ± 8 mm, outside papillate, curved inwards; **Cn** ± 3.5 mm, surpassing the sessile **Gy**, papillose, **Cs** and **Ci** fused for ± ½ of the **Cn** length, divided into an upper and a lower part by a conspicuous constriction; **Ci** as long as the **Cs**, thinner, planar, spreading, bifid, forming a conspicuous concave fold; **Cs** fused with the **Fil**, fleshy, ligule shorter than the **Cs**; **Anth** wings as long as the **Anth**; **Sty** head rose-coloured to purple, capitate or inversely funnel-shaped; **Fr** 9 - 15 cm, fusiform, tip beaked; **Se** ± 7.5 mm, pear-shaped, longitudinally ribbed, winged.

Closely related to *C. bisinuatum*.

C. petignatii Liede & Rauh (Bull. Mus. Nation. Hist. Nat., Sect. B, Adansonia 18(1-2): 126-128, ills., 1996). **T:** Madagascar, Toliara (*Decary* 9011 [P]). – **D:** S Madagascar (Toliara).

Stems ascending, diverging, delicately striate, 2 - 3.5 mm ⌀, roughly tuberculate, strongly wax-covered, glabrous or glabrescent; **L** scaly, ± 1.3 × ± 1 mm, triangular, early caducous; **Inf** 2- to 5-flowered, pseudo-umbellate, almost sessile; **Fl** buds cylindrical; **Cl** green to white, cup-shaped; **Cl** lobes basally united, ± 8 mm, curved inwards; **Cn** ± 5 mm, surpassing the stipitate **Gy**, **Cs** and **Ci** fused for ± ½ of the **Cn** length; **Ci** longer than the **Cs**, planar, oblong-triangular, forming a conspicuous convex fold; **Cs** widely fused with the **Fil**, planar, triangular; **Anth** wings as long as the **Anth**; **Sty** head umbonate.

Related to *C. messeri*.

C. phillipsonianum Liede & Meve (Bull. Mus. Nation. Hist. Nat., Sect. B, Adansonia 18(1-2): 109-111, ills., 1996). **T:** Madagascar, Antsiranana (*Bardot-Vaucoulon* 40 [P]). – **D:** N Madagascar (Antsiranana).

Stems erect, shrub-like, 3 - 4.5 mm ⌀, glabrous; **L** petiole 1.1 - 2 cm, lamina 3 - 5 × 0.9 - 1.5 cm, ovate to elliptic, basally cordate; **Inf** 15- to 30-flowered; **Fl** buds conical to ovoid; **Ped** 1.5 - 3 cm; **Cl** brown-purple with yellowish margins, 4 - 5 mm ⌀, united for ± ¼ of its length; **Cl** lobes slightly papillose, curved inwards; **Cn** pale yellowish-green, ± 2.5 mm, as tall as the stipitate **Gy**, **Cs** and **Ci** fused for > ½ of the **Cn** length; **Ci** planar, forming a conspicuous convex fold; **Cs** not fused with the **Fil**; **Anth** wings as long as the **Anth**; **Sty** head oblong-conical and furcate; **Fr** ± 6 cm, obclavate, tip strongly beaked; **Se** ± 6.5 mm, ovate, with a few scattered **Ha**, winged.

The growth form of this species is exceptional as it possesses succulent stems and leaves. It represents a basal member of Madagascan *Cynanchum*.

C. praecox Schlechter *ex* S. Moore (J. Bot. 40: 256, 1902). **T**: Zimbabwe (*Rund* 512 [BM]). – **D**: Nigeria, Sierra Leone, Cameroon, Malawi, Tanzania, Zaïre, Zimbabwe.

Incl. *Cynanchum pygmaeum* Schlechter (1913).

Rhizome 1.5 - 3 mm ⌀; stems 3 - 10 cm, erect, not twining; **L** 4 - 6 cm × 2 - 8 mm, linear, rarely elliptic to ovate; **Inf** 5- to 15-flowered, bostrychoid; **Fl** scented, buds oblong-conical; **Ped** 4 - 6 (-8) mm; **Cl** yellow to brown; **Cl** lobes ± 6 mm, curved inwards; **Cn** 2 - 4 mm, surpassing the sessile **Gy**; **Cs** and **Ci** fused for ± ½ of the **Cn** length, planar; **Ci** as long as the **Cs**, oblong; **Cs** ovate; **Anth** wings longer than the **Anth**; **Sty** head capitate.

A species growing after fires and closely related to *C. orangeanum*.

C. pycnoneuroides Choux (Ann. Inst. Bot.-Géol. Colon. Marseille, sér. 3, 2: 361, 1914). **T**: Madagascar, Fianarantsoa (*Perrier* 11687 [P]). – **D**: E Madagascar (Fianarantsoa).

Rhizomes ± 1 cm ⌀; stems 40 - 60 cm, creeping to ascending, not twining, 1 - 2 cm ⌀, glabrous, basally with **L** scars; **L** only at the stem tips, 4.5 - 9 cm × ± 2.5 mm, linear; **Inf** 10- to 15-flowered, bostrychoid; **Fl** buds ovoid; **Ped** 2 - 4 mm; **Cl** green to white, cup-shaped, united for ± ¼ of its length; **Cl** lobes ± 5 mm, spreading to recurved; **Cn** ± 2.5 mm, as tall as the stipitate **Gy**, **Cs** and **Ci** fused for ± ½ of the **Cn** length; **Ci** shorter than the **Cs**, planar, ovate; **Cs** planar, triangular, ligule as long as the **Cs**; **Anth** wings longer than the **Anth**; **Sty** head flatly conical.

C. radiatum Jumelle & H. Perrier (Rev. Gén. Bot. 23: 259, 1911). **T** [neo]: Madagascar, Toliara (*Liede & al.* 2744 [P, TAN]). – **D**: S Madagascar (Toliara).

Stems ascending, twining, finely striate, ± 3.5 mm ⌀, inconspicuously wax-covered, becoming glabrous; **L** scaly, ± 1 × ± 0.6 mm; **Inf** 4- to 5-flowered, pseudo-umbellate, sessile; **Fl** buds cylindrical; **Cl** black-green, united for ± ¼ of its length; **Cl** lobes ± 3.5 mm, curved inwards to spreading; **Cn** ± 2.5 mm, surpassing the sessile **Gy**, **Cs** and **Ci** fused for > ¾ of the **Cn** length, divided into a lower and an upper part by a conspicuous constriction; **Ci** shorter than the **Cs**, planar, trifid, spreading, middle lobe triangular and shorter than the ovate lateral lobes, strongly keeled; **Cs** basally fused with the **Fil**, placed against the back of the **St**, planar, oblong; **Anth** wings longer than the **Anth**; **Sty** head flatly conical.

Belonging to the *C. gerrardii* species group.

C. rauhianum Descoings (Bull. Soc. Bot. France 110(3-4): 155-157, ills., 1963). **T**: Madagascar, Toliara (*Rauh* M986 [Herb. Descoings, HEID]). – **D**: S Madagascar (Toliara: Isalo Mts.).

Stems 60 - 80 cm, erect, not twining, 4 - 8 mm ⌀, strongly wax-covered, glabrous; **L** scaly, early caducous; **Inf** 15- to 30-flowered, bostrychoid, almost sessile; **Inf** axis persistent; **Fl** buds globose; **Cl** brown, cup-shaped; **Cl** lobes basally united for 3 - 4.5 mm, spreading, apically recurved; **Cn** ± 3 mm, slightly surpassing the sessile **Gy**, papillose, **Cs** and **Ci** fused for > ¾ of the **Cn** length, divided into a lower and an upper part by a conspicuous constriction; **Ci** shorter than the **Cs**, planar and spreading, pointed, forming a conspicuous convex fold; **Cs** planar, rectangular, ligule shorter than the **Cs**; **Anth** wings as long as the **Anth**; **Sty** head rose-coloured to purple, conical; **Fr** ± 5.5 cm, fusiform; **Se** ± 4 mm, pear-shaped, smooth, winged.

Closely related to *C. macrolobum*.

C. rossii Rauh (CSJA 42(2): 68-72, ills., 1970). Madagascar, Toliara (*Rauh* 21986 [HEID]). – **D**: S Madagascar (Toliara: Cap Ste. Marie). **Fig. XVI.c**

Stems 30 - 40 cm, decumbent, not twining, 4 - 6 mm ⌀, acutely 4-angled; **L** scaly, 1 - 1.5 mm, triangular, early caducous; **Inf** 1- to 2-flowered, almost sessile; **Fl** buds oblong-conical; **Cl** dull olive-green, margins white, later brown, cup-shaped; **Cl** lobes basally united, ± 4 mm, curved downwards; **Cn** ± 3 mm, surpassing and completely hiding the sessile **Gy**, papillose, **Cs** and **Ci** fused for ± ½ of the **Cn** length; **Ci** planar, rectangular, forming a conspicuous convex fold; **Cs** planar, triangular; **Anth** wings as long as the **Anth**; **Sty** head conical; **Fr** ± 6 cm, fusiform, tip strongly beaked, without wing; **Se** ± 4 mm, pear-shaped, strongly hairy, without wing.

Related to *C. descoingsii* and *C. crassipedicellatum*.

C. sessiliflorum (Decaisne) Liede (Bull. Mus. Nation. Hist. Nat., Sect. B, Adansonia 14(3-4): 449, 1993). **T**: Madagascar (*Goudot* s.n. [G-DEL]). – **D**: Madagascar (Antananarivo, Antsiranana, Mahajanga). **Fig. XVI.b**

≡ *Pycnoneurum sessiliflorum* Decaisne (1838).

R tubers 5 - 15 × 1 - 1.5 cm; stems 40 - 60 cm, erect, not twining; **L** 12 - 16 × ± 0.6 cm, linear; **Inf**

15- to 30-flowered, bostrychoid to pseudo-umbellate; basal ½ of the **Fl** buds cylindrical, apical ½ forming an appendage made up by the contorted **Cl** lobes; **Ped** 2 - 4 mm; **Cl** white or rose-coloured, oblong-clavate, united in the lower ½; **Cl** lobes 9 - 11 mm; **Cn** ± 2.5 mm, shorter than the sessile **Gy**; **Cs** and **Ci** completely fused to form a cup-shaped structure; **Anth** wings as long as the **Anth**; **Sty** head flat to umbonate.

Closely related to *C. junciforme*.

C. sigridiae Meve & M. Teissier (Bradleya 14: 10-13, ills., 1996). **T:** Madagascar, Toliara (*Teissier* 135 [K, MSUN]). – **D:** S Madagascar (Toliara). **Fig. XVII.b**

R torulose; stems 1.5 - 2 m, twining, delicately longitudinally striate, basally corky; **L** scaly, 3 - 5 × 0.6 mm; **Inf** 2- to 4-flowered, pseudo-umbellate, almost sessile; **Fl** buds conical; **Ped** 2 - 4 mm; **Cl** green, purple along the main veins; **Cl** lobes basally united, 4 - 5 mm, very strongly rolled back, slightly twisted; **Cn** 4 - 5.5 mm, surpassing the sessile **Gy**, tubular, divided into an upper and a lower part, **Cs** and **Ci** fused for ± ⅔ of the **Cn** length, drawn out into 0.5 - 2 mm long threads; **Anth** wings as long as the **Anth**; **Sty** head green, flatly conical.

C. subtile Liede (Bull. Mus. Nation. Hist. Nat., Sect. B, Adansonia 14(3-4): 440-442, ill., 1993). **T:** Madagascar, Toliara (*Liede & Conrad* 2827 [P, MO, TAN]). – **D:** S Madagascar (Toliara: Cap Ste.-Marie).

R tubers 2 - 3 × 2 - 3 cm; stems 30 - 50 cm, ascending, twining; **L** petiole 3 - 5 mm, lamina 12 - 16 × ± 7 mm, obovate; **Inf** 25- to 35-flowered, pseudo-umbellate; **Fl** unscented, buds oblong-conical; **Ped** ± 4 mm; **Cl** greenish-yellow, united for ± ¼ of its length; **Cl** lobes ± 4 mm, spreading; **Cn** ± 3.5 mm, surpassing the sessile **Gy**, **Cs** and **Ci** fused for > ¾ of the **Cn** length; **Ci** planar, keeled; **Cs** not differentiated; **Anth** wings shorter than the **Anth**; **Sty** head flatly conical; **Fr** ± 5 cm, fusiform, tip strongly beaked, keeled; **Se** ± 5 mm, ovate.

A member of the *C. lineare* species group.

C. toliari Liede & Meve (Adansonia, sér. 3, 23(2): 351, 2001). **T:** Madagascar, Toliara (*Grandidier* s.n. [P]). – **D:** S Madagascar (Toliara: Ambovombe, Ampanihy etc.). **I:** CSJA 64: 31, 1992.

Incl. *Prosopostelma madagascariense* Jumelle & H. Perrier (1911) ≡ *Folotsia madagascariensis* (Jumelle & H. Perrier) Descoings (1961).

Stems 2 - 3 m, ascending, twining, 6 - 8 mm ⌀, glabrous; **L** rudiments petiolate ± 1.5 mm, lamina ± 4 × 2.5 mm, early deciduous; **Inf** 8- to 15-flowered; **Fl** scented; buds oblong-conical; **Cl** yellowish-brown, lobes united for ± ¼ of their length, ± 3.5 mm, tips bent backwards; **Ci** ± 3.5 mm, overtopping the stipitate **Gy**, **Cs** and **Ci** united for > ½ of the **Cn** length, articulated into a lower and an upper part; **Ci** laminar, ovate (with a cleft in the lower ½); **Cs** placed against the back of the **St**; **Anth** wings shorter than the **Anth**; **Sty** head cream-coloured to yellow, flatly conical.

C. verrucosum (Descoings) Liede & Meve (Adansonia, sér. 3, 23(2): 351, 2001). **T:** Madagascar, Toliara (*Descoings* 3999 [P]). – **Lit:** Liede & al. (1993); Liede & Meve (1996). **D:** S Madagascar (Toliara: Cap Ste.-Marie). **Fig. XXXI.e**

≡ *Karimbolea verrucosa* Descoings (1960).

Stems 15 cm, erect, not twining, 6 - 10 mm ⌀, strongly tuberculate, old with strong wax cover, otherwise glabrous; **L** 1 mm petiolate, lamina 3 - 4 × 1 mm, ovate; **Inf** 2- to 3-flowered, pseudo-umbellate; **Fl** with slight honey-like scent, buds conical; **Ped** 2 - 4 mm; **Cl** rose-red to brown, cup-like; **Pet** basally united, 7 - 8 mm, tips bent inwards; **Cn** ± 4 mm, as tall as the sessile **Gy**; **Cs** and **Ci** united for > ½ of the **Cn** length; **Ci** laminar, strongly carinate, shorter than the **Cs**; **Cs** with a massive ligule; **Anth** wings as long as the **Anth**; **Sty** head rose-red, tabular; **Fr** 5 - 6 cm, fusiform; **Se** ± 4.5 mm, pear-shaped, papillate, without wing.

The only *Cynanchum* with ± erect pollinia.

×DERNIA

B. Müller & F. Albers

×Dernia Kimnach & al. (CSJA 65(2): 75, 1993).

= *Duvalia* × *Huernia*. Apart from the combination covered below, the hybrid *D. sp.* × *H. clavigera* is also reported (Leach 1988).

×D. 'Surprise' Kimnach & al. (CSJA 65(2): 75, ill., 1993).

= *D. sp.* (probably *D. reclinata*) × *H. nouhuysii*.

DISCHIDIA

C. Hoffmann, R. van Donkelaar & F. Albers

Dischidia R. Brown (Prodr., 461, 1810). **T:** *Dischidia nummularia* R. Brown. – **Lit:** Rintz (1980); Forster & Liddle (1988). **D:** Bangladesh, Sri Lanka, India, Nepal, China, Taiwan, Vietnam, Camboodia, Laos, Myanmar, Malaysia, Indonesia, Singapore, Philippines, Thailand, Papua New Guinea, Solomon-Islands, Australia. **Etym:** Gr. 'dischides', cleft in two; for the apically bifid staminal Corona.

Incl. *Collyris* Vahl (1810). **T:** not designated.
Incl. *Conchophyllum* Blume (1826). **T:** *Conchophyllum imbricatum* Blume.
Incl. *Leptostemma* Blume (1826). **T:** not designated.
Incl. *Triplosperma* G. Don (1837). **T:** *Stapelia cochinchinensis* Loureiro.
Incl. *Spathidolepis* Schlechter (1905). **T:** *Spathidolepis torricellensis* Schlechter.

Plants epiphytic, rarely epilithic, twining, latex white; **R** at the nodes and/or adventitious; stems 1 - 4 mm ∅, glabrous, tomentose or villose; **L** decussate, rarely alternate, shortly to very shortly petiolate, lamina mesomorphic, leathery or succulent, either (a) narrowly elliptic to ovate or oblanceolate, flat or lenticular, entire, or (b) rounded and shield-shaped, or (c) pitcher-shaped or tubular (pitcher **L**); **Inf** often with part-**Inf**, racemose, lateral, rarely terminal, rachis ± conspicuous; **Fl** 5-merous, actinomorphic; **Ped** short; **Sep** ovate, ± acute; **Cl** narrowly or broadly urceolate, fleshy, outside smooth and glabrous, rarely verrucose or pubescent, inside glabrous or pubescent; petaloid **Cn** absent, annular or 5-lobed and alternating with the **Cl** lobes; **Cn** uniseriate, staminal, **Cn** lobes rarely simple, usually tips bifid and involute; **Gy** sessile or stipitate, usually conical; **Sty** head usually conical, enclosed by the processes of the **Anth**; **Poll** erect, narrowly ovoid to clavate, caudicles broadly triangular; **Fr** narrowly fusiform, terete, flattened or reniform in cross-section, tapering, glabrous and smooth; **Se** ovate, flat, brown, with a tuft of white **Ha** (coma) at the micropylar end.

Many Dischidas are living in symbiosis with ants (Zizka 1990). The ants are inhabiting hollows between the shield-shaped leaves and the wood or stones upon which they are resting like limpets, or the ants are inhabiting pendent or ± spreading pitcher leaves. Detritus collected by the ants accumulates underneath or within the leaves (as does rainwater) and is thus made available to adventitious roots.

The majority of the ± 40 species will be covered here. A revision of the genus is urgently needed, and the following account remains preliminary, and the genus *Dischidiopsis* (see there) with its few species probably belongs here as synonym. In addition, a number of apparently non-succulent species are omitted.

Depending on the pubescence on the inside of the corolla, the genus can be subdivided into 5 groups (Rintz 1980):

[1] **Cl** glabrous throughout.
[2] Only **Cl** lobes pubescent.
[3] Only **Cl** throat pubescent.
[4] Both **Cl** lobes and the throat pubescent.
[5] **Cl** throat with a double circle of pubescence.

D. acutifolia Maingay & Hooker *fil.* (Fl. Brit. India 4: 51, 1883). **T:** Malaysia (*Maingay* 1122 [K]). — **D:** Indonesia (Borneo, Java, Sulawesi, Sumatra), Malaysia (Malaya), S Myanmar, Philippines, S Thailand.

D. acutifolia ssp. **acutifolia** — **D:** Indonesia (Borneo, Java, Sulawesi, Sumatra), S Myanmar, Malaysia (Malaya), Philippines, Thailand. **I:** Rintz (1980).

Incl. *Dischidia zollingeri* Schlechter (1908); **incl.** *Dischidia brachystele* Schlechter (1915); **incl.** *Dischidia hoyoides* Schlechter (1916); **incl.** *Dischidia pedunculata* Schlechter (1916).

[3] **R** at the nodes and adventitious; stems glabrous; **L** to 6 mm petiolate, lamina ovate or elliptic, 5 - 8 × 2 - 3.5 cm, flat, ± leathery, glabrous, acute, base acute; **Inf** 1- to 5-flowered; peduncle erect, 1 - 3 cm; **Ped** ± 1 mm, glabrous; **Sep** ovate, ± 6 mm, acute, membranous; **Cl** pale yellow; **Cl** tube ± 2.5 × 2 mm; **Cl** lobes ± 1 mm, partially revolute, pink; **Cs** lobes stipitate, apex reniform; **Sty** head pyramidal; **Poll** ± 0.25 mm, caudicles narrowly triangular, 0.35 mm, corpuscle 0.25 mm; **Fr** 6 × 0.3 cm, terete, pale green to pale yellow.

D. acutifolia ssp. **klossii** (Ridley) Rintz (Blumea 26: 101, ill., 1980). **T:** Thailand (*Kloss* 6721 [K]). — **D:** Malaysia (Malaya), Thailand.

≡ *Dischidia klossii* Ridley (1951); **incl.** *Dischidia lancifolia* Ridley (1918).

[3] Differs from ssp. *acutifolia*: **L** elliptic, 3 - 5 × 0.4 - 0.8 cm; **Fr** ± 3.5 × 0.2 cm.

D. albida Griffith (Not. Pl. Asiat. 4: 46, 1854). **T:** Malaysia, Malaya (*Griffith* 3781 [K]). — **D:** Indonesia (Borneo, Sulawesi, Sumatra), Malaysia (Malaya), Singapore. **I:** Rintz (1980).

Incl. *Dischidia kutchinensis* Beccari (1886); **incl.** *Dischidia kawengica* Schlechter (1916); **incl.** *Dischidia semperflorens* Schlechter (1916); **incl.** *Dischidia lagenifera* Ridley (1917); **incl.** *Dischidia fultonii* Henderson (1927).

[2] **R** at the nodes and adventitious; stems very thin, glabrous; **L** petiole 2.5 - 4 mm long, faintly pubescent, lamina ovate, 1.5 - 2 × 1 cm, flat or lenticular, glabrous, acute, ± round at the base; **Inf** 1- to 5-flowered, often with part-**Inf**; peduncle 2 - 3 mm; **Ped** ± 2 mm, glabrous; **Sep** ovate, 1 mm, acute; **Cl** tube belly-like, ± 4 × 2 mm, white; **Cl** lobes erect, ± 2 mm, pink; **Cs** lobes stipitate, 2-lobed, apically acute, basally with a pubescent hump; **Poll** elongate, ± flat, caudicles triangular; **Fr** terete, ± 6 × 0.3 cm, pale green.

D. astephana Scortechini *ex* King & Gamble (J. Proc. Asiat. Soc. Bengal 74(2): 582, 1908). **T:** Malaysia, Malaya (*Scortechini* s.n. [not located]). — **D:** Indonesia (Borneo), Malaysia (Malaya). **I:** Rintz (1980); Zizka (1990: 93). **Fig. XVII.c**

Incl. *Conchophyllum angulatum* Schlechter (1908).

[3] Epiphytic, creeping on dead wood and trunks; **R** at the nodes below the **L**; stems thin, villose, ± glabrous with age; **L** basin-shaped and appressed to the substrate, circular, ± 2.5 × 2.5 cm, acute, base with a conical appendage 2 mm long, upper face brown, humped, lower face purple, villose, ± glabrous with age; **Inf** 1- to 4-flowered; peduncle ± 5 mm; **Ped** ± 2 mm, pubescent; **Sep** ± triangular, ob-

tuse, pink, pubescent; **Cl** extremely 5-angular-winged, angles yellow-orange to deep red, remainder deep blue; **Cl** tube ± 5 × 7 mm; **Cl** lobes ± 2 mm, revolute, yellow-orange to deeply red; **Cs** lobes horizontal, deeply divided, broadly spatulate, tips bifid; **Anth** red, white-angled, **Anth** appendage acuminate; **Poll** ± 0.7 mm, caudicles slender, ± 0.2 mm, corpuscle ± 0.3 mm; **Fr** semiterete, ± 3 × 0.5 cm, red.

D. bengalensis Colebrooke (Trans. Linn. Soc. London 12: 357, t. 15, 1817). **T:** [lecto – icono]: l.c. t. 15. – **D:** NE India, Nepal, S Myanmar, Camboodia, Vietnam, Laos, Indonesia (Borneo, Java, Moluccas, Sulawesi, Sumatra), Malaysia (Malaya), Philippines, Singapur, S Thailand. **I:** Rintz (1980). **Fig. XVII.a**

Incl. *Dischidia spatulata* Blume (1826); **incl.** *Dischidia cuneifolia* Wallich (1831); **incl.** *Dischidia loeseneriana* Schlechter (1916).

[3] **R** at the nodes and adventitious; stems sparsely leafy, to 4 mm ⌀, pendent, partly weakly twining, fleshy, pale green to yellowish-green, glabrous; **L** decussate, to 1.5 mm petiolate, lamina oblanceolate to spatulate, 2 - 3 × 0.4 - 0.6 cm, round to blunt, base attenuate, thickened and fleshy, glabrous; **Inf** composed of part-**Inf**, 1- to 6-flowered; peduncle short; **Ped** 2 - 3 mm, pubescent; **Sep** ± reddish, nearly glabrous; **Cl** pale yellow, greenish-white or white; **Cl** tube ± 3 × 4 mm; **Cl** lobes ± 2 mm, partly revolute; **Cs** lobes stipitate, apically obtuse, 2-lobed, free lobes revolute; **Poll** ± 0.38 mm, caudicles broadly triangular, ± 0.25 mm, corpuscle ± 0.2 mm; **Fr** 5 - 7 × 0.3 cm, terete, pale green to pale brown.

Apart from *D. nummularia* perhaps the species of the genus most often seen in cultivation. It may become gigantic.

D. cochleata Blume (Bijdr. Fl. Ned. Ind., 1060, 1826). **T:** Indonesia, Java (*Blume* 1692 [L]). – **D:** Indonesia (Borneo, Java, Sumatra), Malaysia (Malaya). **I:** Rintz (1980).

Incl. *Dischidia coccinea* Griffith (1854).

[1] **R** at the nodes; stems densely leafy, glabrous, faintly hairy when young; **L** circular and shield-shaped, 1.5 - 4.5 × 2 - 5 cm, cordate to reniform at the base, with an appendage of 1 mm, upper face humped, lower face purple, glabrous; **Inf** 1- to 7-flowered; peduncle ± 0.5 - 3 cm, glabrous; **Ped** 1.5 - 3 mm, glabrous; **Sep** minute, acute or obtuse; **Cl** red; **Cl** lobes triangular-ovate, involute, blue; petaloid **Cn** annular; **Cs** lobes stipitate, apically cordate, 2-lobed, free lobes involute; **Sty** head flat; **Poll** narrowly lanceolate, 0.75 mm, caudicles ± 0.1 mm, thin, corpuscle ± 0.3 mm; **Fr** erect, fusiform, ± 0.25 × 0.4 cm, semiterete, green; **Se** obovoid, ± 0.13 mm.

D. dolichantha Schlechter (BJS 40(Beiblatt 92): 8, 1908). **T:** Indonesia, Sumatra (*Schlechter* 13261 [B]). – **D:** Indonesia (Borneo, Sumatra), Malaysia (Malaya). **I:** Rintz (1980). **Fig. XVII.e**

Incl. *Dischidia tubiflora* King & Gamble (1908).

[4] **R** at the nodes and adventitious; stems glabrous; **L** petiole 2 - 6 mm long, pubescent, lamina lanceolate to ovate, 2 - 3 × 1 cm, acute to acuminate, round, flat to lenticular, glabrous or faintly villose; **Inf** 1- to 5-flowered; peduncle erect, 3 - 10 mm; **Ped** ± 3 mm; **Sep** ovate, pubescent; **Cl** pale pink or white; **Cl** tube ± 12 × 3 mm, often inflated ⅔ above the base; **Cl** lobes ovate, 3 - 4 mm, strongly revolute; **Cs** lobes stipitate, apically broadly obtuse, 2-lobed, free lobes revolute; **Anth** appendage rising high; **Sty** head acuminate; **Poll** slender, ± 0.65 mm, caudicles triangular, ± 0.5 mm, corpuscle ± 0.3 mm; **Fr** ± 5 × 0.3 cm, terete; **Se** linear, minute.

D. formosana Maximowicz (Bull. Acad. Imp. Sci. Saint-Pétersbourg 23: 385, 1877). **T:** Taiwan (*Oldham* 332 [K]). – **D:** Taiwan. **I:** Lu & Kao (1978).

[3] Forming runners; stems thin, glabrous; **L** decussate, 2 - 4 mm petiolate, lamina circular or obcordate, 1 - 2 cm, fleshy, glabrous, base narrowed or cuneate; **Inf** 1- to 5-flowered; peduncle 2 - 3 cm; **Sep** ovate, 2.8 mm, membranous, glabrous; **Cl** small, white; **Cl** lobes revolute; **Cs** lobes hastate; **Anth** appendage erect, membranous; **Poll** flat, corpuscle elongate; **Fr** terete, 4 - 5 cm, slender, glabrous.

D. fruticulosa Ridley (J. Straits Branch Roy. Asiat. Soc. 79: 96, 1918). **T:** Malaysia, Malaya (*Robinson* s.n. [K]). – **D:** Malaysia (Malaya). **I:** Rintz (1980). **Fig. XVII.f**

[4] Epiphytic; **R** at the nodes and adventitious; stems robust, woody with age, glabrous; **L** alternate and decussate, with a petiole 2.5 mm long, lamina elliptic, 5 - 7 × 2 - 3 cm, acute, base cuneate, flat, glabrous; **Inf** 1- to 6-flowered; peduncle 0.5 - 2 cm; **Ped** ± 2 mm, glabrous; **Sep** elongate-ovate, obtuse; **Cl** pale yellow or white; **Cl** tube ± 3 × 2.5 mm, ± pear-shaped, throat deep red; **Cl** lobes erect, ± 1 mm, acute, pink or deep red; **Cs** lobes stipitate, hastate; **Anth** appendage obtuse; **Gy** stipitate; **Poll** cylindrical, ± 0.3 mm, caudicles narrowly triangular, ± 0.2 mm, corpuscle linear-elongate, 0.25 mm; **Fr** terete, ± 6 × 0.4 cm, pale green or red.

D. hirsuta (Blume) Decaisne (PSRV 8: 632, 1844). **T:** Indonesia, Java (*Blume* s.n. [L]). – **D:** NE India, S Myanmar, Camboodia, Laos, Vietnam, Singapore, Indonesia (Borneo, Java, Moluccas, Sulawesi, Sumatra), Philippines, S Thailand, Papua New Guinea. **I:** Rintz (1980). **Fig. XVIII.b**

≡ *Leptostemma hirsutum* Blume (1826); **incl.** *Leptostemma fasciculatum* Blume (1826) ≡ *Dischidia fasciculata* (Blume) Decaisne (1844); **incl.** *Dischidia brunoniana* Griffith (1854); **incl.** *Dischidia euryloma* Schlechter (1905); **incl.** *Dischidia sub-*

peltigera Schlechter (1905); **incl.** *Dischidia pulchella* Schlechter (1916); **incl.** *Dischidia verruculosa* Schlechter (1916).

[5] **R** at the nodes and adventitious; stems thin, branched, ± sparsely leafy, ± reddish-brown, densely villose when young, ± glabrous with age; **L** with a glabrous petiole ± 2 - 3 mm long, lamina ovate, 1.5 - 2.5 × 1 - 1.5 cm, acute, base broadly cuneate to round, flat, leathery to fleshy, margins often ± revolute, villose when young; **Inf** 1- to 5-flowered; peduncle ± erect, ± 8 mm; **Ped** 2 - 3 mm, glabrous, reddish; **Sep** ± 1.3 mm, obtuse, reddish; **Cl** outside verrucose, angular at the base; **Cl** tube ± 6 × 5 mm, urceolate, yellowish to deep red; **Cl** lobes ± 3 mm, ascending to revolute, fleshy, pink, alternating with 5 small erect projections; **Cs** lobes stipitate, apically obtuse, 2-lobed, free lobes involute; **Poll** 0.7 mm, caudicles triangular, 0.5 mm, corpuscle 0.4 mm; **Fr** linear-lanceolate, ± 6 × 0.5 cm, reniform in cross-section, green.

D. imbricata (Blume) Steudel (Nomencl. Bot. ed. 2, 1: 519, 1840). **T:** Indonesia, Java (*Blume* s.n. [L]). – **D:** NE India, S Myanmar, Camboodia, Laos, Vietnam, S Thailand, Malaysia (Malaya), Indonesia (Borneo, Java, Sumatra). **I:** Rintz (1980). **Fig. XVII.d**

≡ *Conchophyllum imbricatum* Blume (1826); **incl.** *Dischidia depressa* Clarke *ex* King & Gamble (1908).

[2] Growing on stones or bark; **R** at the nodes and adventitious; **L** sessile, circular, resting upon the substrate like limpets, partly overlapping adjoining **L**, ± 2 × 2.5 cm, outside green, inside deep red with green margin, glabrous; **Inf** with several part-**Inf**, 1- to 4-flowered; peduncle 1 - 2.5 cm, erect; **Ped** ± 2 mm, glabrous; **Sep** ± round, ± reddish, glabrous; **Cl** white or pale yellow; **Cl** tube ± 3 × 4 mm; **Cl** lobes erect to involute, ± 2 mm, tips ± reddish when young; petaloid **Cn** 5-lobed; **Cs** lobes bidentate; **Poll** cylindrical, ± 0.35 mm, caudicles slender, ± 0.2 mm, corpuscle ± 0.15 mm.

Ants are settling below the leaves (Kerr & Craib 1951).

D. lanceolata (Blume) Decaisne (PSRV 8: 631, 1844). **T:** Indonesia, Java (*Blume* s.n. [L?]). – **D:** Indonesia (Java). **I:** Steenis (1972). **Fig. XVIII.a**

≡ *Leptostemma lanceolatum* Blume (1826).

[?] Stems 1 mm ∅, sparsely branched, glabrous; **L** decussate, petiole glabrous, to 1 cm, lamina lanceolate, 5 - 9 × 1.5 - 3 cm, thinly fleshy, glabrous, acute, acuminate or obtuse, base acute to obtuse; peduncle 4 - 8 mm, glabrous; **Ped** 2 - 3 mm, faintly pubescent; **Sep** ovate-elongate, glabrous; **Cl** ± 8 mm, whitish-pink, pale violet, outside with pale blotches; **Cl** tube basally belly-like; **Cl** lobes slender, 1.75 - 2 mm, erect to revolute, outside violet, inside greenish; **Anth** appendage acute, white; **Sty** head obtuse; **Poll** 0.5 - 0.6 mm, caudicles broadly triangular; **Fr** linear-lanceolate, 8.5 - 10.5 cm, acute, glabrous.

D. latifolia (Blume) Decaisne (PSRV 8: 631, 1844). **T:** Indonesia, Java (*Blume* s.n. [L?]). – **D:** Indonesia (Java).

≡ *Leptostemma latifolium* Blume (1826).

[3] Stems twining, robust, ± 3 mm ∅; **L** decussate, petiole glabrous, 0.75 - 1.5 mm long, lamina broadly elliptic or slightly rhombic, 3.5 - 10 × 2 - 5 cm, thickly leathery to fleshy, tip triangular to obtuse, base cuneate; **Inf** 0.25 - 1 cm; peduncle ± 5 mm, thick, glabrous; **Cl** tube ± 2.5 mm; **Cs** lobes minute, broadly triangular-ovate, entire, ± acuminate; **Poll** narrowly ovoid, ± 0.3 mm, corpuscle robust, 0.15 mm.

D. livida Elmer (Leafl. Philipp. Bot. 10: 3558-3560, 1938). **T:** Philippines, Luzon (*Elmer* 16591 [not located]). – **D:** Philippines. **Fig. XIX.a**

[1?] **R** at the nodes; stems very thin or filiform, green, smooth, glabrous; **L** decussate, petiole 5 mm, very thin, faintly canaliculate, glabrous or rarely pubescent, lamina lanceolate, 4 × 1 cm, flat, thin, dark green, slightly shining, glabrous, acuminate, base broadly cuneate; peduncle 1 - 1.5 cm; **Ped** 1 - 2 mm, pale purple, glabrous; **Sep** elongate, somewhat obtuse, glabrous; **Cl** 4 - 6 mm, bluish to violet, becoming paler towards the apex, outside verrucose; **Cl** lobes erect, acute, alternating with small translucent projections, inside whitish; **Cs** lobes translucent, shortly stipitate, tips 2-lobed; **Fr** subterete, 5 cm, straight, glabrous; **Se** linear, flat, brown.

D. major (Vahl) Merrill (Interpr. Herb. Amboin., 437, 1917). **T:** Malaysia, Malaya (*Koenig* s.n. [C]). – **D:** NE India, Myanmar, Camboodia, Laos, Vietnam, Malaysia (Malaya), Indonesia (Borneo, Java, Sulawesi, Sumatra, Timor), Singapore, Thailand, Philippines, Papua New Guinea, Australia (Queensland). **I:** Rintz (1980); Forster & Liddle (1988). **Fig. XVIII.d, XVIII.e**

≡ *Collyris major* Vahl (1810); **incl.** *Dischidia clavata* Wallich (1831); **incl.** *Dischidia rafflesiana* Wallich (1831); **incl.** *Dischidia timorensis* Decaisne (1834); **incl.** *Dischidia merguiensis* Beccari (1886); **incl.** *Dischidia bauerlenii* Schlechter (1908); **incl.** *Dischidia pubiflora* Schlechter (1916).

[2] **R** at the nodes and adventitious, those of the nodes often growing into the pitcher **L**; stems thick, 3 - 4 mm ∅, glabrous; **L** of 2 shapes: *either* with a petiole 2 - 3 × 1 mm, circular, 2.5 - 2.7 × 1.6 - 1.8 cm, acuminate, base cuneate, flat, glabrous or faintly pubescent, *or* with a petiole 8 × 1.5 mm, pitcher-shaped or tubular, 8 - 10 × 3 - 4 cm, inside dark purple, tips obtuse, one side flattened; **Inf** 1- to 8-flowered, often with part-**Inf**; peduncle 15 × 1.5 mm; **Ped** 3.5 - 4 × 1 mm, glabrous or faintly pubescent; **Sep** ovate, 1 - 1.5 × 1 mm, margins minutely denticulate; **Cl** 3 - 4 × 2.5 - 3 mm, yellow-green or striped with yellow and green; **Cl** lobes lanceolate,

1.5 × 0.5 mm, acuminate, revolute, tips reddish, outside faintly pubescent; petaloid **Cn** flat; **Cs** lobes stipitate, apically cordate, 2-lobed, lobes involute, 1 × 1 mm; **Anth** appendage ovate, 0.75 mm; **Sty** head obtuse, 0.25 mm; **Poll** 0.5 - 0.6 × 0.1 - 0.2 mm, translator ± 0.3 × 0.1 mm, caudicles 0.3 × 0.1 mm; **Fr** fusiform, 4.5 - 5 × 0.3 cm, reniform in cross-section, yellow-green, glabrous; **Se** 2 - 2.5 × 1 mm, tuft of **Ha** 1.3 - 1.5 cm.

Also known under the name *D. rafflesiana* (Zizka 1990). *D. major* is particularly characterized by its large pitcher leaves.

D. nummularia R. Brown (Prodr., 461, 1810). **T** [lecto]: Australia, Queensland (*Banks & Solander* s.n. [BM]). − **D:** NE India, Sri Lanka, Bangladesh, Myanmar, China (Hainan), Camboodia, Laos, S Vietnam, Malaysia (Malaya), Indonesia (Borneo, Java, Moluccas, Sulawesi, Timor), Philippines, Singapore, Thailand, Solomon Islands, Papua New Guinea, Australia (Queensland). **I:** Rintz (1980); Forster & Liddle (1988).

Incl. *Collyris minor* Vahl (1810) ≡ *Dischidia minor* (Vahl) Merrill (1939); **incl.** *Dischidia gaudichaudii* Decaisne (1844) ≡ *Dischidia nummularia* var. *gaudichaudii* (Decaisne) Beccari (1886); **incl.** *Dischidia orbicularis* Decaisne (1844); **incl.** *Dischidia nummularia* var. *gracilis* Beccari (1884); **incl.** *Dischidia beiningiana* Schlechter (1905); **incl.** *Dischidia dirhiza* Schlechter (1905); **incl.** *Dischidia ridleyana* Schlechter (1905); **incl.** *Dischidia schumanniana* Schlechter (1905); **incl.** *Dischidia copelandii* Schlechter (1906); **incl.** *Dischidia glabra* Warburg (1907) ≡ *Dischidia nummularia* var. *glabra* (Warburg) Bakhuizen van den Brink *fil.* & Backer (1950); **incl.** *Dischidia microphylla* Schlechter (1908); **incl.** *Dischidia aemula* Schlechter (1913); **incl.** *Dischidia sepikana* Schlechter (1913); **incl.** *Dischidia actephila* Schlechter (1916); **incl.** *Dischidia decipiens* Schlechter (1916); **incl.** *Dischidia immortalis* Guillaumin (1937).

[3] **R** at the nodes and adventitious; stems filiform, glabrous; **L** with a petiole 2 - 3 × 1 mm, lamina circular to ovate, 7 - 14 × 7 - 10 mm, flat, farinose-white, tips with a minute mucro, base round or shortly cuneate; **Inf** 1- to 6-flowered; peduncle 1 - 10 mm; **Ped** 3 - 5 × 1 mm, glabrous; **Sep** ovate, 0.5 - 0.7 × 0.5 mm, glabrous; **Cl** 3 - 4 × 2 - 3 mm, white; **Cl** tube 2 - 3 × 2 - 3 mm, campanulate-urceolate, white; **Cl** lobes ovate, 2 - 2.5 × 0.75 - 1 mm, acuminate, ± revolute; **Cs** lobes stipitate, tip obtuse, 2-lobed, free lobes spatulate, 0.5 × 0.5 mm, involute; **Anth** appendage ovate, 0.5 mm; **Sty** head conical, 0.25 mm; **Poll** elongate, ± 0.2 × 0.1 mm, corpuscle elongate, 0.1 × 0.06 mm, brown, caudicles broadly triangular, ± 0.3 × 0.1 mm; **Fr** fusiform, semiterete, 2.5 - 4 × 0.5 - 0.8 cm, pale green or yellow, glabrous; **Se** 2 - 2.5 × 1 mm, tuft of **Ha** 1 cm.

A widely distributed taxon that is often seen in cultivation despite its inconspicuous flowers. According to Rintz (1980) *D. glaucescens* Elmer and *D. pubicaulis* Schlechter should be included as further synonyms.

D. ovata Bentham (London J. Bot. 2: 226-227, 1843). **T:** Papua-New Guinea (*Hinds* s.n. [K]). − **D:** Papua New Guinea, Australia (Queensland). **I:** Forster & Liddle (1988). **Fig. XVIII.c**

Incl. *Dischidia picta* Blume (1849) ≡ *Leptostemma pictum* (Blume) Zippel & Herbert (1849).

[3] **R** at the nodes only; stems creeping, 1 mm ∅, glabrous; **L** with a petiole 2 - 9 × 0.75 - 1 mm, lamina broadly ovate, 2 - 5 × 0.5 - 3.5 cm, fleshy, pale green to brown with paler venation, finely acuminate, base round or slightly cordate; **Inf** 1- to 8-flowered, often with part-**Inf**; peduncle to 15 × 1 mm; **Ped** 3 × 1 mm; **Sep** ovate, 1 × 0.75 mm, glabrous; **Cl** 6 × 4 mm, yellow, reddish, greenish, striped; **Cl** tube gasteriform-urceolate, angled at the base; **Cl** lobes ovate, 2 × 1 mm, erect, ± blotched with white, outside minutely papillose, usually not striped, yellow-red; **Cs** lobes stipitate, bilobed, free lobes involute, 1.75 × 1.5 mm; **Anth** appendage ovate, 0.75 mm; **Sty** head conical, 0.25 mm; **Poll** ± 0.5 × 0.2 mm, corpuscle ± 0.4 × 0.1 mm, caudicles ± 0.3 × 0.1 mm; **Fr** fusiform, 5.5 - 6.5 × 0.8 - 1 cm, semiterete; **Se** 3 - 4 × 1 - 1.5 mm, pale brown, tuft of **Ha** 8 - 12 mm.

A pretty and easily recognized species that is often cultivated.

D. punctata (Blume) Decaisne (PSRV 8: 631, 1844). **T:** Indonesia, Java (*Blume* s.n. [L]). − **D:** S Myanmar, Indonesia (Borneo, Java, Sumatra), Philippines, S Thailand. **I:** Rintz (1980). **Fig. XVIII.f**

≡ *Leptostemma punctatum* Blume (1826); **incl.** *Dischidia joloensis* Schlechter (1915); **incl.** *Dischidia viridiflora* Ridley (1920); **incl.** *Dischidia punctatoides* Bakhuizen van den Brink *fil.* & Backer (1950).

[3] **R** at the nodes and adventitious; stems glabrous but pubescent to faintly villose when young; **L** decussate, rarely in whorls of 4, with a villose petiole ± 2 mm long, lamina ovate, 2 - 3 × 1.2 - 1.5 cm, ± flat, ± fleshy, glabrous, acute, cuneate at the base; **Inf** 1- to 5-flowered; peduncle bent backwards, ± 5 mm; **Ped** ± 3 mm, faintly villose; **Sep** ovate, 1 mm, obtuse; **Cl** violet or yellow-green, striped with dark green, outside verrucose; **Cl** tube ± 7 × 4 mm; **Cl** lobes ± 2 mm, partially revolute; **Cs** lobes stipitate, tip smooth, obtuse, bilobed, free lobes revolute; **Poll** ± 0.6 mm, caudicles ± 0.4 mm, corpuscle 0.25 mm; **Fr** fusiform, 6 - 7 × 0.4 cm, reniform in cross-section, very acute, green, glabrous; tuft of **Ha** ± 2.5 cm.

D. rhombifolia Blume (Bijdr. Fl. Ned. Ind., 1063, 1826). **T:** Indonesia, Java (*Anonymus* s.n. [L]). − **D:** Indonesia (Java).

≡ *Dischidia nummularia* var. *rhombifolia* (Blume) Bakhuizen van den Brink *fil.* & Backer (1950); **incl.** *Dischidia horsfieldiana* Miquel (1856); **incl.** *Leptostemma truncatum* Zollinger *ex* Miquel (1856) (*nom. illeg.*, Art. 53.1).

[3] Stems twining, thin, glabrous; **L** decussate, rarely in whorls of 4, ± shortly petiolate, lamina rhombic or elliptic, to 1.3 cm, ± flat, glabrous, ± acute when young but obtuse whith age; peduncle short; **Sep** linear-lanceolate; **Cl** unknown; **Cs** lobes linear, obtuse.

Lately treated as a variety of *D. nummularia* by Bakhuizen van den Brink & Backer (1950).

D. ruscifolia Decaisne *ex* Beccari (Malesia 2: 272, 1886). **T:** Philippines (*Cuming* 1086 [not located]). – **D:** Philippines.

[?] Plants pendent; **R** sparse; stems terete, branched, with very dense foliage, green to yellowish-green, finely verrucose, base woody, pubescent; **L** broadly cordate to triangular, ± 1.5 cm, abruptly constricted into a terminal point, very fleshy to succulent, yellowish-green, glabrous; **Inf** 1- to 2-flowered; **Ped** thin, bent; **Cl** pure white; **Fr** semiterete, pale green.

D. sagittata (Blume) Decaisne (PSRV 8: 631, 1844). **T:** Indonesia, Java (*Blume* s.n. [L?]). – **D:** Indonesia (Java). **Fig. XIX.d**

≡ *Leptostemma sagitattum* Blume (1826).

[1] Stems twining, thin, glabrous; **L** decussate or rarely in whorls of 3 - 4, shortly petiolate, lamina narrowly lanceolate to obovate, leathery-fleshy, glabrous, obtuse or acute, obtuse at the base; **Inf** 5 - 15 mm; peduncle short, very thick; **Cl** 2.5 - 3 mm, subglobose, whitish to yellowish; **Cl** lobes erect to involute; **Cs** lobes hastate, acute to obtuse; **Poll** cylindrical, ± 0.2 mm, corpuscle ± 0.2 mm, caudicles thin; **Fr** narrowly ovate to fusiform, 3.5 - 5 cm, ± acute, glabrous; tuft of **Ha** of the **Se** ± 2.5 cm.

D. singularis Craib (BMI 1911: 419, 1911). **T** [syn]: Thailand (*Kerr* 1294 [K]). – **D:** Thailand. **Fig. XIX.e**

[4] Stems thin; **L** linear or lanceolate, 2.5 - 6.5 × 0.3 - 2 cm, acute, base ± narrowed, margins ± revolute; **Inf** many-flowered; **Ped** ± 1 mm; **Sep** ovate-lanceolate, small, obtuse; **Cl** reddish- or greenish-yellow; **Cl** tube with red blotches alternating with the **Cl** lobes, lobes lanceolate, 1.5 mm, strongly revolute, whitish; **Cs** lobes stipitate, erect, nearly round, membranous; **Fr** 5 - 8.5 cm, smooth; **Se** linear, 3 - 3.5 mm.

D. truncata (Blume) Decaisne (PSRV 8: 632, 1844). **T:** Indonesia, Java (*Blume* s.n. [L?]). – **D:** Indonesia (Java, Sulawesi). **Fig. XIX.c**

≡ *Leptostemma truncatum* Blume (1826); **incl.** *Dischidia truncata* var. *celebica* Miquel (1856).

[3,4] Stems sparsely leafy, glabrous when young, smooth; **L** decussate, with a glabrous petiole 2 - 4 mm long, lamina ovate, 1 - 2 × 0.7 - 1.5 cm, thinly leathery to fleshy, glabrous, acute, base cuneate to obtuse-rounded; **Inf** 1.5 - 6 cm; peduncle 0.5 - 5 cm, glabrous; **Ped** very short; **Sep** ciliate; **Cl** 4.5 mm, pale violet or white; **Cl** tube ± 3.5 mm; **Cl** lobes obtuse, often reddish; **Cs** lobes obtusely hastate, margins dentate; **Poll** narrow, ± 0.6 mm, caudicles very slender, ± 0.6 mm.

D. vidalii Beccari (Malesia 2: 272, 1886). – **D:** Philippines. **Fig. XIX.b**

Incl. *Dischidia lanceolata* Fernández-Villar (1880) (*nom. illeg.*, Art. 53.1); **incl.** *Dischidia pectenoides* H. Pearson (1912).

[4] Plants pendent; stems ± twining, glabrous; **R** often growing into the pitcher **L**; **L** decussate, of 2 shapes: *either* shortly petiolate, elliptic-ovate to lanceolate, 1.5 - 2.5 cm, thick, fleshy, acuminate, base narrower, *or* very shortly petiolate, urceolate, 4 - 6 × 4 - 5 cm, 2 cm high; **Inf** 3- to 8-flowered; peduncle 1 - 2 cm; **Ped** ± 6 mm, glabrous; **Sep** ovate, ± 4 mm, obtuse or round, green, margins ciliate; **Cl** 8 mm, red; **Cl** lobes lanceolate, 1.25 mm, acute, erect or involute, inside keeled; petaloid **Cn** 5-lobed, **Cs** lobes involute, bifid; **Fr** 5 - 7 cm, slender.

A very attractive large-flowered species that is easily flowering in cultivation. It is often for sale in flower shops, mostly under the name *D. pectenoides*. The pitcher leaves are inhabited by ants in nature.

DISCHIDIOPSIS

C. Hoffmann & F. Albers

Dischidiopsis Schlechter (in J. R. Perkins, Fragm. Fl. Philipp., 128, 1904). **T:** not designated. – **D:** Philippines. **Etym:** Gr. '-opsis', similar to; and for the genus *Dischidia* (*Asclepiadaceae*).

Pendent or climbing epiphytes, twining, with white latex; stems ± fleshy, ± glabrous, 2 - 3 m, 2 - 5 mm ∅, strongly branched; **L** opposite, shortly to very shortly petiolate, petiole almost as long as broad, lamina narrowly oblong to ovate, 4 - 10 × 1 - 3 cm, ± fleshy, acute to attenuate, entire, glabrous, rarely pubescent, base acute; **Inf** racemose, lateral; peduncle with ± conspicuous short thick rachis; **Fl** 5-merous, actinomorphic; **Ped** short, glabrous; **Sep** oblong-ovate, 1 mm, obtuse, glabrous; **Cl** urceolate, basally ventricose to globose, fleshy, glabrous, glossy, inside whitish; **Cl** tube inside long-hairy, **Ha** erect; **Cl** lobes ovate-lanceolate to triangular, hardly 1 mm; **Cn** petaloid, 5-lobed, short, uniseriate; **Cs** lobes scoop-like, obtuse; **Anth** processes enveloping the **Sty** head, narrowly oblong, acute, transparent; **Poll** cylindrical, erect, corpuscle narrowly oblong; **Fr** fusiform.

The oligotypic genus *Dischidiopsis* can only be distinguished from *Dischidia* on the base of the reduced, slightly scoop-shaped staminal corona, and its status as a separate genus appears unjustified. A formal transfer to *Dischidia* has not been published so far, however. Of the few described taxa, the following species is the only that is somewhat better known:

D. parasitica (Blanco) Merrill (Sp. Blancoan., 317, 1918). **T:** Philippines, Batangas Prov. (*Blanco* s.n. [not located]). – **D:** Philippines (Luzon, Lubang). **I:** Asklepios 67: 12-13, 1996. **Fig. XX.b**

≡ *Marsdenia parasitica* Blanco (1837) ≡ *Dischidia parasitica* (Blanco) Hort. van Donkelaar (s.a.) (*nom. inval.*, Art. 29.1); **incl.** *Dischidiopsis philippinensis* Schlechter (1904); **incl.** *Dischidia merrillii* Schlechter (1906) ≡ *Conchophyllum merrillii* (Schlechter) Merrill (1912).

Inf subsessile; **Cl** slender urceolate-cylindrical, purple, ± 4 mm; **Cl** tube 3.5 mm; **Cl** lobes 0.5 - 1 mm, erect; **Poll** ± 0.4 × 0.11 mm; otherwise description as for the genus.

An attractive and in comparison with *Dischidia* large-flowered species encountered rather frequently in cultivation.

DUVALIA

U. Meve

Duvalia Haworth (Synops. Pl. Succ., 44, 1812). **T:** *Stapelia elegans* Masson. – **Lit:** Bayer (1984); Meve & Albers (1990); Meve (1997). **D:** Disjunct, Arabian Peninsula and Horn of Africa as well as S Africa. **Etym:** For Henri Auguste Duval (1777 - 1814), French physician and botanist.

Small stem succulents with plagiotropic stems and mat-like growth, mostly in the shelter of shrubs; stems globular to cylindrical (or clavate, Sect. *Arabica*), 4-, 5- or 6-angled, 1 - 10 cm long, green to greyish-green, sometimes spotted maroon, glabrous; **L** rudiments triangular, caducous, with stipular **Gl** (Sect. *Duvalia*), or **L** rudiments conical-acuminate and ± persistent, lacking stipular **Gl** (Sect. *Arabica*); **Inf** mostly near stem bases with 1 to several **Fl** appearing successively over a longer period of time; **Ped** 2 - 70 × 1 - 3 mm, (greyish-)green, terete, glabrous; **Sep** green, 1 - 6 × 1 - 2.5 mm, lanceolate, acute, glabrous; **Fl** actinomorphic, 1 - 5 cm ∅, mostly flat on the ground; **Cl** yellow, ochre, maroon, brown to very dark purple, mostly with strong foetid odour; **Cl** tube raised to a ± fleshy annulus, free **Cl** lobes flat, convex or bent backwards along the mid-vein; **Cn** biseriate, **Ci** and dorsal parts of **Cs** fused to a ± flat disc lying on the upper or next to the inner surface of the annulus, thus blocking the entrance of the **Cl** tube, **Cs** with short inner leaf-like appendages placed ± parallel on the **Anth**, and outer ± ovoid to fusiform acuminate appendages horizontally spreading to vertically erect; **Gy** stipitate, guide rails with cartilaginous margins, wing-like enlarged at the base, with glandular **Ha** within; **Anth** lying on the whitish **Sty** head; **Poll** mostly yellow, flattened-ovoid, 0.27 - 0.55 × 0.2 - 0.35 mm, at the apex of the inner side with a massive broad germination mouth on top of a translucent wall; caudicles 0.08 - 0.2 mm long, terminally to subterminally connected to the **Poll**, corpuscle ellipsoid, wings of the translator up to 0.5 mm long, whitish-translucent; **Fr** 5 - 18 × 0.4 - 1 (-1.2) cm, green, partly spotted dark red, glabrous, with ± 40 - 120 **Se**, the latter flat, ellipsoid, 3 - 8 × 2.5 - 5 mm, their rim 0.2 - 1.2 mm broad, brown, with a tuft of **Ha** 10 - 18 mm long.

Duvalia is well defined by its disc-like corona (unique within the family) and the stipitate gynostegium. The genus is most closely related to *Huernia* (Meve 1997), with which hybrids are known (= ×*Dernia*). For another intergeneric hybrid see under ×*Duvaliaranthus*.

Meve & Albers (1990) divide *Duvalia* into 2 sections:

[1] Sect. *Duvalia*: Stems globular to cylindrical, 4- to 6-angled, green, **L** rudiments flattened, ± triangular, mostly with stipular **Gl**. Distribution: Botswana, Malawi, Moçambique, Namibia, Zambia, Zimbabwe, RSA.

[2] Sect. *Arabica* Meve & F. Albers 1990: Stems ± clavate (rarely compactly cylindrical), 4- to indistinctly 5-angled, greyish-green, **L** rudiments ± conical, acute, lacking stipular **Gl**. Distribution: Saudi-Arabia, Yemen, Sudan, Djibouti, Ethiopia, Somalia.

D. anemoniflora (Deflers) R. A. Dyer & Lavranos (FPA 44: t. 1734, in adnot., 1976). **T:** Arabia [S Yemen] (*Deflers* 387 [lecto: Deflers, Ascl. Arab. Trop., t. 6, 1896]). – **D:** S Yemen.

≡ *Stapelia anemoniflora* Deflers (1896) ≡ *Caralluma anemoniflora* (Deflers) A. Berger (1910).

This is a 'nomen dubium'. There is no original material and neither is the lectotype useful to establish the identity of this taxon. Meve & Albers (1990) exclude it from *Duvalia*, and Gilbert (1990) excludes it from *Caralluma*.

D. angustiloba N. E. Brown (Gard. Chron., ser. nov. 20: 230, 1883). **T:** RSA, Western Cape (*Dickson* s.n. in *Barkly* 33 [K]). – **D:** RSA (Western Cape, Estern Cape: C Great Karoo); on sand, 800 - 1000 m. **Fig. XX.c**

[1] Stems 0.8 - 3 × 0.8 - 2 cm, 4- to 5-angled, ovoid-cylindrical, dark green (to greyish-green), **L** rudiments ± 2 mm, very sharply pointed; **Inf** dichasia with 5 - 20 procumbent **Fl**; **Ped** 2 - 4 cm; **Sep** 2 - 3 mm; **Cl** 1.5 - 2.2 cm ∅, very delicate, spider-like, chocolate-brown (rarely light green spotted brown), scentless or with slight foetid odour; free **Cl** lobes 6 - 9 × 4 mm, lamina entirely reflexed, margins glabrous or hairy (to papillate) at the base, **Ha** 0.05 - 0.1 mm; annulus round to 5-angled, 3.5 - 4.5 × 0.4 - 0.8

mm, slightly constricted at the base, hairy (to papillate), **Ha** 0.03 - 0.06 mm; **Cn** white (spotted maroon), disc narrow, fringe-like, 5-angled, ± 3 mm ∅, leaf-like **Cs** lobes 1.2 - 1.5 × 1.5 mm; **Poll** ± 0.27 × 0.18 mm, translator wings ± 0.22 mm long; **Fr** 8 - 10 cm; **Se** 4 - 5 × 2.5 - 3 mm.

D. caespitosa (Masson) Haworth (Synops. Pl. Succ., 45, 1812). **T**: [lecto – icono]: Masson, Stapel. Nov. t. 29, 1797. – **D**: RSA.

≡ *Stapelia caespitosa* Masson (1797); **incl.** *Stapelia fasciculata* Thunberg (1794) ≡ *Piaranthus fasciculatus* (Thunberg) Schultes (1820).

This very variable complex can easily be divided into 2 varieties by morphological characters. Furthermore, *D. caespitosa* var. *compacta* is restricted to winter-rainfall areas of the Western Cape, and is replaced by var. *caespitosa* in summer-rainfall areas and transitional regions.

D. caespitosa var. **caespitosa** – **D**: RSA (Northern Cape, Western Cape, Eastern Cape: Great and Little Karoo; Free State).

Incl. *Stapelia barbata* Hort. ex Salm-Dyck (s.a.); **incl.** *Stapelia jacquinii* Loudon (s.a.); **incl.** *Duvalia marlothii* N. E. Brown *in sched.* (s.a.) (*nom. inval.*); **incl.** *Duvalia propinqua* A. Berger (s.a.) (*nom. inval.*); **incl.** *Stapelia reclinata* Masson (1796) ≡ *Duvalia reclinata* (Masson) Haworth (1812); **incl.** *Stapelia radiata* Sims (1803) ≡ *Duvalia radiata* (Sims) Haworth (1812); **incl.** *Stapelia caespitosa* var. *hirtella* Loudon (1805); **incl.** *Stapelia hirtella* Jacquin (1806) ≡ *Duvalia hirtella* (Jacquin) Sweet (1827) ≡ *Duvalia radiata* var. *hirtella* (Jacquin) A. C. White & B. Sloane (1937); **incl.** *Stapelia replicata* Jacquin (1806) ≡ *Duvalia replicata* (Jacquin) Sweet (1827); **incl.** *Duvalia glomerata* Haworth (1812) ≡ *Stapelia glomerata* (Haworth) Schultes (1820); **incl.** *Duvalia laevigata* Haworth (1812) ≡ *Stapelia laevigata* (Haworth) Schultes (1820); **incl.** *Duvalia tuberculata* Haworth (1812) ≡ *Stapelia tuberculata* (Haworth) Schultes (1820); **incl.** *Stapelia cymosa* hort. *ex* Schultes (1820); **incl.** *Stapelia concolor* Salm-Dyck (1834) ≡ *Duvalia concolor* (Salm-Dyck) Schlechter (s.a.); **incl.** *Stapelia barbata* Salm-Dyck (1834) (*nom. nud.*); **incl.** *Duvalia hirtella* var. *minor* N. E. Brown (1908) ≡ *Duvalia radiata* var. *minor* (N. E. Brown) A. C. White & B. Sloane (1937); **incl.** *Duvalia hirtella* var. *obscura* N. E. Brown (1908) ≡ *Duvalia radiata* var. *obscura* (N. E. Brown) A. C. White & B. Sloane (1937); **incl.** *Duvalia reclinata* var. *angulata* N. E. Brown (1908); **incl.** *Duvalia reclinata* var. *bifida* N. E. Brown (1908); **incl.** *Duvalia emiliana* A. C. White (1933).

[1] Stems 1.5 - 10 (-13) × 1 - 2.2 cm, 4- to 5-angled, ± ovoid to cylindrical, dark green; **L** rudiments ± 2 mm, stipular **Gl** distinct; **Inf** with 1 to several **Fl** near the stem base; **Ped** 1 - 2.5 cm; **Sep** 2 - 6 mm; **Cl** 2 - 3.5 cm ∅, (pale) maroon, chocolate-brown, brownish-purple to dark purple, with strong foetid odour; free **Cl** lobes 9 - 15 × 6 - 9 mm, lamina ± entirely reflexed, margins basally or ± entirely hairy, **Ha** ± 0.2 - 0.4 mm (rarely longer) or clavate **Ha** (0.2-) 0.6 - 2.6 mm; annulus circular to pentagonal, sometimes constricted basally, 7 - 11 mm ∅, 2 - 5 mm high, often spotted, papillate or hairy, **Ha** 0.04 - 1 mm; **Cn** yellow to (reddish-) brown, disc circular to pentagonal, 5 - 7 mm ∅; **Cs** lobes 1.5 - 2.2 × 1.5 - 2.5 mm; **Poll** 0.35 - 0.5 × 0.23 - 0.3 mm; **Fr** 6 - 17 cm; **Se** 4 - 6 × ± 3 mm.

D. caespitosa var. **compacta** (Haworth) Meve (Pl. Syst. Evol., Suppl. 10: 59, 1997). **T**: RSA, Cape Prov. (*Anonymus* s.n. [OXF]). – **D**: RSA (Northern Cape, Western Cape, Eastern Cape: Namaqualand, Great Karoo).

≡ *Duvalia compacta* Haworth (1812) ≡ *Stapelia compacta* (Haworth) Schultes (1820); **incl.** *Stapelia mastodes* Jacquin (1806) ≡ *Duvalia mastodes* (Jacquin) Sweet (1827).

[1] Differs from var. *caespitosa*: Stems 1.5 - 7 (-9) × 1 - 2.5 cm, 4- to 5-angled, compact; **L** rudiments 1.5 - 4 mm; **Inf** mostly close to the stem tip; **Cl** 1.6 - 2.4 (-2.8) cm ∅; free **Cl** lobes 8 - 13 × 4 - 6 mm, margins mostly glabrous or basally with simple 0.05 - 0.1 (-0.6) mm long **Ha**; annulus 5.5 - 8 mm ∅, 1.2 - 2.5 mm high, margin glabrous or papillate, **Ha** papillae 0.03 - 0.1 mm; **Cn** disc 3.5 - 6 mm ∅; **Fr** 7 - 12 cm.

D. corderoyi (Hooker *fil.*) N. E. Brown (CBM 102: t. 6245, in adnot., 1876). **T**: RSA, Cape Prov. (*Burke* s.n. [K]). – **D**: RSA (Northern Cape: Great Karoo, Upper Karoo; Free State).

≡ *Stapelia corderoyi* Hooker *fil.* (1874).

[1] Stems 1.2 - 3 (-4.5) × 1.2 - 2.5 cm, 6-angled, globular to ovate, dark green; **L** rudiments ± 1 × 1 mm; **Inf** with 1 to several **Fl** near the stem base; **Ped** 1.5 - 2.5 cm; **Sep** ± 4 mm; **Fl** with strong foetid odour; **Cl** 3 - 4.5 cm ∅, pale maroon to dark purple, rarely pale greenish-brown; **Cl** lobes with longitudinal furrows, 12 - 16 × 9 - 12 mm, lamina widely but not entirely reflexed, margins with purple clavate **Ha** 2 - 4 mm long, angles between the **Cl** lobes basally papillate to hairy, **Ha** 0.3 - 0.6 mm; annulus fleshy, spotted, 10 - 15 mm ∅, 2.5 - 4 mm high, flattened, densely pubescent, **Ha** simple, white to purple, 3 mm; **Cn** reddish-brown, disc 6.5 - 9 mm ∅; **Cs** lobes cream to maroon, 2 - 3 × 2 - 3 mm; **Poll** 0.45 - 0.55 × 0.3 - 0.35 mm; **Fr** 10 - 15 cm; **Se** 5 - 6 × 3 - 3.5 mm.

The species is easily discernible by its thick 6-angled stems and its fleshy and very hairy flowers.

D. eilensis Lavranos (CSJA 44(6): 260-261, ills., 1972). **T**: Somalia (*Lavranos* 7223 [FT]). – **D**: Somalia (Basaso Region, near Eil).

[2] Stems 2 - 4 × 1 - 1.5 cm, 4-angled, ovoid to cylindrical, greyish- (to yellowish-) green spotted dark green or dark maroon; **L** rudiments to 5 mm,

conical, acute; **Inf** near the stem base with 1 to several procumbent **Fl** with foetid odour; **Ped** 2 - 5 cm; **Sep** 3 - 4 mm; **Cl** 2 - 3 cm ⌀, pale yellow to cream, spotted and speckled deep (reddish-) brown (sometimes entirely pale yellow or reddish-brown); **Cl** lobes 6 - 8 × 7 mm, lamina reflexed to a large extent, apically with finer spots than basally, basally papillate, **Ha** papillae 0.1 - 0.2 mm, apically with coarse **Ha** to 1.2 mm; annulus round, thick, 12 - 15 mm ⌀, 1.5 - 2 mm high, rim with smooth transition into the **Cl** lobes, papillose, papillae 0.05 mm; **Cn** cream to very dark brown, partly spotted, disc 5.5 - 6.5 mm ⌀ with undulate cartilaginous margin; **Cs** lobes ± 1.5 × 1.5 mm; **Poll** 0.4 × 0.27 mm; **Se** pale brown, 5 × 3.5 mm.

According to its stem morphology, *D. eilensis* clearly belongs to Sect. *Arabica*. Flower-morphologically however, it shows resemblance to the species native to the Cape.

D. elegans (Masson) Haworth (Synops. Pl. Succ., 44, 1812). **T:** [lecto – icono]: Masson, Stapel. Nov. t. 27, 1797. – **D:** RSA (Western Cape: Little Karoo, Worcester-Robertson-Karoo).

≡ *Stapelia elegans* Masson (1797); **incl.** *Stapelia radiata* Jacquin (1806); **incl.** *Stapelia jacquiniana* Schultes (1820) ≡ *Duvalia jacquiniana* (Schultes) Sweet (1826); **incl.** *Duvalia jacquinii* Loudon (1841); **incl.** *Duvalia elegans* var. *seminuda* N. E. Brown (1908); **incl.** *Duvalia elegans* fa. *magnicorona* A. C. White & B. Sloane (1937).

[1] Stems 2 - 6 × 0.8 -1.8 cm, bluntly 4- to 5-angled, (ovoid to) cylindrical, dark green; **L** rudiments 1 - 1.5 mm, stipular **Gl** minute, rarely absent; **Inf** near stem base with several procumbent **Fl** with foetid odour; **Ped** 1.2 - 2.5 cm; **Sep** ± 3 × 1 mm; **Cl** 1.5 - 2.2 cm ⌀, very dark purple, shiny, hairy, **Ha** simple, purple, ± curved, 1 - 3 mm; free **Cl** lobes 6.5 - 10 × 3.5 - 6 mm, lamina flat to convex, margins slightly reflexed; annulus weakly expressed, circular, 6 - 8 mm ⌀, 0.8 - 1.5 mm high, densly pubescent; **Cn** reddish-brown to purple, disc ± 5 - 7 mm ⌀; **Cs** lobes 1.2 - 1.8 × 1.2 - 1.6 mm; **Poll** 0.28 - 0.34 × 0.2 mm; **Fr** 8 - 13 cm; **Se** ± 5 × 3 mm.

D. galgallensis Lavranos (CSJA 46(4): 184-185, ill., 1974). **T:** Somalia, Bosaso Distr. (*Lavranos & Horwood* 10260 [FT, ZSS]). – **D:** Somalia; known only from the type.

[2] Stems 0.8 - 2 × 0.7 - 1.2 cm, bluntly 4-angled, pale (yellowish-) green; **L** rudiments 3 mm, acute, hard; **Inf** arising from lower stem regions with several **Fl**, odour unknown; **Ped** 1.5 - 2 cm; **Sep** 4 - 5 mm; **Cl** 3 - 3.6 cm ⌀, pale mahogany, shiny; **Cl** lobes 10 - 13 × 6 - 7 mm, lamina slightly convex, margins slightly bent outwards, hairy, **Ha** whitish or purple, to 3 mm; annulus ± circular, weak, 10 - 12 mm ⌀, ± 1 mm high, densely pubescent; **Ha** ± 1 mm; **Cn** cream (spotted maroon), disc ± 10-angled, slightly emarginate, covering the annulus, 10 - 11 mm ⌀; **Cs** lobes ± 1.8 mm × 2 mm; **Poll** 0.45 × 0.28 mm; **Fr** and **Se** unknown.

This little-known species is closely related to the *D. sulcata*-complex.

D. gracilis Meve (Pl. Syst. Evol., Suppl. 10: 75-77, ills., 1997). **T:** RSA, Eastern Cape (*Meve* 346 [K, MSUN, NBG]). – **D:** RSA (Eastern Cape). **Fig. XX.d**

[1] Stems 1 - 3.5 × 0.7 - 2 cm, 4- to 5-angled, ovoid to cylindrical, dark green to greyish-green; **L** rudiments 2.5 - 4 mm, very acute; **Inf** with 1 to several procumbent **Fl** from near the stem base, with slight foetid odour; **Ped** 1 - 2 cm; **Sep** 3 - 4 mm; **Cl** 1.6 - 2.2 cm ⌀, (reddish-) brown tinged greenish; **Cl** lobes 7 - 9 × 4 - 5 mm, lamina reflexed to a large extent, basal ⅓ papillose to hairy, **Ha** papillae 0.2 mm, margins of the **Cl** lobes basally with **Ha** papillae 0.1 - 0.2 mm long; annulus pentagonal, flattened, 6 - 7 mm ⌀, 0.6 - 0.8 mm high, ± spotted green or greenish-brown and covered with **Ha** papillae 0.05 - 0.2 mm long; **Cn** greenish-yellow to greenish-brown, disc ± 4 mm ⌀; **Cs** lobes ± 1.5 × 1.6 mm, compact; **Poll** ± 0.37 × 0.21 mm; **Fr** 10 - 16 cm; **Se** 4.5 × 3 mm.

This rare species has to be placed in the group of closely related species with small flowers and very acute leaf rudiments (*D. angustiloba, D. maculata, D. modesta*) from the Eastern Cape.

D. immaculata (C. A. Lückhoff) M. B. Bayer *ex* L. C. Leach (SAJB 55(2): 268, 1989). **T:** [lecto – icono]: White & Sloane, The Stap. ed. 2, 2: 766, fig. 769, 1937. – **D:** RSA (Western Cape); in relatively moist coastal Fynbos.

≡ *Duvalia maculata* var. *immaculata* C. A. Lückhoff (1937).

[1] Stems 2 - 5 (-12) × 0.6 - 1.2 cm, oblong-ovoid to cylindrical, rhizomatous, green; **L** rudiments < 1 mm, obtuse, stipular rudiments minute, only microscopically discernible; **Inf** several in the lower ½ of the stems, with mostly erect **Fl** with slight foetid odour; **Ped** 0.5 - 2.5 cm; **Sep** ± 2 mm; **Cl** 2 - 3 cm ⌀, with dull surface, dark chocolate-brown; **Cl** lobes 8 - 12 × 6 - 8 mm, lamina entirely reflexed; annulus pentagonal, 8 - 11 mm ⌀, 2 - 3 mm high, bowl-shaped, raised considerably above the disc, rim undulate; **Cn** straw-yellow, disc circular, 3.2 - 5 mm ⌀; **Cs** lobes compact, 1 - 1.8 × 2 mm, outer appendages ovoid, inner ones bluntly fusiform; **Poll** 0.45 × 0.2 mm; **Fr** 8 - 15 cm; **Se** 6 × 3.5 mm.

D. maculata N. E. Brown (FC 4(1): 1033, 1908). **T:** RSA, Cape Prov. (*Pillans* 31 [K]). – **D:** Namibia (Karasberge), RSA (Northern Cape, Eastern Cape: Great and Upper Karoo).

Incl. *Duvalia minuta* Nel (1937).

[1] Stems 1 - 4 (-6) × 0.8 - 1.8 cm, 4-angled, ± ovoid to cylindrical, dark green to greyish-green; **L** rudiments 2 - 3 mm, acute; **Inf** from the lower ½ of

the stems, with 1 to several procumbent ± scentless **Fl**; **Ped** 1 - 2.5 cm; **Sep** 3 mm; **Cl** 1.5 - 2 cm ⌀, maroon; **Cl** lobes 5 - 7 × 5 - 6.5 mm, lamina almost entirely reflexed, basally with simple **Ha**, **Ha** 0.2 - 2.5 mm, basal angles between the **Cl** lobes papillose; annulus ± pentagonal, white to cream, regularly spotted and dotted maroon (at least) on its inner surface, bowl-shaped, basally constricted, 5 - 8 mm ⌀, 1.5 - 2 mm high, papillose, **Ha** papillae 0.05 - 0.25 mm; **Cn** canary-yellow, disc 3 - 3.6 mm ⌀, fleshy, convex; **Cs** lobes 1 × 1.2 mm; **Poll** yellowish-green, 0.3 × 0.2 mm; **Fr** 6 - 12 cm; **Se** ± 4 - 5 × 2.5 - 3 mm, wings slightly undulate, sometimes notched.

A species with a wide and partly disjunct distribution.

D. modesta N. E. Brown (FC 4(1): 1028, 1908). **T**: RSA, Eastern Cape (*Pillans* 35 [K]). — **D**: RSA (Eastern Cape).

[1] Stems delicate, 1 - 4 × 0.8 - 1.5 (-2) cm, 4- to indistinctly 5-angled, ovoid (to cylindrical), dark green; **L** rudiments 1.5 mm, stipular **Gl** small, sometimes absent; **Inf** about midway on the stems, with a large number of procumbent **Fl** with slight foetid odour; **Ped** 1 - 2 cm; **Sep** 3 - 4 mm; **Cl** 1.5 - 2.5 cm ⌀, brown to brownish-purple; **Cl** lobes 7 - 9 × 3 - 6 mm, lamina basally spreading, apically ± completely reflexed, margins basally ciliate, **Ha** thin, often curved, purple, 1.5 - 2.5 mm; annulus circular or pentagonal, 5.5 - 8 mm ⌀, 1 - 1.5 mm high, rim papillose, **Ha** papillae 0.05 - 0.1 mm; **Cn** ivory, maroon, brown to dark purple, disc 4 - 5.5 mm ⌀; **Cs** lobes 1 - 2 × 1 - 1.5 mm; **Poll** 0.3 × 0.2 mm; **Fr** 8 - 13 cm; **Se** 4 × 2.5 mm.

D. parviflora N. E. Brown (FC 4(1): 1034, 1908). **T**: RSA, Cape Prov. (*Pillans* 621 [K]). — **D**: RSA (Western Cape, Eastern Cape: Little Karoo). **Fig. XX.e**

[1] Stems 1 - 2.5 × 1 - 1.5 cm, globular to ovoid, green to greyish-green; **L** rudiments < 1 mm; **Inf** near the stem base with 2 - 6 essentially scentless **Fl**; **Ped** 3 - 8 mm; **Sep** 2 - 3 mm; **Cl** 1 - 1.5 cm ⌀, cream, pale yellow to yellowish-olive; **Cl** lobes often with tips tinged brownish, 3.5 - 6 × 1.5 - 2 mm, lamina convex, not completely reflexed, with longitudinal furrows, glabrous; annulus slender, rounded to pentagonal, 3.5 - 5 mm ⌀, 1 - 2 mm high; **Cn** ivory to straw-coloured, disc 3.5 - 4 mm ⌀, to 1 mm high, ± pentagonal, convex, rising above the annulus; **Cs** lobes 1 - 1.4 × 1 - 1.5 mm, blunt; **Poll** ± 0.3 × 0.2 mm; **Fr** 8 - 12 cm; **Se** 3.5 × 2.5 mm.

D. pillansii N. E. Brown (FC 4(1): 1026, 1908). **T**: RSA, Eastern Cape (*Pillans* 42 [K, BM]). — **D**: RSA (Eastern Cape).

Incl. *Duvalia pillansii* var. *albanica* N. E. Brown (1908).

[1] Stems 1 - 3 × 0.7 - 1.4 cm, 4- (rarely 5-) angled, ovoid-cylindrical, slender, green; **L** rudiments ≤ 1 mm, stipular **Gl** very small, sometimes absent; **Inf** about midway on the stems with few slightly erect **Fl** with faint foetid odour; **Ped** 7 - 15 mm; **Sep** 3 - 4 mm; **Cl** 2 - 3.5 cm ⌀, red to brownish-purple; **Cl** lobes 7 - 14 × 8 mm, lamina glabrous, basally spreading, towards the tip increasingly reflexed, margins ciliate, **Ha** simple, curved, purple, 2 - 3 mm; annulus circular to slightly pentagonal, margin with smooth transition into the **Cl** lobes, 7 - 10 mm ⌀, 2 - 2.5 mm high, margin white to cream, papillose, **Ha** papillae 0.05 - 0.1 mm; **Cn** cream to straw-coloured, disc ± 5 mm ⌀; **Cs** lobes 1 - 1.5 × 1.5 mm; **Poll** 0.35 × 0.24 mm; **Fr** 8 cm; **Se** 4.5 - 5.5 × 3 mm, wings often emarginate.

D. polita N. E. Brown (Gard. Chron., ser. nov. 6: 130, 1876). **T**: K. — **D**: Botswana, Malawi, Moçambique, Namibia, Zambia, Zimbabwe, RSA. **I**: CBM 102: t. 6245, 1876. **Fig. XX.f**

Incl. *Stapelia echinata* hort. *ex* N. E. Brown (s.a.); **incl.** *Duvalia dentata* N. E. Brown (1895); **incl.** *Duvalia transvaalensis* Schlechter (1895) ≡ *Duvalia polita* var. *transvaalensis* (Schlechter) A. C. White & B. Sloane (1937); **incl.** *Duvalia transvaalensis* var. *parviflora* L. Bolus (1915); **incl.** *Duvalia polita* fa. *intermedia* A. C. White & B. Sloane (1937); **incl.** *Stapelia polita* hort. *ex* N. E. Brown (s.a.).

[1] Above-ground stems 2 - 10 × 0.7 - 1.5 cm, 6-angled, cylindrical, green; **L** rudiments narrow, acute, 4 - 6 mm, stipular **Gl** ± 0.5 mm, oblong; subterranean stems 1 - 10 × 0.5 - 0.8 cm, cylindrical to pencil-like, ivory; **Inf** from the basal stem parts, with a large number of procumbent **Fl** with strong foetid odour; **Ped** 1.5 - 2.5 cm; **Sep** 4 - 5 mm; **Cl** 2 - 3.5 cm ⌀, maroon to very dark brown or deep purple or spotted this way on yellowish-green ground; **Cl** lobes 10 - 15 × 7 - 10 mm, lamina only slightly curved back along the mid-rib, apically ± erect, shiny, glabrous but margins mostly ciliate between the lobes, clavate **Ha** vibratile, purple, 0.5 - 2.5 mm; annulus with or without spots, circular, margin with smooth transition to the **Cl** lobes, 8 - 12 mm ⌀, 2.5 - 5 mm high, upper rim papillate, **Ha** papillae 0.05 - 0.2 mm; **Cn** pale maroon, disc 4.5 - 6.5 mm ⌀; **Cs** lobes 1.5 - 2.5 × 2 mm, outer appendages spoon-shaped in top view; **Poll** 0.4 × 0.27 mm; **Fr** to 20 cm; **Se** 5 - 7.5 × 4 - 5 mm.

This species is most unusual within Sect. *Duvalia* not only because of its habitat requirements (savannah, *Acacia*-thornscrub) and distribution area. *D. polita* is also easily recognizable vegetatively on account of its large leaf rudiments and the strong tendency to produce rhizomes.

D. pubescens N. E. Brown (FC 4(1): 1029, 1908). **T**: RSA, Cape Prov. (*Ayres* s.n. in *Pillans* 94 [K]). — **D**: S Namibia, RSA (Northern Cape, Western Cape). **Fig. XX.g**

Incl. *Duvalia elegans* var. *namaquana* N. E.

Brown (1908); **incl.** *Duvalia pubescens* var. *major* N. E. Brown (1908).

[1] Stems 1 - 5 × 0.7 - 2.2 cm, 4- to 5-angled, ovoid to cylindrical, dark green; **L** rudiments 1 - 2 mm; **Inf** near the stem base with 1 to several procumbent to slightly erect **Fl** with foetid odour; **Ped** 1 - 2.5 cm; **Sep** 3 - 4 mm; **Cl** 1.8 - 3.2 cm ∅, (pale) maroon, brown or deep purple; **Cl** lobes 7 - 13 × 5 - 7 mm, lamina reflexed to a large extent, upper surface (in some cases only basally) finely hairy, **Ha** simple, erect, white or purple, 0.5 - 1.4 mm; annulus circular to clearly pentagonal, occasionally spotted, 6.5 - 12 mm ∅, 1.5 - 3 mm high, finely hairy, **Ha** 0.3 - 1 mm; **Cn** pale orange-brown or maroon (also white in Namibia), disc 4.5 - 6 mm ∅; **Cs** lobes 1 - 2.4 × 1.2 - 2 mm; **Poll** 0.35 - 0.45 × 0.25 - 0.3 mm; **Fr** 8 - 14 cm; **Se** 4.5 - 5.5 × 3 mm.

Regarded as a form of *D. caespitosa* by some authors, *D. pubescens* is clearly distinguishable by its unique indumentum of the corolla and its allopatric distribution within the N winter-rainfall area.

D. sulcata N. E. Brown (BMI 1910: 193, 1910). **T:** S Arabia (Yemen ?) (*Bent* s.n. [K]). – **D:** Saudi Arabia, Yemen, Sudan, Ethiopia, Djibouti, Somalia.

A very variable complex. 3 subspecies can be recognized as geographically and morphologically (to a wide extent) clearly distinct entities.

D. sulcata ssp. **seminuda** (Lavranos) Meve (Pl. Syst. Evol., Suppl. 10: 116, 1997). **T:** Yemen, Subaihi Country (*Rauh & Lavranos* 3157 [K]). – **D:** N Yemen, Saudi Arabia.

≡ *Duvalia sulcata* var. *seminuda* Lavranos (1967).

[2] Stems 2 - 6 × 1 - 2.2 cm, bluntly 4-angled, clavate, greyish-green spotted dark green to maroon; **L** rudiments ± 5 - 8 mm; **Inf** with 1 to several procumbent **Fl**, often long-lasting and producing a short rachis; **Fl** scentless or with foetid odour; **Ped** 1.5 - 4 cm; **Sep** 3 - 6 mm; **Cl** 3 - 4.5 cm ∅, ochre, pale maroon, reddish- or dark brown; **Cl** lobes 10 - 16 × 7 - 12 mm, lamina slightly convex, furrowed, glabrous, apex and basal angles between the lobes finely tuberculate, papillate or tomentose, margins with vibratile clavate **Ha**, these white or purple, 2 - 3 mm; annulus 9 - 12 mm ∅, 2 - 4 mm high, massive, glabrous to slightly hairy; **Cn** cream to pale maroon, partly spotted reddish-brown, disc 6.5 - 10 mm ∅, ± round, 5- to 10-angled; **Cs** lobes sometimes white, 2.2 - 4 × 1.4 - 2.2 mm, outer appendages obtuse to acuminate; **Poll** 0.35 - 0.5 × 0.25 - 0.3 mm.

Ssp. *seminuda* is mainly found at medium altitudes in the W Asir and the Mountain-Tihama between 250 and 700 m.

D. sulcata ssp. **somalensis** (Lavranos) Meve (Pl. Syst. Evol., Suppl. 10: 118, 1997). **T:** Somalia (*Lavranos* 6838 [FT]). – **D:** Ethiopia, Djibouti, N Yemen, Somalia.

≡ *Duvalia somalensis* Lavranos (1971).

[2] Differs from ssp. *seminuda*: Stems 3 - 5 × 1 - 1.5 cm; **Fl** with foetid odour; **Ped** 2.5 - 3 cm; **Cl** 3 - 3.5 cm ∅, brown or yellowish and spotted, dotted and striped maroon, glabrous, shiny; **Cl** lobes ± 15 × 7 mm, lamina convex, half reflexed, apex papillate; annulus ± circular, 10 - 11 mm ∅, 2 - 3 mm high, robust, with few **Ha** papillae; **Cn** cream-coloured, partly spotted maroon, 6.5 - 7.5 mm ∅; **Cs** lobes 2.2 - 3 × 2 mm, outer appendages very acute; **Poll** 0.4 - 0.45 × 0.25 - 0.3 mm.

This little-known taxon presumably has a wider distribution at the Horn of Africa than presently known.

D. sulcata ssp. **sulcata** – **D:** S Yemen, Sudan. **Fig. XXI.b**

[2] Differs from ssp. *seminuda*: Stems ± 2 - 10 × 1 - 2 cm; **Fl** with foetid odour; **Ped** 1 - 4 cm; **Cl** 3 - 4.5 cm ∅, (greenish-) ochre, pale maroon, reddish-brown or pale chestnut-brown, shiny; **Cl** lobes 10 - 18 × 7.5 - 12 mm, lamina ± flatly spreading, with 5 longitudinal furrows on the upper surface, apex finely papillate, margins slightly reflexed, with vibratile clavate **Ha**, **Ha** 1.5 - 2.5 mm; annulus ± circular, weak, 7 - 14 mm ∅, 1 - 1.5 (-2) mm high, densely hairy, **Ha** white (pale pink), simple, 5 - 8 mm; **Cn** disc 6 - 8 mm ∅, round, 5- to 10-angled, covering the annulus almost entirely, **Cs** lobes white or cream, 2.2 - 4 × 1.4 - 2.2 mm, outer appendages very acute, erect (slightly curved inwards); **Poll** 0.35 - 0.5 × 0.3 mm; **Fr** 12 - 16 cm × 7 - 8 mm; **Se** 6 × 3 - 4 mm, greyish-brown.

This subspecies is characterized by the flat, densely hairy annulus; it has a disjunct distribution in southern Yemen and in Sudan.

D. velutina Lavranos (CSJA 55(1): 24-26, ills., 1983). **T:** Saudi Arabia (*Lavranos & Collenette* 18243 [E, MO]). – **D:** Saudi Arabia, N Yemen. **Fig. XXI.c**

[2] Stems 2.5 - 6 (-9) × 1 - 2 cm, bluntly 4- (to 5-) angled, clavate, greyish-green spotted dark-green to dark maroon; **L** rudiments 5 - 11 mm; **Inf** with 1 to several procumbent **Fl**, appearing near the stem base, with up to 3 cm long rachis; **Fl** with foetid odour; **Ped** 3 - 7 cm; **Sep** ± 5 mm; **Cl** 2 - 4 cm ∅, green, pale maroon or (pale) reddish-brown; **Cl** lobes 10 - 14 × 10 - 12 mm, lamina ± flatly spreading, convex, upper surface with 5 - 7 longitudinal furrows, velvety with **Ha** 0.1 - 0.15 mm (at the tip to 0.4 mm), margins of the lobes not more than halfways reflexed, glabrous or hairy, **Ha** simple, whitish, to 1 mm; annulus convex, with smooth transition to the **Cl** lobes, glabrous or papillate, papillae to 0.08 mm; **Cn** cream to pale maroon, often spotted with dark maroon, disc 7 - 8.5 mm ∅; **Cs** lobes mostly white, 2.5 - 5 × 1 - 2 mm, outer appendages vertical, sometimes curved, very acute; **Poll** 0.4 - 0.45 × 0.24 - 0.28 mm; **Fr** ± 10 × 1.2 cm.

Typical for the lower altitudes of the SW Arabian Peninsula (50 - 400 m).

D. vestita Meve (KuaS 39(9): 194-197, ills., 1988). **T**: RSA, Western Cape (*Meve & Liede* 397 [K, MSUN]). – **D**: RSA (Western Cape).

[1] Stems 1.5 - 7 (-10) × 0.8 - 1.8 cm, 4-angled, cylindrical, procumbent to almost erect, stoloniferous, dark green; **L** rudiments ≤ 1 mm, stipular **Gl** minute or absent; **Inf** successively appearing near the stem base; **Fl** 1 to several, procumbent to slightly ascending, with strong foetid odour; **Ped** 1 - 2 cm; **Sep** 2 - 3 mm; **Cl** 2 - 3 cm ⌀, dark maroon to very dark purple, ± shiny; **Cl** lobes 7 - 13 × 5 - 8 mm, lamina reflexed to a great extent, upper surface basally, halfways or entirely covered with fine purple **Ha** to 2.2 mm long; annulus flat and broad, 7 - 10 mm ⌀, 1 - 3 mm high, with purple **Ha** 1.2 - 2 mm long; **Cn** maroon to purple, disc 6 - 7 mm ⌀, slightly covering the rim of the annulus; **Cs** lobes 1.5 - 2 × 1.6 - 2.2 mm; **Poll** 0.4 × 0.27 mm; **Fr** to 16 cm; **Se** 5 - 6 × 4 - 4.5 mm, with very broad margin.

Most closely related to *D. elegans* and *D. caespitosa*.

DUVALIANDRA

U. Meve

Duvaliandra M. G. Gilbert (Cact. Succ. J. Gr. Brit. 42(4): 99-101, 1980). **T**: *Caralluma dioscoridis* Lavranos. – **D**: Socotra. **Etym**: For the similarity to the genus *Duvalia* (*Asclepiadaceae*).

Clustering stem succulents; stems decumbent to ascending, 4-angled, quadrangular in cross-section, 2 - 10 × 1 - 1.7 cm, pale grey-green, smooth; **Tu** small, hardly protruding, **L**-less; **Inf** basal to almost apical, sessile or shortly and broadly pedunculate, with few minute **Bra**, 1- to 3-flowered, only 1 **Fl** open at a time, scent strong and foetid; **Ped** 5 - 30 × 1.5 - 2.5 mm; **Sep** triangular, acute, 2 - 4 mm, glabrous; **Cl** outside pale green, inside flesh-coloured to red-brown, base of tube and tips of the lobes often yellowish to flesh-coloured; **Cl** rotate, 4 - 5 cm ⌀, fleshy, coarse, outside glabrous, inside glossy, with red-brown **Ha**, these ± 2 mm, simple, erect; **Cl** tube 2 - 4 × 8 - 12 mm, bowl-shaped, basally **Ha**-less but papillate; **Cl** lobes triangular, acute, 15 - 20 × 10 - 14 mm, ± reflexed, margin basally with purple clavate 3 mm long **Ha**; **Cn** apparently uniseriate, sessile, purple, lentil-shaped, 6 - 8 mm ⌀; **Cs** ± trapezoid, with short obtuse **Ci** lobes from a thickened and broadened base, ascending at a 45° angle, surrounding the **Sty** head; **Anth** small, transversely reniform, placed almost vertically on the short **Sty** head; **Poll** ovoid, ± 0.4 × 0.3 mm, with very strong germination mouths spreading at right angles, caudicles ± 0.1 mm, apically broadened disc-like, corpuscle ± rectangular, ± 0.2 × 0.15 mm, basally winged, ± 0.3 mm; **Fr** cylindrical, acute, 3 - 6 cm.

The diagnosis of the genus reports the uniseriate corona as characteristic. However, the corona indeed consists of 2 series as in all members of the stapeliads, as there is a flat projecting band of tissue at interstaminal position which unites the staminal **Cn** lobes. This formation can be interpreted as interstaminal corona.

The morphology of pollinaria and fruits indicates, as does the name, a relationship of *Duvaliandra* with *Duvalia*, while in stem morphology and biogeography, *White-sloanea* appears to be much closer. On the other hand, corona, anthers and style head indicate that this Socotran endemic occupies an isolated position.

D. dioscoridis (Lavranos) M. G. Gilbert (Cact. Succ. J. Gr. Brit. 42(4): 101, ill. (p. 100), 1980). **T**: Socotra (*Radcliffe-Smith & Lavranos* 438 [K, PRE, Z]). – **D**: Socotra. **Fig. XX.a**

≡ *Caralluma dioscoridis* Lavranos (1971).
Description as for the genus.

×DUVALIARANTHUS

U. Meve

×**Duvaliaranthus** Bruyns (JSAB 42: 365, 1976).
= *Duvalia* × *Piaranthus*.

×**D. albostriatus** Bruyns (JSAB 42: 365-367, ill., 1976). – **D**: RSA (Western Cape: Little Karoo).

Naturally occurring intergeneric hybrid from the Little Karoo with intermediate characters between the putative parents *Duvalia caespitosa* and *Piaranthus sp*.

ECHIDNOPSIS

B. Müller & F. Albers

Echidnopsis Hooker fil. (CBM 97: t. 5930 + text, 1871). **T**: *Echidnopsis cereiformis* Hooker *fil*. – **Lit**: Bruyns (1988a); Plowes (1993). **D**: SW Oman, Yemen, Socotra, E Africa (Djibouti to Tanzania, N Kenya). **Etym**: Gr. 'echidna', snake, adder; and Gr. '-opsis', looking like; for the often creeping stems.

Incl. *Virchowia* Vatke *ex* K. Schumann (1893) (*nom. illeg.*, Art. 53.1).

Incl. *Pseudopectinaria* Lavranos (1971). **T**: *Pseudopectinaria malum* Lavranos.

Stem succulents branching from the base, with few to numerous prostrate or erect shoots, forming mats or clusters; stems 2 - 20 (-60) cm, 0.8 - 2.5 cm ⌀, **Tu** ± hexagonal, arranged in (5-) 6 - 20 blunt **Ri**, apically with mostly short-lived small ± triangular **L** rudiment, lamina and midrib clearly distinct, stipular **Gl** often present; **Inf** developing successively 1 or several **Fl**, without peduncle (rarely developing again from the same synflorescence); **Ped** (0.5-) 1 - 5 (-15) mm, glabrous; **Sep** usually < 2 mm, ovoid or acutely triangular; **Cl** (0.3-) 0.6 - 1.5 (-2) cm ⌀, campanulate to urceolate, or globose or cylindrical, outside glabrous and smooth, inside glabrous or scattered with **Ha**, rarely papillose; **Cn** biseriate; **Ci** mostly cup-shaped of variable depth (rarely consisting of 5 separate lobes); **Cs** lobes dorsiventrally flattened, incumbent on the **Anth**, often meeting each other above the centre of the **Fl**, rising up above the **Sty** head; **Anth** ± elongate-rectangular; **Poll** ± globose, very small, yellow (rarely orange), 0.16 - 0.22 mm ⌀, corpuscle very small, caudicles simple; **Fr** as pairs of narrowly fusiform follicles, uniformly coloured; **Se** ± flattened pear-shaped, ventrally ± concave, rim conspicuous, brittle, with a tuft of simple white **Ha** at one end.

There are 2 competing classification schemes available, namely Bruyns (1988a) and Plowes (1993). Their concepts of delimiting species and subspecies differ widely and the authors represent 2 extreme positions as "lumper" and "splitter". A compromising approach is attempted here by using suggestions of both classifications.

The genus is divided into the following 4 sections (Bruyns 1988a):

[1] Sect. *Glabra* Bruyns 1988: Stem surface rugose (plaited); **Fl** never erect; **Ci** shallowly cup-shaped.

[2] Sect. *Echidnopsis*: Stem surface rugose and papillose; **Fl** erect; **Ci** not cup-shaped.

[3] Sect. *Vadosicorona* Bruyns 1988: Stem surface smooth; **Fl** never erect; **Ci** shallowly cup-shaped with dorsal appendages; **Cs** lobes longer than the **Anth** and ± rising up.

[4] Sect. *Profundicorona* Bruyns 1988: Stem surface smooth or in part hispid; **Fl** never erect; **Ci** deeply cup-shaped without dorsal appendages, inside with long white **Ha**; **Cs** lobes shorter than the **Anth**.

As yet the following natural hybrids have been recorded: *E. dammanniana* × *E. virchowii*, *E. scutellata* × *E. sharpei* (Gilbert 2293) and *E. watsonii* × *E. rubrolutea* (Plowes 5434). In addition, *E. cereiformis* × *E. dammanniana* is known from cultivation.

The following names are of unresolved application but are referred to this genus: *Stapelia multangula* Forsskål (1775) ≡ *Echidnopsis multangula* (Forsskål) Chiovenda (1923).

E. angustiloba E. A. Bruce & P. R. O. Bally (CSJA 13: 180, 1941). **T** [lecto]: Kenya (*Copley* s.n. in *Bally* S26 [K, ZSS]). – **D**: Kenya. **I**: Bruyns (1988a). **Fig. XXI.e**

[3] Stems 5 - 10 cm, 1 - 2 cm ⌀; **Tu** conical, arranged in 10 - 11 **Ri**; **L** rudiments 2 - 3 mm, with a broad base tapering towards the apex, withering to remain as white **Sp**; **Ped** < 1.5 mm; **Sep** to 2 mm, narrowly ovate-lanceolate, glabrous; **Cl** inside yellow or reddish with yellow **Cl** lobes, 1 - 1.5 cm ⌀, shallow to almost campanulate, tube campanulate, ± 2 mm ⌀, 1 mm deep, **Cl** lobes 3 - 5 mm long, ovoid-triangular at the base, tapering into a pointed apex, margins of upper ⅔ strongly revolute, inside with scattered white **Ha**; **Cn** pale to bright yellow, 2.3 - 2.8 mm ⌀, bluntly pentagonal; **Ci** weakly to conspicuously dentate, inside rarely with a few short **Ha**; **Cs** lobes narrow, tip linear, base broad.

E. archeri P. R. O. Bally (Cact. Succ. J. Gr. Brit. 19: 59, 60, 63, 1957). **T**: Kenya (*Archer* s.n. in *Bally* S235 [K, ZSS]). – **D**: Kenya, Somalia.

Incl. *Echidnopsis similis* Plowes (1993).

[4] Stems decumbent, to 5 cm, 1 - 1.5 cm ⌀; **Tu** arranged in 8 **Ri**; **L** rudiments 2 - 3 mm, deltoid; **Ped** 1 mm; **Sep** 1 - 2 mm, ovate-lanceolate, pointed, apex recurved; **Cl** outside purple to purplish-red (sometimes pink), inside dark purple to crimson, 5 - 6 mm ⌀, 5 - 8 mm deep, cup-shaped, tube 3 - 4 mm ⌀, 3 - 4 mm deep, **Cl** lobes erect, with revolute margins, sometimes finely ciliate, apex revolute, pointed, outside sometimes with a small tuft of **Ha**, 2 - 2.8 mm long, basally 3.6 - 4 mm broad, outside sparsely papillate or smooth, inside sparsely hairy, velvety; **Cn** 2 - 2.2 mm ⌀, pentagonal; **Ci** yellow at the base, sometimes purple along the margin, sometimes dentate, inside with a few long **Ha**; **Cs** lobes broad at the base, with a small tip, triangular, glabrous.

E. ballyi (Marnier-Lapostolle) P. R. O. Bally (KuaS 14(10): 190, ills., 1963). **T**: Somalia (*Bally* 11854 [P, ZSS]). – **D**: N Somalia.

≡ *Stapeliopsis ballyi* Marnier-Lapostolle (1959).

[4] Stems decumbent, grey-green to green, 5 - 12 cm, 1.5 - 2.5 cm ⌀; **Tu** to 7 mm long, laterally slightly flattened, ± 4-angular, arranged in 6 - 8 **Ri**; **L** rudiments fleshy, broad at the base, pointed at the apex; **Ped** 1 - 1.5 cm, **Fl** nodding; **Sep** 2 mm, lanceolate, apex recurved; **Cl** outside mostly dark purple, **Cl** lobes yellow, dotted with red, tube 1 - 1.5 cm long, pear-shaped to urceolate, inside dark purplish-red, at the widest part 0.8 - 1 cm ⌀, mouth only 2 mm ⌀; **Cl** lobes erect or spreading, deltoid, with weak folds merging into the tube, margins and tip revolute, outside shining, with few longitudinal folds, inside glabrous, with longitudinal folds; **Cn** 2.25 mm ⌀, pentagonal; **Ci** deeply cup-shaped, mouth slightly constricted, inconspicuously divided into 5 lobes, 2 mm deep, outside glabrous, inside hairy, **Ha** stiff, translucent, purplish-red at the base of the **Ci**, yellow towards its mouth; **Cs** yellowish with purplish-red spots, lobes (spatulate-) pointed, to 1 × basally 0.4 mm.

E. bentii N. E. Brown *ex* Hooker *fil.* (CBM 127: t. 7760 + text, 1901). **T**: Yemen, Socotra (*Bent* s.n. [K]). – **Lit**: Meve & Wolf (2001). **D**: Socotra.

[4] Stems cylindrical, 7- to 8-ribbed, procumbent-ascending, 3 - 15 × 1 cm, soft, brown-green or green, glabrous; **Tu** ovoid to rhomboid; **L** rudiments deltoid-lanceolate, ± 1 mm, acute, glabrous; **Inf** sessile, 2-flowered; **Bra** subulate, 1 mm; **Ped** 1.5 mm; **Sep** ovate-lanceolate, ± 1.2 mm; **Fl** fleshy; **Cl** rotate-campanulate, ± 1 cm ∅, outside pale green with red-brown dots, inner side dark purple, finely papillate; **Cl** tube campanulate, 3 × 4 - 5 mm; **Cl** lobes deltoid, ± 3.5 × 3.5 mm, horizontally spreading, slightly convex, margins recurved; **Cn** biseriate, (bright) purple, ± 2.5 × 3 mm, urceolate; **Ci** as 5 flattened tips originating at the upper rim of the urceolate part; **Cs** lobes cylindrical-clavate, 3.5 × 1 mm, ± erect, apically often yellowish; **Poll** ± ovoid, ± 0.3 × 0.2 mm; **Fr** slenderly fusiform, glabrous; **Se** unknown.

Recently rediscovered on Socotra after more than 100 years of uncertainty regarding its status and origin. (Meve & Wolf 2001). – [U. Meve]

E. bihendulensis P. R. O. Bally (Cact. Succ. J. Gr. Brit. 19: 58, 60, 1957). **T:** Somalia (*Bally* S125 [K, ZSS]). – **D:** Somalia.

[4] Stems grey-green to green, to 12 cm, 1.5 - 2.5 cm ∅; **Tu** 2 - 3 mm, conical, arranged in 16 **Ri**; **L** rudiments ascending, 3 - 3.5 mm, subulate, withering to persist as acute **Sp**; **Ped** < 1 mm; **Sep** 1.5 - 3 mm, lanceolate and pointed, finely papillose; **Cl** outside pale green, inside yellow or reddish-brown, 0.6 - 0.75 cm ∅, tube campanulate, 1 mm deep; **Cl** lobes erect or spreading, 2.5 - 4 mm, deltoid, margins and apex revolute, inside scattered with ascending **Ha**, margins often ciliate with small thick acute translucent **Ha**; **Cn** 1.5 - 3.3 mm ∅, 1 - 1.5 mm high, bluntly pentagonal; **Ci** yellow with brown margin, in the zone of connation with the **Cs** outside deeply incised; **Cs** yellow with red margin, basally broad, apically obtuse.

E. cereiformis Hooker *fil.* (CBM 97: t. 5930 + text, 1871). **T:** K. – **D:** Ethiopia, Sudan. **Fig. XXI.a**

Incl. *Piaranthus fascicularis* hort. (s.a.) (*nom. inval.*, Art. 29.1); incl. *Stapelia cylindrica* hort. (s.a.) (*nom. inval.*, Art. 29.1); incl. *Boucerosia cylindrica* Brongniart (1860); incl. *Apteranthes cylindrica* Decaisne (1871) ≡ *Echidnopsis cylindrica* (Decaisne) K. Schumann (1895); incl. *Apteranthes tessellata* Decaisne (1871) ≡ *Echidnopsis tessellata* (Decaisne) K. Schumann (1895); incl. *Echidnopsis cereiformis* var. *brunnea* A. Berger (1902); incl. *Echidnopsis cereiformis* var. *obscura* A. Berger (1902).

[2] Stems green to brownish, 5 - 15 (-60) cm, 1.2 - 2 cm ∅; **Tu** ± 6-angled, flat, arranged in 8 **Ri**, initially with thick **L** rudiment; **Ped** < 1 mm; **Sep** finely papillose; **Cl** inside brownish-purple, reddish-brown or bright yellow, 5 - 8 mm ∅, campanulate, tube saucer-shaped, 1 mm deep; **Cl** lobes erect or slightly spreading, ± deltoid, 2 - 3 × basally 2 - 2.5 mm, upper ½ with weak longitudinal folds, with scattered **Ha** on both sides or inside glabrous, **Ha** pointed, prickly; **Cn** light yellow; **Ci** lobes reduced to a linear to deltoid tooth; **Cs** apically often with a chestnut-brown edge, lobes broadly triangular, apically truncate, often steeply raising towards the **Sty** head, usually overtopping the **Anth**.

Closely related to *E. nubica*.

E. chrysantha Lavranos (CSJA 43(2): 65-66, 1971). **T:** Somalia (*Lavranos* 7325 [FI, K, ZSS]). – **D:** Somalia.

This species belongs to the *E. scutellata*-complex and is closely related to *E. planifolia*.

E. chrysantha ssp. **chrysantha**

[3] Stems erect or decumbent in small groups, with underground runners, dark green, 3 - 10 cm, 0.7 - 1.1 cm ∅; **Tu** arranged in 8 (-10) prominent **Ri**, apically with **L** rudiments, these 1.5 - 2.5 mm, (narrowly) triangular, acute, with finely dentate margin, withering to remain as white scarred **Bri**; **Inf** with 2 **Fl** arising from the stem tips; **Ped** 2 mm, finely rugose; **Sep** 1.8 mm, narrowly triangular, densely finely rugose; **Cl** golden-yellow, sometimes with a weak greenish tinge, 9 mm ∅, shallowly campanulate, tube broadly campanulate, 2 mm long; **Cl** lobes ascending-spreading, broadly triangular, 3 mm, on both sides rather weakly rugose; **Cn** golden-yellow, 2 mm ∅; **Ci** cup-shaped, indistinctly pentagonal, notched at the apex; **Cs** lobes broadly triangular at the base, thin at the apex, erect.

E. chrysantha ssp. **filipes** (Lavranos) Plowes (Haseltonia 1: 75, 1993). **T:** Somalia (*Lavranos & Bavazzano* 8484 [FI, E, ZSS]).

≡ *Echidnopsis chrysantha* var. *filipes* Lavranos (1974).

[3] Differs from ssp. *chrysantha*: **Ped** 1 cm, slender; **Cl** lobes narrower.

E. dammanniana Sprenger (Cat. Dammann & Co., 4, fig. 5, 1892). **T** [neo]: Ethiopia, Harerge Region (*Gilbert* 2374 [K, ETH]). – **D:** Ethiopia, Kenya, Somalia. **I:** Plowes (1993).

Incl. *Echidnopsis somalensis* N. E. Brown (1903).

[2] Stems brownish-green, 5 - 15 (-60) cm, 1.2 - 2 cm ∅, **Tu** 4- or 6-angled, arranged in 8 **Ri**, **L** rudiments ascending, thick; **Ped** < 1 mm; **Sep** finely papillose; **Cl** outside greenish-brown, inside purplish- or greenish-brown, dotted with dull greenish-yellow, 0.7 - 1.1 cm ∅, curved outwards in an umbrella-like manner; **Cl** lobes ovate-deltoid or -lanceolate, considerably recurved and pressed against the stems, 2.5 - 3.2 × basally 2.3 - 2.5 mm, outside scattered with **Ha**, inside with pointed **Ha**; **Cn** bright blackish-purple, sometimes mottled with yellow; **Ci** lobes 0.5 mm, ascending, lanceolate with longitudinal furrow; **Cs** lobes truncate-deltoid, cov-

ering the **Anth**, rising up slightly towards the **Sty** head, outer dorsal-marginal zones forming a small limb together with the **Ci**.

E. ericiflora Lavranos (Nation. Cact. Succ. J. 27: 70-71, ills., 1972). **T:** Kenya (*Lavranos* 9305 [EA, PRE, ZSS]). – **D:** SE Kenya.

[3] Stems rooting over the whole length, to 20 cm, 4 - 8 mm ⌀; **Tu** arranged in 6 - 8 **Ri**; **L** rudiments ovate-deltoid; **Ped** 1 - 1.5 mm; **Sep** 1.5 mm; **Cl** outside wine-red, inside reddish to dark red, urceolate, 5 - 8 mm long; **Cl** lobes basally reddish, apically yellowish, tube max. 4 - 4.5 mm ⌀, mouth only 2 mm ⌀; **Cl** lobes erect, deltoid, 1 - 2 mm, slightly spreading when fully expanded, bottom of the tube inside with scattered **Ha**; **Cn** 2.8 - 3 mm ⌀, pentagonal; **Ci** yellowish-reddish-black, basally yellow with ± black margins, lobes slightly divided in the middle, enclosing the nectar cavity; **Cs** yellow at the base, purplish-red and papillose at the apex.

E. globosa Thulin & Hjertson (Nordic J. Bot. 15(3): 261-262, ills., 1995). **T:** Yemen, Hadhramaut (*Thulin & al.* 8248 [UPS, K]). – **D:** Yemen (Hadhramaut).

[3] Stems with **R** over the whole length, ≤ 16 × 0.6 - 1.2 cm ⌀; **Tu** 4- to 6-angled, arranged in 8 **Ri**; **L** rudiments not persistent; **Inf** 1- to 2-flowered; **Ped** 1.5 - 2.5 × ± 1 mm; **Sep** 1.2 - 2 mm, lanceolate, apex curved outwards; **Cl** globose, 5.5 - 8 × 5.5 - 7 mm, outside purplish-brown, spotted with dull yellowish-white towards the apex, inside dark purple; **Cl** lobes dull yellowish-white, basally spotted with purple, 1.5 - 2.5 × 2 - 2.8 mm, triangular, covering the mouth of the tube and thus only the tips curved outwards, inside glabrous; **Cn** dark purple, ± 2.2 - 2.4 mm ⌀, pentagonal; **Ci** shallowly cup-shaped, margins of the lobes flat, ± entire; **Cs** lobes broad at the base, narrow at the apex.

To be placed in the *E. scutellata*-complex; closely related to *E. ericiflora* and *E. squamulata.*

E. insularis Lavranos (CSJA 2: 136-138, ills., 1970). **T:** Socotra (*Radcliffe-Smith & Lavranos* 310 [K]). – **D:** Socotra.

[3] Stems brown or brownish-green, 2 - 6 cm, ± 5 mm ⌀; **Tu** arranged in 6 **Ri**, apically with minute **L** rudiments; **Ped** 2.5 - 3 mm, ± 1 mm ⌀; **Sep** green, 1 - 1.5 mm, apex recurved, glabrous; **Cl** outside greenish-yellow, urceolate, basally 4 mm ⌀, very fleshy, tube inside greenish-yellow, with fine purple-coloured longitudinal lines, 7 - 8 mm deep, mouth 3 mm ⌀; **Cl** lobes erect, attenuate-deltoid, 3 × basally 1.75 mm; **Cn** yellow, 3 mm ⌀, pentagonal; **Ci** fleshy, each lobe deeply incised in the middle, teeth slightly revolute; **Cs** lobes basally broadly rectangular, tapering towards the apex, long, ± cylindrical, vertical, touching the **Anth** near the base and distinctly overtopping them to rise above the centre without meeting each other.

E. jacksonii P. R. O. Bally *ex* Plowes (Haseltonia 1: 78, 1993). **T:** Ethiopia, Bale Region (*Jackson* s.n. in *Bally* S112 [K]). – **D:** Ethiopia.

[3] Stems prostrate-decumbent, slender, ± 9 cm, ± 1.5 cm ⌀; **Tu** arranged in 12 **Ri**; **Cl** campanulate, tube purple-chestnut-brown, shortly campanulate, 5 mm ⌀, 4 mm deep; **Cl** lobes 3 mm, yellow, broadly triangular; **Cn** sessile, broadly cup-shaped; **Ci** apically ± fused to give a disc-like structure, margin slightly dentate or serrate; **Cs** rising above the **Sty** head for about ½ the hight of the **Cl** tube.

Closely related to *E. urceolata.*

E. leachii Lavranos (Nation. Cact. Succ. J. 27: 69-70, ills., 1972). **T:** Tanzania (*Leach & Brunton* 10143 [EA, K, ZSS [status ?]]). – **D:** Tanzania.

[1] Stems greyish-green, brownish with age, to 15 cm, 0.8 - 1 cm ⌀; **Tu** rugose, arranged in 6 **Ri**; **L** rudiments broadly deltoid, pointed; **Ped** < 1 mm; **Sep** to 1.5 mm, lanceolate, pointed; **Cl** outside dark purple to pink, 4 - 5 mm ⌀, 4 mm long, shortly campanulate, tube inside yellowish, shallowly cup-shaped, 1 - 2 mm deep; **Cl** lobes bright purple to pink, ovate-deltoid, ascending, tips slightly revolute, towards the apex laterally slightly 2-layered, 2 - 2.5 × basally 1.5 - 1.8 mm; **Cn** yellow, towards the outside suffused with purple; **Ci** cup-shaped, margin strongly curved inwards, 2 - 2.5 mm ⌀; **Cs** lobes very short, apically rounded or triangular.

E. malum (Lavranos) Bruyns (Bradleya 6: 43, ills. (p. 44), 1988). **T:** Somalia (*Lavranos* 6721 [FI]). – **D:** Somalia.

≡ *Pseudopectinaria malum* Lavranos (1971).

[4] **Br** cylindrical, 5- to 6-ribbed, procumbent-repent, 3 - 20 × 0.8 - 1 cm, dark grey-green, hispid; **Tu** narrow, elongate; **L** rudiment broadly triangular-cordate, ± 1 - 1.5 mm, acute, hispid; **Inf** sessile, 1-flowered; **Ped** 10 - 15 mm, papillose; **Sep** lanceolate, 2 - 3 mm, papillose; **Fl** erect, fleshy; **Cl** globose-urceolate, 14 - 25 × 18 - 22 mm, outer side bright green with red streaks, papillose, inner side dark purple, rugulose, with white to purple **Ha**, 1 - 1.5 mm, **Cl** tube occupying the whole length of the **Cl**, apically with a circular opening; **Cl** lobes deltoid, purple, pointing inwards, apically connate, ± 4 × 4 mm; **Cn** 2-seriate, red, ± 3.5 - 5 mm, globose-urceolate, with a circular opening and a recurved dentate margin, inside hairy; **Ci** lobes shortly deltoid, yellow, developed from the base of the urn, decumbent on the **Anth** base; **Poll** globose-ovoid, ± 0.3 × 0.25 mm.

Because of its highly derived stem and flower morphology, some authors prefer a classification of this species in the monotypic genus *Pseudopectinaria.* – [U. Meve]

E. mariae Lavranos (CSJA 54(5): 215-216, ills., 1982). **T:** Kenya, Northern Prov. (*Lavranos &*

Bleck 19527 [E, K, EA, ZSS [status ?]]). – **D:** N Kenya.

Incl. *Echidnopsis ethiopica* Keller (1971) (*nom. inval.*, Art. 36.1, 37.1); **incl.** *Echidnopsis scutellata* ssp. *australis* Bruyns (1988).

[3] Stems prostrate, profusely branching, vividly green or brownish-purple in full sunlight, 15 cm, ± 1.5 cm ∅; **Tu** conical, apically rounded, arranged in (8-) 10 (-11) **Ri**, apically with green **L** rudiment, this thick, spatulate, 1.5 × basally 0.8 mm, slightly papillose, after drying persistent for some time; **Inf** 1- to 2-flowered, arising from near the stem tips; **Ped** 1.5 mm, glabrous; **Sep** 1 mm, lanceolate; **Cl** outside pale green, inside pale yellowish-brown, ± with reddish spots, 6 - 9 mm ∅, ± flat to campanulate; **Cl** lobes 1.5 - 2 × basally 2.5 - 3 mm, deltoid, margins and apex revolute, marginally sometimes with small cylindrical **Ha**, otherwise glabrous; **Cn** pale yellowish-brown, basally yellow, ± with reddish spots, 2 - 2.5 mm ∅, ± pentagonal; **Ci** cup-shaped, bluntly 5-parted, lobes slightly dentate; **Cs** elongate-subulate, rising above the **Sty** head.

Bruyns places this species in the synonymy of *E. scutellata*.

E. mijerteina Lavranos (CSJA 43(2): 64-65, ills., 1971). **T:** Somalia, Mijerteina Prov. (*Lavranos* 7268 [FI]). – **D:** N Somalia, SW Kenya. **I:** Bruyns (1988a).

Incl. *Echidnopsis mijerteina* var. *marchandii* Lavranos (1974).

[4] Stems creeping, rooting almost over their whole length, tips usually ending just below the soil surface, dull green, to 20 cm, 0.8 - 1.5 cm ∅, scabrous; **Tu** indistinct, arranged in 8 - 10 **Ri**; **L** rudiments subulate, scabrous; **Ped** 3 - 6 mm, scabrous; **Sep** 2 - 3 mm, lanceolate, scabrous; **Cl** outside white to pale brownish, inside purplish-red, cylindrical, curved, tube 1.5 - 2.5 cm long, basally 4 - 6 mm ∅, constricted upwards, enlarged at the mouth, transversely furrowed ± over its entire length; **Cl** lobes erect to spreading, 1 - 2.5 × basally 3 - 3.5 mm, outside scabrous, inside scattered with short **Ha**; **Cn** yellow, 3 mm ∅, pentagonal; **Ci** fused into a deep tube, 2 - 2.5 mm, apically 5-lobed, margins spreading, sometimes weakly dentate, inside hairy, **Ha** erect; **Cs** lobes rectangular, truncate, touching the **Anth** bases.

E. milleri Lavranos (CSJA 65(6): 294-295, ills., 1993). **T:** Socotra (*Miller & al.* 10145 [E]). – **D:** Socotra.

[3] Stems creeping or decumbent, ± blackish-green, 3 - 8 cm, 0.6 - 1 cm ∅; **Tu** longer than broad, arranged in 6 **Ri**, apically with a narrow ovate-deltoid **L** rudiment which is curved outwards and fugacious; **Inf** usually arising from lower stem parts; **Ped** 0.5 - 1 mm; **Sep** basally fused, free parts subulate, 1 mm; **Cl** outside greenish, inside flesh-coloured, 1 - 1.2 cm ∅, flat to ± campanulate, tube 1 - 2 mm deep; **Cl** lobes first spreading, later slightly ascending, deltoid, 4 × basally 4 mm, outside and inside glabrous; **Cn** ± 4.5 mm ∅, pentagonal in top view; **Ci** white, deeply lobed, cup-shaped in outline, lobes ovate-deltoid, irregularly 3-dentate; **Cs** white, apically pink, lobes filiform, 3 mm, in a column rising above the **Gy**, apically divergent.

E. montana (R. A. Dyer & E. A. Bruce) P. R. O. Bally (Cact. Succ. J. Gr. Brit. 26: 89, ills., 1964). **T:** Ethiopia (*McLoughlin* 835 [PRE, K, ZSS]). – **D:** Ethiopia.

≡ *Caralluma montana* R. A. Dyer & E. A. Bruce (1947).

[4] **R** suckers present; stems to 20 cm, 0.7 - 1.5 cm ∅; **Tu** prominent, 1 - 3 mm long, 4-angled, pointed, arranged in 6 **Ri**; **L** rudiments to 1.5 mm, ovate-deltoid; **Ped** 1 - 2 mm; **Sep** 1 - 2 mm, ovate to lanceolate, finely papillose; **Cl** inside greenish-yellow or purplish-brown with greenish-yellow, tube short, to 2 mm; **Cl** lobes 9 - 10 × basally 2 mm, linear, margins folded almost to the middle in 2 layers, tips slightly thickened, glabrous; **Cn** ± 2.5 mm ∅, indistinctly pentagonal or rotate; **Ci** cup-shaped with crenate margin, apically triangular, entire or dentate, inside ± hairy, **Ha** pointing inwards or erect, whitish, dotted with purple; **Cs** lobes basally broad, truncate to triangular, apically dorsally flattened, not overtopping the **Anth**.

E. nubica N. E. Brown (BMI 1895: 263, 1895). **T:** Sudan, "Nubia" (*Schweinfurth* 228 [K, NAP]). – **D:** SE Sudan, Ethiopia. **I:** Plowes (1993).

[2] Stems decumbent or erect, basally branching, to > 20 cm, 1.5 - 2.5 cm ∅; **Tu** 4- to 6-angled, separated by deep furrows, arranged in (6-) 8 (-10) **Ri**, apically with **L** rudiment, this drying and rapidly caducous; **Inf** 1- to 3-flowered; **Ped** ± 2 mm; **Sep** (ovate-) lanceolate, slightly scabrous; **Cl** outside dull purple, inside green marked with brownish-red spots, ± 3 mm ∅, flat or slightly campanulate; **Cl** lobes apically brownish-red, broadly ovate, pointed, spreading outwards; **Ci** lobes reduced to a rounded deltoid tooth; **Cs** lobes broadly triangular, apically truncate, steeply rising up towards the **Sty** head, longer than the **Anth**; **Fr** reddish with dark spots, > 10 cm, 0.8 cm ∅, beaked, borne on **Ped** elongated to 5 mm.

Closely related to *E. cereiformis*.

E. planiflora P. R. O. Bally (Cact. Succ. J. Gr. Brit. 18: 109, 1956). **T:** Ethiopia (*Mitford-Barberton* s.n. in *Bally* S105 [ZSS]). – **D:** Djibouti, Ethiopia, Somalia.

≡ *Echidnopsis scutellata* ssp. *planiflora* (P. R. O. Bally) Bruyns (1988); **incl.** *Trichocaulon somaliense* Guillaumin (1938); **incl.** *Echidnopsis flavicorona* Plowes (1993); **incl.** *Echidnopsis hirsuta* Plowes (1993).

[3] Stems ± erect; **Tu** arranged in 10 - 16 **Ri**, with **L** rudiments to 3× as long as broad, ± erect, persistent as dry **Sp**; **Inf** 2-flowered, arising from apical stem parts; buds broadly conical; **Ped** ± 2 mm; **Sep** lanceolate; **Cl** outside pale green, inside pale yellowish-green, margins tinged with reddish-brown or completely reddish-brown, 0.7 - 1 cm ∅, ± flat to saucer-shaped, sometimes with a bulging non-thickened part around the **Cn** ('annulus'), finely papillose, papillae weakly rounded; **Cl** lobes 2.5 - 3.2 × basally 1.6 - 2.5 mm, triangular or ovate-lanceolate, margins and tip revolute, margins glabrous or ± densely ciliate, towards the apex often with cylindrical **Ha**, these translucent, reddish- or greenish-white; **Cn** dark brick-red, reddish-yellow or light yellow, ± sessile, in top view rotate or bluntly pentagonal; **Ci** sometimes with a narrow shiny dark purple margin, ± cup-shaped, margin entire or slightly serrate; **Cs** lobes apically less coloured, meeting above the **Sty** head, ± column-like ascending; **Poll** ± globose.

Belonging to the *E. scutellata*-complex.

E. radians Bleck (CSJA 49(6): 263-264, 1977). **T:** Kenya (*Powys* s.n. in *Lavranos* 12554 [MO ?, E ?, EA ?, ZSS [status ?]]). – **D:** Kenya, Somalia.

Incl. *Echidnopsis adamsonii* P. R. O. Bally (1956) (*nom. inval.*, Art. 36.1); incl. *Echidnopsis modesta* P. R. O. Bally *ex* Plowes (1993).

[3] Stems prostrate, to 20 cm; **Tu** conical, arranged in 10 - 12 **Ri**, with **L** rudiments, 0.8 mm, deltoid, fugacious without leaving a scar; **Ped** to 1 mm; **Sep** 3 mm; **Cl** outside purple, inside yellowish or white, ± campanulate, tube inversely pear-shaped, 1.5 - 5 mm long, mouth 3 mm ∅; **Cl** lobes ± elongate-rectangular, 2 - 5 mm long, ± erect to spreading or slightly curved downwards (outwards); **Cn** to 3 mm ∅, ± sessile; **Ci** cup-shaped, margin dentate, lobes hardly noticeably divided in the middle; **Cs** basally broad, apically linear, rising above the **Sty** head to form a column.

Plowes (1993) segregates *E. modesta* as a separate species.

E. repens R. A. Dyer & I. Verdoorn (CSJA 11: 68, 1939). **T:** Tanzania (*Pole-Evans & Erens* 1020 [PRE]). – **D:** N Tanzania, S Kenya.

≡ *Echidnopsis sharpei* ssp. *repens* (R. A. Dyer & Verdoorn) Bruyns (1988).

[4] Stems creeping, rooting almost over their whole length, 6 - 9 mm ∅; **Tu** flat, much longer than broad, arranged in 8 - 10 **Ri**; **L** rudiments to 1.5 mm, lanceolate; **Ped** 2 mm; **Cl** outside green, inside deeply wine-red, 7 - 9 mm ∅, tube campanulate and touching the **Gy**; **Cl** lobes sometimes tipped yellow, ovate-deltoid or ovate, ascending-spreading or spreading, occasionally recurved outwards, **Cl** tube and margin of the **Cl** lobes inside with scattered **Ha**; **Cn** ± 1 × 2.5 mm, cup-shaped, prominently pentagonal; **Ci** basally light red, margin purplish-red; **Cs** lobes rectangular-ovate, ± 0.7mm.

Belonging to the *E. sharpei*-complex.

E. rubrolutea Plowes (Haseltonia 1: 76, ill. (p. 73), 1993). **T:** Somalia (*Lavranos & Horwood* 10421 p.p. [SRGH, ZSS]). – **D:** Somalia.

[3] Stems decumbent, 3 - 8 mm, ± 1 cm ∅; **Tu** arranged in 8 **Ri**, apically with **L** rudiment, this narrowly lanceolate, spreading, quickly caducous; **Inf** 1- to 2-flowered, arising from upper stem parts; **Ped** 2 mm; **Sep** narrowly triangular; **Cl** outside purplish-red, inside greenish-yellow or dull brownish-yellow, 8 mm ∅, campanulate, tube tapering towards the base; **Cl** lobes triangular, sometimes longer than broad, margins glabrous, inside finely papillose, papillae arranged in longitudinal furrows towards the tube; **Cn** yellow, ± sessile, pentagonal in top view; **Ci** forming pouches between the **Cs** lobes; **Cs** lobes touching each other above the **Sty** head, sometimes ascending, apically rounded.

Belonging to the *E. scutellata*-complex, and closely related to *E. chrysantha*.

E. scutellata (Deflers) A. Berger (Stapel. & Klein., 26, 1910). **T:** Yemen (*Deflers* 1167 [G]). – **D:** Oman, Yemen.

≡ *Caralluma scutellata* Deflers (1896).

E. scutellata ssp. **dhofarensis** Bruyns (Bradleya 6: 18-19, ills., 1988). **T:** Oman, Dhofar (*Miller* 2811 [E, ZSS [status ?]]). – **D:** Oman. **Fig. XXI.d**

[3] Differs from ssp. *scutellata*: **Tu** arranged in 8 **Ri**, not papillose; **L** rudiments thick, ± as long as broad, glabrous, not papillose; **Cl** inside pale yellow, 0.9 - 1 cm ∅, tube short, cup-shaped; **Cl** lobes deltoid-ovate, tip and margins revolute, margins not hairy; **Cn** yellow or red.

E. scutellata ssp. **scutellata** – **D:** N Yemen.

[3] Stems prostrate to erect, dark green, 3 - 10 cm, 0.6 - 2 cm ∅; **Tu** 6-angled to conical, arranged in 8 **Ri**; **L** rudiments less than 2× as long as broad; **Ped** 1 - 10 mm; **Sep** to 2 mm, lanceolate, finely papillose; **Cl** outside yellowish-green, spotted with purplish-red, inside pale to bright yellow, > 10 mm ∅, ± campanulate; **Cl** lobes (ovate-) deltoid, margins and tips curved outwards, apically thickened, outside and inside glabrous; **Cn** yellow or red, ± 2 mm ∅, rotate to pentagonal, campanulate to very shallow; **Ci** campanulate, lobes indistinctly divided in the middle, inside flattened, apex and margin dentate and crenulate, straight or curved outwards; **Cs** lobes basally broad, apically narrowed, covering and overtopping the **Anth**, tips often meeting above the **Sty** head; **Poll** of rather ovate shape.

E. seibanica Lavranos (JSAB 30: 88, ill., 1964). **T:** Yemen, Hadhramaut (*Lavranos* 1934 [K, ZSS]). – **D:** S Yemen.

[4] Stems prostrate or deflexed, sometimes with **R** suckers, green to bluish-green, 3 - 15 cm, 6 - 9 mm ⌀; **Tu** arranged in 6 - 8 **Ri**; **L** rudiments erect, ovate, to 0.8 mm; **Ped** 1 mm, finely papillose; **Sep** 1 mm, deltoid, appressed to the **Cl**; **Cl** outside cream-coloured, inside sulphur-yellow, tube and lower ⅓ of the **Cl** lobes with tiny red dots, **Cl** 6 - 7 mm ⌀, shallowly campanulate, tube 2 mm ⌀, 0.75 mm long; **Cl** lobes initially ascending, later spreading, 2.5 × basally 1.5 mm, apically rounded, margins revolute, outside finely papillose, inside scattered with **Ha**; **Cn** 1.5 mm ⌀, rotate; **Ci** pink dotted with purplish-red, apically reddish-purple, steeply rounded-cup-shaped, lobes weakly bipartite, inside hairy, **Ha** translucent, stiff; **Cs** whitish with numerous tiny purplish-red dots, lobes < 0.3 mm, shorter than the **Anth**, basally broad, apically narrow.

E. sharpei A. C. White & B. Sloane (CSJA 11: 67, 1939). **T:** Kenya (*Sharpe & Jex-Blake* s.n. [K, ZSS [status ?]]). – **D:** Ethiopia, Somalia, Kenya.

E. sharpei ssp. **ciliata** (P. R. O. Bally) Bruyns (Bradleya 6: 38-39, ills., 1988). **T** [lecto]: Somalia (*Bally* 7167 [icono: Cact. Succ. J. Gr. Brit., 19: 59, top fig., 1957]). – **D:** Somalia.

≡ *Echidnopsis ciliata* P. R. O. Bally (1957); **incl.** *Echidnopsis bavazzanoi* Lavranos (1974); **incl.** *Echidnopsis lavraniana* Plowes (1993).

[4] Differs from ssp. *sharpei*: Stems creeping; **Tu** slightly longer than broad, arranged in 8 **Ri**; **L** rudiments to 1.2 mm, stout, lanceolate-deltoid; **Ped** 3 mm; **Cl** outside green, spotted with purple, inside dark purple, 1 - 1.5 cm ⌀, tube broadly campanulate, not touching the **Gy**; **Cl** lobes deltoid, 5 - 6 × 5 mm, inside ± densely hairy; **Cn** pentagonal; **Ci** basally white, reddish above, margins purplish-black; **Cs** reddish.

Plowes (1993) separates *E. bavazzanoi* and *E. lavraniana* and ssp. *ciliata* is likewise treated as a species of its own.

E. sharpei ssp. **sharpei** – **D:** Ethiopia, Somalia, Kenya. **I:** Bruyns (1988a).

[4] Stems creeping to decumbent, to 15 cm, 0.8 - 1.5 cm ⌀; **Tu** prominent, rounded, arranged in 8 - 10 **Ri**; **L** rudiments ± 2 mm, deltoid, curved downwards, sometimes marcescent; **Ped** to 1 mm; **Sep** 1 - 1.5 mm, ovate-lanceolate; **Cl** outside green or green mottled with purple, inside red, sometimes yellow or white, 0.6 - 1.2 cm ⌀, slightly campanulate, tube shortly campanulate, pentagonal, enclosing and laterally touching the lower ½ of the **Gy**; **Cl** lobes ± deltoid, 3 - 6 mm long, margins ± strongly revolute, tips revolute and therefore blunt, outside glabrous, inside velvety-hairy, **Ha** to 1 mm, white or red, sparse or ± dense; **Cn** 2 - 2.5 mm ⌀, pentagonal, ± sessile; **Ci** yellow, sometimes pale yellow or white with red margins and reddish stripes, cup-shaped, margins weakly dentate, outside sometimes, inside always with long white **Ha**; **Cs** basally broad, apically narrowly triangular, not overtopping the **Anth**, with some erect dorsal **Ha**.

E. socotrana Lavranos (CSJA 65(6): 293-294, ills., 1993). **T:** Socotra (*Miller & al.* 10139 [E]). – **D:** Socotra.

[3] Stems usually decumbent, vividly green, 3 - 7 cm, 0.6 - 1 cm ⌀; **Tu** arranged in 6 **Ri**, apically with **L** rudiment, this ovate-deltoid, short, dark green, soon withering and caducous; **Inf** with 1 - 2 **Fl** successively arising from apical stem parts; **Ped** 1 mm; **Sep** deltoid, 1.5 mm; **Cl** yellow, 5 mm ⌀, campanulate, tube rather flat; **Cl** lobes ± erect, deltoid, 3 mm long, basally > 3 mm broad, margins revolute, glabrous on both sides; **Cn** pentagonal, ± 3 mm ⌀; **Ci** dark reddish-brown, glossy, deeply 5-partite, ± cup-shaped, lobes ascending, deltoid, apically blunt, in the middle with longitudinal furrow, margins laterally bulging, glabrous; **Cs** yellow, lower part covered by the **Ci**, lobes ribbon-shaped, apically rounded, lying ± vertically on the **Anth**, not meeting above the **Sty** head.

Probably better placed under *E. scutellata* as a further subspecies.

E. squamulata (Decaisne) P. R. O. Bally (KuaS 14(9): 173, 1963). **T:** Yemen (*Botta* s.n. [P]). – **D:** Yemen. **I:** Bruyns (1988a).

≡ *Ceropegia squamulata* Decaisne (1838).

[3] Stems creeping, rooting almost over their whole length, green to brownish-green, to 45 cm, 5 - 8 mm ⌀; **Tu** flat, arranged in 6 - 8 (-9) **Ri**; **L** rudiments ovate-deltoid, revolute; **Ped** 5 - 8 mm, ± 1 mm ⌀, dotted with reddish-brown; **Sep** to 3 mm, lanceolate, pointed, glabrous; **Cl** outside reddish-brown, margins sometimes with pale yellow spots, urceolate, pentagonal, 1.1 - 1.8 cm long, max. 5 - 8 mm ⌀, mouth 2 - 3 mm ⌀; **Cl** lobes deltoid, erect, spotted with pale yellow, inside reddish-brown to dark purple or yellow spotted with red, tips abruptly curved outwards, 1.5 - 2.5 × basally 1.25 - 2.5 mm, outside glabrous, inside velvety-papillose, papillae fine, acute, scattered, **Cl** tube margins finely papillose, otherwise shiny; **Cn** 3.5 - 5 mm ⌀, pentagonal; **Ci** red spotted with yellow or reddish-brown, cup- or basin-shaped, margins irregularly dentate, often inflexed, inside rugose, covered with nectar; **Cs** pinkish-red, erect-connivent.

A very variable species, which may perhaps be divided into 2 subspecies.

E. urceolata P. R. O. Bally (Candollea 18: 341-343, ills., 1963). **T:** Kenya, Northern Frontier Prov. (*Williams* s.n. in *Bally* B8008 [K, ZSS]). – **D:** Ethiopia, NE Kenya.

Incl. *Echidnopsis urceolaris* P.R.O.Bally (1956) (*nom. inval.*, Art. 36.1).

[3] Stems prostrate-decumbent, to 9 cm, 2.5 cm

∅; **Tu** conical, densely arranged in 18 - 20 **Ri**; **L** rudiments 3 - 3.5 mm, spreading, lanceolate, rapidly drying off to persist as sharp **Sp**; **Ped** 1 - 1.5 mm; **Sep** 1.5 - 3 × basally 0.7 - 1.5 mm, lanceolate with tips often curved downwards; **Cl** outside pale yellow, towards base purple, 1 - 1.8 cm long, tubular to urceolate, tube distinctly pentagonal in cross-section, basally 4 - 7 mm ∅, mouth 3 - 4 mm ∅; **Cl** lobes pale greenish-yellowish, inside dark purplish-red, 1.2 - 3 mm, deltoid, margins and tips slightly reflexed; **Cn** 2.5 - 3 mm ∅, 1 - 2 mm high, shortly cylindrical, pentagonal, sometimes shortly stipitate; **Ci** large, cup-shaped, margins dentate, erect, enclosing the nectar cavities; **Cs** lobes basally broad, tapering towards the apex.

The corolla is very polymorphic and varies from almost cylindrical-tubular to urceolate-globose.

E. virchowii K. Schumann (Monatsschr. Kakt.-kunde 3: 98-101, ill., 1893). **T:** Somalia (*Hildebrandt* s.n. [K]). – **D:** Somalia.

Incl. *Virchowia africana* Vatke ex K. Schumann (1893) (*nom. nud.*).

E. virchowii var. **stellata** (Lavranos) Plowes (Haseltonia 1: 69, 1993). **T:** Somalia, Erigavo Distr. (*Lavranos* F342 [FI]). – **D:** Somalia.

≡ *Echidnopsis stellata* Lavranos (1974).

[2] Differs from var. *virchowii*: Stems prostrate-decumbent, slender, to 15 cm, 1.5 cm ∅, stem surface finely papillose, **Tu** arranged in 6 **Ri**, apically with **L** rudiments, these triangular, papillose and fugacious; buds pear-shaped or acutely pyramidal; **Cl** 1.3 cm ∅, stellate, yellow, densely spotted with red-brown, spots merging at the tips of the **Cl** lobes, these apically slightly reflexed, 5.5 × basally 2.5 mm, densely papillose on both sides; **Cn** whitish with dense purple-brown markings, 2.75 mm ∅.

E. virchowii var. **virchowii** – **D:** N Somalia. **Fig. XXI.f**

[2] Stems prostrate, brownish-grey to green, 5 - 20 cm, 1.5 - 2 cm ∅; **Tu** 4-angled, arranged in 8 **Ri**; **L** rudiments (ovate-) triangular, thick, ascending; **Inf** 1- to 3-flowered; **Ped** < 1 mm; **Sep** 1 mm (ovate-) triangular, acute, green, brown towards the apex, finely papillose; **Cl** outside grey-green, inside purple-brown speckled with pale yellow-green, (0.6-) 0.8 - 1.1 cm ∅, tube steeply cup-shaped, < 1 mm deep; **Cl** lobes ovate- to lanceolate-triangular, reflexed and appressed to the stem, 3 - 4.5 × basally 2 - 3 mm, outside scattered with **Ha**, inside densely papillose, papillae globose, stalked, often translucent-white; **Cn** purple-brown, towards the nectar cavities pale yellow; **Ci** basin-shaped, with crenulate or dentate limb; **Cs** ± deltoid, with the emarginate back confluent with the **Ci**.

E. watsonii P. R. O. Bally (Candollea 18: 343-345, ills., 1963). **T:** Somalia, Northern Prov. (*Bally & Watson* 9997 [K]). – **D:** N Somalia, Kenya.

[2] Stems creeping, creeping-prostrate or prostrate, to 20 cm; **Tu** depressed to conical, 4- to 6-angled, arranged in 8 - 12 **Ri**; **L** rudiments 0.8 - 2 mm, deltoid to subulate, reflexed, very thick; **Ped** to 1 mm; **Sep** to 3 mm, lanceolate with ovate base; **Cl** outside dark purplish-red, inside yellow to whitish, tube inversely pear-shaped, globose or very shortly cup-shaped, to 11 mm long, mouth 2 mm ∅: **Cl** lobes subulate, to 11 mm, margins strongly reflexed, spreading, outside glabrous, inside with scattered **Ha**; **Cn** dark purplish-red or dark yellow, 1.8 mm ∅, ± sessile, circular or indistinctly pentagonal; **Ci** cup-shaped, limb ± dentate, lobes slightly divided in the middle, inside flattened; **Cs** basally broad, apically linear.

E. yemenensis Plowes (Haseltonia 1: 71, 1993). **T:** Yemen (*Plowes & Barad* 7764 [SRGH]). – **D:** W Yemen.

[3] Stems cylindrical, prostrate, creeping, 5 - 12 cm, 5 - 8 mm ∅; **Tu** 6-angled, arranged in 8 **Ri**, with fugacious **L** rudiment; **Inf** 1- to 2-flowered; **Ped** 2 - 3 mm; **Cl** reddish, yellowish or bicoloured, > 1 cm ∅, campanulate, tube flat, broad; **Cl** lobes erect, inflexed, (ovate-) deltoid, margins and tips slightly revolute, glabrous on both sides; **Cn** cup-shaped, ± sessile, not in contact with the wall of the **Cl** tube; **Ci** united to form a flat serrate ring; **Cs** meeting each other above the **Sty** head and rising up; **Poll** globose.

Belonging to the *E. scutellata*-complex.

EDITHCOLEA

B. Müller & F. Albers

Edithcolea N. E. Brown (BMI 1895: 220, 1895). **T:** *Edithcolea grandis* N. E. Brown. – **D:** Ethiopia, Somalia, Kenya, Uganda, Tanzania, Socotra. **Etym:** For Miss Edith Cole (1859 - 1940) who collected plants during a botanical expedition led by E. Lort-Phillips into N Somalia 1894 - 1895.

Perennial stem succulents; stems prostrate, richly branched, ≥ 10 × 1 - 1.5 cm, brown-olive, bluntly 4- to 5-angled, glabrous; **Ri** sometimes spirally twisted; **Tu** conical, apically with brown acute hard persisting **Sp**; **Inf** mostly 1-flowered, on upper stem parts; **Ped** 18 - 23 × 3.5 mm, glabrous; **Sep** 5 - 9 × 1.5 - 3 mm, ovate-lanceolate to subulate; **Cl** ± flat, incised to ½, 8 - 12.5 cm ∅, outside olive-green, inside pale yellow with scattered to densely velvety purplish brown dots or spots, spots sometimes raised; **Cl** tube saucer-shaped, 6 × 6 mm, mouth raised slightly like an annulus, inside finely papillate with narrow raised purplish-brown concentric rings; **Cl** lobes 3.5 - 5 × 2.5 - 3 cm, ovate-tailed, upper ½ dark brown to olive-green, smooth, glabrous, margin of tube with 5 narrow radial bands with clavate **Ha**, these purplish-brown, ≤ 1 cm, further clavate **Ha** arching from margin to margin over the centre of the **Cl** lobes; **Cn** biseriate, of the C(is)+Cs-type, yellowish; **Ci** segments ± square or transversally rectangular, basally saccate, inside scatteredly hairy; **Cs** segments lying on the **Anth**, apically broadened and triangular, marginally overlapping, dorsally densely spinulose or papillate; **Anth** rectangular; **Poll** large in relation to the **Cn**, ± 0.5 × 0.2 mm, elongate-ovate, apically transparent, caudicles short, broad, corpuscle black; **Fr** fusiform, spreading; **Se** broadly ovate, broadly winged, pale brown, ± 6 × 4 mm; **Ha** crown 25 - 30 mm.

A fairly isolated genus with a single species only. Its position within the *Ceropegieae* is difficult to ascertain; similarities exist primarily with *Caralluma socotrana*.

E. grandis N. E. Brown (BMI 1895: 220, 1895). **T:** Somalia (*Cole* s.n. [K ?]). – **D:** Ethiopia, Somalia, Kenya, Uganda, Tanzania, Socotra.
  Incl. *Edithcolea grandis* var. *baylissiana* Lavranos & D. S. Hardy (1963); **incl.** *Edithcolea sordida* N. E. Brown (1903).

Description as for the species.

FANNINIA

U. Meve

Fanninia Harvey (Gen. South Afr. Pl., ed. 2, 325, 1868). **T:** *Fanninia caloglossa* Harvey. – **D:** RSA. **Etym:** For George Fannin, owner of the farm in RSA where the taxon was discovered.
  Incl. *Panninia* Baillon (1888) (*nom. inval.*, Art. 61.1).

Geophytic perennials with deciduous herbaceous above-ground parts, villous; **R** tuberously thickened; stems erect, sparsely leafy; **L** decussate, ± sessile, lanceolate-elliptic, 3 - 5 × 1 - 2 cm; **Inf** terminal extra-axillary pedunculate 4- to 8-flowered pseudo-umbels; **Ped** filiform; **Fl** ± nutant; **Cl** rotate; **Cl** lobes only slightly fused basally, oblong-elliptic, 10 - 12 mm, very delicate, whitish with purple or green midrib, inside and outside long white-hairy; **Cn** uniseriate; **Cs** lobes erect, 7 - 9 mm, reddish to purple, flat, lingulate, apically ± crenate, midrib thickened, basally-laterally with erect subulate processes; **Gy** stipitate; **Anth** appendages fimbriate.

A monotypic genus from the broader relationship around *Asclepias*. The corolla lobes, which are long-hairy on both faces, are characteristic, as well as the long-stipitate gynostegium and the erect lingulate corona segments.

F. caloglossa Harvey (Gen. South Afr. Pl., ed. 2, 325, 1868). **T:** RSA, KwaZulu-Natal (*Anonymus* s.n. [not located]). – **D:** RSA (KwaZulu-Natal, Eastern Cape); grasslands. **I:** Hilliard & Burtt (1987: t. 24E).

Description as for the genus.

FOCKEA

U. Meve

Fockea Endlicher (in Endlicher & Fenzl, Nov. Stirp. Dec. 3: 17, 1839). **T:** *Fockea capensis* Endlicher. – **Lit:** Marloth (1913-1932); Court (1987). **D:** S Africa. **Etym:** For Charles Focke (1802 - 1856), Dutch botanist, collecting esp. in Surinam.
  Incl. *Chymocormus* Harvey (1842). **T:** *Pergularia edulis* Thunberg.

Perennial slightly woody geophytes with **R** tubers, latex white; stems erect, sometimes decumbent, mostly twining, terete, 1 - 8 mm ⌀ (to the thickness of an arm in *F. multiflora*), normally perennial; **L** sessile to shortly petiolate, linear-elliptic to broadly ovate, short-lived or perennial, stipular rudiments mostly present, subulate, membranous; **Inf** extra-axillary, lateral, sessile or shortly pedunculate, several- to many-flowered, clustered in pseudo-umbels; **Ped** short; **Sep** triangular to lanceolate; **Cl** contorted in bud, ± 5 - 15 mm long, with a cylindrical to campanulate **Cl** tube shorter than the **Cl** lobes; **Cl** lobes ± contorted, inside densely papillose or pubescent, yellow or green, margins revolute; **Cn** biseriate, sessile, ± cylindrically fused, with copious nectar, longitudinally grooved or caniculate; **Ci** united fringe-like, margin ± toothed; **Cs** differentiated into a 3-lobed outer appendage ± enclosing a more deeply inserted inner appendage (= ligula); **Gy** sessile; **Anth** triangular to ovoid, erect, apically with an erect ovoid sterile appendage, appendages hyaline, inflated, considerably overtopping and covering the **Sty** head; pollinaria without caudicles, **Poll** directly inserted at the corpuscle, ovate to broadly ellipsoid, 0.2 - 0.3 mm long, flat; **Fr** solitary, fusiform, rostrate, smooth, glabrous, sometimes keeled; **Se** ellipsoid to ovate, wingless or rarely winged.

Fockea and the closely related genus *Cibirhiza* are isolated and basal groups within the *Asclepiadoideae* and belong to the separate tribe *Fockeeae* (Kunze & al. 1994). Both form root tubers or caudices, have a complicate corona structure (see generic description), large anther appendages and pollinia directly inserted at the corpuscle. The last-mentioned character points to a narrow relationship with subfamily *Secamonoideae*.

The 6 species of *Fockea* can be simply distinguished on the base of vegetative characters (tuber, leaves).

F. angustifolia K. Schumann (BJS 17: 146, 1893). **T:** RSA, Northern Cape (*Marloth* 1008 [not located]). – **D:** Kenya, Tanzania, Namibia, Zambia, Botswana, Zimbabwe, RSA (Free State, Northern Prov., North-West Prov., Northern Cape, KwaZulu-Natal, Mpumalanga), Swaziland. **I:** FPA 43: t. 1711, 1976.
  Incl. *Fockea sessiliflora* Schlechter (1895); **incl.** *Fockea lugardii* N. E. Brown (1903); **incl.** *Fockea*

dammarana Schlechter (1905); **incl.** *Fockea mildbraedii* Schlechter (1908); **incl.** *Fockea tugelensis* N. E. Brown (1908); **incl.** *Fockea monroi* S. Moore (1914); **incl.** *Cynanchum omissum* Bullock (1956); **incl.** *Fockea angustifolia* var. *volkii* G. Court (1987) (*nom. inval.*, Art. 34.1c).

Stems erect to decumbent or twining, 30 - 150 cm; underground **R** tuber globose to turnip-shaped or irregularly shaped, to 40 × 25 cm; **Br** 2 - 5 mm ⌀, somewhat woody, bark somewhat rough, young pubescent; **L** ± sessile, linear to elliptic, 1 - 10 × 0.4 - 1.2 cm, ± pubescent; **Inf** 3- to 6-flowered; **Ped** 0.5 - 3 mm, pubescent; **Sep** triangular, ± 2 mm, pubescent; **Cl** greenish to brownish-yellow, outside tomentose, inside mostly glabrous; **Cl** tube campanulate, 1 - 1.5 mm; **Cl** lobes linear-lanceolate, 7 - 15 × 1.5 - 2.5 mm; **Cn** white to yellowish, **Cn** tube 3 - 4 mm, central appendage of the outer **Cs** lobes erect to reflexed, 2 - 2.5 mm, subulate, inner **Cs** lobes ± 1 mm; **Fr** ± 8 × 0.8 cm; **Se** 10 × 3 mm.

A widespread and polymorphic taxon, esp. in respect to leaf morphology. Its habit is always purely geophytic, and the tubers are always placed underground. Most closely related to *F. comaru* (see also there). Material encountered in cultivation under the name *F. tugelensis* (here treated as synonym) is normally identifiable as *Petopentia natalensis*.

F. capensis Endlicher (in Endlicher & Fenzl, Nov. Stirp. Dec. 3: 17, t. 91, 1839). **T:** [icono]: l.c. t. 91. – **D:** RSA (Western Cape, Eastern Cape); Karoo.

≡ *Fockea edulis* var. *capensis* (Endlicher) G. D. Rowley (1998); **incl.** *Cynanchum crispum* Jacquin (1800) (*nom. illeg.*, Art. 53.1) ≡ *Fockea crispa* (Jacquin) K. Schumann (1895); **incl.** *Brachystelma crispum* E. Meyer (1837) (*nom. illeg.*, Art. 53.1).

Stems 20 - 60 cm, divaricately branched, hardly twining; **R** tuber partly above-ground, turnip-shaped, to ± 50 × 30 cm, bark grey-brown, strongly pustulate; **Br** 2 - 4 mm ⌀, somewhat woody, pubescent to densely felty-hairy; **L** shortly petiolate, ± elliptic and obtuse, 8 - 25 × 5 - 14 mm, coriaceous, strongly undulate, both faces ± densely pubescent; **Inf** 3- to 5-flowered; **Ped** 2 - 5 mm, pubescent; **Sep** lanceolate, ± 2 mm, pubescent; **Cl** yellow-green, outside and inside pubescent; **Cl** tube campanulate, 2 - 3 mm; **Cl** lobes linear-lanceolate, ± 7 - 11 × 2 mm, lamina folded back along the midrib; **Cn** white; **Cn** tube 3 - 4 (-5) mm, central appendage of the outer **Cs** 1 - 2 mm, subulate, inner **Cs** lobe ± 2 mm, lanceolate-subulate.

F. crispa is very similar to *F. edulis*. It can be easily distinguished, however, on the base of the dense hairiness and the undulate leaves. *F. crispa* can be interpreted as the vicariant Karoo species versus the coastal *F. edulis*.

F. comaru (E. Meyer) N. E. Brown (FC 4(1): 781-782, 1908). **T:** RSA, Eastern Cape (*Drège* s.n. [B †]). – **D:** RSA (Western Cape, Eastern Cape). **I:** Rowley (1987: 70, 78, as *Microloma*).

≡ *Brachystelma comaru* E. Meyer (1837); **incl.** *Fockea gracilis* R. A. Dyer (1933).

Stems erect to decumbent, hardly twining; **R** tuber mostly underground, turnip-shaped or irregularly shaped, 10 - 25 cm ⌀; **Br** 10 - 30 cm, ± 2 - 5 mm ⌀, basally becoming woody, young pubescent; **L** ± sessile, linear-subulate, 20 - 30 × 1 - 2 mm, margins revolute, delicately pubescent; **Inf** ± sessile, 2- to 6-flowered; **Ped** 1 - 2.5 mm, pubescent; **Sep** narrowly lanceolate, ± 2 mm, pubescent; **Cl** green-yellow, green-brown to brown, outside pubescent, inside mostly glabrous; **Cl** tube campanulate, 3 - 4 mm; **Cl** lobes linear, 5 - 10 × 1 mm; **Cn** white; **Cn** tube 4.5 - 5.5 mm, central appendage of the outer **Cs** lobes reflexed, 1.5 - 2 mm, lanceolate, inner **Cs** lobes hardly 1 mm.

Most closely related to *F. angustifolia* and perhaps only representing the karroid ecotype within this complex (Court 1987).

F. edulis (Thunberg) K. Schumann (BJS 17: 146, 1893). **T:** RSA, Western Cape (*Thunberg 6227* [UPS]). – **D:** RSA (Western Cape, Eastern Cape), Swaziland; almost exclusively near coasts. **I:** Eggli (1994). **Fig. XXII.b**

≡ *Pergularia edulis* Thunberg (1794) ≡ *Chymocormus edulis* (Thunberg) Harvey (1842); **incl.** *Ceropegia suberosa* Spv. (s.a.) (*nom. inval.*, Art. 29.1); **incl.** *Brachystelma macrorrhizum* E. Meyer (1837); **incl.** *Fockea glabra* Decaisne (1844); **incl.** *Fockea cylindrica* R. A. Dyer (1933).

Stems climbing and twining to 2 m, ± glabrous; **R** tuber under- or halfway above-ground, massive, turnip-shaped or irregularly shaped, to 1 × 0.5 (-1) m, edible, bark grey-brown, scatteredly papillate-pustulate; **Br** 2 - 5 mm ⌀, somewhat woody, young pubescent; **L** ± sessile to shortly petiolate, often densely clustered on short shoots, ovate-lanceolate or elliptic, obtuse or acute, 1 - 4 × 0.5 - 1.5 cm, coriaceous, often slightly undulate, glabrous; **Inf** very shortly pedunculate, 2- to 6-flowered; **Ped** 2 - 6 mm, pubescent; **Sep** lanceolate, ± 2 mm, pubescent; **Cl** pale to green-yellow, outside pubescent, inside glabrous or velvety-hairy; **Cl** tube campanulate, 2 - 4 mm; **Cl** lobes linear-lanceolate, 8 - 12 × 2 mm; **Cn** white; **Cn** tube ± 4 mm, outer **Cs** ± divergent backwards, central appendage 2 - 3 mm, subulate, inner **Cs** lobe ± 3 mm, subulate, erect.

This species is widely cultivated. Cultivation is easy and the plants are floriferous.

F. multiflora K. Schumann (BJS 17: 145-146, 1893). **T:** Tanzania (*Stuhlmann 848* [B †?]). – **D:** Angola, Botswana, Moçambique, Kenya, Tanzania, Zambia, Namibia, Zimbabwe. **I:** Schumann (1895: 294).

Incl. *Fockea schinzii* N. E. Brown (1895).

Stems climbing and twining to 10 m tall, basally

with massive **R** tuber, tuber irregularly shaped, partly above-ground, mostly tapering into ascending twining stems to the thickness of an arm, bark brown, ± smooth; flowering **Br** 7 - 10 mm ⌀, fleshy, becoming woody, young tomentose; **L** shortly petiolate, often densely clustered on short shoots, ovate to elliptic, ± acute, 2 - 6 × 1 - 3 cm, ± hairy, lower face white-felted with raised midrib; **Inf** 1 - 2 cm pedunculate, many-flowered; **Ped** 1 - 2 cm, densely hispid; **Sep** lanceolate, ± 2.5 mm, hispid; **Cl** yellow to green, tube campanulate, 1.5 - 2.5 mm; **Cl** lobes oblong-ovate, ± 6 - 10 × 2 mm, outside glabrous, inside pubescent; **Cn** white; **Cn** tube ± 3 mm, central appendage of the outer **Cs** erect to reclining, ± 3 mm, narrowly triangular-subulate, inner **Cs** lobes ± 2 mm, subulate, erect.

A widespread and sometimes frequent species in Mopane or *Brachystegia* woodland. The tubers are extremely diverse in shape. The leaves with white-felted lower faces are significant for this species.

F. sinuata (E. Meyer) Druce (Bot. Soc. Exch. Club Brit. Isles 1916: 623, 1917). **T:** RSA, Eastern Cape (*Drège* s.n. [B †?]). – **D:** Namibia, RSA (Northern Cape, Western Cape, Eastern Cape, Free State). **I:** Rowley (1987: 70).

≡ *Brachystelma sinuatum* E. Meyer (1837); **incl.** *Fockea undulata* N. E. Brown (1895).

Stems erect, not twining, few-branched; underground **R** tuber turnip-shaped, 20 - 35 × 10 - 20 cm; stems 5 - 20 cm × 2 - 5 mm, basally becoming woody, pubescent; **L** sessile, linear, 2 - 3.5 cm × 2 - 3 mm, longitudinally regularly undulate, upper face pubescent, lower face glabrous except the midrib, margins mostly not revolute; **Inf** sessile, 3- to 6-flowered; **Ped** 1 - 2.5 mm, pubescent; **Sep** lanceolate, 2.5 mm, pubescent; **Cl** green-brown, outside pubescent, inside glabrous; **Cl** tube campanulate, ± 2.5 mm; **Cl** lobes linear-lanceolate, 4 - 6 × 2 mm, ascending; **Cn** white; **Cn** tube ± 3 mm, central appendage of the outer **Cs** lobes spreading backwards, ± 3 mm, lanceolate, inner **Cs** lobe subulate; **Se** without clear tuft of **Ha** but hairy along the entire margin.

This is the smallest species of the genus, easily recognizable by the extremely undulate leaves, the small flowers and the seeds which are hairy along the entire margin.

GLOSSOSTELMA

F. Albers

Glossostelma Schlechter (BJS 33: 321, 1895). **T:** *Glossostelma angolense* Schlechter. – **Lit:** Bullock (1952); Goyder (1995). **D:** S tropical Africa from Angola and Zaïre to Moçambique and S Tanzania. **Etym:** Gr. 'glossa', tongue; and Gr. 'stelma', crown, garland, wreath; for the structure of the corona.

Slender or robust erect perennial herbs with a short, often stout, vertical rhizome and several fusiform tuberous lateral **R**; stems usually solitary, unbranched, uniseriately pubescent, rarely glabrous; **L** opposite, petiolate, slightly fleshy, lamina linear to ovate or spatulate; **Inf** extra-axillary, terminal or near the stem tips, umbellate, with or without peduncle; **Cl** campanulate, deeply lobed; **Cn** uniseriate, staminal; **Cs** lobes often fleshy, sometimes dorsally flattened, united shortly at the base and attached to the **Gy** near the base of the **Anth**; **Gy** usually with a conspicuous stalk; **Anth** with sterile appendages; **Poll** pendent, flattened, corpuscle ovoid, brown or black, caudicles flattened, broadened and geniculate near the attachment of the **Poll**, held ± at right angles to the axis of the corpuscle; **Fr** erect, usually only 1 of a pair developing, lanceolate or ovate-lanceolate, with an attenuate tip, smooth, glabrous or pubescent; **Se** flattened, ovate to suborbicular.

Glossostelma species often have large attractive flowers. The genus is closely related to *Pachycarpus*. *Glossostelma* is generally shorter and less robust in habit than *Pachycarpus*. A subhispid indumentum is common on the leaves of *Pachycarpus* but normally absent in *Glossostelma*. Although the coronal structure is diverse in *Glossostelma*, the typical *Pachycarpus* corona, consisting of a horizontal plate spreading out of the gynostegium, with or without appendages, is not known to occur in the former. All 12 species of the genus are here treated.

The following names are of unresolved application but are referred to this genus: *Schizoglossum lividiflorum* K. Schumann (1900); *Schizoglossum macroglossum* K. Schumann (1903); *Schizoglossum violaceum* K. Schumann (1893).

G. angolense Schlechter (J. Bot. 33: 322, t. 352B, 1895). **T:** Angola (*Welwitsch* 4190 [BM]). – **D:** Angola; known only from the type.

Incl. *Xysmalobium grande* N. E. Brown (1902).

Erect herbs to 1 m; **L** petiolate, petiole 12 - 15 mm, lamina ± 10 × 5.5 - 6.2 cm, broadly spatulate, glabrous, with closely parallel lateral veins at ± 90° to the prominent midrib; **Inf** with 4 - 6 **Fl**; **Ped** 18 - 23 mm; **Sep** 6 - 7 × 4 - 5 mm, broadly ovate, obtuse or rounded; **Cl** green tinged with purple, lobes 15 - 18 × 7 - 9 mm, ovate-oblong, rounded at the tip; **Cs** lobes 0.5 × 2 - 3 mm, oblong with rounded tip, dorsally compressed but slightly fleshy, erect, almost reaching the top of the **Anth**; **Anth** appendages conspicuous, 2 - 3 mm, broadly ovate; **Fr** not known.

Closely related to *G. spathulatum*. The most obvious difference is the form of the corona lobes, which are dorsally compressed in *G. angolense* rather than subglobose.

G. brevilobum Goyder (KB 50(3): 551-553, ills., 1995). **T:** Malawi (*Pawek* 5954 [K]). – **D:** Zaïre,

Burundi, Tanzania, Malawi. **Fig. XXII.a, XXII.c, XXII.e**

Erect herbs to 12 - 40 cm; **L** sessile or petiolate, lower **L** 1.3 - 2 × 0.4 - 1 cm, upper **L** 3 - 7 × 1 - 3.6 cm, spatulate, obovate or narrowly oblong to ovate, tip rounded, obtuse or subacute, apiculate, glabrous or sparsely hairy on the lower face, midrib channelled on the upper face; **Inf** with up to 9 **Fl**; **Ped** 8 - 13 mm, pubescent on one side; **Sep** 2 - 3 × 1 - 1.5 mm, ovate or triangular, acute; **Cl** green or brown, campanulate, lobes 5 - 8 × 4 - 5 mm, ovate, tip acute or subacute, glabrous; **Cs** lobes 0.5 × 1 mm, forming fleshy outward-pointing pegs shortly united into an annulus at the base; **Anth** appendages ± 1.5 × 2 mm, broadly ovate; **Poll** obtriangular, ± 0.75 × 0.5 - 0.6 mm, corpuscle 0.5 - 0.8 mm, ovoid, brown, caudicles 0.4 - 0.7 mm; **Fr** not known.

G. cabrae (De Wildeman) Goyder (KB 50(3): 541, ills. (pp. 548-549), 1995). **T:** Zaïre (*Cabra-Michel 52* [BR]). – **D:** Zaïre, Tanzania, Angola, Zambia, Moçambique.

≡ *Asclepias cabrae* De Wildeman (1904); **incl.** *Xysmalobium speciosum* S. Moore (1909).

Erect herbs to 50 cm; **L** petiolate, lower **L** 1.5 × 0.2 cm, upper **L** 3.5 - 8 × 1.5 - 3.1 cm, oblong, obovate or rarely linear, tip obtuse, rounded or retuse, glabrous or with short sparse **Ha** on the lower face, midrib channelled on the upper face; **Inf** with 2 - 4 **Fl**; **Ped** 18 - 25 mm, pubescent on one side; **Sep** 5 - 6 × 1 - 1.5 mm, triangular, acute; **Cl** green or cream with reddish-brown markings within, campanulate, lobes 23 - 26 × 9 - 14 mm, obovate, tip acute or obtuse, recurved, glabrous except on the margins; **Cs** lobes green, 2 × 9 - 11 mm, lower ½ somewhat compressed laterally, falcate, arched over the **Sty** head and narrowed into the base of the erect, dorsally flattened, slightly fleshy, spatulate upper ½, tip rounded or truncate, with 1 to occasionally 2 narrowly triangular downward-pointing flaps up to 2 mm long on the ventral face; **Anth** appendages ± 1.5 × 2 mm, broadly ovate; **Poll** ± 1 × 0.25 mm, falcate-oblong, corpuscle ± 0.3 mm, caudicles ± 0.5 mm; **Fr** not known.

Closely related to both *G. ceciliae* and *G. spathulatum* but to be distinguished by the shape of the corona lobes.

G. carsonii (N. E. Brown) Bullock (KB 7: 415, 1952). **T:** Zambia (*Carson s.n.* [K]). – **D:** Zaïre, Rwanda, Burundi, Tanzania, Angola, Zambia, Zimbabwe, Malawi, Moçambique. **I:** Plowes & Drummond (1976); Goyder (1995).

≡ *Xysmalobium carsonii* N. E. Brown (1895) ≡ *Schizoglossum carsonii* (N. E. Brown) N. E. Brown (1902) ≡ *Asclepias carsonii* (N. E. Brown) Schlechter (1916); **incl.** *Gomphocarpus chlorojodinus* K. Schumann (1901) ≡ *Schizoglossum chlorojodinum* (K. Schumann) N. E. Brown (1902); **incl.** *Schizoglossum kassneri* S. Moore (1912).

Erect herbs to 25 - 85 cm; **L** linear or narrowly linear-lanceolate, lowest **L** 1 - 6 cm, upper **L** 5.5 - 20 cm, tip acute, glabrous; **Inf** with 3 - 5 (-9) spreading or erect **Fl**; **Ped** 10 - 20 mm, pubescent on one side; **Sep** ovate-triangular, acute; **Cl** very variable in colour, usually green or brown with purple stippling, stippling denser on the outside, but sometimes entire **Cl** cream or maroon, campanulate, lobes 12 - 17 × 4.5 - 11 mm, obovate or obovate-oblong, tip subacute but appearing truncate because of the recurved apex, glabrous except the papillose apical region and margins; **Cs** lobes creamy-white, orange-yellow or occasionally suffused with purple, dorsally compressed, thin, erect, oblong or obovate-oblong, tip rounded, subacute or variously toothed, apical portion erect or inflexed over the **Sty** head for up to ½ of its length, upper margins reflexed or not, basal portion of the lobes with incurved margins, glabrous or minutely verrucate-papillate; **Anth** appendages 1 - 2 mm, sometimes toothed at the margin; **Poll** 1 × 0.3 mm, falcate, attached on the caudiculum near the tip, corpuscle 0.3 mm, caudicles 0.3 mm; **Fr** 11 - 13 × 1.1 cm, glabrous or whitish-pubescent.

This is the most common and widespread species of the genus.

G. ceciliae (N. E. Brown) Goyder (KB 50(3): 541, ills. (pp. 547-548), 1995). **T:** Zimbabwe (*Cecil 60* [K]). – **D:** Zaïre, Tanzania, Angola, Zambia, Zimbabwe.

≡ *Xysmalobium ceciliae* N. E. Brown (1902).

Robust erect herbs to 20 - 90 cm; **L** petiolate or sessile, lower **L** 2.5 - 3.5 × 0.7 - 0.9 cm, upper **L** 5 - 8.5 × 0.9 - 2.1 cm, narrowly oblong or oblanceolate, tip acute, obtuse or rounded, glabrous, midrib channelled on the upper face; **Inf** with (1-) 3 - 5 **Fl**; **Ped** 16 - 25 mm, pubescent on one side; **Sep** 6 - 10 × 2 - 4 mm, ovate or triangular, acute; **Cl** green or cream within, outside tinged brown or reddish-brown, campanulate, lobes 23 - 28 × 9 - 19 mm, obovate, tip acute or obtuse, recurved, glabrous; **Cs** lobes green or white, 7 - 10 mm, 3 - 4 mm wide at the slightly swollen base, laterally compressed but not flattened above, falcate, upper ½ arching over the **Sty** head, dilated slightly at the truncate-clavate tip; **Anth** appendages 2 - 3 × 2 - 3 mm, broadly ovate; **Poll** 1.25 × 0.5 mm, falcate-oblong, corpuscle ± 0.3 mm, ovoid, translator arms ± 0.5 mm; **Fr** 11 - 24 × 1 - 1.3 cm, lanceolate to long-fusiform, attenuate at both ends, puberulent; **Se** ± 5 × 4 mm, ovate.

G. erectum (De Wildeman) Goyder (KB 50(3): 545, 1995). **T:** Zaïre (*Gillet s.n.* [BR]). – **D:** Zaïre; known from only 2 collections.

≡ *Asclepias erecta* De Wildeman (1904).

Erect herbs to 20 - 35 cm; **L** petiolate, lowest **L** 2 × 0.6 cm, upper **L** 3.5 - 5.5 × 0.7 - 1 cm, oblong or oblong-elliptic, tip acute or obtuse, glabrous except the midrib; **Inf** subsessile, with 4 - 7 **Fl**; **Ped** 7 - 10 mm, pubescent on one side; **Sep** 4 × 1 mm, narrowly triangular, acute; **Cl** green, campanulate, lobes 6 - 8 × 3 mm, ovate, acute; **Cs** lobes 2.5 × 1 mm, base ovoid but with flattened inner face, tapering above into an erect or inflexed fleshy tongue which slightly overtops the staminal column, **Cs** lobes linked at the base by a minute pair of acute teeth; **Poll** ± 0.8 × 0.3 mm, falcate-oblong, corpuscle ± 0.5 mm, ovoid, brown, caudicles 0.4 - 0.5 mm, very slender near the corpuscle, broader and contorted distally; **Fr** not known.

G. lisianthoides (Decaisne) Bullock (KB 7: 416, 1952). **T:** Angola (*Anonymus* s.n. [P, K]). – **D:** Gabon, Zaïre, Angola, Zambia, Malawi.

≡ *Gomphocarpus lisianthoides* Decaisne (1838) ≡ *Asclepias lisianthoides* (Decaisne) N. E. Brown (1902); **incl.** *Gomphocarpus chironioides* Decaisne (1844); **incl.** *Xysmalobium dissolutum* K. Schumann (1893) ≡ *Asclepias dissoluta* (K. Schumann) Schlechter (1900); **incl.** *Xysmalobium fritillarioides* Rendle (1894); **incl.** *Asclepias congolensis* De Wildeman (1904); **incl.** *Asclepias nemorensis* S. Moore (1909).

Slender erect herbs to 70 cm; **L** linear, tip acute, glabrous; **Inf** with 2 - 4 nodding **Fl**; **Ped** (10-) 16 - 24 (-28) mm, glabrous or pubescent on one side only; **Sep** narrowly triangular, acute; **Cl** greenish-yellow or cream within, outside suffused with pink or brown, spreading-campanulate, lobes 10 - 13 × 5 - 7 mm, ovate-oblong, obtuse; **Cs** lobes 4 - 5 mm, lower ½ of the lobes bulbous and ± orbicular when fresh (appearing channelled on the inner surface in dried material), upper ½ of the lobes narrowed abruptly into an oblong tongue, truncate or shortly bifid at the tip and inflexed over the **Sty** head; **Anth** appendages membranous, ovate; **Poll** 0.7 × 0.3 mm, apically attached on the caudicle, corpuscle 0.3 mm, caudicles 0.3 mm, extremely slender; **Fr** 8 - 11 × 0.8 - 1.2 cm, glabrous or puberulent; **Se** ovate.

Closely related to *G. carsonii*. Dried material is very difficult to distinguish. See Goyder (1995) for nomenclatural notes.

G. mbisiense Goyder (KB 50(3): 543-544, ills., 1995). **T:** Tanzania (*Sanane* 1426 [K]). – **D:** Tanzania (known only from the Mbisi Forest and Sumbawanga).

Erect herbs 15 - 35 cm tall; **L** petiolate or subsessile, petiole to 6 mm, lamina (3.5-) 5 - 7 × 2.1 (-2.6) cm, lanceolate-oblong to ovate-oblong, tip acute or subacute, slightly fleshy; **Inf** with 2 - 6 **Fl**; **Ped** 10 - 18 mm, pubescent; **Sep** 5 - 7 × 1.5 mm, oblong or lanceolate to triangular; **Cl** green or cream, lobes 13 - 20 × 4 - 7 mm, ovate or oblong, tips sometimes recurved; **Cs** lobes 1.5 - 2 × 1 - 1.5 mm, swollen at the base, with a slender horn at the top of the inner face, projecting upwards or inwards for up to 1 mm, tip of the **Cs** ± at the same level as the tip of the **Anth** or sometimes slightly shorter; **Poll** ± 0.7 × 0.5 mm, broadly oblong but distally rounded, corpuscle ± 6 mm, ovoid, black, caudicles ± 0.8 mm, slender, flattened, contorted distally; immature **Fr** densely rufous-pubescent, mature **Fr** not known.

G. nyikense Goyder (KB 50(3): 545-547, ills., 1995). **T:** Malawi (*Robson* 452 [K, LISC, PRE, SRGH]). – **D:** Malawi, Zambia.

Erect herbs with 1 or 2 stems, 10 - 20 cm tall; **L** sessile, 2.5 - 7 × 0.2 - 0.3 cm, linear, tip acute, slightly fleshy, lateral veins obscure; **Inf** with 4 - 10 **Fl**; **Ped** 5 - 12 mm, pubescent on one side; **Sep** 2 - 3 × 1 mm, triangular-ovate, tip attenuate; **Cl** greenish or brown, lobes ± 5 × 2 mm, ovate-oblong, tip acute or subacute; **Cs** lobes ± 2 mm, ovoid, with no teeth or projections on the inner face, slightly exceeding the top of the head of the **Anth**; **Poll** 0.5 × 0.3 mm, broadly oblong and basally rounded, corpuscle ± 0.4 mm, subcylindrical, brown, caudicles ± 0.3 mm, flattened and contorted distally; very immature **Fr** densely rufous-pubescent, mature **Fr** not known.

G. rusapense Goyder (KB 50(3): 540-543, ills., 1995). **T:** Zimbabwe (*Drummond* 5069 [K, SRGH]). – **D:** Zimbabwe.

Erect herbs to 20 cm; **L** petiolate or subsessile, petiole to 2 mm, lamina 2.5 - 5 × 0.5 - 1 cm, lanceolate or oblong, tip acute, glabrous except for the midrib and margins beneath; **Inf** with 2 - 4 **Fl**; **Ped** 8 - 17 mm; **Sep** ± 4 × 1.5 mm, ovate or triangular; **Cl** green speckled with reddish-brown markings within, campanulate, lobes 13 × 5 - 7 mm, glabrous except for the puberulous tip and margins; **Cs** lobes ± 2 × 5 mm, swollen and appearing slightly pouched at the base, with 2 short lateral teeth on the inner surface, upper part narrowed gradually into a dorsally flattened tongue arched over the **Sty** head; **Anth** appendages ± 1.5 × 2 mm, broadly ovate; **Poll** ± 1 × 0.5 mm, falcate-oblong, corpuscle ± 0.5 mm, ovoid, black, caudicles 0.5 cm, broadly flattened; **Fr** not known.

G. spathulatum (K. Schumann) Bullock (KB 7: 414, 1952). **T** [lecto]: Angola (*Mechow* 539a [K]). – **D:** Zaïre, Tanzania, Angola, Zambia, Zimbabwe, Malawi, Moçambique. **I:** Cribb & Leedal (1982); Goyder (1995).

≡ *Schizoglossum spathulatum* K. Schumann (1893) ≡ *Gomphocarpus spathulatus* (K. Schumann) Schlechter (1895) ≡ *Xysmalobium spathulatum* (K. Schumann) N. E. Brown (1902); **incl.** *Xysmalobium bellum* N. E. Brown (1895).

Erect herbs to 50 cm; **L** petiolate or occasionally sessile, petiole 5 - 15 mm, lamina spatulate, obovate, oblanceolate, lanceolate or oblong, lowest **L** 2 - 3.5 × 0.3 - 1.2 cm, upper **L** 2.5 - 8 × 2 - 4 cm, obtuse, round or retuse, glabrous or sparsely hairy beneath, midrib channelled on the upper face, lateral veins clearly visible, parallel to each other and ± at 90° to the midrib; **Inf** with 2 - 4 (-8) **Fl**; **Bra** often of 2 types, one filiform, the other narrowly triangular with ciliate margins; **Ped** 11 - 25 mm, pubescent on one side; **Sep** 6 - 10 × 4 - 5 mm, ovate, acute or subacute; **Cl** green or maroon with reddish-brown, yellow or white markings within, campanulate, lobes 13 - 20 × 8 - 14 mm, obovate, tip rounded or obtuse, recurved, glabrous; **Cs** lobes white or yellowish, ± 3 × 2 mm, reaching the top of the staminal column or exceeding it slightly, subglobose but with flattened inner face and 3 short inward-pointing teeth near the top (see Goyder (1995) for variation of the teeth); **Anth** appendages ± 1.5 × 2 mm, broadly ovate; **Poll** 0.7 - 1 × 0.5 mm, falcate-oblong, corpuscle 0.5 mm, caudicles 0.5 mm; **Fr** 7 × 2 cm, ovate-lanceolate, puberulent; **Se** suborbicular.

G. xysmalobioides (S. Moore) Bullock (KB 7: 417, 1952). **T:** Angola (*Gossweiler* 4009 [BM]). – **D:** Angola; known only from the type.

≡ *Asclepias xysmalobioides* S. Moore (1912).

Erect herbs to 30 cm, stem shortly pubescent with longer **Ha** in stripes; **L** petiolate, lamina ovate, ± 3.5 - 4 × 2.5 - 3 cm, tip obtuse to subacute, glabrous; **Inf** with ± 4 **Fl**; **Ped** 7 - 8 mm; **Sep** 5 mm, lanceolate, acute; **Cl** probably green with red streaks, campanulate, lobes ± 14 × 6.5 mm, ovate-oblong, obtuse; **Cs** lobes 8.5 mm, consisting of an erect ligulate portion ± 6.5 mm long with an obtuse tip, and a pair of rounded lateral lobes ± 2.5 mm long forming a cucullate basal pouch; **Poll** 1 mm, oblong pear-shaped, corpuscle ± 1 mm, narrowly oblong-ovate; **Fr** not known.

HOODIA

B. Müller & F. Albers

Hoodia Sweet *ex* Decaisne (PSRV 8: 664, 1844). **T:** *Stapelia gordonii* Masson. – **Lit:** Plowes (1992); Bruyns (1993). **D:** SW Angola, NW Namibia (border regions of the Namib Desert), RSA, Botswana, Zimbabwe. **Etym:** For a Mr. Hood, succulent plant grower in England around 1830.

Incl. *Monothylaceum* G. Don (1838) (*nom. inval.*, Art. 32).

Incl. *Scytanthus* Hooker (1844). **T:** *Scytanthus currorii* Hooker.

Incl. *Trichocaulon* N. E. Brown (1878). **T:** *Trichocaulon flavum* N. E. Brown [Lectotype, designated by P. V. Bruyns, Bot. Jahrb. Syst. 115(2): 210, 1993.].

Basally branched and often lignified stem succulents with few to many erect (rarely procumbent) cylindrical stems; stems 0.3 - 2.2 m tall, 2.5 - 6 (-11) cm ⌀; podaria in 11 - 31 blunt vertical **Ri**, apically tapering into a soft or hard **Sp**, glabrous; **Inf** of single or numerous **Fl** opening successively (rarely simultaneously); **Ped** 0.05 mm to 6 cm, ± 1 - 4 mm ⌀, glabrous; **Sep** 5, overlapping at the broadened bases, ovoid-lanceolate to deltoid, tapering, glabrous; **Cl** 0.8 - 17 cm ⌀, small and deeply divided into single **Cl** lobes, or large and dish-like to shallowly cup-shaped, outside glabrous and smooth, inside glabrous, papillate or softly hairy, annulus absent, rudimentary or conspicuous; **Cn** biseriate, mostly glabrous; **Ci** basally cup-shaped with apically retuse or ± deeply divided appendages; **Cs** of dorsiventrally flattened lobes, placed against the back of the **Anth**, dorsally fused with the **Ci**; staminal column rising close to the base of the **Cl** tube; **Anth** incumbent on the **Sty** head, almost square; **Sty** head not prolonged beyond the **Anth**, apically convexly compressed; **Poll** ± ascending-horizontal, with short caudicles; **Fr** acutely tapering to fusiform, slender, the 2 follicles of a pair spreading at an angle of 30 - 60°, uniformly reddish-green, glabrous, smooth; **Se** round or pear-shaped, ventrally ± concave, winged, brittle, apically with a tuft of simple **Ha**.

Bruyns (1993) divides the genus into 2 sections:
[1] Sect. *Hoodia*: **Ped** 2 - 4 mm ⌀; **Cl** never dark maroon to purplish-black, (2-) 2.5 - 17 cm ⌀, mostly dish-like to broadly and shallowly cup-shaped, rarely funnel-shaped (*H. parviflora*), with a central depression just enclosing the **Gy**; **Cl** lobes ovoid-deltoid, more than 2× as broad as long (excl. the narrow tip) and shorter than the width of the fused **Cl** area outside the central depression.
[2] Sect. *Trichocaulon* (N. E. Brown) Bruyns 1993: **Ped** to 1 mm ⌀; **Cl** 0.8 - 2 cm ⌀, or larger (2 - 4 cm ⌀) and then dark maroon to purplish-black, when dish-like without central depression; **Cl** lobes < 2× as broad as long and much longer than the width of the fused **Cl** area outside the central depression.

Hybrids are known with *Orbea* (see ×*Hoodialluma*) and with *Tromotriche* (see there).

The following names are of unresolved application but are referred to this genus: *Hoodia similis* Dinter (1928) (*nom. inval.*); *Trichocaulon karasmontanum* Dinter (1927) (*nom. inval.*).

H. alstonii (N. E. Brown) Plowes (Asklepios No. 56: 7, 1992). **T:** RSA, Northern Cape (*Alston* s.n. in *MacOwan* 2017 [K, SAM]). – **D:** S Namibia, RSA (Northern Cape); winter-rainfall regions. **Fig. XXII.d**

≡ *Trichocaulon alstonii* N. E. Brown (1906); **incl.** *Trichocaulon halenbergense* Dinter (1931) (*nom. inval.*, Art. 32).

[2] Stems 0.5 - 1 m tall, whitish-grey-green, 4 - 8 cm ⌀; **Tu** arranged in 20 - 22 **Ri**, apically with a stout acute pale brown **Sp** (6-) 8 - 10 mm long; **Inf** shortly pedunculate with groups of 1 - 8 **Fl** with very foetid odour; **Ped** 1 - 2 (-4) mm; **Sep** 2 - 2.5 × basally ± 1 mm; **Cl** yellow becoming whitish towards the bottom of the tube, 10 - 18 mm ⌀, tube

campanulate to funnel-shaped, 2 - 3 (-4) mm deep, slightly thickened near the **Gy** and touching its sides; **Cl** lobes ovate and normally tapering very acutely, (4-) 6 - 8 × basally 4 - 5 mm, inside glabrous and smooth; **Cn** pale yellow; **Ci** 1 - 1.5 mm, deeply divided down to the level of the **Cs** into 2 erect blunt teeth; **Cs** ± 0.75 mm, rectangular, blunt, surpassing the **Anth**; **Fr** 3 - 4 cm.

H. currorii (Hooker) Decaisne (PSRV 8: 665, 1844). **T:** Angola (*Curror* s.n. [K]). – **D:** Namibia, Angola, Botswana, Zimbabwe, RSA.

≡ *Scytanthus currorii* Hooker (1844).

H. currorii ssp. **currorii** – **D:** Namibia, Angola; semi-arid zones of the coastal Namib Desert (13° S - 23° S).

Incl. *Scytanthus burkei* Hooker (1844); **incl.** *Adenium namaquarium* Henslow (1901); **incl.** *Hoodia macrantha* Dinter (1914); **incl.** *Hoodia gibbosa* Nel (1937); **incl.** *Hoodia montana* Nel (1937); **incl.** *Hoodia currorii* var. *minor* R. A. Dyer (1966).

[1] Stems 0.15 - 1 m tall, grey to brownish-green, 4 - 6 (-8) cm ∅; **Tu** arranged in 11 - 16 (-24) **Ri**, laterally flattened, apically with an acute **Sp** 0.6 - 1 cm long; **Inf** with groups of 1 - 4 **Fl**; **Ped** 1.2 - 5 (-6) cm, 4 - 6 mm ∅, elliptic in cross-section; **Sep** appressed to the **Cl**, 4 - 8 × basally 3 mm; **Cl** outside pale maroon, centre usually marked with narrow (pale) reddish spots, inside dark red, light maroon or yellowish-pink, usually with darker veins, with a glossy pale orange area around the **Gy** and annulus, with ± distinct central depression, **Cl** (5-) 6 - 17 (-18) cm ∅, ± circular or distinctly 5-lobed; **Cl** lobes ± 1 - 2.5 cm, narrowed tip area 0.6 - 2 cm, basally ± 5.5 - 7.5 cm broad, inside with pink to purple **Ha**, **Ha** 0.5 - 3.5 mm, each from a flattened-conical papilla, annulus 6 - 9 mm ∅; **Cn** sunken for (2.8-) 3 - 6 mm, purple-red or brown, glossy, 2 - 3 × 3.8 - 5 mm, outside sometimes sparsely hairy; **Ci** forming a 5-lobed cup, slightly higher than the **Sty** head or up to 2× as high, appendages split above the middle into upright blunt or deltoid teeth, outer margin folded inwards; **Cs** linear, blunt, apically curved towards each other; **Fr** 15 - 22 cm.

H. currorii ssp. **lugardii** (N. E. Brown) Bruyns (BJS 115(2): 205, 1993). **T:** Botswana (*Lugard* 303 [K]). – **D:** C Botswana, S Zimbabwe, RSA (Northern Prov.).

≡ *Hoodia lugardii* N. E. Brown (1903).

[1] Plants like ssp. *currorii*; **Ped** 3 - 7 × 2 - 2.5 mm, green; **Cl** outside dark red, centre pale pinkish or pale maroon, inside brick-red to pale maroon, with pale red area around **Gy** and annulus, **Cl** 4 - 7.5 cm ∅, *either* circular and with or without concave indentations between the free **Cl** lobe areas, *or* broadly 5-lobed with a central depression; **Cl** lobes 0.5 - 1.1 cm, narrowed tip area 5 - 7 mm, basally 2 cm broad, inside with erect **Ha**, **Ha** 0.5 - 2.5 mm, each from a small reddish conical papilla, annulus 5 - 6 mm ∅; **Cn** sunken for ± 4 mm, ± 2.5 × 4 mm, outside usually weakly hairy; **Ci** not fused as far at the base; otherwise as ssp. *currorii*.

H. dregei N. E. Brown (FC 4(1): 897, 1909). **T:** RSA, Cape Prov. (*Drège* 5616 [K]). – **D:** RSA (Northern Cape, Western Cape: Great Karoo).

[1] Forming dense cushions; stems 20 (-50) × 2.5 - 6 cm; **Tu** arranged in 16 - 24 **Ri**, laterally flattened, apically with a weak **Sp**, **Sp** dark purplish-brown when young, 5 - 7 mm; **Inf** in groups of 1 - 5 **Fl**; **Ped** 0.7 - 1.2 cm, 2 - 3 mm ∅; **Cl** dark maroon or greenish-yellow, 2.8 - 4.8 cm ∅, flat or dish-like, ± distinctly 5-lobed; **Cl** lobes 4 - 7 mm, narrowed tip area 3 - 6 mm, basally 1.2 - 1.8 cm broad, inside densely hairy with soft white **Bri**, **Bri** 1 - 2 mm, originating from small red columnar papillae, annulus appearing as 5 pinkish slightly glossy swellings, 3.5 - 4 mm ∅; **Cn** sunken for 1 mm, dark purplish-black, 3 - 4 mm ∅, surpassing the annulus; **Ci** basally cup-shaped, horizontally spreading, appendages transversally rectangular, bluntly retuse or ± deeply divided, < 0.5 × basally 1 - 1.7 mm; **Cs** lobes broadly linear, blunt, 1 mm, not surpassing the **Anth**, dorsally connected to the **Ci** by a blunt appendage.

H. flava (N. E. Brown) Plowes (Asklepios No. 56: 8, 1992). **T:** RSA, Northern Cape (*Bain* s.n. [K]). – **D:** Namibia (Karas Mts.); RSA (Northern Cape). **Fig. XXIII.c**

≡ *Trichocaulon flavum* N. E. Brown (1878).

[2] Stems 0.2 - 0.5 m tall, 2 - 7 cm ∅; **Tu** arranged in 18 - 31 **Ri**, apically with a brownish **Sp** 4 - 7 mm long; **Inf** in groups of 1 - 3 **Fl**, scent reminiscent of old fish; **Ped** 0.5 - 1 mm; **Sep** appressed to the **Cl**, 2.5 - 3 × basally 1 - 1.5 mm; **Cl** *either* greenish-yellow or yellowish-green and **Cl** lobes with brown tips, *or* entirely brown, 1.1 - 1.3 cm ∅, dish- to bowl-shaped; **Cl** lobes broadly ovate-deltoid, 2.5 - 4 × 3.5 - 5 mm, occasionally with fusiform tip area, inside smooth or finely papillate, papillae semiglobose with fine apical **Bri**, **Cl** lobes glabrous; **Cn** yellow; **Ci** 1.6 - 2.2 mm, divided from below the middle into 2 broad weakly dorsiventrally flattened linear horn-like appendages (appendages strongly spreading, approaching and touching each other in **Cs** position); **Cs** lobes linear, blunt, 0.6 - 1 mm, occasionally longer than the **Anth** and touching each other above the **Sty** head; **Fr** 7.5 - 18 cm.

H. gordonii (Masson) Sweet *ex* Decaisne (PSRV 8: 665, 1844). **T:** [lecto – icono]: Masson, Stapel. Nov. t. 40, 1797. – **D:** Namibia, RSA (Northern Cape); absent from winter-rainfall regions. **Fig. XXIII.a, XXIII.b**

≡ *Stapelia gordonii* Masson (1797) ≡ *Gonostemon gordonii* (Masson) Sweet (1826) ≡ *Monothylaceum gordonii* (Masson) Don (1837) ≡ *Scytanthus gordonii* (Masson) Hooker (1844); **incl.** *Hoodia barklyi* Thiselton-Dyer (1876); **incl.** *Hoodia bainii* Thiselton-Dyer (1878); **incl.** *Hoodia albispina* N. E. Brown (1909); **incl.** *Hoodia burkei* N. E. Brown (1909); **incl.** *Hoodia pillansii* N. E. Brown (1909); **incl.** *Hoodia dinteri* Schlechter *ex* Dinter (1922) (*nom. nud.*); **incl.** *Hoodia rosea* Obermeijer & Letty (1936); **incl.** *Hoodia husabensis* Nel (1937); **incl.** *Hoodia langii* Obermeijer & Letty (1937); **incl.** *Hoodia whitesloaneana* Dinter (1937) (*nom. nud.*).

[1] Stems 0.5 - 1 m tall, greyish-green or -brown, strictly cylindrical, 2.5 - 5 cm ⌀; **Tu** protruding, blunt, arranged in 11 - 17 **Ri**, apically with an acute **Sp** 6 - 12 mm long; **Inf** sometimes very slightly pedunculate, with groups of 1 - 4 **Fl** with foetid odour; **Ped** 0.8 - 3 cm, 2 - 3 mm ⌀, elliptic in cross-section; **Sep** placed against the **Cl**, 5 - 6 × basally 2 - 4 mm; **Cl** outside pale maroon with darker venation, inside light maroon or purple-red, usually with darker venation, (4-) 5 - 10 cm ⌀, ± circular or distinctly 5-lobed, with a central depression; **Cl** lobes to 1.5 mm, narrowed tip area 3 - 6 mm, basally 5 cm broad, inside glabrous and smooth or finely papillate, papillae conical, apically with a 2 - 2.5 mm long **Bri**, papillae partly coloured red along the margin of the **Cl**-depression; **Cn** sunken for 1 - 1.5 mm, purplish-black, 1.5 - 2 × 4 - 6 mm, touching the surrounding **Cl** depression along its margin; **Ci** *either* undivided and apically blunt *or* divided from the middle into 2 spreading-ascending blunt processes, < 1 mm; **Cs** lobes 1 mm, linear, with blunt slightly compressed tips, touching each other, dorsally attached to the **Ci** by a broad crest; **Fr** 9 - 11.5 cm.

H. juttae Dinter (Neue Pfl. Deutsch-SWA, 34, 1914). **T:** Namibia (*Dinter* 3203 [SAM, S]). – **D:** Namibia (Karas Mts.). **Fig. XXIV.a**

≡ *Hoodia bainii* var. *juttae* (Dinter) H. Huber (1961).

[1] Forming dense cushions, 30 - 50 cm; stems pale green-grey, 3 - 5 cm ⌀; **Tu** arranged in 15 - 17 **Ri**, apically with a 8 - 11 mm long hard **Sp**; **Inf** with groups of 1 - 4 **Fl**; **Ped** 1 - 3 cm, 2.5 - 4 mm ⌀, slightly elliptic in cross-section; **Cl** outside pale yellowish-brown, inside pale yellowish-brown or dark maroon with darker venation, 2 - 5.5 cm ⌀, ± flatly dish-like or ± distinctly 5-lobed, with a central cup-shaped depression, the latter thickened annulus-like along its margin; **Cl** lobes 4 - 8 mm, narrowed tip area 2 - 5 mm, basally 1.5 - 2.5 cm broad, inside glabrous and smooth; **Cn** sunken for 1 mm, dark purplish-black, 3 - 4.5 mm ⌀; **Ci** spreading, transversally oblong-rectangular, bluntly retuse or shortly bifid, to 1 × basally ± 1.5 mm, incumbent on and slightly surpassing the annulus-like thickening; **Cs** lobes broadly linear, blunt, slightly surpassing the **Anth**, slightly overlapping with their tip area, attached to the **Ci** by a blunt dorsal appendage; **Fr** to 14 cm.

H. longispina Plowes (Brit. Cact. Succ. J. 11(2): 56-58, ills., 1993). **T:** Namibia (*Plowes* 5321 [SRGH]). – **D:** Namibia.

[1] Plants with 10 - 12 stems, ± 50 cm tall; stems grey-green; **Tu** arranged in 14 - 16 **Ri**, apically with a slender **Sp** 1.6 - 1.8 cm long; **Inf** groups of 5 **Fl**; **Ped** 1.8 - 2 cm; **Cl** pale maroon, 7 - 8 cm ⌀, saucer-shaped, hairy, with a weak central depression, margin glossy, ascending, bulging; **Cl** lobes hardly distinguishable, 0.8 - 1 cm, margin reflexed, tip drawn out in a bristly manner; **Cn** glossy red; **Ci** 5-lobed, cup-shaped, appendages short, truncate, retuse; **Cs** lobes rectangular, blunt, weakly bulging where meeting the **Ci**.

According to own observations, this taxon is only a variation with hairy flowers of the polymorphic *H. gordonii*.

H. mossamedensis (L. C. Leach) Plowes (Asklepios No. 56: 9, 1992). **T:** Angola, Moçamedes Distr. (*Leach & Cannell* 14690 [LISC, SRGH]). – **D:** SW Angola.

≡ *Trichocaulon mossamedense* L. C. Leach (1974).

[2] Stems green-grey, 30 × 4 - 5 cm ⌀; **Tu** 4 - 7 mm long, conically compressed, arranged in 16 - 20 **Ri**, apically with a brown **Sp** to 6 mm long; **Inf** groups of 1 - 4 **Fl**, from button-like peduncles; **Ped** 1 - 1.8 cm, 1 mm ⌀, spreading from the stem or ascending parallel to the stem; **Sep** 2.5 - 3 × basally ± 1 mm; **Cl** dark chestnut-brown, 0.9 - 1.8 cm ⌀, campanulate, circular or 5-lobed, with a cup-shaped central depression; **Cl** lobes slightly recurved outwards, often slightly thickened along their margins, ± 4 × basally 3.5 mm, margins slightly curved outwards, inside glabrous, papillate, papillae small, with a horizontally spreading thick **Bri**; **Cn** sunken for ± 1.5 mm, chestnut-brown-black, slightly cup-shaped, ± 1.5 × 3.5 mm; **Ci** spreading, apically slightly curved outwards, ± 0.75 × 1.25 mm, ± bifid, laterally attached to the **Cs**; **Cs** lobes rectangular, blunt, almost as long as the **Anth**, attached to the **Ci** by a small blunt upright appendage.

H. officinalis (N. E. Brown) Plowes (Asklepios No. 56: 9, 1992). **T:** Botswana ? (*Anonymus* s.n. [K]). – **D:** Namibia, RSA.

≡ *Trichocaulon officinale* N. E. Brown (1895).

H. officinalis ssp. **delaetiana** (Dinter) Bruyns (BJS 115(2): 216, 1993). **T** [neo]: Namibia (*Merxmüller & Giess* 32150 [WIND, M]). – **D:** SW Namibia (Klinghardt Mts.).

≡ *Trichocaulon delaetianum* Dinter (1923) ≡ *Hoodia delaetiana* (Dinter) Plowes (1992).

[2] Stems green-grey, to 40 × 4 - 7 cm ⌀; **Tu** arranged in 19 - 23 **Ri**, apically with an acute brown **Sp** to 1.2 cm long; **Ped** 1 - 2 mm; **Sep** (1-) 2 - 2.5 × basally 1.5 mm; **Cl** outside reddish-brown, inside

brownish-yellow to yellow, (1.2-) 1.4 - 2 cm ⌀, broadly campanulate with a broadly dish-shaped 2 - 3 mm deep depression; **Cl** lobes ovate, acuminate, 4 - 6 × basally 5 - 7 mm, inside smooth or weakly papillate, papillae apically with a short **Bri**; **Cn** dark maroon to reddish, 3 - 3.5 mm ⌀; **Cs** lobes ± ½ as long as the **Anth**, blunt; otherwise as ssp. *officinalis*.

H. officinalis ssp. **officinalis** − **D:** S Namibia (excl. SW winter-rainfall regions and the Kalahari), RSA (Free State).
 Incl. *Trichocaulon rusticum* N. E. Brown (1909) ≡ *Hoodia rustica* (N. E. Brown) Plowes (1992); **incl.** *Trichocaulon pubiflorum* Dinter (1932).
 [2] Plants to 0.3 × 0.5 m, often smaller; stems green-grey, 3.5 - 6.5 cm ⌀; **Tu** arranged in (14-) 17 - 22 **Ri**, apically with a brown **Sp** 6 mm long; **Inf** groups of 1 - 3 **Fl**; **Ped** mostly < 1 mm; **Sep** 2.5 - 3.5 mm; **Cl** outside pale green with brownish venation, inside maroon or yellowish-brown, tube paler, below the **Cn** yellowish, 1 - 1.4 cm ⌀, flatly circular or slightly campanulate with a broadly dish-shaped depression; **Cl** lobes ovate-deltoid, acuminate, 3 - 5 × basally 3.5 - 5.5 mm, tips curved outwards, inside papillate, papillae small, apically with a small **Bri**; **Cn** yellow, sometimes finely dotted with red (then **Cs** brownish), 1.5 - 2 × 4 mm, usually outside weakly pubescent; **Ci** forming pouches between the **Cs** lobes, divided into 2 blunt upright teeth down to the middle, teeth laterally fused with the **Cs**, not surpassing the dorsal appendage of the **Cs**; **Cs** lobes mostly < 0.5 mm, shorter than the **Anth**, dorsally with a broad blunt erect appendage; **Fr** 11.5 - 12.5 cm.

H. parviflora N. E. Brown (BMI 1895: 265, 1895). **T:** Angola, Moçamedes Distr. (*Welwitsch* 4265 [K, G]). − **D:** SW Angola, NW Namibia.
 [1] Plants 0.3 - 1 m ⌀; stems conspicuously dull bluish- or violet-green, 30 - 220 × 3.5 - 11 cm; **Tu** laterally flattened, arranged in 14 - 18 **Ri**, apically with a strong **Sp** 0.6 - 1 cm long; **Inf** groups of 1 - 4 **Fl**; **Ped** 2 - 4 mm, 3 - 4 mm ⌀, elliptic in cross-section; **Sep** appressed to the **Cl** for their lower ⅓, 5 - 6 mm, 3 mm broad at the almost cordate base, spreading, tips reflexed; **Cl** outside whitish with darker venation, inside yellow or brownish-orange with dark (reddish) veins, 3 - 5.5 cm ⌀, conically campanulate, ± 5-lobed, with a broadly cup-shaped central depression; **Cl** lobes ± 1 cm, narrowed tip area 5 - 7 mm × basally 2 - 2.7 cm, inside papillate, papillae conical, apically with a **Ha**-like **Bri** to 3.5 mm, annulus of 5 bulges along the margin of the central depression, ± 7 mm ⌀; **Cn** sunken for 3 - 4 mm, purplish-black, 2 - 2.5 × 4 - 4.5 mm; **Ci** basally cup-shaped, ascending, divided into 2 deltoid or blunt teeth down to the middle, 1 mm; **Cs** linear, blunt, shorter than the **Anth**, dorsally attached to the **Ci** by a flat tissue crest.

H. pedicellata (Schinz) Plowes (Asklepios No. 56: 9, 1992). **T:** Namibia (*Stapff* s.n. [K]). − **D:** SW Angola, Namibia (80 km broad coastal fog belt of the Namib Desert). **Fig. XXIII.e**
 ≡ *Trichocaulon pedicellatum* Schinz (1888).
 [2] Plants with ± 20 stems; stems grey-green, 10 - 25 (-50) × 2.5 - 5 cm; **Tu** arranged in 11 - 20 **Ri**, apically with a delicate dark ± caducous **Sp** 1.5 - 3 mm long; **Inf** groups of 1 - 4 **Fl** curved downwards; **Ped** 4 - 12 mm, pendent; **Sep** 1.5 - 2 mm; **Cl** inside chestnut-brown to dark purplish-brown, 0.8 - 1.4 cm ⌀, flat; **Cl** lobes lanceolate, acute, reflexed along the margin, 3 - 6 × basally 2.5 - 3 mm, often broadened towards the middle, inside glabrous, finely papillate, with weak swellings within the fusion area of the **Cl** lobes; **Cn** only slightly sunken with the base, purplish-brown or yellow, 3 - 3.5 mm ⌀; **Ci** horizontally spreading starting from the middle of the **Gy**, almost halfway divided into 2 spreading very short teeth, laterally attached to the dorsal side of the **Cs**; **Cs** lobes ascending from the level of the **Ci**, very variable in length, blunt, often touching each other above the **Sty** head.

H. pilifera (Linné *fil.*) Plowes (Asklepios No. 56: 10, 1992). **T:** RSA, Cape Prov. (*Thunberg* 6332 [UPS]). − **D:** RSA.
 ≡ *Stapelia pilifera* Linné *fil.* (1781) ≡ *Piaranthus piliferus* (Linné *fil.*) Sweet (1830) ≡ *Trichocaulon piliferum* (Linné *fil.*) N. E. Brown (1873).

H. pilifera ssp. **annulata** (N. E. Brown) Bruyns (BJS 115(2): 235, 1993). **T:** RSA, Eastern Cape (*Lee* s.n. in *Pillans* 135 [BOL]). − **D:** RSA (Eastern Cape: Great Karoo).
 ≡ *Trichocaulon annulatum* N. E. Brown (1909) ≡ *Hoodia annulata* (N. E. Brown) Plowes (1992).
 [2] Plants to 2 m ⌀; stems grey-green, to 45 × 3 - 5 cm; **Tu** conical, arranged in 20 - 22 (-30) **Ri**, apically with a strong brown **Sp** 3 - 5 (-6) mm long; **Inf** groups of 1 - 3 **Fl**; **Ped** < 1 mm; **Sep** 2 - 2.5 × basally 1.5 mm; **Cl** reddish-purple, turning to greenish towards the **Fl** bottom, (1.5-) 2 - 3 cm ⌀, circular, with a prominent central annulus; **Cl** lobes broadly deltoid, acuminate, appressed to the stem, marginally slightly curved towards the outside, 5 - 7 × basally 8 - 9 mm, outside smooth except the weakly prominent venation, inside glabrous and densely papillate, papillae conical, to 0.5 mm long, apically with an erect or horizontal dark purplish-black **Bri**, annulus 5 mm high, cup-shaped, enclosing the **Gy**; **Cn** dark purplish-black, 4 × 5 - 6 mm, cup-shaped; **Ci** basally erect, divided into 2 spreading teeth down to the middle, ± lanceolate, 1.5 mm long, recurved, with a longitudinal furrow; **Cs** lobes linear, blunt, ± 1 mm, occasionally surpassing the **Anth**.

H. pilifera ssp. **pilifera** − **D:** RSA (Western Cape, Eastern Cape: Little and S Great Karoo). **Fig. XXIII.d**

[2] Stems dark green or pale grey-green, to 80 × 3 - 6 cm; **Tu** laterally slightly flattened, arranged in 21 - 34 **Ri**, apically with a strong grey or brown **Sp** 5 - 9 mm long; **Inf** groups of 1 - 3 **Fl** with strong foetid odour; **Ped** 0.5 - 1.5 mm; **Sep** 2 - 3 mm; **Cl** outside reddish-green, inside dark purplish- or pinkish-brown, 1.6 - 2 cm ∅, campanulate, with an ascending annulus around the central cup-shaped depression; **Cl** lobes broadly deltoid, acuminate, marginally slightly curved outwards, 4 - 6 × basally 6 - 7 mm, inside glabrous and papillate, papillae blunt, apically with an acutely prolonged horizontally spreading **Bri**, central depression often smooth, annulus 5-angled, 4 - 5 mm ∅; **Cn** sunken for 2 - 2.5 mm, dark purplish-brown, basally reddish, 4 - 5 mm ∅; **Ci** basally cup-shaped, below the middle divided into 2 teeth, 1 - 1.5 mm, dorsiventrally flattened, curved outwards or ± erect, spreading; **Cs** lobes straightly rectangular, 0.5 - 1.5 mm, partly ascending over the **Sty** head and touching each other.

H. pilifera ssp. **pillansii** (N. E. Brown) Bruyns (BJS 115(2): 238, 1993). **T:** RSA, Western Cape (*Pillans* 9 [K, BOL]). – **D:** RSA (Western Cape: Great Karoo). **Fig. XXIII.f**
≡ *Trichocaulon pillansii* N. E. Brown (1904); **incl.** *Trichocaulon pillansii* var. *major* N. E. Brown (1904); **incl.** *Trichocaulon grande* N. E. Brown (1909) ≡ *Hoodia grandis* (N. E. Brown) Plowes (1992); **incl.** *Hoodia colei* Plowes (1992).
[2] Plants to 50 cm ∅; stems grey-green, 30 (-60) × 3 - 6 cm; **Tu** laterally slightly flattened, arranged in 25 - 34 **Ri**, apically with a grey **Sp** 5 - 6 mm long; **Inf** groups of 1 - 3 **Fl** with unpleasant slightly fish-like odour; **Ped** 1 - 1.5 mm, 1 - 1.5 mm ∅; **Sep** 1.5 - 2.5 mm; **Cl** outside pale reddish-green, inside pale yellow, greenish-yellow or pale reddish, 0.8 - 2 cm ∅, with a cup-shaped central depression with 3.5 - 5 mm ∅, with slightly thickened margin, sometimes with a weak annulus; **Cl** lobes broadly deltoid, very acute, marginally often strongly reflexed, 3.5 - 9 × basally 3 - 7 mm, inside glabrous and papillate, papillae blunt, columnar, apically with a small horizontally spreading **Bri**; **Cn** sunken for 2 - 3 mm, yellow (paler than the **Cl**), 3 - 4 mm ∅; **Ci** 1 - 2 mm long, otherwise as ssp. *pilifera*; **Cs** lobes 0.5 - 1 (-1.5) mm long, otherwise as ssp. *pilifera* (however, only rarely touching each other above the **Sty** head and ascending over the middle).

H. ruschii Dinter (RSN 30: 192, 1932). **T:** Namibia (*Rusch jr.* s.n. in *Dinter* 7976 [BOL, B, G, PRE, S, Z]). – **D:** Namibia (E flank of the Tiras Mts.).
[2] Stems brownish or grey-green, very stout, 45 - 50 × 4 - 6 cm; **Tu** conical, laterally flattened, arranged in 22 - 28 **Ri**, apically with a stiff **Sp** 6 - 8 mm long; **Inf** groups of 4 - 10 **Fl** with extremely foetid odour, on button-like persistent peduncles, opening simultaneously; **Ped** 2 - 4 mm, 2 mm ∅; **Sep** 2 - 4 × 1 - 2 mm; **Cl** outside pale green, reddish towards the tube, inside maroon, 2 - 4 cm ∅, broadly campanulate, tube broadly conical, 6 - 8 mm deep, 0.8 - 1 cm ∅, slightly thickened around and laterally touching the **Gy**; **Cl** lobes deltoid-acuminate, basally slightly ovate, with slightly outwards-curved tips, lobes 0.8 - 1.4 × basally 0.9 - 1.4 cm, inside glabrous and papillate, papillae with a slender spreading **Bri**; **Cn** dark purplish-black, ± 1 × 2.2 mm; **Ci** divided ± from the base into 2 upright lobes, blunt, along their entire length laterally fused with and not surpassing the **Cs**; **Cs** lobes blunt, rectangular, ± ½ as long as the **Anth**.

H. triebneri (Nel) Bruyns (BJS 115(2): 222, 1993). **T:** Namibia (*Triebner* s.n. in *SUG* 6020 [BOL]). – **D:** W Namibia.
≡ *Trichocaulon triebneri* Nel (1935); **incl.** *Hoodia foetida* Plowes (1992).
[2] Plants to 45 cm ∅; stems grey-green, to 30 × 2.5 - 4 cm; **Tu** conical, arranged in 12 - 14 (-16) **Ri**, apically with a pale **Sp** 5 - 6 mm long; **Inf** groups of 6 - 12 **Fl**, some simultaneously opening on button-like persistent peduncles; **Ped** 3 - 4 mm, ± 1 mm ∅; **Sep** 2.2 - 2.5 × basally ± 1 mm; **Cl** outside reddish-green, inside blackish to reddish-purple (paler at the rim of the tube), 1.1 - 1.5 cm ∅, campanulate, with a conical tube 3.5 - 4 mm deep, tube slightly thickened around the **Gy** and touching its sides; **Cl** lobes deltoid-acuminate, sometimes with fine erect tips, lobes 3 - 4.5 × basally 4 - 5 mm, **Cl** inside glabrous, covered with obconical papillae apically ending in a slender spreading **Bri**; **Cn** as in *H. ruschii*; **Fr** 10.5 cm.

×HOODIALLUMA

B. Müller & F. Albers

×**Hoodialluma** G. D. Rowley (Repert. Pl. Succ. 27: 4, 1976).
Incl. *Hoodiopsis* C. A. Lückhoff (1933). **T:** *Hoodiopsis triebneri* C. A. Lückhoff.
= *Hoodia* × *Orbea* (*Caralluma*). The only hybrid which was formerly placed here, ×*H. triebneri* (C. A. Lückhoff) G. D. Rowley, represents the cross *Hoodia gordonii* × *Orbea* (*Orbeopsis*) *lutea* ssp. *vaga* (Bruyns 1993: 265-266).

HOYA

C. Hoffmann, R. van Donkelaar & F. Albers

Hoya R. Brown (Prodr., 459, 1810). **T:** *Asclepias carnosa* Linné. – **Lit:** Rintz (1978); Forster & Liddle (1990); Forster & Liddle (1992); Forster & al. (1998). **D:** Asia from Pakistan, India, Bangladesh and Bhutan to tropical SE Asia, Papua New Guinea, Australia, Japan and numerous Pacific Islands.

Etym: For Thomas Hoy (†1821), gardener at Syon House, England.
Incl. *Sperlingia* Vahl (1810). **T:** *Sperlingia verticillata* Vahl.
Incl. *Schollia* J. Jacquin (1811). **T:** *Schollia crassifolia* J. Jacquin.
Incl. *Physostelma* Wight (1834). **T:** *Physostelma wallichii* Wight.
Incl. *Pterostelma* Wight (1834). **T:** *Pterostelma acuminata* Wight.
Incl. *Centrostemma* Decaisne (1838). **T:** *Hoya multiflora* Blume.
Incl. *Cyrtoceras* Bennett (1838). **T:** *Cyrtoceras reflexum* Bennett.
Incl. *Cystidianthus* Hasskarl (1842). **T:** *Hoya campanulata* Blume.
Incl. *Plocostemma* Blume (1849). **T:** *Plocostemma lasiantha* Korthals *ex* Blume.
Incl. *Acanthostemma* Blume (1850). **T:** *Hoya rumphii* Blume [Lectotype, designated by Pfeiffer 1871-74 (cf. P. V. Heath, Calyx 2(4): 135, 1992, & l.c. 3(2): 50, 1993).].
Incl. *Cathetostemma* Blume (1850). **T:** *Hoya laurifolia* Decaisne.
Incl. *Otostemma* Blume (1850). **T:** *Hoya lacunosa* Blume.
Incl. *Eriostemma* (Schlechter) Kloppenburg & Gilding (2001). **T:** *Hoya coronaria* Blume.

Plants epiphytic, epilithic, rarely rooting in the ground, creeping, climbing, pendent, left-twining, rarely ± shrubby, latex white, rarely clear; **R** fibrous; stems terete, ± sparsely branched, glabrous to pubescent; **L** decussate or very rarely alternate, with a ± long petiole, lamina lanceolate to obcordate, entire, leathery, fleshy to succulent; **Inf** racemose, occasionally with part-**Inf**, lateral or rarely terminal, 1- to many-flowered, globose to flat, rarely with a peduncle > 10 cm long; **Fl** 5-merous, actinomorphic; **Ped** all of equal size and ± straight or length variable within an **Inf** and ± bent; **Cl** star-shaped and spreading, campanulate to almost broadly urceolate, outside glabrous, inside rarely glabrous; **Cl** lobes ± connate, often ± strongly reflexed, wax-like, fleshy; **Cn** uniseriate, staminal; **Cs** lobes ± horizontal, fleshy, canaliculate on the lower face, with outer and inner appendages, the latter ± incumbent on the **Anth**; **Gy** sessile or stipitate; **Sty** head ± flat, covered by the **Anth** appendages; **Poll** erect, elongate to ovate, outer margin with or without germination crest, caudicles not winged or broadly winged, corpuscle elongate or rhombic; **Fr** narrowly fusiform to ovoid, smooth or rough; **Se** linear to ovate, flattened, with a tuft of white **Ha** (coma).

The genus *Hoya* is among the best-known asclepiads, owing to *H. lanceolata* ssp. *bella* and *H. carnosa* ("Wax Flower", "Wax Plant"), both cultivated all over the world. The distribution area of the genus stretches from Pakistan to New Caledonia and the Fiji Islands with a centre in the Malesian sector. Of the over 300 species described, not more than 200 species may be recognized. A generic revision has never been published. Typical of these easily cultivated plants are their often large, leathery to succulent leaves with a shining surface, as well as the large flowers in pseudo-umbellate inflorescences, which make them popular among plant-lovers. The available infrageneric concept with some 24 sections is insufficient and problematic, and a generic subdivision is therefore omitted here. In the following account, only those ± succulent species are treated that are most worth cultivating.

H. anulata Schlechter (Nachtr. Fl. Schutzgeb. Südsee, 362, 1905). **T:** Papua New Guinea, Madang Prov. (*Schlechter* 14185 [B ?, BM]). – **D:** Indonesia (Irian Jaya), Papua New Guinea, Australia (Queensland). **I:** White (1928); Forster & Liddle (1990). **Fig. XXIV.b**
Incl. *Hoya schlechteriana* S. Moore (1916); **incl.** *Hoya poolei* C. T. White & Francis (1928); **incl.** *Hoya pseudolittoralis* C. Norman (1937); **incl.** *Hoya alata* K. D. Hill (1988).

Epiphytic, rarely epilithic; stems twining, terete, to 5 mm ⌀, glabrous or rarely scattered hairy; **L** petiole 2 - 12 × 2 - 4 mm, lamina ovate, obovate or rhombic, 3 - 9.5 × 2 - 5 cm, fleshy, pale green to bronze, glabrous, acuminate, base cordate to round, margins revolute; **Inf** to 12-flowered, pendent, peduncle 4 - 9 × ± 0.2 cm, glabrous; **Fl** 3 - 4 × 12 - 13 mm ⌀; **Ped** 16 - 25 × 1 mm, glabrous or sparsely hairy; **Sep** triangular-ovate, 1 mm, acute; **Cl** star-shaped and spreading, 1.5 cm ⌀, pale pink to almost white, basally ± reddish; **Cl** lobes triangular, 4 - 5 × 5 mm, acute, slightly revolute, outside glabrous, inside densely pubescent; **Cn** 2 × 6 - 7 mm, yellow, pale pink to dark red; **Cs** lobes elongate-linear, 3 × 1 - 1.5 mm, outer appendage acute, inner appendage acuminate, ascending, often darker than the outer appendage; **Anth** appendage ± 0.5 × 0.5 mm, acute; **Sty** head ± 1 mm ⌀; **Poll** erect, elongate, 0.35 - 0.4 × ± 0.15 mm, corpuscle elongate, 0.13 - 0.14 × 0.06 - 0.1 mm, tan-coloured, caudicles winged, 0.1 - 0.13 × 0.1 - 0.13 mm; **Fr** fusiform, 7 - 14 × 0.7 - 1.2 cm.

H. archboldiana C. Norman (Brittonia 2: 328, 1937). **T:** Papua New Guinea, Central Prov. (*Brass* 3621 [NY]). – **D:** Indonesia, Papua New Guinea. **I:** Forster & al. (1995).

Stems terete, up to several m long, corky with age, glabrous; **L** petiole 1.4 - 2 mm long, lower face canaliculate, glabrous, ± 4 mm ⌀, lamina lanceolate-ovate to lanceolate-elliptic, 16 × 7 cm, ± fleshy, upper face shining dark green, lower face pale green, glabrous, acute, with a cordate base; peduncle 2.5 - 3 × 3 mm, glabrous; **Fl** 1.8 - 2 × 4 - 4.7 cm ⌀; **Ped** 4.5 - 5.5 × 1.8 - 2 mm, glabrous; **Sep** lanceolate-ovate, 3 - 4.5 × 4 - 4.1 mm, glabrous; **Cl** lobes triangular, 13 - 14 × 18 - 19 mm, revolute, margin slightly involute; **Cn** 10 - 11 × 17 - 18 mm, pink; **Cs** lobes ± placed against the **Cl**, elongate-lanceolate, 2.5 - 2.7 × 3.5 - 3.6 mm, outer appendage ± obtuse, inner appendage acute, ascending; **Anth** appendage lanceolate, ± 2 × 1.5 mm; **Sty** head globose-depressed, 2.5 - 3 mm ∅; **Poll** narrowly elongate, ± 1.6 × 0.4 mm, corpuscle ovate, ± 0.75 × 0.4 mm, caudicles ± 0.4 × 0.2 mm, winged.

Closely allied to *H. macgillivrayi*. Both present large red campanulate flowers and narrowly lanceolate coronal lobes.

H. australis R. Brown *ex* Traill (Trans. Hort. Soc. London 7: 28, 1830). **T:** Australia, Queensland (*Banks & Solander* s.n. [BM (Herb. R. Brown), NSW]). – **Lit:** Forster (1991). **D:** Indonesia, Australia, Fiji, New Caledonia, Papua New Guinea, Solomon Islands, Samoa, Tonga, Vanuatu.

H. australis ssp. **australis** – **D:** Australia (Queensland, New South Wales), Samoa, Vanuatu. **I:** Forster (1991). **Fig. XXIV.d**

Incl. *Hoya dalrympleana* F. Mueller (1861); **incl.** *Hoya pilosa* Seemann (1861); **incl.** *Hoya keysii* L. H. Bailey (1884); **incl.** *Hoya pubescens* Reinecke (1898); **incl.** *Hoya oligotricha* K. D. Hill (1988).

Plants climbing; stems twining, terete, succulent, ± woody when old, glabrous or hairy when young; **L** 2 cm petiolate, lamina elliptic, narrowly ovate, elongate, ovate or rounded, to 15 × 12 cm, succulent, fleshy to leathery, sparsely to densely hairy, acute, shortly acuminate or acuminate, base rounded, cordate, obtuse or cuneate, margin weakly revolute; **Inf** to 50-flowered, pendent; peduncle 1 - 3 cm, glabrous or hairy; **Ped** 2 - 4 cm; **Sep** ovate to triangular, 1 - 5 × 1 - 3 mm, sparsely to densely hairy; **Cl** 1 - 2.5 cm ∅, white to cream-coloured, red at the base; **Cl** lobes ovate, 5 - 10 × 3 - 7 mm, acute, margin revolute; **Cn** cream-coloured; **Cs** lobes ovate, 1.2 - 3.5 × 1 - 2.5 mm, upper face concave, lower face canaliculate, outer appendage ± rounded, inner appendage acute; **Sty** head conical; **Poll** elongate, ± 0.7 × 0.3 mm, corpuscle ovate-oblong, ± 0.35 × 0.2 mm, caudicles winged; **Fr** fusiform, 9 - 13.5 × 1 - 1.5 cm; **Se** elongate, 5 - 7 × 2 - 3 mm, tan-coloured, tuft of **Ha** 2.5 - 3 cm.

H. australis ssp. **oramicola** P. I. Forster & Liddle (Austrobaileya 3(3): 516, 1991). **T:** Australia, Northern Territory (*Russell-Smith & Lucas* 5812 [DNA]). – **D:** Australia (Northern Territory). **I:** Forster (1991).

Differs from ssp. *australis*: **L** succulent, > 5 cm long, margin strongly revolute, sparsely to densely hairy.

H. australis ssp. **rupicola** (K. D. Hill) P. I. Forster & Liddle (Austrobaileya 3(3): 514, 1991). **T:** Australia, Northern Territory (*Fox* 2548 [NSW, CANB, DNA, NT]). – **D:** Australia (Western Australia, Northern Territory). **I:** Forster (1991).

≡ *Hoya rupicola* K. D. Hill (1988).

Differs from ssp. *australis*: Stems not climbing; **L** very succulent, margin not strongly revolute, densely hairy.

H. australis ssp. **sanae** (L. H. Bailey) K. D. Hill (Telopea 3(2): 251, 1988). **T:** Australia, Queensland (*Anonymus* s.n. [BRI (holo ?)]). – **D:** Australia (Queensland). **I:** Forster (1991).

≡ *Hoya sanae* L. H. Bailey (1897).

Differs from ssp. *australis*: **L** succulent, < 5 cm long, margin strongly revolute, sparsely to densely hairy.

H. australis ssp. **tenuipes** (K. D. Hill) P. I. Forster & Liddle (Austrobaileya 3(3): 512, 1991). **T:** Australia, Queensland (*Wallace* 83252 [NSW, BRI, K, L]). – **D:** Indonesia (Irian Jaya), Papua New Guinea, Australia (Queensland), Fiji, Solomon Islands, Tonga.

≡ *Hoya oligotricha* ssp. *tenuipes* K. D. Hill (1988); **incl.** *Gymnema recurvifolium* Blume (1850); **incl.** *Hoya bicarinata* A. Gray (1862); **incl.** *Hoya barrackii* Horne (1881); **incl.** *Hoya papillantha* K. Schumann (1898); **incl.** *Hoya lactea* S. Moore (1911).

Differs from ssp. *australis*: **L** leathery, margin not strongly revolute, glabrous or scatteredly to sparsely hairy.

H. benguetensis Schlechter (Philipp. J. Sci. 1: 301, 1906). **T:** Philippines, Luzon (*Elmer* 15979 [not located]). – **D:** Philippines. **Fig. XXIV.c**

Stems twining, filiform, terete, branched, sparsely leafy, glabrous; **L** with a fleshy petiole 5 - 8 mm long, lamina elliptic or ovate-elliptic, 6 - 10 × 2.5 - 4 cm, thickly leathery, glabrous, acuminate; **Inf** many-flowered, pendulous; peduncle either short or to 7.5 cm; **Ped** filiform, 1 - 1.3 cm, thin, glabrous; **Sep** ovate, ± 2 mm, obtuse, glabrous; **Cl** star-shaped and spreading, ± 1 cm ∅, yellowish or reddish, outside glabrous, inside minutely mealy-papillose; **Cl** lobes ovate, 3 mm, acute; **Cs** lobes reddish, outer appendage obtuse, weakly ascending, inner appendage obtusely rostrate, upper face keeled in ± the upper ½; **Poll** obliquely clavate, translator very small, corpuscle rhombic.

H. bilobata Schlechter (Philipp. J. Sci. 1: 301-302, 1906). **T:** Philippines, Mindanao (*Copeland* 420 [B]). – **D:** Philippines.

Stems branched, terete, densely leafy, faintly pubescent; **L** with a fleshy petiole ± 3 mm long, lamina broadly elliptic to subcircular, 1.7 - 2.2 × 1.3 - 1.8 cm, obtuse, leathery, glabrous; **Inf** ± 20-flowered, pendent; peduncle 1 - 3 cm, terete, glabrous; **Fl** very small; **Ped** to 8 mm, filiform, ± yellowish, glabrous; **Sep** elongate, to 1 mm, obtuse, glabrous; **Cl** button-shaped, dark pink, outside glabrous, inside densely shortly papillose; **Cl** lobes ovate, ± 1.5

mm, obtuse, strongly revolute; outer appendage of the **Cs** lobes whitish, bifid, inner appendage obtuse, weakly ascending, dark pink, depressed in the middle; **Poll** cylindrical, elongate, translator very short, corpuscle minute, rhombic.

H. bordenii Schlechter (Philipp. J. Sci. 1: 302, 1906). **T:** Philippines, Luzon (*Borden* 1213 [B]). – **D:** Philippines. **Fig. XXIV.e**

Stems twining, filiform, branched, sparsely leafy, glabrous; **L** with a fleshy petiole 1.5 - 2 cm long, lamina flat, lanceolate-elongate or narrowly elliptic, 11 - 18 × 2.5 - 4.5 cm, acuminate, leathery, glabrous; **Inf** pendent; peduncle 3 - 5 cm, terete, glabrous; **Ped** filiform, 2.7 cm, very slender, glabrous; **Sep** ovate, ± 1.5 mm, obtuse, basally faintly pubescent; **Cl** star-shaped, ± 1 cm ∅, deeply red, outside glabrous, inside sparsely minutely tuberculate-papillose; **Cl** lobes ovate, acute, ± strongly revolute; **Cs** lobes narrowly elliptic, acute, flat, with an elongate hump along the middle; **Anth** appendage narrowly falcate, margin cartilaginous; **Poll** cylindrical, obliquely elongate, translator very short, corpuscle rhombic, minute.

H. calycina Schlechter (BJS 50: 125, 1913). **T:** Papua New Guinea, Madang Prov. (*Schlechter* 17510 [B, BRI [photo]]). – **D:** Indonesia (Irian Jaya), Papua New Guinea.

H. calycina ssp. **calycina** – **D:** Papua New Guinea. **I:** Forster & Liddle (1992).

Stems climbing, twining, terete, several m long, 6 mm ∅, little branched, corky when old, glabrous or sparsely to densely hairy; **L** petiole 12 - 30 × 1.5 - 3.5 mm, lower face faintly grooved, lamina flat, elliptic to elliptic-ovate, 16 - 20 × 9 cm, ± finely acuminate, base rounded, cuneate or weakly cordate, upper face glabrous or sparsely pubescent, lower face shortly tomentose to velvety; **Inf** ± 7 cm long, 10-flowered, erect; **Bra** triangular to lanceolate, 0.8 - 1 × 0.5 - 1 mm; peduncle 6 - 50 × 3 - 7 mm, ± hairy; **Fl** 9 - 10 × 18 - 28 mm; **Ped** 24 - 44 × 1.6 - 2 mm, ± hairy; **Sep** lanceolate to lanceolate-ovate, 2.8 - 5 × 1.2 - 3 mm, outside sparsely to densely hairy; **Cl** campanulate to spreading and star-shaped, white, basally ± red to purple, outside sparsely to densely hairy; **Cl** lobes lanceolate-ovate, acute, 7 - 13 × 4 - 9 mm, along margin and outside ± hairy; **Cl** tube 3 - 5.4 × 8 - 14 mm, glabrous; **Cn** 4 - 4.3 × 7 - 13 mm; **Cs** lobes ovate, outer appendage round, inner appendage acuminate, ascending; **Anth** appendage lanceolate, ± 0.8 - 1.7 mm; **Sty** head depressed-globose, 1.5 - 1.9 mm ∅; **Poll** oblong, ± 1.1 × 0.4 mm; caudicles cylindrical, unwinged, ± 0.3 × 0.07 mm, corpuscle ovate, ± 0.6 × 0.4 mm.

According to Forster & Liddle (1992) closely related to *H. albiflora* (here not treated) as well as to *H. australis*. *H. calycina* differs from *H. albiflora* by the spreading corolla and from *H. australis* by the corolla being more or less glabrous inside.

H. calycina ssp. **glabrifolia** P. I. Forster & Liddle (Austrobaileya 3(4): 633, 1992). **T:** Indonesia, Irian Jaya (*Brass* 13465 [BRI, A, BO, L]). – **D:** Indonesia (Irian Jaya), Papua New Guinea.

Differs from ssp. *calycina*: **L** glabrous or scatteredly hairy on the lower face; **Sep** and **Cl** outside glabrous or scatteredly hairy.

H. carnosa (Linné *fil.*) R. Brown (Prodr., 459, 1810). – **D:** India, S China, Japan, Taiwan, Australia (Queensland), Fiji. **I:** Eggli (1994: 160).

≡ *Asclepias carnosa* Linné *fil.* (1781) ≡ *Schollia carnosa* (Linné *fil.*) Schrank *ex* Steudel (s.a.) ≡ *Cynanchum carnosum* (Linné *fil.*) hort. *ex* Decaisne (1844) (*nom. inval.*, Art. 34.1c); **incl.** *Hoya carnosa* var. *compacta* hort. (s.a.) (*nom. inval.*, Art. 32.1c); **incl.** *Hoya carnosa* var. *marmorata* hort. (s.a.) (*nom. inval.*, Art. 32.1c); **incl.** *Hoya carnosa* var. *variegata* de Vries (s.a.) (*nom. inval.*, Art. 32.1c); **incl.** *Stapelia chinensis* Loureiro (1790) ≡ *Schollia chinensis* (Loureiro) Jacquin (1811) ≡ *Hoya chinensis* (Loureiro) Traill (1830); **incl.** *Schollia crassifolia* Jacquin (1811) ≡ *Hoya crassifolia* (Jacquin) Haworth (1812); **incl.** *Hoya rotundifolia* Siebold (1840); **incl.** *Hoya variegata* Siebold *ex* Morren (1846); **incl.** *Hoya picta* Hort (1853); **incl.** *Hoya motoskei* Teijsmann & Binnendijk (1855); **incl.** *Hoya carnosa* var. *japonica* Siebold *ex* Maximowicz (1870); **incl.** *Hoya intermedia* A. C. Smith (1942); **incl.** *Hoya carnosa* var. *gushanica* W. Xu (1989).

Stems weakly succulent, pale grey, smooth, glabrous; **L** 1 - 1.5 cm petiolate, lamina broadly ovate-cordate, ovate or ovate-elongate, 3.5 - 13 × 3 - 5 cm, thick, fleshy, obtuse or acuminate, base rounded to slightly cordate; **Inf** to 30-flowered, pendent or ± erect; peduncle to 4 cm; **Ped** 2 - 4 cm, reddish, pubescent; **Sep** elongate; **Cl** 1.5 - 2 cm ∅, fleshy, whitish; **Cl** lobes broadly ovate or triangular, margin revolute, inside densely papillose; **Cn** white; **Cs** lobes ovate-lanceolate, upper face convex, outer appendage acute, inner appendage acute, reddish; **Fr** fusiform, 6 - 10 × 0.5 - 1.5 cm; **Se** narrow, tuft of **Ha** 2.5 cm.

Widely cultivated. There are numerous named cultivars, e.g. showing leaves spotted with silver-grey or leaves variegated with gold, or irregularly undulate leaves (see Clark (1996) for a list).

H. caudata Hooker *fil.* (Fl. Brit. India 4: 60, 1883). **T:** Malaysia (*Maingay* 1128 [K]). – **D:** Malaysia (Malacca), S Thailand. **I:** Rintz (1978). **Fig. XXIV.f**

Incl. *Hoya crassifolia* Ridley (1912) (*nom. illeg.*, Art. 53.1); **incl.** *Hoya flagellata* Kerr (1940).

Stems twining, 0.25 - 0.5 mm ∅, deep red when young and densely hairy; **L** with a thick petiole ± 0.6 mm long, villose, becoming glabrous with age, lamina ovate to elongate-ovate, 4 - 15 × 2.5 - 7 cm, thick, rigid, acuminate, base cordate, margin often strongly wrinkled; **Inf** 1- to 10-flowered, concave, pendent; peduncle bent, 2 - 3 cm, thin; **Ped** bent, 0.5 - 3 cm, thin, rigid, glabrous; **Sep** linear-lanceolate, 0.13 - 0.19 mm, glabrous; **Cl** spreading, ± 1.3 cm ⌀; **Cl** lobes ovate, caudate, pale pink, inside pubescent, with long felt-like marginal **Ha**; **Cs** lobes elliptic-ovate, outer appendage ± rounded, deep or pale red, inner appendage acute, white or deep red; **Anth** appendage long and whip-like, 4 - 5 mm, white; **Poll** cylindrical, elongate, ± 0.5 mm, caudicles broadly winged; **Fr** terete, 7.5 - 10 × ± 0.4 cm, apically narrower, faintly striped, smooth; **Se** flat, ± 5 × 0.75 mm, tuft of **Ha** white.

A striking feature are the very long appendages of the anthers (but see also *H. imbricata*).

H. cinnamomifolia Hooker (CBM 1848: t. 4347 + text, 1848). – **D:** Indonesia (Java). **Fig. XXV.b**

Stems twining, terete, to 3 m, glabrous; **L** rather shortly and thickly petiolate, lamina ovate or ovate-elongate, 5 - 7 × 3 cm, fleshy to leathery, acuminate, margin slightly revolute; **Inf** many-flowered, semiglobose to globose; peduncle ± 3 cm; **Cl** thickly fleshy, yellow-green, glabrous; **Cl** lobes broadly ovate, acute, strongly revolute; **Cn** ± flat, deep purple to dark red; **Cs** lobes ovate, acute, upper face ± keeled along the middle.

H. crassicaulis Elmer *ex* Kloppenburg (Fraterna 1995(3): 10-13, ills., 1996). **T:** Philippines, Luzon (*Elmer* 14440 [BO, UC]). – **D:** Philippines.

Epiphytic; stems twining, terete, 5 mm ⌀, robust, branched, slightly woody, yellowish-brown, glabrous; **L** with a thick petiole 1.5 - 5 cm long, fleshy, canaliculate on the lower face, brownish, glabrous, lamina elliptic to ovate-elongate, 10 - 17 × 5 - 8 cm, leathery, upper face ± shining, glabrous, acute to acuminate, strongly revolute at the tip, base broadly cuneate, rounded or weakly cordate, margin often slightly revolute; **Inf** 60- to 125-flowered, globose; peduncle ascending, ± 5 cm, with minute papillae or **Gl**, ochre-brown; **Ped** filiform, 1.5 - 2.5 cm, pale yellowish-brown; **Sep** narrowly elongate, 1.5 mm, acute, apically reddish-brown, outside sparsely ciliate or glabrous; **Cl** ± 1 cm ⌀, deep yellow; **Cl** lobes rhombic-ovate, acute, strongly revolute, reddish, glabrous; **Cn** ± 4 × 2 mm, flat, pale yellow; **Cs** lobes ovate-elongate, outer appendage acute, inner appendage acute, weakly keeled on the upper face; **Fr** terete, elongate, 9 - 12 × 0.4 cm; **Se** flat, 3 - 5 mm, pale brown, tuft of **Ha** dull white.

H. diptera Seemann (Bonplandia 9: 257, 1861). **T:** Fiji Islands (*Seemann* 320 [BM, BU, GH]). – **D:** Fiji (Viti Levu, Taviuni).

Stems thin, glabrous or ± faintly hairy; **L** with a rugose petiole 5 - 15 mm long, lamina elliptic, ovate-elliptic or narrowly elongate, 3.5 - 8 × 1 - 3.2 cm, slightly fleshy, ± acuminate, base obtuse; **Inf** 5- to 10-flowered; peduncle 1 - 4.5 cm, glabrous or faintly pubescent; **Ped** 0.8 - 2 cm, thin, glabrous or sparsely pale-pubescent; **Sep** triangular, ± 1 mm, membranous, margin ciliate; **Cl** 1.1 - 1.6 cm ⌀, yellowish-green, rarely reddish at the base, inside densely velvety; **Cl** lobes triangular to ovate, 4 - 6 × 4 - 6 mm, acute, thin, margin often revolute; **Cs** lobes elongate, 3 - 4.2 × 1.6 - 1.8 mm, upper face flat, lower face rounded, outer appendage obtuse, inner appendage acuminate, ascending, dark pink; **Poll** 0.5 - 0.6 mm.

H. diversifolia Blume (Bijdr. Fl. Ned. Ind., 1064, 1826). **T:** Indonesia, Java (*Zollinger* 2581 [KW, P]). – **D:** Cambodia, Laos, Myanmar, S Vietnam, Malaysia, Singapore, S Thailand, Indonesia (Borneo, Java, Sumatra). **I:** Rintz (1978). **Fig. XXV.c**

Incl. *Sussuela esculenta* Rumphius (1747) ≡ *Hoya esculenta* (Rumphius) Tsiang (1936); **incl.** *Hoya orbiculata* Wallich *ex* Wight (1834); **incl.** *Hoya crassipes* Turczaninow (1848); **incl.** *Hoya zollingeriana* Miquel (1856).

Stems twining, robust, brown, smooth; **L** with a thick petiole 0.7 - 2 cm long, lamina elliptic to obovate, ± 13 × 5 cm, dull, very fleshy, glabrous, acute or obtuse, base cuneate; **Inf** 1- to 20-flowered, convex, erect; peduncle 3.5 cm, rigid; **Ped** ± 2 cm; **Sep** elliptic, obtuse, usually ciliate at the tip; **Cl** ± 1.3 cm ⌀, pale pink, inside densely pubescent; **Cl** lobes ovate, obtuse; **Cn** flat, dark pink; **Cs** lobes broadly ovate, depressed in the middle, lower face canaliculate, outer appendage rounded, inner appendage acute; **Anth** appendage acute, membranous; **Poll** elongate, ± 0.75 mm, winged; **Fr** ± 14 × 0.6 cm.

H. eitapensis Schlechter (BJS 50: 109, 1913). **T:** Papua New Guinea (*Schlechter* 19964 [B]). – **D:** Papua New Guinea.

Stems twining, branched, filiform, sparsely leafy, glabrous; **L** with a petiole ± 2 mm long, erect and flat, lamina elliptic, 3 - 5 × 1.2 - 2 cm, fleshy, glabrous, ± obtuse, base cuneate; **Inf** 20- to 35-flowered; peduncle ± 5 cm; **Ped** to 2 cm, very thin, apically ± very shortly hairy; **Sep** ovate-triangular, 1.75 mm, ± obtuse, ± shortly hairy; **Cl** disc-shaped, 8 mm ⌀, yellowish-white; **Cl** lobes ovate, acute with revolute tips, outside glabrous, inside ± very shortly papillose to weakly pubescent; **Cs** lobes flat, elongate, 2 mm, inner appendage slightly ascending, acuminate, outer appendage ± obtuse; **Poll** cylindrical, oblong, apically weakly falcate, translator very short, corpuscle rhombic, minute.

H. elliptica Hooker *fil.* (Fl. Brit. India 4: 58, 1885). **T:** Malaysia, Malacca (*Maingay* 3286 [K]). – **D:** Malaysia, Singapore, Thailand. **I:** Rintz (1978).

Stems twining, terete, pale, glabrous; **L** petiole 5 - 7 mm, flat, lamina elliptic, 4.5 - 10 × 2.5 - 4.5 cm,

± fleshy, pale, glabrous, obtuse, base obtuse, margin firm, slightly revolute; **Inf** 1- to 30-flowered, convex, erect; peduncle ± 2.5 cm, thick, pink; **Ped** 2.5 - 5 cm, thin, pink, slightly pubescent; **Sep** ovate-elongate, ± 2 mm, obtuse, pale pink, slightly pubescent; **Cl** 1.5 - 2 cm ⌀, white, outside glabrous, inside sparsely pubescent; **Cl** lobes obcordate, finely acuminate; **Cs** lobes very narrowly ovate-elongate, obtuse, white, outer appendage bladder-like, ascending, apically narrowly keeled, inner appendage shortly subulate; **Anth** appendage acute, membranous; **Sty** head flat, humped in the centre; **Poll** elongate-falcate, winged, caudicles thick, corpuscle very large, lower face flat, 2-winged; **Fr** 15 - 20 × 0.5 cm, often pale pink.

H. engleriana Hosseus (Notizbl. Königl. Bot. Gart. Berlin 4: 315, ill., 1907). **T:** Thailand (*Hosseus* s.n. [not located]). – **D:** Cambodia, Laos, Vietnam, Thailand.

Stems partially pendent, filiform, branched, ± densely leafy, pubescent; **L** shortly or very shortly petiolate, petiole with a leaf cushion, lamina narrowly ovate-lanceolate, 1.5 × 0.4 cm, convex, slightly fleshy, upper face waxy, dark green, white-pubescent, margins recurved; **Inf** terminal, usually 4-flowered, shortly or very shortly pedunculate; **Fl** pleasantly perfumed; **Ped** ± 5 mm, hairy; **Cl** star-shaped and spreading, 1.5 cm ⌀, fleshy, white, outside glabrous, inside densely shortly tuberculate-pubescent; **Cl** lobes broadly ovate-triangular, ± acuminate, margin slightly revolute; **Cn** pink to violet; outer appendage of the **Cs** lobes obtuse, becoming darker towards the tips, inner appendage acuminate, bent upwards, apically darker.

Closely related to *H. linearis*, which exhibits a similar white corolla and linear-lanceolate though longer leaves.

H. erythrostemma Kerr (BMI 1939: 460, 1939). **T:** Thailand, Surat (*Kloss* 6909 [K]). – **D:** Myanmar, Malaysia, S Thailand. **I:** Rintz (1978).

Stems twining; **L** elliptic, 8 - 9 × 2.5 - 3 cm, fleshy, acute, base cuneate; **Inf** 1- to 30-flowered, convex, ± erect; peduncle horizontal, rigid, 3 - 4 cm; **Ped** rigid, ± 2 cm; **Cl** ± 1 cm ⌀, white to pale pink, inside long-hairy, **Ha** long; **Cl** lobes strongly revolute; **Cn** flat, deep red; **Cs** lobes very narrowly elliptic, thin, outer and inner appendages acute; **Poll** ± 0.5 mm, winged, caudicles ± 0.13 mm, corpuscle ovoid, ± 0.2 mm; **Fr** ± 14 × 0.4 cm.

H. finlaysonii Wight (Contr. Bot. India, 38, 1834). **T:** Malaysia, Malacca (*Wight* s.n. [K]). – **D:** Myanmar, S Thailand, Malaysia, Indonesia (Borneo, Sumatra). **I:** Rintz (1978).

Stems twining, robust, pale brown; **L** with a thick petiole 0.6 - 13 cm long, lamina elliptic, 21 × 6 cm, thick, rigid, venation conspicuously prominent, upper face greenish-red, lower face dark red, minutely papillose when young, acute, base cuneate; **Inf** 1- to 40-flowered, convex, erect; peduncle 2 - 3 cm; **Ped** 2.5 cm, very thin, rigid, ± blotched with red; **Cl** 6 mm ⌀, white to yellow, lobes tipped deep red, inside finely pubescent; **Cl** lobes ovate, acute, strongly revolute; **Cn** white; **Cs** lobes ovate, outer appendage ovate-acute, ascending, inner appendage short, acute; **Anth** appendage rounded, short, membranous; **Poll** elongately falcate, obtuse, winged, caudicles very short, thick, corpuscle conical; **Fr** 12 × 0.6 cm; **Se** cylindrical, 0.75 cm.

H. fuscomarginata N. E. Brown (BMI 1910: 278, 1910). – **D:** Only known from cultivation. **Fig. XXV.a**

Stems robust, ± woody with age; **L** petiole 2.5 - 4.5 × 0.7 cm, ash-grey, lamina ovate-lanceolate, 17 - 22 × 7 - 8 cm, thickly fleshy, acute or abruptly acuminate, base obtuse or slightly cuneate, green with darker margins, glabrous; **Inf** many-flowered; peduncle 3.5 - 4.5 × 0.25 cm, sparsely minutely pubescent; **Sep** ovate, 1 mm, acute, sparsely minutely pubescent; **Cl** star-shaped and spreading, 1.2 - 1.3 cm ⌀, yellowish-brown, yellowish-green or pale pink, glabrous; **Cl** lobes ovate, 5 × 5 mm, acuminate, apically revolute; **Cs** lobes narrowly ovate, 3.5 mm, outer appendage narrowly acute, darker, inner appendage acute, ascending.

A widely cultivated species of dubious origin.

H. globulosa Hooker *fil.* (Gard. Chron., ser. nov. 7: 732, fig. 115, 1882). **T** [lecto]: India, Sikkim (*Hooker* Hoya no. 32 [BM, K]). – **D:** India (Sikkim-Himalaya). **Fig. XXV.e**

Stems robust, woody, ± long-hairy; **L** 1.3 - 2.5 cm petiolate, lamina elliptic or elongate, 12 - 17.5 × 5 - 9 cm, leathery, hairy, acuminate, base rounded; **Inf** globose, pendent; peduncle 7.5 - 10 cm; **Ped** 2.5 cm, whitish, villose; **Sep** round; **Cl** 1.3 cm ⌀, yellowish or cream-coloured; **Cl** lobes short, revolute, inside glabrous; **Cn** basally pink; **Cs** lobes broadly elliptic, outer appendage round, inner appendage erect, acute; **Anth** appendage large, broad; **Fr** cylindrical, 30 - 40 cm, slender; **Se** ± 7 mm, thin.

H. heuschkeliana Kloppenburg (Hoyan 11(1: Part 2): i-iii, ills., 1989). **T:** Philippines, Luzon (*Pancho* 2175 [CAHP, UC]). – **Lit:** Meve (2001). **D:** Philippines (Luzon). **Fig. XXV.f**

Epiphytic, climbing on tree trunks, latex white; stems procumbent-creeping, green, cylindrical, wiry, 1 - 2 mm ⌀, with scattered decumbent **Ha**, with adventitious **R**, richly branching; **L** 3 - 5 mm petiolate, broadly ovate to elliptic, 2 - 5 × 1 - 2.2 cm, upper face dark green, lower face paler, usually slightly convex, base obtuse to cuneate, succulent to leathery, marginally with scattered **Ha**; **Inf** globose, 2- to 7-flowered, inserted extra-axillary on the lower face of the stems; peduncle 2 - 3 mm, with scattered appressed **Ha**; **Ped** 2 - 4 mm; **Sep** deltoid,

± 1.2 mm; **Fl** with ample nectar, with dull-sweetish scent esp. in late afternoon; **Cl** depressedly globose-urceolate, ± 3.5 - 4 × 5 - 7 mm, outside glabrous, inside finely papillose; **Cl** tube cream-coloured; **Cl** lobes triangular, ± 2.5 × 2.5 mm, acute, erect-connivent except for the completely recurved pinkish tips, margins slightly recurved, pinkish; **Cn** yellowish; **Cn** lobes lanceolate, ± 3 × 1.2 mm, ascending at 45°, lower face canaliculate, basally winged, bifid, dorsal projection apron-shaped, inner projection acute, erect, ± connivent above the conical **Sty** head; **Poll** rounded-rectangular, ± 0.25 × 0.14 mm, corpuscle rhomboid, ± 0.25 × 0.1 mm, caudicles short, ± cylindrical; **Fr** and **Se** not known.

This small-leaved *Hoya* possesses an urceolate corolla that is otherwise typical for the genus *Dischidia*. Because of the corona, which has the typical morphology of *Hoya* species, it cannot be confused with that genus. – [U. Meve]

H. hypolasia Schlechter (BJS 50: 123-124, ills., 1913). **T**: Papua New Guinea (*Schlechter* 18075 [B]). – **D**: Papua New Guinea. **Fig. XXVI.a**

Stems twining, sparsely branched, narrowly filiform, sparsely leafy, with numerous verrucose lenticels, minutely softly pubescent, ± glabrous with age; **L** petiole ± 1 cm long, canaliculate on the upper face, very shortly pubescent, lamina flat, lanceolate, 12 - 20 × 3.3 - 5.3 cm, leathery, upper face shining, glabrous, lower face very shortly and faintly pubescent, acuminate, base slightly cordate; **Inf** ± 10-flowered; peduncle 5 cm, shortly and faintly pubescent; **Ped** ± 5 cm, thin, glabrous; **Sep** ovate, ± 2 mm, obtuse, margin shortly hairy; **Cl** ± 2.1 cm ∅, yellowish-white, outside faintly suffused with red, glabrous, inside with minute papillae; **Cl** lobes broadly rhombic-ovate, acute; **Cs** lobes obliquely 4-angled in lateral view, 3 × 3.5 mm; **Anth** small; **Poll** obovoid-oblong, translator very small, corpuscle minute, oblong.

H. imbricata Decaisne (PSRV 8: 637, 1844). – **D**: Indonesia (Sulawesi), Philippines. **I**: Koorders (1919). **Fig. XXV.d**

Incl. *Hoya imbricata* fa. *typica* De Candolle (1846) (*nom. inval.*, Art. 24.3); **incl.** *Hoya maxima* Teijsmann & Binnendijk (1863); **incl.** *Collyris major* Neves (1877) (*nom. illeg.*, Art. 53.1); **incl.** *Conchophyllum maximum* H. Karsten (1895) ≡ *Dischidia maxima* (H. Karsten) Koorders (1898) ≡ *Hoya maxima* (H. Karsten) Warburg (1907) (*nom. illeg.*, Art. 53.1); **incl.** *Hoya imbricata* var. *basi-subcordata* Koorders (1919); **incl.** *Hoya pseudomaxima* Koorders (1919).

Stems climbing, rooting at the nodes, stems and **L** densely appressed to the bark; **L** imbricate, very shortly petiolate, lamina circular, 8 - 10 cm ∅, raised like a watch-glass, lower face purple, sparsely ciliate or glabrous, tip obtuse to rounded, base rounded to cordate; **Inf** many-flowered, erect, concave; peduncle ± 10 cm, glabrous; **Ped** variable in length, ± bent, green, glabrous; **Sep** narrowly elliptic, very acute, glabrous; **Cl** 8 - 10 mm ∅, cream-coloured; **Cl** lobes thin, very finely acuminate, strongly revolute, outside glabrous, inside very densely and finely hairy; outer appendages of the **Cs** lobes narrowly rounded, ± flat, inner appendages elliptic, steeply ascending, white; **Anth** appendage far overtopping the **Cn**, acuminate; **Fr** fusiform, straight to slightly bent, acute, glabrous; **Se** narrow, 7 mm, brown, tuft of **Ha** < 14 mm.

See note under *H. caudata*.

H. inconspicua Hemsley (BMI 1894: 213, 1894). **T**: Solomon Islands (*Anonymus* s.n. [K]). – **D**: Australia (Queensland), Solomon Islands, Papua New Guinea. **I**: Forster & Liddle (1990). **Fig. XXVI.b**

Incl. *Hoya litoralis* Schlechter (1905); **incl.** *Hoya dodecatheiflora* Fosberg (1940).

Epiphytic; stems ± twining, to 4 mm ∅, glabrous to finely hairy; **L** with a cylindrical curved petiole 6 - 15 × 1.5 mm, lamina ovate, broadly or narrowly lanceolate, 3 - 10 × 1.5 - 3 cm, thickly leathery to fleshy, finely hairy, acute to acuminate, base cuneate to rounded; **Inf** to 20-flowered, pendent; peduncle 7 cm × 1 - 2 mm, red-brown, finely hairy; **Sep** ± 1 × 1 mm, acute, pale pink, outside finely hairy; **Cl** button-shaped; **Cl** lobes ovate, ± 4 × 4 mm ∅, strongly revolute, pale pink, outside glabrous, inside very densely hairy, **Ha** to 0.5 mm; **Cn** ± 2.5 × 5 - 7 mm, pale pink; **Cs** lobes ± 3 × 1 mm, outer appendage bifid, ± 1 × 0.5 mm, inner appendage subulate; **Anth** appendage ovate, ± 1 × 0.75 mm, margin translucent; **Sty** head ± 1.5 mm; **Poll** elongate, ± 0.4 × 0.17 mm, corpuscle elongate, ± 0.15 × 0.1 mm, caudicles ± 0.16 × 0.13 mm, winged; **Fr** fusiform, ± 10 × 0.6 cm, glabrous.

Forster & Liddle (1992) treat *H. inconspicua*, *H. litoralis* and *H. dodecatheiflora* as synonyms of *H. revoluta*.

H. kerrii Craib (BMI 1911: 418, 1911). **T**: Thailand (*Kerr* 1810 [BM]). – **D**: China, Cambodia, Laos, S Vietnam, NW Thailand, Indonesia (Java). **I**: Costantin (1912); Pham-Hoang (1972). **Fig. XXVI.c**

≡ *Hoya obovata* var. *kerrii* (Craib) Costantin (1912).

Stems climbing, to 1 m, ± 7 mm ∅, pale, glabrous; **L** with a thick petiole 0.5 - 2 cm long, lamina obcordate, 4 - 12 × 5 - 9.5 cm, thickly fleshy, ± glabrous, tip deeply 2-lobed to 1 - 1.75 cm, base rounded or broadly cuneate, margin revolute; **Inf** many-flowered, 4 - 5 cm ∅; peduncle 2 - 6 cm, 3 mm ∅, glabrous; **Ped** 1.3 - 1.8 cm, thin, pubescent; **Sep** ovate-elongate, 2.5 mm, obtuse, outside pubescent; **Cl** 9 - 13 mm ∅, whitish, inside papillose to densely pubescent; **Cl** lobes ovate-triangular, 4 - 5 × 5 mm, acute, basally connate, strongly revolute;

Cn pink to purple; **Cs** lobes broadly ovate, 2.5 mm, upper face depressed in the middle, outer appendage broadly rounded, inner appendage acute; **Poll** 1 mm.

Closely allied to *H. obovata* (see there for a note).

H. lacunosa Blume (Bijdr. Fl. Ned. Ind., 1063, 1826). **T:** Indonesia, Java (*Horsfield* s.n. [K]). — **D:** India, China, Malaysia, Singapore, Thailand, Indonesia (Borneo, Java, Sumatra). **I:** Rintz (1978).

≡ *Otostemma lacunosum* (Blume) Blume (1848); **incl.** *Hoya suaveolens* Miquel (1856); **incl.** *Hoya lacunosa* var. *pallidiflora* Hooker *fil.* (1861).

Stems thin, glabrous, rooting at the nodes; **L** petiole 4 mm long, thick, nearly glabrous, lamina ovate or lanceolate, 3 - 7 × 2.5 - 3 cm, fleshy, shining, glabrous, ± acuminate, base rounded or narrow, sections between the veins slightly sunken; **Inf** 1- to 30-flowered, convex, pendent; peduncle < 5 cm, rigid; **Fl** scented; **Ped** ± 0.4 - 2.5 cm, ± bent, rigid; **Sep** minute, obtuse; **Cl** button-shaped, ± 8 mm ⌀, whitish, inside pubescent, **Ha** long, thick; **Cl** lobes triangular, strongly revolute; **Cn** whitish; **Cs** lobes elliptic, outer appendage obtuse, slightly ascending, inner appendage acute, ascending; **Sty** head conical; **Anth** appendage very thinly membranous, acute; **Poll** elongate-falcate, caudicles calycinate, thick, broadly winged; **Fr** 5 - 6 × 0.5 cm, smooth.

H. lanceolata Wallich *ex* D. Don (Prodr. Fl. Nepal., 130, 1825). **T:** Nepal (*Wallich* s.n. [BM]). — **D:** Bhutan, India, Myanmar, Nepal.

H. lanceolata ssp. **bella** (Hooker) D. H. Kent (Asklepios No. 23: 27, 1981). — **D:** India, Nepal, S Myanmar.

≡ *Hoya bella* Hooker (1848); **incl.** *Hoya paxtonii* hort. (s.a.) (*nom. inval.*, Art. 29.1).

Differs from ssp. *lanceolata*: Stems curved, densely pubescent; **L** ovate-lanceolate or elongate-trapeziform, base rounded, dark green, paler on the lower face, becoming purple under strong light, midrib prominent; **Inf** 8- to 11-flowered; **Fl** with a strong sweet scent; **Cl** 1.8 cm ⌀, whitish; **Cl** lobes 7 × 7 mm, acuminate; **Cs** lobes ovate to narrowly ovate, inner appendage ± acute, slightly ascending, upper face concave, lower face paler.

H. lanceolata ssp. **lanceolata** — **D:** India (Assam, W Himalaya, Sikkim), Bhutan, Nepal, S Myanmar.

Incl. *Hoya sikkimensis* hort. (s.a.) (*nom. inval.*, Art. 29.1).

Stems pendent, limp, densely leafy, shortly pubescent; **L** 2.5 mm petiolate, lamina lanceolate to rhombic-lanceolate or narrowly trapeziform, 2 - 5 cm, ± fleshy, glabrous, acuminate, base narrowly rounded to acute, green becoming pale yellow in strong light; **Inf** 3- to 10-flowered, flat, pendent, often terminal; peduncle 0.5 - 2 cm, sparsely pubescent; **Fl** with a strong fresh scent; **Ped** ± 1.5 cm; **Sep** oblong-lanceolate, pubescent; **Cl** 1.3 cm ⌀, pure white; **Cl** lobes star-shaped and spreading, 6 × 4 mm, obtuse, inside pubescent; **Cn** pink to dark purple; **Cs** lobes ± cylindrical, outer appendage ± acute, inner appendage ± obtuse, slightly ascending, upper face convex; **Fr** 12 - 15 cm, very slender; **Se** 2 mm, tuft of **Ha** ± 2.5 cm.

H. latifolia G. Don (Gen. Hist. 4: 127, 1838). **T:** Malaysia (*Wallich* s.n. [K ?]). — **D:** Myanmar, Malaysia, S Thailand, Indonesia (Borneo, Java, Sumatra). **I:** Rintz (1978).

Incl. *Hoya macrophylla* Wight (1834) (*nom. illeg.*, Art. 53.1); **incl.** *Hoya polystachya* Blume (1849); **incl.** *Hoya latifolia* ssp. *kinabaluensis* C. M. Burton (1991); **incl.** *Hoya loyceandrewsiana* T. Green (1995).

Stems deep red when young; **L** with a robust petiole 6 - 12 mm long, lamina ovate, 10 - 25 × 7 - 17 cm, fleshy, upper face bright green, lower face deep red, acute to acuminate, base rounded to cordate; **Inf** 1- to 40-flowered, convex, erect, part-**Inf** paired, 3 - 4 cm; peduncle ± 5 cm, very thick, rigid, rachis thick; **Ped** ± 2 cm, thin, shortly pubescent; **Sep** ovate, minute, acute; **Cl** 8 mm ⌀, white to pale brownish-yellow, inside finely pubescent; **Cl** lobes ovate, acute; **Cn** white; **Cs** lobes ovate, outer appendage acute, ± flat, inner appendage acute, slightly ascending; **Poll** winged, ± 0.35 mm, corpuscle angled, ± 0.15 mm; **Fr** 7.5 × 0.5 cm, weakly pubescent.

Vegetatively very similar to *H. verticillata* var. *citrina* (Rintz 1978).

H. limoniaca S. Moore (JLSB 45: 368, 1921). **T:** New Caledonia, Isle of Pines (*Compton* 2253 [BM]). — **D:** New Caledonia. **Fig. XXVI.e**

Stems twining, subterete, thickened at the nodes, subglabrous; **L** to 15 mm petiolate, lamina ovate, 7 - 8 × 3 - 3.5 cm, thickly leathery, apically narrower, obtuse, base rounded-obtuse; **Inf** few-flowered; peduncle subterete, sparsely hairy; **Fl** faintly scented; **Ped** filiform, 12 - 15 mm, glabrous; **Sep** ovate, 1.5 mm, obtuse, outside papillose, glabrous; **Cl** 1 cm ⌀, white to pale yellow; **Cl** lobes ovate, obtuse, inside glabrous or finely minutely papillose; **Cn** nearly flat; **Cs** lobes ovate, 4 mm, acute, outer appendage white, inner appendage reddish; **Anth** appendage ovate-lanceolate, obtuse; **Poll** elongate and pear-shaped, 0.6 mm.

Forming a species complex together with *H. nicholsoniae*.

H. linearis Wallich *ex* D. Don (Prodr. Fl. Nepal., 130, 1825). **T:** Nepal (*Wallich* s.n. [BM]). — **D:** India (Sikkim), Nepal, China (Yunnan). **I:** CBM 1883: t. 6682. **Fig. XXVI.d**

Incl. *Hoya linearis* var. *nepalensis* Hooker *fil.* (1883); **incl.** *Hoya linearis* var. *sikkimensis* Hooker *fil.* (1883).

Epiphytic; stems pendent, to 1.5 m, thin, soft, greyish-green, ± villose; **L** 2 mm petiolate, lamina linear, 2.5 - 5 × 0.5 cm, fleshy, dark green, ± villose, acute, base cuneate, lower face canaliculate; **Inf** many-flowered, terminal; **Ped** 2.5 cm; **Sep** linear-lanceolate, villose; **Cl** ± 1 cm ⌀, ± white, inside glabrous or papillose, margin ciliate; **Cl** lobes acute, revolute; **Cs** lobes nearly cylindrical, very pale pink or yellow, outer appendage incised, furrowed, inner appendage slightly darker, acute; **Fr** fusiform, 6.5 × 0.5 cm; **Se** 3 mm, tuft of **Ha** 2 cm, white.

See also under *H. engleriana*.

H. longifolia Wallich *ex* Wight (Contr. Bot. India, 36, 1834). **T:** Nepal (*Wallich* s.n. [K, E]). – **D:** India, Bhutan, Nepal, Pakistan, China, Singapore, Thailand. **Fig. XXVI.f**

Epiphytic or epilithic; stems pendent to twining, to 1 m, glabrous or pubescent; **L** with a thick petiole 8 - 13 mm long, lamina narrowly lanceolate, 5 - 10 × 1.2 - 2.6 cm, lower face canaliculate, upper face convex, thickly fleshy, dull green, ± acuminate, base narrowly acute or rounded; **Inf** many-flowered; peduncle 1.2 - 5 cm, stout, glabrous; **Ped** 1.2 - 2.6 cm; **Sep** ovate, 1 - 1.5 mm, acute, glabrous; **Cl** star-shaped and spreading, 1 cm ⌀, white; **Cl** lobes nearly triangular, acute, inside glabrous or pubescent, margin ciliate; **Cn** white or pink; **Cs** lobes broadly ovate, depressed in the middle, lower face convex, outer appendage broadly rounded, slightly ascending, inner appendage acute, ascending; **Fr** terete, 10 - 30 × 0.5 - 0.8 cm, smooth; **Se** ovoid, 1 - 3 mm, tuft of **Ha** 1.2 - 1.5 cm.

According to Hara (1971), the leaves of *H. longifolia* are very variable in shape and size (see also *H. shepherdii*), and its flowers resemble those of *H. thomsonii*.

H. macgillivrayi F. M. Bailey (Queensland Agric. J., n.s. 1: 190, fig. 14, 1914). **T:** Australia, Queensland (*Macgillivray* s.n. [BRI AQ333104]). – **D:** Australia (Queensland). **I:** Forster & Liddle (1990). **Fig. XXVII.a**

Epiphytic, rarely on rocks; stems twining, terete, to 5 mm ⌀, glabrous; **L** 2 - 3.5 × 0.2 - 0.5 cm petiolate, lamina ovate to lanceolate, 15 - 20 × 2.5 - 8 cm, copper-coloured when young, dark green with age, glabrous, acute, base cordate; **Inf** to 12-flowered; peduncle 4 - 20 × 0.1 - 0.2 cm, green to yellowish; **Fl** 5.5 - 8 cm ⌀; **Ped** 5.4 - 8.5 mm, green, glabrous; **Sep** 3 - 4 × ± 2 mm, acute; **Cl** cup-shaped, uniformly dark red or with a white centre; **Cl** lobes 2 - 2.5 × 1.9 - 2.3 cm, acute, margins extremly recurved; **Cn** ± 1 × 1.8 - 2.4 cm, dark red, glabrous; **Cs** lobes linear, 10 - 12 × 2 - 3 mm, inner appendage acute, 5 - 10 mm, ascending, outer appendage broadly ovate or acute, slightly ascending; **Anth** appendage ± 2.5 mm, acute; **Sty** head ± 3 mm ⌀; **Poll** elongate, ± 1.4 × 0.45 mm, corpuscle ± 0.8 × 0.5 mm; caudicles ± 0.4 × 0.2 mm.

The corona of *H. macgillivrayi* is rather variable (Forster & Liddle 1990). The species is closely allied to *H. archboldiana*.

H. macrophylla Blume (Bijdr. Fl. Ned. Ind., 1063, 1826). **T:** Indonesia, Java (*Anonymus* s.n. [L, P]). – **D:** Indonesia (Borneo, Java). **I:** Pham-Hoang (1972).

Incl. *Hoya clandestina* Blume (1848); **incl.** *Hoya browniana* Koorders (1911).

Stems subterete; **L** with a thick petiole to 2.5 cm long, lamina ovate or ovate-elongate, 14 - 25 × 4.5 - 11 cm, fleshy, pale green, glabrous, acute to acuminate, base acute, rarely nearly rounded-obtuse; **Inf** many-flowered, ± globose; peduncle 6 - 10 cm, ± violet, glabrous; **Ped** 1 - 3 cm, purple, glabrous or very sparsely hairy; **Sep** triangular, ± 0.6 mm, red, outside very sparsely hairy; **Cl** reddish-white to violet; **Cl** lobes ovate, acute, revolute, fleshy, outside glabrous, inside densely white-papillose; **Cn** ± 3.5 mm ⌀, white, centre white or pink; **Cs** lobes keeled on the upper face, outer appendage acute, ascending, inner appendage acute; **Poll** linear-lanceolate.

H. magnifica P. I. Forster & Liddle (Austrobaileya 3(4): 629-630, fig. 1, 1992). **T:** Papua New Guinea, Morobe Prov. (*Streimann & Kairo* NGF39381 [CANB, A, BO, BRI, K, L, LAE]). – **D:** Papua New Guinea. **Fig. XXVII.d**

Stems climbing, twining, terete, several m long, 5 mm ⌀, densely hairy, becoming sparsely hairy with age; **L** petiole 14 - 43 × 2.3 - 2.5 mm, densely hairy, erect, lamina elliptic-ovate to narrowly ovate, 19 × 10 cm, sparsely hairy on the upper face, densely hairy on the lower face, ± finely acuminate, base cordate to rounded; **Inf** horizontal to pendent, ± 9 cm ⌀; **Bra** triangular, 1.4 - 1.5 × 1.2 - 1.3 mm, ± hairy; peduncle 9 - 10 × 5 - 6 mm, densely hairy; **Fl** 1 - 1.5 × 4 - 4.5 cm; **Ped** 2 - 5 cm, densely hairy; **Sep** lanceolate-ovate, 13 - 16 × 7 - 10 mm, outside hairy; **Cl** campanulate, white; **Cl** lobes spreading or revolute, triangular, 15 - 18 × 15 - 18 mm, glabrous; **Cl** tube 10 - 14 × 22 - 25 mm, glabrous; **Cn** white, 4 - 5 × 10 - 16 mm; **Cn** lobes rising high, outer appendage round, bent upwards, inner appendage lanceolate; **Anth** appendage lanceolate, 1.7 - 2 × 1.3 - 1.7 mm; **Sty** head depressed-globose, 4.5 - 5 mm ⌀; **Poll** elongate, 1.1 × 0.4 mm, corpuscle ovate, ± 0.8 × 0.5 mm, caudicles cylindrical, 0.3 × 0.1 mm, not winged; **Fr** fusiform, ± 19 × 1.2 cm, glabrous.

According to the protologue closely related to *H. albiflora*, but distinguishable by the larger flowers.

H. meliflua (Blanco) Merrill (Sp. Blancoan., 318, 1918). – **D:** Philippines. **Fig. XXVII.b**

≡ *Stapelia meliflua* Blanco (1837); **incl.** *Hoya carnosa* Blanco (1845) (*nom. illeg.*, Art. 53.1); **incl.** *Hoya diversifolia* Fernández-Villar (1880) (*nom. illeg.*, Art. 53.1); **incl.** *Hoya parasitica* Fernández-

Villar (1880) (*nom. illeg.*, Art. 53.1); **incl.** *Hoya luzonica* Schlechter (1904).

Stems terete, 3 - 4 m, slightly fleshy, branched, glabrous; **L** with a slightly fleshy petiole 1.5 - 2.5 cm long, glabrous, lamina elongate-ovate, 9 - 13 × 5 - 6 cm, slightly obtuse or acute, leathery to fleshy, glabrous on both faces; **Inf** many-flowered; peduncle terete, 6 - 8 cm, fleshy, glabrous; **Ped** 1 - 2.5 cm, thin; **Sep** broadly ovate-elongate or nearly circular, 3 mm, glabrous; **Cl** button-shaped, ± 15 mm ⌀, lilac to purple, below the **Cs** lobes dark red; **Cl** lobes broadly triangular, 6 mm, inside densely papillose to densely velvety; **Cn** ± 0.75 mm ⌀, glabrous; **Cs** lobes broadly elliptic to triangular, outer appendage ± broadly ovate, depressed in the middle, white, inner appendage slightly ascending, acute, dark purple; **Sty** head conical; **Poll** keeled, translator minute, corpuscle acute; **Fr** ± 12 × 0.7 cm; **Se** lanceolate in top view, 6 × 1.5 mm, flat, tuft of **Ha** 1.5 mm.

H. meredithii T. Green (Phytologia 64(4): 304-306, ills., 1988). **T:** Malaysia, Sarawak (*Meredith* s.n. in *Green* 80-05 [BISH, NY]). – **D:** E Malaysia. **Fig. XXVII.c**

Epiphytic or rooting in the ground; stems twining, filiform, sparsely leafy and rooting; **L** petiole 2.5 - 3 × 0.8 cm, stout, twisted, lamina ovate, 12 - 30 × 7.5 - 18 cm, acuminate, base obtuse, margin undulate; **Inf** many-flowered, semiglobose; peduncle 10 × 0.2 cm; **Fl** 1 cm ⌀, pale yellow to yellow-green; **Ped** erect; **Sep** triangular, 1.5 mm, acute, glabrous; **Cl** dark yellow; **Cl** lobes ovate, acute, strongly revolute; **Cn** 5 mm ⌀, flat, fleshy, white to greenish-white; **Cs** lobes elliptic, outer appendage acute, inner appendage acute; **Poll** elongately flattened, translator very short, bent, corpuscle minute.

H. nicholsoniae F. Mueller (Fragm. Phytogr. Austral. 5: 159, 1866). **T:** Australia, Queensland (*Dallachy* s.n. [MEL]). – **D:** Australia (Queensland), Papua New Guinea, Solomon Islands. **I:** Forster & Liddle (1990). **Fig. XXVII.e**

Incl. *Hoya hellwigii* Warburg *ex* K. Schumann & Lauterbach (1901); **incl.** *Hoya hellwigiana* Warburg (1907); **incl.** *Hoya sogerensis* S. Moore (1911).

Epiphytic or rarely on rocks; stems twining, ± 5 mm ⌀, glabrous; **L** petiole 2 - 3 × 0.3 - 0.8 cm, lamina narrowly ovate to ovate, 4.5 - 21 × 4.5 - 10 cm, leathery to fleshy, (brown-) green to purple, acute, base rounded, cuneate or slightly cordate; **Inf** to 40-flowered; peduncle ± 12 × 0.2 cm, green to ± purple, glabrous; **Ped** ± 18 - 30 mm, pink to cream-coloured, glabrous; **Sep** triangular, ± 1.5 - 3 × 2 mm; **Cl** 10 - 18 mm ⌀, pale yellow, cream-coloured, green or pale pink, minutely sparsely pubescent; **Cl** lobes ovate to ovate-lanceolate, 6 - 7 × 4.5 - 6 mm, revolute; **Cn** 2 - 3 × 7 - 8 mm; **Cs** lobes ovate-lanceolate, outer appendage acute, rarely rounded, inner appendage more narrowly acute, concolorous with the **Cl** or paler; **Anth** appendage triangular, ± 1 mm; **Sty** head ± 1.5 mm ⌀; **Poll** erect, elongate, ± 0.7 × 0.23 mm, caudicles ± 0.19 mm, winged, corpuscle elongate, ± 0.25 × 0.15 mm; **Fr** fusiform, 8 - 15 × 0.7 - 1.2 cm; **Se** 6 - 7 × 3 - 4 mm, tuft of **Ha** white, 2 - 2.5 cm.

The SE Asian *H. nicholsoniae* is rather variable in relation to size and shape of the leaves and colour of the corolla and corona (Forster & Liddle 1990). Forster & Liddle (1992) treat *H. nicholsoniae* as a synonym of *H. pottsii* from continental Asia.

H. nummularioides Costantin (in Lecomte, Fl. Indo-Chine 4: 129-130, ills., 1912). **T** [syn]: Cambodia (*Geoffray* s.n. [not located]). – **D:** Cambodia, Laos, Vietnam, Thailand. **Fig. XXVII.f**

Stems pubescent; **L** petiole 2 - 5 mm long, thick, weakly pubescent, lamina rounded to ovate, 12 - 25 × 9 - 16 m, fleshy, acute, weakly pubescent; **Inf** 12- to 18-flowered; peduncle 1.5 - 3 cm, weakly pubescent; **Fl** 4 mm ⌀, fragrant; **Ped** 6 - 10 mm, thin, weakly pubescent; **Sep** ovate, obtuse, weakly pubescent; **Cl** white to pink; **Cl** lobes triangular, tips strongly revolute, inside weakly pubescent; **Cn** dark pink; **Cs** lobes ovate, outer appendage ± acute, paler, inner appendage acuminate, tips white; **Anth** appendage very short; **Sty** head conical, minute; **Poll** rather flat, caudicles small; **Fr** cylindrical, acute, slightly bent.

H. obovata Decaisne (PSRV 8: 635, 1844). **T:** Indonesia, Sulawesi (*Labillardière* s.n. [not located]). – **D:** India, Indonesia, Thailand, Fiji. **Fig. XXVIII.b**

Stems twining, robust, green; **L** ± 1 cm petiolate, lamina obovate to obcordate, 15 × 8 - 10 cm, very thickly leathery, glabrous, tip 2-lobed, base narrow, margin revolute; **Inf** ± flat, 1.5 cm ⌀; peduncle short; **Cl** 1.5 cm ⌀, cream-coloured, yellowish-orange to pink or dark pink, inside densely papillose; **Cl** lobes ovate, apically revolute; **Cn** purple-red, centre ± darker; **Cs** lobes broadly ovate, depressed in the middle, 4 mm, outer appendage ± rounded, inner appendage acute.

H. obovata differs from *H. kerrii* by a larger corona, indeed by an overall larger flower, as well as by a slightly different leaf shape and the general absence of any pubescence.

H. pachyclada Kerr (BMI 1939: 462, 1939). **T:** Thailand (*Kerr* 20007 [not located]). – **D:** Thailand. **Fig. XXVIII.c**

Epiphytic; stems robust, slightly succulent, 6 - 8 mm ⌀, dark when young, weakly pubescent; **L** petiole 6 - 12 mm long, thick, canaliculate on the upper face, minutely weakly pubescent, lamina obovate, 7.8 - 10.5 × 5.2 - 6 cm, thickly leathery, sparsely and shortly hairy, shortly obtuse, base cuneate or slightly rounded; **Inf** many-flowered; peduncle 1.5 - 3.5 cm, minutely weakly pubescent; **Ped** 2 - 3 cm, thin, weakly pubescent; **Sep** ovate, ± 2 mm, obtuse, outside minutely weakly pubescent; **Cl** ± 17 mm ⌀, yellowish-white, inside glabrous; **Cl** lobes ovate, 7 × 5 mm, acute, strongly revolute; **Cn** flat, white; **Cs** lobes elliptic, ± 4 mm, depressed in the middle, outer appendage acute, inner appendage acuminate, upper face keeled along the middle; **Anth** appendage membranous, finely pointed, tip slightly split; **Poll** very short, corpuscle with subulate tip.

According to the protologue most closely allied to *H. diversifolia*, which differs by smaller leaves and flowers as well as blunter corona lobes.

H. parviflora Wight (Contr. Bot. India, 37, 1834). **T:** Myanmar (*Wallich* 8156A [K]). – **D:** India, Myanmar, Indonesia, Malaysia, Thailand. **I:** Rintz (1978).

Stems twining, terete, thin, young often red; **L** with a thick petiole 4 mm long, lamina lanceolate, 6 - 11.5 × 1 - 2.2 cm, fleshy, glabrous, acute, base acute, margins slightly revolute; **Inf** 1- to 40-flowered, concave, pendent; peduncle recurved, 4.4 - 5.1 cm, thick; **Ped** 0.3 - 3 cm, thin, ± bent; **Sep** ovate, very small, membranous; **Cl** button-shaped, 5 mm ⌀, white to pink, inside villose at the base; **Cl** lobes triangular, revolute; **Cn** broadly conical; **Cs** lobes obovate or obcordate, longitudinally folded, outer appendage 3-fid, inner appendage acute; **Anth** appendage very thin, membranous; **Sty** head conical; **Poll** elongately truncate, minute, flat, winged, caudicles minutely cup-shaped, winged; corpuscle triangular; **Fr** 11.5 - 16 × 0.6 cm, smooth; **Se** elongate, ± 7 mm, pale brown, tuft of **Ha** 2.5 cm, white.

H. parvifolia Schlechter (BJS 40(Beiblatt 92): 15, 1908). **T:** Indonesia, Sumatra (*Schlechter* 13307 [B]). – **D:** Indonesia (Sumatra).

Stems twining, filiform, thin, sparsely leafy, ± densely and shortly villose, with aerial **R**; **L** with a villose petiole 3 mm long, lamina elongate to elongate-elliptic, 1.2 - 1.7 × 0.8 - 1 cm, rather obtuse, densely and shortly villose; **Inf** 15- to 20-flowered; peduncle 4 - 5 cm, thin, villose; **Ped** 0.7 - 1.2 cm, glabrous; **Sep** ovate, 2 mm, obtuse, revolute, margin minutely hairy, otherwise glabrous; **Cl** 7 mm ⌀, pale yellow; **Cl** lobes ovate, acute, tips revolute, outside glabrous, inside densely white-villose; outer appendage of the **Cs** lobes elliptic, inner appendage ascending, acuminate, depressed in the middle; **Anth** trapeziform, **Anth** appendage triangular, translucent; **St** head conical; **Poll** erect, obliquely elongate, outside basally with a horn-shaped process, translator minute, corpuscle elliptic, ± 5× smaller than the **Poll**.

Flower colour and shape are reminiscent of *H. lacunosa*, but vegetatively, the taxon rather resembles *Dischidia*.

H. pauciflora Wight (Icon. Pl. Ind. Orient. t. 1269, 1848). **T:** Sri Lanka (*Wight* 511 [K]). – **D:** India (Malabar, Kerala), Sri Lanka. **Fig. XXVIII.d**

Incl. *Hoya wightiana* Thwaites (1860).

Often epiphytic; **R** adventitious; stems terete, thin, slightly twining when young, glabrous; **L** 2 - 6 mm petiolate, lamina oblong-lanceolate or oblong-linear, 2 - 5.5 × 0.5 - 1 cm, thickly fleshy, shining, lower face often blotched with red, glabrous, tip acute to rounded, base acute or rounded, margin often revolute; **Inf** 2- to 12-flowered; peduncle very short; **Ped** thin, glabrous; **Sep** lanceolate, small, glabrous; **Cl** 2 cm ⌀, white; **Cl** lobes broadly obtuse or weakly acute, inside glabrous or minutely densely pubescent; **Cn** pink to dark purple; **Cs** lobes ovate, fleshy, depressed in the middle, outer appendage rounded, inner appendage spur-like ascending; **Fr** fusiform, ± 10 cm, very slender.

See note under *H. retusa*.

H. pottsii Traill (Trans. Hort. Soc. London 7: 25, fig. 1, 1827). **T:** [lecto – icono]: l.c. fig. 1. – **D:** China. **I:** CBM 62: t. 3425, 1835; Tsiang & Li (1977).

Incl. *Hoya angustifolia* Traill (1830) ≡ *Hoya pottsii* var. *angustifolia* (Traill) Tsiang & P. T. Li (1974); **incl.** *Hoya trinervis* Traill (1830); **incl.** *Hoya obscurinerva* Merrill (1923).

Stems twining, terete, 3 - 4 mm ⌀, branched; **L** petiole 2 - 2.5 cm long, terete, very thick, lamina 5 - 20 × 5 - 10 cm, ovate or ovate-cordate, leathery to fleshy, upper face pale green, lower face dark green, tip acute to acuminate, base acute, margin slightly revolute; **Inf** to 25-flowered, ± globose; peduncle 5 - 8 cm; **Ped** 2 - 2.5 cm, thin, glabrous; **Sep** broadly ovate, short; **Cl** 1 cm ⌀, faintly yellowish-green to white; **Cl** lobes broadly ovate, acute, revolute, inside minutely weakly pubescent; **Cn** flat, white, with a ± yellowish centre; **Cs** lobes ovate, depressed in the middle, outer appendage ± acuminate, inner appendage acute; **Fr** fusiform, acute; **Se** linear-lanceolate in top-view.

This species is extremely variable with regard to flower colour as well as size and shape of the leaves, depending on the habitat and origin (Forster & Liddle 1990). See also the note under *H. nicholsoniae*.

H. purpureofusca Hooker (CBM 76: t. 4520 + text, 1850). **T:** Indonesia, Java (*Loob* s.n. [K?]). – **D:** Indonesia (Java). **Fig. XXVIII.a**

≡ *Hoya cinnamomifolia* var. *purpureofusca* (Hooker) Kloppenburg (2001).

Stems terete, glabrous; **L** petiole ± 15 mm long, very thick, brownish, lamina ovate, ± 10 - 12.5 × 7.5 - 10 cm, fleshy, acute to ± acuminate, base rounded; **Inf** many-flowered, semiglobose; peduncle to ± 8 cm; **Ped** thin; **Sep** usually subulate; **Cl** ± 1.5 cm ⌀, dark pink to dark purple, inside glabrous or pubescent to villose; **Cl** lobes revolute, margins slightly involute; **Cn** flat; **Cs** lobes ovate, acute, dark pink to strongly purple-brown, upper face keeled and depressed in the middle, lower face convex.

Closely allied to *H. cinnamomifolia*. The 2 species are hardly separable vegetatively.

H. retusa Dalzell (Hooker's J. Bot. Kew Gard. Misc. 4: 294, 1852). **T:** India (*Dalzell* s.n. [K]). – **D:** India (Assam, Bombay Presidency), Indonesia (Sulawesi). **I:** Talbot (1909-1911).

Epiphytic, pendent; stems terete, thin, green when young, blotched with purple, grey with age, nodes sparsely pubescent, otherwise glabrous; **L** petiole 5 mm long, terete, green, blotched with purple, ± glabrous, lamina linear, 4.5 × 0.5 cm, thickly fleshy, pale green, lower face canaliculate, sparsely blotched with purple, glabrous, apically obcordate, base acute; **Inf** 1- to 3-flowered; **Bra** triangular, minute, obtuse, margin ciliate; **Ped** terete, 7 - 8 mm, ± glabrous; **Sep** elliptic-oblong, 1.5 × 1 mm, weakly acute, lower face weakly pubescent, upper face glandular; **Cl** star-shaped and spreading, 1.2 - 1.4 cm ⌀, white; **Cl** lobes triangular, 5 mm, acute, velvety-pubescent; **Cn** 5 mm ⌀, pink to dark purple; **Cs** lobes ovate-elongate, outer appendage ± broadly acute, ascending, inner appendage acute; **Sty** head flattened; **Poll** elongate.

Closely allied to *H. pauciflora*, which likewise exhibits a purple corona contrasting with a white corolla. However, both are well-separable by the leaf characters.

H. revoluta Wight *ex* Hooker *fil.* (Fl. Brit. India 4: 55, 1883). **T** [lecto]: Malaysia, Malacca (*Maingay* 1127 [K]). – **D:** Cambodia, Laos, Vietnam, Indonesia (Borneo, Java, Sumatra), Malaysia (Malacca), S Thailand. **I:** Rintz (1978).

Incl. *Hoya ovalifolia* Wallich (s.a.) (*nom. inval.*, Art. 32.1c).

Stems twining, terete, thin, glabrous; **L** petiole thick, 4 mm long, lamina elliptic, 8 × 4 cm, glabrous, acuminate, base acuminate, margin strongly revolute; **Inf** 1- to 30-flowered, concave, pendent; peduncle to 5 cm, rigid; **Ped** 0.2 - 5 cm, ± strongly bent; **Sep** elongate, very small, obtuse; **Cl** ± 5 mm ⌀, pale pink, inside finely weakly pubescent; **Cl** lobes triangular, weakly keeled towards the base; **Cs** lobes elliptic, outer appendage bifid, white or pale pink, inner appendage acuminate, ascending, deep red, whitish or pale pink; **Anth** appendage membranous, acuminate; **Sty** head conical-acute; **Poll** elongate, obtuse, winged, caudicles short, thick, corpuscle small; **Fr** ± 25 × 0.2 cm.

See note under *H. inconspicua*.

H. serpens Hooker *fil.* (Fl. Brit. India 4: 55, 1883). **T:** India, Sikkim (*Griffith* s.n. [K]). – **D:** India (E Himalaya), Nepal, Australia (Queensland). **Fig. XXVIII.e**

Stems twining, nodes rooting, terete, to 1 mm ⌀, sparsely hairy; **L** petiole ± 4 × 1 mm, cylindrical, lamina nearly circular, ± 15 × 12 mm, upper face dull green, lower face paler, ± sparsely hairy; **Inf** pendent, to 8-flowered; peduncle 28 - 35 × ± 1 mm; **Ped** 19 - 20 mm, glabrous or very sparsely hairy; **Sep** broadly ovate, ± 1.5 mm, outside sparsely hairy; **Cl** ± 15 mm ⌀, white; **Cl** lobes 5 × 3 mm, acute, margins revolute, inside densely white-hairy; **Cn** ± 3 - 7 mm, white; **Cs** lobes elliptic, 3.5 × 2 mm, outer appendage ovate, acuminate, ascending, inner appendage acute, dark pink; **Anth** appendage triangular, 0.5 mm; **Sty** head ± 1 mm ⌀; **Poll** ± 0.7 × 0.25 mm, apically narrowly winged, corpuscle ± 0.35 × 0.15 mm, tan-coloured, caudicles ± 0.17 × 0.07 mm.

In Australia only a single locality is known. According to Forster & Liddle (1990) the taxon has probably been introduced.

H. shepherdii Hooker (CBM 87: t. 5269 + text, 1861). **T:** India, Sikkim (*Hooker & Thomson* s.n. [K?]). – **D:** India, SW China.

Lithophytic; stems terete, to 1 m, papillose; **L** shortly teretly petiolate, lamina linear-lanceolate, ± 15 × 1 cm, ± acuminate, revolute to pendent, fleshy to leathery, dark green, upper face very dark and canaliculate, lower face paler, semicircular in transverse view; **Inf** 5 cm ⌀, with several **Fl**; peduncle 1 - 4 cm, clavate; **Fl** white and pink; **Ped** clavate; **Sep** broadly triangular; **Cl** ± 1.3 cm ⌀; **Cl** lobes broadly triangular, acute, tips weakly revolute, basally fused, white, densely villose; **Cn** white; **Cs** lobes broadly ovate, outer appendage acuminate, ascending, inner appendage acute, dark violet.

Hara (1971) places *H. shepherdii* in the synonymy of *H. longifolia*, yet it is separable by longer, narrower and succulent leaves, smaller inflorescences, a shorter peduncle, and by the flat corona.

H. siamica Craib (BMI 1911: 419, 1911). **T:** Thailand (*Kerr* 724 [K?]). – **D:** India, Cambodia, Laos, Vietnam, NW Thailand.

Epiphytic; stems to 1.5 m, glabrous; **L** with a glabrous petiole 5 - 15 mm long, lamina lanceolate, elongate-lanceolate or ovate-lanceolate, 3.5 - 9.5 × 1.5 - 2.2 cm, fleshy, glabrous, somewhat acute, base obtuse, margin slightly revolute; **Inf** 10- to 14-flowered; peduncle 1 - 5 cm, glabrous; **Ped** 2 cm, glabrous; **Sep** small, ovate-lanceolate, acute, outside ± hairy; **Cl** 1.5 cm ⌀, white; **Cl** lobes ovate-triangular, outside glabrous, tip and margin inside weakly pubescent; **Cs** lobes somewhat erect, broadly ovate, 3 - 3.5 mm, margin convex and depressed in the middle.

According to the protologue closely allied to *H. longifolia*.

H. thailandica Thaithong (Nordic J. Bot. 21(2): 143-145, ills., 2001). **T:** Thailand, Chiang Mai (*Thaithong* 1488 [BK, AAU]). – **D:** Thailand (Chiang Mai).

Plants scrambling, twining; stems cylindrical, ± succulent, glabrous; **L** 4 - 6 cm petiolate, lamina ovate to elliptic, 8 - 12 × 6 - 8 cm, finely acuminate, coriaceous, base rounded cuneate, glabrous; **Inf** 10- to 15-flowered, peduncle 4 - 6 cm, glabrous; **Ped** 2.5 - 4 cm; **Sep** boat- to hood-shaped, 6 - 8 × 6 - 7 mm, ciliate; **Cl** ± 2 cm ∅, white, upper face pubescent; **Cl** lobes broadly ovate, ± 5 × 6 mm, abruptly acuminate, margins recurved; **Cn** purple-red; **Ci** lobes ovate, ± 3 × 2.5 mm, horizontal, concave above, grooved below, outer appendage ± rounded, inner appendage acuminate, ascending; **Sty** head conical; **Poll** oblong, ± 1 × 0.3 mm, caudicles winged.

Belonging to the *H. australis* complex and most closely related to the similarly little-succulent ssp. *tenuis*. The long peduncles and the large boat- to hood-shaped sepals are significant for *H. thailandica*. – [U. Meve]

H. thomsonii Hooker *fil.* (Fl. Brit. India 4: 61, 1883). **T:** India, Assam (*Hooker & Thomson* s.n. [K?]). – **D:** India (Assam), Tibet, China (Xizang). **Fig. XXVIII.f**

Stems twining, to 2 m, thin; **L** 5 - 13 mm petiolate, lamina elongate or obovate-oblong, 5 - 7.5 × 1.5 - 3.8 cm, ± fleshy, lower face sparsely pubescent, tips acuminate, base obtuse or rounded; **Inf** pendent; peduncle 1.3 - 2.5 cm, weakly pubescent or ± glabrous; **Ped** sparsely weakly pubescent to glabrous; **Sep** ovate, acute, glabrous; **Cl** ± 1.3 cm ∅, white; **Cl** lobes acute, margin ciliate, inside weakly pubescent or glabrous; **Cn** white; **Cs** lobes obovate, outer appendage broad and rounded, inner appendage spur-like ascending, depressed in the middle.

Another species from the complex around *H. longifolia*.

H. tsangii C. M. Burton (Hoyan 13(1: Part 2): 5, 1991). **T:** Philippines, Mindanao (*Elmer* 13372 [US]). – **D:** China, Philippines.

Incl. *Hoya angustifolia* Elmer (1938) (*nom. illeg.*, Art. 53.1).

Stems terete, to 3 mm ∅, branched, smooth, glabrous; **L** petiole ± 1 cm, thick, canaliculate, lamina narrowly lanceolate, 5 - 12 × 1 - 1.5 cm, acute, base narrowly cuneate, margin revolute, papillose; **Inf** pendent, concave to flat; peduncle terete, 3 - 5 cm, glabrous, rachis very thick; **Ped** filiform, 1 - 2 cm, strongly bent; **Cl** button-shaped, ± 1 cm ∅, dark pink to deep red; **Cl** lobes elongate-ovate, 2 - 3.5 mm, acute, strongly revolute, inside densely greyish-pubescent; **Cn** dark pink to red; **Cs** lobes narrowly elliptic, outer appendage bifid, inner appendage acute, highly ascending, often with a yellow centre; translator winged; **Fr** 1 cm pedicellate, subterete, 8 - 14 cm, acute, bent; **Se** in top view linear, 6 mm, flat, brown.

H. verticillata (Vahl) G. Don (Gen. Hist. 4: 128, 1837). **T:** India, Assam (*Anonymus* s.n. [not located]). – **D:** E India, Myanmar, Cambodia, Laos, Vietnam, Indonesia (Borneo, Java, Sulawesi, Sumatra), Brunei, Malaysia, Thailand, Singapore. **I:** Rintz (1978).

≡ *Sperlingia verticillata* Vahl (1810); **incl.** *Asclepias parasitica* Wallich *ex* Hornemann (1819); **incl.** *Hoya acuta* Haworth (1821); **incl.** *Asclepias parasitica* Roxburgh (1832) (*nom. illeg.*, Art. 53.1) ≡ *Hoya parasitica* (Roxburgh) Wallich *ex* Wight (1834); **incl.** *Hoya hookeriana* Wight (1834); **incl.** *Hoya pallida* Lindley (1839); **incl.** *Hoya albens* Miller *ex* Steudel (1841); **incl.** *Hoya ridleyi* King & Gamble (1903); **incl.** *Hoya parasitica* var. *geoffrayi* Costantin (1912); **incl.** *Hoya parasitica* var. *spirei* Costantin (1912); **incl.** *Hoya globifera* Ridley (1915); **incl.** *Hoya citrina* Ridley (1922) ≡ *Hoya acuta* var. *citrina* (Ridley) D. H. Kent (1989) ≡ *Hoya verticillata* var. *citrina* (Ridley) Veldkamp (1996); **incl.** *Hoya amoena* Bakhuizen van den Brink *fil.* & Backer (1950); **incl.** *Hoya parasitica* var. *citrina* Rintz (1978); **incl.** *Hoya parasitica* var. *hendersonii* Kiew (1995) ≡ *Hoya verticillata* var. *hendersonii* (Ridley) Veldkamp (1996).

Epiphytic; stems creeping, twining, robust and fleshy, pale, glabrous; **L** petiole very thick, 5 - 25 mm long, lamina ovate-elliptic to oblong-elliptic, ± 10 × 4 cm, fleshy, glabrous, acute to acuminate, base cuneate; **Inf** 1- to 40-flowered, convex, erect; peduncle 3 - 5 cm, rigid; **Ped** ± 3 cm, thin, glabrous; **Sep** elongate-obtuse; **Cl** ± 1.5 cm ∅, white, inside glabrous or weakly pubescent; **Cl** lobes ovate, acute, revolute; **Cs** lobes ovate, acute, outer appendage white, inner appendage deep pink; **Anth** appendage membranous, acute; **Sty** head conical-acute; **Poll** elongate, obtuse, winged, caudicles short, thick, corpuscle conical; **Fr** ± 12 × 0.4 cm, often numerous.

In contrast to the meaning of the epithet, the leaves are not arranged in whorls.

H. vitellina Blume (Mus. Bot. 1: 45, 1849). **T:** Indonesia, Java (*Anonymus* s.n. [not located]). – **D:** Indonesia (Java). **Fig. XXVIII.g**

Stems twining, glabrous; **L** shortly petiolate, ovate, obovate or oblong, ± 15 cm, thickly fleshy, glabrous, acute or rounded, base rounded or obtuse; **Inf** many-flowered; peduncle like the petioles; **Cl** < 1 cm ∅, yolk-coloured, shining, glabrous; **Cl** lobes ovate-oblong, acute, revolute; **Cs** lobes narrowly elliptic, flat, outer appendage acute, inner appendage acute, upper face keeled.

×HUERNELIA

B. Müller & F. Albers

×**Huernelia** Barad (CSJA 67(4): 255, 1995).
= *Huernia* × *Stapelia*. The hybrid *H. zebrina* × *S. glanduliflora* was created in cultivation by G. Barad.

HUERNIA

B. Müller & F. Albers

Huernia R. Brown (Mem. Wern. Nat. Hist. Soc. 1: 22, 1810). **T:** *Stapelia campanulata* Masson. – **Lit:** Leach (1988). **D:** Arabian Peninsula, Africa. **Etym:** For Justus Heurnius [van Heurne] (1577 - 1652), Dutch missionary and first European to collect plants at the Cape of Good Hope, RSA.

Dwarf perennial stem succulents; stems matforming or creeping, rarely pendulous; **Tu** arranged in 4 - 5 (rarely 6 or more) **Ri**, apically with acute small **L** rudiment, becoming hard and persistent (sometimes as a soft **Bri**), without stipular **Gl**, glabrous; **Inf** developing successively several (rarely 1) **Fl** mostly at or near the base of the stems; peduncle short, strong, mostly bracteate; **Ped** ± short, glabrous; **Sep** 5, rather variable in size, usually (narrowly) ovoid to elongate-acuminate, glabrous; **Cl** rarely > 5 cm ∅, tubular, campanulate, bicampanulate or ± flat, with or without annulus, with 5 **Cl** lobes and 5 intermediate lobes; **Cl** lobes spreading or concave in longitudinal direction, apically sometimes with a longitudinal furrow, outside glabrous or slightly scabrous, inside ± papillate or verrucose (rarely smooth), sometimes with reticulate folds, more rarely irregularly transversely plaited, often with simple **Bri**, clavate and / or more strongly inflated **Ha**; **Cn** biseriate, sessile; **Ci** 5-lobed, sometimes fused into a flattened or ± conical or convex disc, in front of the guide rails with small fleshy humps, papillose; **Cs** 5 simple lobes incumbent on the backs of the **Anth**, glabrous, usually with a warty transverse hump at the base, apically often finely verrucose; **Anth** incumbent on the **Sty** head, ± quadrangular; **Poll** elongate-elliptic, with short caudicles; **Fr** 8 - 9 × 1 cm, (narrowly) fusiform, the follicles of a pair ± erect, glabrous; **Se** ± ovate, ± flat, winged, with apical tuft of **Ha**.

The genus is divided into 3 sections:
[1] Sect. *Huernia*: **Cs** overtopping the **Anth** by far, erect, with a ± slight hump at the back.
[2] Sect. *Plagiostelma* K. Schumann 1897: **Cs** not or slightly overtopping the **Anth**, with a distinct hump at the back, broad, sometimes erect, often transversely furrowed and of contrasting colour.
[3] Sect. *Fallacistelma* L. C. Leach 1988: **Cs** distinctly overtopping the **Anth**, within one and the same species variable in length, with a ± distinct hump at the back.

In addition to hybrids with *Duvalia* (see ×*Dernia*), *Stapelia* (see ×*Huernelia*), and *Stapelianthus* (see ×*Huernianthus*), the intrageneric combinations *H. hislopii* ssp. *hislopii* × *H. verekeri* and *H. hislopii* ssp. *robusta* × *H. verekeri* have been confirmed (Leach 1969: 52). A further hybrid has formally been named (see *H. ×distincta*).

The following names are of unresolved application but are referred to this genus: *Huernia albomaculata* A. C. White & B. Sloane (1937); *Huernia crispa* Haworth (1812) ≡ *Huernia barbata* var. *crispa* (Haworth) Loudon (1830); *Huernia transmutata* White & Sloane (1937); *Huernia venusta* (Masson) Haworth (1812); *Stapelia venusta* Masson (1796).

H. andreaeana (Rauh) L. C. Leach (JSAB 40(1): 21, ills., 1974). **T:** Kenya (*Rauh* Ke867 [HEID, PRE [photos]]). – **D:** Kenya.
≡ *Duvalia andreaeana* Rauh (1961).

[2] Stems creeping, few-branched, mat-forming, rooting over the whole length, dark green, 4- to 5-angled; **Tu** small, obtuse; **L** rudiments marcescent; **Inf** 1-flowered, ± decumbent, upwards-directed; **Ped** ± 2 cm, rarely with basal additional bud; **Sep** ± 10 × 3 mm, protruding between the **Cl** lobes; **Cl** inside and outside pale ochre, 2.5 - 2.7 cm ∅, annulus pale wine-red; **Cl** lobes shiny, ± 10 × 6 mm, narrowly triangular, acute, with apical longitudinal furrow, margins inflexed, irregularly dentate, teeth black-purple, inside scattered with weak black-purple tubercles, ± 1 mm, sturdy; **Ci** lobes fused into a thickened disc; **Cs** yellowish, lobes as long as the **Anth**, dorsal hump ± erect, conspicuously enlarged, blunt, reaching above the annulus (similar to *H. tanganyikensis*).

H. archeri L. C. Leach (Excelsa Tax. Ser. 4: 88-90, ills., 1988). **T:** Kenya (*Archer* s.n. in *Leach* 15561 [PRE]). – **D:** Kenya (only known from the type locality).

[1] Stems deflexed, mat-forming, to 32 × ± 0.4 - 0.7 cm; **Tu** blunt, arranged in numerous **Ri**; **L** rudiments ± 2 mm, narrowly triangular, erectly spreading; **Inf** 1-flowered, buds with long beak; **Bra** ± 4 mm; **Ped** ≤ 1.8 cm, basally with small additional bud; **Sep** ± 9 mm, protruding between the **Cl** lobes; **Cl** inside yellow, mottled with maroon, ± 3.6 cm ∅, tube with fine concentric streaks or spots, ± flat, 8 - 9 mm ∅, outside finely scabrous, inside densely papillose, papillae narrowly conical, ± 0.5 - 2 mm, limb of the tube finely verrucose, warts dark, conical; **Cl** lobes ± 14 × 6 mm, elongate-triangular, pointed, ± caudate, margins dark; **Ci** black, lobes ± 2.5 × 2 mm, apically slightly crenate, spreading widely; **Cs** cream-coloured with maroon patterning, 2.5 × ± 0.75 mm, erect, ± capitate, 1 mm ∅, verruculose, warts small, acute; **Poll** yellow.

H. aspera N. E. Brown (Gard. Chron., ser. 3, 2: 364, 1887). **T:** Tanzania (*Kirk* s.n. [K]). – **D:** Kenya, Tanzania, Malawi.

[2] Stems prostrate or sometimes decumbent, intertwined, rooting, to 20 × 1 - 1.5 cm, 5- to 6-angular; **Tu** small, blunt, spreading; **L** rudiments acute, often recurved; peduncle very short; **Bra** 2 - 3.5 mm, acute; **Fl** solitary, slightly nodding; **Ped** 0.5 - 1.2 cm; **Sep** ≤ 7 × 1.5 mm; **Cl** inside and outside red-brown to black-purple, ± 2 - 2.5 cm ∅, broadly

campanulate or bowl-shaped; **Cl** lobes spreading, densely papillose on either side, papillae ± conical; **Ci** pale yellow, lobes transversely rectangular; **Cs** yellowish, apically blunt, slightly erect; **Poll** yellow-brown.

H. barbata (Masson) Haworth (Synops. Pl. Succ., 31, 1812). **T:** [lecto – icono]: Masson, Stapel. Nov. t. 7, 1796. – **D:** RSA (Northern Cape, Free State, Eastern Cape), Lesotho.

≡ *Stapelia barbata* Masson (1796); **incl.** *Stapelia tubulosa* hort. *ex* Steudel (s.a.); **incl.** *Stapelia duodecimfida* Jacquin (1806) ≡ *Huernia duodecimfida* (Jacquin) Sweet (1830); **incl.** *Stapelia tubata* Jacquin (1806) ≡ *Huernia tubata* (Jacquin) Haworth (1812) ≡ *Huernia barbata* var. *tubata* (Jacquin) N. E. Brown (1909); **incl.** *Huernia barbata* var. *griquensis* N. E. Brown (1909); **incl.** *Stapelia crispata* hort. *ex* Haworth (s.a.).

[1] Stems forming clumps, 4- to 5-angular; **Tu** triangular, pointed; peduncle short and thick, pointed; **Fl** facing outwards or upwards; **Ped** ≤ 1 cm, ± S-shaped; **Sep** ≤ 8 mm; **Cl** outside whitish, greenish-creamy or pale yellowish, base and above the veins rarely finely dotted with maroon, inside whitish to pale yellow, 2 - 6 cm ∅, margin of tube and lobes dotted with reddish; **Cl** tube basally red, glabrous or with short prickles, with maroon or cream-coloured concentric lines, 1.7 - 5.5 × 1.5 - 6 cm, broadly tubular-campanulate; **Cl** lobes mostly longer than broad, narrowly triangular, pointed, spreading widely or curved inwards, inside ± densely papillose, papillae dark maroon, bluntly ± dome-shaped, with a short **Ha** at the tip, **Ha** reddish to black, long, stiff, simple or clavate, often inflated, globose and pointed, particularly in the upper part of the tube; **Ci** lobes often fused into a crenate disc, or deeply bifid; **Cs** black, rarely brown, cream-coloured or pink, then with darker dots, tips black, lobes narrowly subulate, erect, tips vertically rising; **Poll** brownish-yellow, germination mouth region brown.

H. bayeri L. C. Leach (Excelsa Tax. Ser. 4: 187-190, ills., 1988). **T:** RSA, Eastern Cape (*Leach & Bayliss* 15662 [NBG, PRE]). – **D:** RSA (Eastern Cape).

[3] Stems 3 - 6 cm, tufted, distinctly 4- to 5-angular; **Tu** triangular; peduncle erect; **Fl** facing outwards; **Ped** ≤ 3.5 cm, erect, often overtopping the stems; **Sep** 6 - 7 × 2 mm, keeled; **Cl** outside greenish-creamy, inside cream-coloured, pale greenish or brightly pink-brownish, rarely weakly lined with reddish towards the apex, 2.5 - 3 cm ∅, campanulate, mouth of the tube occasionally annulus-like, glabrous; **Cl** lobes 5 - 7 × 6 - 9 mm, triangular, erect or ascending, sometimes revolute, outside partly finely scabrous, inside finely scabrous, scabrous parts elevated, often with short stiff **Ha**, inside ± densely papillose, papillae conical-truncate, apically with a **Ha**, **Ha** colourless or white, ≤ 2 mm, slender, linear; intermediate lobes protruding particularly in the bud; **Cn** black-purple, 4 × ± 4.5 - 5 mm; **Ci** lobes ± 1 × 1.5 mm, bifid; **Cs** lobes ± 3 mm, subulate, finely pointed, meeting each other above the **Anth**, above divaricate, basally with a prominent verrucose hump; **Poll** brownish.

H. boleana M. G. Gilbert (CSJA 47(1): 9-11, ills., 1975). **T:** Ethiopia (*Gilbert & Gilbert* 2431 [K, ADAB, NBG]). – **D:** Ethiopia.

[1] Stems erect or occasionally creeping or pendulous, to 9 (pendulous stems to 35) × 1.2 cm, distinctly 5-angular; **Tu** ± 6 mm, conically compressed, subulate; peduncle very short, humped; **Bra** ≤ 5 mm; **Ped** ≤ 1.2 mm; **Sep** 16 × 0.75 mm; **Cl** outside white, inside pale cream-coloured, dotted with red, narrowly campanulate, tube 1.2 × 1.25 cm, ± cylindrical, basally rounded, ± glabrous; **Cl** lobes 15 × 7 mm, tapering to a long narrow point, spreading, margins dark, outside glabrous or finely scabrous, inside densely papillose, papillae weakly dotted with red, 1 - 2.5 mm, cylindrical or slightly compressed, ± obtuse, **Cl** lobes verrucose, warts dark, ± 0.25 mm, ± acute; intermediate lobes ± 3 mm, spreading; **Cn** black, ± 6 × 8.5 mm; **Ci** lobes ± ovate or ± square, ± truncate, irregularly crenate, at the base of the **Anth** with a verrucose-conical hump; **Cs** 3.5 - 4 mm, basally convergent, just below the middle slightly diverging, apically ± truncate, verruculose; **Poll** yellowish.

H. brevirostris N. E. Brown (Gard. Chron., ser. nov. 7: 780, 1877). **T:** RSA, Eastern Cape (*Bolus* 575 [K]). – **D:** SE RSA (Eastern Cape).

H. brevirostris ssp. **baviaana** L. C. Leach (Excelsa Tax. Ser. 4: 166-168, ills., 1988). **T:** RSA, Eastern Cape (*Bruyns* 1605 [NBG, GRA, K, M, MO, PRE]). – **D:** RSA (Eastern Cape).

[3] Differs from ssp. *brevirostris*: **Cl** darker, markings denser, inside more densely papillose, papillae of the tube apically with a short stiff acute **Bri**; **Cs** whitish or cream-coloured, rarely light yellow or brown, margin dark maroon, basally with maroon markings, lobes longer, ± curved, divaricate, subulate.

H. brevirostris ssp. **brevirostris**

[3] Stems erect or decumbent, compact, forming clumps, occasionally creeping, dull to dark grey-green, partly suffused with purple, to ± 5 × 1.5 - 2 cm, distinctly 4- to 5-angular; **Tu** triangular-conical; **L** rudiments deciduous; **Fl** facing outwards, often bent upwards; **Bra** very acute; **Ped** 1 - 2.5 cm; **Sep** ± 5 mm; **Cl** outside greenish, brownish-creamy or light yellow, often finely stippled with red, inside cream-coloured to yellowish, ± densely mottled with brown-red or maroon, 2.5 - 4 (-4.5) cm

⌀, ± flat; **Cl** tube often paler, mottled or streaked, sometimes without markings, shallowly bowl-shaped, with a pentagonal mouth, margin abruptly reflexed, sometimes raised in an annulus-like manner, base of tube glabrous, sometimes with some **Bri** around the **Cn**, **Bri** basally crimson or maroon, short, straight; **Cl** lobes a little broader than long, acutely triangular, spreading or slightly revolute, often weakly convex in longitudinal direction, inside ± densely papillose, papillae apically often crimson to black, blunt to ± broadly conical, apically sometimes pointed or with a short **Ha**; **Ci** dark crimson, black or whitish to light yellow to brown, with maroon markings, lobes elongate, ± deeply bifid; **Cs** mostly brownish, tips darker, lobes narrowly triangular, 1.5 - 3.25 mm, mostly dorsiventrally compressed, meeting each other above the **Anth**, rarely decumbent on the **Anth**, apical part erect, tips slightly divergent or subulately acute, apex ± obtuse; **Poll** brownish-yellow.

H. brevirostris ssp. **intermedia** (N. E. Brown) L. C. Leach (Excelsa Tax. Ser. 4: 164, 1988). **T**: RSA, Eastern Cape (*Pillans* 72 [BOL, K]). – **D**: RSA (Eastern Cape).

≡ *Huernia brevirostris* var. *intermedia* N. E. Brown (1909); **incl.** *Huernia scabra* N. E. Brown (1909) ≡ *Huernia brevirostris* var. *scabra* (N. E. Brown) A. C. White & B. Sloane (1937); **incl.** *Huernia scabra* var. *immaculata* N. E. Brown (1909) ≡ *Huernia brevirostris* var. *immaculata* (N. E. Brown) A. C. White & B. Sloane (1937); **incl.** *Huernia scabra* var. *longula* N. E. Brown (1909) ≡ *Huernia brevirostris* var. *longula* (N. E. Brown) A. C. White & B. Sloane (1937); **incl.** *Huernia scabra* var. *pallida* N. E. Brown (1909) ≡ *Huernia brevirostris* var. *pallida* (N. E. Brown) A. C. White & B. Sloane (1937); **incl.** *Huernia brevirostris* var. *histrionica* A. C. White & B. Sloane (1937); **incl.** *Huernia brevirostris* var. *parvipuncta* A. C. White & B. Sloane (1937).

[3] Differs from ssp. *brevirostris*: **Cl** darker, rarely with a low annulus, markings less prominent, sometimes plainly cream-coloured to yellowish.

H. campanulata (Masson) Haworth (Synops. Pl. Succ., 28, 1812). **T**: [lecto – icono]: Masson, Stapel. Nov. t. 6, 1796. – **D**: RSA (Western Cape, Eastern Cape).

≡ *Stapelia campanulata* Masson (1796); **incl.** *Huernia campanulata* var. *denticoronata* N. E. Brown (1909).

[1] Plants forming clumps; stems dull green, occasionally with reddish spots, usually ± 6 × 1 - 1.5 cm, prominently 4- to 5-angled; **Tu** pointed, laterally compressed, spreading; **Fl** facing upwards; **Ped** ≤ 1.5 cm, curved upwards; **Sep** conspicuously furrowed along the middle; **Cl** outside greenish- to brownish-creamy, towards the limb often spotted with crimson to reddish, inside whitish to pale yellow, 2.5 - 4.5 cm ⌀, variably bicampanulate, mouth of the tube and **Cl** lobes ± coarsely spotted with crimson, tube basally black, above with irregular concentric lines of dark crimson, margin abruptly curved away from the mouth; **Cl** lobes a little broader than long, acutely triangular, ± erect or spreading, ± revolute, inside ± densely papillose, papillae dark crimson, small, ± blunt, apically pubescent, **Ha** ± large, translucent, globosely inflated, **Ha** within the upper part of the **Cl** tube reddish to black, long, stiff, clavate, pointing inwards; **Ci** dark crimson to black, lobes transversely rectangular or ± quadrangular, emarginate, crenate or deeply lobed; **Cs** very variable, lobes subulate, apically divaricate; **Poll** yellow, corpuscle sometimes malformed.

H. clavigera (Jacquin) Haworth (Suppl. Pl. Succ., 10, 1819). **T**: [lecto – icono]: Jacquin, Stapel. Hort. Vindob. Cult., t. 5. – **D**: RSA (Northern Cape, Western Cape).

≡ *Stapelia clavigera* Jacquin (1806); **incl.** *Huernia clavigera* var. *maritima* N. E. Brown (1909); **incl.** *Huernia decemdentata* N. E. Brown (1909); **incl.** *Huernia ingeae* Lavranos (1982).

[1] Stems erect or decumbent, occasionally ± creeping, forming loose clumps, dull to dark green, sometimes suffused with purple, 5 - 6 × 2 cm, 4- to 5-angular; **Tu** acutely deltoid, ± curved downwards; **Fl** facing upwards; peduncle very short, acute; **Ped** ≤ 1.5 cm, ± S-shaped; **Cl** outside greenish-creamy or pale brown (tube spotted with maroon), inside greenish-creamy or pale yellow, ± narrowly bicampanulate, to 5 cm ⌀, mouth of the tube and **Cl** lobes ± densely stippled with red or maroon; **Cl** tube dark maroon to black, rarely with concentric white or yellow lines, 1.8 - 3.5 × 1.5 - 3 cm ⌀, margin of tube abruptly spreading, ± annulus-like; **Cl** lobes ± erect or flat, sometimes revolute, inside densely finely papillose, papillae dark purple, apically pubescent, **Ha** ± globose or narrowly elliptic, finely pointed, inflated, rarely simple, **Ha** of the upper part of the tube dark flesh-coloured, long, clavate, stiff, erect; **Ci** very variable, crimson to black, rarely with a white band circling the **Gy**; **Cs** brown to black, occasionally reddish with cream-coloured spots, apically black, lobes narrowly subulate, tips ± erect; **Poll** brownish-yellow.

H. concinna N. E. Brown (FTA 4(1): 497, 1903). **T**: Somalia (*Lort-Phillips* s.n. [K]). – **D**: Somalia.

≡ *Huernia macrocarpa* ssp. *concinna* (N. E. Brown) M. G. Gilbert (1975).

[2] Stems tufted, to 4 (-6) × 1 cm, prominently 5-angular; **Tu** ± 6 mm, conical-subulate, spreading; **Inf** few-flowered; **Ped** ± 5 mm; **Sep** 6 - 8 mm, thin towards the tip; **Cl** outside whitish, inside pale sulphur-yellow, spotted with brownish-crimson, ± 2.5 cm ⌀; **Cl** tube broadly campanulate, ± 6 × 14 mm; **Cl** lobes 8 × 9 mm, acuminate, slightly spreading, margins dark purple-brown, inside papillose, papil-

lae slender; **Ci** purple-brown, lobes 0.5 × ± 1.5 mm, transversely rectangular, emarginate; **Cs** yellow with purple-brown markings, lobes ± 1 × 0.66 mm, broadly ovate or ± triangular, ± blunt, borne in a cone above the **Anth** and meeting each other with their tips, ± covering the **Gy**.

H. ×distincta N. E. Brown (*pro sp.*) (FC 4(1): 910, 1909). **T:** RSA, Cape Prov. (*Pillans* 83 [K, BOL]).
= *H. clavigera* × *H. pillansii*.

H. echidnopsioides (L. C. Leach) L. C. Leach (Excelsa Tax. Ser. 4: 51, ill. (p. 52), 1988). **T:** RSA, Eastern Cape (*Leach & Bayliss* 13612 [PRE, K, SRGH]). – **D:** RSA (Eastern Cape).
≡ *Huernia pillansii* ssp. *echidnopsioides* L. C. Leach (1968) ≡ *Huernia longii* ssp. *echidnopsioides* (L. C. Leach) Bruyns (1984).
[1] Stems ± cylindrical, rhizome-like runners of up to 3 m forming single stems at intervals (rarely forming clumps), stems 4 - 6 × ± 1 cm, sides deeply furrowed; **Tu** 1 - 2 mm, arranged densely in (6-) 7 - 9 (-10) vertical or spiral **Ri**; **L** rudiments acute, spreading or curved, eventually drying; peduncle ≤ 7 mm; **Ped** 4 - 7 mm; **Sep** 5 - 7 mm; **Cl** inside (pale) yellowish, labyrinth-like markings composed of small red to brown spots, 2 - 3 (-3.5) cm ∅; **Cl** tube 9 × 9 mm, campanulate, basally glabrous; **Cl** lobes diverging, revolute, inside densely papillose, papillae narrowly cylindrical, blunt, rarely very slightly clavate, often tipped with **Bri**, **Ha** irregularly marked red, ≤ 1 mm, delicate; **Ci** lobes fused into a ± round crenate disc, or free and ± quadrangular, blunt, variably emarginate or bifid; **Cs** lobes erect, ribbon-like, tips swollen, meeting each other, acutely warty, basally with a prominent hump; **Poll** yellow to greenish-brown.

H. erectiloba L. C. Leach & Lavranos (Kirkia 3: 38-40, ills., 1963). **T:** Moçambique (*Leach & Rutherford-Smith* 10914 [SRGH, MO, PRE]). – **D:** N Moçambique.
[1] Stems prostrate to decumbent, forming lax clumps, to 20 cm, prominently 4- to 5-angular; **Tu** ≤ 5 mm, acute, spreading; **Fl** facing upwards; **Ped** ≤ 2 cm; **Sep** 5 - 7 mm; **Cl** outside cream-coloured with fine red spots, inside cream-coloured with blood-red spots and broken or basally confluent concentric lines, 2.5 cm ∅, bicampanulate, basally rounded; **Cl** tube ± 10 × 7 - 8 mm, usually with a weak edge just above the base, mouth slightly constricted, lateral walls thickened, annulus low, slightly shining, limb of the tube abruptly rising or slightly revolute; **Cl** lobes ± 7 - 8 × < 9 mm, broadly triangular, acuminate, erect or tips revolute, laterally ± fringed, crisped, outside slightly rugose, basally sometimes pubescent, **Ha** straight, stiff, inside pustulate-papillate, ± pubescent, **Ha** ≤ 2 mm, simple, stiff, erect; **Cn** ± 5 × 6 mm; **Ci** cream-coloured, margins and tips dark, lobes ± quadrangular, bluntly notched or slightly bifid; **Cs** cream-coloured with blood-red markings, apically dark, lobes ± 4 mm, subulate, mucronate, rising up above the **Anth** and meeting each other, then revolute, basally weakly bulging; **Poll** greenish-brown, germination mouth region darker.

H. erinacea P. R. O. Bally (FPA 31: t. 1206 + text, 1956). **T:** Kenya, Northern Frontier Prov. (*Gillett* 12629 [K, PRE]). – **D:** Kenya, border to Ethiopia.
Fig. XXIX.b
[1] Stems prostrate (rarely decumbent), to 6 × 1 cm, 5-angular; **Tu** blunt; **Fl** upright; **Ped** 0.7 - 2 cm; **Sep** ± 6 × 1.75 mm; **Cl** outside dull yellow, inside yellow, lined or spotted with red-brown, 4 - 5 cm ∅, flat; **Cl** tube ± flat, basally glabrous, mouth 1.2 - 1.5 cm ∅, slightly constricted; **Cl** lobes to 2 cm, narrowly triangular, margins purple-red and irregularly dentate-papillose, inside ± densely papillose, papillae stippled with purple, ≤ 1.5 mm, ± obtusely conical, glabrous; **Cn** ± 9 mm ∅; **Ci** black-purple, lobes ± transversely rectangular, centrally with a short appendix or ± broadly quadrangular, deeply emarginate; **Cs** whitish, basally purple, apically weakly spotted, lobes 3.5 mm, ± cylindrical, blunt, slightly divergent; **Poll** yellowish, caudicles very short.

H. formosa L. C. Leach (Excelsa 12: 95-97, ills., 1986). **T:** Somalia (*Lavranos* 9330 [NBG]). – **D:** N Somalia.
[2] Stems tufted, to 4.5 × 1.5 mm, prominently 5-angular; **Tu** ≤ 6 mm, conical-subulate; **Inf** few-flowered, **Fl** bent upwards; **Bra** ≤ 5 mm, pointed; **Ped** 1 - 1.25 cm; **Sep** ± 8 × 2 mm; **Cl** outside greenish, inside cream-coloured, spotted with brown-red, 3.5 - 4 cm ∅, shallowly saucer-shaped; **Cl** tube 3 - 4 × 15 mm, bowl-shaped, basally glabrous, limb of the tube and tips of the **Cl** lobes spreading outwards, lobes 1 × 1.4 cm, acutely triangular, with dark margins, outside finely roughened in part, inside densely papillose, papillae ± 1 mm, narrowly conical, ± finely pointed, inconspicuous towards the tip; intermediate lobes inconspicuous; **Cn** 4 × 9 - 10 mm, dome-shaped; **Ci** dark brown, lobes ± 4.5 × 1 mm, broadly bidentate; **Cs** red-brown, whitish towards the tip, lobes ± 3 × 3 mm, ovate or obtusely triangular, usually not meeting each other, ± covering the **Gy**; **Anth** truncate, notched; **Poll** brown, corpuscle reddish.

H. guttata (Masson) Haworth (Synops. Pl. Succ., 30, 1812). **T:** [lecto – icono]: Masson, Stapel. Nov. t. 4, 1796. – **D:** RSA (Western Cape, Eastern Cape).
≡ *Stapelia guttata* Masson (1796); **incl.** *Stapelia lentiginosa* Sims (1801) ≡ *Huernia lentiginosa* (Sims) Haworth (1812); **incl.** *Stapelia ocellata* Jacquin (1806) ≡ *Huernia ocellata* (Jacquin) Schultes (1820); **incl.** *Stapelia venusta* var. *minor* Jacquin

(1806) (*incorrect name*, Art. 11.4); **incl.** *Stapelia venusta* Jacquin (1806) (*nom. illeg.*, Art. 53.1).

H. guttata ssp. **calitzdorpensis** L. C. Leach (Excelsa Tax. Ser. 4: 38-39, 1988). **T:** RSA, Western Cape (*Leach & Roussouw* 16147 [NBG, STE]). – **D:** RSA (Western Cape, Eastern Cape).

[1] Differs from ssp. *guttata*: **Cl** darker, to 7 cm ∅, tube dark maroon, occasionally marked with some vague yellowish lines, **Ha** within the tube much longer and more strongly clavate.

H. guttata ssp. **guttata** – **D:** RSA (Eastern Cape).

[1] Stems deflexed or ± creeping to erect, forming clumps, to 7 × 1.5 cm, 4- to 5-angular; **Tu** triangular, acute; **Fl** facing upwards; **Ped** 1 - 3 cm; **Sep** ≤ 7 mm; **Cl** outside greenish, inside white to yellow, stippled with maroon to crimson (becoming confluent on the annulus), 2 - 3 cm ∅; **Cl** tube marked with transverse lines, campanulate, basally slightly rugose around the **Cn**, annulus broad, dull, pointing into the tube, mouth of the tube hairy (**Ha** stiff, sometimes ± clavate); **Cl** lobes broadly triangular, acuminate, ascending, above widely spreading, ± pustulate-papillate; **Cn** 4.5 - 5.5 × 6 - 8 mm; **Ci** with variable colours, lobes ± square, notched to incised or elongate and deeply bifid; **Cs** mostly cream-coloured with maroon markings, apically dark, lobes subulate, basally geniculate, narrowed, meeting each other above the **Anth**, apically erect, diverging, often with nectar droplets; **Poll** yellow, germination mouth region brown.

H. hadhramautica Lavranos (JSAB 29: 99-101, ills. (pl. 10: 1-2), 1963). **T:** Yemen, Hadhramaut (*Lavranos* 1935 [K]). – **D:** S Yemen.

[2] Stems erect, tufted, ± 7 × 0.8 cm, 5-angular; **Tu** ≤ 1.2 mm, basally triangularly compressed, apically subulate, widely spreading; **Fl** facing outwards; **Ped** ± 5 mm; **Sep** 7 - 9 mm; **Cl** outside pinkish-creamy, inside dark maroon, ± 2.5 cm ∅, broadly campanulate, tube prominent, basally pink, glabrous; **Cl** lobes broader than long, outside densely finely papillose, inside finely papillose, papillae maroon, conical; **Cn** ± 5 mm ∅, disc-shaped; **Ci** lobes convex, apically rounded; **Cs** lobes simple, meeting each other above the **Sty** head.

H. hallii E. & B. M. Lamb (Nation. Cact. Succ. J. 13(3): 57, ills., 1958). **T** [neo]: Namibia (*Hall* s.n. [NBG]). – **D:** S Namibia. **Fig. XXIX.c**

≡ *Huernia namaquensis* ssp. *hallii* (E. & B. M. Lamb) Bruyns (1982).

[3] Stems prostrate to decumbent, clump-forming, pale green, 1.5 - 2.5 (-3) × 0.6 - 1.2 (-1.5) cm, prominently 4- to 5-angular; **Tu** 0.5 - 1.5 (-2) mm; **Inf** with or without peduncle, nodding or pointing outwards, borne at the base or middle of the stems or irregularly; **Bra** ± 4 mm, acute; **Ped** tipped with 5 wart-like convexities, glabrous; **Sep** 4 - 7 mm; **Cl** outside greenish-creamy, inside cream-coloured, occasionally ± bright pink, marked with maroon to brown-red spots and ± entire concentric lines, to ± 3 cm ∅, campanulate, tube basally glabrous, mouth not or hardly constricted, margin spreading; **Cl** lobes acuminate, triangular, erect or widely spreading and reflexed, margins maroon, inside papillose, papillae ± 0.2 mm, ± conical, finely pointed, occasionally larger, blunt; **Cn** 2.5 - 3.5 × 4.5 mm; **Ci** cream-coloured marked with red and crimson, fused into a ± circular disc showing 10 notches, ± strongly convex; **Cs** yellow marked with maroon, lobes ± as long as the **Anth**, if longer then meeting each other, ± erect, ± obtuse, basally obtusely thickened, with a conspicuous transverse furrow; **Anth** yellow, ± acute, with weak dorsal concave notch; **Poll** yellowish.

H. hislopii Turrill (BMI 1922: 30, 1922). **T:** Zimbabwe (*Hislop* s.n. [K]). – **D:** Zimbabwe.

H. hislopii ssp. **hislopii** – **D:** Zimbabwe, Moçambique. **Fig. XXIX.d**

[1] Stems erect, tufted, occasionally forming large clumps, to 5 × 0.8 cm, prominently 5-angular; **Fl** facing upwards, rarely prostrate; peduncle ± rudimentary; **Ped** ≤ 2.5 cm; **Sep** 6 - 10 × ± 2 mm; **Cl** outside cream-coloured, inside ± cream-coloured, stippled and mottled with red-brown, 4 - 5 (-6) cm ∅, campanulate, tube ± globose, marked with transverse red-brown lines, mouth abruptly reflexed; **Cl** lobes ≤ 2.2 × 1.8 cm, acuminate-triangular, sometimes ± caudate, often reflexed, inside ± densely papillose, papillae cream-coloured, spotted with red-brown, ≤ 1 mm, obtuse, ± compressed, basally broadened, mucronate; **Cn** ± 5.5 mm; **Ci** dark maroon to black, lobes rectangular; **Cs** cream-coloured with red-brown markings, lobes ± 4 - 4.5 mm, obtuse, divergent, rarely erect and only slightly divergent; **Poll** yellow.

H. kirkii in the sense of White & Sloane (1937) belongs here.

H. hislopii ssp. **robusta** L. C. Leach & Plowes (JSAB 32(1): 53, ill., 1966). **T:** Zimbabwe, Lupani Distr. (*Leach* 11628 [SRGH, BM, G, K, LISC, PRE]). – **D:** NW Zimbabwe.

[1] Differs from ssp. *hislopii*: Stems more robust; **Tu** arranged in 5 - 7 **Ri**; **Cl** tube longer, weakly inflated, basally without or with weak concentric lines; **Cl** lobes shorter, rarely acuminate.

H. humilis (Masson) Haworth (Synops. Pl. Succ., 30, 1812). **T:** [lecto – icono]: Masson, Stapel. Nov. t. 5, 1796. – **D:** RSA (Eastern Cape). **Fig. XXIX.a**

≡ *Stapelia humilis* Masson (1796); **incl.** *Huernia simplex* N. E. Brown (1909).

[2] Stems stout, ± pyramid-shaped or ± globose, forming clumps, green to greyish, marked with maroon, 2.5 × 2.5 cm, 4-angular; **Tu** ≤ 2 mm, triangu-

lar; **Inf** few- or 1-flowered, pointing outwards; **Ped** ± 1 cm, ascending; **Sep** 2.5 - 4 × ± 2 mm; **Cl** outside bright green, inside pale yellow, with evenly arranged maroon dots, 2.5 - 3 cm ⌀; **Cl** tube ± 6 mm ⌀, basally bright maroon, above and on the annulus spotted with reddish; **Cl** lobes broadly triangular, divergent, inside finely papillose, papillae maroon, very variable in size, acute; **Ci** dark maroon to black, lobes ± shortly bifid; **Cs** lobes slender, pointed, ± as long as the **Anth**; **Poll** yellowish.

H. venusta in the sense of Retief 1980 (Flow. Pl. Afr. 46: t. 1812) belongs here.

H. hystrix (Hooker *fil.*) N. E. Brown (Gard. Chron., ser. nov. 5: 795, 1876). **T:** [icono]: Curtis's Bot. Mag. 95: t. 5751, 1869. – **D:** Zimbabwe, Moçambique, RSA, Swaziland.

≡ *Stapelia hystrix* Hooker *fil.* (1869); **incl.** *Huernia appendiculata* A. Berger (1910) ≡ *Huernia hystrix* var. *appendiculata* (A. Berger) A. C. White & B. Sloane (1937).

H. hystrix var. **hystrix** – **D:** SE Zimbabwe, W Moçambique, RSA (Northern Prov., Mpumalanga, KwaZulu-Natal, Eastern Cape), Swaziland. **Fig. XXIX.e**

[1] Stems decumbent-erect, spreading, sometimes creeping, ± 6 - 7 cm, prominently 5-angular, laterally deeply furrowed; **L** rudiments persistent as hard **Sp**; **Fl** lying flat on the ground; **Ped** ≤ 6 cm; **Sep** 7 - 10 mm, obtusely keeled; **Cl** outside greenish, inside cream-coloured dotted and streaked with brown-red, flat, 3 - 5 cm ⌀; **Cl** tube short, broad, ± cup-shaped, basally often only weakly roughened, mouth of the tube and **Cl** lobes abruptly divaricate; **Cl** lobes broadly triangular, occasionally shortly pointed, outside slightly roughened, inside densely papillose, papillae 3.5 - 5 (-5.5) mm, fleshy, subulate, sometimes finely pointed, flattened, basally broadened; **Ci** very variable in shape and colour, sometimes much reduced or absent; **Cs** lobes erect, dorsiventrally strongly compressed, with ± parallel sides, terminating in a fleshy process, ± invertedly foot-shaped, glabrous or scattered with fine **Ha**, outside flat, glabrous, tuberculate-rugose and/or furrowed, basally with a broad wart-like convexity; **Poll** yellowish with darker germination mouth region.

H. hystrix var. **parvula** L. C. Leach (JSAB 42(4): 450, 1976). **T:** RSA, KwaZulu-Natal (*Strey* 9730 [PRE, K, MO, SRGH]). – **D:** RSA (KwaZulu-Natal).

[1] Plants smaller than in var. *hystrix*; stems rarely > 3 cm; **Ped** ≤ 1.75 cm; **Cl** 3 - 4.5 cm ⌀, bud longer conically pointed, papillae rarely > 2.5 mm; **Cs** lobes apically hoof-like, obtusely clavate, more densely puberulous and tuberculate-furrowed.

H. insigniflora Maass (Möllers Deutsche Gärtn.-Zeit. 1928(7): 79, 1928). **T** [neo]: RSA (*Anonymus* s.n. [K]). – **D:** RSA (Northern Prov., Mpumalanga).

Incl. *Huernia confusa* E. Phillips (1932).

[2] Stems basally branching, erect, forming small clumps, grey-green, ± quadrangular in cross section, 4- (rarely 5-) angular; **Tu** small, disappearing on older stems; **Inf** few-flowered, pointing outwards; **Ped** ± 1 cm; **Sep** ± 6 mm; **Cl** outside light green, inside greenish-white, ivory-coloured or light yellow, at times pink, 2.5 - 3.5 cm ⌀, tube and the glabrous annulus crimson to dark purple or brown, tube ± 7 × 6 mm; **Cl** lobes triangular, shortly pointed, divaricate, irregularly transversely streaked with light red or purple, outside partly finely roughened, inside puberulous, **Ha** maroon or streaked with reddish, short, acute, basally sometimes swollen; **Ci** cream-coloured to light yellow, rarely black-purple, sometimes with maroon markings, lobes ± shortly lobed, fused into a disc, margin sometimes with 10 notches; **Cs** yellowish with dark maroon margin, black or rarely dark maroon, basal hump ivory to orange, lobes acute, as long as the **Anth** or exceeding them; **Poll** yellow to dark greenish, ± elliptic to elongate-elliptic.

H. humilis in the sense of Schlechter (1895a: 54) belongs here.

H. keniensis R. E. Fries (Acta Horti Berg. 9(3): 79-80, t. 2, 1929). **T:** Kenya (*Fries* 1024 [UPS [not preserved?]]). – **D:** Kenya, Tanzania.

H. keniensis var. **globosa** L. E. Newton (Asklepios No. 74: 23-24, ills., 1998). **T:** Kenya, Rift Valley Prov. (*Newton* 2927 [K, EA, MSUN]). – **D:** Kenya.

[2] Differs from var. *keniensis*: **Cl** globose-urceolate, ± 20 × 25 mm, inside deep purple with minute papillae of uniform size. – [U. Meve]

H. keniensis var. **grandiflora** P. R. O. Bally (FPA 38: t. 1511B + text, 1967). **T:** Kenya, Samburu Distr. (*Hennings* s.n. in *Bally* S204 [G, EA]). – **D:** Kenya.

[2] Differs from var. *keniensis*: **Cl** outside parchment-white or dark with pale venation, ± 5 cm ⌀, lobes prominently 3-veined with 2 inconspicuous additional veins, outside ± glabrous, inside densely papillose.

H. keniensis var. **keniensis** – **D:** Kenya, N Tanzania. **Fig. XXX.a**

Incl. *Huernia keniensis* var. *quintitia* Leach (1988) (*nom. inval.*, Art. 36.1, 37.1).

[2] Stems prostrate-decumbent or erect, irregularly branching, ± 12 × 1 cm, 5-angular; **Tu** small, acute or larger and more prominent; **Ped** ≥ 5 mm; **Sep** ± 6 mm; **Cl** outside rarely cream-coloured, inside dark purple, 2 - 2.5 cm ⌀, deeply campanulate; **Cl** lobes broadly triangular with revolute tips, 5-veined, outside papillose-roughened, inside densely

papillose, papillae ± conical or ± cylindrical, at the mouth of the tube and on the lobes obtuse, in the tube more delicate; **Cn** dark, often black-purple; **Ci** lobes fused into a ± crenate disc; **Cs** lobes apically slightly pointing upwards, narrow, with obtuse tips; **Gy** tuberculate-papillose below the basal convexity.

H. keniensis var. **molonyae** A. C. White & B. Sloane (Stapelieae, ed. 2, 3: 832-833, 1144, ills., 1937). **T:** Kenya (*Molony* s.n. [not preserved]). – **D:** Kenya.

[2] As var. *keniensis*, but **Cl** ± glabrous inside and outside, inside with few scattered **Ha**.

Possibly an abnormal form.

H. keniensis var. **nairobiensis** A. C. White & B. Sloane (Stapelieae, ed. 2, 3: 837, 1937). **T:** Kenya (*Molony* s.n. [K alc 6754]). – **D:** Kenya, Tanzania.

[2] As var. *keniensis*, but stems erect, more robust (specimens from Tanzania prostrate, to 18 cm); **Tu** more distinct; **Cl** larger, indistinctly bicampanulate, broadly funnel-shaped, tube broad, bowl-shaped; **Cl** lobes 3-veined, outside papillose-roughened, inside densely papillose, papillae small; **Cn** erect for more than ½ of its height; **Cs** lobes ± 1.3 mm, apically obtuse, shortly rising up, slightly overlapping above the **Sty** head.

H. kennedyana Lavranos (JSAB 31(4): 313-315, ills., 1965). **T:** RSA, Eastern Cape (*Kennedy* s.n. in *Lavranos* 2356 [PRE]). – **D:** RSA (Eastern Cape). **Fig. XXIX.f**

[1] Stems erect, dwarf, in densely caespitose mats, green, in part suffused with purple; **Tu** short, compact, ± globose, arranged in (6-) 7 - 9 (-10) **Ri**; **L** rudiments ± 0.75 mm, soft, occasionally deciduous; **Fl** facing upwards, peduncle compact; **Bra** ≤ 2 mm, distinctly keeled; **Ped** 4 - 9 (-15) mm; **Sep** 3 - 4 mm, finely tuberculate; **Cl** outside pale yellow, basally pink, inside cream-coloured to dull yellow, 2 - 2.5 cm ∅; **Cl** tube with red-brown to dark maroon sometimes elongate spots, basally dotted, above streaked, campanulate or semiglobose, 5.5 × ± 10 mm; **Cl** lobes 6 - 7 × 6 - 6.5 mm, triangular, acuminate, outside slightly roughened, inside densely papillose, papillae whitish or pale yellow, ± 3 mm, cylindrical, ± obtuse, often ± clavate, sometimes finely acute, margin of lobes finely crenate-papillose; **Ci** pale pink, lobes transversely rectangular, shortly bifid or finely crenate; **Cs** yellow, tipped whitish, lobes ± 2 mm, subulate, dorsiventrally compressed, basally broad, apically blunt, often slightly clavate, sometimes shortly reflexed, basal convexity much reduced; **Poll** yellow.

H. kirkii N. E. Brown (FC 4(1): 920, 1909). **T:** RSA, Transvaal (*Kirk* s.n. [K]). – **D:** SE Zimbabwe, W Moçambique, RSA (Northern Prov., Mpumalanga, E KwaZulu-Natal). **I:** Leach (1988).

Incl. *Huernia bicampanulata* I. Verdoorn (1932).

[1] Stems erect, tufted, often forming large clumps, ± 8 cm, distinctly 4- to 5-angular; **Tu** 5 mm, laterally compressed; **L** rudiments acute, hardened; **Fl** facing upwards, **Inf** first without peduncle, later rudimentary; **Bra** ≤ 5 mm; **Ped** ≤ 3 cm; **Sep** ± 8 × 1.5 mm; **Cl** outside pale yellow, inside cream-coloured, with irregular reddish or maroon stipples and spots, 3 - 5 cm ∅; **Cl** tube black-maroon, bicampanulate, basally glabrous; **Cl** lobes 8 - 10 × 11 - 13 mm, triangular, acuminate, erect to divaricate, outside sometimes very weakly roughened, inside ± densely papillose, papillae dark maroon, ≤ 2 mm, ± cylindrical to conical, apically with short a acute **Bri**; **Cn** black-maroon, ± 6 × 7.5 mm; **Ci** lobes narrowly transversely rectangular or ± square, crenate, dentate or bifid; **Cs** lobes erect, meeting each other, apically clavate, reflexed-divaricate; **Poll** brownish-yellow, germination mouth region darker.

H. laevis J. R. Wood (KB 39(1): 128-130, ills., 1984). **T:** Yemen (*Wood* 3037 [K]). – **D:** N Yemen.

[2] Stems robust, prostrate-decumbent, forming clumps, to 4 cm, prominently 5-angular; **Tu** 5 - 6 mm; **Fl** facing upwards; **Bra** ± 5 mm, linear; **Sep** 1 - 1.2 cm; **Cl** outside dotted with white-red, inside shining with markings composed of irregular, often elongate maroon stipples and spots, ± 3 cm ∅; **Cl** tube broadly funnel-shaped, ± 5 × 6 - 7 mm, margin weakly bulging like an annulus; **Cl** lobes ± 10 × 8 mm, long-triangular with a long point, weakly revolute, inside glabrous, smooth; intermediate lobes inconspicuous; **Cn** black-brown, ± 5-angular; **Ci** lobes shortly bluntly bifid; **Cs** lobes ± 1.6 mm, apically ± acute, slightly overtopping the **Anth** but not meeting each other.

H. lavrani L. C. Leach (Excelsa 12: 94, 96-97, ills., 1986). **T:** Somalia (*Lavranos & Bavazzano* 8534 [NBG, K]). – **D:** N Somalia. **Fig. XXIX.g**

[2] Stems erect, tufted, to 4 × 1.25 cm, prominently 5-angular; **Tu** ≤ 6 mm, conical, subulate, spreading; **Fl** facing outwards, bent upwards; peduncle rudimentary; **Bra** 3 mm, pointed, distinctly keeled; **Ped** 5 - 10 mm; **Sep** 5 - 7 × 1.5 mm; **Cl** inside cream-coloured, 1.5 × ± 2 cm, deeply bowl-shaped, ± glabrous; **Cl** tube basally with irregular concentric red-brown streaks; **Cl** lobes mottled or stippled, lower ⅔ with dark margin, 8 mm, triangular, (very shortly) pointed, erect or spreading, outside partly finely roughened, inside finely acutely tuberculose; intermediate lobes 1.5 mm, divaricate; **Cn** 4 × ± 8 mm, dome-shaped; **Ci** dark purple-brown, lobes ± 3 × 1 mm, rather broadly bidentate or finely crenate; **Cs** basally dark to red-brown, lobes ± 3 × 2.5 mm, ± oval, basally truncate, meeting each other above the **Sty** head to slightly overlapping, ± covering the **Gy**; **Anth** sometimes shortly obtusely bifid; **Poll** brown, germination mouth region darker, corpuscle red-orange.

H. leachii Lavranos (JSAB 25: 311-313, pl. 29-30, 1959). **T**: Moçambique, Manica e Sofala (*Leach* 5641 [PRE]). – **D**: Moçambique, Malawi. **Fig. XXX.b**

[2] Stems creeping or prostrate, ± cylindrical, to 150 × 0.5 - 0.8 cm; **Tu** ± 2 mm, acute, fleshy, strongly spreading, irregularly arranged (sometimes arranged in 4 indistinct **Ri**); **Inf** few- to many-flowered, peduncle ± persistent; **Fl** nodding; **Ped** ≤ 8 mm; **Sep** ± 5 mm; **Cl** outside cream-coloured, inside cream-coloured to brownish, bowl-shaped, ± 1.7 × ± 2.5 cm ∅, tube pink at the base, above with fine concentric (partly broken) purple lines; **Cl** lobes ± erect or slightly spreading, slightly reflexed, papillose-roughened, inside densely papillose, margin dark purple, finely papillose, papillae ≤ 0.4 mm, semiglobose at the ground of the tube, above cylindrical, blunt, rarely finely pointed; **Ci** purple, lobes 0.5 - 1.5 mm, fused into a crenate disc, 3.5 mm ∅, round or emarginate; **Cs** yellow, margin dark purple, lobes as long as the **Anth**, ± 1.5 mm broad, ± subulate, apically obtuse; **Poll** dark greenish-brown, germination mouth region ± black, ± elliptic, very compact.

H. lenewtonii Plowes (Asklepios No. 64: 21-22, pl. 5-6, 1995). **T**: Kenya, Eastern Prov. (*Newton & Powys* 3703 [K, EA]). – **D**: N Kenya. **Fig. XXX.c**

[2] Stems prostrate-decumbent, green, often mottled with purple, 10 × 1.5 cm, 5-angular; **Tu** 3 - 4 mm, spreading; **Inf** with 2 - 3 **Fl** near the base of the plant, nodding; **Bra** light pink, 2 × 1 mm, broadly triangular, acute; **Ped** 15 - 25 mm, ± erect; **Sep** pale green, 4 - 5 × 1.5 - 2 mm; **Cl** outside greenish-yellow, inside ruby-red, 3.5 cm ∅, broadly funnel-shaped, tube slightly darker, 1 × 1 cm; **Cl** lobes 1 × 1.5 cm, triangular, pointed, margins sometimes greenish-yellow, outside scattered with papillae being round or slightly pointed, inside densely velvety-papillose; **Ci** ruby-red, lobes 2 × 2 mm, ± rectangular, borne tightly together, forming a crenate circle; **Cs** basally ruby-red, yellowish above, tipped dark red, lobes 1.5 mm, basally hump-backed, apically obtuse, covered with fine blisters; **Poll** pale brown, corpuscle dark reddish-brown.

According to the protologue closely related to *H. keniensis* var. *nairobiensis* and *H. penzigii* (= *H. macrocarpa*) and belonging to the *H. keniensis* complex.

H. levyi Obermeyer (FPSA 16: t. 616 + text, 1936). **T**: Zimbabwe (*Levy* s.n. [PRE, NH]). – **D**: Botswana, Zambia, Zimbabwe.

[1] Stems erect or prostrate-decumbent, forming clumps, to 10 × ≤ 1.5 cm, 4- to 5-angular; **Tu** ≤ 8 mm, triangular, often wing-like compressed; peduncle (if present) with many acute **Bra**, these ≤ 3 × 0.75 mm; **Ped** ≤ 8 mm; **Sep** ≤ 6 × 2 mm; **Cl** outside greenish-brown, inside basally dark maroon, tubular to campanulate, to 4 × 3 cm ∅, 3 - 5 mm above the base thickened annulus-like, base of the tube often smooth; **Cl** lobes yellowish dotted with brownish-red, ± 5 × 5 (-8) mm, triangular, shortly pointed, spreading, revolute, outside roughened, inside ± densely papillose, papillae dark maroon, ≤ 1.25 mm, elongate conical-cylindrical, obtuse, often tipped with an erect **Bri** of 2 mm; **Cn** 4 - 5 mm; **Ci** black-maroon, lobes fused into a ± circular disc; **Cs** brownish, lobes ≤ 3.5 × 1 mm, erect or rising up and meeting each other, strong, obtuse, dorsiventrally compressed; **Poll** yellowish; **Fr** yellow-brown, streaked with purple, 15 cm, **Ped** 3 - 4 mm ∅, strong, ± woody.

H. lodarensis Lavranos (JSAB 30(2): 87, 1964). **T**: Saudi Arabia (*Lavranos* 1900 [K]). – **D**: Saudi Arabia.

[1] Stems erect, compact, forming clumps, 3 - 4 cm, 5-angular; **Tu** 4 mm, basally compact-conical, apically finely subulate, divaricate; **Bra** ≤ 6 mm, acute; **Ped** ≤ 1.2 cm; **Sep** 8 - 10 × 10 - 15 mm; **Cl** inside and outside cream-coloured, inside irregularly spotted and streaked with purple-brown, ± 3.6 cm ∅; **Cl** lobes ± as long as broad, triangular, strongly revolute, apically caudate, outside slightly roughened, margins erect, dark purple-brown, inside ± densely papillose, papillae stippled and streaked with purple-brown, ≤ 1.5 (-1.75) mm, strong, ± conical, compressed; **Ci** dark purple, lobes as long as broad, tips appressed to the base of the tube; **Cs** whitish, finely dotted with purple, lobes ± 3 mm, erect-divaricate, dorsally flattened, basally widened, apically finely roughened, obtuse.

H. loeseneriana Schlechter (BJS 20(Beiblatt 51): 55, 1895). **T** [neo]: RSA, Transvaal (*Rossouw* 92 [NBG]). – **D**: N RSA (North-West Prov., Gauteng, Free State). **I**: White & Sloane (1937).

[1] Stems erect, tufted, often forming mats, ± 3.5 (-8) cm, compact, 4-angular; **Tu** acute; **Inf** 1- or few-flowered, pointing outwards; **Bra** ± 3 mm, straight, subulate; **Ped** ± 7 mm, basally sometimes thickened (then with several **Bra**), often with an additional bud; **Sep** ≤ 7 mm; **Cl** outside pale yellow to brownish, inside densely marked with brown to dark crimson concentric lines or stipples, ± 2 - 2.6 cm ∅, campanulate; **Cl** tube 7 - 9 × ± 10 mm, ± cup-shaped, mouth a little wider, mouth of the tube and **Cl** lobes ± erect or divaricate; **Cl** lobes ± 6 × > 6 mm, triangular, outside slightly roughened, inside ± papillose, papillae ≤ 1 mm, bluntly conical, compressed, basally broadened, often finely pointed; **Cn** 3.5 × ± 5 mm; **Ci** lobes ± square; **Cs** lobes 1.5 - 2 mm, ± erect, dorsiventrally compressed, tips knobby, meeting each other above the **Anth** or slightly divergent; **Poll** yellowish, germination mouth region darker.

H. longii Pillans (JSAB 5: 65, 1939). **T**: RSA, Eastern Cape (*Long* 1154 [BOL]). – **D**: RSA (Eastern Cape: Uitenhage Distr.). **I**: Leach (1988).

[1] Stems ± erect, tufted, ± 4 × ≤ 0.8 cm; **Tu** obtuse, indistinct, arranged in 6 - 8 **Ri** or borne in a weak spiral; **L** rudiments soon becoming a **Sp**, small, acute, ± spreading; **Fl** facing upwards; **Ped** 2 - 5 mm; **Sep** ≤ 5 mm; **Cl** both sides cream-coloured, inside densely stippled with brown-red, 2 - 3 (-3.5) cm ⌀, campanulate; **Cl** tube 9 × ≤ 10 mm, glabrous; **Cl** lobes 5.5 - 10.5 × 6 - 7.5 mm, ± triangular, divergent, revolute, inside densely papillose, papillae marked with red-brown, cylindrical or weakly clavate, obtuse, apically often finely hairy; **Cn** dark brown; **Ci** lobes ± square or ± triangular; **Cs** lobes 2 - 2.5 mm, subulate, basally dorsiventrally compressed, apically ± cylindrical, tip sometimes slightly clavate, convergent above the **Anth**, apically revolute-divergent, basally with a pointed convexity; **Poll** yellow.

H. longituba N. E. Brown (FC 4(1): 912, 1909). **T:** RSA, Cape Prov. (*Pillans* 609 [BOL, GRA]). – **D:** Zimbabwe, Botswana, RSA.

H. longituba ssp. **cashelensis** L. C. Leach & Plowes (JSAB 32(1): 49, ill., 1966). **T:** Zimbabwe, Melsetter Distr. (*Leach* 5405 [PRE, K, SRGH]). – **D:** Zimbabwe.

[1] Differs from ssp. *longituba*: Stems 5- to 6-angular; **Fl** smaller, base of the tube often slightly widened, papillae **Bri**-tipped.

H. longituba ssp. **longituba** – **D:** Botswana, RSA (Northern Prov., North-West Prov., Free State, Northern Cape).

[1] Stems erect, tufted, ± 5 × 0.6 cm, prominently 4- to 5-angular; **Tu** acute; **Fl** facing upwards or ± erect; **Bra** ≤ 6 mm; **Ped** ± 1 (-1.8) cm; **Sep** ≤ 1 cm; **Cl** inside cream-coloured, ± 2.8 × 4 cm, tubular or campanulate; **Cl** tube basally with concentric red-brown or maroon lines, streaked above, then dotted, ≤ 2 × 1.5 cm, basally flat and glabrous; **Cl** lobes 8 - 10 × 8 - 10 mm, acutely triangular, erect or divaricate, rarely revolute, outside sometimes partly finely roughened, inside papillose, papillae stippled with reddish, ≤ 1.5 mm, slender, conical, often finely mucronate; **Cn** 5 × ± 8 mm; **Ci** black-maroon, lobes ± square, lobed, dentate, crenate or emarginate; **Cs** lobes ± 1 mm, erect and touching each other, dorsiventrally ± compressed, apically clavate, revolute; **Poll** yellow, germination mouth region brown.

H. macrocarpa (A. Richard) Sprenger (Cat. Dammann & Co., 59: 4, 7, fig. 6, 1892). **T:** Ethiopia (*Penzig* s.n. [K [epi]]). – **Lit:** Brodie (1998). **D:** Sudan, Eritrea, Ethiopia, Saudi Arabia, Yemen. **I:** Leach (1988: 108, as *H. penzigii*). **Fig. XXX.d**

≡ *Stapelia macrocarpa* A. Richard (1851); **incl.** *Huernia penzigii* N. E. Brown (1892) ≡ *Huernia macrocarpa* var. *penzigii* (N. E. Brown) A. C. White & B. Sloane (1937); **incl.** *Huernia arabica* N. E. Brown (1895) ≡ *Huernia penzigii* var. *arabica* (N. E. Brown) A. Berger (1910) ≡ *Huernia macrocarpa* var. *arabica* (N. E. Brown) A. C. White & B. Sloane (1937); **incl.** *Huernia penzigii* var. *schimperi* A. Berger (1910) ≡ *Huernia macrocarpa* fa. *schimperi* (A. Berger) S. Brodie (1998); **incl.** *Huernia penzigii* var. *schweinfurthii* A. Berger (1910) ≡ *Huernia macrocarpa* var. *schweinfurthii* (A. Berger) A. C. White & B. Sloane (1937); **incl.** *Huernia macrocarpa* var. *cerasina* A. C. White & B. Sloane (1937) (*nom. inval.*, Art. 37.1); **incl.** *Huernia macrocarpa* var. *flavicoronata* A. C. White & B. Sloane (1937) (*nom. inval.*, Art. 37.1); **incl.** *Stapelia macrocarpa* Martelli (s.a.).

[2] Stems erect or shortly prostrate and then rising, tufted, to 9 × 1.5 cm, 4- to 5-angular; **Tu** ≤ 1 cm, conical-subulate, divaricate; **Fl** facing outwards or nodding; peduncle short, very robust; **Bra** ≤ 6 mm; **Ped** 0.5 - 1 cm; **Sep** 6 - 10 (-12) × 0.75 - 1.5 mm; **Cl** inside brownish-red or black-purple, with purple transverse striping on yellowish ground, ± 1.5 - 2 cm ⌀, campanulate; **Cl** lobes 5 - 7 mm, triangular, erect or slightly spreading, outside sometimes finely roughened, inside weakly papillose, papillae small, obtuse, rarely slightly pointed; **Cn** brown or black-purple; **Ci** lobes transversely rectangular to square, shallowly emarginate or shortly broadly bidentate, at the **Anth** bases with a prominent conical convexity; **Cs** basally sometimes bright yellow, darker towards the tip, tip ± black, lobes ± 1.4 × ≥ 0.5 mm, apically obtuse; **Poll** yellowish; **Fr** cream-coloured, striped black-purple, 4 - 17 × 0.4 - 1 cm, erect with woody **Ped** to 6.5 cm.

Size and colouring of the corolla are variable in this complex.

H. marnieriana Lavranos (JSAB 29: 97-99, ills. (pl. 9: 1-2), 1963). **T:** Yemen (*Lavranos* 1958 [K]). – **D:** S Yemen.

[2] Stems erect, tufted, 5 × ± 0.8 cm, 5-angular; **Tu** ≤ 8 mm, basally compressed into a deltoid shape, apically acutely subulate, much spreading; **Fl** facing outwards; **Bra** ≤ 5 mm; **Ped** ≤ 1 cm; **Sep** ≤ 8 mm; **Cl** outside cream-coloured to brownish, inside whitish, basally pink, ± 3 cm ⌀, ± flat, mouth of the tube and lobes divaricate; **Cl** lobes ≤ 1 × > 1 cm, broadly triangular, inside ± densely finely papillose, papillae red, ≤ 1 mm, obtusely conical; intermediate lobes indistinct or absent; **Cn** ± 5 mm ⌀; **Ci** whitish, apically pink, lobes ± transversely rectangular; **Cs** pale pink, lobes 1.25 × 0.75 mm, attenuate, basally with a dorsal hump, apically ± obtuse.

H. namaquensis Pillans (J. Bot. 68: 102, 1930). **T:** RSA, Northern Cape (*Pillans* 5155 [BOL]). – **D:** S Namibia, RSA (Northern Cape).

Incl. *Huernia herrei* A. C. White & B. Sloane (1937); **incl.** *Huernia herrei* var. *immaculata* A. C. White & B. Sloane (1937); **incl.** *Huernia owamboensis* R. A. Dyer (1980).

[3] Stems prostrate-decumbent, with sympodial

branching near the base, attenuate, often slightly twisted, tufted, (dark) green, sometimes suffused with purple, to 4 (-6) × 1.5 - 2 cm, 4- to 6-angular; **Tu** ± 4 mm, triangular, ± compressed, spreading, usually reflexed; **L** rudiments fine, acute, ± hard, persistent; **Fl** ± nodding; peduncle 1.2 (-2) cm; **Bra** ± 3 mm; **Ped** brownish or purple, 1.5 (-2.5) cm; **Sep** 3 - 5 (-6) × 1 - 1.5 mm; **Cl** outside yellowish to brown, inside whitish-creamy to pale yellow, stippled with reddish or purple or irregularly marked with ± concentric lines (lines short, broken, occasionally without markings), (1.6-) 2 - 2.5 (-3.5) cm ∅; **Cl** tube ± flat, basally rounded and glabrous, sometimes ± obtuse, mouth constricted, margin spreading; **Cl** lobes triangular, weakly acuminate, slightly revolute, inside ± densely papillose, papillae mostly obtuse, semiglobose, obtusely cylindrical or ± conical, apically shortly pointed or **Bri**-tipped; intermediate lobes ± 2 mm; **Ci** crimson or purple, lobes deeply bifid, weakly 2-notched, irregularly dentate or fused into a disc (specimens from the Hellskloof); **Cs** yellow with crimson markings, lobes swollen, attenuate, tip obtuse, at least as long as the **Anth**, meeting each other above the **Sty** head, rarely with the tips shortly bent upwards, sometimes ± erect, rarely basally with a prominent transverse furrow; **Poll** yellowish.

H. nigeriana Lavranos (JSAB 27: 233, ills., 1961). **T:** Nigeria (*Lavranos* 1058 [PRE]). – **D:** Nigeria.

≡ *Huernia volkartii* var. *nigeriana* (Lavranos) Lavranos (1964).

[1] Stems erect or prostrate-decumbent, tufted or ± creeping, to 8 × 1 cm, 5-angular; **Tu** small, acute and spreading; **Ped** 0.8 - 1 cm; **Sep** 6 - 7 mm; **Cl** outside cream-coloured, inside cream-coloured with blood-red spots and broken lines, 2 - 2.5 cm ∅; **Cl** tube ± deeply cup-shaped, 7 - 9 × 9 - 10 mm; **Cl** lobes 6.5 - 7 × 5.5 - 6.5 mm, triangular, spreading, margin dark red, denticulate, outside finely roughened, sometimes ± glabrous, inside papillose, papillae 1 - 1.5 mm, ± slender, ± conical, sometimes finely mucronate; **Ci** reddish-brown, lobes as long as broad; **Cs** lobes 2.5 - 3 mm, meeting each other above the **Sty** head, above upper ½ spreading, basally broadened, rarely humped, apex small, obtuse, finely papillose.

Nigerian collections cited under *H. volkartii* by various authors belong here.

H. nouhuysii I. Verdoorn (FPSA 11: t. 412 + text, 1931). **T:** RSA, Northern Prov. (*Van Nouhuys* s.n. [lecto: l.c. t. 412]). – **D:** RSA (Northern Prov.).
Fig. XXX.e

[3] Stems erect, forming clumps, to 9 (-20) × 1 - 1.5 cm, 4- to 6-angular or spirally twisted; **Tu** acute, spreading, often curved; **L** rudiments whitish, acute, hardening; **Inf** few-flowered, often 2 **Fl** open at the same time; peduncle very short, compact; **Bra** ± 2 mm; **Ped** 0.5 - 1.5 cm, ± erect; **Sep** ± 4 × 1.5 mm, tips slightly revolute; **Cl** inside greenish, dotted and streaked in concentric circles with brown-red, campanulate, 3 cm ∅; **Cl** tube ± 5 × 10 - 13 mm, basally glabrous; **Cl** lobes 5 - 6 mm, triangular, spreading, indistinctly revolute, inside ± densely papillose, papillae obtuse or ± acutely conical, larger papillae terminated by a fine red **Sp** or by a small reddish point; **Cn** 3 - 4 × 5 - 6 mm; **Ci** lobes ± square or fused into a disc showing 10 notches of ± even distribution, erect, obtuse, compressed; **Cs** lobes 2 - 2.5 mm, ± attenuate, obtuse, on the back of the **Anth** slightly reflexed, towards the apex connivent-erect; **Poll** brownish.

H. occulta L. C. Leach & Plowes (JSAB 32(1): 56-58, ills., 1966). **T:** Zimbabwe (*Leach* 11661 [SRGH, K, PRE]). – **D:** Zimbabwe.

[1] Stems slender, prostrate, to 30 × 0.5 cm, 5-angular, sides deeply furrowed; **Tu** small, acute, spreading; **Fl** facing upwards; **Ped** ≤ 2.5 cm; **Sep** 6 - 8 × 1.5 mm; **Cl** inside cream-coloured, stippled and mottled with brownish-red, 3 cm ∅; **Cl** tube shining, black-maroon, bicampanulate, ± 0.75 × 10 mm, basally glabrous, margin of the tube abruptly divaricate, dish- or shallowly bowl-shaped; **Cl** lobes ± as long as broad, triangular, ± erect or divergent, sometimes apically weakly revolute, inside ± densely papillose, papillae ≤ 1.25 mm, conical, often terminated by a short point; **Cn** black-maroon, ± 8 mm ∅; **Ci** lobes ± square, emarginate, 2-lobed or shortly toothed; **Cs** lobes ± 3.5 mm, ± cylindrical, touching each other, basally with indistinct blunt convexity, tapering off towards the tip, tip obtuse, finely obtusely tuberculate or plaited, tips slightly divergent; **Poll** yellow, elongate-ovoid, pollinarium ± 0.375 × 0.25 mm.

A close ally of *H. hislopii*.

H. oculata Hooker *fil.* (CBM 108: t. 6658 + text, 1882). **T:** Namibia, Damaraland (*Een* s.n. [K]). – **D:** Angola, Namibia.

Incl. *Huernia rogersii* R. A. Dyer (1927).

[2] Stems erect, compact, forming clumps, to 10 × 0.8 cm, prominently 5-angular; **Tu** acute, wing-like compressed; **Inf** many-flowered, pointing outwards; peduncle compact; **Bra** 7 - 8 mm; **Ped** 4 - 5 mm; **Sep** ≤ 1.5 cm, apically very acute; **Cl** outside greenish with purple, basally paler, inside basally white, sometimes slightly suffused with pink, ± 2 cm ∅, shallowly bowl-shaped, lobes and margin of the tube black-purple; **Cl** lobes rarely green, 4 × ± 7 - 8 mm, broadly triangular, ± acuminate, spreading, slightly revolute, both sides roughened-papillose, papillae of the inside terminated by fine acute **Ha**; **Cn** pale yellow, red or purple, 3.5 × 4.5 mm; **Ci** reduced, with variable margins; **Cs** lobes ± 2 × 1.25 mm, meeting each other above the **Anth**, straight, often curved, then partly with overlapping tips, attenuate, tip obtuse; **Poll** brownish or greenish-brown.

H. pendula E. A. Bruce (FPA 28: t. 1108 + text, 1951). **T:** RSA, Eastern Cape (*Phillips* 1 [PRE]). – **D:** RSA (Eastern Cape).

[2] Stems ± cylindrical, pendulous from rock faces, on level ground first erect, then curved, rooting when in contact with soil, to 90 (-150) × 0.5 - 0.8 cm, indistinctly 4-angular; **Tu** small, acute; base of **L** rudiments sterile, tuberculate; **Fl** nodding, borne on basal short lateral **Br**, occasionally spread along the stems; **Ped** ≤ 8 mm; **Sep** 3 - 4 mm; **Cl** outside purple, inside dark maroon, 1.4 × ≤ 2 cm, bowl-shaped; **Cl** lobes ± erect or slightly spreading, revolute, inside densely papillose, papillae dark maroon, ± 0.2 mm ∅, mucronate; **Cn** ± 3 mm; **Ci** black-maroon, disc-shaped, 3 - 3.5 mm ∅, crenate, weakly conical; **Cs** dark maroon, lobes ± 1.5 mm, ± tongue-shaped, apically obtuse; **Poll** yellowish.

H. piersii N. E. Brown (FC 4(1): 909, 1909). **T:** RSA, Eastern Cape (*Piers* s.n. in *Pillans* 622 [K, BOL]). – **D:** RSA (Eastern Cape). **I:** Leach (1988).

[1] Stems erect, compact, crowded, dull green, dotted with purple, to 4.5 (-5) × 1 - 1.5 cm, 4-angular, **Ri** ± distant from each other; **Tu** 1 - 2 mm, spreading; **Fl** facing outwards or upwards; **Ped** to 2.5 (-4) cm; **Sep** 2.5 - 5 × 1.5 mm, very variable; **Cl** outside greenish-creamy to pale yellow, inside cream-coloured to yellowish, ≤ 3.5 cm ∅, ± bicampanulate; **Cl** tube basally bright red-brown, above with concentric narrow red-brown lines, otherwise stippled with red-brown, cylindrical or broadly bowl-shaped, ≥ 1.2 × ≤ 1.2 cm, margin abruptly spreading; **Cl** lobes ≤ 1 × 1 cm, triangular, acuminate, ± erect or revolute, inside densely papillose, papillae columnar, apically often hairy, **Ha** dark crimson (rarely white), long, erect, mostly ± clavate, rigid, **Ha** of the lobes inflated, globose or globose-acute; intermediate lobes 1.5 mm, prominent, acute; **Cn** ≤ 4.5 mm; **Ci** dark crimson to black, lobes ± square or rectangular, ± deeply bifid; **Cs** brownish, apically darker, lobes subulate, meeting each other above the **Anth**, divergent, apically ± obtuse, ± verruculose, first moist with ample nectar; **Poll** brownish-yellow.

H. pillansii N. E. Brown (Gard. Chron., ser. 3, 35: 50, 1904). **T:** RSA, Cape Prov. (*Pillans* 23 [K, BOL, GRA]). – **D:** RSA (Western Cape, Eastern Cape). Fig. XXXI.c
  Incl. *Huernia pillansii* ssp. *pillansii*.

[1] Stems ± cylindrical or narrowly ovate, erect, tufted, to 18 cm (in cultivation); **Tu** small, crowded, arranged in (9-) 10 - 16 (-24) **Ri** or spirally organized; **L** rudiments 2 - 8 mm, as soft **Bri**, ± divaricate, usually marcescent; **Fl** ± facing upwards; **Ped** 2 - 8 mm; **Sep** 8.5 - 12.5 mm; **Cl** outside yellow or brownish, inside cream-coloured or yellowish, stippled with reddish, 3 - 5 cm ∅, campanulate, tube with labyrinth-like reddish dots, ± cylindrical (at times cup-shaped), 5 - 9 mm; **Cl** lobes 1.2 - 2.2 × 0.5 - 1 cm, triangular, spreading or revolute, inside densely papillose, papillae ± 1 mm, cylindrical, obtuse, apically partly hairy, **Ha** streaked red, delicate, sometimes slightly clavate; intermediate lobes 1.5 mm, subulate, spreading to revolute; **Ci** lobes ± trapeziform; **Cs** lobes 2 - 2.5 mm, erect, dorsiventrally compressed and flattened, slowly attenuate, apex prominent, acutely verrucose, knobby, basally with a ± prominent hump; **Poll** yellowish to greenish, germination mouth region darker.

H. plowesii L. C. Leach (Excelsa Tax. Ser. 4: 134-136, ills., 1988). **T:** Namibia (*Plowes* 6761 [NBG, K, PRE]). – **D:** SW Namibia.

[2] Stems erect or decumbent, ± pyramid-shaped, forming small clumps, to 3 cm, 4-angular; **Tu** acute; **Fl** pointing outwards; **Ped** ≤ 1 cm; **Sep** 5 - 6 × 2 mm; **Cl** inside dark purple-brown, 2.5 - 3.5 cm ∅, tube ± 6 mm ∅, inside 5-angular, mouth thickened, annulus sometimes magenta, ± 5 mm ∅; **Cl** lobes cream-coloured, irregularly spotted with maroon, somewhat broader than long, triangular, acuminate, divaricate, inside basally glabrous, scattered with **Ha** above, **Ha** ≤ 3 mm, simple, rigid, borne at the apex of small tubercles, lobes finely pubescent, **Ha** ≤ 0.25 mm, acute; **Ci** lobes ± 5 × 2 mm, ± bifid; **Cs** ± black, rarely brownish, lobes narrowly triangular, more elongate forms more pointed and meeting each other above the **Sty** head, with a basal dorsal hump; **Anth** orange; **Poll** yellow, germination mouth region paler, caudicles orange.

Belongs to *H. guttata* in the sense of Huber (1967: 59).

H. praestans N. E. Brown (FC 4(1): 914, 1909). **T:** RSA, Western Cape (*Pillans* 667 [K, BOL]). – **D:** RSA (Western Cape). **I:** Leach (1988).

[1] Stems ± erect (rarely creeping), forming small clumps, green, spotted purple, to 5 × 1.5 cm, 4- to 5-angular; **Tu** ≤ 5mm, elongate and triangularly pointed; **Fl** pointing outwards; **Ped** ≤ 1.2 cm; **Sep** 3 - 6 × ± 1.5 mm; **Cl** outside cream-coloured, weakly stippled with reddish, inside cream-coloured or pale greenish-yellow, to 4.5 (-5.5) cm ∅, shallowly bowl-shaped, margin of the tube and lobes irregularly stippled with red to maroon, base of the tube bright maroon, above with concentric lines, sometimes coarsely spotted; **Cl** tube bowl-shaped, basally usually ± obtuse, inside slightly constricted, mouth 5-angular, margin bulging annulus-like, abruptly divaricate; **Cl** lobes shorter than broad, oblong-triangular, erect to ± horizontally spreading, radially often strongly convex, inside ± densely papillose, papillae ± column-shaped, apically hairy, **Ha** dark purple, ± clavate or ± globose to globose-acute; **Cn** cream-coloured; **Ci** lobes with maroon spots and edges, ≤ 2 mm, ± square to transversely rectangular, emarginate, finely crenate or bifid; **Cs** apically dark, lobes ± 3

mm, subulate, meeting each other above the **Anth** or connivent with divergent tips; **Poll** yellow.

H. procumbens (R. A. Dyer) L. C. Leach (BT 10(1): 54, 1969). **T:** RSA, Northern Prov. (*Van der Schiff* 3618 [PRE]). – **D:** Zimbabwe, RSA (Northern Prov.).

≡ *Duvalia procumbens* R. A. Dyer (1956).

[2] Stems creeping, sometimes pendulous from rock faces, or semi-underground, to ± 30 × 1 - 1.5 cm, 5-angular, sides deeply furrowed; **Tu** obtuse; **L** rudiments ± 3 mm, very acute, dry persistent as a curved white point; **Inf** few-flowered, upright; peduncle rudimentary, emerging from the base of the l. **Ped**; **Ped** ≤ 1.5 cm; **Sep** ≤ 1 cm, prominent between the **Cl** lobes; **Cl** inside cream- or putty-coloured, 4 - 5 cm ⌀, shallow, annulus maroon, prominent, glabrous or weakly rugose; **Cl** lobes edged with red, 15 - 24 × 4 - 5 mm, narrowly triangular, acuminate, often ± revolute, convex towards the tip, inside apically hairy, **Ha** reddish, short, acute; **Cn** 3 × 3.5 mm; **Ci** lobes fused into a thickened obtusely pentagonal disc; **Cs** lobes ± 1.5 × 0.75 mm, shorter than the **Anth**, dorsal hump erect (rarely spreading), strongly enlarged, triangular in cross-section, ending ± bluntly in the lower part; **Poll** greenish-brown.

H. quinta (E. Phillips) A. C. White & B. Sloane (Stapelieae, ed. 2, 3: 885, 1937). **T:** RSA (*Anonymus* s.n. [PRE 10134]). – **D:** RSA.

≡ *Huernia scabra* var. *quinta* E. Phillips (1932).

H. quinta var. **blyderiverensis** L. C. Leach (Excelsa Tax. Ser. 4: 178-179, 1988). **T:** RSA, Mpumalanga (*Percy-Lancaster* 466 [NBG]). – **D:** RSA (Mpumalanga).

[3] Differs from var. *quinta*: **Cl** larger, with transverse lines and dots, margin of the tube and lobes spreading, mouth of the tube conspicuously constricted; **Bri** ± 0.25 mm or absent.

H. quinta var. **quinta** – **D:** RSA (Northern Prov., Mpumalanga).

[3] Stems erect, robust, weakly twisted, forming clumps, to 7 × 2 cm, distinctly 4- to 5-angular; **Tu** ≤ 4 mm; **L** rudiments white, hard and persistent; **Fl** facing upwards, sometimes 2 - 3 flowering simultaneously; peduncle sometimes branching; **Bra** ± 2 mm; **Ped** < 1 cm; **Sep** 5 × ± 1.5 mm; **Cl** inside white to whitish-creamy, (2-) 2.5 - 3 (-3.5) cm ⌀; **Cl** tube stippled with reddish, mouth with few concentric partly broken red lines, 0.7 - 1 cm ⌀, bowl-shaped, mouth pentagonal, margin of the tube and **Cl** lobes spreading, often slightly revolute; **Cl** lobes triangular, slightly acuminate, inside ± densely papillose, papillae ± prominent, ± conically compressed or ± semiglobose, sometimes tipped with a rigid **Bri** up to ± 1 mm; **Cn** cream-coloured or pale brown, rarely with maroon markings, 3 - 4 × ± 4.5 - 5.5 mm, fused into a flat or convex disc; **Ci** lobes very short, emarginate or sometimes crenate; **Cs** lobes ± 1 mm broad, tapering to an obtuse tip, furnished with very fine and obtuse warts, meeting each other above the **Sty** head, conical, basally with a transverse furrow; **Poll** brownish, germination mouth region darker.

H. recondita M. G. Gilbert (CSJA 47(1): 6-13, ills., 1975). **T:** Ethiopia (*Gilbert & Gilbert* 1729 [K, ADAB, EA, NBG [ex cult.]]). – **D:** W Ethiopia, NW Kenya.

[1] Stems prostrate, ± cylindrical, to 50 × 1 - 1.25 cm, 4- (rarely 5- to 7-) angular; **Tu** obtuse; **L** rudiments deciduous; peduncle (if present) compact; **Ped** 15 - 30 × ± 2 mm; **Sep** 7 - 10 × 1.5 - 2 mm; **Cl** inside yellowish, irregularly dotted with red, 3.5 - 4.5 cm ⌀; **Cl** tube basally red-brown, above with concentric lines, ± 8 × 10 - 12.5 mm, shallowly campanulate, mouth weakly constricted, margin spreading; **Cl** lobes 12 - 15 × 9 - 12 mm, triangular, shortly pointed, weakly revolute, margins thickened, dark red, outside ± densely and finely roughened, inside ± densely papillose, papillae streaked with red, ± 0.75 - 2.8 mm, either narrow and acute or cylindrical and slightly clavate; intermediate lobes ± 2 - 3 mm, acute, spreading; **Cn** ± 5 mm; **Ci** black, lobes ± 2 mm broad, apically a little narrower, ± square, irregularly obtusely bidentate, basally with a small ± conical hump; **Cs** dark red, the middle paler, lobes 3 - 3.5 mm, attenuate, apically ± elongate, abruptly revolute, finely papillose, obtuse, erectly connivent above the **Anth**, meeting each other below the tips, basally with a ± rectangular dorsal hump; **Poll** yellowish.

H. reticulata (Masson) Haworth (Synops. Pl. Succ., 28, 1812). **T:** [lecto – icono]: Masson, Stapel. Nov. t. 2, 1796. – **D:** RSA (Western Cape).

≡ *Stapelia reticulata* Masson (1796); **incl.** *Stapelia crassa* Donn *ex* Haworth (1812); **incl.** *Stapelia reticulata* var. *deformis* Jacquin (s.a.).

[1] Stems robust, decumbent-erect, forming loose clumps, prominently 5-angular; **Tu** ± 5 mm, acute; **Ped** 0.5 - 1.5 cm; **Sep** ± 6 mm; **Cl** outside greenish-creamy with pale purple-red dots, tube inside dark reddish-maroon, annulus red-black, broad, shining, sometimes pale yellow dotted with maroon, margin of the tube and lobes cream-coloured to pale yellow, with irregular dense maroon dots and spots, spots large, ± square, arranged in a net-like pattern; **Cl** 2.5 - 5 cm ⌀, tube campanulate, 0.8 - 1 (-1.2) × 1 - 1.2 cm, with a broad spreading margin; **Cl** lobes ≤ 1.5 × ≥ 1.5 cm, broadly triangular, steeply ascending or revolute, inside ± densely papillose, papillae delicate, conical, apically hairy, **Ha** purple, 1.5 - 4 mm, short with a rigid point or longer, robust, clavate, glass-like, **Cl** lobes ± granular, wrinkled, wrinkles flat, irregular; intermediate lobes pointing outwards or revolute; **Cn** 5.5 - 6 × 0.5 - 0.75 mm; **Ci** black-crimson, lobes ± 2.5 mm, ± square; **Cs**

whitish with crimson markings, sometimes crimson, more rarely brownish, lobes ± 4 mm, subulate, basally geniculate, apically finely pointed, meeting each other above the **Anth**, tips divaricate (sometimes erect); **Anth** alternating with humps; **Poll** yellow or pale brown.

H. rosea L. E. Newton & Lavranos (CSJA 65(6): 279-280, ills., 1993). **T:** Yemen (*Lavranos & Newton* 13071 [E, ZSS]). − **D:** Yemen.

[2] Stems erect, tufted, 3 - 6.5 cm, distinctly 5-angular; **Inf** 3-flowered; **Ped** pink, 7 × 1.5 mm; **Sep** pink, 10 × 2 mm; **Cl** either side pink, 2.5 - 3 cm ⌀, campanulate, tube 1 × 1 cm, fleshy, mouth 2 mm across; **Cl** lobes paler, 1 × 1 cm, triangular, spreading, very finely acuminate, outside verruculose, inside densely verrucose, warts pale pink, ≤ 0.5 mm; intermediate lobes very short; **Ci** pink, lobes 2.5 × 2 mm, sides ± parallel, bifid, rounded, hairy, **Ha** red, fine; **Cs** ± pink, dorsal hump yellow, lobes 1.5 × 1.5 mm, erect, basally flattened, spreading in a slightly oblique direction above the dorsal hump, apically connivent, tip delicately papillose. Otherwise as *H. hadhramautica*.

H. rubra Plowes (Asklepios No. 64: 20-21, pl. 1-3, 1995). **T:** Yemen (*Plowes & Barad* 7797 [K, SRGH]). − **D:** N Yemen.

[2] Stems erect, forming compact clumps, ± 5 - 7 cm, 5-angular; **Tu** ± 7 mm, tapering, horizontally spreading; **Inf** 3-flowered, horizontal or slightly nodding; **Bra** 8 mm, very narrow; **Ped** 5 mm; **Sep** 14 mm, very narrow; **Cl** outside pale greenish-white, inside deeply maroon or ruby-red, glabrous, 3 cm ⌀; **Cl** tube 3 mm, annulus shallow; **Cl** lobes occasionally slightly paler, ± strongly revolute, lobes and annulus weakly papillose, papillae cushion-shaped, deciduous; **Cn** ivory, 3 × 5.5 mm; **Ci** lobes very short and broad, fused into a ± circular disc; **Cs** lobes acuminate, basally weakly humped, apically meeting each other; **Poll** rusty brown, corpuscle dark reddish-brown.

Closely related to *H. hadhramautica*.

H. saudi-arabica D. V. Field (KB 35(4): 754, ill., 1981). **T:** Saudi Arabia (*Collenette* 549 [K, NBG]). − **D:** S Saudi Arabia. **Fig. XXX.f**

[1] Stems prostrate-decumbent or erect, tufted, 3 - 6 × ± 1 cm, 5-angular; **Tu** ± 8 mm, triangular-subulate; **Ped** 0.8 - 1.6 cm; **Sep** 8 - 13 × 1.5 mm; **Cl** inside dark purple, campanulate, ± 4 cm ⌀, sometimes with few small whitish spots or basally with fine golden lines, margin of the tube and lobes occasionally weakly spotted with white, margins of lobes darker; **Cl** tube broad, ± bowl-shaped, basally rounded; **Cl** lobes triangular, acuminate, strongly revolute, outside rarely finely roughened, inside glabrous, rarely with fine acute tubercles; **Cl** tube inside densely papillose, papillae basally whitish-yellow, apically dark, ≤ 1.5 mm, ± columnar, obtuse, often ± compressed; **Cn** 5.5 - 6 mm; **Ci** black-purple, lobes ± 2 mm, ± square, apically ± dentate, ± finely crenate; **Cs** lobes 4 - 4.5 mm, slightly connivent, upper ½ divergent, apically ± glabrous, subulate; **Poll** yellow.

H. schneideriana A. Berger (Monatsschr. Kakt.-kunde 23: 177, 1913). **T:** Tanzania, Rungwe Distr. (*Stolz* 1407 [B †]). − **D:** Tanzania. **I:** Leach (1988).

[2] Stems erect, creeping or pendulous, to 40 × 1.2 - 1.4 cm, (6- to) 7-angular; **Tu** small, obtuse; **Inf** many-flowered, **Fl** facing outwards, sometimes 2 flowering simultaneously; peduncle very strong, ± persistent; **Ped** ≤ 1 cm; **Sep** ± 5 × 2 mm; **Cl** inside and outside pale brown, 2.6 - 3 cm ⌀, tube deep dark brown, ± 5 × ± 9 mm, ± semiglobose, margin of the tube and lobes horizontally spreading; **Cl** lobes ± 8 × 8 mm, triangular, acute, inside partly very finely papillose, margin of the tube and lobes densely pubescent, **Ha** dark brown; intermediate lobes small, revolute, not visible in top-view; **Cn** ± 2.5 × 5 mm, basally broadly ± conical; **Ci** brown, lobes short, very slightly and irregularly crenate, basally with a deep furrow; **Cs** yellowish, lobes as long as the **Anth**, tapering to an obtuse point, slightly raised, basally with an obtuse dorsal hump; **Anth** obtuse, fleshy; **Poll** brownish, germination mouth region distinct.

H. similis N. E. Brown (BMI 1895: 265, 1895). **T:** Angola, Luanda Distr. (*Welwitsch* 4264 [K, LISU]). − **D:** Angola. **I:** Leach (1988).

[2] Stems prostrate or ± erect, branching freely, ± rampant, to 7 × 1.2 cm, 4-angular; **Tu** ± 2 mm, obtuse, spreading; **Fl** nodding; peduncle 6 - 8 mm; **Sep** ± 2 × 1.25 - 1.5 mm; **Cl** outside dull purple, inside basally white, 9 mm ⌀, broadly bowl-shaped, towards the tips and margins of the lobes dark purple, tube 5 mm deep; **Cl** lobes ± 3 × 5 mm, acute, slightly spreading and revolute, inside densely whitish-papillose, papillae of the **Cl** tube and the lobes dark purple, ≤ 0.15 mm, ± semiglobose or ± cylindrical, ± acuminate to mucronate; **Cn** 2 - 2.5 × ± 3.5 mm; **Ci** lobes apically obtuse, emarginate, fused into an obtusely angled pentagon; **Cs** lobes ± 1 × 0.6 mm, attenuate, with obtuse apex; **Poll** yellow-brown, germination mouth region brown.

H. somalica N. E. Brown (BMI 1898: 309, 1898). **T** [lecto]: Somalia (*Lort-Phillips* s.n. [K]). − **D:** E Ethiopia, NW Somalia. **I:** Leach (1988).

[2] Stems very compact, forming clumps, ± 4.5 × ± 1 cm, prominently 5-angular; **Tu** ≤ 8 mm, very acute; **Fl** facing outwards; peduncle compact; **Ped** ± 5 mm; **Sep** ± 7 × 2 mm; **Cl** outside cream-coloured, inside dark purple-brown, ± 4 cm ⌀, margin of the tube ± shining, lobes dull yellowish; **Cl** tube ± 1.5 cm ⌀, ± bowl-shaped, outside sometimes weakly

roughened, margin revolute, convex and annulus-like, glabrous or with weak pustules; **Cl** lobes 7 × ± 5 - 8 mm, acute or oblong-triangular, spreading or revolute, strongly papillose, papillae apically dark red or purple, ≤ 0.5 mm, cylindrical, obtuse; **Cn** 4.5 × 5 mm; **Ci** dark maroon, margin sometimes yellow, lobes obtusely and shortly bifid, longitudinal furrow reaching from the point of insertion of the **Ci** to the base of the **Cs**; **Cs** yellow or dark purple, dorsal hump yellow, lobes ± 2 mm, tapering to a blunt tip, meeting each other above the **Anth**; **Poll** brown, germination mouth region dark.

H. stapelioides Schlechter (BJS 20(Beiblatt 51): 55, 1895). **T**: RSA, Transvaal (*Schlechter* 4487 [B †]). – **D**: RSA (Northern Prov., Mpumalanga, Gauteng, North-West Prov.), Swaziland.

Incl. *Huernia vogtsii* E. Phillips (1932).

[1] Stems short, square, mat-forming, or longer and ± tufted, 3 - 10 × ≤ 1.5 cm, 4- (to 5-) angular; **L** rudiments sometimes as hard **Sp**; **Fl** facing outwards; **Ped** ± 1 (-1.8) cm; **Sep** 0.8 - 1 cm; **Cl** inside and outside (pale) yellow, inside dotted and streaked with maroon or brown-red, 3 - 4.2 cm ⌀, tube ± cup-shaped, glabrous; **Cl** lobes 1.5× as long as broad, oblong-triangular, sometimes ± caudate, outside weakly roughened, inside densely papillose, papillae ≤ 2.5 mm, narrowly conical, strongly compressed, flattened; **Ci** lobes ± square; **Cs** lobes 2.5 - 3 mm, erect, clavate, convex, apically ± foot-shaped, setaceous-verrucose, basally with a hump, broad, obtuse, verrucose; **Poll** yellow, germination mouth region darker.

H. tanganyikensis (E. A. Bruce & P. R. O. Bally) L. C. Leach (BT 10(1): 54, 1969). **T**: Tanzania, Arusha Distr. (*Bally* S19 [K, PRE [photos]]). – **D**: NE Tanzania. **Fig. XXX.g**

≡ *Duvalia tanganyikensis* E. A. Bruce & P. R. O. Bally (1941).

[2] Stems creeping, poorly branching, ± cylindrical, forming dense mats, 20 (-85) × 1 - 1.8 cm, obtusely 5-angular; **Tu** 1 - 2 cm, acute; **Inf** few-flowered, pointing upwards; peduncle partly persistent, curved when elongate, knobby; **Ped** ≤ 2 cm; **Sep** ± 1.2 cm; **Cl** outside greenish-creamy, inside pink, 3 - 3.5 cm ⌀, shallow, annulus bright dark maroon, ± 1 cm ⌀, prominent; **Cl** lobes ± 12 × 8 mm, triangular, acuminate, smooth, glabrous; intermediate lobes delicate; **Cn** ± 3 mm; **Ci** 3.5 mm ⌀, lobes fused into a thickended ± circular or obtusely pentagonal disc; **Cs** yellow, lobes ± acute or bidentate, ± as long as the **Anth**, dorsal hump 1 - 2 × 0.75 mm, ± ovate, obtuse, ± erect; **Poll** brownish-orange, germination mouth region amber, ± quadrangular-elliptic.

H. thudichumii L. C. Leach (Excelsa Tax. Ser. 4: 132-134, ills., 1988). **T**: RSA, Western Cape (*Thudichum* 214A [NBG]). – **D**: RSA (Western Cape).

[2] Stems erect, with age ± pyramid-shaped, forming clumps, to 4 × 1.5 cm; **Tu** arranged in 4 - 5 **Ri**, sometimes disappearing; **Inf** few-flowered; peduncle knobby, persistent; **Ped** ≤ 1.5 cm; **Sep** 4 × 2 mm; **Cl** inside ivory, often with pale orange, sometimes with greenish, tube basally pink, lobes partly indistinctly stippled with pale pink, (3.5-) 4 (-4.5) cm ⌀, margins maroon; **Cl** tube shallow, sides sometimes rugose, annulus prominent; **Cl** lobes broadly triangular, acute or shortly tapered, ± transversely convex, slightly curved outwards, finely papillose, papillae maroon, ± inflated, semiglobose, obtuse or rather acute; **Ci** brown-red or dark crimson, lobes short, irregularly emarginate, with flat irregular notches, 2-crenate, sometimes fused into a 10-notched or ± circular disc; **Cs** orange-red, lobes ovate, obtuse, very fleshy, attenuate, apex ± acute, sometimes meeting each other above the **Sty** head, ± ½ as long as the **Anth**, sometimes overtopping these, without conspicuous dorsal edge or basally with a prominent obtuse convexity; **Poll** yellow or orange, germination mouth region darker, compressed.

H. thuretii Cels (Hort. Franç., sér. 3, 73: t. 3 + text, 1866). **T**: [lecto – icono]: l.c. t. 3. – **D**: RSA, Namibia.

≡ *Stapelia thuretii* (Cels) Croucher (1877); **incl.** *Huernia inornata* Obermeyer (1937); **incl.** *Huernia striata* Obermeyer (1937).

H. thuretii var. **primulina** (N. E. Brown) L. C. Leach (Excelsa Tax. Ser. 4: 185, 1988). **T** [lecto]: RSA, Eastern Cape (*Barkly* 13 [K]). – **D**: RSA (Eastern Cape).

≡ *Huernia primulina* N. E. Brown (1890); **incl.** *Huernia primulina* var. *rugosa* N. E. Brown (1909); **incl.** *Huernia thuretii* var. *rugosa* N. E. Brown (1909).

[3] Differs from var. *thuretii*: Stems longer; **Cl** uniformly pale yellow, a little larger, often with a low annulus, margin of the tube and lobes more papillose, papillae sometimes rather dense, prominent.

H. thuretii var. **thuretii** – **D**: Namibia, RSA (Eastern Cape).

[3] Stems erect, crowded in ± large clumps, ± 5 × 0.8 - 1 cm, 4- (to 5- to 6-) angular; **Fl** facing upwards or outwards, opening in rapid succession, sometimes ± irregularly borne along the stems, often with several **Inf** on single stems; peduncle (if present) long, slender, erect; **Ped** ≤ 2.5 cm, basally thickened; **Sep** ± 5 mm; **Cl** outside cream-coloured or greenish pale yellow, often dotted with red-brown, inside cream-coloured to (pale) yellowish or rarely brown, to 2.5 cm ⌀, with concentric pink to dark brown transverse lines, lines broken, partly composed of dots, tube basally bright maroon, globose (in very small **Fl** rarely tubular), mouth slightly constricted, margin abruptly spreading; **Cl** lobes tri-

angular, acute or weakly acuminate, often longitudinally convex, ± erect to divaricate, sometimes revolute, outside sometimes weakly roughened, inside sometimes very weakly papillose, apex often ± densely furnished with fine acute tubercles to papillae; **Ci** dark crimson to black, lobes narrowly 3- or 4-angular, emarginate, bifid or deeply lobed; **Cs** orange-brown to ± black, lobes 1.5 - 2 mm, connivent-erect, ± ribbon-like, acuminate, often with nectar droplets; **Poll** brownish-yellow.

H. transvaalensis Stent (BMI 1914: 249, 1914). **T:** RSA, Transvaal (*Pole Evans* s.n. [PRE]). – **D:** RSA (Northern Prov., North-West Prov., Gauteng). **I:** Leach (1988).

[1] Stems prostrate to decumbent, compact, forming loose clumps, to > 10 × ± 1.1 cm, prominently 4- to 5-angular; **Tu** often wing-like compressed; **Fl** facing ± upwards; **Ped** ≤ 2.5 cm; **Sep** ± 1 cm; **Cl** 4.5 - 5.5 cm ⌀, outside greenish-creamy with maroon, tube and annulus inside dark maroon, margin of the tube and lobes cream-coloured to yellowish, with irregular zebra-like markings; **Cl** tube short, basally 7 - 8 mm ⌀, to 3 mm above the ground smooth and glabrous, above with (± scattered) papillae, these conical or obtuse, apically hairy, **Ha** 1.5 - 4 mm, rigid, needle-like, sometimes cylindrical, obtuse; annulus prominent, shining; **Cl** lobes as broad as or broader than long, triangular, acuminate, divergent, finely pubescent, **Ha** short, acute; **Cn** 4.5 - 5 mm; **Ci** lobes cream-coloured, margin maroon, 1.5 × ± 3 mm, shortly obtusely bilobed; **Cs** lobes ± 3 mm, erect, acuminate, apically narrow, convergent, meeting each other or slightly divergent; **Poll** yellow, germination mouth region darker.

H. urceolata L. C. Leach (FPA 39: t. 1550 + 4 pp. text, 1969). **T:** Angola, Moçamedes Distr. (*Leach & Cannell* 14025 [PRE, K, LISC]). – **D:** Angola, Namibia.

[2] Stems erect, very robust, forming clumps, grey-green, 14 × ≤ 2.5 cm, prominently 5-angular; **Tu** ≤ 1.5 cm, oblong-triangularly pointed; **Inf** many-flowered, **Fl** nodding, peduncle (if present) knobby; **Ped** 0.8 - 1 cm; **Sep** ≤ 1 cm; **Cl** outside purple, inside darker, 1 - 1.4 cm ⌀, globose-urceolate, mouth of the tube 6 - 8 mm ⌀, bulging annulus-like; **Cl** lobes narrowly acuminate, spreading and revolute, inside glabrous and smooth; **Cn** dark crimson; **Ci** much reduced or absent, ± 6 mm ⌀, fused into a finely crenate or ± entire disc; **Cs** with a dark apical margin, lobes ± ovate, very fleshy and swollen, dorsal hump shorter than the **Anth**; **Anth** yellow with brownish margins, ascending and curved, meeting each other ± above the centre, forming a small dome above the **Cs**; **Poll** ± elliptic or often ± ovate.

H. verekeri Stent (BMI 1933: 145, 1933). **T:** Zimbabwe (*Vereker* 5427 [K, PRE]). – **D:** Zambia, Zimbabwe, Malawi, Moçambique, Botswana, Namibia.

Incl. *Huernia verekeri* var. *stevensonii* A. C. White & B. Sloane (1937).

H. verekeri var. **angolensis** L. C. Leach (JSAB 40(1): 18-20, ills., 1974). **T:** Angola, Huila (*Leach & Cannell* 14650 [LISC, BM, LUAI, PRE, SRGH]). – **D:** Angola.

[2] Differs from var. *verekeri*: Stems deflexed; **Ri** blunt; **Tu** small; **Cl** tube light red; **Cn** reddish.

H. verekeri var. **pauciflora** L. C. Leach (BT 10(1): 49, 51, ills., 1969). **T:** Moçambique (*Leach & Bayliss* 11899 [SRGH, K, LISC, NBG, PRE]). – **D:** Moçambique.

[2] Differs from var. *verekeri*: Stems prostrate, much longer; **Ri** blunt; **Tu** smaller, more distant; **Inf** few-flowered or **Fl** solitary.

H. verekeri var. **verekeri** – **D:** Zambia, Zimbabwe, S Malawi, Moçambique, Botswana, Namibia. **Fig. XXXI.d**

[2] Stems erect, mat-forming, ± 7 cm, distinctly (5- to) 6- (to 7-) angular; **Tu** ≤ 1.5 cm, strongly compressed; **Inf** profusely flowering, pointing outwards; **Ped** ± 1 cm; **Sep** 5 - 8 mm, usually prominent between the **Cl** lobes; **Cl** ± 3.5 cm ⌀, inside basally white, mouth of the tube dark maroon, otherwise pale greenish-creamy or yellow; **Cl** tube ± 3 × 6.5 - 8 mm, ± semiglobose, circular or obtusely pentagonal in top-view; **Cl** lobes ≤ 1.4 cm, acute, horizontally spreading, convex, both sides glabrous, margin of the tube and lobes hairy, **Ha** maroon, short, acute; intermediate lobes small, strongly revolute; **Ci** ivory, lobes reduced or fused into a disc ± 3 mm ⌀; **Cs** lobes ± acute, shorter than the **Anth**, dorsal hump much enlarged, horizontally spreading, ± round; **Poll** yellow-brown.

H. volkartii Peitscher *ex* Werdermann (Gartenflora 85: 78, 1936). **T:** Angola (*Gossweiler* s.n. [not conserved]). – **D:** Angola, Namibia, Zimbabwe, Moçambique

Incl. *Huernia montana* Kers (1969).

H. volkartii var. **repens** (Lavranos) Lavranos (JSAB 38: 43, 1972). **T:** Moçambique, Manica e Sofala (*Schweickerdt* 3469 [PRE, SRGH]). – **D:** E Zimbabwe, W Moçambique.

≡ *Huernia repens* Lavranos (1961).

[1] Differs from var. *volkartii*: Stems long, creeping; **L** rudiments ± deciduous.

H. volkartii var. **volkartii** – **D:** Angola, Namibia, Zimbabwe, Moçambique. **I:** Leach (1988).

[1] Stems (shortly decumbent to) erect, tufted, sometimes forming mats, 3 - 5 cm, (4- to) 5-angular; **Fl** facing outwards or upwards; **Ped** to > 1 cm; **Sep** 5.5 - 8 mm; **Cl** outside cream-coloured to

brownish, inside cream-coloured, streaked irregularly with dull crimson, ± 2 - 3 cm ⌀, campanulate; **Cl** tube ± 5 × ± 1 cm, glabrous, mouth sometimes weakly constricted; **Cl** lobes 4.5 - 9 × 7 - 9.5 mm, triangular, divergent and revolute, outside roughened, inside ± densely papillose, papillae ≤ 2.5 mm, ± compressed, basally slightly broadened, apically sometimes finely mucronate; intermediate lobes 2 - 2.5 mm; **Cn** 3.5 × ≤ 7 mm; **Ci** very variable in shape and length; **Cs** lobes ± 2.5 mm, dorsiventrally ± strongly compressed, tips meeting each other, apex obtuse, short or abruptly revolute, occasionally slightly foot-like, ventrally finely tuberculate, at the base of the **Anth** with a conical wart-like convexity; **Poll** yellow.

H. whitesloaneana Nel (CSJA 8: 9, 1936). **T:** RSA, Northern Prov. (*Nel* s.n. [STE]). – **D:** RSA (Northern Prov.).

[1] Stems erect, tufted, to 5 cm; **Tu** arranged in 4 - 5 prominent **Ri**; **Ped** 4 - 8 mm; **Sep** 5 - 7 mm; **Cl** outside spotted with deep purple, inside cream-coloured, basally with irregular concentric purple-red lines, above broken, otherwise mottled or dotted, 1.2 - 2.2 cm ⌀, campanulate or ± tubular; **Cl** tube 6 - 8 mm ⌀, basally glabrous; **Cl** lobes 6 - 8 mm broad, triangular, erect to divaricate, ± strongly revolute, inside ± densely papillose, papillae ≤ 1.25 mm, conical or compressed, obtuse, often with a short mucro; **Cn** 5 × 4 mm; **Ci** fused with the base of the **Cl** tube, lobes broadly triangular, obtusely emarginate or bidentate, margin thickened; **Cs** lobes ± 3 mm, ± cylindrical-truncate, basally broadened, apex sometimes weakly clavate, dorsiventrally compressed, connivent above the **Anth**, above ± 1.5 mm divergent, towards the tip finely rugose, basally with a blunt hump; **Poll** yellowish.

H. witzenbergensis C. A. Lückhoff (in A. C. White & B. Sloane, Stapelieae, ed. 2, 3: 890, fig. 934, 1937). **T:** [lecto – icono]: l.c. fig. 934. – **D:** RSA (Western Cape).

[3] Stems erect, tufted, 7 × 2 cm, prominently 4- to 5-angular; **Tu** 5 - 6 mm, acute, compressed-triangular; **Fl** facing outwards; **Ped** 1.5 - 2 cm; **Sep** ± 6 mm; **Cl** inside bright sulphur-yellow, ± 4 cm ⌀; **Cl** tube ± 6 × 12 mm, ± glabrous, mouth thickened, weakly constricted, margin abruptly spreading; **Cl** lobes ± 1.3 × ≥ 1.3 cm, dull black, triangular, weakly revolute, inside ± densely papillose, papillae conical, acute, terminated by a short **Ha**; **Ci** dull crimson, lobes 2 mm, ± square, shortly bidentate; **Cs** lobes 2.5 mm, attenuate, apex ± obtuse, meeting each other above the **Anth**, apically slightly divergent.

H. zebrina N. E. Brown (FC 4(1): 921, 1909). **T:** RSA, KwaZulu-Natal (*Saunders* s.n. [K]). – **D:** Namibia, Botswana, Zimbabwe, Moçambique, RSA, Swaziland.

H. zebrina ssp. **magniflora** (E. Phillips) L. C. Leach (Excelsa Tax. Ser. 4: 139, 1988). **T:** RSA, Northern Prov. (*Ralston* s.n. [PRE 2058]). – **D:** Namibia, Botswana, Zimbabwe, RSA (Northern Prov.). **Fig. XXXI.a**

≡ *Huernia zebrina* var. *magniflora* E. Phillips (1936); **incl.** *Huernia blackbeardiae* R. A. Dyer *ex* H. Jacobsen (1933).

[2] Differs from ssp. *zebrina*: Stems frequently forming clumps, longer, more robust, usually 4-angled; **Cl** ≤ 8.5 cm ⌀, more variable in size, colour and markings.

H. zebrina ssp. **zebrina** – **D:** Zimbabwe, Moçambique, RSA (Mpumalanga, KwaZulu-Natal), Swaziland.

Incl. *Huernia zebrina* var. *zebrina*.

[2] Stems deflexed-decumbent and irregularly branching, ± creeping, occasionally forming mats, ± 5 × 1.2 cm, ± prominently 5- (to 6-) angled; **Inf** few-flowered; **Ped** 1.2 - 1.4 cm; **Sep** 6 - 8 mm; **Cl** inside yellowish-creamy, (2.5-) 3.5 - 4.5 (-5) cm ⌀, both sides glabrous, annulus shining, densely stippled with red to purple-brown, stipples often merging, otherwise yellowish, with irregular 'zebra' markings of reddish to red-brown or purple transverse stripes; **Cl** tube ± 7 × 6 mm; **Cl** lobes triangular, shortly tapering, divaricate, pubescent, **Ha** purple; **Ci** cream-coloured or light yellow, margin dark purple, very variable in shape; **Cs** dark, basal hump cream-coloured to yellow, lobes attenuate, apex ± obtuse, not meeting above the **Sty** head, shorter than the **Anth** or overtopping them a little.

×HUERNIANTHUS

B. Müller & F. Albers

×**Huernianthus** Barad (CSJA 67(4): 255, 1995).
= *Huernia* × *Stapelianthus*.

×**H. 'Alexis'** Barad (CSJA 67(4): 255, ill., 1995).
= *H. longituba* × *S. keraudreniae*.

HUERNIOPSIS

U. Meve, B. Müller & F. Albers

Huerniopsis N. E. Brown (JLSB 17: 171, 1878). **T:** *Huerniopsis decipiens* N. E. Brown. – **Lit:** Land (1992). **D:** Botswana, Namibia, RSA (Northern Cape, North-West Prov.). **Etym:** For the genus *Huernia* (*Asclepiadaceae*); and Gr. '-opsis', like.

Mat-forming perennials; stems clavate, 3 - 7 × 1 - 1.5 cm, apically ascending, basally branched, coarse, dull green, flushed or spotted with purple, bluntly 4- (to 5-) angled; **Tu** conical, raised, **L** rudiments subulate, ± 2 mm, basally with large stipular **Gl**, deciduous; **Inf** in the upper ½ of the stems, few-flowered, **Fl** opening only once in succession during late afternoon, with ample nectar and strong foetid odour; **Ped** short, thick; **Sep** (ovate-) lanceolate, glabrous, spreading; **Cl** reddish-brown to dark carmine, sometimes spotted with yellow, slightly rugose, 1 - 4.5 cm ⌀, coarse; **Cl** tube shortly cup-like or deeply campanulate, 5-angled; **Cl** lobes ovate-deltoid, basally depressed, sometimes centrally furrowed, outside greenish, glabrous, inside velvety; **Cn** biseriate with ample nectar; **Ci** fringe-like, strongly reduced or absent; **Cs** massive, lanceolate to linear, basally incumbent on the **Anth** and far surpassing the **Sty** head; **Sty** head small, 5-angled, truncate; **Anth** ± rectangular, surpassing the **Sty** head or incumbent on the top of the **Sty** head; **Poll** ± ovoid, caudicles short, apically broadened and discoid, basally united with the corpuscle by an oval broadened part; **Fr** narrowly fusiform, 10 - 18 cm.

Bruyns (1999a) included *Huerniopsis* in *Piaranthus*, a concept that is not applied here. *Huerniopsis* is in sister-group position to *Piaranthus* (Meve 1994). The 2 groups are each monophyletic, and there is no cogent reason to subsume them in a single genus.

In cultivation, a hybrid with *White-sloanea* has been produced (see ×*Whitesloaniopsis*).

H. atrosanguinea (N. E. Brown) A. C. White & B. Sloane (Stapelieae, ed. 2, 3: 967, 1937). **T**: Botswana (*Lugard* 263 [K]). – **D**: SE Botswana, RSA (Northern Cape, North-West Prov.). **I**: Asklepios No. 57: 18, fig. 3. **Fig. XXXI.b**

≡ *Stapelia atrosanguinea* N. E. Brown (1901) ≡ *Caralluma atrosanguinea* (N. E. Brown) N. E. Brown (1904) ≡ *Piaranthus atrosanguineus* (N. E. Brown) Bruyns (1999); **incl.** *Huerniopsis gibbosa* Nel (1937) ≡ *Huerniopsis atrosanguinea* var. *gibbosa* (Nel) hort. (s.a.) (*nom. inval.*, Art. 29.1); **incl.** *Huerniopsis papillata* Nel (1937).

Stems 3 - 7.5 × 0.5 - 1.5 cm; **Inf** 1- to 3-flowered, **Bra** ± 2 mm, subulate; **Ped** 3 × 2 mm; **Sep** 4 × 2 mm, acute; **Cl** outside grey-green, dotted with purple, inside dark blood-red, sometimes spotted with greenish-yellow, 3 - 4.5 cm ⌀, odour strongly foetid; **Cl** tube shallow, ± 8 × 3 mm, 5-angled; **Cl** lobes 15 - 20 × 8 - 12 mm, radiating stellately, with 1 (-3) longitudinal furrows, basally with 2 semiglobose swellings, margins slightly bent outwards, very finely papillate, sometimes hairy towards the tip, **Ha** white, short, stiff; **Cn** white to cream, 5-angled, massive; **Ci** ± 1 × 1 mm, ovate-deltoid, apically sometimes emarginate, upper face canaliculate, horizontally spreading, compact; **Cs** lobes ± 6 - 9 × 1 - 3 mm, lanceolate to linear, blunt, erect with tips bent inwards or outwards; **Gy** ± 3 × 5 mm, guide rails shortly triangular, 0.5 mm long; **Poll** ± 0.7 × 0.4 mm, translator shortly winged.

H. decipiens N. E. Brown (JLSB 17: 171, 1878). **T**: RSA (*MacOwan* 2246 [K]). – **D**: Botswana, Namibia, RSA (Northern Cape). **I**: Asklepios No. 57: 18, figs. 4-5.

≡ *Piaranthus decipiens* (N. E. Brown) Bruyns (1999); **incl.** *Piaranthus grivanus* N. E. Brown (1890) ≡ *Caralluma grivana* (N. E. Brown) Schlechter (1898).

Stems 2 - 7 × 0.8 - 1.5 cm; **Inf** 2- to 4-flowered; **Ped** 3 - 6 × 2 mm; **Sep** 5 - 8 mm, subulate; **Cl** outside pale green-grey, often patterned with purple, inside brown to red-brown, mostly spotted or banded with yellowish, 2 - 2.5 cm ⌀, odour intensely sweetish-foetid; **Cl** tube ± 5 × 8 mm, shallow to deeply campanulate, 5-angled; **Cl** lobes 9 - 12 × 6 - 8 mm, triangular, acute, basally depressed, basally almost erect, then increasingly reflexed, inside velvety, margins basally with coarse purple vibratile clavate **Ha** 1 - 2.5 mm long; **Cn** cream to pale brown, sometimes spotted with red-brown, conical, fleshy; **Ci** fringe-like, raised, shallowly emarginate to incised in the middle, rarely almost completely absent; **Cs** lobes 5 - 6 × 2 - 3 mm, basally rectangular, towards the tip lanceolate, raised at 45°, apically subulate, slightly spreading outwards, papillate; **Gy** ± 4 × 4 mm, guide rails raised, ± 1.5 mm long; **Poll** 0.8 × 0.5 mm, translator with ± semiglobose projections at the base of the massive corpuscle; **Fr** 12 - 18 × 0.5 - 0.6 cm; **Se** ± 6 × 4 mm, pale brown.

ISCHNOLEPIS

U. Meve

Ischnolepis Jumelle & H. Perrier (Rev. Gén. Bot. 21: 53, 1909). **T**: *Ischnolepis tuberosa* Jumelle & H. Perrier. – **D**: Madagascar. **Etym**: Gr. 'ischnos', dry; and Gr. 'lepis', scale; perhaps for the filiform slender corona segments.

Glabrous geophytic shrubs with white viscous latex; **R** tuberous, numerous, ± 5 cm ⌀, with a total weight of 50 - 100 kg; stems 5 - 100 (-150) cm tall, erect, usually single and little branched, perennial, woody, bark red-brown, glabrous, smooth, shining, nodes thickened, often several clustered because of lost **Int**; **L** nearly sessile, in whorls of 3, linear to very narrowly elliptic, ± pendent, 2 - 20 cm × 1.5 - 4 mm, acute, glabrous, midrib prominent on the lower face; **Inf** lateral, 1 - 2 cm pedunculate, 1- to 4-flowered; **Bra** 2 - 3 mm; **Ped** 1 - 2 cm; **Sep** free, 2 - 3 mm, acute; **Fl** buds conical; **Fl** rotate, 1.5 - 2.5 cm ⌀, glabrous; **Cl** tube 1 - 2 mm; **Cl** lobes spreading, triangular-lanceolate, 6 - 10 × 3 mm, acute, yellow; **Cn** petaloid; **Cn** lobes arising from the **Cl** tube in front of the **Fil**, yellow, filiform-subulate, 4 - 5 mm, erect-connivent; **Fil** basally connate with the **Cl** lobes; **Anth** white, lanceolate, arranged into a cone, connective triangular-lanceolate, apically free, ± clavate; translator spoon ovate, ± 1 mm, tip incised, yellowish-brown, stalk none, adhesive disc convex, 3-lobed with 1 apical and 2 lateral concave adhesive areas; **Sty** head short, bulging; **Fr** in pairs, narrowly fusiform, 10 - 14 × 0.3 - 0.4 cm; **Se** green, ovate, ± 6 × 2 mm, glabrous, tuft of **Ha** whitish, ± 15 mm.

This monotypic genus is a member of subfamily *Periplocoideae* and is possibly closely related to *Petopentia* and *Pentopetia*. A striking feature are the leaves arranged in whorls of 3. When 1 or 2 internodes are lost, up to 9 leaves may be counted per condensed "node".

I. graminifolia (Costantin & Gallaud) Klackenberg (Candollea 54(2): 332, 1999). **T:** Madagascar (*Deans Cowan* s.n. [P, BM]). – **D:** C and NE Madagascar.

≡ *Pentopetia graminifolia* Costantin & Gallaud (1907); **incl.** *Ischnolepis tuberosa* Jumelle & H. Perrier (1909).

Description as for the genus.

LARRYLEACHIA

B. Müller, F. Albers & U. Meve

Larryleachia Plowes (Excelsa 17: 5, 1997). **T:** *Stapelia cactiformis* Hooker. – **Lit:** Bruyns (1993: as *Lavrania*); Plowes (1997). **D:** Namibia, RSA, Botswana. **Etym:** For Leslie (Larry) C. Leach (1909 - 1996), English-born electrical engineer and self-taught botanist in Zimbabwe and later in RSA, specialist of succulent Asclepiads and Euphorbias.
Incl. *Leachia* Plowes (1992) (*nom. illeg.*, Art. 53.1). **T:** *Stapelia cactiformis* Hooker.
Incl. *Leachiella* Plowes (1992) (*nom. illeg.*, Art. 53.1). **T:** *Stapelia cactiformis* Hooker.
Basally branched and often lignified small stem succulents with few to numerous, erect or procumbent unarmed stems to 15 (rarely 30) cm; stems usually grey-green, 2 - 6 cm ⌀, cylindrical or clavate, glabrous; **Tu** flat, rounded, polygonal, arranged in 12 - 20 **Ri**; **L** rudiments < 1 mm, thickish, conical, in a depression near the **Tu** tip, persistent; **Inf** concentrated near the stem tips, with solitary or numerous successively (rarely simultaneously) opening **Fl**; **Ped** 0.5 - 3 × 1 mm, glabrous; **Sep** 5, slightly overlapping at the broadened bases, acuminate, occasionally ciliate; **Cl** (0.5-) 0.7 - 1.6 cm ⌀, flatly campanulate or deeply divided; **Cl** lobes flatly spreading, ovate-deltoid, ± acuminate, strongly curved outwards, outside smooth, glabrous, inside papillate or glabrous; **Cn** 2-seriate, of the **C**(**is**) +**Cs**-type, glabrous; **Ci** basally fused cup-like, lobes apically retuse or ± deeply divided; **Cs** of 5 lobes, dorsiventrally flattened, lying on the dorsal side of the **Anth**, often dorsally fused with the **Ci** via an appendage; **Anth** ± square; **Sty** head truncate-depressed; **Poll** ± horizontal, caudicles short; **Fr** 2 - 9 × 0.5 - 1 cm, acutely tapering to fusiform, glabrous, smooth, follicles of a pair diverging at an angle of 30 - 180°; **Se** brown or grey, 3 - 7 mm, orbicular or pear-shaped, ventrally ± concave, winged, brittle, with a crown of **Ha** on one side.

The species classified here correspond to the smooth-stemmed species of the former genus *Trichocaulon* (= *Lavrania* Sect. *Cactoidea* in the sense of Bruyns (1993)). Recent molecular studies (Meve & Liede 2001b) have shown that *Larryleachia* has to be separated from *Lavrania*. The generic treatment by Plowes (1997) is therefore adopted, but the species treatments are predominantly based on Bruyns (1993).

The following names are of unresolved application but are referred to this genus: *Stapelia clavata* Willdenow (1798) ≡ *Trichocaulon clavatum* (Willdenow) H. Huber (1961); *Trichocaulon sociarum* A. C. White & B. Sloane (1937) ≡ *Leachia sociarum* (A. C. White & B. Sloane) Plowes (1992) (*incorrect name*, Art. 11.3) ≡ *Leachiella sociarum* (A. C. White & B. Sloane) Plowes (1992) (*incorrect name*, Art. 11.3) ≡ *Larryleachia sociarum* (A. C. White & B. Sloane) Plowes (1997) ≡ *Hoodia sociarum* (A. C. White & B. Sloane) Halda (1998).

L. cactiformis (Hooker) Plowes (Excelsa 17: 5, 1997). **T:** [lecto – icono]: Curtis's Bot. Mag., t. 4127, 1845. – **D:** Namibia. **Fig. XXXII.b**

≡ *Stapelia cactiformis* Hooker (1845) ≡ *Trichocaulon cactiforme* (Hooker) N. E. Brown (1890) ≡ *Leachia cactiformis* (Hooker) Plowes (1992) (*incorrect name*, Art. 11.3) ≡ *Leachiella cactiformis* (Hooker) Plowes (1992) (*incorrect name*, Art. 11.3) ≡ *Lavrania cactiformis* (Hooker) Bruyns (1993) ≡ *Hoodia cactiformis* (Hooker) Halda (1998); **incl.** *Trichocaulon simile* N. E. Brown (1909) ≡ *Leachia similis* (N. E. Brown) Plowes (1992) (*incorrect name*, Art. 11.3) ≡ *Leachiella similis* (N. E. Brown) Plowes (1992) (*incorrect name*, Art. 11.3) ≡ *Larryleachia similis* (N. E. Brown) Plowes (1997) ≡ *Hoodia similis* (N. E. Brown) Halda (1998); **incl.** *Trichocaulon felinum* D. T. Cole (1985) ≡ *Leachia felina* (D. T. Cole) Plowes (1992) (*incorrect name*, Art. 11.3) ≡ *Leachiella felina* (D. T. Cole) Plowes (1992) (*incorrect name*, Art. 11.3) ≡ *Larryleachia felina* (D. T. Cole) Plowes (1997) ≡ *Hoodia felina* (D. T. Cole) Halda (1998).

Stems often single, to 15 (rarely 20) cm, whitish-green, 3 - 6 cm ⌀, ± 12- to 16-ribbed; **Inf** 1- to 5-flowered; **Ped** 1 mm; **Sep** 1.5 mm, margin occasionally hairy; **Cl** inside white or pale yellow striped with purple, rarely spotted, or ± uniformly purple or wine-red, 0.6 - 1.5 cm ⌀, tube flat or deeply cup-shaped, enclosing the **Cn**; **Cl** lobes 2 - 3.5 × 2.5 - 5.5 mm, tips curved outwards, inside papillate, apically with a small **Bri**; **Cn** yellow, spotted or striped with red, 3 - 4.5 mm ⌀; **Ci** lobes bifid to more than ½, appendages horn-like, blunt, spreading-ascending; **Cs** margin occasionally red, lobes acuminate or truncately retuse, shorter than the **Anth**, sometimes slightly overlapping above the **Sty** head, dorsal appendage ± deltoid, upright or slightly spreading, laterally fused with the **Ci**; **Fr** 2 - 3 cm, diverging at an angle of 180°. − [B. Müller & F. Albers]

L. marlothii (N. E. Brown) Plowes (Excelsa 17: 7, 1997). **T:** [neo − icono]: Flow. Pl. South Afr. 18: t. 681, 1938. − **D:** Namibia, Angola?, RSA (Northern Cape). **Fig. XXXII.c**

≡ *Trichocaulon marlothii* N. E. Brown (1909) ≡ *Leachia marlothii* (N. E. Brown) Plowes (1992) (*incorrect name*, Art. 11.3) ≡ *Leachiella marlothii* (N. E. Brown) Plowes (1992) (*incorrect name*, Art. 11.3) ≡ *Lavrania marlothii* (N. E. Brown) Bruyns (1993) ≡ *Hoodia marlothii* (N. E. Brown) Halda (1998); **incl.** *Trichocaulon dinteri* A. Berger (1910) ≡ *Leachia dinteri* (A. Berger) Plowes (1992) (*incorrect name*, Art. 11.3) ≡ *Leachiella dinteri* (A. Berger) Plowes (1992) (*incorrect name*, Art. 11.3) ≡ *Larryleachia dinteri* (A. Berger) Plowes (1997) ≡ *Hoodia dinteri* (A. Berger) Halda (1998); **incl.** *Trichocaulon keetmanshoopense* Dinter (1914); **incl.** *Trichocaulon sinus-luederitzii* Dinter (1914).

Plants to 15 cm; stems 2 - 5.5 (-6.5) cm ∅, ± 12- to 19-ribbed; **Inf** 1- to 5-flowered, predominantly apical; **Ped** to 1 mm; **Sep** 1.5 - 2 × 1 mm; **Cl** outside reddish spotted with green, inside cream-coloured, irregularly speckled with red or dark maroon, occasionally ± uniformly dark purple, 0.8 - 1.6 cm ∅, ± campanulate, central depression 1 - 3 mm, ± shallowly saucer- or cup-shaped, margins not or only slightly thickened; **Cl** lobes 2 - 5 × 3 - 5 mm, tips curved outwards, margins more weakly curved; **Cn** cream-coloured, irregularly marked with pinkish or chestnut-brown, 3.5 - 4.5 mm ∅, often with nectar droplets; **Ci** basally cup-shaped, lobes bifid for more than ½, appendages 1 - 1.5 mm, ± cylindrical, blunt, erect or strongly diverging; **Cs** lobes ± 2 mm, linear, blunt, erectly meeting above the **Sty** head, basally sometimes with a blunt dorsal appendage; **Fr** 5 - 9.5 cm, follicles diverging at an angle of 30 - 60°. − [B. Müller & F. Albers]

L. perlata (Dinter) Plowes (Excelsa 17: 9, 1997). **T:** Namibia (*Dinter* s.n. [B]). − **D:** Namibia, RSA (Northern Cape), both sides of the Orange River. **Fig. XXXI.h**

≡ *Trichocaulon perlatum* Dinter (1923) ≡ *Leachia perlata* (Dinter) Plowes (1992) (*incorrect name*, Art. 11.3) ≡ *Leachiella perlata* (Dinter) Plowes (1992) (*incorrect name*, Art. 11.3) ≡ *Lavrania perlata* (Dinter) Bruyns (1993) ≡ *Hoodia perlata* (Dinter) Halda (1998); **incl.** *Trichocaulon cinereum* Pillans (1928); **incl.** *Trichocaulon kubusense* Nel (1933); **incl.** *Trichocaulon truncatum* Pillans (1937).

Plants to 30 cm; stems 2.5 - 6 cm ∅, 12- to 14-ribbed; **L** rudiments later ± median on the **Tu**; **Inf** 3- to 6- (to 12-) flowered; **Fl** with strong foetid odour, often several simultaneously open; peduncle short; **Ped** 0.5 - 2 mm; **Sep** 1 - 1.5 × 0.5 - 1 mm, margins hairy, **Ha** translucent, thick; **Cl** outside yellowish-green, inside greenish-white dotted with red, or dark red, (0.5-) 0.7 - 1 cm ∅, central depression ± 1 mm, shallowly cup-shaped, basally enclosing the **Cn**; **Cl** lobes 3 - 4.5 × 2 - 2.5 mm, strongly curved outwards and pressed against the **Ped**, margins strongly curved outwards, fused **Cl** area thickened in an annulus-like manner, inside densely papillate, papillae blunt, with horizontally spreading appendage, cylindrical; **Cn** yellow or white, irregularly dotted with red or chestnut-brown, 2 - 3.5 mm ∅; **Ci** lobes *either* bifid for ½, appendages 0.5 mm, dorsiventrally flattened, diverging, *or* retuse, truncate; **Cs** lobes broadly rectangular, truncate, often retuse, sometimes surpassing the **Anth**, tips sometimes erect, dorsal appendage broadly truncate, ± upright; **Fr** ± 6 cm, follicles diverging at an angle of 180°. − [B. Müller & F. Albers]

L. picta (N. E. Brown) Plowes (Excelsa 17: 9, 1997). **T:** [neo − icono]: Flow. Pl. South Afr. 16: t. 620, 1936. − **D:** SW Namibia (E margin of the winter-rainfall region) to C Botswana. **Fig. XXXII.d**

≡ *Trichocaulon pictum* N. E. Brown (1909) ≡ *Leachia picta* (N. E. Brown) Plowes (1992) (*incorrect name*, Art. 11.3) ≡ *Leachiella picta* (N. E. Brown) Plowes (1992) (*incorrect name*, Art. 11.3) ≡ *Lavrania picta* (N. E. Brown) Bruyns (1993) ≡ *Hoodia picta* (N. E. Brown) Halda (1998); **incl.** *Lavrania picta* ssp. *picta*; **incl.** *Trichocaulon meloforme* Marloth (1912) ≡ *Leachia meloformis* (Marloth) Plowes (1992) (*incorrect name*, Art. 11.3) ≡ *Leachiella meloformis* (Marloth) Plowes (1992) (*incorrect name*, Art. 11.3) ≡ *Larryleachia meloformis* (Marloth) Plowes (1997) ≡ *Hoodia meloformis* (Marloth) Halda (1998); **incl.** *Trichocaulon engleri* Dinter (1914).

Stems mostly single, 3 - 6 × 2 - 5 cm, 12- to 16-ribbed; **Ped** 0.5 - 1 × 1 mm; **Sep** pale green with reddish hue, 2 × 1 mm, apically keeled; **Cl** outside green dotted with red, inside yellow marked with ± orbicular purplish-brown spots, (8-) 12 - 16 mm ∅, central depression cup-shaped, pentagonal, with 5 local swellings; **Cl** lobes apically strongly reflexed, inside papillate, papillae weakly rounded, rarely with an apical **Bri**; **Cn** purplish-black speckled with whitish; **Ci** lobes ≤ 2 mm, deeply bifid, appendages horn-like, widely diverging; **Cs** lobes erectly meeting above the **Sty** head, ascending column-like, column 2 mm, tips often diverging, dorsal appendage < 1 mm, conical, spreading; **Fr** 4.5 cm, follicles diverging at an angle of 30 - 60°; otherwise like *L. cactiformis*. − [B. Müller & F. Albers]

L. tirasmontana (Plowes) Plowes (Excelsa 17: 15, 1997). **T:** Namibia (*Plowes* 4306 [SRGH]). − **D:** Namibia.

≡ *Leachiella tirasmontana* Plowes (1993) (*incorrect name*, Art. 11.4) ≡ *Hoodia tirasmontana* (Plowes) Halda (1998); **incl.** *Leachia tirasmontana* Plowes (1992) (*nom. inval.*, Art. 32.1, 36.1); **incl.** *Lavrania picta* ssp. *parvipunctata* Bruyns (1993).

Differs from *L. picta*: **Cl** inside cream-coloured, finely dotted with red, 8 - 9 mm ⌀, central depression ± 2 mm, shallowly cup-shaped; **Cl** lobes 2 × 3 mm, tips darker, inside papillate, papillae apically with a fine **Bri**; **Cn** pale yellow, densely dotted with red; **Ci** ± 1.5 mm; **Cs** ± 2 mm, linear, apically rounded, apically meeting, column ≤ 0.5 mm; **Fr** 5.5 cm.

Closely related to *L. picta* and treated as subspecies by Bruyns (1993). – [B. Müller & F. Albers]

LAVRANIA

U. Meve

Lavrania Plowes (CSJA 58(3): 122-123, 1986). **T:** *Lavrania haagnerae* Plowes. – **Lit:** Bruyns (1993); Meve & Liede (2001b). **D:** Namibia. **Etym:** For John J. Lavranos (*1926), Greek-born well-known collector of succulents throughout S and E Africa.

Branched stem succulents, stems procumbent-ascending, stems pale green, cylindrical, 10 - 25 × 2 - 3 cm, with 10 - 12 regular rows of flat **Tu**; **L** rudiments very short, conical, acute; **Inf** ± sessile, bracteate, from basal stem parts, 3- to 15-flowered; **Ped** 2 - 3 mm; **Fl** strongly urine-scented; **Sep** 2.5 mm; **Cl** yellowish with coarse red-brown spots, 13 - 16 mm ⌀, rotate, densely papillose with coarse obtuse papillae; **Cl** tube shallowly cup-shaped, ± 1 mm deep; **Cl** lobes broadly triangular, 3 × 6 mm, spreading horizontally, margins slightly recurved; **Cn** 2-seriate, pink-red, ± 2 × 5 mm, with ample nectar; **Ci** basally shortly fused to form a shallow cup, **Ci** lobes rectangular, obtuse or ± fused with the wing-like extended **Cs** base, ascending; free **Cs** lobes hardly 1 mm, rectangular, obtuse, decumbent on the rectangular **Anth**; **Sty** head white; **Poll** yellow, broadly drop-shaped, ± 0.3 × 0.22 mm, corpuscle rectangular-fusiform, translator wings > 1 mm, acuminate; **Fr** slenderly fusiform, 6 - 7 cm × 3 - 4 mm, divergent at an angle of 30 - 60°.

Whether *Lavrania*, here treated as monotypic genus, should be merged with *Larryleachia* or not, was controversially discussed for many years (see Meve & Liede (2001b) and comments for *Larryleachia*).

L. haagnerae Plowes (CSJA 58(3): 123, 119 (ill.), 1986). **T:** Namibia (*Haagner* s.n. in *Plowes* 5046 [PRE]). – **D:** NW Namibia. **Fig. XXXII.e**

≡ *Hoodia haagnerae* (Plowes) Halda (1998).
Description as for the genus.

MADANGIA

F. Albers

Madangia P. I. Forster & al. (Austrobaileya 5(1): 53, 1997). **T:** *Madangia inflata* P. I. Forster & al. – **D:** Papua New Guinea. **Etym:** For its occurrence in Madang Prov., Papua New Guinea.

Wiry twiners; stems to several m long; **L** petiolate, lamina elliptic to elliptic-ovate, fleshy, to 12 × 6 cm, base cordate with overlapping lobes; **Inf** to 9-flowered, pendent; peduncle 8 - 22 mm, with scattered **Ha**; **Ped** 3.2 - 4.5 cm; **Sep** triangular, 1 × 2 mm, glabrous; **Cl** globose, 17 - 18 mm ⌀, white; **Cl** lobes 4 - 6 × 6 - 7 mm, completely fused except the slightly recurved tips, inside papillate; **Cn** 1-seriate, staminal, 10 mm ⌀; **Cs** lobes membranous, contiguous to each other for almost their entire length, lower edge of **Cs** lobes strongly recurved forming a membranous frill; **Poll** 0.77 - 0.78 × 0.29 - 0.31 mm, corpuscle 0.38 - 0.43 × 0.21 - 0.24 mm; **Fr** not known.

This monotypic genus is closely allied to *Hoya*. The species differs markedly by the globose corolla and the staminal corona. Omlor (1998) is uncertain about the generic rank.

M. inflata P. I. Forster & al. (Austrobaileya 5(1): 54, ills. (p. 56), 1997). **T:** Papua New Guinea, Madang Prov. (*Liddle* 1076 [BRI]). – **D:** Papua New Guinea (Madang Prov.); only known from the type locality.

Description as for the genus.

MARSDENIA

U. Meve

Marsdenia R. Brown (Prodr., 460, 1810). **T:** *Marsdenia tinctoria* R. Brown. – **Lit:** Rothe (1915); Forster (1995); Omlor (1998). **D:** Worldwide in all largely frost-free regions. **Etym:** For William Marsden (1754 - 1836), Irish orientalist and traveller in Sumatra, secretary to the British Admirality.

Incl. *Stephanotis* Thouars (1806) (*nomen rejiciendum*, Art. 56.1). **T:** *Stephanotis thouarsii* Brongniart.

Incl. *Harrisonia* Hooker (1826) (*nom. illeg.*, Art. 53.1). **T:** *Harrisonia loniceroides* Hooker.

Incl. *Baxtera* Reichenbach (1828). **T:** *Harrisonia loniceroides* Hooker.

Incl. *Dregea* E. Meyer (1838). **T:** *Dregea floribunda* E. Meyer.

Incl. *Pterophora* Harvey (1838). **T:** *Pterophora dregei* Harvey [*nom. illeg.*, ≡ *Dregea floribunda* E. Meyer].

Incl. *Pterygocarpus* Hochstetter (1843). **T:** *Pterygocarpus abyssinicus* Hochstetter.

Incl. *Leichardtia* R. Brown (1849). **T:** *Marsdenia leichardtiana* F. Mueller.

Incl. *Jasminanthes* Blume (1850). **T:** *Jasminanthes suaveolens* Blume.
Incl. *Ruehssia* H. Karsten *ex* Schleidel (1853). **T:** *Ruehssia purpurea* Schleidel.
Incl. *Tetragonocarpus* Hasskarl (1857). **T:** *Tetragonocarpus teysmannii* Hasskarl.
Incl. *Chlorochlamys* Miquel (1869). **T:** *Chlorochlamys celebica* Miquel.
Incl. *Stephanotella* E. Fournier (1885). **T:** *Stephanotella glaziovii* E. Fournier.
Incl. *Verlotia* E. Fournier (1885). **T:** not typified.
Incl. *Pseudomarsdenia* Baillon (1890). **T:** *Pseudomarsdenia bourgeana* Baillon.
Incl. *Traunia* K. Schumann (1895). **T:** *Traunia albiflora* K. Schumann.
Incl. *Dalzielia* Turrill (1916). **T:** *Dalzielia oblanceolata* Turrill.
Incl. *Loniceroides* Bullock (1964) (*nom. illeg.*, Art. 52.1). **T:** *Harrisonia loniceroides* Hooker.

Geophytes, lianas, subshrubs or shrubs with white, in part viscous latex; **R** fibrous, irregularly thickened or forming water-storing tubers; stems erect, occasionally decumbent, usually twining, perennial; **L** ± sessile or petiolate, linear, elliptic, lanceolate or ovate, usually leathery; **Inf** extra-axillary, lateral, ± sessile, 1- to many-flowered; **Ped** short; **Cl** usually campanulate or urceolate or rotate, **Cl** tube often longer than the **Cl** lobes; **Cl** lobes triangular, ovate or lanceolate, usually robustly fleshy, inside often hairy; **Cn** uniseriate; **Cs** simple, ± incumbent on the back of the **Anth**, partly auriculate or winged, or ± reduced; **Gy** sessile or shortly stipitate, **Anth** appendages acuminate or obtuse; **Poll** erect, cylindrical, ellipsoid or ovoid, corpuscle often fairly massive; **Fr** single, usually beaked, sometimes striate, keeled or winged; **Se** usually slightly winged.

This large genus (> 200 species) is widely distributed throughout the warmer regions of the world. *Marsdenia* species are usually woody; some species additionally show subsucculent stems and leaves as well as various types of root succulence. More or less woody tuberous roots are occasionally found (so-called xylopodia), particularly in Australia ("group 4" of Forster (1995)). The delimitation of genera like *Dregea*, *Wattakaka* or *Gymnema* is still in debate, even though Forster (1995) united all relevant taxa into a (now rather variable) collective genus *Marsdenia*, which even includes the famous ornamental *Stephanotis* ("Madagascar Jasmine"). Omlor (1998) has recently again recognized most of these genera as separate taxa. A division of *Marsdenia* s.str. into sections was published by Rothe (1915).

M. australis (R. Brown) Druce (Bot. Soc. Exch. Club Brit. Isles 1916: 634, 1917). **T:** Australia, New South Wales (*Sturt* s.n. [BM]). – **D:** Australia, esp. C Australia; usually in open *Acacia* woodland. **I:** Forster (1995).

≡ *Leichardtia australis* R. Brown (1845).

Delicate twining plants with ± globose **R** tubers lined up in a row; stems to 3 mm ∅, densely hairy; **L** petiolate, linear to lanceolate, to 12 × 3 cm, grey-green; **Inf** pedunculate, umbellate, many-flowered; **Ped** 5 - 9 mm; **Sep** ovate-lanceolate, ± 3 × 2.5 mm; **Cl** urceolate, 5 - 9 × 4 - 6 mm, white to pale yellow, glabrous; **Cl** tube 4 - 6 × 3.5 - 6 mm; **Cl** lobes ± ovate, erect, concave, 2 - 4 × 1.5 - 2 mm; **Cn** 1 - 3 × 2.5 - 3 mm; **Cs** lobes erect, 1 - 2 mm, acuminate, dorsally-laterally ± winged; **Anth** appendages lanceolate, 1 - 1.5 mm; **Poll** elongate, ± 0.95 × 0.25 mm, erect; **Fr** 4 - 10 × 2 - 3 cm.

This species is suggestive of the S African genus *Microloma*.

M. brevis P. I. Forster (Austral. Syst. Bot. 8: 781-782, 1995). **T:** Australia, Queensland (*Lockyer* s.n. [BRI]). – **D:** Australia (Queensland).

Differs from *M. coronata*: **Cl** tube urceolate; **Cn** lobes very much shorter, crenulate.

Closely allied to *M. coronata*.

M. coronata Bentham (Fl. Austral. 4: 341-342, 1869). **T:** Australia, Queensland (*Mueller* s.n. [MEL]). – **D:** Australia (SE Queensland); in open *Eucalyptus* forests. **I:** Forster (1995: 880).

Delicate twining plants with ± globose **R** tubers lined up in a row; stems 1 - 2 mm ∅, biseriately hairy; **L** petiolate, linear-lanceolate, to 5.5 × 1 cm, upper face bright green, lower face pale green; **Inf** pedunculate, umbellate, many-flowered; **Ped** 6 - 8 mm; **Sep** ovate-lanceolate, ± 2.5 × 1.5 mm; **Cl** shallowly campanulate, yellow to yellow-green, glabrous; **Cl** tube 2.5 - 3 × 3 mm; **Cl** lobes triangular, revolute, ± 1.8 × 1.7 mm; **Cn** ± 1.5 × 1.3 mm; **Cs** lobes erect, 1 mm, shield-shaped, auriculate, acuminate; **Anth** appendages triangular-ovate, ± 0.7 mm ∅; **Poll** narrowly reniform, ± 0.25 × 0.25 mm; **Fr** ± 10 × 1.5 cm.

Forming a group of closely allied species together with *M. brevis*, *M. rara* and *M. microlepis*.

M. gillespieae Morillo (Ernstia, ser. 2, 1(3): 111-112, ills. (p. 118), 1991). **T:** Guyana (*Gillespie* 1782a [US, VEN]). – **D:** Guyana; in rock fissures close to rivers in rainforest.

Perennial woody geophytic twining lianas with white latex; **R** tuber (xylopodium) 40 - 60 × 3 - 5 cm (young plants); stems twining to 2 m tall, thickened to 2 cm ∅ above the tuber, branched, glabrous, rapidly becoming corky, bark brown, finely fissured, ± pimpled; **L** 1 - 2 cm petiolate, lamina elliptic, basally cuneate, apically cuspidate, ± 4 - 16 × 2 - 8 cm, soft, elastic, subsucculent, glabrous; **Inf** lateral, 8 - 12 mm pedunculate, pseudo-umbellate, many-flowered; **Ped** 4 - 5 mm; **Sep** ovate, ± 2 mm, ciliate; **Cl** campanulate, ± 10 - 13 mm ∅, purple-brown; **Cl** tube ± 2.5 mm deep, inside with 2 stripes of **Ha**; **Cl** lobes narrowly ovate, 3 - 4 × ± 2 mm, inside with scattered delicate **Ha**; **Cn** ± sessile, 1.2 - 1.4 × 0.9 - 1.1 mm; **Cs** lobes ± 1.3 mm, erect, elliptic, basal ½ winged, dorsally depressed, apically convex, cuspidate; **Poll** narrowly ovoid, 0.22 × 0.1 mm, erect on upwardly bent caudiculae; **Fr** fusiform, ± 15 cm × 1.5 - 2 mm, glabrous; **Se** ± ellipsoid, 12 × 6 mm, broadly winged, dark brown, coma ± 5 cm.

The delicate small flowers attractively constrast the imposing growth of these easily cultivated plants.

M. guanchezii Morillo (Ernstia 37: 10-12, ills., 1986). **T:** Venezuela (*Guanchez & Melgueiro* 4130 [TFA, F, VEN]). – **D:** Venezuela (Amazonas Region); fissures of granite rocks.

Robust twining lianas with a massive **R** tuber (xylopodium), tubers 50 - 90 × 2 - 10 cm; stems twining to 3 m high, basally to 1 cm ∅, thickened, branched, glabrous; **L** 3 - 10 mm petiolate, lamina narrowly to broadly elliptic, robust, subsucculent, acute, basally cuneate, 3.2 - 5.5 × 1 - 2 cm, upper face glabrous, lower face ± hairy; **Inf** 5 - 9 mm pedunculate, many-flowered; **Ped** 4 - 6 mm; **Sep** ovate, ± 1.3 mm, ciliate; **Cl** campanulate, ± 6 mm ∅, olive-green; **Cl** tube ± 1 × 2 mm; **Cl** lobes ovate, ± 2 × 1.75 mm, inside slightly papillose; **Cn** sessile, 1.5 × 1.2 mm; **Cs** triangular, ± 1.75 mm long, acute, erect, flat, fleshy; **Poll** ± ovoid, ± 0.2 × 0.14 mm, erect on upwards-bent caudicles; **Fr** ovoid-fusiform, 9 - 10 × 1.5 - 2 mm, glabrous.

In habit similar to *M. megalantha*, yet with very small inconspicuous flowers.

M. megalantha Goyder & Morillo (Asklepios No. 63: 18-22, ills., pl. 1-2, 1994). **T:** Brazil, Bahia (*Taylor, Zappi & Eggli* 1557 [CEPEC, K, MO, MSUN, NY, RB, SPF, VEN, ZSS]). – **D:** NE Brazil (Bahia); in humus-filled rock fissures. **Fig. XXXIII.a**

Perennial subsucculent subshrubs to 60 cm; **R** irregularly thickened, fleshy; stems erect, apically twining, basally thickened, 5 - 10 (basally to 15) mm ∅, bark silver-grey; **L** shortly petiolate, ovate-elliptic to broadly ovate, basally cuneate, 1 - 5 × 0.5 - 2.5 cm, glabrous or sparsely pubescent, robust, subsucculent, often crowded on short shoots; **Inf** 1-flowered; **Ped** 5 - 10 mm, sparsely hairy; **Sep** ovate, 4 - 5 × 2 - 3 mm, pubescent; **Cl** flatly rotate to shallowly campanulate, 3 - 3.5 cm ∅, outside pale green, inside yellow to green or brown, margins paler; **Cl** tube 15 - 17 mm ∅; **Cl** lobes ± ovate, 10 - 13 × 11 - 14 mm, ciliate; **Cn** robust-fleshy, ± 9 × 6 mm, pale green; **Cs** terminating with 5 erect-connivent ellipsoid lobes of ± 5 mm length, each with wing-like appendages at the base, these incumbent on the **Cl**; **Anth** appendages semicircular, ± 0.5 × 1 mm; **Poll** broadly obovoid, 0.35 × 0.25 mm, **Sty** head conical, pale green, overtopping the **Cn**.

Pretty and magnificent plants with flowers attaining an unusual size for *Marsdenia*; readily flowering in cultivation. Vegetatively similar species are restricted to S America (see *M. guanchezii*); less fleshy taxa are found in the Old World Sect. *Dregea* (e.g. *D. arabica* (Decaisne) Omlor).

M. microlepis Bentham (Fl. Austral. 4: 342, 1869). **T:** Australia, Queensland (*Dallachy* s.n. [K, MEL]). – **D:** Australia (Queensland); forests. **I:** Forster (1995: 881).

Differs from *M. coronata*: **L** broader; **Fl** urceolate; segments of the **Cn** peltate-ovate.

Closely related to *M. coronata*.

M. pumila P. I. Forster (Austral. Syst. Bot. 8: 780-781, ills. (p. 879), 1995). **T:** Australia, Queensland (*Forster* 14880 [BRI, QRS]). – **D:** Australia (Queensland).

Delicate twining plants with ± globose **R** tubers lined up in a row; stems to 1 mm ∅, sparsely hairy; **L** petiolate, linear to lanceolate, to 9.5 × 3 cm, upper face bright green, lower face pale green; **Inf** pedunculate, umbellate, few-flowered; **Ped** 3 - 4 mm; **Sep** lanceolate, ± 2 × 1 mm; **Cl** funnel-shaped, ± 6 × 2 mm, pale green; **Cl** tube ± 4 × 2 mm, inside with tufts of **Ha** below the sinuses of the **Cl** lobes, these ± ovate, erect or slightly revolute, ± 1.4 × 1.4 mm; **Cn** 1.5 mm long; **Cs** lobes erect, ± 1.5 × 0.8 mm, acuminate, basally widening into a subsquare shape; **Anth** appendages lanceolate, 0.5 mm; **Poll** obovoid, ± 0.2 × 0.11 mm, on long upcurving caudicles.

Quite similar to *M. australis*.

M. rara P. I. Forster (Austral. Syst. Bot. 8: 782-783, 1995). **T:** Australia, Queensland (*Lyons* 149 [BRI]). – **D:** Australia (NE Queensland: neighbourhood of Cairns); *Eucalyptus* forests.

Differs from *M. brevis*: **Inf** with longer peduncle; **Cl** tube narrower.

M. rara is given as sister species of *M. brevis*. Both possess a crenulate staminal corona and are related to *M. coronata*.

M. viridiflora R. Brown (Prodr., 461, 1810). **T:** Australia, Queensland (*Brown* s.n. [BM, CANB]). – **D:** Australia (Northern Territory, Western Australia, Queensland), W Papua New Guinea; in open woodland. **I:** Forster (1995: 882-883).

Incl. *Bidaria leptophylla* F. Mueller (1859) ≡ *Marsdenia leptophylla* (F. Mueller) Bentham (1869); **incl.** *Marsdenia viridiflora* ssp. *tropica* P. I. Forster (1995).

Delicate twining plants with ± globose **R** tubers lined up in a row; stems to 4 mm ∅, sparsely hairy; **L** petiolate, linear, lanceolate or ovate, to 7 × 2 cm, slightly hairy, upper face dark green, lower face pale green; **Inf** pedunculate, umbellate, few- to many-flowered; **Ped** 5 - 9 mm; **Sep** ovate-lanceolate; **Cl** campanulate, 3 - 5 × 5 - 8 mm, green to yellow, glabrous or with dense tufts of **Ha** around the entrance; **Cl** tube 2 - 4 × 3 - 5 mm; **Cl** lobes ovate to triangular, ± revolute, glabrous, 2 - 3 mm ⌀; **Cn** up to 1.5 × 3 mm, greenish-yellow; **Cs** lobes erect, 1 - 1.5 mm, acuminate, triangular-ovoid, slightly auriculate; **Anth** appendages ovate, 0.5 - 1 mm; **Poll** oblong, ± 0.33 × 0.16 mm, obliquely erect; **Fr** ± 9 × 2.5 - 3 cm.

Very variable species, which has been subdivided by Forster (1995) into the narrow-leaved and small-flowered typical subspecies and the broad-leaved and large-flowered ssp. *tropica*.

MATELEA

U. Meve

Matelea Aublet (Hist. Pl. Guiane, 277, 1775). **T:** *Matelea palustris* Aublet. – **Lit:** Stevens (1988). **D:** Tropical and subtropical America. **Etym:** Not explained.
Incl. *Dictyanthus* Decaisne (1844). **T:** *Dictyanthus pavonii* Decaisne.

Geophytes, subshrubs, shrubs, lianas or herbaceous twining plants with white latex; **R** fibrous, occasionally forming underground or halfways above-ground ± woody tubers (caudices), which include the hypocotyl; stems usually twining, rarely decumbent or erect, occasionally forming rhizomes; **L** petiolate, elliptic, lanceolate, ovate or subcircular; **Inf** extra-axillary, lateral, ± shortly pedunculate, 1- to many-flowered; **Cl** rotate to campanulate, **Cl** tube well-developed, **Cl** lobes narrowly lanceolate, triangular to ovate; **Cn** very sophisticated and variable, uni- to biseriate, often connate to the **Cl**, partially also petaloid; **Gy** ± sessile or stipitate; **Anth** appendages ± absent; **Poll** horizontal to pendent, variable, with cartilaginous margin where attached to the caudicle; **Sty** head ± 5-angled, apically flat; **Fr** single, usually beaked, slender to thickly fusiform, often very large, striate, keeled, winged, hairy, verrucose or hispid; **Se** usually winged.

Matelea is currently defined in a very broad sense, thus with ± 250 species constituting the largest genus within the *Asclepiadeae* – *Gonolobinae*. Furthermore, it is poorly delimited against *Gonolobus* (± 110 species). The absence of anther appendages is regarded as an important character defining the genus *Matelea*. Leaf or stem succulence is absent, the occurence of forms showing root succulence seems to be restricted to *Matelea* s.str. Not all of these species will be covered here (for many species the subterranean organs are unknown anyway), but rather a choice of particularly attractive taxa. These mainly belong to Subgen. *Dictyanthus* (Decaisne) R. E. Woodson (1941), which is again classified at generic rank by Stevens (2000).

The list of synonyms (here omitted) of *Matelea* contains more than 30 names that either represent synonyms of the genus or of infrageneric units. A revision would be needed to elaborate a correct and usable list of synonyms as well as an infrageneric classification, which is presently wanting.

M. cyclophylla (Standley) R. F. Woodson (Ann. Missouri Bot. Gard. 28: 233, 1941). **T:** Mexico, Guerrero (*Rose & al.* 9355 [US]). – **D:** Mexico. **I:** Rowley (1987: 79); CSJA 67: 48, 1995, both as *Gonolobus*. **Fig. XXXIII.d, XXXIII.e, XXXIII.f**
≡ *Vincetoxicum cyclophyllum* Standley (1924) ≡ *Gonolobus cyclophyllus* (Standley) hort. (s.a.) (*nom. inval.*, Art. 29.1).

Robust twining climbers with globose to turnip-shaped mostly underground caudex with coarsely fissured corky bark; stems to 3 - 8 mm ⌀ (often thicker at the base), glabrous; **L** long-petiolate, reniform to cordate, 6 - 12 × 5 - 10 cm, upper face green, lower face pale green, glabrous or sparsely pubescent; **Inf** ± sessile, few-flowered; **Ped** short, robust, ± pubescent; **Sep** linear-lanceolate, acute, 1 - 1.5 cm, glabrous or pubescent; **Cl** shallowly rotate, 2 - 3 cm ⌀, green or purple, ± hairy; **Cl** tube ± ½ as long as the **Cl**; **Cl** lobes triangular-ovate, ± 1 cm; **Cn** biseriate, ± 3 × 7 mm, flat, lentiform with thickened and raised margin, yellow-green or purple (-red); **Ci** fused into a disc, connate to the broadened base of the **Cs**; free **Cs** lobes obliquely erect, barely 1 mm, acuminate; **Anth** appendages fringe-like, white; **Fr** broadly fusiform, 12 - 20 × 3 - 4 cm.

A fairly vigorous plant with striking caudices and conspicuous flowers, more frequently seen in cultivation.

M. decumbens W. D. Stevens (Phytologia 53: 403-404, 1983). **T:** Mexico, Hidalgo (*Rzedowski* 24011 [MSC, ENCB, F, WIS]). – **D:** Mexico, Guatemala, El Salvador.

Closely related to *M. hemsleyana*, but stems decumbent; **Inf** few-flowered; **Fl** only half as large, greenish; **Cl** lobes erect.

M. dictyantha R. F. Woodson (Ann. Missouri Bot. Gard. 28: 236, 1941). **T:** Mexico, Oaxaca (*Jürgensen* 692 [KW, K]). – **D:** Mexico (Guerrero, Puebla, Jalisco, Oaxaca); mountain regions, 1500 - 2500 m. **I:** Stevens (1988).
Incl. *Rytidoloma reticulatum* Turczaninow (1852) ≡ *Dictyanthus reticulatus* (Turczaninow) Bentham & Hooker *ex* Hemsley (1882).

Erect, creeping or climbing-twining herbaceous perennials; caudex irregularly turnip-shaped, to 5 × 3 cm, with corky bark, partly with additional rhizomes; stems 15 - 70 (-150) cm, glandular-hairy, becoming woody at the base; **L** 1 - 6 cm petiolate, lamina broadly ovate with cordate base, 2.5 - 6 (-10) × 2.5 - 5 (-7.5) cm, ± hairy, often also glandular; **Inf** sessile or to 1 cm pedunculate, 1-flowered; **Ped** 5 - 15 mm; **Sep** narrowly ovate, 6 - 11 mm; **Cl** shallowly campanulate, 3 - 4 cm ⌀, inside with purple-brown reticulate markings; **Cl** tube 7 - 15 × 15 - 25 mm, inside with ridges becoming broader towards the base; **Cl** lobes 8 - 15 mm, margins revolute; **Cn** uniseriate, shortly stipitate, ± 4 × 10 mm; **Cs** lobes linear, 5 - 8 mm, semi-erect, tips slightly narrowed, 1 - 1.5 mm, dorsally placed against the bulging **Cl** tube; **Poll** narrowly drop-shaped, 1.2 - 1.45 × 0.3 - 0.38 mm; **Fr** broadly fusiform, 4.5 - 7 × 1 - 2 cm, shortly glandular-hairy, softly prickly.

A spectacular plant whose flowers are only surpassed in size by those of *M. standleyana*.

M. hemsleyana R. F. Woodson (Ann. Missouri Bot. Gard. 28: 237, 1941). **T:** Mexico, Chiapas (*Ghiesbreght* 663 [K, GH, MO, NY]). – **D:** Mexico, Guatemala, El Salvador; mostly in grassland. **I:** Stevens (1988).

Incl. *Dictyanthus prostratus* Brandegee (1920); **incl.** *Dictyanthus parviflorus* Hemsley (1924); **incl.** *Matelea diffusa* R. F. Woodson (1941).

Erect herbaceous perennials, rarely slightly twining; caudex irregularly turnip-shaped, ± 4 × 2 cm, with thick corky bark; stems 20 - 90 cm, glandular-hairy, becoming woody at the base; **L** 0.7 - 2.5 cm petiolate, lamina ovate to broadly ovate with cordate base, 1.5 - 3.5 × 1.3 - 3.5 cm, ± hairy, often glandular; **Inf** 1 - 4 mm pedunculate, 1-flowered; **Ped** 3 - 5 mm; **Sep** ovate-lanceolate, 4 - 6 mm; **Cl** shallowly campanulate, 1 - 1.8 cm ∅, inside purple-brown, with reticulate markings; **Cl** tube 3 - 6 × 5 - 8 mm, inside with prominent ridges in **Cs** position; **Cl** lobes broadly triangular, 3 - 6 mm; **Cn** uniseriate, ± 2.5 × 3 mm; **Cs** lobes broadly triangular-spatulate, 1 - 1.5 mm, dorsally placed against the bulging **Cl** tube; **Poll** drop-shaped, 0.6 - 0.9 × 0.25 - 0.35 mm; **Fr** broadly fusiform, 5 - 7 × 1 - 1.8 cm, softly prickly.

M. macvaughiana W. D. Stevens (Ann. Missouri Bot. Gard. 75: 1545-1548, ills., 1988). **T:** Mexico, Jalisco (*Pringle* 8629 [MSC, ENCB, F, G, GH, L, MEXU, MO, NY, P, PH, POM, UC, US, VT, W]). – **D:** Mexico (Jalisco); in (temporarily) moist grassland.

≡ *Dictyanthus macvaughianus* (W. D. Stevens) W. D. Stevens (2000).

Erect herbaceous perennials, occasionally twining; caudex ± fusiform, 4 × 2 cm, with slightly corky bark; stems 20 - 85 cm, densely hairy; **L** 1 - 5 cm petiolate, lamina narrowly to broadly ovate, base cordate (auriculate), 3 - 9.5 × 2 - 7 cm, ± hairy, often glandular; **Inf** sessile or to 16 mm pedunculate, usually 1-flowered; **Ped** 5 - 20 mm; **Sep** ovate-lanceolate, 8 - 12 mm; **Cl** campanulate, 3 - 4 cm ∅; **Cl** tube ± 5 × 12 mm, inside with weakly prominent ridges in **Cs** position, with purple-brown centrifugal linear markings; **Cl** lobes triangular, acute, 9 - 17 mm, margins revolute, inside with purple-brown reticulate markings; **Cn** uniseriate, stipitate; **Cs** lobes linear-spatulate, semi-erect, dorsally placed against the bulging **Cl** tube, 6 - 10 mm; **Poll** 1.45 - 1.65 × 0.47 mm; **Fr** broadly fusiform, slightly flattened, ± 8.5 × 2 cm, softly prickly.

In terms of floral morphology most similar to *M. standleyana*.

M. pavonii (Decaisne) R. F. Woodson (Ann. Missouri Bot. Gard. 28: 237, 1941). **T:** Mexico (*Sessé & al.* s.n. [FT, P]). – **D:** Mexico (Sinaloa to Oaxaca); mountain regions. **I:** Stevens (1988).

≡ *Dictyanthus pavonii* Decaisne (1844); **incl.** *Tympananthe suberosa* Hasskarl (1847); **incl.** *Dictyanthus campanulatus* Reichenbach (1850); **incl.** *Dictyanthus stapeliiflorus* Reichenbach (1850) ≡ *Matelea stapeliiflora* (Reichenbach) R. F. Woodson (1941); **incl.** *Stapelia campanulata* Pavón *ex* Sessé & Moçiño (1888) (*nom. illeg.*, Art. 53.1).

Stems twining, herbaceous, sparsely hairy, becoming woody and corky at the base, often forming rhizomes; **L** 2 - 8 cm petiolate, lamina ovate with ± cordate base, 5 - 12 × 3 - 10 cm, ± ciliate; **Inf** 1 - 6 (-9) cm pedunculate, 1-flowered; **Ped** 7 - 30 mm; **Sep** narrowly ovate, 9 - 18 mm; **Cl** campanulate, 6 - 7 cm ∅, pale grey-green with concentric purple-brown lines; **Cl** tube ± 1.5 × 3.5 cm, basal ⅔ inside bulging in **Cs** position; **Cl** lobes triangular, 11 - 25 mm, acute, ± folded, margins revolute; **Cn** uniseriate, shortly stipitate, ± 10 mm ∅; **Cs** lobes linear-falcate, 8 - 13 mm, ascending, dorsally placed against the bulging **Cl** tube; **Poll** narrowly drop-shaped, 1.5 - 1.6 × 0.4 - 0.5 mm; **Fr** broadly fusiform, ± 7 - 10.5 × 2.2 cm, shortly hairy, softly prickly.

The most common *Matelea* species with *Stapelia*-like flowers. Closely allied to *M. standleyana*.

M. sepicola W. D. Stevens (Phytologia 32: 387-392, ills., 1975). **T:** Mexico, Jalisco (*Stevens* 1436 [MSC, ENCB, F, MO]). – **D:** Mexico (Sinaloa, Nayarit, Jalisco); roadsides, scrub.

≡ *Dictyanthus sepicola* (W. D. Stevens) W. D. Stevens (2000).

Herbaceous climbers; caudex irregularly shaped, with coarsely fissured corky bark; stems to 2 - 3 mm ∅ (thicker at the base), weakly hairy; **L** 3 - 8 cm petiolate, lamina broadly ovate, 3.5 - 10 × 2.5 - 8.5 cm, ± verrucose and hairy; **Inf** shortly pedunculate, few-flowered, ± globose; **Ped** 1.5 - 3.5 mm, hairy; **Sep** acute, 3 - 5 mm; **Cl** rotate with urceolate **Cl** tube, 9 - 12 mm ∅, inside green to green-brown with red-brown stripes, glabrous; **Cl** tube ± 4 mm ∅; **Cl** lobes triangular-ovate, slightly revolute, 2.5 - 4.5 mm; **Cn** uniseriate, subglobose, completely filling the **Cl** tube; **Cs** lobes ± erect, 1.5 - 3 mm, acuminate; **Poll** obliquely ovoid-triangular, ± 0.65 × 0.37 mm; **Fr** broadly fusiform, 5 - 7.5 × 1.2 - 2 cm, shortly hairy, softly prickly.

M. standleyana R. F. Woodson (Ann. Missouri Bot. Gard. 28: 236, 1941). **T:** Mexico, Oaxaca (*Conzatti*

3760 [US, GH]). – **D:** Mexico (Oaxaca, Chiapas, Veracruz). **I:** Stevens (1988).

Incl. *Dictyanthus tigrinus* Conzatti & Standley (1924).

Twining, herbaceous, often forming rhizomes; **L** petiolate, ovate with cordate base, 5 - 10 × 3.5 - 10 cm, ± glabrous; **Inf** ± 2 cm pedunculate, 1-flowered; **Ped** 7 - 15 mm; **Sep** narrowly ovate, 9 - 18 mm; **Fl** with faint revolting smell; **Cl** campanulate, 6 - 7 cm ∅, pale grey-green with concentric purple-brown lines; **Cl** tube ± 12 × 30 mm, basal ½ inside bulging in **Cs** position; **Cl** lobes triangular, 17 - 28 mm, acute, margins revolute; **Cn** uniseriate, shortly stipitate, ± 8 mm ∅; **Cs** lobes linear-falcate, 9 - 13 mm, ascending, dorsally placed against the bulging **Cl** tube; **Poll** squarely drop-shaped, 1.5 - 1.9 × 0.45 - 0.63 mm; **Fr** broadly fusiform, ± 8.5 × 3 cm, shortly hairy, softly prickly.

This species apparently possesses the largest flowers among the New World *Asclepiadaceae* being pollinated by flies (presumably like all species of Subgen. *Dictyanthus*). It is therefore very similar to African *Stapelia* or *Brachystelma* flowers.

M. tuberosa (Robinson) R. F. Woodson (Ann. Missouri Bot. Gard. 28: 237, 1941). **T:** Mexico, Jalisco (*Pringle* 3568 [GH, F, VT]). – **D:** Mexico (S Sonora to Jalisco); open forests, grassland. **I:** Stevens (1988).

≡ *Dictyanthus tuberosus* Robinson (1892).

Erect herbaceous perennials, caudex irregularly turnip-shaped, ± 5 × 3 cm, with thick corky bark; stems 10 - 70 (-100) cm, glandular, hairy, becoming woody at the base; **L** 0.7 - 3 cm petiolate, lamina ovate to broadly ovate, base cordate, 1.5 - 4.5 × 1.5 - 4 cm, ± hairy, often glandular; **Inf** 1 - 9 mm pedunculate, 1- to 2-flowered; **Ped** 4 - 5 mm; **Sep** ovate-lanceolate, 5 - 9 mm, hairy; **Cl** deeply campanulately funnel-shaped, 1 - 1.8 cm ∅, inside marked with purple-brown streaks and reticulations; **Cl** tube 6 - 8 × 5 - 7 mm, inside with ridges becoming broader towards the base; **Cl** lobes triangular, 3 - 6 mm, margins revolute; **Cn** uniseriate, stipitate, ± 2.5 × 3 mm; **Cs** lobes arrow-headed, semi-erect, 1 - 1.5 mm, dorsally placed against the bulging **Cl** tube; **Poll** drop-shaped, 0.65 - 0.85 × 0.35 mm; **Fr** broadly fusiform, 5.5 - 6.5 × 1.1 - 1.9 cm, softly prickly.

MICHOLITZIA

U. Meve & C. Hoffmann

Micholitzia N. E. Brown (BMI 1909: 358, 1909). **T:** *Micholitzia obcordata* N. E. Brown. – **Lit:** Maxwell (1991); Li (1992); Goyder & Kent (1994). **D:** China, India, Myanmar, Thailand. **Etym:** For Wilhelm Micholitz (1854 - 1932), plant collector for Sander & Co., England.

Incl. *Antiostelma* (Tsiang & P. T. Li) P. T. Li (1992). **T:** *Hoya lantsangensis* Tsiang & P. T. Li.

Erect, prostrate or ± pendent epiphytes with white latex; stems ± 30 cm, terete, branched, green and weakly pubescent when young, becoming grey and glabrous with age; **L** decussate, 3 - 4 mm petiolate, lamina obcordate to obovate, 1.8 - 3.2 × 1.4 - 2.3 cm, pale grey or dark green, weakly pubescent, ± truncate to obtuse with a minute mucro, cuneate to ± attenuate at the base; **Inf** racemose, lateral or terminal, 4- to 9-flowered; peduncle ± 2 × 1 mm ∅; **Ped** 1 - 2 mm, weakly pubescent; **Sep** ovate, 1 mm, green, becoming brownish towards the tip, weakly pubescent; **Cl** narrowly urceolate, basally with 5 humps, olive-green, outside weakly pubescent; **Cl** tube ± 6 × basally 3 mm ∅, pale green, inside bearded; **Cl** lobes spreading, triangular-ovate, 1.5 × 1 mm, acute, inside reddish-brown to yellowish, villose; **Cn** uniseriate, staminal, orange; **Cs** lobes fleshy, apically obtuse, lower face dorsally-medianly with a vertical groove, upper face broadly humped; **Gy** pyramidal, guiding rails protruding; **Sty** head acutely conical, covered by the **Anth** appendages; **Anth** appendages lanceolate, acute, pale orange; **Poll** broadly D-shaped, 0.25 × 0.25 mm, germination crest lateral on the outside, but becoming horizontally oriented by an outwards-twist of the **Poll**; **Fr** single, narrowly ellipsoid, erect.

This monotypic genus of the *Marsdenieae* belongs to the alliance of *Dischidia* and *Hoya*. Although very similar in habit to *Dischidia*, it is presumably most closely related to *Hoya*, as particularly suggested by the similar morphology of the corona (fleshy, dorsally grooved).

M. obcordata N. E. Brown (BMI 1909: 358, 1909). **T:** Myanmar (*Micholitz* s.n. [K]). – **D:** China, India, Myanmar, Thailand. **I:** Goyder & Kent (1994). **Fig. XXXIII.b**

≡ *Dischidia obcordata* (N. E. Brown) J. F. Maxwell & R. van Donkelaar (1991); **incl.** *Hoya yunnanensis* hort. (s.a.) (*nom. inval.*, Art. 29.1); **incl.** *Hoya manipurensis* Deb (1955) ≡ *Antiostelma manipurense* (Deb) P. T. Li (1992); **incl.** *Hoya lantsangensis* Tsiang & P. T. Li (1974) ≡ *Antiostelma lantsangense* (Tsiang & P. T. Li) P. T. Li (1992).

Description as for the genus.

MIRAGLOSSUM

B. Müller, P. Stegemann & F. Albers

Miraglossum Kupicha (KB 38(4): 625, 1984). **T:** *Schizoglossum pulchellum* Schlechter. – **Lit:** Kupicha (1984). **D:** E RSA. **Etym:** Lat. 'mirus', wonderful, remarkable; and Gr. 'glossa', tongue; for the conspicuous corona segments.

Erect herbaceous perennials with deciduous above-ground parts and tuberous **R**; stems 17 - 60 cm, unbranched, with soft **Ha** evenly distributed or rarely distichous; **L** irregularly arranged, within the **Inf** usually whorled, shortly petiolate, lamina simple, entire, linear to narrowly triangular, margins revolute, usually hairy; **Inf** sessile, from nodes in upper stem sections (terminal in species with reduced upper **L**); **Bra** filiform, inconspicuous; **Ped** softly hairy; **Sep** triangular, usually hairy on both faces, margins ciliate; **Cl** campanulate or inversely campanulate or bowl-shaped; **Cl** lobes elliptic, usually revolute, outside variously hairy, margins inside densely soft-hairy on the left side, ciliate on the right side; **Cs** lobes thickly fleshy, variable in shape; **Sty** head flat or shield-shaped; **Anth** ± flat to undulate, folded-compressed, incumbent on the **Sty** head; **Poll** sausage-shaped, dorsiventrally slightly compressed, without distinct germination crest, translators ribbon-like, flattened, subapically attached to the **Poll**; **Fr** usually with prickles, these somewhat appressed, woolly, **Ha** usually white; **Se** numerous, dark brown, narrowly ovate, rugose, tipped with a tuft of **Ha**.

Closely related to *Schizoglossum*, yet distinguishable by the sessile clustered inflorescences as well as by the pollinia lacking a distinct germination crest.

M. anomalum (N. E. Brown) Kupicha (KB 38(4): 629, 1984). **T:** RSA, Eastern Cape (*Flanagan* 396 [K, BOL, GRA, K, NBG, SAM, PRE]). – **D:** RSA (Eastern Cape, SW KwaZulu-Natal); grassland, rocky slopes, 450 - 1560 m.

≡ *Schizoglossum anomalum* N. E. Brown (1907).

Plants 20 - 35 cm; **Int** ≤ 2 cm, **Ha** long, soft; **L** clustered or lax, not reduced within the **Inf**, lamina 25 - 47 × 2 - 9 mm, linear to triangular, obtuse at the base, hairy, **Ha** long, soft; **Inf** 3- to 9-flowered; **Ped** ≤ 12 mm; **Sep** 4 - 4.5 × 1.5 mm, outside densely soft-hairy, inside glabrous or apically soft-hairy; **Cl** green, inversely campanulate; **Cl** lobes 4 - 5 × 2 - 3 mm, linear to lanceolate, laterally overlapping, margins revolute, inside ± densely soft-hairy towards the tip; **Cs** green, apically white, 1 - 1.5 mm, lobes simple, ± narrowly ovate, tips acute, sides weakly bulging outwards; **Sty** head greenish-beige; **Anth** 0.8 mm, flat; **Poll** 0.63 - 0.75 mm, corpuscles 0.33 - 0.43 mm; **Fr** ± 5 cm, prickles 2 mm; **Se** 2 mm.

M. davyi (N. E. Brown) Kupicha (KB 38(4): 632, 1984). **T:** RSA, Mpumalanga (*Burtt Davy* 964 [K, NH]). – **D:** RSA (Mpumalanga); grassland, sandy ground or black loam, ± 1700 m.

≡ *Schizoglossum davyi* N. E. Brown (1907).

Plants 30 - 60 cm; **Int** 2 - 3 cm; **L** densely arranged, not reduced within the **Inf**, petiole 1 - 2 mm, lamina 24 - 47 × 2 - 6 mm, linear, acute, basally obtuse, both faces woolly, particularly on the midrib of the lower face; **Inf** 3 - 4; **Ped** ≤ 2 cm; **Sep** ± 11 × 3.5 mm, both faces densely woolly, **Ha** either transparent, very fine, erect, or brown and appressed; **Cl** ± flat; **Cl** lobes ± 14 × 7.5 mm, ovate-lanceolate, outside soft-hairy, **Ha** transparent, erect, or brown and appressed; **Cs** ± 5 mm, lobes square at the base with bifid tips, appendages triangular, fleshy, laterally spreading at 180°, inner appendage triangular at the base, pointing towards the centre of the **Fl**, apically filiform, bent outwards into a horizontal position; **Anth** 1 mm, ± flat; **Poll** 1.1 - 1.15 mm, corpuscles 0.9 mm; **Fr** 6 - 7 cm, prickles 6 - 8 mm; **Se** 2 mm, tuft of **Ha** white, ± 15 mm.

M. laeve Kupicha (KB 38(4): 630, 1984). **T:** RSA, Mpumalanga (*Acocks* 21011 [PRE]). – **D:** RSA (Mpumalanga); scrub, ± 1500 m.

Plants 17 - 25 cm; stems with curly **Ha**, **Int** ≤ 2 cm; **L** densely arranged, not reduced within the **Inf**, lamina 30 - 37 × 1 - 3 mm, glabrous; **Inf** 6 - 8, with 3 - 4 **Fl**; **Ped** ≤ 8 mm; **Sep** 4 - 4.5 × 1 - 1.5 mm, outside soft-hairy, inside glabrous; **Cl** green, inversely bowl-shaped; **Cl** lobes 6 - 6.5 × 3 mm, margins revolute, outside pubescent, **Ha** brownish, appressed; **Cs** 4.5 - 5 mm, lobes basally transversely rectangular, shoulders rounded, lobes rolled inwards and thus forming ear-conch-shaped vaults, apical appendage ± 3 mm, filiform, bent and ascending above the **Sty** head; **Anth** 0.8 mm, ± flat; **Poll** 0.85 mm, corpuscles 0.35 - 0.43 mm; **Fr** bristly when young, **Ha** yellowish.

Closely allied to *M. verticillare*.

M. pilosum (Schlechter) Kupicha (KB 38(4): 631, 1984). **T:** RSA, KwaZulu-Natal (*Schlechter* 3238 [not located]). – **D:** RSA (S KwaZulu-Natal); grassland, 900 - 1740 m.

≡ *Schizoglossum pilosum* Schlechter (1895).

Plants 12 - 36 cm; stems with long **Ha**; **Int** 3.5 - 10 cm; **L** decussate, usually small and inconspicuous within the **Inf**, petiole 1 - 5 mm, lamina 15 - 37 × 2 - 14 mm, linear to linear-ovate, basally obtuse, soft-hairy, **Ha** long; **Inf** terminal, sometimes also from 1 - 2 nodes below, with 6 - 8 **Fl**; **Ped** ≤ 8 mm; **Sep** 3.5 - 4 × 1 - 2 mm, outer face densely hairy, **Ha** soft, erect, inner face glabrous, apically with a single **Ha**; **Cl** green, inversely campanulate; **Cl** lobes 5 - 6 × 3 mm, ± ovate, margins revolute, outside hairy, **Ha** soft, erect; **Cs** white at the base, appendages purple or greenish-brown, lobes 4.5 mm, square at the base, shoulders rounded, apical appendage slender, horn-like, ± falcately inflexed, inner appendage simple or bifid, tongue-shaped, flat, incumbent on the **Anth**; **Sty** head white; **Anth** 0.7 - 1.2 mm, undulate; **Poll** 0.55 - 0.6 mm, corpuscles 0.2 - 0.28 mm; **Fr** without prickles.

M. pulchellum (Schlechter) Kupicha (KB 38(4): 627, 1984). **T:** RSA, Mpumalanga (*Galpin* 1089 [PRE, K]). – **D:** RSA (E Mpumalanga, KwaZulu-Natal, esp. frontier to Lesotho and Free State), Swaziland; grassland, 600 - 2340 m.

≡ *Schizoglossum pulchellum* Schlechter (1894).

Plants 20 - 25 cm; **Int** within the **Inf** 8 - 70 mm; **L** dense or lax, decussate or irregular, lamina 20 - 50 × 1.5 mm, linear-lanceolate, pubescent; **Inf** terminal or from 1 - 2 nodes below, indistinctly offset, with 3 - 5 **Fl**; **Sep** 4 - 6 × 1.5 - 2 mm, outer face densely hairy, **Ha** transparent, rigid, erect, or **Ha** brown, thin, appressed, **Ha** of the inner face apically brown, appressed; **Cl** brownish-green, inversely campanulate; **Cl** lobes 7 - 8 × 4 - 4.5 mm, narrowly ovate, margins revolute, outside densely hairy, **Ha** transparent, thin, rigid, erect, or **Ha** brown, accumbent; **Cs** (white-) green at the base, appendages purple or brown, lobes erect, ± rounded, 2.5 - 4 mm, inner appendage subulate, stout, basal ½ erect, upper ½ bent outwards into a horizontal position; **Anth** 1.25 - 1.7 mm, undulate; **Poll** 0.63 - 0.93 mm, corpuscles 0.3 - 0.38 mm; **Fr** 3.5 - 7 cm, prickles ± 2 mm; **Se** ± 2.5 mm, tuft of **Ha** creamy-white, ± 25 mm.

M. superbum Kupicha (KB 38(4): 631, 1984). **T:** RSA, KwaZulu-Natal (*Hilliard & Burtt* 7848 [E, K, NU, PRE]). – **D:** RSA (KwaZulu-Natal, frontier to Lesotho); grassland, black soil of the valley bottoms, 1500 - 2100 m.

Plants 25 - 50 cm; stems pubescent; **Int** 1 - 4 cm; **L** densely arranged in the vegetative stem parts, but lax and usually small and inconspicuous within the **Inf**, lamina 15 - 40 × 0.5 - 2 mm, glabrous or weakly soft-hairy; **Inf** terminal, sometimes also from 1 - 2 nodes below, with 3 - 5 **Fl**; **Ped** ≤ 13 mm; **Sep** 4 - 6 × 1.5 - 2 mm, outside soft-hairy, upper ½ of the inside pubescent; **Cl** dull green, streaked with dull purple, inversely bowl-shaped; **Cl** lobes 6 - 9 × 4 mm, narrowly ovate, margins revolute, outside glabrous; **Cs** basally yellow-green, apical appendage (dark) brown, inner appendage brownish-purple, 6 - 7 mm, lobes basally transversely rectangular, erect, apical appendage cross-shaped, inner appendage trifid, lateral segments filiform, bent towards each other, central segment like a shark's fin, joined with the apical appendage; **Sty** head white; **Anth** white, ± 2 mm, papery, conspicuous; **Poll** 0.68 - 0.78 mm, corpuscles 0.45 - 0.5 mm, sagittate.

M. verticillare (Schlechter) Kupicha (KB 38(4): 630, 1984). **T:** RSA, KwaZulu-Natal (*Schlechter* 3242 [BOL, Z]). – **D:** RSA (Eastern Cape, S KwaZulu-Natal); rocky grassland, 900 - 1800 m.

≡ *Schizoglossum verticillare* Schlechter (1895).

Plants 10 - 35 cm; stems pubescent with porrect or curly **Ha**; **Int** 8 - 18 mm; **L** clustered, rarely laxly arranged, lamina 17 - 45 × 1 - 2 mm, ± linear, hairy, **Ha** soft, accumbent; **Inf** 3 - 5, with 2 - 6 **Fl**; **Ped** ≥ 10 mm; **Sep** 3 - 5 × 1 - 2 mm, hairy on both faces, **Ha** transparent, soft, erect, or brown and accumbent; **Cl** green, bowl-shaped; **Cl** lobes 4.5 - 6.5 × 2.5 - 3.5 mm, narrowly ovate, tips involute, outside soft-hairy, **Ha** transparent, erect or brownish and accumbent; **Cs** white, base and margins dark green, lobes basally 1.5 mm, transversely rectangular, involute, apical appendage long, filiform, bent clockwise to form a loose crown above the white **Sty** head; **Anth** 1 mm, undulate; **Poll** 0.85 - 1.1 mm, corpuscles 0.28 - 0.35 mm.

NOTECHIDNOPSIS

U. Meve

Notechidnopsis Lavranos & M. B. Bleck (CSJA 57(6): 255-256, ills., 1985). **T:** *Caralluma tessellata* Pillans. – **Lit:** Bruyns (1999d); Meve & Liede (2001b). **D:** RSA. **Etym:** Gr. 'notos', South, and for the genus *Echidnopsis* (*Asclepiadaceae*); for the similarity to that genus and the more southern distribution.

Sparsely branched stem succulents forming loose cushions, stoloniferous; stems ± erect or decumbent and creeping underground, respectively, 6- to 10-angled, 2 - 20 × 0.8 - 3 cm, cylindrical, (blue-) green, **Ri** consisting of ± rectangular convex **Tu** in a regular pattern; **Tu** 2 - 5 × 2 - 8 mm, conical and terminating in a minute apiculus or convex with a minute central elevation (**Tu** of old stems becoming more and more flat); **L** rudiments absent; **Inf** 1- to 10- (to 15-) flowered fascicles, laterally arising near the stem tips, partly on persistent cushion-shaped pseudopeduncles; **Fl** with copious nectar and strong sweet foetid odours; **Ped** 2 - 4 × 1 mm, glabrous; **Sep** subulate-lanceolate, 1 - 2 mm, glabrous; **Cl** yellowish-green blotched with red or blood-red with yellowish blotches, 4 - 11 (-15) mm ∅, rotate, outside glabrous; **Cl** tube short, shallow, 3 - 5.5 mm ∅; **Cl** lobes triangular, 2 - 5 × 2 - 4 mm, fully expanding, inside densely short-haired, **Ha** ± 0.1 mm; **Cn** biseriate, 2 - 3 × 3 - 4 mm; **Ci** fused dish-like or to form a shallow cup, free **Ci** lobes < 1 mm, triangular or rectangular with an apical notch; **Cs** ± 1mm, linear-lanceolate, incumbent on the **Anth** or erect; **Anth** ± rectangular; **Poll** D-shaped, caudicles short, corpuscle fusiform, basally shortly winged; **Sty** head convex; **Fr** narrowly fusiform, 3 - 4 cm; **Se** brown, ± 4 × 2.5 mm, tuft of **Ha** ± 14 mm.

Notechidnopsis – if at all a good genus – is related to the genus *Lavrania*, which is likewise of S African origin. Both are characterized by a combination of multi-angled stems tessellated by small ± leafless tubercles and with small rotate flowers with shallowly cup-shaped coronas.

N. columnaris (Nel) Lavranos & M. B. Bleck (CSJA 57(6): 256-257, ills., 1985). **T:** RSA, Northern Cape (*Herre* s.n. in *SUG* 6023 [BOL]). – **D:** RSA (Northern Cape: Richtersveld). **Fig. XXXII.a, XXXII.g (left)**

≡ *Trichocaulon columnare* Nel (1933) ≡ *Echidnopsis columnaris* (Nel) R. A. Dyer & D. S. Hardy (1968).

Stems 8-angled, erect, rarely prostrate-decumbent, 12 - 20 × 2 - 3 cm, sparsely branched, glaucous-green, regularly tessellated with always rectangular **Tu**, these ± 4 × 8 mm, conical with a short rigid tip, projecting horizontally or slightly curved downwards; **Inf** near the stem tips, 1- to 10- (to 15-) flowered, on cushion-shaped pseudopeduncles; **Ped** 2 - 4 mm; **Fl** opening in succession, with copious nectar and a foetid-fruity scent; **Cl** rotate, outside pale green, inside yellowish-green (-ochre) finely spotted with red-brown, 5 - 11 (-15) mm ⌀, inside densely short-haired; **Cl** tube 4 - 5 mm ⌀; **Cl** lobes triangular, 2 - 5 × 2 - 4 mm, fully expanding; **Cn** pale yellow blotched with purple, ± 2.5 × 2.5 - 4 mm; **Ci** fused into a dish, lobes triangular, ± 0.7 × 0.7 mm, tips slightly erect; **Cs** linear to lanceolate (spoon-shaped), apically ± notched, 1 - 1.5 mm, connivent above the **Sty** head; **Poll** D-shaped, ± 0.35 × 0.25 mm; **Fr** 4 - 5 cm.

A fairly rare species, which is solely known from the Richtersveld in the NW corner of RSA.

N. tessellata (Pillans) Lavranos & M. B. Bleck (CSJA 57(6): 256, ill., 1985). **T:** RSA, Western Cape (*Ross-Frames* s.n. in *NBG* 198/31 [BOL, K]). – **D:** RSA (Northern Cape, Western Cape). **I:** Eggli (1994: 166). **Fig. XXXII.f, XXXII.g (right)**

≡ *Caralluma tessellata* Pillans (1933); **incl.** *Caralluma serpentina* Nel (1935) ≡ *Echidnopsis serpentina* (Nel) A. C. White & B. Sloane (1937); **incl.** *Echidnopsis framesii* A. C. White & B. Sloane (1937).

Stems (6- to) 10-angled, creeping to prostrate-decumbent, 2 - 20 × 0.8 - 1.8 cm, sparsely branched, green, regularly tessellated with always rectangular (sometimes irregular owing to interspersed **Inf**) **Tu**, these ± 2 × 4 mm, convex with a minute central elevation; **Inf** near the stem tips, 1- to 8- (to 15-) flowered, **Fl** opening in succession (rarely simultaneously), with foetid-fruity scent; **Ped** 3 - 5 mm; **Cl** rotate, outside pale green, inside red (-brown), often blotched with yellow, 6 - 10 mm ⌀, (densely) short-haired; **Cl** tube 3 - 4.5 mm ⌀; **Cl** lobes triangular, 3 - 4 × 2 - 3 mm, fully expanding; **Cn** yolk-yellow, 1.5 - 2 × 2.5 - 3 mm; **Ci** basally fused into a cup, lobes rectangular, apically emarginate, ± 0.7 × 0.7 mm, placed against the **Cl** tube or slightly erect; **Cs** lanceolate to triangular, entire, < 1 mm, incumbent on the back of the **Anth** or (almost) upright; **Poll** narrowly D-shaped, ± 0.38 × 0.20 mm; **Fr** ± 3 cm.

ODONTOSTELMA

U. Meve

Odontostelma Rendle (J. Bot. 32: 161, 1894). **T:** *Odontostelma welwitschii* Rendle. – **D:** Angola, Zimbabwe. **Etym:** Gr. 'odous, odontos', tooth; and Gr. 'stelma', crown, garland, wreath; for the nature of the corona.

Geophytic herbaceous perennials to 15 cm tall; **R** ± fusiform, tuberous; **L** decussate, ± sessile, narrowly linear; **Inf** terminal or lateral, borne near the stem tips, axillary, pedunculate, few-flowered, pseudo-umbellate; **Cl** shallowly campanulate; **Cl** lobes elongate-spatulate; **Cn** 1- (to 2-) seriate; **Cs** lobes erect, flat, small, basally fused into a ring-like structure; **Gy** sessile, basally fused to the **Cl**; appendages of the **Anth** hyaline, bent inwards; **Poll** narrowly ellipsoid; **Fr** very slenderly fusiform, to 9 cm, acute.

A bitypic genus from the group around *Xysmalobium*. A striking feature is the corona, which is basally fused into a ring.

O. minus S. Moore (J. Bot. 50: 363, 1912). **T:** Angola (*Gossweiler* 2332 [BM]). – **D:** Angola.

Habit as *O. welwitschii*, but **Fl** very much smaller.

O. welwitschii Rendle (J. Bot. 32: 161-162, t. 344, 1894). – **D:** Angola, Zimbabwe; bushland and savanna.

≡ *Schizoglossum welwitschii* (Rendle) N. E. Brown (1902).

Description as for the genus.

OPHIONELLA

U. Meve

Ophionella Bruyns (Cact. Succ. J. Gr. Brit. 43(2/3): 70-73, ills., 1981). **T:** *Pectinaria arcuata* N. E. Brown. – **Lit:** Bruyns (1999c). **D:** RSA. **Etym:** Diminutive of Gr. 'ophis', snake, serpent; referring to the snake-like sinuous stems.

Stems rounded 4-angled, creeping, with the tips always ending underground, 4 - 25 cm × 3 - 8 mm; **Tu** flatly ellipsoidal; **L** minute, deltoid, slightly reflexed; **Cl** bud-like to campanulate, somewhat thickened at the mouth; **Cl** tube narrowly to broadly cup-shaped, 3 - 6 × 2 - 7 mm; **Cl** lobes deltoid to lanceolate, ascending and apically joined to spreading, 3.5 - 9 × 2 - 4.8 mm; **Cn** biseriate, shortly stipitate, shallowly bowl-shaped, 1 - 1.5 × 2 - 4 mm; **Ci** lobes spreading, somewhat ascending, truncate with a thickened tooth, 0.5 mm; **Cs** lobes ± 0.25 mm, deltoid to subquadrate, with a dorsal projection, appressed to the backs of the **Anth**; **Poll** ellipsoidal, translator wings minute; **Fr** terete-fusiform, ± 3 cm.

O. arcuata (N. E. Brown) Bruyns (Cact. Succ. J. Gr. Brit. 43(2/3): 72, ills., 1981). **T:** RSA, Eastern Cape (*Pillans* 182 [K, BOL]). – **D:** RSA (Eastern Cape).

≡ *Pectinaria arcuata* N. E. Brown (1909).

Stems forming densely interwoven mats of up to 30 cm ⌀, grey to green-brown, 4 - 25 × 0.4 - 0.8 cm, slightly 4-angled, glabrous, papillose, papillae dome-shaped; **Tu** obtuse, broad, flattened, on young stems at the base with dark drop-like markings; **L** rudiments minute, **Sc**-like; **Inf** with 1 - 4 **Fl** opening in succession, borne basally on the sides of the stems, pedunculate or with a rudimentary peduncle; **Ped** 1 - 4 mm, glabrous; **Sep** 0.8 - 2 mm, (ovate-)lanceolate, glabrous; **Cl** outside pale flesh-coloured, inside creamy-white, tube purple-red at the base, short and ovoid or broadly cup-shaped, ± 4 × 3 - 7 mm, mouth and base of the **Cl** lobes, respectively, slightly bulging; **Cl** lobes elongate, bent upwards and tips often joined, 3.5 - 8 × 2 - 3.5 mm, at the base usually with a distinct appendage, this pointing outwards, pleated, inside densely and finely prickly; **Cn** pale yellow, dorsal process of the **Cs** occasionally blotched with reddish-purple, ≤ 1.5 × 2 - 3 mm; **Ci** lobes < 0.5 mm, trapeziform, truncate, ± at the base joined to the dorsal process of the **Cs**; **Cs** lobes broad at the base, dorsal process conspicuous, ≤ 1 mm, fleshy, often verrucose, ± clavate, horizontal to ascending, apical section ≤ 0.5 mm, narrower, incumbent on the **Anth** and usually overtopping them; **Anth** rectangular, ± covered by the **Cs**; **Poll** ± 0.25 × 0.16 mm, ovoid-elliptic. – [B. Müller & F. Albers]

O. arcuata ssp. **arcuata** – **D:** RSA (Eastern Cape: between Somerset East and Bedford). **Fig. XXXIII.c**

Cl long; **Cl** lobes 5 - 8 × 2 - 2.5 mm, lanceolate, bent upwards and inwards, tips joined (occasionally divergent); **Cl** tube ± 4 × 3 mm, ± ovoid; **Cs** lobes apically ± triangular, dorsal process pointing upwards. – [B. Müller & F. Albers]

O. arcuata ssp. **mirkinii** (Pillans) Bruyns (Bot. J. Linn. Soc. 131(4): 395-396, ills., 1999). **T:** RSA, Eastern Cape (*Mirkin* s.n. [BOL 22432]). – **D:** RSA (Eastern Cape: between Willowmore and Addo).

≡ *Pectinaria mirkinii* Pillans (1939) ≡ *Ophionella arcuata* var. *mirkinii* (Pillans) Bruyns (1981).

Cl shorter than in ssp. *arcuata*, tube ≤ 4 × 5 - 7 mm, broadly cup-shaped; **Cl** lobes ≤ 4 × 3 - 3.5 mm, triangular, bent upwards and inwards, tips free; **Cn** 1 - 1.2 × 2.8 - 3.2 mm; **Cs** lobes apically ± rectangular, truncate or tips slightly concave, dorsal process rather horizontal. – [B. Müller & F. Albers]

O. willowmorensis Bruyns (Bot. J. Linn. Soc. 131(4): 396-397, ills., 1999). **T:** RSA, Eastern Cape (*Bruyns* 4966 [BOL]). – **D:** RSA (SE Eastern Cape).

Stems often subterranean, 3 - 15 cm × 3 - 5 mm; **Inf** subsessile, 1- (to 4-) flowered; **Bra** subulate, 1 mm; **Ped** 1.5 - 6 × 0.6 mm, glabrous, ascending; **Sep** ovate-acuminate, ± 1 × 0.4 mm; **Cl** bud-like, broadly ellipsoid to nearly (depressed-) globose, 3.5 - 7 × 4 - 6 mm, outer face white, sometimes spotted pale red, papillate, inner face coarsely red-spotted, glabrous except for 1 mm long stiff **Ha** in the median regions; **Cl** tube shallowly bowl-shaped, 4 - 6 mm ⌀; **Cl** lobes deltoid to lanceolate, ascending, often becoming descending towards the tips, apically fused, 3 - 5 × 2 - 3.5 mm; **Cn** shortly stipitate, pale yellow, shallowly bowl-shaped, ± 1.2 - 2 mm ⌀; **Ci** lobes spreading, somewhat ascending, deltoid-lanceolate, acute, ± 0.5 mm; **Cs** lobes ± 0.25 mm, deltoid to subquadrate, emarginate, flat; **Poll** 0.18 × 0.09 mm, translator wings ellipsoid.

A distinct tiny species endemic to sandstone slabs of the Witteberge in the SE Eastern Cape.

ORBEA

B. Müller, J. Kiel, F. Albers & U. Meve

Orbea Haworth (Synops. Pl. Succ., 37, 1812). **T:** *Stapelia variegata* Linné [Lectotype, designated by L. C. Leach, Trans. Rhodesia Sci. Assoc. 59(1): 2, 1978.]. – **Lit:** Leach (1975); Leach (1978a); Plowes (1994b); Bruyns (2001). **D:** Arabien, Afrika. **Etym:** Lat. 'orbis', circle; for the thickened corolla part (annulus) surrounding the flower centre.

Incl. *Podanthes* Haworth (1812). **T:** *Podanthes pulchra* Haworth [Lectotype, designated by Rehder & al., Bull. Misc. Inform. (Kew) 1935: 451.].
Incl. *Diplocyatha* N. E. Brown (1878). **T:** *Stapelia ciliata* Thunberg.
Incl. *Stapeliopsis* E. Phillips (1932) (*nom. illeg.*, Art. 53.1). **T:** *Stapelia cooperi* N. E. Brown.
Incl. *Stultitia* E. Phillips (1933). **T:** *Stapelia cooperi* N. E. Brown.
Incl. *Orbeopsis* L. C. Leach (1978). **T:** *Caralluma lutea* N. E. Brown.
Incl. *Pachycymbium* L. C. Leach (1978). **T:** *Caralluma keithii* R. A. Dyer.
Incl. *Angolluma* Munster (1990). **T:** *Boucerosia decaisneana* Lemaire.
Incl. *Ballyanthus* Bruyns (2001). **T:** *Stapelia prognatha* P. R. O. Bally.

Basally branched stem succulents with erect or ascending stems, clump-forming to caespitose, often with subterranean runners, latex clear, colourless or yellowish; stems 4-angled or 4- to 5-angled, 2 - 25 × 1 - 2 (-3) cm, green or grey-green, almost always spotted with red, glabrous; **Tu** conical, spreading, decussate, ± tapering without differentiation into the acute **L** rudiment, usually with stipular **Gl**; **Inf** (sub-) sessile to pedunculate, basal, lateral to subapical, 1- to many-flowered; **Fl** opening in succession or simultaneously; **Ped** 0.1 - 30 (-90) mm, apicall usually narrowing, glabrous; **Sep** green, acute, glabrous; **Cl** inside whitish, cream-coloured, yellow, red or brown, often spotted or banded in 2 contrasting colours, ± rotate, 1 - 7 cm ⌀, smooth, verrucose or transversely rugose, glabrous, papillate or hairy; **Cl** tube flat, cup-shaped or campanulate, often bulging ring-like with a thickened and/or fleshy disc (annulus) enclosing the **Cn**; **Cl** lobes broadly triangular (-ovate) to lanceolate, flatly spreading or margins recurved; **Cn** biseriate, the 2 series usually distinctly differentiated (rarely fused to a ring-like structure, or with the interstaminal series reduced), but never completely separated on the staminal column, occasionally urceolate, often stalked, glabrous; **Ci** lobes entire, 2-dentate to 2-lobed, flat and spreading to erect and pouch-forming; **Cs** lobes polymorphic, often with additional apical or basal appendages, erect or decumbent on the **Anth**; **Anth** whitish, linear, rectangular or trapezoid; guide rails ± erect, not embedded in gynostegial tissue; **Poll** bean- or D-shaped or ellipsoid, germination crests bent on the dorsal (upper) face of the **Poll**, translator ± winged; **Fr** fusiform, 4 - 16 × 0.6 - 1.5 cm, erect, arranged at an acute angle, often spotted with red; **Se** brown, elliptic, margins ± voluminous.

The separation of the "Ango-Group" from *Caralluma* and its classification under *Pachycymbium* by Gilbert (1990) was replaced by splitting the complex into *Pachycymbium* s.str. and *Angolluma* (Plowes 1994b). The change of *Angolluma miscella* to *Orbea* (Meve 2000b) was followed by Bruyns (2001) with a taxonomical round-up. He subsumed the closely related genera *Angolluma, Orbeanthus, Orbeopsis* and *Pachycymbium* completely under *Orbea*. Apart, Bruyns (2001) created the new genus *Ballyanthus* for *Orbea prognatha*, which is marked by an atypical pollinia structure. In contrast to this concept, *Ballyanthus* is not accepted here as an independent taxon, and the bitypic genus *Orbeanthus* continues to be maintained as a separate genus (see there).

O. abayensis (M. G. Gilbert) Bruyns (Aloe 37(4): 73, 2001). **T**: Ethiopia, Shewa Region (*Gilbert 3891* [K, ETH]). – **D**: C Ethiopia. **I**: Gilbert (1978). **Fig. XXXIV.a**

≡ *Caralluma abayensis* M. G. Gilbert (1978) ≡ *Pachycymbium abayense* (M. G. Gilbert) M. G. Gilbert (1990) ≡ *Angolluma abayensis* (M. G. Gilbert) Plowes (1994); **incl.** *Angolluma gilbertii* Plowes (1994).

Stems erect, in compact clumps, to 14 cm; **Tu** 1.5 - 2 cm, robust, spreading-ascending, stipular rudiments rare; **Inf** with up to 3 erect **Fl**; **Ped** 8.5 - 9 × 2.5 mm, conspicuously bent outwards; **Sep** 4.5 - 5 × 2.5 - 3 mm; **Cl** inside pale brown, ± 4 cm ∅, flat, annulus 1 × 1 mm, minutely tomentose, basally closely surrounding the **Cn**; **Cl** lobes ± 15 × 9 mm, ovate-triangular, apically bent outwards, inside minutely velvety-papillate, glabrous; **Cn** 2.5 × 8 - 9 mm; **Ci** bowl-shaped, lobes forming 5 shallow pouches, apically thickly fleshy, slightly incised in the middle; **Cs** lobes ovate-lanceolate, surpassing the **Anth**, region where united with the **Ci** with 2 flat dorsal bulges; **Poll** 0.5 × 0.4 mm, D-shaped.

Closely related to *O. laticorona*.

O. albocastanea (Marloth) Bruyns (Aloe 37(4): 73, 2001). **T**: Namibia (*Marloth 5110* [PRE]). – **D**: Namibia. **I**: White & Sloane (1937).

≡ *Stapelia albocastanea* Marloth (1913) ≡ *Caralluma albocastanea* (Marloth) L. C. Leach (1970) ≡ *Orbeopsis albocastanea* (Marloth) L. C. Leach (1978); **incl.** *Stapelia caroli-schmidtii* Dinter & A. Berger (1914).

Stems decumbent, rooting when touching the ground, dull green, stippled with dull red, 6 - 8 × ± 1 cm, ± 4-angled; **Ri** separated up to 1.4 cm; **Tu** 3 - 5 mm; **Inf** 3- to 6-flowered, **Fl** opening in succession, peduncle short; **Ped** 4 - 6 cm, bent upwards; **Cl** ± white, cream-coloured, blotched or stippled with pale grey or dull green, purple-brown, reddish or ± black-purple, 2.5 cm ∅, central depression 8 mm ∅, shallowly cup-shaped; **Cl** lobes ± 11 × 5 mm, narrowly ovate-lanceolate, acuminate, margins and tips revolute, outside glabrous, inside coarsely rugose, margins hairy with vibratile clavate **Ha**; **Cn** dark brown; **Ci** lobes 2 × 1.5 mm, rectangular, tips deeply crenate, convex, divaricate, revolute, closely appressed to the **Cl**; **Cs** apically sometimes whitish, lobes 4 - 5 mm, complex, apical appendage ± 3 mm, subulate-filiform, erect-connivent, tips ± revolute, basal appendage ± 2.5 mm, acute, ± erect, divaricate; **Poll** 0.3 × 0.4 mm, broadly D-shaped, widely distant.

A member of the *O. lutea*-complex and closely related to *O. knobelii*. – [B. Müller & F. Albers]

O. araysiana (Lavranos & Bilaidi) Bruyns (Aloe 37(4): 73, 2001). **T**: Yemen (*Lavranos & al. 8602* [FT]). – **D**: S Yemen.

≡ *Stultitia araysiana* Lavranos & Bilaidi (1971) ≡ *Pachycymbium araysianum* (Lavranos & Bilaidi) M. G. Gilbert (1990) ≡ *Angolluma araysiana* (Lavranos & Bilaidi) Plowes (1994).

Stems ascending, in dense clusters, grey-green, spotted with brown, 3 - 8 cm; **Tu** ± 0.5 - 1 cm, horizontally spreading, sometimes slightly bent upwards; **Inf** with 1 - 2 erect **Fl** at the stem tips; **Bra** 2 - 4, ± 0.5 × 0.25 mm, triangular, acute; **Ped** ± 7 × ≤ 2 mm, terete; **Sep** 3 - 4 × ± 2 mm, sometimes keeled, alternating with colleters; **Cl** outside grey-green, spotted with brown, inside pale red-brown, glossy, 2 - 4 cm ∅, flat, deeply incised, fleshy, tube very short, surrounding the **Cn** basally, annulus low; **Cl** lobes ± 15 × 6 - 9 mm, ovate-triangular, acute, margins slightly revolute, outside minutely bumpy-papillate, inside papillate, densely hairy, **Ha** white, short, bristly; **Cn** glossy black, 6 - 7 mm ∅; **Ci** open and cup-like, lobes forming wide pouches, ± rectangular when seen from above, apically bulging; **Cs** lobes ± 4.5 mm, linear-triangular with 2 shallow depressions, geniculate, basally erect, apical portion ± horizontal or erect, ± 2 mm, linear, obtuse, overlapping, ± strongly surpassing the **Sty** head, with longitudinal grooves where united with the **Ci**; **Poll** bright orange, 0.45 × ± 0.35 mm, D-shaped.

O. baldratii (A. C. White & B. Sloane) Bruyns (Aloe 37(4): 73, 2001). **T:** [lecto − icono]: A. C. White & B. Sloane, Stapelieae, ed. 2, 1: figs. 200-201, 1937. − **D:** Ethiopia, Eritrea, Somalia, Kenya, Tanzania.

≡ *Caralluma baldratii* A. C. White & B. Sloane (1937) ≡ *Pachycymbium baldratii* (A. C. White & B. Sloane) M. G. Gilbert (1990) ≡ *Angolluma baldratii* (A. C. White & B. Sloane) Plowes (1994).

O. baldratii ssp. **baldratii** − **D:** Ethiopia, Eritrea, Kenya, N Tanzania.
  Incl. *Pachycymbium baldratii* ssp. *baldratii*.

Stems mostly solitary, erect, 5 - 10 cm, green spotted with red, sides shallowly grooved; underground runners short; **Tu** ± 1 cm, conically triangular, horizontally spreading; **L** rudiments bent slightly upwards, with stipular rudiments; **Inf** with 1 - 3 sessile **Fl**; **Sep** ± 2.5 × 1 mm, alternating with colleters; **Cl** inside pale mahogany- or cream-coloured, minutely spotted with red, 2.8 cm ⌀, deeply incised; **Cl** lobes ± 12 × 2 mm, narrowly triangular or ovate-lanceolate, long attenuate, horizontally spreading, apically darker, margin revolute, inside densely downy-tomentose, **Ha** white, short, stiff, acute; **Cn** ± 5 mm ⌀; **Ci** black-purple, cup-shaped, lobes 0.5 mm, forming ovate pouches, apically sinuately notched; **Cs** cherry-red, lobes ± 1.5 × 1 mm, apically variously toothed, with short basal appendage where united with the **Ci**, also with 2 dorsal appendages, 1.5 mm, narrow; **Poll** 0.45 × 0.25 mm, ovoid.

O. baldratii ssp. **somalensis** Bruyns (Aloe 37(4): 73, ill., 2001). **T:** Somalia (*Bally* 16017 [ZSS]). − **D:** Somalia.

Differs from ssp. *baldratii*: **Ci** lobes much smaller; **Cs** lobes square, basally slightly emarginate. − [U. Meve]

O. carnosa (Stent) Bruyns (Aloe 37(4): 73, 2001). **T:** RSA, Gauteng (*Pole-Evans* 11020 [PRE]). − **D:** RSA (Northern Prov., Gauteng). **Fig. XXXIV.b**

≡ *Caralluma carnosa* Stent (1916) ≡ *Pachycymbium carnosum* (Stent) L. C. Leach (1978); **incl.** *Caralluma keithii* R. A. Dyer (1935) ≡ *Pachycymbium keithii* (R. A. Dyer) L. C. Leach (1978); **incl.** *Caralluma fosteri* Pillans (1937); **incl.** *Caralluma schweickerdtii* Obermeyer (1937); **incl.** *Pachycymbium lancasteri* Lavranos (1984).

Stems apically slightly tapering, basally branching, grey-green, blotched with reddish-brown, 6 - 20 × 1 - 2 cm; **Tu** ≤ 1.5 cm, conical; **L** rudiments acute; **Fl** buds ± globose, laterally at the bases of the **Cl** with 5 tooth-like protrusions; **Ped** 1 - 3 mm, conical; **Sep** greenish-yellow with pale purple, 2.5 - 7 mm, margin membranous; **Cl** outside greenish-purple, weakly marbled with purple or stippled with reddish, inside cream-coloured and spotted with dark carmine, 7 - 10 × 8 - 15 mm, tube 4.5 × 5.5 mm, annulus cushion-shaped, conspicuously grooved; **Cl** lobes 3 - 5 × ± 5 mm, inside wrinkled, densely tuberculate-papillose, hairy, **Ha** ± 0.6 × 0.15 mm, blunt, margins of the **Cl** lobes delicately hairy, clavate **Ha** 0.5 - 1 mm; **Cn** cream-coloured, purple-spotted; **Ci** basally fused into an urn- or cup-like structure, 3 × 3.5 - 4.5 mm, free part of the lobes 1.4 mm, deeply bidentate, teeth lanceolate, obtuse, ± erect, basally fused with the **Cs**; **Cs** sometimes purple along the margin, lobes 1 mm, strap-like, dorsally bulging, not surpassing the **Anth**; **Poll** ± 0.45 - 0.6 × ± 0.45 mm. − [B. Müller, J. Kiel & F. Albers]

O. caudata (N. E. Brown) Bruyns (Aloe 37(4): 73, 2001). **T:** Malawi (*Cameron* 25 [K]). − **D:** Tanzania, Zambia, Malawi, Moçambique, Zimbabwe, Botswana.

≡ *Caralluma caudata* N. E. Brown (1903) ≡ *Orbeopsis caudata* (N. E. Brown) L. C. Leach (1978).

O. caudata ssp. **caudata** − **D:** S Tanzania, Zambia, Malawi, Moçambique, Zimbabwe.
  Incl. *Caralluma caudata* ssp. *caudata* ≡ *Orbeopsis caudata* ssp. *caudata*; **incl.** *Caralluma longecornuta* Croizat *ex* Gomes e Sousa (1935) (*nom. nud.*); **incl.** *Caralluma caudata* var. *fusca* C. A. Lückhoff (1937); **incl.** *Caralluma praegracilis* Obermeyer (1937).

Stems erect or ascending, forming ± lax mats 10 - 20 cm ⌀, brownish-green or dark olive-green or grey-green, sometimes ± densely blotched with brownish or purple, ± shining, 8 - 10 × 0.65 cm; **Tu** 9 - 13 × 5 - 8 mm, **Sp**-like, ± widely apart, ± horizontally spreading; **Inf** sessile pseudo-umbels with 3 - 6 **Fl** opening ± simultaneously, with strongly foetid odour; **Fl** buds 3.8 cm, 5-angled, acuminate, terminating in a slight spiral; **Bra** ± 2.5 mm, acute; **Ped** 5 - 20 × 2.5 - 3 mm, terete, erect; **Sep** 4 - 9 × 1.5 - 2 mm, projecting from between the bases of the **Cl** lobes; **Cl** outside whitish, finely stippled with red, inside pale or dark yellow, stippled or blotched with purple or crimson (rarely completely dark purple), 5.5 - 9 cm ⌀, deeply incised, central depression purple, ± saucer-shaped, embracing the **Cn**; **Cl** lobes 22 - 40 × 6 - 9 mm, acuminate to caudate, horizontally spreading, margins of the upper ½ revolute, inside velvety, finely papillose, papillae dark purple with fine transversal excrescences, central depression basally weakly setulose, margins of the **Cl** lobes slightly hairy, clavate **Ha** purple to dark violet, long, vibratile; **Cn** ± yellowish-red, 4.5 - 7 mm ⌀; **Ci** apically paler, bowl-shaped, lobes 1.5 - 3 × ± 1.6 mm, bidentate, segments short, divergent, sometimes slightly undulate or ± truncate, basally partly with verrucose excrescences; **Cs** lobes 2 - 2.5 × 0.5 - 0.8 mm, narrowly to broadly ribbon-shaped, apically tapering or obtusely lobed, occasionally joining or laterally overlapping, sometimes revolute and slightly raised above the **Sty** head, at the base fused with the **Ci** by hump-like excrescences; **Poll** 0.5 × ± 0.35 mm, bean- or D-shaped. – [B. Müller & F. Albers]

O. caudata ssp. **rhodesiaca** (L. C. Leach) Bruyns (Aloe 37(4): 74, 2001). **T:** Zimbabwe, Belingwe Distr. (*Leach & Bullock* 13145 [SRGH, BM, BOL, BR, K, LISC, PRE, ZSS]). – **D:** Angola, Namibia, Zimbabwe, Botswana. **Fig. XXXIV.f**

≡ *Caralluma caudata* ssp. *rhodesiaca* L. C. Leach (1973) ≡ *Orbeopsis caudata* ssp. *rhodesiaca* (L. C. Leach) L. C. Leach (1978); **incl.** *Caralluma chibensis* C. A. Lückhoff (1935) ≡ *Caralluma caudata* var. *chibensis* (C. A. Lückhoff) C. A. Lückhoff (1937); **incl.** *Caralluma caudata* var. *milleri* Nel (1937); **incl.** *Caralluma caudata* var. *stevensonii* Obermeyer (1937).

Differs from ssp. *caudata*: **Tu** 1 - 2.2 cm, very acute, strongly compressed, acuminate, tips bent upwards; **Fl** opening simultaneously or with long intervals, often only 2 open at any time, buds conical, attenuate; **Cl** paler, occasionally dark yellow as in ssp. *caudata*, blotches pale brownish-red, purple, violet, ± ovate, of variable size, central depression shallowly bowl-shaped or broadly cup-shaped, ± 4 × 8 - 12 mm, 5-angled, finely pubescent, **Ha** white; **Ci** crimson or reddish, margins paler, apically slightly undulate or truncate, inside with a ± distinct median longitudinal furrow; **Cs** crimson, tips white, plaited at the transition towards the **Ci**. – [B. Müller & F. Albers]

O. chrysostephana (Deflers) Bruyns (Aloe 37(4): 74, 2001). **T:** Yemen (*Deflers* 1071 [P]). – **D:** S Yemen. **I:** Deflers (1896).

≡ *Stapelia chrysostephana* Deflers (1896) ≡ *Caralluma chrysostephana* (Deflers) A. Berger (1910) ≡ *Pachycymbium chrysostephanum* (Deflers) M. G. Gilbert (1990) ≡ *Angolluma chrysostephana* (Deflers) Plowes (1994).

Stems erect, 8 - 10 cm, in clusters, grey- to green-blue, delicately marbled with dark green; **Tu** 1 cm, thickly fleshy, horizontally spreading; **Inf** with 1 - 2 erect scentless **Fl**; **Ped** green, minutely dotted or lineate with red, 15 × 4 mm; **Sep** 2 - 3 mm; **Cl** outside grey-green, lineate with red, inside dark purple, 2 - 2.5 cm ⌀, flat, deeply incised, fleshy, tube 2.5 × 2.5 mm, cup-shaped; **Cl** lobes 1.2 - 1.5 cm, ovate-elliptic, acute, ascending, margins reflexed and meeting on the backside, inside hairy (only the tip glabrous), **Ha** clavate, white, long; **Cn** golden yellow, 3 - 3.5 × ± 1.5 mm ⌀, with copious nectar; **Ci** cup-shaped, lobes forming ± tubular pouches, apically bulging, slightly spreading; **Cs** basally purple, lobes 1 mm, linear, surpassing the **Anth**, apically obtuse, erect or slightly bent outwards; **Poll** 0.45 × 0.25 mm, ovoid; **Fr** green, lineate with red, 8 - 10 × 0.8 cm.

O. ciliata (Thunberg) L. C. Leach (Kirkia 10(1): 291, 1975). **T:** RSA, Western Cape (*Thunberg* 6327 [UPS]). – **D:** RSA (Western Cape, S of 31° S). **Fig. XXXIV.g**

≡ *Stapelia ciliata* Thunberg (1794) ≡ *Podanthes ciliata* (Thunberg) Haworth (1812) ≡ *Tromotriche ciliata* (Thunberg) Sweet (1830) ≡ *Diplocyatha ciliata* (Thunberg) N. E. Brown (1878).

Stems short, decumbent-erect, to ± 6 × ± 1.5 cm ⌀, distinctly 4-angled; **Tu** ≤ 8 mm, acute, strongly divergent; **L** rudiments acute, hard; **Inf** 1-flowered, rarely with an additional bud at the **Ped** base; **Ped** 10 - 12.5 × ± 2 mm; **Sep** 6 - 8 × 3 mm, convex; **Cl** outside pale green, finely stippled with reddish-purple, inside pale greenish-yellow, uniformly dark purple at the base, 7 - 8 cm ⌀, bowl-shaped; **Cl** tube basally flattened, densely hairy, **Ha** purple, short, acute, rigid; annulus ± 8 mm high, solid, ± broadly funnel-shaped, limb thickened, spreading, revolute, sometimes indistinctly obtusely 5-angled; **Cl** lobes 3 - 3.2 × 1.4 - 1.7 cm, divergent, (strongly) revolute, outside glabrous, inside ± densely verrucose-papillose, papillae with purple tips, margins of the **Cl** lobes hairy, **Ha** white, to 3 mm, simple or flattened-clavate, strongly vibratile; **Cn** 1.5 × ± 4 mm ⌀, base faintly 5-angled; **Ci** pale yellow, weakly stippled with purple, deeply bowl-shaped, lobes ± 2 × 2 mm, ± square, tips irregularly obtusely emarginate, sometimes bidentate, divaricate-ascending; **Cs** leather-coloured, stippled with purple, lobes shorter than the **Anth**, ± acute, dorsally ± humped, erect at the base and then geniculate; nectar slit large, shallow, drop-shaped; **Poll** ± 0.8 × 0.5 mm, very broadly bean- or ± D-shaped.

Formerly placed in the separate genus *Diplocyatha*, this species belongs to the *O. variegata*-complex. – [B. Müller, K. Seidler & F. Albers]

O. commutata (A. Berger) Bruyns (Aloe 37(4): 74, 2001). **T:** [lecto – icono]: A. Berger, Stapel. & Klein., fig. 23. – **D:** Yemen, Ethiopia, Uganda, Kenya.

≡ *Caralluma commutata* A. Berger (1910) ≡ *Pachycymbium commutatum* (A. Berger) M. G. Gilbert (1990) ≡ *Angolluma commutata* (A. Berger) Plowes (1994); **incl.** *Angolluma commutata* ssp. *commutata*; **incl.** *Caralluma commutata* ssp. *eu-commutata* Maire (1934) (*nom. illeg.*, Art. 26.1); **incl.** *Angolluma commutata* ssp. *sheilae* Plowes (1994).

Stems erect, coarsely branched, in open tufts, pale green, glossy, with reddish spots and streaks, 10 - 11 cm; **Tu** 5 - 7 mm, conical-subulate, spreading horizontally, apically turned upwards; **Inf** with 1 - 2 erect **Fl**; buds spreading, broad, pointed; **Ped** 3 - 4 × 2 - 2.5 mm; **Sep** 2.5 × 2 - 2.5 mm; **Cl** inside brown, 2.3 cm ⌀, flat, ± divided for ½, fleshy, glabrous; tube 1 mm, closely surrounding the base of the **Cn**; **Cl** lobes 6 - 10 (-18) × 7 - 9 mm, triangular-ovate, horizontally spreading, apically pointed, often slightly reflexed, inside finely pitted-papillate, glabrous; **Cn** dark brown, glossy, 4.5 × 7 mm ⌀; **Ci** basin- or bowl-shaped, lobes forming deep saccate pouches, apically somewhat emarginate in the middle; **Cs** lobes 1.5 mm, spatulate, apically bidentate, narrowing, not touching, dorsally with a deep longitudinal furrow, where united with the **Ci** laterally with obtuse appendages; **Poll** 0.5 × 0.35 mm, D-shaped.

O. cooperi (N. E. Brown) L. C. Leach (Kirkia 10(1): 291, 1975). – **D:** RSA (Northern Cape, Eastern Cape). **Fig. XXXV.b**

≡ *Stapelia cooperi* N. E. Brown (1909) ≡ *Stapeliopsis cooperi* (N. E. Brown) E. Phillips (1932) (*incorrect name*, Art. 11.3) ≡ *Stultitia cooperi* (N. E. Brown) E. Phillips (1933).

Stems squat, in lax clumps, rhizome-like runners often ≥ 2× as long as the aerial stems; stems 2 - 3 (-5) × ± 1 cm ⌀, distinctly 4-angled; **Tu** 1 cm; **Ped** ± 2.5 cm; **Sep** ± 3 mm; **Cl** sand-coloured, finely striped with purple, 3.2 - 3.5 cm ⌀, flat; annulus cream-coloured, 8 mm ⌀, verrucose-papillose, cushion-like, depressed in the centre; **Cl** lobes 11 - 14 × 8 - 9 mm, horizontally spreading, inside transversely rugose, margins revolute, basally hairy, clavate **Ha** purple-violet, ≤ 1 mm, vibratile; **Cn** sand-coloured or brown, ± 4.5 mm ⌀, sessile, overtopping the **Cl**; **Ci** bowl-shaped, lobes 0.9 × ≤ 0.8 mm, ± square, obtusely bidentate, margins plaited; **Cs** lobes 3.5 × 0.9 mm, broadly or narrowly ovate, apically tapering, weakly clavate, with fine parallel longitudinal folds, raised above the **Sty** head, tip revolute, dorsally at the intersection to the **Ci** with numerous humps; nectar slit ± triangular; **Poll** ± 0.6 × ± 0.4 mm, bean-shaped.

Closely allied to *O. tapscottii*. – [B. Müller, K. Seidler & F. Albers]

O. decaisneana (Lemaire) Bruyns (Aloe 37(4): 74, 2001). **T:** [lecto – icono]: Herb. Gén. Amateur, sér. 2, 4: t. 21, 1844. – **D:** Algeria, Morocco, Mauretania, Senegal, Sudan.

≡ *Boucerosia decaisneana* Lemaire (1844) ≡ *Desmidorchis decaisneana* (Lemaire) Kuntze (1891) ≡ *Caralluma decaisneana* (Lemaire) N. E. Brown (1892) ≡ *Stapelia decaisneana* (Lemaire) A. Chevalier (1934) ≡ *Pachycymbium decaisneanum* (Lemaire) M. G. Gilbert (1990) ≡ *Angolluma decaisneana* (Lemaire) Munster *ex* L. E. Newton (1993); **incl.** *Caralluma hesperidum* Maire (1922) ≡ *Caralluma commutata* var. *hesperidum* (Maire) Font Quer (1924) ≡ *Caralluma commutata* ssp. *hesperidum* (Maire) Maire (1934) ≡ *Caralluma decaisneana* ssp. *hesperidum* (Maire) Raynaud (1991) ≡ *Angolluma hesperidum* (Maire) Plowes (1994); **incl.** *Caralluma venenosa* Maire (1931) ≡ *Angolluma venenosa* (Maire) Plowes (1994).

Stems erect, ± cylindrical, slender, apically tapering, whitish green, marbled or spotted with brown, 10 - 40 × 1.5 cm ⌀, 3- to 6-ribbed, sides conspicuously furrowed; **Tu** 0.7 - 1.5 cm, strong, conical-subulate, horizontal to ascending; **Inf** with 1 - 3 ± erect **Fl**; peduncle cushion-like, 1- to 20-flowered, sometimes with 1 mm long filiform **Bra**; **Ped** 1 - 5 mm, ± conical; **Sep** 4 - 5 mm; **Cl** dark purple or reddish-brown, 1.5 - 2.5 cm ⌀, flat, tube 4 (-6) × 8 mm ⌀, campanulate, embracing the **Cn**; **Cl** lobes 7 - 12 × 5 - 8 mm, elongate-ovate, acute, horizontally spreading, inside glabrous to finely papillate, papillae whitish, surface appearing frosted, apically sometimes with a short white **Ha**, margins at times with purple clavate **Ha**; **Cn** flesh-coloured to purple, ± 5-angled; **Ci** narrowly cup-shaped, lobes forming pouches, apically often emarginate; **Cs** lobes triangular to broadly lanceolate, 1 mm broad, ± surpassing the **Sty** head; **Poll** ovoid, 0.5 × ± 0.3 mm, large in relation to **Fl** size.

A widely distributed and variable complex.

O. deflersiana (Lavranos) Bruyns (Aloe 37(4): 74, 2001). **T:** Yemen (*Lavranos* 1788 [K, PRE]). – **D:** S Yemen. **Fig. XXXV.a**

≡ *Caralluma deflersiana* Lavranos (1963) ≡ *Pachycymbium deflersianum* (Lavranos) M. G. Gilbert (1990) ≡ *Angolluma deflersiana* (Lavranos) Plowes (1994).

Stems ascending-erect, densely clustered or in cushions, grey-green, conspicuously brownish spotted, 5 - 7 × 1.2 - 1.5 cm; **Tu** 0.8 - 1.2 cm, strong, horizontally spreading, apically slightly upturned; **Inf** 1-flowered; **Fl** erect; **Bra** 2 mm, triangular, acute; **Ped** grey-green, 10 × 3 mm; **Sep** 5 - 7 × 2 mm, apically slightly bent outwards; **Cl** outside grey-green, densely brownish spotted, inside dark brownish-purple, ± 1.5 - 3 cm ⌀, campanulate, thickly fleshy; **Cl** tube basally rose-coloured, 1.3 × 0.8 - 1 cm ⌀, cylindrical, basally rounded and glabrous, embracing the **Cn**; **Cl** lobes 18 × 9 mm, ovate-lanceolate, weakly ascending, margins slightly revolute, inside densely minutely warty-papillate, densely short-tomentose; **Cn** ± 8 × 6 - 8 mm ⌀; **Ci** rose, margins dark purplish-brown, cup-shaped, lobes forming concave pouches, apically depressed in the middle, towards the margins ascending, rounded, margin entire; **Cs** dark purplish-brown, lobes 3 - 4 × 0.75 mm, linear, apically rounded or emarginate, thickened, touching each other, basally with a transverse groove, bulging where united with the **Ci**, laterally bordered by 2 longitudinal grooves; **Poll** ± 0.75 × 0.45 mm, ovoid.

O. denboefii (Lavranos) Bruyns (Aloe 37(4): 74, 2001). **T** [neo]: Kenya, Rift Valley Prov. (*Foresti* 824 [K]). – **D:** S Kenya, N Tanzania. **I:** Meve (1999).

≡ *Caralluma denboefii* Lavranos (1983) ≡ *Pachycymbium denboefii* (Lavranos) M. G. Gilbert (1990) ≡ *Angolluma denboefii* (Lavranos) Plowes (1994).

Stems unbranched, pistachio-green, spotted with dark or brownish-green, to > 20 × 0.8 - 1.2 (-1.5) cm ⌀; **Tu** ≤ 1.5 cm, ascending, apically turned slightly upwards; **Fl** solitary, occasionally ± nutant; **Ped** ± 12 × 1.7 mm; **Sep** 8 mm, ascending; **Cl** outside yellowish-cream with short brown longitudinal streaks, inside bright golden-yellow, ± 3.5 cm ⌀, campanulate, tube 10 - 12 × 8 mm ⌀, cylindrical, basally rounded, embracing the **Cn**; **Cl** lobes 11 × 7 mm, narrowly triangular, weakly ascending, margins strongly revolute, inside smooth, glabrous; **Cn** bright or golden-yellow, ± 6 mm ⌀; **Ci** ± semiglobose, lobes forming pouches, apically somewhat emarginate and crenate; **Cs** lobes narrowly triangular, apically overlapping; **Poll** ± 0.6 × 0.4 mm, ovoid.

O. distincta (E. A. Bruce) Bruyns (Aloe 37(4): 74, 2001). **T**: Tanzania (*Greenway* 4260 [K]). – **D**: E Tanzania, E Kenya.

≡ *Caralluma distincta* E. A. Bruce (1940) ≡ *Pachycymbium distinctum* (E. A. Bruce) M. G. Gilbert (1990) ≡ *Angolluma distincta* (E. A. Bruce) Plowes (1994).

Stems prostrate, ± cylindrical, basally tapering, (grey-) olive (-green), streaked with brownish, 5 - 8 mm ⌀; **Tu** 3 - 5 × 1 mm, subulate; **Inf** with 1 - 2 erect **Fl**; **Ped** ± 9 × 3 mm; **Sep** 5 - 6 × 2 - 3 mm, spreading; **Cl** outside cream-coloured or whitish with dark brownish longitudinal streaks, inside cream-coloured, 3.5 - 4 cm ⌀, campanulate, tube 1.5 - 1.7 × 1 cm ⌀, embracing the **Cn**, mouth somewhat narrower, 5 mm above the base hairy, **Ha** 1.5 - 1.7 mm; **Cl** lobes intensely carmine to dark flesh-coloured, 1.6 × 1 cm, elongate-ovate, apically shortly attenuate, erect or horizontally spreading, margins strongly revolute, inside wrinkled and with folds, with 5 longitudinal depressions from base to tip, inside glabrous; **Cn** ± 2 × 1.7 mm; **Ci** cup-shaped, lobes forming shallow ovate to reniform or cordate pouches, apically thickened, incised in the middle; **Cs** lobes basally triangular, apically linear and laterally compressed, with conical dorsal appendage, with small bulges where united with the **Ci**; **Poll** ± 0.5 × 0.45 mm, broadly D-shaped.

O. dummeri (N. E. Brown) Bruyns (Aloe 37(4): 74, 2001). **T**: Uganda (*Dummer* s.n. [K]). – **D**: S Uganda, S Kenya, N Tanzania, E Zaïre.

≡ *Stapelia dummeri* N. E. Brown (1917) ≡ *Caralluma dummeri* (N. E. Brown) A. C. White & B. Sloane (1940) ≡ *Pachycymbium dummeri* (N. E. Brown) M. G. Gilbert (1990) ≡ *Angolluma dummeri* (N. E. Brown) Plowes (1994); **incl.** *Caralluma circes* M. G. Gilbert (1978) ≡ *Pachycymbium circes* (M. G. Gilbert) M. G. Gilbert (1990) ≡ *Angolluma circes* (M. G. Gilbert) Plowes (1994) ≡ *Orbea dummeri* ssp. *circes* (M. G. Gilbert) Bruyns (2001); **incl.** *Caralluma dummeri* fa. *colorata* Lodé (1992).

Stems prostrate-ascending, ± cylindrical or indistinctly 4-ribbed, 6 - 9 cm, in small lax mats, pale grey-green with purple spots and stripes; **Tu** 8 - 15 × 3 mm, conical-subulate, ± slender, horizontally spreading; **Inf** with 1 - 4 (-6) **Fl**; peduncle 1.5 mm; **Bra** ± 1.8 mm, lanceolate-triangular, pointed; **Ped** 8 - 12 × ± 1.5 mm ⌀; **Sep** 4 - 5 × 2 mm; **Cl** inside yellow-green to olive-green, 2.5 - 4 cm ⌀, campanulate, deeply incised, tube 4 - 5 × 8 - 9 mm, cup-shaped, embracing the **Cn**; **Cl** lobes 13 - 15 × 4 mm, triangular, long attenuate, horizontally spreading or erect, convex, inside papillate, papillae cylindrical, ≤ 0.5 mm, apically with white 1 - 1.5 mm long bristly **Ha**; **Cn** white or yellowish, 4.5 × 4.3 mm ⌀; **Ci** tubular, lobes ± 1.5 mm, forming deep pouches, rectangular when seen from above, apically with 2 lateral teeth, sometimes with additional small tooth-like appendages; **Cs** lobes 1.7 - 2 × ± 1.3 mm, rectangular, apically 2- to 3-toothed, central tooth elongate, apically overlapping; **Poll** 0.5 × 0.3 mm, ovoid.

O. gemugofana (M. G. Gilbert) Bruyns (Aloe 37(4): 74, 2001). **T**: Ethiopia, Gemu Gofa Region (*Gilbert* 1731 [K, EA, ETH]). – **D**: Ethiopia, Kenya, Uganda. **Fig. XXXV.c**

≡ *Caralluma gemugofana* M. G. Gilbert (1978) ≡ *Pachycymbium gemugofanum* (M. G. Gilbert) M. G. Gilbert (1990) ≡ *Angolluma gemugofana* (M. G. Gilbert) Plowes (1994).

Stems ± erect, rarely decumbent, never rhizomatous, 10 - 20 × 0.5 - 1 cm ⌀; **Tu** ≤ 2.2 cm, narrow, long attenuate, ascending; **Inf** with ≤ 3 erect **Fl** near stem tips, unscented; **Bra** ± 1.5 × 1 mm, lanceolate, acute, ± caducous; buds erect, acute; **Ped** (2-) 3.5 - 5 × ± 2 mm; **Sep** 3 - 5 × 1 - 1.5 mm; **Cl** inside bright yellow, greenish-yellow or brown-yellow (tube and lobes basally sometimes brighter), 2 cm ⌀, campanulate or flat, fleshy, without annulus, tube (1-) 2 - 3.5 × 8 - 8.5 mm ⌀, embracing the **Cn**; **Cl** lobes 8 - 9 × 4 - 5 mm, ovate-triangular, convex, horizontally spreading or slightly reflexed, inside glabrous, smooth; **Cn** ± 2.5 × 5 - 6 mm ⌀; **Ci** dull rose, brown or shiny dark red, flatly bowl- or disc-shaped, obtusely 5-angled, strongly reduced, apical margin overtopping, thin, indented, somewhat depressed below the guiding rails, glandular-tomentose; **Cs** dark red, lobes ± 1.2 mm, elliptic, rarely narrowly triangular, convex, apically irregularly toothed, surpassing the **Anth**, apically sometimes overlapping; **Poll** yellow, 0.55 × 0.3 mm, ovoid to D-shaped; **Fr** grey, spotted, 11 - 16 × 1 - 1.5 cm.

The characters of the corona are intermediate between *O. sprengeri* and *O. schweinfurthii*.

O. gerstneri (Letty) Bruyns (Aloe 37(4): 74, 2001). **T**: RSA, KwaZulu-Natal (*Gerstner* 740 [PRE]). – **D**: RSA.

≡ *Caralluma gerstneri* Letty (1936) ≡ *Orbeopsis gerstneri* (Letty) L. C. Leach (1978).

O. gerstneri ssp. **elongata** (R. A. Dyer) Bruyns (Aloe 37(4): 74, 2001). **T**: RSA, Mpumalanga (*Phillips* s.n. [PRE 30367]). – **D**: RSA (Mpumalanga, KwaZulu-Natal). **Fig. XXXV.d**

≡ *Caralluma gerstneri* ssp. *elongata* R. A. Dyer (1969) ≡ *Orbeopsis gerstneri* ssp. *elongata* (R. A. Dyer) L. C. Leach (1978).

Rhizom-like runners absent; stems to 10 cm, narrowly rectangular; **Tu** ≤ 0.5 cm, stipular rudiments none; **Fl** opening simultaneously, with strongly foetid odour; **Sep** 4 mm; **Cl** red-brown, blotched with yellowish, campanulate; **Cl** tube deeper and narrower than in ssp. *gerstneri*, 4.6 - 4.8 mm ⌀, mouth bulging annulus-like; **Cl** lobes 18 × 7 mm, inside with excrescences arranged in transverse and longitudinal lines; **Cn** 7 mm ⌀; **Ci** lobes 6.5 × 1.5 (apically 0.5) mm, erect, at the base broadly rectangular, lateral margins fleshily bulging, upper ½ narrow, ribbon-like, bidentate, bent outwards above the limb of the **Cl** tube; **Cs** lobes 5 × 0.75 mm, very thickly fleshy, complex, with 3 filiform appendages, apical appendage longest, basal appendage shortest, first ± vertically raised above the **Sty** head, then revolute, laterally with some coarsely hump-shaped excrescences; **Anth** yellow; **Poll** broadly D-shaped.

This taxon should possibly better be treated as a separate species.

O. gerstneri ssp. **gerstneri** – **D**: RSA (KwaZulu-Natal).

Incl. *Caralluma gerstneri* ssp. *gerstneri* ≡ *Orbeopsis gerstneri* ssp. *gerstneri*.

Stems slender, in dense clumps, with stout rhizome-like runners, bluish-green, partially stippled with purple, to 7 × 0.5 - 1 cm, distinctly 4-angled; **Tu** ≤ 1.3 cm, conical, divaricate; **Inf** 2- to 6-flowered, **Fl** opening in succession, sometimes 2 - 3 **Fl** open at the same time, with faint foetid odour; peduncle 2 mm, stout; **Fl** buds conical, obtuse, base of the **Cl** lobes with 5 small teeth; **Ped** 10 × 2 mm; **Sep** apically brownish, 8 × 1.5 mm; **Cl** outside pale cobalt-green, inside dark purple-red, 3.5 cm ⌀; **Cl** tube cup-shaped, 5-angled, embracing the **Cn**, inside densely velvety-papillose, papillae fine, below the **Cn** white; **Cl** lobes parchment-coloured apically, densely stippled and blotched with purple, 15 × 8 mm, narrowly triangular, horizontally spreading, rugose-velvety, crystal-like, margins hairy in the lower ⅓, clavate **Ha** purple, 1.5 mm, vibratile, occasionally terminating in a **Bri**; **Cn** darker red, basin-shaped or shallowly cup-shaped, 4 × 9 mm, stalk slender; **Ci** lobes 2 × 2 mm, 2-keeled, tips 3-dentate, middle tooth pale yellow, elongate-ovate, ± acute, lateral teeth narrower; **Cs** lobes complex, apical appendage ± cylindrical, filiform, tip clavate, overtopping the **Anth**, apically revolute, basal appendage subulate, lower ⅔ incumbent on the lobe, then spreading, rugose, basally crested; **Poll** ± 0.5 × 0.4 mm broad, D-shaped.

O. halipedicola L. C. Leach (Excelsa Tax. Ser. 1: 40-43, ills., 1978). **T**: Moçambique (*Leach* 12396 [SRGH]). – **D**: Moçambique.

O. halipedicola ssp. **halipedicola** – **D**: Moçambique.

Stems slender, ± prostrate, sometimes creeping, partially with subterranean runners; stems to 15 × ± 0.5 cm ⌀; **Ri** 4, ± widely spaced, sides furrowed; **Tu** ± 1 cm, narrow, attenuate, usually ± bent upwards or erect; **Inf** few-flowered; buds broadly ovoid-pentagonal, apically acute, basally truncate; **Bra** 1.5 - 2 × 1 - 1.5 mm, ovate, acute, concave; **Ped** 2 - 3 × ± 0.2 cm ⌀, horizontally spreading; **Sep** 5 - 6 × 2.5 mm, alternating with **Gl**, these ≤ 2 mm, subulate or filiform, fleshy; **Cl** purple-brown, striped or blotched with shades of yellow or white, 3.7 - 4 cm ⌀, flat; annulus 1.5 - 2.5 × 8 - 9 mm ⌀, ± rounded, depressed in the centre, sunken in the middle, limb spreading, obtuse, sometimes slightly revolute, ± convex, with concentric rings; **Cl** lobes 14 - 16 × 9 - 10 mm, spreading, outside glabrous, along the middle partly ± conspicuously furrowed, inside glabrous, very shortly pubescent, margins slightly revolute, hairy, clavate **Ha** purple, 2 mm; **Cn** yellowish-purple, 6 - 6.5 mm ⌀; **Ci** inversely shallowly bowl-shaped, lobes ± 1.5 × 2 mm, ± transversely rectangular, apically truncate, irregularly dentate, central margin distinctly raising, narrowly triangular, apically with a fleshy appendage reaching beyond the margin of the lobes, divaricate, slightly revolute; **Cs** 3.5 - 4 × ± 1 mm, narrowly ovoid, erect, apical appendage ± filiform, erect, tip obtuse, slightly revolute, sometimes faintly clavate or indistinctly rugose or weakly irregularly dentate, basally with 2 fleshy ± acute teeth, dorsally humped, much overtopping the **Anth**, occasionally meeting each other above the **Sty** head; nectar slit ± triangular, deeply sunken; **Poll** ± 0.6 × 0.6 mm, ± D-shaped, caudicles apically much broadened.

Closely allied to *O. umbracula*. – [B. Müller, K. Seidler & F. Albers]

O. halipedicola ssp. **septentrionalis** L. C. Leach (Excelsa Tax. Ser. 1: 43-44, 1978). **T**: Moçambique (*Tinley* s.n. [SRGH]). – **D**: Moçambique.

Differs from ssp. *halipedicola*: Stems erect, more compact, **Tu** shorter, straight; annulus lower; **Ci** lobes ± flat, only faintly ribbed. – [B. Müller, K. Seidler & F. Albers]

O. huernioides (P. R. O. Bally) Bruyns (Aloe 37(4): 74, 2001). **T**: Somalia, Northern Prov. (*Bally* 11166 [G [alc], K, ZSS]). – **D**: NE Somalia.
≡ *Caralluma huernioides* P. R. O. Bally (1965) ≡ *Pachycymbium huernioides* (P. R. O. Bally) M. G. Gilbert (1990) ≡ *Angolluma huernioides* (P. R. O. Bally) Plowes (1994).

Stems decumbent or ascending, pale green, sometimes spotted dark green or brownish-violet, ± 12 × 1.4 cm, thickly fleshy; **Tu** ≤ 10 × 4 mm, spaced 0.5 - 2.5 cm from each other, fleshy, narrowly subulate, apically spreading or ascending; **L** rudiments Sc-like, early caducous; **Inf** with 3 - 7 nutant **Fl**; **Bra** 2

- 2.5 mm, delicate, subulate; **Ped** pale green, spotted with brown, 1.8 × 2.5 mm, bent outwards; **Sep** green, ± 4 × 7 mm, apically filiform, sometimes bent outwards; **Cl** outside pale green with few green or brown longitudinal stripes, inside yellow, spotted reddish-brown, 3.5 cm ⌀, campanulate, tube 1.5 × 1.6 cm ⌀, embracing the **Cn**, dots at the mouth of the tube confluent and forming a stellate pattern; **Cl** lobes 8.5 - 9 × 8 mm, broadly triangular, ascending-erect, alternating with 1 mm long intermediate lobes, acute, reflexed, inside warty-papillate, ± densely hairy, **Ha** reddish-brown, 0.1 - 0.15 mm, conical, apically thickened; **Cn** glossy, ± 5.5 × 6.5 mm, ± 5-angular; **Ci** dark carmine, lobes pouch-like, 2 × 2 mm, inversely triangular, obtuse, thick, entire; **Cs** lobes 3.5 - 4 × 1.3 - 1.8 mm, ovate-lanceolate, fleshy, basally erect, apically ± meeting; **Poll** 0.5 × ± 0.4 mm, ellipsoid.

O. huillensis (Hiern) Bruyns (Aloe 37(4): 74, 2001). **T:** Angola (*Welwitsch* 4266 [BM, LISU]). – **D:** Angola, Namibia, Zambia, Zimbabwe, Botswana.

≡ *Caralluma huillensis* Hiern (1898) ≡ *Orbeopsis huillensis* (Hiern) L. C. Leach (1978); **incl.** *Caralluma gossweileri* S. Moore (1912) ≡ *Orbeopsis gossweileri* (S. Moore) L. C. Leach (1978).

O. huillensis ssp. **flava** Bruyns (Aloe 37(4): 75, ill. (p. 73), 2001). **T:** Namibia (*Bruyns* 5522 [BOL]). – **D:** Namibia.

Differs from ssp. *huillensis*: **Br** more slender; **Fl** yellow. – [U. Meve]

O. huillensis ssp. **huillensis** – **D:** Angola, Zambia, Zimbabwe, Botswana.

Stems stumpy, ± prostrate, bright green, faintly mottled with purple, to 10 × 1.5 cm, sides distinctly furrowed; **Tu** ≤ 2 × 0.75 cm, conical-triangular, acute, stout, spreading; **L** rudiments ± 5 mm, sometimes with a single stipular rudiment only; **Inf** pseudo-umbellate, occasionally 2 together, with 6 - 15 **Fl** with revolting foetid odour; peduncle short, stout; **Bra** numerous, 3 - 8 × 2 mm, lanceolate, acuminate, tips narrowed, base with 1 - 2 lateral teeth; **Ped** 15 - 45 × 3.5 mm; **Sep** 5.5 - 9 × ± 3 mm, alternating with minute **Gl**; **Cl** maroon or dark brown, ± 10 cm ⌀, deeply bowl-shaped, deeply incised; **Cl** lobes alternating with short obtuse appendages; **Cl** tube 7 × 8 - 12 mm, broadly tubular or campanulate, mouth bulging annulus-like; **Cl** lobes to 4.5 × 1.25 cm, lanceolate, divaricate, margins slightly revolute, outside glabrous, inside sometimes very finely pubescent, at the base with coarse transverse foldings (sometimes ± down to the base of the tube), towards the tip coarsely granular and transversely folded; **Cn** darker than the **Cl**, ± 8 mm ⌀; **Ci** shallowly bowl-shaped, lobes ± 2.5 × 1.5 mm, tips 4-dentate, 3-keeled, convex, at the base ± concave, ascending-divergent; **Cs** lobes ± 3.5 × ± 1 mm, ± cylindrical, attenuate, tips obtuse, base sometimes weakly humped, geniculate, first very shortly erect, then bent inwards, meeting above the **Sty** head, apically ± revolute; **Poll** 0.6 × 0.5 mm, D-shaped.

Closely related to *O. valida*. – [B. Müller & F. Albers]

O. knobelii (E. Phillips) Bruyns (Aloe 37(4): 75, 2001). **T:** Botswana (*Knobel* s.n. [PRE 8308]). – **D:** Botswana, RSA (Northern Cape, North-West Prov., Free State).

≡ *Stapelia knobelii* E. Phillips (1930) ≡ *Caralluma knobelii* (E. Phillips) E. Phillips (1935) ≡ *Orbeopsis knobelii* (E. Phillips) L. C. Leach (1978); **incl.** *Caralluma kalaharica* Nel (1937); **incl.** *Caralluma langii* A. C. White & B. Sloane (1937) ≡ *Caralluma knobelii* var. *langii* (A. C. White & B. Sloane) A. C. White & B. Sloane (1937).

Stems erect or shortly decumbent at the base, with long rhizome-like runners, green or grey-green, blotched with purple, *either* to 10 × ± 0.8 cm and **Tu** ≤ 1 cm *or* ± 6 cm, stumpy and **Tu** ≤ 6 mm; **Tu** stout, (shortly) conical, acute, ± horizontally spreading; **L** rudiments whitish, acute; **Inf** with 6 - 11 **Fl**, on young stems near the apex, **Fl** opening ± simultaneously; **Bra** 1 - 2 mm, lanceolate; **Ped** 0.5 - 3.5 cm; **Sep** ≤ 5 mm; **Cl** outside pale green, inside white, 2.5 - 3.5 cm ⌀, flat, central depression 1 cm ⌀, 5-angled; **Cl** lobes in the upper ⅓ yellowish-green, blotched with black-purple, or pale yellow, irregularly blotched with brown-purple, to 1.5 × ≤ 1 cm, horizontally spreading, margins revolute, tips revolute in older **Fl**, inside rugose, ± granulate, at the base with ± concentric foldings around the **Cn**, margins weakly hairy, clavate **Ha** brown-purple, long, vibratile; **Cn** 8 mm ⌀; **Ci** black-purple, margins flesh- to rose-coloured, pentagonally basin- or bowl-shaped, lobes 3 × ± 2.5 mm, ± square, tips ± truncate, obtusely denticulate, in the middle with a broad strip of longitudinal furrows; **Cs** lobes 4 mm, apical appendage ± 2.5 mm, filiform, apically humped, connivent-erect, sometimes with an additional conical appendage, incumbent on the **Sty** head, basal appendage 1.5 mm, ± horizontally spreading, sometimes bidentate, ± square or subulate; **Poll** 0.3 × 0.4 mm, broadly D-shaped, densely grouped.

Closely related to *O. albocastanea* and to the *O. lutea*-complex. – [B. Müller & F. Albers]

O. laikipiensis (M. G. Gilbert) Bruyns (Aloe 37(4): 75, 2001). **T:** Kenya (*Bally* 12376 [K]). – **D:** Kenya. **Fig. XXXIV.c**

≡ *Pachycymbium laikipiense* M. G. Gilbert (1990) ≡ *Angolluma laikipiensis* (M. G. Gilbert) Plowes (1994).

Stems ascending, to 6 cm, in small clumps or clusters, with short rhizomatous runners; **Tu** ≤ 9 mm, broad, horizontally spreading; **L** rudiments ± 3 mm, sometimes with stipular rudiments; **Inf** to 3-flowered, sometimes with rudimentary peduncle; **Ped** 3 - 3.5 mm; **Sep** ± 2.5 × 1.5 mm; **Cl** inside pale wine-red, rarely pale yellow, ± 2.5 cm ⌀, incised ± to the base; **Cl** lobes 11 - 12 × 3 - 3.5 mm, linear-lancoleate, ascending, margins and tip revolute, inside velvety; **Cn** bright yellow, ± 1.5 × 4.5 mm; **Ci** cup-shaped, lobes pouch-like; **Cs** lobes 1 × ± 0.5 mm, apically slightly toothed, meeting, basally with short erect appendages where united with the **Ci**, apically acute or obtuse and then shallowly toothed; **Poll** 0.45 × ± 0.25 mm, ovoid.

Very similar to *O. wissmannii* from the Yemen. Here also belongs *O. baldratii*.

O. laticorona (M. G. Gilbert) Bruyns (Aloe 37(4): 75, 2001). **T:** Ethiopia, Shewa Region (*Gilbert 1383* [K, ETH]). – **D:** Ethiopia, E Sudan, NW Kenya. **Fig. XXXV.e**

≡ *Caralluma sprengeri* ssp. *laticorona* M. G. Gilbert (1978) ≡ *Pachycymbium laticoronum* (M. G. Gilbert) M. G. Gilbert (1990) ≡ *Angolluma laticorona* (M. G. Gilbert) Plowes (1994).

Stems ascending, to 10 × 0.8 cm ⌀, never rhizomatous; **Tu** ± 5 mm, horizontally spreading; **L** rudiments slightly ascending, with stipular rudiments; **Inf** to 6-flowered, erect; buds horizontal to nutant, apically mostly blunt; **Ped** ± 3 × 2 mm; **Sep** ± 5 × 2 mm; **Cl** inside pale to medium brown or orange, ± 3 cm ⌀, flat, deeply incised, tube very short, basally embracing the **Cn**, annulus rose, narrow, covered with glossy mucilaginous nectar; **Cl** lobes 11 - 16 × 7 - 9 mm, ovate, horizontally spreading or more strongly bent outwards, tip and margins slightly revolute, inside warty-papillate, papillae conical-cylindrical; **Cn** dark brown, 6.5 - 8 mm ⌀, flatly bowl-shaped, covered with glossy mucilaginous nectar during anthesis; **Ci** bulging ring-like, ± 1.4 mm broad; **Cs** lobes ± 1.8 mm, ± rectangular, overtopping the **Anth**, very smooth, basally sometimes with short lateral appendages, apically obtuse or toothed, apically partly overlapping; **Poll** greenish or (orange-) yellow, ± 0.4 × 0.3 mm, ovoid.

Closely related to *O. sprengeri*. The corolla is similar esp. to *O. sprengeri* ssp. *foetida*.

O. longidens (N. E. Brown) L. C. Leach (Kirkia 10 (1): 290, 1975). **T:** Moçambique (*Tillett s.n.* [K]). – **D:** S Moçambique, neighbouring RSA.

≡ *Stapelia longidens* N. E. Brown (1895).

Stems 2 - 8 (-15) cm, sometimes ± decumbent-erect, tufted; **Tu** ± 1.5 cm, tapering; **Inf** 1- to 3-flowered, occasionally ± prostrate; **Ped** ± 2 cm; **Sep** ± 8 × 3 mm; **Cl** cream-coloured or yellowish, 3 - 5 cm ⌀, ± shallow, deeply divided into separate lobes; **Cl** tube densely stippled with maroon, broadly campanulate, 5 × 10 mm ⌀, basally with an annulus, this 2 × ± 6 mm ⌀, shallowly dome-shaped, sometimes indistinctly clump-like; **Cl** lobes inside glabrous, often finely pubescent at the tip, margins hairy, clavate **Ha** ≤ 2.5 mm, very slender; **Cn** ± 3 × 4.5 - 5 mm; **Ci** lobes ± rectangular, fleshy, bent dome-like; **Cs** lobes with tips indistinctly erect, ± subulate, dorsally sometimes slightly humped, irregularly obtusely verrucose; nectar slit ± narrowly triangular, beneath the **Anth** wings, partially covered by the **Ci**; **Poll** 0.6 - 0.75 × ± 0.5 mm, ± broadly D-shaped.

Closely allied to *O. speciosa*. – [B. Müller, K. Seidler & F. Albers]

O. lugardii (N. E. Brown) Bruyns (Aloe 37(4): 75, 2001). **T:** Botswana (*Lugard 74* [K]). – **D:** Namibia, Botswana, Zimbabwe, RSA (Northern Cape). **Fig. XXXIV.d**

≡ *Caralluma lugardii* N. E. Brown (1904) ≡ *Pachycymbium lugardii* (N. E. Brown) M. G. Gilbert (1990) ≡ *Angolluma lugardii* (N. E. Brown) Plowes (1994); **incl.** *Caralluma longicuspis* N. E. Brown (1909).

Stems erect or decumbent, 5 - 7.5 (-15) × 0.6 - 1.9 cm ⌀, densely clustered, grey-green or -blue, spotted with brown; **Tu** 3 - 5 mm, strong, conical-subulate, upper face flattened, horizontally spreading; **Inf** with 3 - 10 erect **Fl**; **Bra** 1 - 2 mm, broadly ovate, acute; **Ped** 3 - 6 × 1.6 mm; **Sep** (2-) 4 - 5 × 1 - 1.4 mm; **Cl** outside grey-green, inside wine-red or dark brown (lobes sometimes olive, or reciprocal), 3.8 - 4.5 cm ⌀, deeply incised, tube 4 - 6 × ± 8 mm ⌀, campanulate, basally embracing the **Cn**; **Cl** lobes 20 - 25 × 3 - 5 mm, linear, acute, erect, ascending or horizontally spreading, margins strongly revolute, inside warty-papillate, minutely tomentose; **Cn** ± 5 mm ⌀; **Ci** black-purple, basal parts of **Ci** and **Cs** united cup-shaped, ± 3 mm, free lobes 1.5 mm, bifid to the base, segments triangular, apically elongate, obtuse, diverging at 90°; **Cs** reddish-yellow or dull orange, apically brown, lobes 4.5 - 5.2 mm, basally erect, triangular-subulate with 2 lateral wings each, ± closely touching the guide rails, central part lanceolate, apically filiform and much surpassing the **Sty** head, touching each other, sometimes intertwined, or apically slightly bent outwards, basally with triangular appendages pointing outwards where united with and obscured by the **Ci**; **Poll** 0.75 × 0.3 mm, ovoid-ellipsoid.

This aberrant species is probably better placed in *Tridentea* according to Gilbert (1990).

O. luntii (N. E. Brown) Bruyns (Aloe 37(4): 75, 2001). **T:** Yemen (*Lunt 209* [K]). – **D:** S Yemen. **I:** Plowes (1994b).

≡ *Caralluma luntii* N. E. Brown (1894) ≡ *Pachycymbium luntii* (N. E. Brown) M. G. Gilbert (1990) ≡ *Angolluma luntii* (N. E. Brown) Plowes (1994).

Stems ascending, to 15 - 20 × 1.2 - 1.8 cm, pale green, marbled with purple, thickly fleshy; **Tu** 1 - 1.1 cm, horizontally spreading, apically **Sp**-like; **Inf** with 1 - 3 **Fl**; **Ped** 1 - 2.5 × 0.6 mm; **Sep** 2.5 mm, recurved; **Cl** inside yellowish-green, ± 2.5 cm ⌀, deeply incised; **Cl** lobes ± 20 × 1.5 mm, linear, slender, ± erect, margins and towards the tip purplish, tips bent inwards and sometimes meeting each other, margins revolute, inside very minutely tomentose; **Cn** purple, ± 3.8 mm ⌀; **Ci** cylindrical, lobes forming small pouches; **Cs** lobes ± 0.7 mm, oblong-linear, acute, dentate, slightly overtopping the **Anth**; **Poll** 0.4 × 0.25 mm, ovoid-triangular.

O. lutea (N. E. Brown) Bruyns (Aloe 37(4): 75, 2001). **T** [lecto]: RSA, Northern Cape (*Tuck* s.n. in *MacOwan* 2240 [K]). – **D:** Namibia, Angola, Moçambique, Botswana, Zimbabwe, RSA.
≡ *Caralluma lutea* N. E. Brown (1891) ≡ *Orbeopsis lutea* (N. E. Brown) L. C. Leach (1978).

O. lutea ssp. **lutea** – **D:** Namibia, Moçambique, Botswana, Zimbabwe, RSA (Northern Cape, North-West Prov., Northern Prov., Gauteng, Mpumalanga, Free State). **Fig. XXXV.f**
Incl. *Caralluma lutea* ssp. *lutea* ≡ *Orbeopsis lutea* ssp. *lutea*; **incl.** *Caralluma lateritia* N. E. Brown (1903) ≡ *Caralluma lutea* var. *lateritia* (N. E. Brown) Nel (1937); **incl.** *Caralluma vansonii* Bremekamp & Obermeyer (1935) ≡ *Caralluma lutea* var. *vansonii* (Bremekamp & Obermeyer) C. A. Lückhoff (1952); **incl.** *Caralluma lateritia* var. *stevensonii* A. C. White & B. Sloane (1937).

Stems erect or decumbent, (pale) green, with dull purple blotches, *either* 8 - 20 × 1 - 2 cm, slender, distinctly 4-angled, *or* 4 - 10 × 1 - 2 cm, stumpy, ± 4-angled; **Tu** 1 - 1.5 cm, conical, partly acuminate, occasionally tips slightly curved upwards, laterally compressed, ± horizontally spreading; **L** rudiments acute, hardened, with or without stipular rudiments; **Inf** with 3 - 26 **Fl**; peduncle to 20 × 8 mm; **Bra** 5 - 7 mm, filiform; **Ped** 1 - 4 × ≤ 0.3 cm; **Sep** 4 - 10 × 2 mm, sometimes projecting between the bases of the **Cl** lobes; **Cl** yellow, sometimes mustard-coloured, brownish, greenish, rarely orange or ± black, red or maroon, sometimes in shaded hues, 3 - 13 cm ⌀; **Cl** tube 7 - 10 mm, shallowly bowl-shaped, embracing the **Cn**; **Cl** lobes ≤ 5 × ± 1 cm, divaricate, margins revolute, outside glabrous, inside transversely finely verrucose-papillose, rugose or furrowed, margins ± densely hairy, clavate **Ha** purple, ≤ 3 mm, ± strongly vibratile; **Cn** yellow to pale brown, often streaked, to 1.5 cm ⌀, stalk slender or broad; **Ci** sometimes red in the middle, bowl-shaped or campanulate, apically revolute, lobes to 5 × 4 mm, tips finely crenulate or obtusely dentate, with a ± broad strip of transverse foldings in the middle; **Cs** lobes to 7 × 0.3 mm, complex, laterally strongly compressed, apical appendage filiform, extending far above the **Anth**, sometimes meeting, ascending, partially with a small subulate projection, incumbent on the **Sty** head, basal appendage subulate, ⅓ as long, appendage falcate, bent outwards, brown-flowered variant more robust; **Sty** head yellowish-white; **Poll** 0.4 × 0.45 mm, broadly D-shaped; **Fr** ± 9 cm. – [B. Müller & F. Albers]

O. lutea ssp. **vaga** (N. E. Brown) Bruyns (Aloe 37(4): 75, 2001). **T**: Namibia (*Schinz* 2047 [Z, K]). – **D:** Namibia, Angola, RSA (Northern Cape).
≡ *Stapelia vaga* N. E. Brown (1895) ≡ *Caralluma vaga* (N. E. Brown) A. C. White & B. Sloane (1937) ≡ *Caralluma lutea* ssp. *vaga* (N. E. Brown) L. E. Leach (1970) ≡ *Orbeopsis lutea* ssp. *vaga* (N. E. Brown) L. C. Leach (1978); **incl.** *Caralluma nebrownii* A. Berger (1906); **incl.** *Caralluma brownii* Dinter & A. Berger (1909); **incl.** *Caralluma pseudonebrownii* Dinter (1914) ≡ *Caralluma nebrownii* var. *pseudonebrownii* (Dinter) A. C. White & B. Sloane (1937); **incl.** *Caralluma hahnii* Nel (1937); **incl.** *Caralluma nebrownii* var. *discolor* Nel (1937).

Differs from ssp. *lutea*: **Tu** 1.5 - 2 (-3) × 3 cm, on older stems shorter; **Bra** 2 - 3 mm; **Ped** 3 - 9 × ± 0.5 cm; **Cl** ± dark flesh-coloured, wrinkles sometimes stippled with yellow; **Cl** tube broadly campanulate, ± 8 × ≤ 15 mm; **Cl** lobes much broader, upper ⅓ furrowed, margins less revolute, less hairy, clavate **Ha** sometimes black-red, to 7 mm; **Cn** brown or reddish, more robust; **Ci** sometimes reddish or yellowish in the middle, bowl-shaped, lobes 4 (apically 5) mm broad, margins ± furrowed; **Cs** lobes complex, with 3 filiform appendages, 4 mm, falcate, revolute, dorsal tooth present, 1.5 - 2 mm, erect, acute, forming a connection to the **Ci**; **Anth** yellow. – [B. Müller & F. Albers]

O. macloughlinii (I. Verdoorn) L. C. Leach (Kirkia 10(1): 291, 1975). **T**: RSA, Eastern Cape (*McLoughlin* s.n. [PRE 26384]). – **D:** RSA (Eastern Cape).
≡ *Stapelia macloughlinii* I. Verdoorn (1941).

Stems to 10 × 0.8 cm ⌀, broadly 4-angled; **Tu** ≤ 2 cm, falcately bent towards the stem axis; **Inf** 1- to 3-flowered; **Ped** ± 2 cm; **Sep** 6 mm; **Cl** dark red, irregularly finely white-mottled, ± 5.5 cm ⌀, inversely shallowly bowl-shaped; annulus ± 1 cm ⌀, fairly inconspicuous; **Cl** lobes 1.7 × 1.1 cm, spreading, revolute, glabrous on both sides, margins densely hairy, clavate **Ha** purple, ≤ 1.2 mm, vibratile; **Cn** reddish-yellow, 5 mm ⌀, overtopping the **Cl**; **Ci** inversely bowl-shaped, lobes 1.9 × 1.8 mm, ± square, in the middle crenulate, basally continuing in a longitudinal furrow, laterally bulging; **Cs** dorsally with an ovate red spot, 2.5 × 1 mm, broadly ovate, overtopping the **Anth**, apically not joined, connected to the **Ci** by a stipe-like basal section; **Anth** ± shortly bifid or apically emarginate, hardly touching each other; nectar slit ± triangular; **Poll** ± 0.7 × 0.6 mm, bean-shaped.

Closely allied to *O. speciosa*. – [B. Müller, K. Seidler & F. Albers]

O. maculata (N. E. Brown) L. C. Leach (Excelsa

Tax. Ser. 1: 49, 1978). **T:** Botswana (*Lugard* 297 [K]). – **D:** Botswana, Zimbabwe, N RSA, Namibia.
≡ *Caralluma maculata* N. E. Brown (1904).

O. maculata ssp. **kaokoensis** Bruyns (Aloe 37(4): 75, ill. (p. 73), 2001). **T:** Namibia (*Bruyns* 4083 [BOL]). – **D:** Namibia (Kaokoveld).

Br 4 - 10 × 0.8 - 1.5 cm; **Tu** 0.8 - 2 cm, broadly triangular; **Cl** upper face pale cream-yellow with red-brown spots, 5 - 6.5 cm ⌀, margins of the lobes with white clavate **Ha**; **Cn** reddish to yellow, 6 - 7 mm ⌀, pentagonal, 1.5 - 2 mm stipitate; **Cs** lobes short, extending to below the guide rails.

Like the typical subspecies, but differing by stronger tubercles and the short interstaminal corona that expands into the areas below the guide rails. – [U. Meve]

O. maculata ssp. **maculata** – **D:** S and SW Botswana, SW Zimbabwe, N RSA (Northern Prov.).

Incl. *Caralluma grandidens* I. Verdoorn (1933) ≡ *Stapelia grandidens* (I. Verdoorn) P. V. Heath (1992).

Stems squat, to 7 cm, distinctly 4-angled, tufted, rarely forming large mats, rhizome-like runners long, fleshy; **Tu** ≤ 1.5 cm, tapering into a conical-acute point or flattened-triangular, acute, widely divaricate; peduncle short, fleshy; buds ± 5-angled, elongate, narrowed towards the tip, basally truncate, sides concave; **Ped** ± 5 × ± 0.2 cm; **Sep** ± 6 × ± 1.5 mm; **Cl** greenish-yellow, ± densely stippled with reddish-brown, ± 6 cm ⌀, flat, deeply divided into lobes; annulus very small, thickened, abruptly raised; **Cl** lobes to 25 × 9 mm, elongate, ± acute, occasionally with a small mucro, revolute, inside slightly rugose, margins revolute, hairy, clavate **Ha** whitish, ± 3 mm, apically elongate, spreading, vibratile; **Cn** 5.5 - 6.6 mm ⌀, ± round, dome-shaped, ± fleshy, ± sunken, shortly and broadly stipitate, outer margin resting on the annulus and only raised above it for ± 2 mm; **Ci** absent; **Cs** lobes ± 3 × (in the middle) ± 1.5 mm, apically abruptly narrowed, often irregularly formed, dorsally slightly verrucose-rugose, slightly longer than the **Anth**; nectar slit ± elliptic; **Poll** ± 0.5 × 0.5 mm, ± D-shaped. – [B. Müller, K. Seidler & F. Albers]

O. maculata ssp. **rangeana** (Dinter & A. Berger) Bruyns (Aloe 37(4): 76, 2001). **T:** Namibia (*Dinter* 1226 [not located]). – **D:** Namibia. Fig. XXXV.g
≡ *Caralluma rangeana* Dinter & A. Berger (1914) ≡ *Orbea rangeana* (Dinter & A. Berger) L. C. Leach (1978) ≡ *Stapelia rangeana* (Dinter & A. Berger) P. V. Heath (1992); **incl.** *Caralluma rangei* Dinter & A. Berger (1914); **incl.** *Piaranthus streyianus* Nel (1949); **incl.** *Caralluma maculata* var. *brevidens* H. Huber (1961).

Br squat, ± 5 × ± 1 cm; **Tu** 4 - 7 mm, triangular-compressed, acuminate; **Cl** greenish-yellow, with brownish-red cross bands and spots, ± 5.5 cm ⌀, clavate **Ha** purple, ± 2 mm; **Cn** ± 2 × 7 mm, reddish; **Cs** lobes 2 - 2.5 × ± 3 mm; **Anth** apically shortly bifid or emarginate, **Nec** cavity triangular to drop-shaped, inserted high up, ± hidden by the **Anth**.

Easily distinguished from the other 2 subspecies on account of the cross-banded rather than spotted or blotched corolla. – [U. Meve]

O. melanantha (Schlechter) Bruyns (Aloe 37(4): 76, 2001). **T:** RSA, Transvaal (*Schlechter* 4694 [not located]). – **D:** Moçambique, Zimbabwe, RSA (Northern Prov., Mpumalanga).

≡ *Stapelia melanantha* Schlechter (1905) ≡ *Caralluma melanantha* (Schlechter) N. E. Brown (1909) ≡ *Orbeopsis melanantha* (Schlechter) L. C. Leach (1978); **incl.** *Caralluma leendertziae* N. E. Brown (1909); **incl.** *Stapelia furcata* N. E. Brown (1909); **incl.** *Caralluma rubiginosa* Werdermann (1932) ≡ *Caralluma melanantha* fa. *rubiginosa* (Werdermann) A. C. White & B. Sloane (1937) (*nom. inval.*, Art. 34.1a); **incl.** *Caralluma melanantha* var. *sousae* Gomes e Sousa (1935); **incl.** *Caralluma australis* Nel (1937).

Stems stumpy, 6 - 10 × 1.5 - 2 cm, distinctly 4-angled; **Tu** 0.6 - 1 cm, acute; **Inf** with 3 - 5 **Fl**; peduncle 10 - 15 × 5 - 8 mm; **Ped** 20 - 45 × ≤ 3.5 m, erect; **Sep** reddish at the base, 4 - 9 × ≤ 3.5 mm; **Cl** dark black-purple or very dark purple-red, ± lilac in the middle, 3.5 - 5 cm ⌀, conically tapering-bowl-shaped, with or without a tube, tube at the base ± distinctly bulging annulus-like when present, embracing the **Cn**; **Cl** lobes 1.1 - 1.8 × 1.1 - 1.4 cm, triangular-ovate, acute, outside smooth and glabrous, inside densely rugose, faintly hairy, **Ha** black, < 2 mm, erect, margins ± densely hairy, clavate **Ha** purple, to 2 mm, vibratile; **Cn** brownish-purple, 7.5 mm ⌀; **Ci** with a yellow blotch in the middle, shallowly bowl-shaped, lobes 3 - 5 × ≤ 3 mm, weakly 2-keeled, spreading, tips coarsely 3-lobed, central lobe dome-shaped or acute, lateral lobes ± acuminate and divergent; **Cs** lobes ≤ 4 × 0.75 (at the base 3) mm, ovate-lanceolate, laterally compressed, apically bifid, with a longitudinal furrow in the middle, partially with hump-shaped excrescences, ± ascending, sometimes meeting above the **Sty** head; **Poll** 0.55 × 0.45 mm, D-shaped.

An extremely variable species, in particular regarding its staminal corona. – [B. Müller & F. Albers]

O. miscella (N. E. Brown) Meve (KuaS 51(7): 186, ill. (p. 183), 2000). **T:** RSA, Cape Prov. (*Pillans* 657 [K]). – **D:** RSA (Western Cape, Eastern Cape). Fig. XXXIV.e
≡ *Stapelia miscella* N. E. Brown (1909) ≡ *Stultitia miscella* (N. E. Brown) C. A. Lückhoff (1938) ≡ *Pachycymbium miscellum* (N. E. Brown) M. G. Gilbert (1990) ≡ *Angolluma miscella* (N. E. Brown)

Plowes (1994); **incl.** *Caralluma bredae* R. A. Dyer (1964); **incl.** *Caralluma bredae* var. *thomallae* R. A. Dyer (1964).

Stems erect, apically tapering, 4 - 6.5 × 0.4 - 0.85 cm ∅, green, rhizomatous runners ≤ 6 - 7 × 0.7 - 0.8 cm ∅; **Tu** 1 - 1.5 mm, triangular, upper face flattened, horizontally spreading; **Inf** with 1 - 4 **Fl**, ± erect, basal or from the middle of young **Br**; **Ped** 4 - 15 × ± 1 mm; **Sep** 1 - 1.5 × ± 1 mm; **Cl** inside dark red-brown, sometimes slightly glossy, centre somewhat paler, 1.3 - 1.5 cm ∅, flat, deeply incised, tube very short, basally embracing the **Cn**, with an annulus, 5 mm ∅, thickly pillow-shaped; **Cl** lobes 3.5 - 5 × ± 3 mm, ovate-lanceolate or ovate-triangular, acute, horizontally spreading or somewhat reflexed, margins revolute, inside wrinkled, glabrous; **Cn** brownish-purple or ± black, ± 4 mm ∅, with abundant nectar; **Ci** disc-shaped, lobes spreading, ± 1 mm broad, transversely rectangular, apically shallowly emarginate, horizontally spreading; **Cs** darker than the **Ci**, lobes ± 1 × 0.4 mm, elliptic-lanceolate, erect, shorter than the **Anth**, obtuse; **Poll** 0.3 × ± 0.15 mm, ellipsoid-rectangular.

A miniature *Orbea*.

O. namaquensis (N. E. Brown) L. C. Leach (Kirkia 10(1): 290, 1975). – **D:** RSA (Northern Cape: Little Namaqualand).

≡ *Stapelia namaquensis* N. E. Brown (1882); **incl.** *Stapelia ciliolata* Rüst (s.a.); **incl.** *Stapelia namaquensis* var. *minor* N. E. Brown (s.a.); **incl.** *Stapelia tridentata* Rüst (s.a.); **incl.** *Stapelia namaquensis* var. *ciliolata* N. E. Brown (1882); **incl.** *Stapelia namaquensis* var. *tridentata* N. E. Brown (1882); **incl.** *Stapelia namaquensis* var. *bidens* N. E. Brown (1909).

Stems ± 6 × 1.5 cm ∅, squat, shortly prostrate and then decumbent, distinctly 4-angular; **Tu** 1 cm, stout, acute; **L** rudiments hardened; **Inf** 1- to 4-flowered; **Ped** purple, ± 2 × 0.5 mm; **Sep** 4 - 6 × ± 2.5 mm; **Cl** pale greenish-yellow, vertically broadly streaked with dark red-brown, partly marked with confluent spots or reticulate, 7.5 - 10 cm ∅, basin-shaped, tube embracing the **Cn**, shortly hairy at the base, **Ha** red-brown, erect; annulus large, fleshy, margin extremely recurved or revolute; **Cl** lobes 3 × 2.5 cm, divaricate, revolute, inside plicate-rugose or papillose, papillae reddish; **Cn** yellow with red-brown spots; **Ci** elongate-rectangular, lobes 6 × 1.5 mm, ribbon-like, ascending, attenuate; **Cs** lobes 3.5 × 1 mm, narrowly ovate, apical appendage filiform, clavate, overtopping the **Sty** head, sometimes tips revolute, partly with a dorsal hump; nectar slit drop-shaped; **Poll** 0.8 × 0.5 mm, bean-shaped; **Fr** to 15 cm.

Belonging to the *O. variegata*-complex. – [B. Müller, K. Seidler & F. Albers]

O. paradoxa (I. Verdoorn) L. C. Leach (Excelsa Tax. Ser. 1: 55, 1978). **T:** Moçambique (*Blignaut & van der Merwe* 403 [PRE]). – **D:** S Moçambique, RSA (KwaZulu-Natal), Swaziland. **Fig. XXXVI.b**

≡ *Stultitia paradoxa* I. Verdoorn (1937) ≡ *Stapelia paradoxa* (I. Verdoorn) P. V. Heath (1992).

Stems slender, 6 - 8 cm, obtusely 4-angled, rhizome-like runners fleshy; **Tu** prominent, pointed, with 1 - 3 pairs of (flattened) stipular rudiments; **Inf** few-flowered, **Fl** usually facing upwards; **Ped** to ± 1 cm; **Sep** ± 4 × 1.5 mm; **Cl** greenish-white, blotched with pink, purple or dark red, ± 2 cm ∅, campanulate, stiffly fleshy; **Cl** tube at the base and the annulus dark red to violet, tube 6.5 × 9 mm ∅, ± semi-orbicular, inside weakly (sometimes densely) hairy, **Ha** 0.75 mm, acute, rigid, erect; annulus 9 mm ∅, prominent, fleshy; **Cl** lobes ± 6 × 6 mm, ± triangular, divaricate, alternating with acute spine-like processes, tips very shortly pubescent, **Ha** white, margins densely hairy, clavate **Ha** purple, 1.5 mm, vibratile; **Cn** 2.5 × ± 5 mm, sessile or slightly raised; **Ci** dark red, campanulate, lobes 1.5 - 2 mm, ± broadly ovate or ± triangular, below the **Anth** wings deeply concave and forming 5 distinct indentations or pouches; **Cs** yellowish, lobes 1.5 - 2 mm, simple, ± ribbon-like, geniculate, tip not ascending, dorsally sometimes weakly humped; nectar slit narrowly drop-shaped; **Poll** 0.5 × 0.3 mm, elongate-elliptic.

Stapelia conjuncta in the sense of Lückhoff (1952) belongs here. – [B. Müller, K. Seidler & F. Albers]

O. prognatha (P. R. O. Bally) L. C. Leach (Kirkia 10(1): 290, 1975). **T:** Somalia, Northern Prov. (*Glover* s.n. in *Bally* S147 [K]). – **D:** Somalia.

≡ *Stapelia prognatha* P. R. O. Bally (1963) ≡ *Ballyanthus prognathus* (P. R. O. Bally) Bruyns (2001).

Stems to 6 × ± 1 cm, squat, shortly prostrate-erect, prominently 4-angled; **Tu** short; **L** rudiments 1 cm, hardened; **Inf** few-flowered; **Ped** ± 2 × 0.3 mm; **Sep** 4 mm; **Cl** red-brown or yellow-green, 1.8 cm ∅, inversely bowl-shaped; annulus red-brown, pedestal-like; **Cl** lobes ± 5 × 4 mm, ± triangular, divaricate, strongly revolute, margins revolute, hairy at the base, clavate **Ha** purple, ± 1 mm, vibratile; **Cn** red-brown, ± 4 mm ∅, far overtopping the **Cl**, sessile; **Ci** ± inversely bowl-shaped, lobes 1 × 1 mm, square, fleshy, tips slightly lobed, not joined with the **Cs**; **Cs** lobes 3 × 0.5 mm, narrowly ribbon-shaped, apical appendage filiform, basally extended into an awn, in the middle with deep longitudinal furrows, margins forming prominent ledges and divergent at the base, apical appendages joined above the **Sty** head; nectar slit ± triangular or narrowly drop-shaped; **Poll** ± 0.4 × ± 0.2 mm, ± elongate-elliptic, germination crests parallel to the **Poll** margins, not outwards-curving.

An atypical species that could probably have originated from hybridization with a taxon of *Duvalia*.

O. pulchella (Masson) L. C. Leach (Kirkia 10(1): 290, 1975). **T:** [lecto – icono]: Masson, Stapel. Nov. t. 36, 1797. – **D:** RSA (Eastern Cape: near Port Elizabeth).

≡ *Stapelia pulchella* Masson (1797) ≡ *Podanthes pulchellus* (Masson) Haworth (1812).

Stems (prostrate-) erect, 5 - 10 × ± 1 cm, green; **Tu** acute, stout; **Inf** with 3 and more **Fl**; **Ped** 2.5 - 3 cm; **Sep** 3 - 6 mm; **Cl** sulphur-yellow, 3.1 - 6 cm ∅, slightly saucer-shaped; **Cl** tube usually pale purple at the base, densely stippled and blotched with purple-brown; annulus pale purple at the base, to 2 × 10 - 18 mm ∅, ± 5-angled; **Cl** lobes 1 - 2.6 × 1 - 2 cm, ovoid or triangular-ovoid, very acute or acuminate, divaricate or revolute, margins revolute, sometimes purple-brown, inside ± densely finely verrucose-rugose; **Cn** 6 mm ∅; **Ci** dark purple-brown, below the tip with a pale yellowish dot or ± A-shaped marking, margins occasionally yellowish, weakly inversely bowl-shaped, lobes 3 - 4 × 1.5 mm, (narrowly) rectangular, obtuse, truncate, emarginate or apically finely 3-dentate, divergent; **Cs** pale yellow, stippled with dark maroon or vice versa, apical appendage 2 - 3 mm, filiform, erect-joining or tips revolute, ± clavate or not thickened, dorsal appendage 0.6 - 1 mm, conical or conical-subulate, obtuse; nectar slit ± drop-shaped; **Poll** 0.6 × 0.5 mm, D-shaped. – [B. Müller, K. Seidler & F. Albers]

O. rogersii (L. Bolus) Bruyns (Aloe 37(4): 76, 2001). **T:** Botswana (*Rogers* 6298 [BOL]). – **D:** Botswana, RSA.

≡ *Stapelia rogersii* L. Bolus (1915) ≡ *Caralluma rogersii* (L. Bolus) E. A. Bruce & R. A. Dyer (1934) ≡ *Pachycymbium rogersii* (L. Bolus) M. G. Gilbert (1990) ≡ *Angolluma rogersii* (L. Bolus) Plowes (1994).

Stems erect or ascending, to 10 × 0.8 cm, sides grooved; **Tu** 1 - 1.8 × ≤ 0.5 cm, subulate, very acute, ascending or horizontally spreading, with stipular rudiments; **Inf** 3 - 4 per stem, several-flowered, often 2 **Fl** of the same **Inf** open simultaneously; **Ped** 1.3 - 15 × 1.5 mm; **Sep** 3 - 4 × ± 1 mm; **Cl** outside and inside pale yellow, 3 - 3.5 cm ∅, flat, deeply incised, united part 5 - 6 mm ∅; **Cl** lobes ± 14 × 4 mm, linear, acute, bent inwards or ascending, rarely horizontally spreading, margins in the upper ½ revolute, inside basally ciliate, club-shaped **Ha** transparent ≤ 1.7 mm, directed inwards, otherwise glabrous, minutely papillose, papillae transparent, short, globose; **Cn** yellowish, ± 3.5 mm ∅, margins purple, **Ci** and **Cs** only basally shortly united; **Ci** lobes ± 2 × 0.7 mm, rectangular, inside with longitudinal groove, apically toothed, basally spreading horizontally, apically bent upwards; **Cs** lobes broadly triangular when seen from above, divided into 2 filiform appendages ± to the base, apical appendage ≤ 7 mm, erect, basally ovate, ascending over the **Sty** head and intertwined, basal appendage ≤ 5 mm, filiform, erect, intertwined with itself; **Poll** 0.6 × 0.45 mm, broadly D-shaped.

O. sacculata (N. E. Brown) Bruyns (Aloe 37(4): 76, 2001). **T:** Ethiopia, Shewa Region (*Drake-Brockman* 129 [K]). – **D:** Ethiopia, Djibouti, NE Somalia.

≡ *Caralluma sacculata* N. E. Brown (1909) ≡ *Pachycymbium sacculatum* (N. E. Brown) M. G. Gilbert (1990) ≡ *Angolluma sacculata* (N. E. Brown) Plowes (1994); **incl.** *Caralluma kochii* Lavranos (1971) ≡ *Pachycymbium kochii* (Lavranos) M. G. Gilbert (1990) ≡ *Angolluma kochii* (Lavranos) Plowes (1994).

Stems erect or basally decumbent, often strongly rhizomatous, 4 - 10 cm; **Tu** 0.8 - 1.1 cm, subulate, very acute, firm; **L** rudiments ± 5 mm, caducous; **Inf** with 2 - 4 erect **Fl**; **Ped** grey-green-blue, 2 - 8 × 2 mm; **Sep** 3 - 6 mm; **Cl** outside grey-green-blue, striped with purple or brown, inside red or yellowish-green, spotted with purple, 1.5 - 3 cm ∅, campanulate or tubular, tube 7 - 12 × 6 - 9 mm ∅, apically narrower, embracing the **Cn**; **Cl** lobes (5-) 7 - 12 × 3 - 5 mm, triangular-ovate, purple, ± horizontally spreading, margins revolute, densely minutely tomentose, upper face of the lamina scatteredly tuberculate-papillate, papillae with a short white apical **Bri**; **Cn** 5 - 6 mm ∅; **Ci** cup-shaped, 3.5 - 5 mm, lobes 3.5 mm, forming pouches, apically elongating into 2 short tips; **Cs** lobes 1 - 3 mm, triangular-lanceolate or linear, obtuse, apically sometimes erect, basally united with the **Ci**; **Poll** yellow, 0.6 - 0.7 × 0.4 - 0.5 mm, ovoid to D-shaped.

O. schweinfurthii (A. Berger) Bruyns (Aloe 37(4): 76, 2001). **T** [neo]: Zaïre (*De Witte* 1023 [BR]). – **D:** Zaïre, Tanzania, Zambia, Zimbabwe, Rwanda.
Fig. XXXVI.c

≡ *Caralluma schweinfurthii* A. Berger (1910) ≡ *Pachycymbium schweinfurthii* (A. Berger) M. G. Gilbert (1990) ≡ *Angolluma schweinfurthii* (A. Berger) Plowes (1994); **incl.** *Caralluma piaranthoides* Obermeyer (1935).

Stems mostly decumbent, basally slightly tapering, in flat mats, 8 - 10 cm, 4- to 5-ribbed; **Tu** 0.8 - 1.5 cm, spreading-ascending; **L** rudiments slightly bent upwards, with stipular rudiments; **Inf** with 2 - 5 erect **Fl** at the stem tips; buds rounded, obtuse; **Bra** ± 2.5 × 1.5 mm, lanceolate-triangular, obtuse; **Ped** 3 - 4 × ± 1.5 mm, pointed; **Sep** 2 - 3 × 1 - 1.2 mm; **Cl** outside green with few red blotches, inside yellow to brown, spotted with wine-red, beige or purple, 1 - 1.5 cm ∅, flat, incised for ± ½, tube very shallow, embracing the **Cn** basally, surrounded by a thickly fleshy annulus; **Cl** lobes 6 - 12 × ≤ 10 mm, triangular-ovate, acute, ± horizontally spreading, convex, apically somewhat bent outwards, inside papillate-warty; **Cn** 4 - 5.5 mm ∅; **Ci** cream-coloured, minutely red-dotted, ± 1 mm, shallowly bowl-shaped; **Cs** lobes ± 1 × 0.7 mm, triangular, overtopping the **Anth**, apically obtuse or 2-toothed, dorsally depressed, margin strongly toothed, free **Cs** parts sometimes with 2 shallow lateral depressions; **Poll** 0.3 × 0.3 mm, D-shaped.

O. semitubiflora (L. E. Newton) Bruyns (Aloe 37(4): 76, ill. (p. 72), 2001). **T:** Tanzania, Arusha Prov. (*Newton* 3419 [K, EA]). – **D:** N Tanzania. **I:** Meve (1999). **Fig. XXXVI.d**

≡ *Angolluma semitubiflora* L. E. Newton (1993).

Stems erect-ascending, in ± dense clusters to 50 cm ⌀ with rhizomatous runners to 14 cm; stems pale green, spotted with brown, 5 × 0.8 cm ⌀; **Tu** 4 - 8 × 2 - 4 mm, ± spreading horizontally; **L** rudiments ± 1 mm; **Inf** with 2 - 3 (-8) erect **Fl** and 2 - 5 caducous buds; **Ped** pale rose, 5 mm; **Sep** pale green, 4 × 1.5 mm; **Cl** outside pale green with few brown stripes, inside golden-yellow or dark purple, **Cl** lobes apically pale yellow, with faint maroon stripes, or **Cl** uniformly maroon, 2 cm ⌀, campanulate, tube 4 × 7 mm ⌀, embracing the **Cn**; **Cl** lobes 7 × 4 mm, erect-ascending, margins revolute, minutely tomentose; **Cn** bright yellow or chestnut-brown; **Ci** cup-shaped, 3 - 4 × centrally 1 mm, lobes forming shallow pouches, 3 - 4 mm broad where united with the **Cs**; **Cs** lobes triangular, short, apically rounded, not touching each other; **Poll** dark red, 0.5 × ± 0.3 mm, ovoid.

O. semota (N. E. Brown) L. C. Leach (Kirkia 10(1): 290, 1975). **T:** Tanzania (*Burtt* 1450 [K]). – **D:** Rwanda, Kenya, Tanzania. **Fig. XXXVI.e**

≡ *Stapelia semota* N. E. Brown (1933); **incl.** *Stapelia discoidea* Obermeyer (1937); **incl.** *Stapelia molonyae* A. C. White & B. Sloane (1937); **incl.** *Stapelia kagerensis* Lebrun & Taton (1948).

Stems single or sometimes branched to form compact cushions, prostrate-decumbent, squat, 5 - 8 × ± 1 cm ⌀, grey-green or dark green, sometimes blotched with dark reddish-brown, distinctly 4-angled; **Tu** 5 - 10 mm, conically triangular, acuminate; **Inf** 1- to 3-flowered; **Bra** ± 1 mm, lanceolate; **Ped** 2 - 3.5 × 0.2 - 0.3 cm; **Sep** 4 - 7 × ± 3 mm, divaricate, reaching the bases of the **Cl** lobes, alternating with small **Gl**; **Cl** dark reddish-brown (then lobe tips stippled with yellow) or yellow, 3.5 - 5 cm ⌀, shallow, deeply incised; annulus 0.9 - 1 cm ⌀, cushion-like, 5-angled; **Cl** lobes 1.5 - 1.9 × 0.9 - 1 cm, divaricate or slightly revolute, inside transversely ± rugose, margins hairy, **Ha** red or silvery-white, ≤ 3 mm, simple, fusiform or clavate, vibratile; **Cn** ± 7.5 mm ⌀, overtopping the **Cl**; **Ci** dark brown or maroon, inversely bowl-shaped, lobes 2 - 2.5 × 2 mm, ± square, obtusely bidentate, ventrally with few plaited outgrowths; **Cs** yellowish, lobes 1.7 × 1 mm, spatulate, slightly overtopping the **Anth**, tips shortly bifid, in the middle broadly furrowed in longitudinal direction, transitional section to the **Ci** dorsally protruding and coarsely folded;

nectar slit ± triangular or drop-shaped; **Poll** ± 0.6 × ± 0.4 mm, D-shaped.

Closely related to *O. woodii*. – [B. Müller, K. Seidler & F. Albers]

O. speciosa L. C. Leach (Excelsa Tax. Ser. 1: 33-36, ills., 1978). **T:** RSA, KwaZulu-Natal (*Lubbers* s.n. in *Leach* 14742 [PRE]). – **D:** RSA (KwaZulu-Natal). **Fig. XXXVI.a**

Stems shortly postrate-erect, ± 9 × 0.6 - 0.8 cm ⌀; **Tu** ≤ 1 cm, acute; **Inf** 1- to 3-flowered, sometimes prostrate; **Fl** buds very broadly 5-angular-ovoid, ± acute, basally truncate; **Bra** ± 6 × 1.5 mm, narrowly ovate, acute, fleshy, strongly concave; **Ped** purple, finely blotched with white, (15-) 20 - 25 × ± 2 mm, ± horizontally spreading; **Sep** 5 - 12 × 2 - 3 mm, alternating with delicate fleshy **Gl**; **Cl** outside green, with fine lines consisting of magenta dots, inside pale yellow spotted and dotted with magenta, 3.5 - 5 cm ⌀; annulus ± 1.5 × ± 7 mm ⌀, obtusely 5-angled; **Cl** lobes 1.25 - 2 × 0.9 - 1.35 cm, divaricate, revolute, outside glabrous, inside finely verrucose, indistinctly obtusely rugose, furrowed along the midrib, margins densely hairy, clavate **Ha** dark magenta, ± 2.5 mm, vibratile; **Cn** 3 × ± 7 mm ⌀, stalk 1.5 × ± 2.5 mm, obtusely 5-angled; **Ci** inversely shallowly bowl-shaped, lobes ± 2 × ± 1.9 mm, ± square, apically obtusely 3- to 4-dentate, longitudinally 2-ribbed, laterally bulging; **Cs** pale yellow, finely stippled with red, lobes ± 2.5 × 1 mm, ± ovate, obtusely ± tapering, straightly bent inwards, slightly overtopping the **Anth** but not joining each other, dorsally weakly rugose, occasionally faintly humped, margins irregularly weakly dentate; **Anth** completely hiding the **Poll**; nectar slit very narrowly triangular, partially covered by the base of the **Ci**; **Poll** ± 0.7 × 0.6 mm, ± D-shaped, caudicles short, strongly broadened at the **Poll**, ± 0.3 mm, winged.

Closely allied to *O. longidens* and *O. macloughlinii*. – [B. Müller, K. Seidler & F. Albers]

O. sprengeri (Schweinfurth) Bruyns (Aloe 37(4): 76, 2001). **T:** [lecto – icono]: Berger, Stapelien & Kleinien, fig. 22: 1-4, 1910. – **D:** Ethiopia, Eritrea, Somalia, Sudan, Yemen.

≡ *Huernia sprengeri* Schweinfurth (1893) ≡ *Caralluma sprengeri* (Schweinfurth) N. E. Brown (1895) ≡ *Pachycymbium sprengeri* (Schweinfurth) M. G. Gilbert (1990) ≡ *Angolluma sprengeri* (Schweinfurth) Plowes (1994).

O. sprengeri ssp. **foetida** (M. G. Gilbert) Bruyns (Aloe 37(4): 76, 2001). **T:** Ethiopia, Harerge Region (*Gilbert* 2297 [K, ETH]). – **D:** Ethiopia.

≡ *Caralluma sprengeri* ssp. *foetida* M. G. Gilbert (1978) ≡ *Pachycymbium sprengeri* ssp. *foetidum* (M. G. Gilbert) M. G. Gilbert (1990) ≡ *Angolluma foetida* (M. G. Gilbert) Plowes (1994).

Differs from ssp. *sprengeri*: Stems prostrate-ascending, to 7 cm; **Inf** to 5-flowered; **Fl** with strong evil scent; buds truncate or broadly pointed, horizontally spreading or bent towards the stems; **Cl** inside pale or dark brown, ± 3 cm ⌀, annulus sometimes with a narrow paler ring, flat, deeply incised, tube very short, embracing the **Cn** basally, annulus narrow; **Cl** lobes 11 - 13.5 × 6 - 7 mm, ovate, ± strongly reflexed, inside finely roughened, papillose, papillae delicate, conical; **Cn** dark brown, 4.5 - 6 mm; **Poll** orange, ± 0.45 × 0.4 mm, broadly ovoid.

The corolla has the same characteristics as *O. laticorona*.

O. sprengeri ssp. **ogadensis** (M. G. Gilbert) Bruyns (Aloe 37(4): 76, 2001). **T:** Ethiopia, Harerge Region (*de Wilde & Gilbert* 6481 [WAG, ETH]). – **D:** Ethiopia.

≡ *Caralluma sprengeri* ssp. *ogadensis* M. G. Gilbert (1978) ≡ *Pachycymbium sprengeri* ssp. *ogadense* (M. G. Gilbert) M. G. Gilbert (1990) ≡ *Angolluma ogadensis* (M. G. Gilbert) Plowes (1994).

Differs from ssp. *sprengeri*: Stems decumbent, sometimes solitary and erect, 5 - 10 cm, rhizomatous runners to 20 cm; buds ascending, ± acute; **Cl** inside leather-brown, to 2 cm ⌀, with a conspicuous white ring in the annulus; **Cl** lobes 9.5 - 11 × 5 - 5.5 mm, lanceolate, horizontally spreading to ascending, margins revolute, inside conspicuously wrinkled, glabrous; **Cn** 4.5 - 9 mm ⌀; **Cs** yellowish, apically dark rose.

O. sprengeri ssp. **sprengeri** – **D:** Ethiopia, Eritrea, E Sudan, N Yemen, W Somalia.

Stems ascending-erect, ± strongly branched, robust, to 15 cm, green, spotted with reddish-purple or dark green, sometimes 5-ribbed, rhizomatous runners short; **Tu** 0.7 - 1.25 cm, strong, spreading-ascending; **Inf** with 5 - 6 **Fl**; buds ± nutant, very acute; **Ped** green, striped with reddish, 3 mm, robust; **Sep** 3 - 4 mm; **Cl** outside pale green with pale brown spots and stripes, inside grey-brown to tan-coloured, 2 - 2.5 cm ⌀, flat, deeply incised, tube very shallow, basally embracing the **Cn**, mouth with narrow slightly ascending annulus, which is covered by the **Cn**; **Cl** lobes 13 - 20 × 5.7 - 7 mm, ± narrowly lanceolate, horizontally spreading, margins slightly revolute, inside velvety-papillate, papillae with apical white ± short **Ha**; **Cn** (reddish-)yellow, 4.3 - 5.2 mm ⌀; **Ci** yellow, ± 1.5 mm, ring-like, not conspicuously thickened; **Cs** lobes oblong-elliptic or rectangular, at least as long as the **Anth**, apically obtuse, lobed or toothed; **Poll** yellow, 0.3 × 0.25 mm, ovoid.

O. subterranea (E. A. Bruce & P. R. O. Bally) Bruyns (Aloe 37(4): 76, 2001). **T:** Kenya (*Boy Joanna* s.n. in *Bally* S4 [K]). – **D:** Kenya, N Tanzania; savannas.

≡ *Caralluma subterranea* E. A. Bruce & P. R. O. Bally (1941) ≡ *Pachycymbium baldratii* ssp. *subterraneum* (E. A. Bruce & P. R. O. Bally) M. G. Gilbert (1990) ≡ *Angolluma subterranea* (E. A. Bruce & P. R. O. Bally) Plowes (1994); **incl.** *Angolluma lenewtonii* Lavranos (1998).

Plants 2 - 10 cm tall, rhizomatous runners to 15 cm, branched; stems green or grey-green, sometimes spotted with dark brown; **Tu** 0.5 - 1 cm, thickly fleshy, spreading-ascending; **Inf** with 1 - 3 **Fl**; **Bra** 1 mm, linear-lanceolate; **Ped** 1 (-5) × 1.7 mm; **Cl** black-carmine, reddish-orange or pale yellow, 1.5 - 2 cm, flat, hairy, tube 2 mm, basally embracing the **Cn**; **Cl** lobes 5 - 7 × ± 3 mm, horizontally spreading, margins near tip revolute; **Cn** brown (-reddish), ± 1.5 × 4 - 4.5 mm ⌀; **Ci** lobes united with the **Cs** for ± ¾ of their length, apically smooth, entire or strongly crenate; **Cs** lobes oblong-rectangular, sometimes with a dorsal appendage lying on the back of the segment, apically blunt or dentate; **Poll** ± 0.5 × 0.25 mm, ovoid.

Stout plants, frequently seen throughout its range. *Angolluma lenewtonii* is preliminarily subsumed here. However, its more intensely hairy corolla makes varietal status for it more appropriate.

O. tapscottii (I. Verdoorn) L. C. Leach (Kirkia 10(1): 291, 1975). **T:** Botswana (*Tapscott* s.n. [K, KMG]). – **D:** S Botswana, N RSA (North-West Prov., Northern Prov.).

≡ *Stapelia tapscottii* I. Verdoorn (1927) ≡ *Stultitia tapscottii* (I. Verdoorn) E. Phillips (1933).

Stems slender, forming ± lax mats, to 12 × 0.8 cm ⌀, with 4 projecting **Ri**; **Tu** ≤ 2 cm, strongly porrect, falcately curved towards the stem; **Ped** ± 2.5 cm; **Sep** 1 cm; **Cl** red-brown, 6 cm ⌀, flat; annulus ± 1 cm ⌀, cushion-like, with a central depression, verrucose-papillose; **Cl** lobes 2 × 1.5 cm, inside transversely rugose, margins slightly revolute, ± hairy at the base, clavate **Ha** purple-violet, ≤ 1 mm, vibratile; **Cn** pale brown, ± 5 mm ⌀, raised above the **Cl**, sessile; **Ci** bowl-shaped, lobes 2.5 × 2 mm, ± rectangular, bidentate, margins plaited; **Cs** lobes 3.2 × 0.9 mm, ovate-lanceolate, apical appendage filiform, conspicuously raised above the **Sty** head, meeting each other, tips clavate, revolute, laterally each with an ascending subulate appendage, transitional section towards the **Ci** bulging; nectar slit ± triangular; **Poll** ± 0.6 × 0.5 mm, D-shaped.

Closely related to *O. cooperi*. – [B. Müller, K. Seidler & F. Albers]

O. tubiformis (E. A. Bruce & P. R. O. Bally) Bruyns (Aloe 37(4): 76, 2001). **T:** Kenya (*Copley* s.n. in *Bally* S33 [K]). – **D:** Kenya. **Fig. XXXVI.f**

≡ *Caralluma tubiformis* E. A. Bruce & P. R. O. Bally (1941) ≡ *Pachycymbium tubiforme* (E. A. Bruce & P. R. O. Bally) M. G. Gilbert (1990) ≡ *Angolluma tubiformis* (E. A. Bruce & P. R. O. Bally) Plowes (1994).

Stems unbranched, ascending-erect, 12 - 15 × 1.5

- 2 cm, pale green, blotched with red-green or dark brown; **Tu** 1.5 - 1.8 cm, basally conical-triangular, very long attenuate-pointed, ascending; **Inf** 1-flowered, erect, subterminal or lateral; **Ped** pale green, striped with brown, 5 - 8 × 2.5 mm; **Sep** ± 5 × 1.5 mm; **Cl** outside green, inside dark flesh-coloured, tube ± 1.2 × 1 cm ∅, embracing the **Cn**; **Cl** lobes ± 10 × 6 mm, triangular, acute, erect or horizontally spreading, margins revolute, inside ± strongly hairy, **Ha** whitish, ≤ 3 mm, apically capitate; **Cn** dark red, cup-shaped, ± 6 mm ∅; **Ci** ± 2 mm, lobes forming widely protruding deep pouches; **Cs** lobes 1.5 - 2 mm, ± as long as the **Anth**, basally obtriangular, laterally with large wing-like appendages closely pressed to the guide rails, middle part ± rectangular, apically acute or slightly rounded, slightly bent upwards; **Poll** 0.7 × 0.35 mm, ellipsoid.

Closely related to *O. sacculata* and *O. kochii*. – [B. Müller, J. Kiel & F. Albers]

O. ubomboensis (I. Verdoorn) Bruyns (Aloe 37(4): 76, 2001). **T:** RSA, KwaZulu-Natal (*Pole-Evans* s.n. [PRE 8764]). – **D:** E Zimbabwe, RSA (KwaZulu-Natal), Swaziland.

≡ *Caralluma ubomboensis* I. Verdoorn (1932) ≡ *Pachycymbium ubomboense* (I. Verdoorn) M. G. Gilbert (1990) ≡ *Angolluma ubomboensis* (I. Verdoorn) Plowes (1994).

Stems ascending, ± 4 cm, dull green; **Tu** ascending, apically erect; **Inf** with 2 - 3 erect **Fl** at or near the stem tips; peduncle short, cylindrical; **Bra** ± 1 × 0.5 mm, linear, acute; buds ± 4 × 4 mm, basally rounded, apically bluntly conical; **Ped** 3 - 5 × ± 0.8 mm; **Sep** 1.5 - 2 × 1 mm; **Cl** inside black- or dark purple, 0.9 - 1.2 cm ∅, flat, deeply incised, tube shallow, ± 4.5 mm ∅, basally embracing the **Cn**; **Cl** lobes 3 - 5 × ≤ 3.5 mm, ovate, horizontally spreading or more strongly recurved, slightly concave, inside wrinkly-velvety, glabrous; **Cn** dark purple, ± 4 mm ∅; **Ci** flatly disc-like, lobes flat, broad, apically deeply bifid with diverging appendages; **Cs** lobes 0.5 - 1 mm, linear-elliptic, basally with 2 wing-like lateral appendages, overtopping the **Anth**, apically rounded, basally united with the **Ci** for a short distance; **Poll** 0.25 × 0.2 mm, ± D-shaped.

O. umbracula (M. D. Henderson) L. C. Leach (Kirkia 10(1): 291, 1975). **T:** Zimbabwe, Bikita Distr. (*Leach & Pienaar* 5584 [PRE]). – **D:** SW Zimbabwe.

≡ *Stultitia umbracula* M. D. Henderson (1962) ≡ *Stapelia umbracula* (M. D. Henderson) P. V. Heath (1992).

Stems squat, branched 2 cm above the ground and forming small lax mats, to ± 7.5 × 0.8 cm ∅, distinctly 4-angled, sides ± deeply furrowed; **Tu** ≤ 2 cm, acute, ascending; **L** rudiments acute; **Inf** 1- to 8-flowered, usually erect, rarely oblique or pointing outwards; peduncle often ± as thick as the stems, ± terete, stumpy, erect; **Fl** buds 5-angled-ovoid, ± acute, sides concave; **Ped** 3 - 4 × ± 0.2 cm, rarely spreading; **Sep** ± 6 × 2 mm; **Cl** outside yellow-green, inside dull maroon, ± 5 cm ∅, inversely campanulate; annulus 4 × 7.5 mm ∅, abruptly ascending, glabrous, inside finely pubescent, **Ha** white, stout, simple, rigid; **Cl** lobes blotched with yellow, ± 16 × 9 mm, strongly revolute and with the tips embracing the **Ped**, inside slightly transversely rugose, margins cartilaginous, **Ha** stiff, white; **Cn** ± 7.5 mm ∅, convex, stalk 2 - 3 mm; **Ci** dull maroon, sides darker, lobes ± 2 × 1.5 - 2 mm, ± square, fleshy, apically irregularly crenulate, lateral margins bulging in a rib-like manner; **Cs** yellow with reddish markings, lobes 4 - 5 × ± 1 mm, elongate, ovate, apical appendage filiform, erect, meeting each other above the **Sty** head, tips clavate, verrucose-rugose, basally with 2 divergent edges merging into the **Ci**; nectar slit narrowly triangular, below the **Anth** wings, basally with a small hump; **Poll** ± 0.6 × 0.6 mm, ± D-shaped.

Closely allied to *O. halipedicola*. – [B. Müller, K. Seidler & F. Albers]

O. valida (N. E. Brown) Bruyns (Aloe 37(4): 76, 2001). **T:** Botswana ? (*Holub* s.n. [K]). – **D:** Namibia, Zambia, Zimbabwe, Botswana.

≡ *Caralluma valida* N. E. Brown (1895) ≡ *Orbeopsis valida* (N. E. Brown) L. C. Leach (1978).

O. valida ssp. **occidentalis** Bruyns (Aloe 37(4): 76, ill. (p. 74), 2001). **T:** Botswana (*Bruyns* 6465 [BOL]). – **D:** S Botswana.

Stems 4 - 15 × 1.5 - 2 cm, with subterranean runners to 20 cm long; **Tu** 5 - 15 mm; **Cl** upper face pinkish-red, 4 - 4.5 cm ∅; **Cl** tube 2 - 3 mm, containing > ½ of the **Cn**; margins of the **Cl** tips with dark red **Ha**; **Ci** lobes spreading.

Like the typical subspecies with the exception of the pale-coloured corolla and the deep corolla tube. – [U. Meve]

O. valida ssp. **valida** – **D:** N Namibia, S Zambia, Zimbabwe, Botswana.

Incl. *Caralluma tsumebensis* Obermeyer (1937) ≡ *Orbeopsis tsumebensis* (Obermeyer) L. C. Leach (1978).

Stems prostrate, apically ascending, stumpy, rooting when touching the ground, green, mottled with purple, to 10 (rarely to 50) × ± 1.5 cm, prominently 4-angled; **Tu** ≤ 1.5 × 0.5 cm, robust, triangular; **L** rudiments ± 5 mm; **Inf** with 20 - 40 **Fl** with revoltingly foetid odour; **Ped** 10 - 15 × 3 mm; **Sep** 8 - 9 mm, tips spreading or bent ± outwards, alternating with tiny **Gl**; **Cl** dark flesh-coloured to ± black, ± 6 cm ∅; **Cl** tube ± 7 × 10 - 12 mm, campanulate, embracing the **Cn**, margin of the tube abruptly spreading; **Cl** lobes 15 - 25 × 5 - 9 mm, lanceolate, divaricate, margins slightly revolute, at the base alternating with small fleshy obtuse appendages, inside strongly rugose, very finely pubescent, **Ha** 0.4 mm, rigid, margins ± densely hairy, clavate **Ha** pale, to 3 mm, vibratile; **Cn** ± 3.5 × ± 5.5 mm; **Ci** conically bowl-shaped, lobes 3.5 × 1.75 mm, thickly fleshy, slightly spreading or ± erect, inside weakly ± irregularly keeled, apically rectangular-acute, sometimes truncate, dentate or acutely 4-lobed; **Cs** lobes ± 4 × ± 1 mm, broadly rectangular, very thickly fleshy, at the base with a pair of prominent dorsal humps, apically flattened, erect, or the tip ± cylindrical, acuminate, or ± bifid (occasionally all 3 forms represented in a single **Cn**), slightly raised above the **Sty** head, meeting, tips revolute; **Poll** ± 0.5 × 0.5 mm, D-shaped.

O. variegata (Linné) Haworth (Synops. Pl. Succ., 40, 1812). **T:** South Africa (*Anonymous* s.n. [LINN 311/1]). – **D:** RSA (Western Cape). **Fig. XXXVI.g**
≡ *Stapelia variegata* Linné (1753); **incl.** *Orbea trisulca* hort. (s.a.); **incl.** *Stapelia anguinea* Jacquin (s.a.); **incl.** *Stapelia bufonia* Sims (s.a.) ≡ *Orbea bufonia* (Sims) Haworth (1812); **incl.** *Stapelia bufoniana* Jacquin (s.a.); **incl.** *Stapelia ciliolata* Todaro (s.a.); **incl.** *Stapelia conspurcata* Willdenow (s.a.) ≡ *Stapelia variegata* var. *conspurcata* (Willdenow) N. E. Brown (s.a.) ≡ *Orbea conspurcata* (Willdenow) Sweet (1827); **incl.** *Stapelia hispida* Horn. *ex* Rüst (s.a.); **incl.** *Stapelia inodora* Decaisne (s.a.); **incl.** *Stapelia marginata* Willdenow (s.a.) ≡ *Stapelia variegata* var. *marginata* (Willdenow) N. E. Brown (s.a.) ≡ *Orbea marginata* (Willdenow) G. Don (1837) ≡ *Orbea planiflora* var. *marginata* (Willdenow) G. Don (1837); **incl.** *Stapelia marmorata* Jacquin (s.a.) ≡ *Stapelia variegata* var. *marmorata* (Jacquin) N. E. Brown (s.a.) ≡ *Orbea marmorata* (Jacquin) G. Don (1837); **incl.** *Stapelia monstrosa* Hort. *ex* Steudel (s.a.); **incl.** *Stapelia normalis* Jacquin (s.a.) ≡ *Stapelia variegata* var. *normalis* (Jacquin) A. Berger (s.a.) ≡ *Orbea normalis* (Jacquin) G. Don (1837); **incl.** *Stapelia ophiuncula* Haworth (s.a.); **incl.** *Stapelia picta* N. E. Brown (s.a.) ≡ *Stapelia variegata* var. *picta* (N. E. Brown) N. E. Brown (s.a.); **incl.** *Stapelia rugosa* Donn *ex* Jacquin (s.a.) ≡ *Orbea rugosa* (Donn *ex* Jacquin) Sweet (1826); **incl.** *Stapelia variegata* var. *atrata* (Todaro) N. E. Brown (s.a.); **incl.** *Stapelia orbiculata* Donn (s.a.) (*nom. nud.*); **incl.** *Stapelia mixta* Masson (1797) ≡ *Stapelia variegata* var. *mixta* (Masson) N. E. Brown (s.a.) ≡ *Orbea mixta* (Masson) Haworth (1812); **incl.** *Stapelia clypeata* J. Donn (1804) ≡ *Orbea variegata* var. *clypeata* (J. Donn) hort. (s.a.) ≡ *Orbea clypeata* (J. Donn) Haworth (1819) ≡ *Stapelia variegata* var. *clypeata* (J. Donn) N. E. Brown (1909); **incl.** *Stapelia lepida* Jacquin (1806) ≡ *Podanthes lepida* (Jacquin) Haworth (1812) ≡ *Orbea lepida* (Jacquin) Haworth (1819); **incl.** *Orbea anguinea* Haworth (1812); **incl.** *Orbea bisulca* Haworth (1812); **incl.** *Orbea curtisii* Haworth (1812) ≡ *Stapelia curtisii* (Haworth) Schultes (s.a.); **incl.** *Orbea picta* Haworth (1812); **incl.** *Orbea quinquenervia* Haworth (1812); **incl.** *Orbea retusa* Haworth (1812); **incl.** *Orbea woodfordiana* Haworth (1812) (*nom. nud.*) ≡ *Stapelia woodfordiana* (Haworth) Schultes (s.a.); **incl.** *Orbea inodora* Haworth (1819); **incl.** *Stapelia atropurpurea* Salm-Dyck (1834) ≡ *Stapelia variegata* var. *atropurpurea* (Salm-Dyck) N. E. Brown (s.a.); **incl.** *Stapelia limosa* Salm-Dyck (1834); **incl.** *Stapelia horizontalis* N. E. Brown (1890); **incl.** *Stapelia orbicularis* Loddiges (s.a.) ≡ *Orbea orbicularis* (Loddiges) Haworth (1812); **incl.** *Stapelia planiflora* Jacquin (s.a.) ≡ *Stapelia variegata* var. *planiflora* (Jacquin) N. E. Brown (s.a.) ≡ *Orbea planiflora* (Jacquin) Haworth (1819); **incl.** *Stapelia retusa* Schultes (s.a.) ≡ *Stapelia variegata* var. *retusa* (Schultes) N. E. Brown (s.a.); **incl.** *Stapelia bufonia* Sims (s.a.) ≡ *Stapelia variegata* var. *bufonia* (Sims) N. E. Brown (s.a.); **incl.** *Stapelia rugosa* Donn *ex* Jacquin (s.a.) ≡ *Stapelia variegata* var. *rugosa* (J. Donn *ex* Jacquin) N. E. Brown (s.a.) ≡ *Tridentea rugosa* (Jacquin) G. Don (s.a.); **incl.** *Stapelia horizontalis* N. E. Brown (1890) ≡ *Stapelia variegata* var. *horizontalis* (N. E. Brown) N. E. Brown (s.a.); **incl.** *Stapelia atrata* Todaro (s.a.); **incl.** *Stapelia beffoniana* Schultes (s.a.); **incl.** *Stapelia bidentata* Hort. *ex* Salm-Dyck (s.a.); **incl.** *Stapelia bisulca* Schultes (s.a.); **incl.** *Stapelia buffoniana* G. Don (s.a.); **incl.** *Stapelia buffonis* Loddiges (s.a.); **incl.** *Stapelia obliqua* Willdenow (s.a.); **incl.** *Stapelia orbicularis* Loddiges (s.a.); **incl.** *Stapelia planiflora* Jacquin (s.a.); **incl.** *Stapelia planiflora* var. *marginata* Willdenow (s.a.); **incl.** *Stapelia quinquenervis* Schultes (s.a.); **incl.** *Stapelia retusa* Schultes (s.a.); **incl.** *Stapelia scutellata* Todaro (s.a.); **incl.** *Stapelia scylla* Sprenger (s.a.); **incl.** *Stapelia variegata* var. *asparagensis* Jacob. (s.a.); **incl.** *Stapelia variegata* var. *brevicornis* N. E. Brown (s.a.); **incl.** *Stapelia variegata* var. *laeta* N. E. Brown (s.a.); **incl.** *Stapelia variegata* var. *pallida* N. E. Brown (s.a.); **incl.** *Stapelia variegata* var. *prometheus* Rüst (s.a.); **incl.** *Stapelia bifolia* Schultes (s.a.) (*nom. nud.*); **incl.** *Stapelia trisulca* J. Donn (s.a.) (*nom. nud.*) ≡ *Stapelia variegata* var. *trisulca* (J. Donn) N. E. Brown (s.a.) (*nom. nud.*).

Stems slender or squat, basally decumbent or mostly erect, green, blotched with purple, slender stems to 15 × ≤ 1 cm, squat stems 4 - 5 × ≤ 1 cm, obtuse or conspicuously 4-angled; **Tu** ≤ 1.5 cm, strongly spreading, stipular rudiments sometimes absent; **Inf** 1- to 5-flowered; **Fl** buds apically attenuate-pointed to pointed, flattish or rounded; **Ped** 2 - 6 × 0.2 - 0.3 mm; **Sep** ≤ 9 mm; **Cl** 5 - 9 cm ⌀, basin-shaped, outside green, lobes and venation suffused with reddish, inside pale greenish-yellow or pale lemon-yellow, variably marked with ± large black-purple, crimson- or purple-brown dots, annulus usually paler yellow, more finely dotted, fleshy-rounded or 5-angled, margin revolute, acutely edged, verrucose-granular; **Cl** tube embracing the **Cn**; **Cl** lobes to ± 2.5 × ± 2.3 cm, widely spreading or slightly revolute, inside densely transversely rugose; **Cn** pale yellow, dark violet or stippled with dark violet, 6 - 7 mm ∅; **Ci** deeply bowl-shaped, lobes 2.3 × 2 mm, elongate-rectangular, bifid down to ½ or bidentate, ascending-divaricate; **Cs** lobes 3 × 1.2 mm, ovate-lanceolate, tips and base with filiform or subulate ± equally sized appendages, apical appendage raised above the **Sty** head, tips clavate, finely verrucose, both appendages revolute; nectar slit drop-shaped; **Poll** 0.8 × 0.5 mm, ± narrowly D-shaped.

O. variegata is among the most widely cultivated stapeliads and has as neophyte become part of the flora of subtropical countries worldwide. It has been known as *Stapelia variegata* for long. For nomenclatural reasons (the species is the generic type of *Orbea*) it has to be placed in *Orbea*. With time a plethora of minor colour variants and artifical hybrids have been described as separate species, which is reflected in the (incomplete) list of synonyms given above. – [B. Müller, K. Seidler & F. Albers]

O. verrucosa (Masson) L. C. Leach (Kirkia 10(1): 290, 1975). **T:** (*Masson* s.n. [BM, SRGH [photo]]). – **D:** RSA.

≡ *Stapelia verrucosa* Masson (1797) ≡ *Podanthes verrucosa* (Masson) Haworth (1812); **incl.** *Stapelia irrorata* Loddiges (s.a.) (*nom. illeg.*, Art. 53.1); **incl.** *Stapelia rugosa* Wendland (s.a.).

O. verrucosa var. **fucosa** (N. E. Brown) L. C. Leach (Excelsa Tax. Ser. 1: 25, 1978). **T:** RSA, Eastern Cape (*Pillans* 173 [BOL]). – **D:** RSA (Eastern Cape).

≡ *Stapelia fucosa* N. E. Brown (1909) ≡ *Stapelia verrucosa* var. *fucosa* (N. E. Brown) P. V. Heath (1992); **incl.** *Stapelia irrorata* Masson (1796) ≡ *Podanthes irrorata* (Masson) Haworth (1812) ≡ *Orbea irrorata* (Masson) L. C. Leach (1978).

Differs from var. *verrucosa*: Stems with comparatively small and weak **Tu**; **Cl** smaller, with red-brown much denser markings, less basin-shaped; **Cl** lobes divaricate or slightly revolute; annulus more solid and more distinctly set off. – [B. Müller, K. Seidler & F. Albers]

O. verrucosa var. **verrucosa** – **D:** RSA (Western Cape, Eastern Cape).

≡ *Stapelia verrucosa* var. *verrucosa*; **incl.** *Stapelia roriflua* Jacquin (1806) ≡ *Podanthes roriflua* (Jacquin) Sweet (1827) ≡ *Piaranthus rorifluus* (Jacquin) Decaisne (1844) ≡ *Stapelia verrucosa* var. *roriflua* (Jacquin) N. E. Brown (1909); **incl.** *Podanthes pulchra* Haworth (1812) ≡ *Stapelia pulchra* (Haworth) Schultes (1820) ≡ *Stapelia verrucosa* var. *pulchra* (Haworth) N. E. Brown (1909); **incl.** *Stapelia wendlandiana* Schultes (1820) ≡ *Orbea wendlandiana* (Schultes) Schultes (1820); **incl.** *Podanthes pulchra* var. *major* Sweet (1827); **incl.** *Podanthes pulchra* var. *verrucosa* G. Don (1837); **incl.** *Stapelia verrucosa* var. *conspicua* N. E. Brown (1909); **incl.** *Stapelia verrucosa* var. *pallescens* N. E. Brown (1909); **incl.** *Stapelia verrucosa* var. *punctifera* N. E. Brown (1909); **incl.** *Stapelia verrucosa* var. *robusta* N. E. Brown (1909).

Stems prostrate-decumbent, slender, growing in laxly tufted mats, to 8 cm, distinctly 4-angled; **Tu** ≤ 1.3 cm; **L** rudiments hardened; **Ped** ± 2 × 0.5 cm; **Sep** 3.5 × ± 1.2 mm; **Cl** pale yellow, ± regularly finely stippled with dark red (dots partly becoming confluent), 3 - 4 cm ∅, basin- or tube-shaped or campanulate, tube (if present) embracing the **Cn**, basally forming an annulus of 8 - 9 mm ∅; **Cl** lobes 1.6 × 1.3 - 1.6 cm, ± triangular, revolute, inside verrucose-papillose, partly with pairs of transverse folds; **Cn** 4.5 - 6.6 mm ∅; **Ci** dark brown, bowl-shaped, lobes 1.8 × 1.7 mm, ± square, shortly bidentate, along the middle distinctly plicate; **Cs** yellow, margins dark brown, lobes 1.7 × 0.7 mm, broadly ovate, slightly overtopping the **Anth**, at the base forming a hump; **Poll** 0.43 × 0.35 mm, D-shaped; **Fr** 10 - 12 × 1.3 cm. – [B. Müller, K. Seidler & F. Albers]

O. vibratilis (E. A. Bruce & P. R. O. Bally) Bruyns (Aloe 37(4): 76, 2001). **T:** Kenya (*Ritchie* s.n. in *Bally* S35 [K]). – **D:** Ethiopia, Uganda, Kenya, Tanzania.

≡ *Caralluma vibratilis* E. A. Bruce & P. R. O. Bally (1941) ≡ *Pachycymbium vibratile* (E. A. Bruce & P. R. O. Bally) M. G. Gilbert (1990) ≡ *Angolluma vibratilis* (E. A. Bruce & P. R. O. Bally) Plowes (1994).

Stems erect, solitary or rarely clustering, ± 6 × 1.5 cm, (grey-) green, spotted with reddish-brown, rhizomatous runners to 75 cm; **Tu** ± 5 × 2 mm, spreading-ascending; **L** rudiments ± 1 mm, spatulate, conspicuously articulated, with stipular rudiments; **Inf** with several nodding **Fl**; peduncle very short; **Ped** 5 - 6 × ± 2 mm; **Sep** ± 5 × 1.2 mm; **Cl** outside green, inside dark flesh-coloured to black-purple, partly paler transversally striped, ± 2 cm ∅, campanulate, tube ± 8 × 7 mm; **Cl** lobes ± 8 × 7 mm, outside apically purple, horizontally spreading, tips long attenuate and curved outwards, inside wrinkled, margins of the lower ½ with vibratile clavate **Ha**; **Cn** dark purple, ± 5 mm ∅; **Ci** cup-shaped, ± 2 mm, lobes forming pouches, apically 2-toothed; **Cs** lobes ± 4 × 1 mm, linear, acute, meeting apically; **Poll** ± 0.5 × 0.4 mm, ovate.

O. wilsonii (P. R. O. Bally) Bruyns (Aloe 37(4): 76, 2001). **T:** Uganda, Toro Distr. (*Wilson* 13 [K [alc]]). – **D:** SW Uganda.

≡ *Caralluma wilsonii* P. R. O. Bally (1966) ≡ *Pachycymbium wilsonii* (P. R. O. Bally) M. G. Gilbert (1990) ≡ *Angolluma wilsonii* (P. R. O. Bally) Plowes (1994).

Stems decumbent-ascending, grey-green and

blotched with brown, 5 - 6 cm, fleshy, sides grooved; **Tu** ≤ 6 mm, slender, spreading-ascending; **L** rudiments 4 mm, subulate-lanceolate, caducous; **Inf** with 2 - 6 **Fl**; peduncle 2 × 2 mm; **Bra** 0.5 - 1 mm, linear; **Ped** 2 - 3 mm; **Sep** 2.75 - 3 mm; **Cl** outside greenish-yellow, inside lemon-yellow, with inconspicuous violet irregular spots, ± 2 cm ⌀, campanulate, tube shallow, 2 - 2.5 × 6 mm ⌀, embracing the **Cn**; **Cl** lobes 3.5 × 3.2 mm, triangular, acute, horizontally spreading, apically slightly bent outwards, inside with scattered **Ha**, **Ha** purple, 0.2 - 0.8 mm, bristly, acute, stiff; **Cn** dark purple-maroon, 3 - 4 mm ⌀; **Ci** cup-shaped, 2 mm, lobes forming pouches, apically with a central notch; **Cs** lobes 1.7 mm, narrowly triangular, with a dorsal crest, apically obtuse or 2-toothed, basally erect; **Poll** ≤ 0.8 × 0.45 mm, D- or bean-shaped.

O. wissmannii (O. Schwartz) Bruyns (Aloe 37(4): 76, 2001). **T:** Yemen (*von Wissmann* 1207 [HBG]). – **D:** Saudi Arabia, Yemen, Oman.

≡ *Caralluma wissmannii* O. Schwartz (1939) ≡ *Pachycymbium wissmannii* (O. Schwartz) M. G. Gilbert (1990) ≡ *Angolluma wissmannii* (O. Schwartz) Plowes (1994).

The varieties introduced by Bruyns (2001) are quite constant in their character states, which suggests that a higher rank (subspecies, species) could be more appropriate.

O. wissmannii var. **eremastrum** (O. Schwartz) Bruyns (Aloe 37(4): 76, 2001). **T:** Yemen (*Rathjens* 103 [HBG]). – **D:** N Yemen, Saudi Arabia.

≡ *Caralluma eremastrum* O. Schwartz (1939) ≡ *Pachycymbium eremastrum* (O. Schwartz) M. G. Gilbert (1990) ≡ *Angolluma eremastrum* (O. Schwartz) Plowes (1994).

Differs from var. *wissmannii*: **Ped** 6 × ± 2.5 mm, ± curved downwards; **Cl** both faces pale yellow, ± 2.4 cm ⌀, rotate, incised for ± ½; **Cl** lobes ascending, sometimes remaining united at the tips, margins in the upper ½ rolled outwards, apically tuberculate-papillose, ± hairy; **Cn** ≤ 5 mm ⌀, pale yellow; **Ci** cup-shaped, lobes 2.8 × 1.6 mm.

The morphological differences between var. *eremastrum* and var. *wissmannii* are considerably more pronounced than those between var. *wissmannii* and var. *parviloba*. Moreover, no intermediate forms between var. *eremastrum* and var. *wissmannii* are known, and the two do not differ in their range. For these reasons, var. *eremastrum* should more appropriately be regarded as a good biological species. – [U. Meve]

O. wissmannii var. **parviloba** Bruyns (Aloe 37(4): 76, 2001). **T:** Yemen (*Miller & King* 5336 [E]). – **D:** S Jemen, Oman. **I:** Miller & Morris (1988: as *Caralluma luntii*).

Differs from var. *wissmannii*: **Fl** much smaller, often with scatteredly hairy **Cl** lobes and white **Cn**. – [U. Meve]

O. wissmannii var. **wissmannii** – **D:** S Saudi Arabia, N and S Yemen.

Incl. *Caralluma meintjesiana* Lavranos (1962) ≡ *Pachycymbium meintjesianum* (Lavranos) M. G. Gilbert (1990).

Stems ascending-erect, densely clustered, (grey-) green, spotted with dark green or brownish-violet, to 8 cm; **Tu** ± 5 mm, robust, horizontally spreading, apically slightly bent upwards; **Inf** with 1 - 3 erect **Fl**; **Bra** 1 - 3 × 0.6 mm, triangular-linear, acute; buds conical; **Ped** green, marbled with brownish, to 10 × 2 mm; **Sep** 3 - 4 × ± 1 mm, sometimes bent outwards; **Cl** outside grey-green, densely and delicately red-dotted except upper ½ of lobes, ± 2 - 3.5 cm ⌀, incised ± to the base; **Cl** lobes 12 - 15 × 1 - 2 mm, narrowly linear, ascending-erect, apically obtuse or pointed, bent outwards, margins ± strongly revolute, blood-red, inside of **Cl** lobes basally densely papillose with small conical papillae, inside sometimes very minutely tomentose, basal ½ glossy from abundant nectar; **Cn** yellow or white, partly dotted with dark red, 3.5 - 5 mm ⌀, cup-shaped, glossy from nectar; **Ci** 2 - 2.5 mm, lobes forming ± shallow concave pouches, apically broad, somewhat outcurved, inside glandular-hairy; **Cs** lobes 1 - 1.7 × ± 0.6 mm, oblong-rectangular, geniculate, basally erect, ± as long as the **Anth**, apically triangular, indented, lobed or emarginate, with horn-like appendages where united with the **Ci**, these appendages 1 mm, robust, apically truncate, emarginate, erect or spreading outwards; **Poll** 0.4 × 0.25 mm, ovoid to D-shaped; **Fr** grey-green, spotted with purple, ± 5.5 × 0.7 cm.

O. woodii (N. E. Brown) L. C. Leach (Kirkia 10(1): 290, 1975). **T:** RSA, KwaZulu-Natal (*Medley-Wood* 4119 [K]). – **D:** RSA (KwaZulu-Natal).

≡ *Stapelia woodii* N. E. Brown (1892); **incl.** *Stapelia woodii* var. *westii* R. A. Dyer (1941).

Stems shortly prostrate-erect to erect, to 6 × 0.6 - 0.8 cm ⌀, forming tufted mats; **Tu** ± 1 (-1.5) cm, acute, ascending; **Inf** 1- to 3-flowered (rarely more); peduncle (if present) stout, persistent forming several **Inf**; **Fl** buds 5-angled, broadly ovoid, sides slightly concave, base truncate, apically shortly tapering; **Ped** 2 - 5.5 × 0.15 - 0.25 cm; **Sep** 5.5 - 7 × 1.5 - 2 mm; **Cl** pale reddish-brown, black-brown or maroon, 3 - 4.5 cm ⌀, inversely bowl-shaped; annulus ± 2.5 × 8 - 9 mm ⌀, very abruptly ascending, cushion- or dome-shaped, with a central depression, ± glabrous, inside very finely pubescent; **Cl** lobes to 1.8 × 1 cm, margins ± strongly revolute, inside weakly to strongly rugose, hairy, clavate **Ha** ≤ 1.75 mm, vibratile; **Cn** dark brown, sometimes black, 6 - 7 mm ⌀, columnar foot short, robust, obtusely 5-angled; **Ci** inversely saucer-shaped, lobes ± 2 × 1.5 mm, apically ± truncate, irregularly obtusely dentate or crenulate, lateral margins raised, fleshily thickened; **Cs** lobes ± broadly ovate, attenuate-acute, apical appendage ± narrowly subulate, very variable in length, sometimes longer than the **Anth**, shortly erect above the **Anth** and meeting each other, lateral margin often bent upwards, occasionally with 1 - 2 teeth at the sides below the apical appendages, sometimes with a small dorsal hump; **Anth** exposed, not meeting each other; nectar slit ± triangular; **Poll** ± 0.6 × 0.5 mm, ± D-shaped.

Closely related to *O. semota*. – [B. Müller, K. Seidler & F. Albers]

ORBEANTHUS

B. Müller & F. Albers

Orbeanthus L. C. Leach (Excelsa Tax. Ser. 1: 71, 1978). **T:** *Stultitia conjuncta* A. C. White & B. Sloane. – **Lit:** Leach (1978a). **D:** RSA (Northern Prov., Mpumalanga). **Etym:** For the genus *Orbea* (*Asclepiadaceae*); and Gr. 'anthos', flower; for the *Orbea*-like flowers.

Dwarf branched stem succulents; stems creeping, rooting along the whole length and forming lax cushions, subterranean runners short; stems dark green or reddish, partially blotched, to 30 × ± 0.9 cm, obtusely 4-angled, glabrous; **Tu** rarely ≥ 3 mm, acute, revolute; **L** rudiments small, ± triangular, acute, flattened, becoming hard, without stipular rudiments; **Inf** with 1 - 2 **Fl** at the the stem tips; **Fl** opening in succession, odour only slighty foetid; **Ped** greenish, glabrous; **Sep** pale green, ± triangular, glabrous; **Cl** broadly bowl-shaped, ± 6 cm ⌀, annulus broad, stout; **Cl** lobes triangular, ± connate, ± revolute, outside smooth, glabrous, inside ± smooth and glabrous; **Cn** biseriate, of the C(is)+Cs type, shortly stipitate, ± deeply sunken, hairy; **Ci** lobes elongately rectangular, ± deeply bifid, ± erect, basally fused with the **Cs**; **Cs** lobes simple, incumbent on the **Anth**, overtopping these by far, dorsally fused with the **Ci**; **Anth** yellow, ± rectangular, without apical appendages; **Poll** D-shaped; **Fr** narrowly fusiform, glabrous, follicles of a pair at an acute angle; **Se** not described.

Orbeanthus is closely related to *Orbea* but shows also affinities with the Madagascan genus *Stapelianthus* (Leach 1978a). Bruyns (2001) includes *Orbeanthus* in *Orbea*. On the base of the considerable differences in vegetative architecture (creeping stems, minute leaf rudiments) and in the corona (hairy), *Orbeanthus* is better kept separate.

O. conjunctus (A. C. White & B. Sloane) L. C. Leach (Excelsa Tax. Ser. 1: 73, 1978). **T:** RSA, Northern Prov. (*Crundall* s.n. [PRE]). – **D:** RSA (Northern Prov.). **I:** Asklepios 57: 18, figs. 1-2, 1992.

≡ *Stultitia conjuncta* A. C. White & B. Sloane (1938) ≡ *Orbea conjuncta* (A. C. White & B. Sloane) Bruyns (2001).

Stems grey-green, blotched with dark brownish-green, to 15 × 0.8 cm; **Tu** 1 - 2 mm, conical, to 1.6 cm apart; **Inf** 2-flowered; **Fl** buds brownish-green, 4 × 2 mm, ovoid, glabrous; **Ped** brown-green, 10 × 3 mm; **Sep** 5 mm; **Cl** 5 - 5.2 cm ⌀, ± semiglobose or campanulate, outside whitish-cream with pink, base of the lobes laterally with triangular yellowish areas, inside pale maroon, tube 5 × 14 mm, mouth partly cream-coloured, narrowed by the annulus, 8 mm ⌀, 5-angled, fleshy, cup-shaped, outside of the **Cl** lobes furrowed for ± 6 mm at the base; **Cl** lobes 3 × ± 10 mm; **Cn** ± 8 mm ⌀; **Ci** dark maroon, stippled with white, deeply bowl-shaped, lobes 3 mm, elongate-rectangular, ± erect, bifid down to ½, tips divergent, flattened, velvety and densely finely pubescent or finely setulose, **Ha** shining, usually translucent, uncoloured, very short, nearly spine-like, attenuate; **Cs** upper ½ purple, lower ½ white, blotched with purple, lobes ± ovate, flat, rarely ascending above the **Sty** head, dorsally humped, velvety densely finely pubescent, the longest **Ha** dark purple; **Anth** pale yellow; **Poll** yellow, ± translucent, 0.5 - 0.7 × 0.45 - 0.6 mm.

O. hardyi (R. A. Dyer) L. C. Leach (Excelsa Tax. Ser. 1: 74, 1978). **T:** RSA, Northern Prov. (*Hardy & al.* s.n. [PRE 29028]). – **D:** RSA (Northern Prov., Mpumalanga). **I:** FPA 36: t. 1403, 1963; Asklepios 57: 18: figs. 6-9, 1992. **Fig. XXXVII.b**

≡ *Stultitia hardyi* R. A. Dyer (1963) ≡ *Orbea hardyi* (R. A. Dyer) Bruyns (2001).

Stems green, blotched with purple, to 30 × 0.6 - 0.9 cm, young distinctly (with age indistinctly) obtusely 4-angled, sides furrowed; **Tu** ± 1 mm, conical, acute, prominent, spreading or slightly inflexed towards the stem; **Inf** 1- to 2-flowered, slowly developing; peduncle ± 2 mm; **Fl** buds apically acute, abruptly broadening above the middle; **Ped** ± 5 mm; **Sep** 1.5 - 3 mm; **Cl** 6.5 - 7 cm ⌀, cream-coloured, irregularly blotched with dark maroon, annulus paler, tube margin and annulus outside maroon, tube 5 - 6 mm, bowl-shaped, embracing the annulus, this to 4 × ± 18 - 20 mm ⌀, cushion-shaped around the central depression, which comprises the **Cn**; **Cl** lobes 2 - 2.5 × 1.3 - 1.8 cm, ovate, acuminate, outside glabrous, inside finely furrowed, finely papillose, tips ± glabrous, otherwise hairy, **Ha** translucent, coloured like the background, small, globose, bladder- or balloon-like or ± cylindrical (then finely taper-pointed); **Cn** ± 8 mm ⌀; **Ci** deeply bowl-shaped, at the base cream-coloured, blotched with brown-carmine, margins apically dark carmine, lobes 3 mm, narrowly rectangular, deeply bifid, weakly blotched, hairy, **Ha** brown or colourless, fine, acute; **Cs** basal ½ cream-coloured, margins and upper ½ dark carmine, lobes 4 mm, linear, geniculate, apically attenuate, clavate, divergent above the **Sty** head, dorsally humped, hairy, **Ha** dark purple, to 0.5 mm, acuminate or ± cylindrical; **Anth** yellowish; **Poll** 0.5 × 0.4 mm.

×ORBELIA

B. Müller & F. Albers

×**Orbelia** G. D. Rowley (Name that Succulent, 95, 1980).
Incl. ×*Orbeopelia* G. D. Rowley (1980).
Incl. ×*Orbeostemon* P. V. Heath (1992).
 = *Orbea* × *Stapelia*. Apart from the following formally named hybrid, the combinations *O. tapscottii* × *S. gettliffei* and *O. maculata* × *S. kwebensis* have also been reported (Plowes, pers. comm.).

The cross *O. caudata* × *S. gigantea* was named *Stapelia* ×*tarantuloides* R. A. Dyer (*pro sp.*) (≡ ×*Carapelia tarantuloides* (R. A. Dyer) G. D. Rowley ≡ ×*Orbeostemon tarantuloides* (R. A. Dyer) P. V. Heath).

×**O. barklyi** (N. E. Brown *pro sp.*) G. D. Rowley (Cact. Succ. J. Gr. Brit. 44(1): 2, 1982). **T:** RSA, Cape Prov. (*Morris* s.n. in *Barkly* 31 [K?]).
 ≡ *Stapelia* ×*barklyi* N. E. Brown (*pro sp.*) (1890) ≡ ×*Gonostapelia barklyi* (N. E. Brown *pro sp.*) P. V. Heath (1992).
 = *O. namaquensis* × *S. pulvinata*.

PACHYCARPUS

U. Meve

Pachycarpus E. Meyer (Comm. Pl. Afr. Austr., 209, 1837). **T:** *Asclepias grandiflora* Linné *fil.* [Lectotype, designated by N. E. Brown 1908.]. — **Lit:** Smith (1988); Goyder (1998a). **D:** Africa S of the Sahara. **Etym:** Gr. 'pachys', thick; and Gr. 'karpos', fruit; for the mostly very large fruits.

Geophytic shrubs with deciduous above-ground parts with white latex; **R** tuberously thickened, usually fusiform; stems erect, unbranched; **L** decussate, petiolate, coriaceous; **Inf** terminal or laterally near the stem tips, axillary, pedunculate, pseudo-umbellate; **Cl** rotate or campanulate; **Cn** uniseriate; **Cs** lobes dorsiventrally flattened, linear, lanceolate or weakly 3-parted, often with a pair of fleshy rails; **Gy** ± stipitate; **Anth** appendages membranous, usually surpassing the **Sty** head; **Poll** flat, ± drop-shaped, pendent; **Fr** usually solitary, erect, ellipsoid to thickly ovoid, coriaceous, conspicuously winged or keeled.

Usually very large and attractively flowering members of the *Asclepiadinae* from the grasslands of E and S Africa. Closely related to *Glossostelma*, *Xysmalobium* and *Gomphocarpus* as well as *Asclepias*. *Raphionacme* is often similar in habit. Of the ± 37 species, 24 are native to S Africa.

The following names are of unresolved application but are referred to this genus: *Schizoglossum cabrae* De Wildeman (1903); *Schizoglossum gossweileri* S. Moore (1912); *Schizoglossum theileri* S. Moore (1921).

PECTINARIA

B. Müller & F. Albers

Pectinaria Haworth (Synops. Pl. Succ., 14, 1819). **T:** *Stapelia articulata* Aiton. — **Lit:** Bruyns (1981). **D:** RSA. **Etym:** Lat. 'pectinarius', comb- (from Lat. 'pecten', comb); for the comb-like processes of the corona.

Perennials, mostly decumbent, forming mats; stems 2 - 8 cm, 6-ribbed, glabrous; **Tu** conspicuous, **L** rudiments minute, triangular; **Inf** 1- (rarely 2-) flowered, apically from the sides of young stems, successively opening, without peduncle; **Ped** to 1 (rarely 2.5) cm, terete; **Sep** green, ± 2 mm, lanceolate, acute; **Cl** either flat with shallow tube (hardly embracing the **Cn**), or **Cl** lobes apically united with each other and then tube deeply conical; **Cl** lobes deltoid, ± long-attenuate, margins often conspicuously folded, inside ± densely papillose; **Cn** biseriate, of the (Cis-)+Cs-type; **Ci** lobes bifid, segments cylindrically horn-like, erect or ascending, ± laterally diverging, each fused with the dorsal appendage of the **Cs**; **Cs** lobes apically ± deltoid, sometimes reduced to a small swelling, incumbent on the **Anth**, dorsal appendage divided into numerous ascending finger-like processes, which form a crest-like ring together with the lobes of the **Ci**; **Anth** rectangular, laterally-apically without appendage, frequently ± hidden by the **Cs**; **Poll** 0.22 × 0.15 mm, usually ovoid-ellipsoid; **Fr** fusiform, those of a pair arranged at an acute angle; **Se** brownish-ochre, ± 3 mm, flat, ovate, with narrow margin, apically with a tuft of simple **Ha**.

Bruyns (1981) divided the genus as follows into 2 sections:

[1] Sect. *Pectinaria*: Stems ± globose; **Ped** short, robust, closely pressed to the stems; **Cl** lobes apically united with each other; **Cl** tube deeply conical, outside and inside densely papillose, papillae prominent, papillae of the inside apically spine-tipped.
[2] Sect. *Erectiflora* Bruyns 1981: Stems more rectangular; **Ped** usually elongate, slender, erect; **Cl** flatly spreading; **Cl** lobes horizontally spreading; **Cl** tube shallow, hardly embracing the **Cn**, outside and inside delicately papillate with spine-tipped papillae.

Bruyns (l.c.) reports the natural hybrid *P. articulata* ssp. *articulata* × *P. longipes*.

P. articulata (Aiton) Haworth (Suppl. Pl. Succ., 14, 1819). **T:** [neo – icono]: Masson, Stapel. Nov., t. 30, 1796. — **D:** RSA (Northern Cape, Western Cape).
 ≡ *Stapelia articulata* Aiton (1789).

P. articulata ssp. **articulata** — **D:** RSA (Northern Cape: NW Sutherland). **I:** Bruyns (1981: 65). **Fig. XXXVII.c**

[1] Stems in dense fascicles, to 6 × 2 cm; **Tu** ± dome-shaped; **L** rudiments minute, soft; **Fl** erect; **Ped** 5 - 15 mm; **Cl** outside purple-brown, inside dark purple or pale yellow, 5 - 8 mm ⌀, ± bowl-shaped, truncate, apically flattened, tube 2.5 - 3.5 mm deep and ⌀, bowl-shaped to weakly conical; **Cl** lobes 2.5 - 3.5 × 3 - 3.5 mm, margins very weakly folded outwards, apically fused with each other already within the **Cl** tube, papillae of the outside rounded, those of the inside crystal-like, 0.25 - 0.3 mm, columnar, spine-tip transparent, rounded; **Cn** dark purple or pale yellow, ± 2 mm ⌀; **Ci** lobe segments 0.3 - 0.4 mm, erect to ascending, laterally slightly diverging, fused with the **Cs** for ± ½ of their length; **Cs** lobes apically 0.4 mm, deltoid, tip rounded, processes of the dorsal appendage ± 0.4 mm; **Poll** 0.2 - 0.22 × 0.14 mm.

P. articulata ssp. **asperiflora** (N. E. Brown) Bruyns (Cact. Succ. J. Gr. Brit. 43(2/3): 65, 1981). **T**: RSA, Western Cape (*Pillans* 70 [BOL]). – **D**: RSA (Western Cape: E margin of the Great Karoo to the Little Karoo S of Laingsburg). **I**: Bruyns (1981: 66). **Fig. XXXVII.d**

≡ *Pectinaria asperiflora* N. E. Brown (1909).

[1] Differs from ssp. *articulata*: **Fl** pointing downwards; **Ped** down-curved; **Cl** 6 - 8 mm ⌀, conical, tube 2.3 - 3.2 mm deep, conical; **Cl** lobes 3.6 - 6 × ± 3.5 mm, margins sometimes conspicuously strongly folded outwards, apically fused with each other 5 mm above the tube mouth; **Cs** lobes apically 0.3 - 0.4 mm, tip truncate, variably incised, sometimes deltoid, processes of the dorsal appendage erect to ascending, often with fleshy parts between the processes and the margins.

P. articulata ssp. **borealis** Bruyns (Cact. Succ. J. Gr. Brit. 43(2/3): 67, ills., 1981). **T**: RSA, Northern Cape (*Bayer* 1506 [NBG]). – **D**: RSA (Northern Cape: Richtersveld). **Fig. XXXVII.e**

[1] Differs from ssp. *articulata*: **Cl** dark purple-brown, tube 3.5 × 7 mm ⌀, inside more weakly papillose; **Ci** lobes forming rather large conspicuously separate pouches between the dorsal appendages of the **Cs**.

P. articulata ssp. **namaquensis** (N. E. Brown) Bruyns (Cact. Succ. J. Gr. Brit. 43(2/3): 66, 1981). **T**: RSA, Northern Cape (*Templeman* s.n. in *Pillans* 22 [BOL]). – **D**: RSA (Northern Cape: Little Namaqualand). **I**: Bruyns (1981: 67).

≡ *Pectinaria articulata* var. *namaquensis* N. E. Brown (1909).

[1] Differs from ssp. *articulata*: Stems 8 - 12 mm ⌀; **Fl** usually horizontal; **Ped** 2 - 10 mm; **Cl** pale greenish-yellow or purple-red, ± 5 mm ⌀, shortly conical; **Cl** lobes ± 4 × 2.5 mm, triangular, margins esp. apically strongly folded outwards, leaving only a small opening into the **Cl** inside, papillae of the inside 0.15 - 0.2 mm, dome-shaped; processes of the dorsal **Cs** appendage ± horizontally pointing outwards.

P. longipes (N. E. Brown) Bruyns (Cact. Succ. J. Gr. Brit. 43(2/3): 69, ill. (p. 68), 1981). **T**: RSA, Northern Cape (*Marloth* 3799 [BOL, PRE]). – **D**: RSA (Northern Cape: Sutherland Distr.).

≡ *Caralluma longipes* N. E. Brown (1909); **incl.** *Caralluma longipes* var. *villettii* C. A. Lückhoff (1938).

[2] Stems brownish-green, to 6 × 1 - 1.5 cm; **Tu** conical; **L** rudiments acute, rarely hardening; **Fl** horizontal; **Ped** 10 - 35 × 1.2 mm, slender, erect, apically curved outwards; **Cl** yellow, 8 - 12 mm ⌀; **Cl** lobes 2.5 - 3 × 2.5 - 3 mm, deltoid to broadly ovate-deltoid, margins folded outwards, papillae spine-tipped, transparent; **Cn** yellowish-orange, ± 2 mm ⌀, basally shortly cup-shaped; **Ci** lobes ± strongly bifid, segments ascending, 0.3 - 0.4 mm, ± cylindrical, near the tips fused with the dorsal appendage of the **Cs**; **Cs** lobes absent or present, broadly rectangular, tip concave, appendages numerous, small, rounded, or rather ± emarginate.

Caralluma longipes var. *villettii* is only insufficiently known and thus provisionally treated as synonym.

P. maughanii (R. A. Dyer) Bruyns (Cact. Succ. J. Gr. Brit. 43(2/3): 69, 1981). **T**: RSA, Western Cape (*Maughan-Brown* 20 [GRA]). – **D**: RSA (Western Cape: Vanrhynsdorp Distr.). **I**: Bruyns (1981: 70).

≡ *Caralluma maughanii* R. A. Dyer (1931).

[2] Stems procumbent to ascending, (brownish-)green, to 8 × 1 - 1.5 cm ⌀; **Tu** dome-shaped; **L** rudiments hardened; **Fl** erect; **Ped** 5 - 25 mm, slender, erect; **Cl** reddish-purple, 1.2 - 1.6 cm ⌀; **Cl** lobes apically or sometimes entirely dark yellow, 5 - 7 × ± 2.5 mm, lanceolate, margins conspicuously folded outwards; **Cn** dark purple-black; **Ci** lobes ≤ 1 mm, slender, for ± ¾ of their length fused with the dorsal appendage of the **Cs**; **Cs** lobes frequently very broad, overlapping in the **Fl** centre or deltoid and not overlapping, apically usually irregularly emarginate, dorsal appendage usually trifid, appendages ± 1 mm, horizontally porrect (more slender if more numerous); **Poll** 0.22 mm ⌀, ± circular.

PETOPENTIA

U. Meve

Petopentia Bullock (KB 10: 362, 1954). **T**: *Pentopetia natalensis* Schlechter. – **D**: RSA. **Etym**: Anagram of the genus name *Pentopetia* (*Asclepiadaceae*), where the taxon was previously classified.

Lianas with white latex; **R** tuber ± globose, large, to 40 cm ⌀ (or more?), sometimes several lined up in a row, bark red-brown, finely fissured; stems twining to 15 m tall, becoming woody, often reddish or with red-brown bark, glabrous; **L** 5 - 15 mm robustly petiolate, petiole canaliculate on the upper face, lamina broadly oblong, 5 - 13 × 1 - 6 cm, basally cordate, rounded or obtuse, apically rounded, often mucronate or with drip-tip, ± leathery, glabrous, upper face dark green and glossy, lower face usually violet; **Inf** lateral, 2 - 15 mm pedunculate, branched, several-flowered; **Ped** 5 - 12 mm; **Fl** buds acutely conical; **Sep** free, 1 - 2 mm, ovate, acute, glabrous; **Cl** rotate, only basally shortly united; **Cl** lobes spreading, narrowly triangular-lanceolate, 12 - 14 × 4 mm, acute, pale yellow or pale green, sometimes with strong purple tinge, coarse, glabrous; **Cn** petaloid, basally arising from the sinuses of the **Cl** lobes and alternating with them, **Cn** lobes greenish, erect, filiform-subulate; **Anth** lanceolate, acute, whitish, conically arranged; **Fil** basally fused with the **Cl** lobes, filiform from a broadened base; pollen in tetrads, rhomboid to T-shaped, pollen spoon broadly obovate; translator 1.5 × 1 mm, yellow-brown; **Fr** in pairs, divergent, flattened fusiform, 6 - 11 × 1.2 - 1.7 cm, shortly beaked; **Se** oblong, black-brown, dorsally with a median keel, ± smooth.

A monotypic genus of the *Periplocoideae*. Similar character states otherwise only occur in the Madagascan genus *Ischnolepis*, but this appears to be predominantly the result of parallel developments. A direct phylogenetic relationship between the 2 genera does certainly not exist. Accordingly, the synonymization of *Petopentia* with *Ischnolepis* (Venter & Verhoeven 2001) is here not supported.

The handsome large-leaved lianas have a conspicuous halfways above-ground caudex. It is widespread in cultivation and in the trade under the misapplied name *Fockea tugelensis*.

P. natalensis (Schlechter) Bullock (KB 10: 362, 1954). **T:** RSA, KwaZulu-Natal (*Wood* s.n. [K]). – **D:** RSA (KwaZulu-Natal, Eastern Cape). **I:** Rowley (1987: 70, as *Fockea tugelensis*); Venter & al. (1990). **Fig. XXXVII.a**

≡ *Pentopetia natalensis* Schlechter (1894) ≡ *Tacazzea natalensis* (Schlechter) N. E. Brown (1907) ≡ *Ischnolepis natalensis* (Schlechter) Venter (2001); **incl.** *Fockea natalensis* hort. (s.a.) (*nom. inval.*, Art. 29.1).

Description as for the genus.

PIARANTHUS

U. Meve

Piaranthus R. Brown (Mem. Wern. Nat. Hist. Soc. 1: 23, 1810). **T:** *Stapelia punctata* Masson [Lectotype, proposed by N. E. Brown 1909, designated by White & Sloane 1937.]. – **Lit:** Meve (1994). **D:** Namibia, RSA (Northern Cape, Western Cape, Eastern Cape). **Etym:** Gr. 'piar', fat; and Gr. 'anthos', flower; referring to the fleshy corolla.
Incl. *Obesia* Haworth (1812). **T:** *Stapelia geminata* Masson [Lectotype, designated by Farr & al. in ING.].

Stem succulents with plagiotropic growth, mat-forming under host plants; stems 1 - 6 × 0.8 - 2 cm, subglobose to cylindrical, ± distinctly 4- (to 5-) ribbed, green to grey-green; **L** deltoid, ± 1 mm ⌀, deciduous, stipular glands very small; **Inf** extra-axillary, close to the stem tips, in reduced cymes (1- to few-flowered), drepanoid, rarely scorpioid, **Fl** ± simultaneously opening; **Ped** 2 - 25 × 1 - 2 mm, smooth, erect; **Sep** 1 - 4 mm, acute, green, smooth; **Cl** with ± sweetish odour of excrements (radiate group), honey-scented or ± scentless (campanulate group); **Cl** 6 - 42 mm ⌀, **Cl** lobes ± completely free (radiate group), or with campanulate **Cl** tube and short free lobes (campanulate group); **Cl** lobes deltoid, outside green, glabrous, inside hairy with increasing density from base to apex (radiate group), or papillose to shortly hairy (campanulate group); **Cn** yellow, simple, but **Cs** double with ventral lanceolate lobes plus dorsal crest-like projections (**Cs** crest), **Cs** crest compact, rounded or protruding in a wing-like manner, ± tuberculate to deeply furrowed, **Ci** reduced to a basal rim; guide-rails protruding, short, convex; **Anth** yellow, oblong; **Sty** head pentagonal, white; **Poll** flattened-ovate, 0.25 - 0.8 × 0.15 - 0.4 mm, yellow, inner sides apically with nail-like rectangularly bent germination mouth; corpuscle 0.18 - 0.4 mm; **Fr** paired, raised vertically, only slightly spreading, 6 - 14 × 0.4 - 0.8 cm, smooth, with 40 - 130 **Se**; **Se** ellipsoid, flat, brown, 3 - 7 × 2 - 4 mm, tuft of **Ha** 10 - 16 mm.

Vegetatively, *Piaranthus* is almost indistinguishable from the genus *Duvalia*. However, these similarities are obviously due to convergence because of the same habitat preferences. On the contrary, *Piaranthus* is closely related to *Huerniopsis* (see comments there) according to Meve (1994) and Bruyns (1999a).

Piaranthus can be informally divided into 2 groups (Meve 1994).
[1] "radiate group": **Cl** star-shaped.
[2] "campanulate group": **Cl** campanulate.

Intrageneric natural hybrids are rare, but the occurrence of the hybrids *P. punctatus* × *P. framesii*, *P. punctatus* × *P. decorus* ssp. *cornutus* and *P. punctatus* × *P. decorus* ssp. *decorus* are sufficiently well-documented (Meve 1994).

In addition to the names enumerated above, *Podanthes* in the sense of Bentham and Hooker as well as *Stapelia* Sect. *Caruncularia* (Haworth) Decaisne 1844 p.p. are further synonyms of *Piaranthus*.

See ×*Duvaliaranthus* for an intergeneric hybrid between *Duvalia* and *Piaranthus*.

P. barrydalensis Meve (Bradleya 12: 68-71, ills., t. 1 (p. 97), 1994). **T:** RSA, Western Cape (*Meve & al.* 128 [K, MSUN]). – **D:** RSA (Western Cape: Little Karoo).

[1] Stems subglobose-ovoid, 1 - 3 (-5) × 0.8 - 2

cm; **L** rudiments minute, < 0.5 mm; **Inf** 1- to 2-flowered; **Ped** 7 - 12 mm; **Sep** 2 - 3 × 0.8 - 1.2 mm; **Cl** with strong sweetish scent of decaying fruits; **Cl** 20 - 26 mm ⌀, stout; **Cl** lobes 5 - 6 mm broad, white to ivory-coloured banded red-brown, hairy, **Ha** 0.1 - 0.4 mm, acute-conical; **Cn** spotted red-brown, ± 2.5 × 5 mm; **Cs** lobes ± 1.5 × 0.7 - 1 mm, linear, incumbent on the **Anth**, **Sty** head freely visible; **Cs** crest ± 1.2 - 1.5 mm broad; **Poll** 0.45 - 0.55 × 0.28 - 0.35 mm; **Fr** 8 - 12 × 0.6 cm; **Se** 4 × 2.5 mm.

White & Sloane (1937: 811) erroneously treated plants of this species as *P. decorus*. *P. barrydalensis* is closely allied to the *P. geminatus* / *P. foetidus* complex.

P. comptus N. E. Brown (HIP 21(1): t. 1924B + text, 1890). **T:** RSA, Cape Prov. (*Barkly* 58 [K]). – **D:** RSA (Western Cape, Eastern Cape, Northern Cape: Great Karoo). **Fig. XXXVIII.b**

≡ *Caralluma compta* (N. E. Brown) Schlechter (1898); **incl.** *Piaranthus comptus* var. *ciliatus* N. E. Brown (1908).

[1] Stems 1 - 5 (-7) × 1 - 2.2 cm, globose to cylindrical, slightly rough; **Inf** 1- to several-flowered; **Ped** 5 - 15 × 1 - 2 mm; **Sep** 1 - 3 × 1 mm; **Fl** with intensive odour of horse sweat mixed with rotten fruits; **Cl** 14 - 24 mm ⌀ (usually ± 17 mm), radiate; **Cl** lobes 3.5 - 5 mm broad, delicate, sometimes basally slightly fused, whitish with red or red-brown dots, patches or tranverse bands, usually densely hairy, **Ha** slightly conical, 0.4 - 1 mm, whitish to brownish purple, basal ¼ of the **Cl** margins often with clavate **Ha**; **Cn** 2.6 - 3.2 × 3.5 - 4.8 mm; **Cs** lobes 1 - 1.5 × 0.6 mm, ± linear, acute; **Cs** crest nearly unfurrowed, 0.8 - 1.5 mm broad; **Poll** ± 0.5 × 0.3 - 0.35 mm; **Fr** 6 - 11 cm; **Se** 3.5 - 5.5 × 2 - 3 mm.

P. decorus (Masson) N. E. Brown (JLSB 17: 163, 1878). **T:** [lecto – icono]: Masson, Stapel. Nov., t. 26, 1797. – **D:** S Namibia, NW RSA.

≡ *Stapelia decora* Masson (1797) ≡ *Obesia decora* (Masson) Haworth (1812) ≡ *Caralluma decora* (Masson) Schlechter (1898).

The 2 subspecies are easily separated by characters of the corona. The ssp. *decorus* possesses extremely large and deeply furrowed **Cs** crests and **Cs** lobes being incumbent on the anthers. In ssp. *cornutus*, in contrast, the **Cs** crests are rather delicate and the **Cs** lobes erect.

P. decorus ssp. **cornutus** (N. E. Brown) Meve (Bradleya 12: 76, ills., t. 2 (p. 100), 1994). **T:** RSA, Northern Cape (*Barkly* 25 p.p. [K]). – **D:** S Namibia, NW RSA (Northern Cape); Namaqualand Broken Veld. **Fig. XXXVII.f, XXXVII.g**

≡ *Piaranthus cornutus* N. E. Brown (1908); **incl.** *Piaranthus cornutus* var. *grandis* N. E. Brown (1908); **incl.** *Piaranthus pulcher* N. E. Brown (1908); **incl.** *Piaranthus nebrownii* Dinter (1914) ≡ *Piaranthus pulcher* var. *nebrownii* (Dinter) A. C. White & B. Sloane (1937); **incl.** *Piaranthus mennellii* C. A. Lückhoff (1935); **incl.** *Piaranthus pallidus* C. A. Lückhoff (1935); **incl.** *Piaranthus ruschii* Nel (1937).

[1] Stems 1 - 4 (-8) × 1 - 2 cm, ovoid to cylindrical, slightly rough; **Inf** 1- to several-flowered; **Ped** 4 - 12 × 1 - 1.5 mm; **Sep** 2 - 4 mm; **Fl** ± scentless or with sweetish odour of decaying fruits; **Cl** 15 - 28 mm ⌀; **Cl** lobes 4 - 5 mm broad, ± free, often somewhat ascending (at an angle of ± 30°), tips often recurved, whitish to brownish-yellow with reddish to brown dots, patches or tranverse bands (rarely without pattern), densely hairy, **Ha** conical, white to purple-brown, 0.1 - 1 mm; **Cn** 2.5 - 3 (-3.5) × 3.5 - 4.7 mm; **Cs** lobes 1.2 - 2 × 0.5 - 0.8 mm, linear, blunt to acute, tips entire or irregularly serrate, mostly ascending to erect with recurved tip; **Cs** crest delicate, 0.8 - 1.5 mm broad; **Poll** 0.45 - 0.6 × 0.28 - 0.35 mm; **Fr** ± 10 cm; **Se** ± 5 × 3 mm.

This subspecies is typically confined to more arid localities as the typical subspecies.

P. decorus ssp. **decorus** – **D:** RSA (Northern Cape, Western Cape); Western Mountain Karoo, to 1300 m.

Incl. *Stapelia serrulata* Jacquin (1806) ≡ *Obesia serrulata* (Jacquin) Sweet (1827) ≡ *Duvalia serrulata* (Jacquin) Loudon (1841) ≡ *Piaranthus serrulatus* (Jacquin) N. E. Brown (1878) ≡ *Caralluma serrulata* (Jacquin) Schlechter (1898).

[1] Differs from ssp. *cornutus*: Stems 1.5 - 5 × 1 - 2 cm; **Ped** 5 - 20 × 1 - 2 mm; **Sep** ± 3 × 1 - 1.5 mm; **Fl** with sweetish odour of decaying fruits; **Cl** 15 - 30 mm ⌀; **Cl** lobes 4 - 5 mm broad, usually horizontally expanded, apically often recurved, ivory-coloured, cream-yellow or green-yellow with bright to dark red-brown or brown dots, patches or transverse bands, densely hairy, **Ha** 0.1 - 0.5 mm, conical, white to bright purple; **Cn** 2.5 - 3 (-3.5) × 4.2 - 5 mm; **Cs** lobes 1.2 -1.5 (-2) × 0.5 - 1 mm, ± incumbent on the **Anth** (apex sometimes slightly erect); **Cs** crest ± rectangular in top view, 1.4 - 2 mm broad, robust, raised and usually deeply furrowed; **Poll** 0.5 - 0.6 × 0.3 - 0.35 mm; **Fr** ± 8 × 0.5 cm; **Se** ± 4.5 × 2.5 mm.

P. framesii Pillans (South Afr. Gard. 18: 62, ill., 1928). **T:** RSA, Cape Prov. (*Frames* s.n. in *NBG* 1283/26 [BOL, NBG]). – **D:** RSA (Northern Cape, Western Cape); Karoo, 500 - 1100 m.

[2] Stems 1 - 5 (-9) × 1 - 2 (-2.5) cm, ± ovoid to cylindrical, compact, slightly rough; **Inf** 1- to several-flowered; **Ped** 8 - 20 × 1.5 - 2 mm; **Sep** 2 - 4 mm; **Fl** scentless or honey-scented; **Cl** 17 - 23 mm ⌀, fused in the central ½; **Cl** tube campanulate, 5 - 8 mm ⌀, 4 - 7 mm high, ± pentagonal; **Cl** lobes 6.5 - 9 × 4 - 6 mm, basally depressed, white to ivory-coloured with reddish patches or transverse bands,

papillose, papillae globose, ± 0.1 mm; **Cn** 4 - 5.5 × 4.5 - 5.5 mm, ± barrel-shaped, winged in staminal position; **Cs** lobes 0.5 - 1.5 × ± 0.5 mm, linear-subulate, entire, crooking over the backs of the **Anth**; **Cs** crest absent; **Sty** head free; **Poll** 0.6 - 0.7 × ± 0.35 mm; **Fr** ± 10 cm; **Se** ± 5 × 2.5 mm.

P. framesii is a characteristic species of Western Karoo mountains (500 - 1100 m). In contrast, the closely related *P. punctatus* prefers the lower areas of the Namaqualand Succulent Karoo. Both species are easily confused, but the papillae on the corolla (instead of hairs) and the winged corona typically characterize *P. framesii*. Recent collections are also indicative of the presence of hybrid populations.

P. geminatus (Masson) N. E. Brown (JLSB 17: 163, 1878). **T:** [lecto – icono]: Masson, Stapel. Nov., t. 25, 1797. – **D:** S RSA (Western Cape, Eastern Cape).

≡ *Stapelia geminata* Masson (1797) ≡ *Obesia geminata* (Masson) Haworth (1812) ≡ *Podanthes geminata* (Masson) G. Nicholson (1886) ≡ *Caralluma geminata* (Masson) Schlechter (1898).

P. geminatus is an extremely variable species (see Meve (1994)), which forms a taxonomically difficult complex. Characters of the corolla serve best to identify the taxa involved.

P. geminatus var. **foetidus** (N. E. Brown) Meve (Bradleya 12: 84, ills., t. 1 (p. 97), 1994). **T:** RSA, Eastern Cape (*Pillans* 165 [K, BOL]). – **D:** RSA (Eastern Cape); Great Karoo, Little Karoo. **Fig. XXXVIII.e**

≡ *Piaranthus foetidus* N. E. Brown (1908); **incl.** *Piaranthus foetidus* var. *diversus* N. E. Brown (1908); **incl.** *Piaranthus foetidus* var. *multipunctatus* N. E. Brown (1908); **incl.** *Piaranthus foetidus* var. *pallidus* N. E. Brown (1908); **incl.** *Piaranthus foetidus* var. *purpureus* N. E. Brown (1908).

[1] Differs from var. *geminatus*: Stems 1 - 5 (-12) × 7 - 2 cm, often clavate; **Fl** with strong sweetish odour of excrement; **Cl** 18 - 30 mm ⌀; **Cl** lobes 5 - 7 mm broad, sometimes basally fused, stout and fleshy, ivory- to sand-coloured, centrifugally increasingly patterned with (red-) brown transverse streaks or bands (rarely plain red-brown), densely hairy, **Ha** 0.1 - 0.75 mm, white; **Cn** 2.5 - 3.5 × 5 - 6 mm, spotted red-brown; **Cs** lobes 1.5 - 2 × 1 mm, partly overlapping; **Cs** crest 1.3 - 2 mm broad; **Poll** 0.55 - 0.7 × 0.3 - 0.4 mm; **Fr** 6 - 9 cm; **Se** 4 - 5 × 2.5 - 3 mm.

This taxon shows a wide ecological amplitude, and it is probably the most common representative of *Piaranthus*.

P. geminatus var. **geminatus** – **D:** S RSA (Western Cape, Eastern Cape). **Fig. XXXVIII.c**

Incl. *Piaranthus disparilis* N. E. Brown (1908); incl. *Piaranthus pillansii* N. E. Brown (1908); incl. *Piaranthus pillansii* var. *fuscatus* N. E. Brown (1908); **incl.** *Piaranthus pillansii* var. *inconstans* N. E. Brown (1908); **incl.** *Piaranthus disparilis* var. *immaculatus* C. A. Lückhoff (1937); **incl.** *Piaranthus globosus* A. C. White & B. Sloane (1937).

[1] Stems 1 - 6 (-15) × 0.8 - 2 (-2.5) cm, subglobose to cylindrical, smooth or slightly rough; **Inf** 1- to 3-flowered (rarely with more **Fl**); **Ped** 2 - 20 × 1 - 2 mm; **Sep** 2 - 4 mm; **Fl** usually valerian-scented with fruity components; **Cl** 15 - 42 mm ⌀; **Cl** lobes ± 5 mm (3.2 - 6.5) broad, ± free, spreading horizontally, yellowish, bright brown or brown-red, without pattern or with (red-) brown dots or transverse bands (sometimes condensed to a ± plain colouration), ± strongly hairy, **Ha** 0.1 - 0.9 mm, conical, white, brown-red or purple, basal margins of the **Cl** lobes very rarely with clavate **Ha**; **Cn** 2.5 - 3.2 × 3.2 - 6.5 mm, usually yellow and unspotted; **Cs** lobes 1 - 2 × 0.5 - 1.2 mm, linear or lanceolate, acute or blunt, apically mostly dentate, parallel-incumbent on the **Anth**; **Cs** crest 1.2 - 2.2 mm broad, furrowed or rounded and knotty-warty; **Poll** 0.4 - 0.65 × 0.25 - 0.4 mm; **Fr** 6 - 13 cm; **Se** 3.5 - 4.5 × 2 - 3 mm.

P. parvulus N. E. Brown (FC 4(1): 1023, 1908). **T:** RSA, Western Cape (*Pillans* 130 [K, BOL]). – **D:** RSA (Western Cape); Great Karoo, Little Karoo. **I:** Meve (1994: 90, 100).

[1] Stems 1.5 - 5 (-9) × 1 - 2 cm, subglobose (-ovoid); **Inf** many-flowered (1 - 10 **Fl**); **Ped** 4 - 25 mm; **Sep** 1 - 2 mm; **Fl** indistinctly valerian-scented; **Cl** 6 - 13 mm ⌀; **Cl** lobes ± 2 mm broad, free, usually ascending (at an angle of 15 - 40°) with recurved tips, bright yellow, densely hairy, **Ha** 0.05 - 0.2 mm, conical, white; **Cn** 2 - 3.2 × 2 - 3.5 mm; free **Cs** lobes 1 - 1.8 × 0.3 - 0.6 mm, linear to acute or subulate, sometimes several times curved, erect; **Anth** freely visible; **Cs** crest ± 1 mm broad, ± rounded and slightly warty or furrowed; **Poll** 0.23 - 0.35 × 0.15 - 0.23 mm; **Fr** ± 7 cm; **Se** 4.5 - 5 × 2.5 mm.

P. punctatus (Masson) Schultes (Syst. Veg. 6: 9, 1820). **T:** [lecto – icono]: Masson, Stapel. Nov., t. 24, 1797. – **D:** RSA (Northern Cape, Western Cape). **Fig. XXXVIII.d**

≡ *Stapelia punctata* Masson (1797) ≡ *Obesia punctata* (Masson) Haworth (1812) ≡ *Caralluma punctata* (Masson) Schlechter (1898).

[2] Stems 1 - 6 (-15, in cultivation) × 0.8 - 2 cm, ovoid to cylindrical, slightly rough; **Inf** 1- to several-flowered; **Ped** 5 - 15 × 1.5 - 2 mm; **Sep** 2 - 3 mm; **Fl** unscented or honey-scented; **Cl** 18 - 28 mm ⌀, central ½ fused; **Cl** tube 6 - 9 mm ⌀, cup-shaped to campanulate, round, 4 - 6 mm high; **Cl** lobes 6 - 9 × 5 - 7 mm broad, white to ivory-coloured with reddish or brown patches or transverse bands, papillose, **Ha**-papillae blunt to acute, conical, 0.1 - 0.2 mm; **Cn** 3 - 4.5 × 3.5 - 5 mm, ± barrel-shaped; **Cs**

lobes 1 - 2 × 0.5 - 1 mm, linear, incumbent on the **Anth** or erect; **Cs** crest ± 1 mm broad, fringe-like; **Poll** 0.5 - 0.7 × 0.3 - 0.4 mm; **Fr** ± 10 cm; **Se** 5 - 6.5 × 2.5 - 3.5 mm.

PSEUDOLITHOS

U. Meve

Pseudolithos P. R. O. Bally (Candollea 20: 41, 1965). **T:** *Lithocaulon sphaericum* P. R. O. Bally. – **Lit:** Bruyns (1990a). **D:** Somalia, Oman. **Etym:** Gr. 'pseudo-', false; and Gr. 'lithos', stone; for the appearance of the stems.

Incl. *Lithocaulon* P. R. O. Bally (1959) (*nom. illeg.*, Art. 53.1). **T:** *Lithocaulon sphaericum* P. R. O. Bally.

Incl. *Anomalluma* Plowes (1993). **T:** *Caralluma dodsoniana* Lavranos.

Compact low-growing stem succulents, 1- to several-stemmed, rarely forming clumps or cushions; stems ± erect (to decumbent), 4-ribbed, globose, cube-like or clavate, 1.5 - 6.5 × 1 - 6 cm, pale brown or (grey-) green, tessellate, sections polygonal, 0.2 - 6 mm ⌀, irregularly arranged, convex, **Tu** 0.5 - 8 mm ⌀, flatly shield-like (cordate) to cylindrical, often with a median groove; **L** rudiments absent or to 1 mm, obtusely triangular (pyramidal); **Inf** basal to apical, sessile at the main stem or on short shoots, partly shortly pedunculate; **Fl** with nectar, usually with strong scent of excrements; **Ped** 2 - 12 × 1 - 2 mm; **Sep** lanceolate, glabrous; **Cl** yellowish, rose, (red-) brown to purple, 5 - 12 mm ⌀; **Cl** tube very short or cup-shaped to urceolate, to 3 × 5 mm; **Cl** lobes broadly triangular to almost linear, 0.5 - 9 × 0.5 - 2.5 mm, spreading or with recurved margins, glabrous or with **Ha**-papillae; **Cn** biseriate, 1.5 - 2.2 × 2 - 4 mm; **Ci** united flatly bowl- or cup-shaped or urceolate, margin entire or deeply incised; **Cs** delicate, 1 - 2 mm, linearly spoon-shaped, incumbent on the **Anth**; **Anth** ± rectangular to spoon-shaped, horizontally incumbent on the concave or convex **Sty** head or slightly ascending; **Poll** D-shaped or ovoid, caudicles short, attached to the corpuscle with basal shortly triangular broadenings; **Fr** slender-fusiform, 3 - 6 cm × 4 - 6 mm, diverging at ± 180°, glabrous; **Se** (black-) brown.

The tessellation of the stem surfaces is esp. characteristic for this genus. Some of the most succulent stapeliads are found among these "living rocks". The species are rare in nature, and the cultivation is delicate and usually only grafts on *Ceropegia linearis* are successful.

P. caput-viperae Lavranos (CSJA 46(3): 125-130, ills., 1974). **T:** Somalia (*Lavranos & Horwood* 10155 [PRE, K]). – **D:** Somalia. **Fig. XXXVIII.f, XXXVIII.g**

Stems 3 - 12, erect to decumbent, globose to clavate, 1.5 - 4 × 0.8 - 1.5 cm, brown-green, tessellate, sections polygonal, irregular in size, convex, enlarged to conspicuous cordate **Tu** along the **Ri**; **L** rudiments shortly pyramidal, ± 0.3 mm ⌀, erect; **Inf** umbellate, near the stem tips, or also laterally on short shooots, to 25-flowered; **Ped** 2 - 3 mm; **Cl** pale green, outside glabrous; **Cl** tube urceolate, 2.5 × 2.5 mm; **Cl** lobes triangular, erect or somewhat spreading, 0.5 × 0.5 mm, inside with short **Ha**-papillae; **Cn** purple, ± 1.5 - 2 mm; **Ci** fused in an urceolate manner in the basal ½, free lobes slightly incurved, medianly incised for ½, inside with downwards-directed **Ha**; **Cs** linear, ± 0.5 mm; **Poll** narrowly D-shaped, ± 0.3 × 0.14 mm, corpuscle narrowly ellipsoid; **Fr** ± 3 cm; **Se** dark brown.

P. caput-viperae and *P. dodsonianus*, although very different as to flower morphology, are the only two species of the genus that still show conspicuously differentiated leaf rudiments and (richly) branched stems (Bruyns & Meve 1995).

P. cubiformis (P. R. O. Bally) P. R. O. Bally (Candollea 20: 41, 1965). **T:** Somalia (*Bally & Daniels* 11785 [K]). – **D:** Somalia.

≡ *Lithocaulon cubiforme* P. R. O. Bally (1959) (*incorrect name*, Art. 11.3); **incl.** *Pseudolithos cubiformis* var. *viridiflorus* Horwood (1975).

Stems solitary, cube-like, 3 - 5 × 4 - 6 cm, pale green, tessellate, sections 3- to 6-angled, flat, enlarged along the **Ri**, cordate, often with a median groove; **L** rudiments ± absent; **Inf** umbellate, sessile on short shoots, to 30-flowered; **Ped** 6 - 12 mm; **Cl** outside pale (grey-) green, glabrous; **Cl** tube cup-shaped, 3 × 4 - 5 mm, inside dull yellow; **Cl** lobes linear-lanceolate, erectly spreading, 7 - 9 × 2 - 2.5 mm, margins recurved, flesh-coloured to greenish, densely beset with stiff **Ha**-papillae, apically with 2 - 4 vibratile ± 2 mm long clavate **Ha**; **Cn** dark purple, ± 2 - 2.5 × 4 mm; **Ci** fused in cup-shaped manner, with median incisions; **Cs** narrowly spoon-shaped, ± 1.5 mm; **Poll** D-shaped, ± 0.3 × 0.2 mm, corpuscle obovate.

These large plants are close to *P. migiurtinus*.

P. dodsonianus (Lavranos) Bruyns & Meve (Edinburgh J. Bot. 52(2): 202, 1995). **T:** Somalia, Northern Region (*Lavranos* 7326 [FT]). – **D:** Somalia, Oman. **Fig. XXXIX.b**

≡ *Caralluma dodsoniana* Lavranos (1971) ≡ *Anomalluma dodsoniana* (Lavranos) Plowes (1993).

Many-stemmed, irregularly branched; stems erect to decumbent, globose to clavate, 1 - 6 × 0.7 - 1.2 cm, green, irregularly tessellate, sections strongly convex, **Tu** cylindrical, uneven, upper face with median incision; **L** rudiments hardly 1 mm ⌀, pyramidal, concave, at a slight angle; **Inf** near stem tips, 1- to 3-flowered; peduncle ± 1 mm, caducous; **Ped** ± 2 mm; **Fl** opening in succession, with ample nectar, with freshly acid scent; **Cl** rotate, outside pale green, inside purple-brown, ± 6 mm ⌀, glabrous; **Cl** tube hardly 1 mm; **Cl** lobes triangular, ± 2 × 2 mm, flatly spreading; **Cn** purple, ± 2.5 - 3.5 × 1 mm; **Ci** basally fused into a dish-like structure, lobes 0.5 - 0.7 mm, ± placed against the **Cl** tube, ± rectangular, with a median incision extending over the whole length (Somalia) or ± straightly truncate (Oman); **Cs** spoon-shaped, ± 1 mm; **Poll** D-shaped, ± 0.28 × 0.19 mm, obovate; **Fr** ± 3 cm.

The generic placement of this species was in dispute since the original description as *Caralluma*. Apart from general morphological and biogeographical aspects, esp. SEM analyses of the epidermis of both stem and corolla furnished the important clues for a placement of this species in *Pseudolithos* (Bruyns & Meve 1995).

P. horwoodii P. R. O. Bally & Lavranos (CSJA 46(5): 220-222, ills., 1974). **T:** Somalia (*Lavranos & Horwood* 10152 [FT]). – **D:** Somalia.

Stems solitary, ± erect, cube-shaped, ± 4.5 cm ∅, (grey-) green, tessellate; **Tu** shield-like, slightly prominent, with a median groove; **L** rudiments ± absent; **Inf** umbellate, 4- to 6-flowered, usually apically, sessile, on the main stem or on short shoots; **Fl** with ample nectar; **Ped** 5 - 7 × 1.25 mm; **Cl** yellowish with purple spots, ± 7.5 mm ∅, openly campanulate (to rotate); **Cl** lobes triangular, 2.5 × 2 mm, greenish-purple towards the tips, with slightly recurved margins, glabrous; **Cn** ± 2 × 1.2 mm; **Ci** yellow-green, shallowly bowl-shaped, margin ± entire, 10-angled; **Cs** yellow, ± 1 mm, linear; **Poll** ovoid, ± 0.5 × 0.38 mm, corpuscle ellipsoid.

Apart from *P. horwoodii*, *P. dodsonianus* is the only other species with rotate and almost free corolla lobes. The pronounced similarity in the corona structure likewise points to a close relationship between the two taxa.

P. mccoyi Lavranos & Mies (Asklepios No. 82: 28-30, ills., 2001). **T:** Oman, Dhofar (*McCoy* 2305 [MO]). – **D:** Oman, Yemen.

Stems clump-forming, ± erect, rounded 4-angled, 1.2 - 4 (-7) × 0.4 - 0.8 cm, greyish-brown; **Cl** rotate, ± 7 mm ∅, with shallow **Cl**tube, dark olive-green with purple-brown spots or dark red-brown; **Ci** lobes ± 1 mm; **Fr** fusiform, ± 2 - 3 cm × 3.5 mm, acute, erect, diverging at an angle of 30°; otherwise like the closely related *P. dodsonianus*.

P. migiurtinus (Chiovenda) P. R. O. Bally (Nation. Cact. Succ. J. 30: 31-32, 1975). **T:** Somalia (*Bally* 10924 [K]). – **D:** Somalia. **I:** Bally (1959: 54, 57). **Fig. XXXIX.c, XXXIX.e**

≡ *White-sloanea migiurtina* Chiovenda (1937); **incl.** *Lithocaulon sphaericum* P. R. O. Bally (1959) (*incorrect name*, Art. 11.3) ≡ *Pseudolithos sphaericus* (P. R. O. Bally) P. R. O. Bally (1965).

Stems solitary, cube-shaped, 3 - 6 × 4 - 6.5 cm, pale (grey-) green, tessellate, sections polygonal, 2 - 8 mm ∅, flat, enlarged to large shield-shaped **Tu** along the **Ri**, with a slight median groove; **L** rudi-ments flatly scaly, minute or absent; **Inf** umbellate, sessile on short shoots, successively with up to 4 pseudo-umbels with 8 to 10 **Fl**; **Ped** 4 - 10 mm; **Fl** with ample nectar and strong excrement-scent; **Cl** outside pale green, glabrous; **Cl** tube cup-shaped, 3 - 4 × 4 - 5 mm, inside pale green; **Cl** lobes brown-violet, triangular, horizontally spreading to semi-erect, 3 × 1.8 - 2.8 mm, margins revolute, densely beset with stiff **Ha**-papillae, apically with a bunch of ± 2 mm long vibratile clavate **Ha**; **Cn** dark purple, ± 2.5 - 4 × 3.5 - 4.5 mm; **Ci** fused to form a cup-shaped structure, basally conspicuously tapering, emarginate; **Cs** slenderly spoon-shaped, ± 1.5 mm; **Poll** D-shaped, ± 0.3 × 0.2 mm, corpuscle obovate; **Fr** narrowly fusiform, 3.5 - 5.5 cm; **Se** black, ± 5 × 4.5 mm, margin finely crenate.

QUAQUA

B. Müller & F. Albers

Quaqua N. E. Brown (Gard. Chron., ser. nov. 12: 8, 1879). **T:** *Quaqua hottentotorum* N. E. Brown. – **Lit:** Bruyns (1983); Plowes (1994a); Bruyns (1999b). **D:** SW Namibia, RSA (Northern Cape: Namaqualand; Western Cape: Great Karoo to Beaufort West). **Etym:** From the name "Qua-Qua", resp. "Kam-qua-qua" for *Q. hottentottorum* in the Nama language.

Incl. *Sarcophagophilus* Dinter (1923). **T:** *Sarcophagophilus winklerianus* Dinter.

Perennial semi-shrubby stem succulents; stems branched basally or further up, usually in clumps ≤ 90 cm ∅, without subterranean runners; **Br** grey to purple, 10 - 45 × 1.5 - 2.5 cm, rarely apically tapering, 4- to 6-ribbed, glabrous; **Ri** rounded, sides with longitudinal grooves; **Tu** robust, horizontally spreading or pointing downwards, apically with a **L** rudiment, this yellow-brown, **Sp**-like, conical, horizontally spreading or upcurved, laterally usually with small stipular rudiments; **Inf** solitary or in groups from the lateral grooves; peduncle very short, cushion-like; **Ped** 1.5 - 5 (-15) × 1 - 2 mm, often elongate at **Fr** time, glabrous; **Sep** 1 - 3 mm, deltoid to lanceolate, acute, glabrous; **Cl** opening variously (depending on the number of **Fl** per stem), 7 - 25 mm ∅, usually much smaller than the stem ∅; **Cl** tube ≤ 5 mm, bowl-shaped or campanulate to conical, partly only centrally depressed; **Cl** lobes 3 - 15 mm, ovate-lanceolate to linear-acute, margins usually conspicuously curved outwards, outside smooth and glabrous, inside occasionally papillose or with **Ha**-tipped emergences, rarely hairy, lobe margins rarely with clavate **Ha**; **Cn** of the C(is)+Cs-type, lobes of the **Ci** basally usually fused in a cup-like manner with the dorsal regions of the **Cs**, ± sessile, occasionally free (then broadly stipitate); **Ci** lobes shortly to deeply bifid, ascending from near the **Cs** base; **Cs** lobes simple, incumbent on the **Anth** but usually not overtopping them, occasionally surpassing the **Sty** head, dorsally often flattened, usually with ± strongly visible dorsal bulge or dorsal appendage; **Anth** simple, rectangular, without apical appendage; **Poll** ± ovoid to rounded, translators short, point of attachment variable, corpuscle variably shaped; **Fr** follicles 10 × ≤ 0.8 cm, glabrous, paired, erect, diverging at an acute angle, ± parallel to each other; **Se** brown, 5 - 8 × 3 - 5 mm, numerous, ovate, flattened, margin thickened, delicately papillose, apically with a tuft of white **Ha**.

Quaqua does not exhibit a close relationship to any other genus of the stapeliads. This can be the result of an independent development, or may be interpreted as an evolutive level between soft-stemmed and spiny-stemmed forms (e.g. *Caralluma* / *Hoodia*).

The genus can be divided in 3 sections according to Bruyns (1999b):

[1] Sect. *Pseudorhytidocaulon* Bruyns 1999: Stems tapering to the tip; **Cl** surface slightly rugulose, covered with fine crinkly **Ha**.

[2] Sect. *Quaqua*: Stems not tapering apically; **Ped** > 1.5 mm ⌀; **Cl** delicately papillate or with rigid **Ha**; **Cl** lobes folded along the midrib.

[3] Sect. *Pauciflora* Bruyns 1999: Stems not tapering apically; **Ped** at most 1 mm ⌀; **Cl** delicately papillate or with rigid **Ha**; **Cl** lobes not folded along the midrib.

Natural hybrids appear to be rare, and so far, only the combinations *Q. cincta* × *Q. swanepoelii* and (with a question mark) *Q. cincta* × *Q. acutiloba* have been mentioned. Hybrids with other genera are not known.

The following names are of unresolved application but are referred to this genus: *Stapelia intermedia* N. E. Brown (1890) ≡ *Caralluma intermedia* (N. E. Brown) Schlechter (1898).

Q. acutiloba (N. E. Brown) Bruyns (Bradleya 1: 44, ills. (p. 45), 1983). **T**: RSA, Northern Cape (*Templeman* s.n. in *Pillans* 8 [BOL, K]). – **D**: S Namibia, RSA (Northern Cape, Western Cape). **Fig. XXXIX.a**

≡ *Caralluma acutiloba* N. E. Brown (1909); **incl.** *Caralluma wilfriedii* Dinter (1923); **incl.** *Caralluma ortholoba* Lavranos (1972).

[2] Stems 15 - 20 × ≤ 2 cm, 4-ribbed, sides deeply grooved; **Tu** 4 - 5 mm, conical, spreading horizontally; **Inf** 1- to 3-flowered; buds elongate-conical; **Ped** 1.5 mm; **Sep** 2 - 3 mm; **Cl** outside pale green-purple, inside uniformly black or yellow, or yellow with purple-black patterning, basally paler, central depression ≤ 2 mm, embracing the **Cn** basally; **Cl** lobes 3 - 6 × 2 - 3.5 mm, usually deltoid, flatly spreading; **Cn** purple-black, rarely yellow; **Ci** lobes < 1 mm, bifid, appendages deltoid, erect; **Cs** lobes ± 1 mm, linear-rectangular, apically pointed, not touching each other in the middle; **Poll** 0.4 × 0.3 mm, broadly ovoid to roundish, translators attached basally.

This species does not show particularly close relationships to other taxa of the genus; at the most it appears distantly related to members of Sect. *Pauciflora*.

Q. albersii Plowes (Excelsa 16: 94, ill. (p. 101), 1994). **T**: RSA, Western Cape (*Albers* 2252 [MSUN]). – **D**: RSA (Western Cape: Vredendal).

[2] Stems erect, lateral stems ascending-erect and shrub-like; stems 25 - 30 × ≤ 2 cm, irregularly 4- (5-) ribbed; **Tu** 5 × 4 mm, rather roundish, horizontally oriented; **Sp** acute, robust, porrect; **Inf** 2- to 4-flowered, near the stem tips; buds ovoid; **Ped** 2 - 3 mm, porrect; **Sep** ± 2 mm, lanceolate; **Cl** outside greenish, spotted with pale purple, inside dark lemon-yellow, with purple short stripes or dots arranged in 3 - 4 concentric circles, ± 1.5 cm ⌀, central depression ± 2.5 mm, broadly bowl-shaped, pale yellow; **Cl** lobes oblong-linear, slightly ascending, tips abruptly rounded, acute, slightly curved outwards, central depression in the direction of the **Cl** lobes on the inside with a thin ring of **Ha**, **Ha** purple, ≤ 3 mm, margins of the **Cl** lobes weakly clavately hairy; **Ci** dark purple, lobes ascending-erect, narrow, tips emarginate or slightly bifid; **Cs** reddish-brown, lobes as long as the **Anth**, rectangular, apically weakly toothed.

Closely related to *Q. pulchra*, possibly of hybrid origin.

Q. arenicola (N. E. Brown) Plowes (Excelsa 16: 98, ill. (p. 101), 1994). **T**: RSA, Western Cape (*Pillans* 44 [BOL, K]). – **D**: RSA (Western Cape: S Great Karoo).

≡ *Caralluma arenicola* N. E. Brown (1909) ≡ *Quaqua armata* ssp. *arenicola* (N. E. Brown) Bruyns (1983).

[3] Stems basally decumbent, then erect, lateral stems freely branching, forming mats ≤ 60 cm ⌀; stems green to grey-purple, ≤ 15 × 2 - 3 cm, regularly 4- (to 5-) ribbed; **Tu** ≤ 1.5 cm, conical; **Inf** 2- to 12-flowered; **Fl** usually simultaneously opening, buds basally rounded, apically conical; **Sep** 1 × 0.5 mm, ovate-lanceolate; **Cl** inside purple-black, tube for the greater part cream-coloured, 4 × 4.5 mm, campanulate to V-shaped; **Cl** lobes 5.5 × 1.5 mm, ascending to erect, basally bulging, margins for ¾ of their length strongly folded outwards, mouth of the **Cl** tube inside papillose, papillae apically with a short **Ha**; **Cn** purple-brown; **Ci** lobes bifid, appendages rounded to broadly triangular, ± laterally diverging; **Cs** lobes rectangular, surpassing the **Anth**; **Poll** 0.3 × 0.25 mm, rounded-ovoid, translators attached basally; **Fr** 13 - 15 × 0.6 cm.

Q. arida (Masson) Plowes (Excelsa 16: 96, 1994).

T: [lecto – icono]: Masson, Stapel. Nov. t. 33, 1797. – **D:** RSA (Western Cape).

≡ *Stapelia arida* Masson (1797) ≡ *Orbea arida* (Masson) Sweet (1827) ≡ *Obesia arida* (Masson) Sweet (1830) ≡ *Piaranthus aridus* (Masson) G. Don (1838) ≡ *Caralluma arida* (Masson) N. E. Brown (1890); **incl.** *Caralluma marlothii* N. E. Brown (1903) ≡ *Quaqua marlothii* (N. E. Brown) Bruyns (1983); **incl.** *Caralluma simulans* N. E. Brown (1909); **incl.** *Caralluma marlothii* var. *viridis* E. & B. M. Lamb (1956).

[2] Stems densly turf-like, green, spotted with purple, ≤ 9 × 2 - 2.5 cm, 4- (to 5-) ribbed; **Tu** 3 - 5 mm, conical, acute, porrect; **Sp** hard; **Inf** with 1 - 3 erect **Fl** near the stem tips; **Fl** buds rounded-ovoid; **Ped** reddish, 9 - 15 × ± 1 mm, elongating at **Fr** time; **Sep** ≤ 2 mm, lanceolate, acute; **Cl** outside pale green, inside whitish-yellow, sometimes spotted with purple, 7 - 10 mm ∅, ± flat, central depression embracing the **Cn** only basally; **Cl** lobes basally dotted with red (purple-brown), ± 5 × 2 mm, ovate-lanceolate, apically acute and slightly thickened, sometimes strongly curved outwards and toching the **Ped**, inside delicately pubescent-/ bristly-hairy, central depression at the margin with a ring of erect purple **Ha**; **Cn** yellow, **Cs** occasionally brown-speckled; **Ci** lobes ± 1.5 mm, erect, bifid, appendages narrow, horn-like, laterally strongly diverging; **Cs** lobes not surpassing the **Anth**, apically rounded to crenate, dorsal bulge broad, truncate to emarginate; **Poll** 0.25 × 0.25 mm, obliquely ovoid, translators attached in the middle.

Q. armata (N. E. Brown) Bruyns (Bradleya 1: 65, ill. (p. 67), 1983). **T:** RSA, Northern Cape (*Barkly* 47 [K]). – **D:** RSA (Northern Cape).

≡ *Caralluma armata* N. E. Brown (1890) ≡ *Sarcophagophilus armatus* (N.E.Brown) Dinter (1928).

[3] Stems basally decumbent, then erect, freely branching, lateral stems freely rooting, plants ≤ 10 (-20) × 60 cm; stems spotted with grey-purple, ≤ 15 × 2 - 3 cm, irregularly 4- to 6-ribbed; **Tu** ≤ 1.5 cm, very broadly triangular, slightly down-curved; **Inf** 2- to 12-flowered, **Fl** often opening simultaneously, buds ovoid; **Ped** 3 mm; **Sep** 2 - 3 × 2 mm, ovate-lanceolate; **Cl** inside pale yellow; tube 3 - 4 × ± 3 - 4 mm; **Cl** lobes brownish towards the tips, 8 - 15 × 4 mm, ovate-lanceolate, acute, margins of the lower ½ curved outwards, inside smooth, glabrous; **Cn** purple-brown, ± 2 × 2 mm; **Ci** lobes ± 0.5 mm, bifid, appendages deltoid; **Cs** lobes ± 1 mm, broadly linear, apically usually crenate, usually surpassing the **Anth**, (almost) touching each other in the centre; **Poll** 0.3 × 0.25 mm, roundish-ovoid, translators attached basally.

Q. aurea (C. A. Lückhoff) Plowes (Excelsa 16: 90, ill. (p. 101), 1994). **T:** RSA, Northern Cape (*Villet* s.n. in *Lückhoff* 269 [Herb. Lückhoff [†], BOL]). – **D:** RSA (Northern Cape).

≡ *Caralluma aurea* C. A. Lückhoff (1938) ≡ *Quaqua incarnata* ssp. *aurea* (C. A. Lückhoff) Bruyns (1983).

[3] Plants 50 × 50 cm; stems grey-green, 1.5 - 2 cm ∅, 4- to 5-ribbed; **Tu** robust; **Sp** acute; **Inf** with 1 - 6 simultaneously opening **Fl** in the apical ½ of the stems; **Cl** outside whitish, spotted with pale rose (to purple-brown), inside yellow, 2 - 2.2 cm ∅; **Cl** tube whitish, 2.5 × ≤ 2 mm, V-shaped; **Cl** lobes 6 - 9 × 2.5 - 3 mm, (ovate-) lanceolate, ascending, margins folded outwards, inside delicately hairy; **Cn** 2 × 3 mm; **Ci** yellow, lobes bifid, segments deltoid; **Cs** lobes rectangular, apically rounded, somewhat longer than the **Anth**, occasionally overlapping; **Poll** 0.35 × 0.3 mm, roundish, translators attached in the middle, corpuscle oblong-rectangular.

Q. bayeriana (Bruyns) Plowes (Excelsa 16: 93, 1994). **T:** RSA, Northern Cape (*Bruyns* 1464 [NBG]). – **D:** RSA (Northern Cape: near Springbok).

≡ *Quaqua parviflora* ssp. *bayeriana* Bruyns (1983).

[2] Stems grey to purple, 10 - 30 × 1 - 2 cm, 4-ribbed; **Tu** ≤ 5 mm; **Sp** yellow, hard; **Inf** with 1 - 3 slightly nodding **Fl**; peduncular cushion very small; **Ped** 2 - 4 mm; **Cl** inside pale yellow-green, irregularly spotted with pale purple, 6 - 7 mm ∅, central depression 1 mm, embracing the lower ¼ of the **Cn**, apically weakly thickened; **Cl** lobes 2.5 × 2 mm, ovate-deltoid, inside densely hairy, **Ha** pale purple, ≤ 0.7 mm, wavy, lobe margins hairy, clavate **Ha** transparent, spotted and lineate with purple; **Cn** purple-black, **Cs** somewhat paler, ≤ 2 × 2.5 mm; **Ci** lobes ± erect, shortly or up to ½ bifid, appendages deltoid to horn-like, laterally diverging up to 180°; **Cs** lobes dorsally flattened, as long as the **Anth**, apically narrowed, dorsal appendage truncate-emarginate; **Poll** 0.2 × 0.15 mm, roundish.

Q. cincta (C. A. Lückhoff) Plowes *ex* Bruyns (Asklepios No. 65: 17, 1995). **T:** RSA, Western Cape (*Primos* s.n. in *Lückhoff* 161 [Herb. Lückhoff [†], BOL]). – **D:** RSA (Western Cape).

≡ *Caralluma cincta* C. A. Lückhoff (1935) ≡ *Quaqua inversa* var. *cincta* (C. A. Lückhoff) Bruyns (1983).

[2] Stems regularly 4-ribbed; **Tu** 5 mm; **Sp** 1 mm, hard, horizontal; **Inf** 1-flowered in the upper ½ of the stems; **Fl** buds roundish-ovoid; **Ped** 3 - 4 mm, erect; **Cl** whitish with concentric stripes or lines; **Cl** tube 3 × 15 mm; **Cl** lobes dark purple- or reddish-brown, basally sometimes spotted or elongate-striate, 5 - 8 × 2 - 4 mm, oblong-deltoid, margins in the middle part clavately hairy; **Cn** dark purple-brown, centrally yellowish; **Ci** lobes rectangular, with a conspicuous longitudinal groove, bifid, appendages deltoid to horn-like, ± widely diverging; **Cs** lobes irregularly dentate, basally with a dorsal appendage, this large, flat, ascending-erect, some-

times the extreme tip emarginate-crenate; **Poll** obliquely ovoid, translators attached in the middle, corpuscle oblong, basally acute; **Fr** 7 × 0.5 cm.

Closely related to *Q. inversa.*

Q. confusa Plowes (Excelsa 16: 92, ill. (p. 101), 1994). **T:** RSA, Western Cape (*Plowes* 3234 [SRGH]). – **D:** RSA (Western Cape: Vanrhynsdorp Distr.).

≡ *Quaqua parviflora* ssp. *confusa* (Plowes) Bruyns (1999).

[2] Forming clumps with ≤ 20 stems and **Br**, ± 20 - 30 cm; stems 4-ribbed; **Sp** yellow, hard, slightly curved downwards; **Ped** 4 - 5 mm, slightly curved downwards; **Cl** inside cream-coloured, 1 - 1.2 cm ∅, lobes or lobe margins occasionally purple (-brown); **Cl** tube basally dotted with red-brown, hypocrateriform; **Cl** lobes in the upper ½ dark purple to purple- or reddish-brown, basally dotted, somewhat longer than broad, ovate-triangular, lobe margins ± strongly clavate-hairy; **Cn** purple-black, **Cs** also reddish-brown; **Ci** lobes ± short, bifid to the base; **Cs** lobes slightly surpassing the **Anth**, basally with irregularly toothed edge or dorsal bulge.

Closely related to *Q. dependens* and *Q. swanepoelii.*

Q. dependens (N. E. Brown) Plowes (Excelsa 16: 91, ill. (p. 101), 1994). **T:** RSA, Western Cape (*Barkly* 78 [K]). – **D:** RSA (Western Cape: Clanwilliam Distr.). **Fig. XXXIX.d**

≡ *Caralluma dependens* N. E. Brown (1890) ≡ *Quaqua parviflora* ssp. *dependens* (N. E. Brown) Bruyns (1983); **incl.** *Caralluma reflexa* C. A. Lückhoff (1938).

[2] Stems densely fasciculate, ≤ 20 × 1.5 - 2 cm, 4-ribbed; **Tu** roundish, curved downwards; **Inf** with 1 - 2 downwards-directed **Fl**; **Fl** buds ovoid; **Ped** 3 - 5 mm, strongly curved downwards; **Cl** pale yellow, striped with purple, 7 - 11 mm ∅, centrally weakly depressed, embracing the **Cn** only at the very base, usually 4 **Cl** lobes slightly curved outwards, at least 1 lobe appressed to the stem; **Cl** lobes apically purple, 4 × 1.5 mm, equally wide to the middle, apically attenuate into a broad tip, margins slightly curved outwards, hairy, clavate **Ha** soft; **Cn** dark purple; **Ci** lobes bifid, appendages horn-like, widely diverging (45 - 180°); **Cs** lobes linear, dorsally slightly flattened, apically ± rounded, narrowed, rarely overtopping the **Anth**; **Poll** 0.3 × 0.2 mm, ovoid, translators attached in the middle.

Similar to *Q. inversa.*

Q. framesii (Pillans) Bruyns (Bradleya 1: 43, ills. (p. 44), 1983). **T:** RSA, Western Cape (*Ross-Frames* s.n. [BOL, K]). – **D:** RSA (Western Cape).

≡ *Caralluma framesii* Pillans (1928).

[3] Stems densely arranged, grey-green, flushed with purple, ≤ 40 × 2 cm, irregularly spiralling 4- to 6-ribbed; **Tu** hard, slightly curved upwards; **Sp** yellowish, acute; **Inf** with 2 - 10 simultaneously opening **Fl**, usually in the upper stem parts; peduncular cushions flattened; **Fl** buds obovoid; **Cl** outside yellow-green, occasionally with slight rosy hue, inside yellowish, ≤ 1.8 cm ∅; **Cl** tube 3 × 3 mm, cup-shaped; **Cl** lobes ≤ 8 × 2 mm, broadly ensiform, usually ascending, tips obtuse, often incurved, margins folded outwards; **Cl** tube in the upper part inside and basis of the **Cl** lobes hairy, **Ha** transparent, ≤ 0.5 mm, cylindrical, stiff, ascending; **Cn** pentagonal in top-view, angles formed by dorsal bulges of the **Cs**; **Ci** lobes very small, apically bifid, basally forming pouches; **Cs** lobes broadly rectangular, tip broad, truncate, frequently crenate; **Anth** apically with a short deltoid appendage; **Poll** 0.3 × 0.2 mm, ± obliquely ovoid, arms of the translator inserted in the middle, corpuscle 0.25 mm, fusiform.

Q. gracilis (C. A. Lückhoff) Plowes (Excelsa 16: 91, ill. (p. 102), 1994). **T:** RSA, Northern Cape (*Villet* s.n. in *Lückhoff* 264 [lecto – icono: S.A.G. 28: 227, ill., 1938]). – **D:** RSA (Northern Cape).

≡ *Caralluma gracilis* C. A. Lückhoff (1938) ≡ *Quaqua parviflora* ssp. *gracilis* (C. A. Lückhoff) Bruyns (1983); **incl.** *Caralluma ericeta* Nel (1943).

[2] Stems 10 - 30 × ± 1.5 cm; **Tu** rounded, apically curved outwards; **Sp** deltoid, hard; **Inf** 1- to 2-flowered; **Fl** buds with conspicuous constriction in the middle; **Ped** 4 - 5 mm, ascending to erect; **Cl** inside whitish, scatteredly brownish-dotted, lobes apically changing to brown or yellow, usually flat, rarely with connate tips; **Cl** lobes 5 - 7 × ≤ 2 mm, ovate-lanceolate, usually in the middle slightly constricted, apically somewhat thickened, margins very delicately hairy; **Cn** darke purple-brown; **Ci** lobes rectangular, erect, only shortly bifid, appendages horn-like, laterally diverging; **Cs** lobes broadly rectangular, apically undulate-emarginate, dorsally somewhat flattened, surpassing the **Anth**, sometimes touching each other in the centre; **Poll** ± 0.35 × 0.25 mm, roundish, translator arms inserted in the middle.

Q. hottentotorum N. E. Brown (Gard. Chron., ser. nov. 12: 8-9, 1879). **T:** RSA, Northern Cape (*Barkly* 50 [K]). – **D:** RSA (W Northern Cape: Namaqualand).

≡ *Caralluma hottentotorum* (N. E. Brown) N. E. Brown (1890) ≡ *Quaqua incarnata* ssp. *hottentotorum* (N. E. Brown) Bruyns (1999); **incl.** *Caralluma hottentotorum* var. *major* N. E. Brown (1909); **incl.** *Caralluma ausana* Dinter & A. Berger (1918); **incl.** *Caralluma hottentotorum* var. *minor* C. A. Lückhoff (1937); **incl.** *Caralluma hottentotorum* var. *tubata* C. A. Lückhoff (1937).

[3] Stems in mats, grey-green, often flushed with reddish, 10 - 15 × 2.5 cm, bluntly 4-ribbed, sides grooved; **Tu** roundish-conical, acute, horizontal or slightly upcurved; **Sp** brown, hard; **Inf** 6- to > 10-

flowered, **Fl** successively opening (sometimes 2 simultaneously), unscented; **Ped** ≤ 2.6 mm, 1 - 1.5 cm at **Fr** time; **Sep** triangular-ovate; **Cl** pale yellowish-green, 0.5 - 1.2 cm ⌀, campanulate; **Cl** tube 1.25 - 2.6 × ≤ 2.5 mm; **Cl** lobes 3.5 - 5.2 × ≤ 2.5 mm, triangular-ovate, acute, upper face slightly keeled, spreading, inside basally weakly hairy, **Ha** in part as delicate papillae; **Cn** pale yellow, ≤ 2.25 mm ⌀, stipitate; **Ci** lobes very small, spreading, or larger, ± square, emarginate or bifid, with 2 median grooves, basally pouch-like; **Cs** lobes basally semicircular, fleshy, towards the tips broadly linear, truncate, (½ or) as long as the **Anth**; **Poll** clavate, apically fused with a transparent margin, translators inserted in the middle; **Fr** grey-green, lineate with red-brown, 6 - 8 cm.

Belonging to the *Q. incarnata*-complex.

Q. incarnata (Linné *fil.*) Bruyns (Bradleya 1: 39, ills. (pp. 40-41), 1983). **T:** RSA, Northern Cape (*Thunberg* 6330 [UPS, K [fragment]]). — **D:** S Namibia; N RSA (Northern Cape). **I:** Eggli (1994: 173). **Fig. XL.a, XL.b**

≡ *Stapelia incarnata* Linné *fil.* (1782) ≡ *Podanthes incarnata* (Linné *fil.*) Sweet (1830) ≡ *Piaranthus incarnatus* (Linné *fil.*) G. Don (1838) ≡ *Boucerosia incarnata* (Linné *fil.*) N. E. Brown (1878) ≡ *Caralluma incarnata* (Linné *fil.*) N. E. Brown (1892); **incl.** *Piaranthus incarnatus* var. *albus* G. Don (1837) ≡ *Caralluma incarnata* var. *alba* (G. Don) N. E. Brown (1909).

[3] Stems grey- to purple-green, 10 - 30 × 1.5 - 2.5 cm, 4-ribbed; **Tu** robust, conical, spreading, apically hardened; **Inf** usually with 1 - 4 often simultaneously opening **Fl**, mostly near the stem tips; **Fl** buds oblong-ovoid, somewhat constricted in the middle; **Cl** outside pale rose to whitish, inside cream-white to pale yellow; **Cl** tube 2 mm, deeply cup-shaped to campanulate, mouth slightly thickened; **Cl** lobes 5 - 8 × 2 mm, oblong-deltoid to lanceolate, ascending to spreading, margins (weakly) folded outwards; mouth of the **Cl** tube and **Cl** lobes (partly only basally) hairy on the inside, **Ha** stiff, cylindrical; **Cn** (pale) yellow; **Ci** lobes ± deeply bifid, appendages spreading, deltoid to linear; **Cs** lobes ± rectangular, shortly pointed; **Poll** 0.3 × 0.2 mm, ± roundish transversely elliptic, translators attached in the middle; **Fr** 6 - 9 × 0.5 - 0.7 cm.

Closely related to *Q. radiata*.

Q. inversa (N. E. Brown) Bruyns (Bradleya 1: 53, ills. (p. 54), 1983). **T:** RSA, Western Cape (*Ayres* s.n. in *Pillans* 92 [BOL, K]). — **D:** RSA (Western Cape). **Fig. XXXIX.f**

≡ *Caralluma inversa* N. E. Brown (1903); **incl.** *Caralluma villetii* C. A. Lückhoff (1935).

[2] Stems grey-green to purple, ≤ 25 × 1.5 - 2.5 cm, 4-ribbed; **Tu** 1 cm, curved downwards; **Sp** ≤ 1 mm, robust, acute; **Inf** 1-flowered; **Fl** buds ± ovoid; **Ped** 3 - 5 mm, curved downwards; **Sep** ≤ 3 mm, lanceolate; **Cl** outside pale green, 1.2 - 2.1 cm ⌀, lobes in the middle with a purple stripe, inside yellowish to whitish, sometimes striped and dotted with purple or reddish; **Cl** tube 1.5 - 2 mm; **Cl** lobes purple-brown to red in the lower ½, apically green, 3 - 8 × 2 - 4 mm, broadly ovate-lanceolate, horizontally spreading, margins curved outwards, hairy, apically slightly thickened, clavate **Ha** flattened; **Cn** purple-brown to very dark purple; **Ci** lobes very slender, linear, shortly bifid, basally with a longitudinal groove; **Cs** lobes broadly rectangular, sometimes surpassing the **Anth**, tip truncate and incised, dorsal bulge truncate; **Poll** 0.3 × 0.3 mm, rounded-ovoid, translators attached in the middle; **Fr** ± 8 × 0.5 cm.

Similar to *Q. dependens*.

Q. linearis (N. E. Brown) Bruyns (Bradleya 1: 71, ills. (p. 73), 1983). **T:** RSA, Western Cape (*Bain* 8 [K]). — **D:** RSA (Western Cape: Witteberg Region). **Fig. XL.g**

≡ *Caralluma linearis* N. E. Brown (1890).

[3] Stems grey to dark purple-black, 6 - 15 × 1 - 2.5 cm, 4-ribbed; **Ri** very regular; **Tu** rounded, obtuse; **Sp** 1 - 3 mm; **Inf** usually with 2 - 4 **Fl**; **Fl** buds oblong finger-like, basally broadened; **Ped** 1 - 3 mm, ≤ 8 mm at **Fr** time; **Sep** ± 2 mm, acutely ovate; **Cl** inside white, sometimes flushed with purple; **Cl** tube 3 - 4 mm; **Cl** lobes dark purple-brown, 8 - 10 × (basally) 2 - 3 mm, broadly ovate-lanceolate, ascending-spreading, tips usually slightly incurved, margins folded outwards for ¾ of their length, inside glabrous; **Ci** white flushed with purple, lobes 1 mm, bi- or sometimes trifid, appendages deltoid; **Cs** purple, lobes 2 mm, rectangular-filiform, dorsally flattened, ascending over the **Anth**, curved outwards; **Poll** 0.3 × 0.2 mm, ovoid, translators attached basally; **Fr** 4 cm.

Q. mammillaris (Linné) Bruyns (Bradleya 1: 63, ills. (p. 64), 1983). **T:** [lecto – icono]: Burman, Rar. Afric. Pl. 27: t. 11, 1738. — **D:** Namibia, RSA (Northern Cape).

≡ *Stapelia mammillaris* Linné (1771) ≡ *Pectinaria mammillaris* (Linné) Sweet (1830) ≡ *Piaranthus mammillaris* (Linné) G. Don (1838) ≡ *Boucerosia mammillaris* (Linné) N. E. Brown (1878) ≡ *Caralluma mammillaris* (Linné) N. E. Brown (1902); **incl.** *Stapelia pulla* Masson (1797) ≡ *Piaranthus pullus* (Masson) R. Brown (1811); **incl.** *Sarcophagophilus winklerianus* Dinter (1923) ≡ *Caralluma winkleriana* (Dinter) A. C. White & B. Sloane (1937); **incl.** *Sarcophagophilus winkleri* Dinter (1923) (*nom. inval.*, Art. 61.1) ≡ *Caralluma winkleri* (Dinter) A. C. White & B. Sloane (1937) (*nom. inval.*, Art. 61.1).

[3] Stems in dense shrubby patches 12 - 50 × 50 (70) cm, individual stems 2.5 cm ⌀, very variable: short and very compact, sometimes with robust **Tu**, or slender-stemmed, irrregularly 4- to 5- (to 6-)

ribbed; **Tu** 1 - 2 cm, conical, robustly spined; **Inf** 4- to 15-flowered, usually near the stem tips; **Fl** buds oblong-triangular, basally rounded; **Ped** ± 2 mm; **Sep** ± 4 mm, acutely ovate; **Cl** outside pale green, inside pale yellow dotted with purple-black; **Cl** tube 3 - 4 mm; **Cl** lobes purple to reddish-black, 12 - 20 × 5 - 7 mm, lanceolate, tapering into a conspicuously acute tip, erect to spreading, margins strongly folded outwards, inside usually papillose, papillae with an apical horizontally curved thick **Ha**; **Cn** dark purple-brown; **Ci** lobes bi- or trifid, appendages deltoid; **Cs** lobes narrowly rectangular-linear, dorsal bulge short, truncate, apically meeting and ascending in a curved manner over the **Sty** head; **Poll** 0.35 × 0.25 mm, ovoid, translators basally attached, corpuscle oblong.

Q. maritima (Bruyns) Plowes (Excelsa 16: 97, 1994). **T:** RSA, Northern Cape (*Bruyns* 1457 [NBG]). – **D:** RSA (Northern Cape: coastal regions of Namaqualand).

≡ *Quaqua armata* ssp. *maritima* Bruyns (1983).

[3] Stems basally decumbent, then erect, lateral stems freely rooting, forming wide mats of ≤ 60 cm ∅; stems ≤ 15 × 2 - 3 cm, 4- to 5- (to 6-) ribbed; **Tu** ≤ 1.5 cm, triangular to conical; **Inf** usually with 2 - 12 often simultaneously opening **Fl**; **Sep** ovate-lanceolate; **Cl** inside purple-black, ± 1 cm ∅; **Cl** tube whitish, 3 - 4 × 3 - 4 mm, mouth slightly thickened; **Cl** lobes 6 - 8 × 4 mm, margins folded outwards ± along the whole length; **Cl** tube and base of the **Cl** lobes inside densely papillose, papillae apically with a **Ha**; **Cn** dark purple-brown; **Ci** lobes < 0.5 mm, bifid, appendages deltoid, tips rounded; **Cs** lobes ± 1 mm, broadly linear, surpassing the **Anth**, usually touching each other in the centre, dorsal bulge rounded.

Closely related to *Q. armata.*

Q. multiflora (R. A. Dyer) Bruyns (Bradleya 1: 68, ills. (p. 71), 1983). **T:** RSA, Northern Cape (*Hanekom* 2475 [PRE, K]). – **D:** RSA (Northern Cape).

≡ *Caralluma multiflora* R. A. Dyer (1977).

[3] Stems basally decumbent, then erect, lateral stems ascending, freely rooting, forming mats or shrubbily divergent; stems 10 × 2 cm, irregularly 4- (to 5-) ribbed, sides deeply furrowed; **Tu** 5 × 5 mm, laterally compressed, horizontally spreading; **Sp** brownish, ≤ 1 mm, acute; **Inf** in the upper ½ of the stems; **Cl** ± 1 cm ∅, purple, basally white; **Cl** tube 3 - 4 × 3 - 4 mm, mouth thickened: **Cl** lobes ovate-lanceolate, margins folded outwards, erect, tips incurved; **Cl** tube inside papillose, papillae apically with a **Ha**, these **Ha** transparent, thick, oriented at right angles to the papilla axis and thus appearing undulate; **Cn** purple; **Ci** lobes bifid, appendages deltoid, ascending; **Cs** lobes oblong-rectangular, apically rounded, meeting above the **Sty** head and ascending, dorsal appendage undulately crenate, as long as the **Ci** and equally ascending (resulting in a circle of teeth); **Poll** 0.32 × 0.25 mm, ovoid, translators attached in the middle.

Closely related to *Q. armata.*

Q. pallens Bruyns (BJS 121(3): 355-356, ills., 1999). **T:** RSA, Northern Cape (*Bruyns* 6767 [BOL]). – **D:** RSA (Northern Cape: Namaqualand).

[3] Stems 4-angled, erect, often scandent in bushes, 6 - 30 × 0.6 - 1 cm, grey-green, often mottled purplish; **L** thorny, yellow, 2 - 3 (-5) mm; **Inf** several towards the stem tips, sessile, 1- to 3-flowered; **Ped** 2 - 5 mm, ascending, holding the **Fl** upwards; **Sep** ovate-lanceolate, ± 2 × 0.8 mm; **Cl** rotate, 4 - 8 mm ∅, glabrous, inner face pale green, rest of the **Cl** white; **Cl** tube shallow around the base of the **Gy**; **Cl** lobes linear-lanceolate, spreading to reflexed, ± 3 × 1.5 mm; **Cn** sessile, yellow-green to white, cup-shaped, ± 1.2 - 2 mm; **Ci** lobes ascending, bifid into widely diverging appendages, ± 0.6 mm; **Cs** lobes rectangular, swollen at the base, truncate, ± 0.7 mm, appressed to the backs of the **Anth**; **Poll** globose-rectangular, 0.25 × 0.19 mm, translator wings minute, deltoid.

An unpretentious species endemic to the Namaqualand Broken Veld, differentiated from *Q. parviflora* by the slender, more greyish and purplish stems, the lack of corolla hairs and any colour pattern, and a sessile corona. – [U. Meve]

Q. parviflora (Masson) Bruyns (Bradleya 1: 46, ills. (p. 47), 1983). **T:** [lecto – icono]: Masson, Stapel. Nov. t. 35, 1797. – **D:** RSA (W Northern Cape, N Western Cape).

≡ *Stapelia parviflora* Masson (1797) ≡ *Piaranthus parviflorus* (Masson) Sweet (1826) ≡ *Caralluma parviflora* (Masson) N. E. Brown (1892); **incl.** *Caralluma virescens* C. A. Lückhoff (1938).

[2] Stems in part ascending, otherwise lateral stems erect, fasciculate, grey to purple, 10 - 30 × 1 - 2 cm, 4-ribbed; **Tu** 2 - 3 (≤ 5) mm, horizontal or curved downwards; **Sp** yellow, hard; **Inf** with 1 - 3 **Fl**; peduncular cushions small; **Ped** 4 - 10 mm, curved slightly up- or downwards; **Sep** deltoid to lanceolate; **Cl** inside green to yellow-whitish, striped with purple, 8 - 10 mm ∅, central depression embracing less than ¼ of the **Cn**; **Cl** lobes 3 - 5 × ≤ 3 mm, lanceolate, margins at the most slightly curved outwards, finely hairy, clavate **Ha** flattened; **Ci** dark purple, basally white, lobes basally sometimes pouch-like, apically oblong-rectangular (linear), ascending-erect, spreading, bifid for up to ½, appendages horn-like, diverging at an angle of 90 - 180°, forming a ± continuous ring; **Cs** whitish, lobes simple, apically cylindrical, rounded or attenuate, usually much shorter than the **Anth**, without dorsal bulge; **Poll** 0.2 × 0.15 mm, ovoid, translator attached in the middle.

Q. pilifera (Bruyns) Plowes (Excelsa 16: 98, 1994). **T:** RSA, Northern Cape (*Bayer & Bruyns* 826

[NBG]). – **D:** RSA (S Northern Cape, N Western Cape: W Great Karoo).

≡ *Quaqua armata* ssp. *pilifera* Bruyns (1983) ≡ *Quaqua arenicola* ssp. *pilifera* (Bruyns) Bruyns (1999).

[3] Stems greenish-purple, ≤ 10 (-15?) cm, 4- (to 5-) ribbed; **Inf** with 2 - 12 usually simultaneously opening **Fl**; **Sep** ovate-lanceolate; **Cl** outside greenish, inside dark purple-brown; **Cl** tube 2 - 3 × ≤ 4 mm, mouth weakly thickened; **Cl** lobes ≤ 8 × 4 mm, ovate-lanceolate, margins slightly folded outwards, inside papillose, papillae small, usually L-shaped, apically with a thick slightly curved **Ha**; **Cn** dark purple; **Ci** lobes ± shortly bifid, appendages ± 0.5 mm, deltoid; **Cs** lobes ± 1 mm, broadly rectangular (linear), apically crenate, overtopping the **Anth**, raising above the **Sty** head, dorsal bulge rounded; **Poll** 0.3 × 0.2 mm, roundish-elliptic.

Q. pillansii (N. E. Brown) Bruyns (Bradleya 1: 69, ills. (p. 72), 1983). **T:** RSA, Western Cape (*Pillans* 678 [BOL, K]). – **D:** RSA (Western Cape).

≡ *Caralluma pillansii* N. E. Brown (1909).

[3] Bushy, 40 × ≤ 40 cm; stems grey-green, spotted purple-brown, 20 - 45 × 2 - 3 cm, 4-ribbed; **Ri** flattened; **Tu** ≤ 1.5 cm, broadly ± triangular, slightly down-curved; **Sp** yellowish to black, short; **Inf** with 4 - 20 **Fl**, densely massed in the upper ½ of the stems, opening in succession; **Fl** buds obovoid; **Ped** robust; **Sep** 2 - 3 mm, acute-ovate; **Cl** outside greenish dotted with purple, inside purple-brown; **Cl** tube ≤ 6 mm; **Cl** lobes greyish, dotted with purple-brown, 6 - 10 × 6 mm, erect to ascending, oblong-ovate, apically very shortly pointed, margins curved outwards, inside papillose, tube papillose, papillae tipped by a purple slender acute **Ha** 0.5 - 0.9 mm long; **Cn** dark purple-brown; **Ci** lobes bifid to emarginate; **Cs** lobes ± 1 mm, broadly rectangular, apically rounded, erect to ascending (at an angle of 45°) or horizontal; **Poll** 0.2 × 0.3 mm, ± transversely ovoid, translators attached at the narrow side.

Probably most closely related to *Q. mammillaris.*

Q. pruinosa (Masson) Bruyns (Bradleya 1: 74, ills. (p. 76), 1983). **T:** [lecto – icono]: Masson, Stapel. Nov. t. 41, 1797. – **D:** RSA (Northern Cape). **Fig. XL.d**

≡ *Stapelia pruinosa* Masson (1797) ≡ *Tromotriche pruinosa* (Masson) Haworth (1812) ≡ *Caralluma pruinosa* (Masson) N. E. Brown (1892); **incl.** *Caralluma pruinosa* var. *nigra* C. A. Lückhoff (1937).

[1] Stems in large mats, lateral stems apically ± tapering, first erect, then horizontal; stems greyish-green to dark purple-grey, ≤ 50 × 1.5 cm, 4-ribbed; **Tu** rounded; **Sp** < 2 mm, delicate; **Inf** with 1 - 3 successively opening **Fl** in upper stem parts; **Fl** buds broadly rounded-deltoid; **Ped** 4 - 6 mm; **Sep** ≤ 2 mm, acutely ovate; **Cl** outside grey-green, blotched with purple-brown, inside dark brown, 10 - 13 mm ⌀, central depression embracing ± the basal ⅓ of the **Cn**; **Cl** lobes 4 - 5 × 3 mm, ovate-deltoid to deltoid-lanceolate, tip acute, inside ± wrinkled and finely hairy, **Ha** white, simple, crisped; **Cn** black; **Ci** lobes bifid, appendages small, deltoid to horn-like; **Cs** lobes 1 mm, linear-rectangular, apically crenate or rounded, usually overlapping in the middle (but sometimes much shorter), dorsal bulge short, comb-like; **Poll** 0.3 × 0.2 mm, obliquely ovoid, translators attached in the middle.

Not very typical for *Quaqua*. The stems are reminiscent of *Rhytidocaulon* (Bruyns 1983).

Q. pulchra (Bruyns) Plowes (Excelsa 16: 93, ill. (p. 102), 1994). **T:** RSA, Western Cape (*Bruyns* 1397 [NBG]). – **D:** RSA (Western Cape). **Fig. XL.c**

≡ *Quaqua parviflora* ssp. *pulchra* Bruyns (1983).

[2] Stems green to purple, 10 - 30 × 1 - 2.5 cm, 4-ribbed; **Tu** ≤ 5 (usually 2 - 3) × 5 mm, down-curved; **Sp** yellow, hard; **Inf** with 1 - 3 downwards-directed **Fl**; peduncular cushions small; **Fl** buds ovoid; **Ped** 2 - 4 mm; **Sep** deltoid; **Cl** inside dark red to pale reddish, 10 - 16 mm ⌀, central depression apically with a white ring, rarely entirely white, embracing ¼ of the **Cn**; **Cl** lobes pale yellow, basally reddish, 4 - 8 × ≤ 2 mm, ovate-lanceolate, apically ± thickened, towards the tips slowly attenuate and then suddenly pointed, margins slightly curved outwards, laxly hairy on the reddish-coloured parts, **Ha** contorted, **Cl** lobe margins clavately hairy; **Cn** purple-red to reddish; **Ci** lobes deeply bifid, appendages curved horn-like, up to 180° diverging, ± forming a ring; **Cs** lobes broadly rectangular, ± dorsally flattened, basally broadly rounded, sometimes overtopping the **Anth**, apically crenate; **Poll** 0.28 × 0.25 mm, ± ovoid, translators inserted in the middle of the long side.

Closely related to *Q. parviflora.*

Q. radiata Plowes (Excelsa 16: 89, ill. (p. 102), 1994). **T:** RSA, Western Cape (*Plowes* 7526 [SRGH]). – **D:** RSA (N Western Cape).

[3] Stems with a waxy-grey cover, 20 - 30 × 1.5 - 2 cm, 4-ribbed; **Tu** conical, porrect or slightly down-curved; **Sp** yellowish, small, hard, usually without stipular rudiments; **Inf** 1- to 5-flowered in the upper ½ of the stems; **Fl** buds conical; **Cl** outside dark rose-coloured or pale purple-red, inside cream-coloured, 8 - 10 mm ⌀; **Cl** tube shallow; **Cl** lobes apically white, sometimes strongly flushed with rose, flatly spreading, 2× as long as the basal width, apical ⅔ of the margins curved outwards, mouth of the tube and **Cl** lobes basally inside densely hairy, **Ha** white, very short; **Cn** yellow; **Ci** lobes short, deeply bifid, ascending-erect; **Cs** lobes ± ½ as long as the **Anth**, linear, apically rounded.

Closely related to *Q. incarnata* and most probably only a variety of it.

Q. ramosa (Masson) Bruyns (Bradleya 1: 74, ills.

(p. 73), 1983). **T:** [lecto – icono]: Masson, Stapel. Nov. t. 32, 1797. – **D:** RSA (Western Cape: Little Karoo).

≡ *Stapelia ramosa* Masson (1797) ≡ *Piaranthus ramosus* (Masson) Sweet (1830) ≡ *Caralluma ramosa* (Masson) N. E. Brown (1890).

[3] Densely shrubby; stems purple to greyish-green, 12 - 30 × 1.5 - 2.5 cm, 4-ribbed; **Ri** obtuse, roundish; **Tu** < 3 mm, only on young tips, older stem parts very round; **Inf** with 2 - 10 **Fl**; **Sep** 2 - 3 mm, lanceolate; **Cl** inside black-purple, basally white; **Cl** tube 3 - 4 mm; **Cl** lobes 10 × 4 mm, lanceolate, erect, curved outwards, apically very acute, margins conspicuously curved outwards; **Cl** tube around the mouth papillose, papillae small, tipped with a white stiff **Ha** 0.5 mm long; **Cn** dark purple-black; **Ci** lobes bifid, appendages short, deltoid; **Cs** lobes ± 1 mm, laterally flattened, broadly linear, completely hiding the **Anth**, dorsal bulge rounded; **Poll** 0.2 × 0.15 mm, ovoid.

Q. swanepoelii (Lavranos) Plowes (Excelsa 16: 91, ill. (p. 102), 1994). **T:** RSA, Northern Cape (*Swanepoel* s.n. in *Lavranos* 8371 [PRE]). – **D:** RSA (Northern Cape). **Fig. XL.e**

≡ *Caralluma swanepoelii* Lavranos (1972) ≡ *Quaqua parviflora* ssp. *swanepoelii* (Lavranos) Bruyns (1983).

[3] Stems densely arranged, 10 - 30 × ≤ 1.5 cm; **Tu** roundish; **Sp** pointed, hard; **Inf** 1-flowered; **Fl** buds broadly deltoid; **Ped** ≤ 2 mm, **Fl** oriented at right angles to the stem; **Sep** ± 2 mm; **Cl** whitish, dotted purple to reddish-brown, 1 - 1.2 cm ∅, flat; **Cl** lobes 3 - 4 × 2 - 3 mm, ovate-deltoid to ovate-lanceolate, margins only slightly curved outwards, clavately hairy; **Cn** dark purple-black; **Ci** lobes usually deeply bifid, appendages horn-like, diverging at 90 - 145°; **Cs** lobes dorsally flattened, frequently with a truncate emarginate dorsal bulge; **Poll** 0.3 × 0.25 mm, obliquely ovoid, translators inserted quite apically on the long side.

Closely related to *Q. gracilis*.

Q. tentaculata (Bruyns) Plowes (Excelsa 16: 89, 1994). **T:** RSA, Northern Cape (*Bruyns* 1666A [NBG]). – **D:** RSA (S Northern Cape).

≡ *Quaqua incarnata* var. *tentaculata* Bruyns (1983) ≡ *Quaqua incarnata* ssp. *tentaculata* (Bruyns) Bruyns (1999).

[1] Stems greyish to purple-green, 10 - 30 × 1.5 - 2.5 cm, 4-ribbed; **Tu** porrect, robust, conical; **Sp** hard; **Inf** with 4 - 10 often simultaneously opening **Fl**, usually close to the stem tips; **Fl** buds oblong; **Cl** usually cream-coloured; **Cl** tube 2.5 × 4 mm, campanulate, inside glabrous; **Cl** lobes 8 × 2 mm, narrowly (ovate-) lanceolate, ascending, margins curved outwards, mouth of the tube finely papillose, papillae conical; **Cn** yellowish, basally quite bowl-shaped; **Ci** lobes shortly bifid, appendages broadly deltoid; **Cs** lobes deltoid.

RAPHIONACME

U. Meve, B. Willke & F. Albers

Raphionacme Harvey (London J. Bot. 1: 22-23, 1842). **T:** *Raphionacme divaricata* Harvey. – **Lit:** Verhoeven & Venter (1997). **D:** Tropical and subtropical Africa, Oman. **Etym:** Gr. 'r[h]apys, r[h]aphys', beet-root; and Gr. 'akme', sharpness, cutting edge; application obscure.

Incl. *Apoxanthera* Hochstetter (1843). **T:** *Apoxanthera pubescens* Hochstetter.

Incl. *Zucchellia* Decaisne (1844). **T:** *Zucchellia angolensis* Decaisne.

Incl. *Rhaphionacme* C. Mueller (1846) (*nom. inval.*, Art. 61.1).

Incl. *Chlorocyathus* Oliver (1887). **T:** *Chlorocyathus monteiroae* Oliver.

Incl. *Zaczatea* Baillon (1889). **T:** *Zaczatea angolensis* Baillon.

Incl. *Raphiacme* K. Schumann (1895) (*nom. inval.*, Art. 61.1).

Erect or twining perennial geophytes with white latex; **R** tuber turnip-shaped to globose or discoid, sometimes woody; stems ± woody; **L** decussate, with stipules, petiolate or sessile, linear to broadly ovate or obovate, ± hairy; **Inf** terminal and/or lateral, few- to many-flowered, ± dense; **Sep** free or basally shortly united; **Cl** rotate to campanulate; **Cl** tube present, often campanulate, inside with longitudinal grooves, basally saccate and nectariferous; **Cl** lobes oblong to ovate or obovate, white, yellow, green, purple or blue, outside ± hairy, inside glabrous or papillose; **Cn** petaloid; **Cn** lobes originating each in the zone where 2 **Cl** lobes are fused, erect, simple or 3-parted, ovate, filiform to subulate or divided into irregular appendages; **Gy** projecting from the **Cl** tube; **Fil** originating from the inner base of the **Cn** lobes; **Anth** oblong, ovoid or triangular, apically connivent; pollen in tetrads, arranged tetragonally or rhomboidally, translator lanceolately spoon-shaped to spatulate; **Sty** terete, apically broadened into a conical to plug-like **Sty** head; **Fr** solitary or in pairs, erect to divergent or pendent, slender to thickly fusiform; **Se** hairy at the micropylar end or with a fringe of hairs along the margins.

These small-growing erect and only rarely high-climbing geophytes from subfamily *Periplocoideae* enjoy increasing popularity amongst growers. A complete treatment of the ± 35 species of the genus is not yet existing but is awaited for some time. The species concept here adopted is only a ± preliminary summary.

The 2 species of *Schlechterella* are closely related and similar in gross morphology but have pollen united into packets (pollinia).

The non-succulent taxon *Raphionacme volubilis* Schlechter 1895 belongs to the genus *Buckollia*.

The following names are of unresolved application but are referred to this genus: *Raphionacme baguirmiensis* A. Chevalier (1913); *Raphionacme engleriana* Schlechter (1926) (*nom. inval.*, Art. 32.1c); *Raphionacme seineri* Schlechter *ex* Dinter (1926) (*nom. inval.*, Art. 32.1c); *Raphionacme sudanica* A. Chevalier (1920).

R. angolensis (Baillon) N. E. Brown (BMI 1895: 248, 1895). **T:** Angola (*Welwitsch* 4202 [BM, G]). – **D:** Angola, Tanzania.
≡ *Zaczatea angolensis* Baillon (1889); **incl.** *Zucchellia angolensis* Decaisne (1844); **incl.** *Raphionacme kubangensis* S. Moore (1912).

R tuber globose to turnip-shaped; stems shrubby, 20 - 60 cm tall, erect to decumbent, ± branched, fleshy, nodes swollen, basally woody, pubescent; **L** petiolate for 4 - 15 mm, lamina elliptic to oblong-elliptic, 2.5 - 7 × 1.3 - 4.5 cm, obtuse, base obtuse or acutish, subsucculent, grey-green; **Inf** terminal, 3.8 - 5 cm ∅, many-flowered; **Ped** 4 - 13 mm; **Sep** lanceolate, 3 - 4 mm; **Cl** violet-purple, basally yellowish; **Cl** tube campanulate, 4 mm; **Cl** lobes oblong or oblong-lanceolate, 8 mm, acute, outside and along the margins pubescent with short bent **Ha**; **Cn** lobes purple-red, 3-parted, middle appendage erect, subulate, 5 mm, sometimes apically incised, ventrally keeled, hardly overtopping the **Anth**, lateral appendages short, ± triangular; **Fil** broadly linear, 2.5 mm; **Anth** narrowly lanceolate, ± 2.5 mm, glued together at the margins; (unripe) **Fr** tomentose-hairy.

The species has somewhat viscid stems and leaves according to Hiern (1898: 678-680) and is thus probably ± glandular. *R. procumbens* is so similar in flower morphology that the 2 taxa are probably conspecific.

R. arabica A. G. Miller & Biagi (Notes Roy. Bot. Gard. Edinburgh 45: 61-63, ills., 1988). **T:** Oman, Dhofar (*McLeish* 480 [E]). – **D:** Oman (Dhofar); open stony regions on limestone, steep slopes and on mountains.

R tuber globose to ellipsoid, 3 - 10 × 2 - 6 cm, brownish-white, edible, sweet-tasting; stems annual, simple or few-branched, erect, to 20 cm, reddish-brown, whitish-hairy; **L** 4 - 10 mm petiolate, often first in verticils of 4, elliptic to obovate or oblong-ovate, 2.5 - 4.5 × 1.3 - 3.5 cm, basally ± acute, upper face green, lower face paler, veins reddish-brown, both faces pubescent, ciliate; **Inf** terminal, far overtopping the upper **L**, dichotomously branching, apically dense-flowered; **Ped** 1.5 - 2.5 mm; **Sep** oblong-ovate, 1 - 1.8 × 0.6 - 1 mm, pubescent; **Cl** tube 1 mm; **Cl** lobes erect to spreading, oblong-obovate, 5 - 8.5 × 1.6 - 2.5 mm, obtuse, apically white with pale purple midrib, basally green, outside brownish-purple and ± pubescent; **Cn** lobes linear from a thickened base, 4 - 6 mm, connivent over the **Sty** head, apically often incised, brownish-purple, apically green; **Gy** 2 × 1 mm; **Sty** head plug-like; **Fil** ± 0.3 mm; **Anth** ovate, ± 1.4 × 0.6 mm, appendage of the connective acute; translator ± 0.8 mm; **Fr** in pairs, narrowly fusiform.

R. brownii is the most-closely related taxon according to the protologue but differs esp. in the longer corolla tube and the almost horizontal anthers.

R. borenensis Venter & M. G. Gilbert (Bot. J. Linn. Soc. 99: 401-405, ills., 1989). **T:** Ethiopia, Bale Region (*Gilbert & Jones* 66 [K]). – **D:** Ethiopia, Kenya, Somalia; usually in poorly vegetated and often stony regions.
Incl. *Raphionacme moyalica* Venter & Verhoeven (1997).

R tuber turnip-shaped, usually broader than long, voluminous; stems erect, to 10 cm, weakly pubescent to hispid; **L** ± sessile, ± lanceolate, 2.5 - 7.5 × 1.5 - 3 cm, acutely to bluntish aristate, base cuneate to obtuse, margins undulate, upper face green, glossy, ± glabrous, lower face dull green, weakly hairy; **Inf** ± terminal, pedunculate for 2 - 20 mm, 2- to 5-flowered; **Ped** 3 - 5 mm, ± pubescent; **Sep** free, triangular-ovate, 1 - 2 mm, weakly pubescent; **Cl** 14 - 20 mm ∅; **Cl** tube campanulate, 3 - 4 mm, outside weakly pubescent; **Cl** lobes spreading, ovate, 4 - 7 × 2 - 4 mm, longitudinally canaliculate, purple; **Cn** lobes filiform from a broadened semi-ellipsoid base, 5 - 12 mm, whitish, apically contorted; **Fil** short, thin; translator spatulate, ± 2 mm, spoon subcordate, obtuse; **Fr** narrowly fusiform, in pairs, spreading at an angle of 180°.

The undulate leaf margins and the longitudinally canaliculate corolla lobes are significant. Details of the corona and the translators indicate a close relationship to *R. monteiroae*.

R. brownii Scott-Elliot (JLSB 30: 91, 1895). **T:** Sierra Leone (*Scott-Elliot* 5179 [K, BM]). – **D:** Senegal, Guinea Bissau, Guinea, Mali, Ivory Coast, Ghana, Togo, Benin, Nigeria, Chad, Central African Republic, Congo, Sudan, Uganda, Tanzania, Malawi, Moçambique; esp. in dry forests and rocky places. **I:** Berhaut (1971).
Incl. *Raphionacme jurensis* N. E. Brown (1902); **incl.** *Raphionacme bagshawei* S. Moore (1907); **incl.** *Raphionacme brownii* var. *longifolia* A. Chevalier (1920); **incl.** *Raphionacme wilczekiana* R. Germain (1952).

R tuber ± woody, turnip-shaped to ovoid, 5 - 8 (-10) × 3 - 7 (-15) cm; stems erect, branched, 15 - 45 (-70) × 0.2 cm, annual, laxly hispid; **L** sessile or shortly petiolate, mostly ascending, linear to oblong (or oblanceolate), fresh green, 1.5 - 11 cm × 2 - 7 mm, obtuse to acute, basally tapering; **Inf** terminal or laterally-axillary, mostly with 2 - 3 part-**Inf** with 4 - 6 **Fl** each; **Ped** 6 - 20 (-25) mm, papillate or sparsely pubescent; **Cal** campanulate; **Sep** triangular-ovate, 1 - 3 mm, acute; **Cl** 10 - 20 mm ∅, cream, rose to reddish-purple, often with greenish tinge; **Cl** tube campanulate to cylindrical, 2 - 4 mm; **Cl** lobes

± ascending, oblong-spatulate, obtuse, 5 - 9 × 2 - 2.5 mm, twisted, margins revolute, outside ± pubescent; **Cn** lobes cream, erect, filiform to subulate, originating dorsally from centripetally raised roundish **Cn** bulges, 5 - 7 mm, apically ± bent, papillose; **Fil** subulate, inserted on the **Cn** bulges; **Anth** dark brown, ovoid-lanceolate, 1 mm, margins undulate; translator 1 mm, spoon short, broadly elliptic, suddenly contracted into a robust stalk; **Fr** erect, 10 - 30 × 0.4 - 0.7 cm, rostrate, laxly shortly pubescent; **Se** narrowly ellipsoid, to 10 mm, **Ha** tuft 12 mm, white.

A widespread and vegetatively rather variable taxon, which is very easily recognizable from the characteristic corona, the flat gynostegium and the short polliniferous part of the translators.

R. burkei N. E. Brown (FC 4(1): 537, 1907). **T:** RSA, Gauteng (*Burke* 64 [not located]). – **D:** Namibia, Botswana, N RSA. **I:** Bruyns (1982b); Rowley (1987: 83, als *R. sp.*).

Incl. *Raphionacme dinteri* Schlechter (1910); **incl.** *Raphionacme pachyodon* K.Schumann (1910); **incl.** *Brachystelma viridiflorum* Turrill (1924).

R tuber not seen; stems erect, 7 - 20 (-40) cm, erect, basally rather richly branched, sparsely pubescent; **L** 2 - 3 mm petiolate, linear-lanceolate to obovate, 1.3 - 5 × 0.2 - 1.5 cm, acutish, basally tapering, margins ± ciliate, lower face sparsely pubescent; **Inf** lateral, ± sessile, almost from each node along the stems, many-flowered, ± globose; **Ped** 1 - 2 mm, pubescent; **Sep** lanceolate, 1.2 - 1.5 × 0.5 - 0.7 mm, pubescent; **Cl** ± 9 mm ∅, green, outside pubescent, inside ± papillose; **Cl** tube (1.2-) 2 - 2.5 mm; **Cl** lobes spreading with reflexed tips, triangular-ovate, 2.5 - 3.5 × 2 mm, obtuse; **Cn** lobes 3-parted, middle appendage narrowly triangular to subulate, 0.5 - 2 mm, purple, sometimes ciliate, lateral appendages triangular to subulate, 0.5 - 0.8 mm, (yellow-) green; **Anth** narrowly ovoid, acute, flat, ± imbricate; **Fr** erect, solitary, thickly fusiform, glabrous.

Appears to form a complex of very closely related taxa together with *R. zeyheri*, *R. galpinii* and *R. velutina*. The contrastingly bicoloured corona is significant for *R. burkei*.

R. caerulea E. A. Bruce (BMI 1935: 279, 1935). **T:** Sierra Leone (*Deighton* 1246 [K]). – **D:** Guinea, Sierra Leone, Ivory Coast, Angola; mostly in seasonally very wet and sometimes moor-like places.

≡ *Pentagonanthus caeruleus* (E. A. Bruce) Bullock (1962).

R tuber cylindrical to globose, 5 - 15 × few cm, fleshy; stems solitary, erect, 40 - 100 cm, basally woody, glabrous or weakly pubescent; **L** ± sessile, linear, (1-) 9 - 12 (-14) cm, 3 - 5 mm thick, attenuate-pointed, basally obtuse, dull green, upper face rough, margin revolute, midvein robust, prominent on the lower face; **Inf** terminal, lax, with 5 - 10 part-**Inf**, these 2- to 3-flowered, sessile; **Ped** 1 - 5 cm; **Sep** lanceolate to lanceolate-triangular, 5 - 11 mm, acute, basally united and with finger-like colleters, ciliate; **Cl** 1.5 - 2.5 cm long; **Cl** tube broadly campanulate, 2 - 4 × 4 mm, pale violet to blue; **Cl** lobes ± ascending to horizontal, oblong-lanceolate, 15 - 21 × 7 mm, acute, margins recurved, outside glabrous or weakly pubescent; **Cn** lobes erect, white, 6 - 10 mm, basally 2.7 mm broad and ovate, then narrowed and acutely tapering or again broadened, apart from the subulate tip with additional lateral denticles; **Fil** 1 - 4 mm; **Anth** sagittate, 3 - 6 mm; translator oblanceolate, 4 - 5 mm; **Sty** 6 - 8.5 mm, **Sty** head 2.5 mm.

R. dyeri Retief & Venter (SAJB 2(4): 326-328, ills., 1983). **T:** RSA, Free State (*Rawlinson* s.n. [PRE 57731, BLFU]). – **D:** RSA (Free State). **I:** Rowley (1987: 73, erroneously as *Brachystelma*).

Suffruticose, erect or decumbent; **R** tuber ovoid; stems dichotomously branched, reddish-green, hispid; **L** (1-) 5 - 15 mm petiolate, lamina ± ovate, 2 - 3.5 (-6.5) × 0.5 - 1.5 cm, spreading, often folded along the midrib, acute or obtusely spine-tipped, basally cuneate to obtuse, margins undulate, dull green to blue-green, lower face somewhat hispid; **Inf** terminal, rarely axillary, shortly pedunculate, 3- to 5-flowered; **Ped** 2 - 8 mm; **Sep** basally united, 1.5 - 2 × 1 mm, purple-green, outside hispid; **Cl** 6 - 11 mm long; **Cl** tube campanulate, 2.5 - 4 mm; **Cl** lobes ± ovate, 4 - 8 × 2 - 3.5 mm, apical ½ with pale purple to magenta-red inverted V-shaped pattern on whitish-green background; **Cn** lobes purple-brown to green-brown, 3-parted, middle appendage terete, filiform, 9 - 12 mm, placed one above the other, laterally flattened, ± 1.5 mm, bent horn-like, fleshy; **Anth** narrowly ovoid, apically sticking together, greenish-white to pale violet; **Fr** narrowly ovoid, 5 - 8 × 0.8 - 1.4 cm; **Se** 8 × 3 mm, **Ha** tuft 2 - 2.5 cm.

The V-shaped pattern on the corolla lobes as well as the morphology of the corona distinguish this species from all others.

R. elsana Venter & Verhoeven (SAJB 53(2): 177-179, ills., 1987). **T:** RSA, KwaZulu-Natal (*Venter* 9085 [BFLU, K, PRE]). – **D:** RSA (KwaZulu-Natal).

R tuber turnip-shaped, to 20 cm ∅; stems erect, ± divaricately branched, to 40 × 0.2 cm, weakly pubescent; **L** 3 mm petiolate, lamina obovate to elliptic, 3 - 3.5 × 1 cm, obtuse-aristate, basally obtuse, upper face olive-green, glabrous, lower face grey-green, ± hairy; **Inf** axillary, 2 mm pedunculate, lax, 2- to 3-flowered; **Ped** 3 mm; **Sep** free, triangular, 2 mm; **Cl** 7 - 9 mm ∅, green with violet tinge; **Cl** tube campanulate, 2 mm; **Cl** lobes ovate, 4 × 2 mm, obtuse, outside weakly pubescent; **Cn** lobes 3-parted, erect, middle appendage terete-subulate, 5 mm, basally broadened, deep violet, lateral appendages subulate, 2 mm, green; **Fil** 1 mm; **Anth** narrowly lanceolate, 2 mm; translator 1.5 mm, spoon ellipsoid; **Fr** solitary, porrect, thickly cuneate, 3.5 × 1 cm, grey with rosy tinge, weakly pubescent; **Se** spatulate, 7 × 4 mm, **Ha** tuft 1 - 1.5 cm, silvery-white.

Very thick solitary fruits are otherwise only known from *R. madiensis* and *R. namibiana*.

R. flanaganii Schlechter (BJS 18(Beiblatt 45): 2, 1894). **T:** RSA, Eastern Cape (*Flanagan* 118 [not located]). – **D:** RSA (Eastern Cape, KwaZulu-Natal, Mpumalanga). **I:** FPA 40: t. 1599, 1970; Eggli (1994).

Incl. *Raphionacme scandens* N. E. Brown (1895).

R tuber to 20 cm ∅, ± turnip-shaped; stems usually perennial, becoming woody, twining, young weakly pubescent; **L** spreading, 3 - 13 mm petiolate, ovate-elliptic to oblong-obovate, 4 - 9 × 1.3 - 1.8 cm, basally ± cuneate, pubescent, margins undulate, upper face dark green, lower face paler green; **Inf** terminal and lateral, ± long pedunculate, numerous, 10- to 20-flowered; **Ped** 2 - 4 mm, tomentose-hairy; **Sep** basally fused, ovate to lanceolate, ± 1.5 mm, hairy; **Cl** ± 12 mm ∅, outside pubescent; **Cl** tube campanulate, 2 mm; **Cl** lobes oblong to lanceolate, 5 - 8 × 1.5 - 4 mm, obtuse, ± reflexed, margins revolute, green, basally purple-red; **Cn** lobes whitish, ± tinged with purple, 3-parted, middle appendage subulate-filiform from a ventrally roundish-thickened base, 4 - 7 mm, erect, apically incumbent, lateral appendages subulate, 0.5 - 2 mm, acute, spreading centrifugally; **Anth** appendage acute, triangular; translator 2 mm, spoon ovate, 1 mm; **Fr** mostly in pairs, divaricate, ± 4 - 5 × 1 cm, fusiform, weakly pubescent.

A vegetatively somewhat isolated species because of the very large tubers and the woody stems. The intensity of flower colour and indumentum varies geographically according to Bayer (1976) (dark green flowers and weakly pubescent in the Eastern Cape, pale green flowers and more densely pubescent in KwaZulu-Natal).

R. galpinii Schlechter (BJS 18(Beiblatt 45): 14-15, 1894). **T:** RSA, Mpumalanga (*Galpin* 613 [not located]). – **D:** Angola, RSA (Gauteng, Mpumalanga, KwaZulu-Natal), Swaziland. **I:** Bruyns (1982b); Fabian & Germishuizen (1982: 207).

Incl. *Raphionacme macrorrhiza* Schlechter (1895); **incl.** *Raphionacme virgultorum* S. Moore (1912).

R tuber top- to turnip-shaped, 25 × 12 cm, bark pale ochre; stems 5 - 25 cm, erect, unbranched, densely leafy towards the tips, coarse, apically tomentose to velvety; **L** 3 - 6 mm petiolate, spreading, linear-lanceolate to narrowly obovate, coarse, 2.5 - 7.5 × 0.5 - 2 cm, ± acute, basally attenuate, lower face paler, midvein raised, both faces velvety-tomentose; **Inf** extra-axillary laterally and terminal, ± pedunculate, many-flowered; **Ped** very short, pubescent; **Sep** linear to lanceolate, 2 - 4.5 mm, hairy; **Cl** 10 - 12 mm ∅; bright green, outside pubescent, inside glabrous; **Cl** tube campanulate to cylindrical, 2 - 4 mm; **Cl** lobes oblong to ovate, 3.5 - 5.5 × 1.5 - 3 mm, obtuse; **Cn** lobes 3-parted from a rectangular base, greenish-brown to lilac-red, middle appendage erect, ± inclined, subulate-filiform, 3.5 - 4 mm, acute, lateral appendages ± erect, subulate or triangular, 1 - 2 mm, often laterally fused; **Sty** head whitish with violet angles; spoon ovate-lanceolate.

Most closely related to *R. velutina* and probably conspecific with the flower-morphologically similar but vegetatively much more delicate *R. globosa*.

R. globosa K. Schumann (BJS 27: 118-119, 1893). **T:** Angola (*Mechow* 327 [not located]). – **D:** Angola, Zaïre, Tanzania, NW Zambia, N Zimbabwe, N and E RSA, Swaziland. **I:** Bruyns (1982b: as *R. elata*). **Fig. XL.f, XL.h**

Incl. *Raphionacme elata* N. E. Brown (1907); **incl.** *Raphionacme lucens* Venter & Verhoeven (1988).

R tuber turnip-shaped, 5 - 10 cm, to 10 cm ∅; stems erect, 15 - 30 (-90) cm, solitary or basally laxly branched, terete, **Int** sometimes very long, ± pubescent; **L** sessile or 2 - 4 (-8) mm petiolate, lamina very variable, linear to ovate-elliptic, 3 - 14 × 0.4 - 1.5 (-5) cm, upper face grey-green to dark green, sometimes glossy, lower face paler, both faces ± pubescent; **Inf** terminal and lateral, almost sessile or to 4 cm pedunculate, globose, many-flowered; **Ped** 2 - 4 mm, pubescent; **Sep** triangular, 2 - 4 mm, outside pubescent, sometimes ciliate; **Cl** 12 - 16 mm ∅, cream to pale green, white to purple, sometimes bicoloured, outside pubescent to tomentose; **Cl** tube campanulate, 2.5 - 5 mm, inside basally with conspicuous V-shaped nectar pouches; **Cl** lobes spreading, ± reclined, oblong to ovate-oblong, 5 - 7 × 2 - 3.5 mm, ± obtuse; **Cn** lobes whitish to greenish-cream, sometimes purple-tinged, 3-parted, ± erect, middle appendage 4 - 5.5 mm, acute, lateral appendages subulate to filiform, 1 - 2 (-3) mm, acute; **Fil** 0.5 mm; translator spoon-shaped, spoon ovate-lanceolate; **Sty** head apically plug-like.

A taxon variable in habit and best recognized from the globose inflorescences and the 15 long and acute appendages of the corona.

R. grandiflora N. E. Brown (BMI 1895: 111, 1895). **T** [lecto]: Zambia, Northern Prov. (*Carson* 5 [K]). – **D:** Tanzania, Zambia, Malawi, Zimbabwe, Moçambique; in dry forests and on rocks. **Fig. XLI.c**

≡ *Pentagonanthus grandiflorus* (N. E. Brown) Bullock (1962); **incl.** *Raphionacme grandiflora* ssp. *glabrescens* Bullock (1953) ≡ *Pentagonanthus grandiflorus* ssp. *glabrescens* (Bullock) Bullock (1962).

R tuber turnip-shaped, ± depressed, to 7.5 - 12 cm

∅; stems erect, solitary, rarely apically branching, herbaceous, hairy or glabrous; **L** shortly petiolate, lamina linear-lanceolate to obovate, 2 - 12 × 0.5 - 2.5 (-5) cm, acute, basally cuneate-attenuate, rather limp, rough, venation prominent on the lower face, margin undulate, green, lower face paler and glossy, both faces hairy, hairy only along the margin and on the veins or glabrous; **Inf** terminal, 1.2 - 3.8 cm pedunculate, weakly branched, 2- to 7-flowered; **Ped** 8 - 16 mm, hairy; **Sep** narrowly lanceolate to ovate-lanceolate, 5 - 9 mm, hairy; **Cl** ± 3 cm ∅, glabrous; **Cl** tube campanulate, pentagonal, ± 6 mm, outside green or violet-blue, spurred, spurs 3 - 4 mm, keeled in between, keels apically 2-toothed, papillose, yellow; **Cl** lobes ascending to erect, oblong to ovate-oblong, 14 - 20 × 6 - 10 mm, ± obtuse, basally with 2 parallel fleshy keels, outside yellow-green to green, inside deep violet, margins ± recurved; **Cn** whitish, with delicate blue tinge; **Cn** lobes ascending, ± 11 mm, shovel-shaped, trifid, middle appendage linear-subulate, ± 5 mm, lateral appendages shortly triangular; **Anth** ± 5 mm, triangular, acute, arranged conically; **Fil** short, broadened, robust; **Sty** head plug-like conical, whitish, ± 7 mm ∅; translator 5 mm, stalk falcately curved, adhesive disc differentiated in 1 down-curved and 2 upcurved triangular denticles, pollen white; **Fr** linear-fusiform, erect, 17 × 1 cm, glabrous.

The large blue flowers are spectacular.

R. haeneliae Venter & Verhoeven (SAJB 62(6): 316-320, ills., 1996). **T:** Namibia (*Haenel* s.n. [K]). – **D:** Namibia; in pure sand or gravel on flats almost devoid of vegetation.

R tubers few to numerous, cylindrical-fusiform, sometimes heavy up to 1 kg; stems subterranean to 25 cm, above-ground to 40 × 0.4 - 0.7 cm, erect, annual, glabrous, branched, fleshy; **L** 1 - 3 mm petiolate, lamina ovate, 4 - 6 × 1 - 4 cm, succulent, basally roundish, bluish-green to dark green, cuticle thick, wax-like, glabrous; **Inf** appearing ± before the **L**, terminal, with ± 5 part-**Inf**, each ± 3-flowered, peduncle 3 - 5 mm; **Ped** 2 - 3 mm; **Sep** free, broadly ovate, 2 × 1.5 mm; **Cl** 10 mm ∅; **Cl** tube campanulate, ± 2 mm; **Cl** lobes spreading, oblong-ovate, 4 × 2 mm, obtuse, outside greenish, inside cream-greenish with central purple zone; **Cn** lobes undivided, filiform from a thickened base, 4 - 5 mm, inclined, tips tortuous, yellowish-whitish; **Gy** protruding from the **Cl** tube; **Fil** 1 mm; **Anth** elliptic, 1.5 × 1.5 mm; translator spatulate, 2 mm, spoon broadly ovate, stalk very thin; **Sty** head broadly plug-like, greenish; **Fr** in pairs, widely divergent, slender-fusiform, flattened, 6.5 - 7 cm × 5 - 6 mm, rostrate; **Se** narrowly ellipsoid, 10 × 4 mm, surface lineate-papillate, brown, **Ha** tuft 12 - 15 mm, whitish.

The unusually large and sometimes numerous root tubers and the succulent leaves are adaptations to the extremely dry habitat in the Namib Desert.

R. hirsuta (E. Meyer) R. A. Dyer (FPSA 22: t. 853 + text, 1942). **T:** RSA, KwaZulu-Natal (*Drège* s.n. [not located]). – **D:** Namibia, Botswana, RSA, Swaziland. **I:** Bruyns (1982b); Fabian & Germishuizen (1982: 206h, as *R. divaricata*).

≡ *Brachystelma hirsutum* E. Meyer (1837); **incl.** *Apoxanthera pubescens* Hochstetter (1843) ≡ *Raphionacme pubescens* (Hochstetter) Hochstetter (1844); **incl.** *Raphionacme divaricata* Harvey (1843); **incl.** *Raphionacme obovata* Turczaninow (1848); **incl.** *Raphionacme purpurea* Harvey (1859); **incl.** *Raphionacme divaricata* var. *glabra* N. E. Brown (1907).

R tuber to 30 cm ∅, variable in shape and size; stems 5 - 30 cm, basally richly dichotomously branched, divaricate, glabrous to densely pubescent; **L** 1 - 6 mm petiolate, lamina oblong-ovate, broadly elliptic, circular or obovate, 1 - 3 (-5) × 0.7 - 1.3 (-2.6) cm, acute or obtuse, basally rounded or cuneate, conspicuously veined, dark green, upper face glabrous, lower face glabrous or pubescent; **Inf** terminal in the bifurcations of the stems and laterally, pedunculate to 2.5 cm, ± pubescent, 5- to many-flowered; **Ped** 2 - 12 mm, pubescent; **Sep** ovate-lanceolate, 2 - 4 mm, acute, ± pubescent; **Cl** campanulate, 10 - 18 mm ∅, lilac-purple; **Cl** tube 2 - 3.5 mm; **Cl** lobes erect to spreading-reclined, elliptic to oblong, 3 - 8 × 1.5 - 3 mm, obtuse; **Cn** lobes mostly white, undivided or 2-parted, or lobes clearly 3-parted or irregularly divided and drawn out into filiform appendages, variable in size and shape (even within a **Fl**), oblong-lanceolate to obovate, 1.4 - 4 × 0.7 - 2 mm, margin ± denticulate, covering the **Sty** head, ± imbricate; **Fil** short; **Fr** solitary, porrect, fusiform, 2.5 - 6 × 0.7 - 1 cm, smooth, slightly pubescent; **Se** oblong, ± 8 mm, obtuse, glabrous, raphe bulging keel-like, **Ha** tuft reflexed.

The species shows considerable variation as to indumentum, leaf morphology, peduncle length and shape of the corona. The species is widespread in S Africa as well as worldwide in cultivation and is recognized by the broadly elliptic, almost glabrous leaves, the almost leaf-like corona lobes and the seeds which are keeled along the underside (but see also *R. palustris*).

R. inconspicua H. Huber (Mitt. Bot. Staatssamml. München 12: 73-74, 1955). **T:** Namibia, Damaraland (*Volk* 2718 [M, WIND?]). – **D:** Namibia (Damaraland).

R tuber unknown; stems erect, basally woody, ± 25 cm; **L** obovate, 3 - 5 × 0.7 - 1.2 cm, basally long-cuneate, upper face dark green, both faces weakly pubescent; **Inf** terminal or laterally-axillary, 7 - 15 mm pedunculate, 12- to 15-flowered; **Ped** 2 - 10 mm, pubescent; **Sep** basally shortly united and urceolate, ± 1 mm; **Cl** ± 15 mm ∅; **Cl** tube 1 mm; **Cl** lobes spreading, 5 - 6 × 1 - 2 mm, obtuse, outside softly hairy; **Cn** lobes 3-parted, middle appendage filiform, 7 mm, tips bent inwards, lateral appendages shortly subulate, 0.7 mm.

Insufficiently known only from the type, and of uncertain relationships.

R. keayi Bullock (KB 8: 63-64, 1953). **T:** Nigeria, Zaria Prov. (*Keay* 25990 [K]). – **D:** Ivory Coast, Nigeria, Cameroon.

R tuber large, woody; stems erect, to 1 m tall, numerous, ribbed, ± densely crisply hairy; **L** almost sessile, oblong-lanceolate, to 12 × 2 cm, acute, basally attenuate, both faces crisply pubescent; **Inf** laterally-axillary, near the stem tips, many-flowered; **Ped** 4 - 6 mm; **Cal** divided almost to the base, **Sep** triangular, 2 mm; **Cl** aestivation valvate; **Cl** tube cylindrical, 2 mm; **Cl** lobes spreading or finally reflexed, triangular, 5 - 6 × 3.5 - 4 mm, acute, green with deep red basal zone, inside delicately papillose, outside papillose to pubescent; **Cn** lobes undivided (to 3-parted), erect from a 1 × 2.5 mm large rectangular laterally toothed base, middle appendage linearly whip-like, 4.5 mm, greenish, connivent over the **Gy**, papillose; **Fil** subulate, decurrent in the **Cl** tube, 1 mm, basally thickened and fleshy; **Anth** triangular, 2 mm; translator spatulate, with large elliptic spoon; **Sty** 6 mm; **Sty** head plug-like, 3 mm.

Handsome plants from W Africa.

R. lanceolata Schinz (Verh. Bot. Vereins Prov. Brandenburg 30: 263-264, 1888). **T** [syn]: Namibia (*Schinz* 167 [not located]). – **D:** Namibia, Botswana, Zimbabwe, Moçambique, RSA (Northern Cape).

Incl. *Raphionacme lanceolata* var. *latifolia* N. E. Brown (1902).

R tuber head-sized, with copious resinous latex, neck-like portion > 8.5 mm ⌀, persistent; stems annual, branched, terete, (10-) 22 - 40 (-90) cm, pubescent; **L** 2 - 6 mm petiolate, petiole canaliculate above, lamina narrowly oblong or lanceolate to elliptic, 1.9 - 5 × 0.3 - 3.2 cm, basally rounded or cuneate, both faces scattered-hispid; **Inf** axillary, shortly pedunculate, 3- to 7-flowered; **Ped** 3 - 9 mm, hispid; **Sep** lanceolate, 3 mm; **Cl** tube 3 mm, slightly constricted below the throat; **Cl** lobes oblong, 6 × 2 - 3 mm, obtuse, contorted from right to left in the bud stage, greenish-spotted with pale violet tips, outside ± pubescent, inside glabrous; **Cn** lobes lanceolate-subulate, 4 - 7.5 mm, basally 1 - 1.5 mm broad, apically acute, mostly shortly bifid, yellow-green; **Fil** basally shortly broadened and shortly pouch-like from the laterally fused margins, 1 mm; **Anth** overtopping the mouth of the **Cl** tube for ≤ 2 mm; translator shortly stalked; **Sty** 1 mm; unripe **Fr** hispid.

Rather closely related to *R. brownii* from which it differs by the larger root tubers and characters of the gynostegium.

R. loandae Schlechter & Rendle (in Hiern, Cat. Afr. Pl. 3: 679-680, 1898). **T:** Angola, Luanda (*Welwitsch* 4274 [BM]). – **D:** Angola; dry hilly places.

R tuber unknown; stems low-growing, branched, decumbent, > 45 cm, slender, densely shortly pubescent; **L** to 4 mm petiolate, lamina lanceolate to elliptic or obovate, 1.3 - 4 cm × 3 - 9 mm, very slightly succulent, deep green, both faces shortly pubescent, basally cuneate, margins undulate, hairy; **Inf** axillary, 2 - 4 mm pedunculate, alternating near the stem tips, pendent, 3- to 6-flowered; **Ped** 3 - 5 mm, pubescent; **Cal** broadly campanulate, deeply divided, **Sep** ovate-lanceolate, 2.5 - 4 mm, acute, outside densely short-tomentose, **Ha** curved, pale; **Cl** green, reflexed, 6 - 8 mm long; **Cl** tube 4 mm; **Cl** lobes ovate, 4 - 6 mm, contorted in the bud stage, outside pubescent; **Cn** lobes 3-parted, purple, middle appendage subulate, 2× as long as the triangular lateral appendages.

Insufficiently known taxon, probably related to *R. procumbens*, but the latter has a very much shorter corolla tube.

R. lobulata Venter & Verhoeven (SAJB 54(6): 603-606, ills., 1988). **T:** RSA, Eastern Cape (*Dyer* 3381 [PRE]). – **D:** RSA (Eastern Cape).

A woody climber without prooven root succulence. This species is incompatible with the generic diagnosis of *Raphionacme* due to the coriaceous leaves, the corolla lobes with hairy insides and the bipartite semiglobose tips of the corona lobes. *Mafekingia* or *Stomatostemma* are more similar, *Buckollia* is probably most-similar.

R. longifolia N. E. Brown (BMI 1895: 110-111, 1895). **T:** Moçambique (*Kirk* s.n. [not located]). – **D:** Ivory Coast, Zambia, Malawi, Moçambique, Zimbabwe.

R tuber turbinate to broadly turnip-shaped (as far as known), 5 - 8 cm ⌀; stems 10 - 25 cm, erect, unbranched, pubescent with scattered curved **Ha**; **L** 1 - 9 mm petiolate, lamina linear to narrowly lanceolate, 3 - 19 × 0.4 - 1.3 cm, ± hairy; **Inf** 2 - 5 mm pedunculate, alternating from upper nodes, dense- and many-flowered; **Ped** 2 - 6 mm; **Sep** ovate, 1.5 mm, acute; **Cl** tube campanulate, ± 9 mm ⌀; **Cl** lobes oblong-lanceolate, 4 - 5.5 mm, acutish, yellowish to green, outside pubescent, inside glabrous; **Cn** lobes 3-parted, middle appendage erectly incumbent, subulate to filiform, 3 - 5 mm, lateral appendages triangular-lanceolate to subulate, 0.5 - 2.5 mm; **Fil** subulate, basally broadened; **Anth** oblong, acute, apically united.

R. longituba E. A. Bruce (BMI 1937: 419-420, 1937). **T:** Tanzania, Tabora Distr. (*Lloyd* 45 [not located]). – **D:** Tanzania, Zambia, Malawi, N Zimbabwe; in grassland and Miombo formations. **I:** Meve (2000a: as *R. ernstiana*).

Incl. *Raphionacme ernstiana* Meve (2000).

Plants pubescent; **R** tuber beetroot-like discoid to globose, ± 20 cm ∅; stems green, divaricately branching, ± 30 cm × 5 mm ∅; **L** subsessile, ± 3 - 6 × 2 cm, ovate (to obovate), green, apically obtuse, basally truncate, margins undulate; **Inf** terminal, dichasially 3-flowered; peduncle 15 - 25 × 1 - 2.5 mm; **Ped** 6 - 12 × 1.5 mm, bracteate; **Sep** narrowly lanceolate, 4 - 7 mm, acute; **Cl** campanulate; **Cl** tube 7 - 10 × 3 - 3.5 mm, inside ridged, greenish; **Cl** lobes ovate, horizontally spreading, 10 - 12 × 4 - 6 mm, margins recurved, upper face white to greenish-white, partly tinged with rose or (purplish-) pink, basally often greenish; **Cn** white to greenish-white; **Cn** lobes broadly ovate to rectangularly ovate, 7 - 8 × 3 - 4 mm, apically often incised-bidentate, erect-connivent, forming a ± closed dome; translator boat-shaped with rounded tip, ± 2.5 × 0.6 mm, without explicit stalk, adhesive disc deltoid, ± 0.7 × 0.5 mm, pollen in decussate subglobose tetrads, ± 0.35 mm ∅; **Fr** ovoid.

The deep corolla tube and the very broad corona tips, which obscure the gynostegium considerably, are notable. More closely related with *R. hirsuta*. – [U. Meve]

R. madiensis S. Moore (J. Bot. 46: 293-294, 1908). **T:** Uganda (*Bagshawe* 1611 [BM]). – **D:** Uganda, Kenya, Tanzania; wood- and grasslands. **I:** Agnew & Agnew (1994). **Fig. XLI.a, XLI.d**

R tuber broadly turnip-shaped, 7 - 14 cm ∅, bark silvery-grey, smooth; stems 5 - 40 × 0.2 - 0.4 cm, erect, weakly and ± divaricately branched, pubescent or glabrous; **L** 5 - 10 mm petiolate, overtopping the **Inf**, lamina oblong, obovate or elliptic, 5 - 12 × 1.5 - 4 cm, ± obtuse, basally attenuate, dark green, both faces hairy or glabrous; **Inf** lateral and terminal, axillary, shortly pedunculate, ± laxly 3- to 10-flowered; **Ped** ± 5 mm, pubescent; **Sep** free, lanceolate, 2 - 3 mm, acute, mostly somewhat spreading, outside pubescent; **Cl** green, violet or brown-violet, 15 - 18 mm ∅, outside pubescent; **Cl** tube campanulate, 2 - 3 × 3 - 4 mm; **Cl** lobes spreading, triangular-oblong, 6 - 7.5 mm, obtuse; **Cn** lobes purple-brown, (1- to) 3-parted, subulate to almost linear, 6 - 8 mm, basally broadened and laterally with 2 small obtusely triangular appendages; **Fil** rectangular from a broadened base, 1 mm; translator spatulate-saccate, 1.5 mm.

This species with large tubers and attractive flowers is esp. worthy of cultivation.

R. michelii De Wildeman (Ann. Mus. Congo Belge, Bot., sér. 5, 1: 181, 1904). **T:** Zaïre (*De Wildeman* 47 [BR, BM]). – **D:** Congo, Zaïre, Kenya.

R tuber flattened; stems erect, 7 - 20 cm, somewhat branched, pubescent; primary **L Sc**-like, other **L** elliptic to obovate, 1.5 - 7 × 0.7 - 2 cm, both faces (roughly) hairy, venation prominent on the lower face; **Inf** mostly terminal, extra-axillary, shortly pedunculate, 1- to many-flowered; **Ped** 1 - 2 cm, weakly pubescent; **Sep** ovate-lanceolate, ± 3 mm, acute, pubescent, ciliate; **Cl** tube almost absent; **Cl** lobes spreading, 13 - 20 × 4 - 5 mm, ± obtuse, rose to violet, outside laxly pubescent, inside glabrous; **Cn** lobes erect, filiform from a triangular base, ± 20 mm, overtopping the **Gy**, apically somewhat contorted; **Anth** lanceolate-ovate.

These compact dwarf shrublets are probably most closely related to *R. splendens* from which they differ among other characters by the style head, which is only shortly stipitate. – [U. Meve]

R. monteiroae (Oliver) N. E. Brown (FC 4(1): 533, 1907). **T:** Moçambique (*Monteiro* 424 [K]). – **D:** Angola, Namibia, Botswana, Tanzania, Zimbabwe, S Moçambique. **I:** Rowley (1987: 83).

≡ *Chlorocyathus monteiroae* Oliver (1887).

R tuber not seen; stems slender, twining, with long **Int**, scattered-pubescent; **L** 2 - 7 mm petiolate, spreading, lamina oblong-ovate to obovate, 2 - 3.5 (-6.5) × 0.7 - 1.5 cm, ± pubescent; **Inf** axillary, 4 - 6 mm pedunculate, few-flowered; **Ped** 3 - 4 mm, pubescent; **Sep** ovate, 2 mm, acute, pubescent; **Cl** ± 2.5 cm ∅, olive-green, outside sparsely pubescent; **Cl** tube campanulate, ± 6 × 6 mm, inside with 5 decurrent 2-winged nectar pouches; **Cl** lobes ascending, ovate-oblong, 8 - 13 × 5 - 6 mm, obtuse, margins ± widely reflexed and undulate, inside ± papillose; **Cn** lobes basally very broad, 3-partite, middle appendage subulate, 2 - 4 mm, red-brown, lateral appendages subulate, 1 mm, pale yellow, tips curved over or below the middle appendage; **Fil** 1.5 mm, translator 2 mm, spoon broadly ovate or obtusely cordate, stalk rather broad.

A very large-flowered species and easily recognized from the olive-green corolla and the red-brown / pale yellow corona. It is related to the non-twining *R. borenensis*.

R. namibiana Venter & Verhoeven (SAJB 52(4): 332-334, ills., 1986). **T:** Namibia (*Marloth* 5017 [PRE]). – **D:** Namibia. **I:** Bruyns (1995a).

Suffruticose with rather clear latex; **R** tuber ovoid to turnip-shaped, to 50 × 15 cm; stems erect, underground 25 - 50 × 1 cm, above-ground 6 - 10 × 0.3 - 0.5 cm, glabrous; **L** to 2 cm petiolate, lamina lanceolate to narrowly ovate, 4 - 15 × 1 - 3 cm, acute, grey-green with purple venation, densely pubescent, later glabrescent, midrib prominent below; **Inf** terminal, few-flowered; peduncle 1 - 2 mm, weakly pubescent; **Ped** 6 - 8 mm, ± pubescent; **Sep** free, triangular to ovate, 5 × 1.5 mm, weakly pubescent; **Cl** 21 × 28 - 30 mm; **Cl** tube 10 - 11 mm long, inside dark green, outside weakly pubescent; **Cl** lobes obovate, 10 × 3 mm, obtuse, purple-blue to rose-purple, basally pale yellow, glabrous; **Cn** lobes basally broadened, laterally fused with each other and forming pouches, apically 3-parted, middle appendage filiform, 6 - 7 mm, lateral appendages 1 mm, curved, horn-like, rose-purple; **Gy** narrowly oblong-ovoid, 7 × 3 mm; **Fil** 2 mm, terete; **Anth** narrowly ovoid, 5 × 3 mm; translator narrowly spatulate; **Fr** normally solitary, freely pendulous, cuneate, 4 - 6 × 2 cm, acute, mostly terete in cross-section, green to brown, glabrous; **Se** flatly ovoid, each end ± pointed, 15 × 4 mm, without a tuft of **Ha** at the micropylar end, but instead with a crown of **Ha** along the **Se** margin.

The crown-like fringe of hairs along the seed margin (instead of the usual tuft of hairs at the micropylar end) is unique in the whole complex.

R. palustris Venter & Verhoeven (SAJB 52(2): 149-152, ills., 1986). **T:** RSA, KwaZulu-Natal (*Venter* 9005 [BLFU, K, PRE]). – **D:** RSA (KwaZulu-Natal); in swamps and wet meadows.

Identical to *R. hirsuta*, but stems longer, to 50 cm; **Inf** 3-flowered; **Cn** lobes more slender; **Fr** erect, narrowly fusiform, 18 - 27 × 0.8 - 1 cm.

Differing only quantitatively in some characters from *R. hirsuta* and probably only representing an ecotype (from wet habitats) of this taxon.

R. procumbens Schlechter (BJS 20(Beiblatt 51): 11, 1895). **T:** RSA, Mpumalanga (*Schlechter* 3867 [BOL]). – **D:** Zimbabwe, RSA (Mpumalanga), Swaziland. **I:** Bruyns (1995a). **Fig. XLI.e**

R tuber not seen; stems 15 - 30 cm, basally branched, decumbent (in cultivation also twining), long and softly pubescent; **L** 2 - 9 mm petiolate, lamina elliptic to lanceolate, 2.5 - 6.5 × 0.5 - 1.5 cm, basally attenuate, both faces velvety-hairy, **Ha** long, soft, denser on the lower face; **Inf** extra-axillary, ± many-flowered, globose, conspicuously pedunculate, somewhat pendent; **Ped** 2.5 - 3 mm, softly hairy; **Sep** ovate-lanceolate, 2 - 3 mm, acute, shortly soft-hairy; **Cl** campanulate, ± 10 mm ⌀, outside softly hairy, inside glabrous; **Cl** tube 1 mm; **Cl** lobes oblong-lanceolate to ovate, 2.5 - 4 × 1.3 mm, inside pale green, basally purple; **Cn** lobes 3-parted from a ± square base, middle appendage erect or slightly inclined, triangular-lanceolate, 1 mm, acute, lateral appendages erect, triangular, acute, slightly shorter to ¼ as long as the middle appendage; **Anth** fused to form an acute cone; translator hardly 1 mm, spoon broadly ovate, stalk linear, 0.6 mm.

Very similar to *R. angolensis* in flower morphology, esp. because of the red and acutely toothed corona.

R. pulchella Venter & Verhoeven (SAJB 54(1): 72-74, ills., 1988). **T:** Moçambique, Manica e Sofala Distr. (*Williams* RSES89 [SRGH]). – **D:** Moçambique, Zimbabwe; grassland.

Suffruticose; **R** unknown; stems to 25 cm, erect, reddish-brown, laxly pubescent; **L** sessile, lamina linear, 5 - 10 cm × 3 - 4 mm, acute, basally cuneate, ± laxly pubescent, midrib prominent below; **Inf** from upper **L** axils, 8 - 15 mm pedunculate, erect, with 1 - 4 part-**Inf**, each to 10-flowered; **Ped** 3 - 5 mm, densely pubescent; **Sep** free, ± ovate, 2 - 4 mm, brownish, pubescent; **Cl** 6 - 9 mm ⌀; **Cl** tube campanulate, 1 - 2 mm ⌀; **Cl** lobes ovate to broadly ovate, 3 - 4 × 1.5 - 2 mm, purple, glabrous; **Cn** lobes 3-parted, middle appendage subulate, 3 - 4 mm, apically spiralled, lateral appendages curved horn-like, ± 1 mm; **Fil** ± 0.6 mm, massive; translator spatulate-saccate, ± 1 mm.

R. splendens Schlechter (J. Bot. 33: 301-302, 1895). **T:** Uganda (*Scott-Elliot* s.n. [BM]). – **D:** Senegal, Sierra Leone, Mali, Niger, Ivory Coast, Ghana, Benin, Nigeria, Cameroon, Central African Republic, Congo, Sudan, Burundi, Ethiopia, Zaïre, Uganda, Kenya, Tanzania, Angola. **I:** Newton (1974: 59); Rowley (1987: 37, 82); both as *R. daronii*. **Fig. XLI.b**

Incl. *Raphiacme macrostemon* K. Schumann *in sched.* (s.a.) (*nom. inval.*, Art. 29.1); **incl.** *Raphionacme excisa* Schlechter (1895); **incl.** *Brachystelma bingeri* A. Chevalier (1901) ≡ *Raphionacme bingeri* (A. Chevalier) J.-P. Lebrun & Stork (1984); **incl.** *Raphionacme gossweileri* S.Moore (1908); **incl.** *Raphionacme daronii* Berhaut (1954); **incl.** *Raphionacme chimanimaniana* Venter & Verhoeven (1989).

R tuber often flattened turnip-shaped, 5 - 15 × 2 - 25 cm, edible raw; stems erect, 15 - 20 (-30) × 0.2 - 0.3 cm, unbranched, terete, young weakly pubescent; **L** erect, linear, lanceolate, elliptic or obovate, 5 - 12 × 0.2 - 4 cm (primary **L** smaller), almost glabrous or variably short-hairy; **Inf** terminal or subterminal, extra-axillary, ± shortly pedunculate, few-flowered with 2 - 4 **Fl**; **Ped** 7 - 25 mm, weakly pubescent; **Sep** ovate, 2 - 3 × 1.5 - 2 mm, acute, outside pubescent to hispid; **Cl** 20 - 50 mm ⌀; **Cl** tube 1.5 - 5 mm; **Cl** lobes completely spreading, oblong, 10 - 25 × 6 - 7 mm, obtuse, white, rose to purple, outside scattered weakly pubescent, inside glabrous; **Cn** lobes erect, filiform from a broadened base with 2-keeled inside, 5 - 25 mm, far overtopping the **Gy**, apically often incised and contorted; **Fil** 5 mm, conical; **Anth** ± linear; **Gy** completely protruding from the **Cl** tube; translator spatulate, ± 3.5 mm, spoon apically incised; **Fr** erect, linearly fusiform, 8 - 10 cm × 4 - 5 mm.

An exceptionally beautiful and large-flowered species. The W African "*R. daronii*" is a smaller-flowered variant of this extremely widespread complex characterized by the long-stipitate stylar head, which surpasses the corolla considerably.

R. sylvicola Venter & Verhoeven (Novon 10(2): 170-172, ills., 2000). **T:** Zambia, Kaputa Region (*Merello & al.* 962 [MO, K]). – **D:** Zambia; forests; only known from the type collection.

Pubescent lianas; **R** tuber unknown; stems usually perennial, becoming woody, twining, pubescent when young, nodes with dentate interpetiolar stipules; **L** spreading, 2 - 3 cm petiolate, lamina ovate-elliptic, (9-) 11 - 13 × 4.5 - 6.5 cm, basally cuneate, pubescent, upper face green, lower face slightly paler; **Inf** subterminal, pedunculate, 5- to 8-flowered; **Ped** 8 - 12 mm; **Sep** free, ovate to lanceolate, 1.5 mm; **Cl** outside slightly pubescent; **Cl** tube campanulate, ± 7 × 3.5 mm; **Cl** lobes oblong-ovate, 4 - 5 × 1.5 - 2 mm, obtuse; petaloid **Cn** 3-partite, median segment subulate-filiform, 8 mm, erect, deeply bilobed, lateral segments triangular-subulate, 1 mm, acute, spreading; **Anth** appendage ovate, 1 mm; translator spatulate, viscidium ± discoid.

Very closely related to *R. flanaganii* with almost identical habit, from which it differs esp. by the large soft leaves and the bilobed inner segment of the corona. – [U. Meve]

R. utilis N. E. Brown & Stapf (BMI 1908: 214-215, ills., 1908). – **D:** Angola, Zimbabwe. **I:** CBM 134: t. 8221, 1908.

R tuber turnip-shaped, 10 × 5 - 14 cm, bark dark brown, scaly; stems ± erect, solitary, rarely branching, 2 - 10 cm, brown-pubescent; **L** 4 - 17 mm petiolate, lamina spreading, ± circular, elliptic or oblong-ovate, 3 - 8.5 × 3 - 4.5 cm, basally cuneate, rounded to cordate, upper face green, lower face pale purple to purple, both faces scatteredly pubescent; **Inf** sessile, terminal and lateral, axillary, many-flowered, congested; **Ped** 3 - 7 mm, brown-pubescent; **Sep** lanceolate, 3 - 5 mm, acute, pubescent; **Cl** 12 - 15 mm ⌀, deeply 5-parted, pale purple, outside laxly weakly pubescent, inside glabrous; **Cl** tube broadly campanulate, short; **Cl** lobes oblong, 5 - 6 mm, margins revolute; **Cn** lobes 3-parted, middle appendage erect, subulate, 3 - 5 mm, acute, completely erect or inclined, dark purple, lateral appendages oblong, 1.5 - 2 mm, neighbouring appendages basally fused, tips purple; **Fr** fusiform, ± 10 cm; **Se** 4 mm, tuft of **Ha** 8 mm.

The authors of the protologue interpret *R. welwitschii* as the most closely related taxon, but this differs in growth habit and flower structure.

R. velutina Schlechter (BJS 20(Beiblatt 51): 12, 1895). **T:** RSA, Eastern Cape (*Schlechter* 3509 [not located]). – **D:** Namibia, RSA (Gauteng, Eastern Cape). **I:** Fabian & Germishuizen (1982: 219).

R tuber not seen; stems 7 - 12 cm, erect, basally dichotomously branched and divaricate, basally woody, densely leafy, velvety-pubescent; **L** 2 - 6 mm petiolate, spreading or ascending, lamina linear, 2 - 4.5 cm × 5 - 9 mm, V-shaped in cross-section, obtuse or acute, basally attenuate, upper face grey-green, lower face darker, both faces velvety-pubescent; **Inf** 4 - 8.5 mm pedunculate, extra-axillary, lateral and terminal, densely 10- to 25-flowered; **Ped** 2 - 3.5 mm, velvety-pubescent; **Sep** lanceolate-triangular, 2 - 2.5 × 1 mm, acute, velvety-hairy; **Fl** green; **Cl** campanulate; **Cl** tube 2 - 3 mm, cylindrical; **Cl** lobes almost erect, ovate-oblong or oblong, 2 - 4 × 2 mm, obtuse, outside velvety-pubescent, inside glabrous; **Cn** lobes 3-parted, appendages filiform or subulate, middle appendage erect, 2 - 3 mm, lateral appendages subacute, ± 1.5 mm; **Anth** ovoid or lanceolate, margin somewhat undulate; translator spatulate, spoon lanceolate, basally abruptly constricted into the linear stalk; **Fr** narrowly ovoid, apically slowly attenuate and acute, usually 5 - 9 cm × 4 - 6 mm.

Certainly closely related to *R. galpinii* and probably even conspecific.

R. vignei E. A. Bruce (BMI 1936: 477, 1936). **T:** Ghana (*Vigne* 3823 [FH]). – **D:** Ivory Coast, Guinea Bissau, Guinea, Ghana, Chad; grassland, savannas. **I:** Rowley (1987: 37).

R tuber discoid, turnip-shaped or globose, 3 - 8 cm ⌀, fleshy; stems erect, 5 - 20 cm, weakly pubescent, the basal cm often underground and woody; **L** shortly petiolate, lamina narrowly lanceolate to obovate, 4 - 10 × 0.5 - 4 cm, basally cuneate, even the lateral veins raised below, upper face ± glabrous, lower face weakly hispid; **Inf** terminal, congested, sessile, 5- to 6-flowered; **Ped** 1 - 2 mm; **Cal** divided almost to the base, **Sep** imbricate, ovate, 1.5 mm, outside pubescent; **Fl** with citrus-like scent; **Cl** pale green to lemon-yellow; **Cl** tube cylindrical-campanulate to campanulate, 2 mm; **Cl** lobes ovate-lanceolate, 3 - 4 mm; **Cn** lobes simple, erect, compressed, filiform from a rectangularly broadened base, ± 5 mm, yellowish, tinged ± purple-red.

A conspicuous plant with slender spicate inflorescences.

R. welwitschii Schlechter & Rendle (J. Bot. 34: 97, 1896). **T:** Angola (*Welwitsch* 4234 [BM]). – **D:** Angola, Malawi, Tanzania, Kenya. **I:** Cribb & Leedal (1982: 102).

≡ *Chlorocyathus welwitschii* (Schlechter & Rendle) Bullock (1959); **incl.** *Raphionacme denticulata* N. E. Brown (1902); **incl.** *Raphionacme verdickii* De Wildeman (1904).

R tuber to 15 cm ⌀; stems 50 - 120 cm tall, twining, velvety-pubescent; **L** spreading, 0.5 - 5 cm petiolate, petiole rather thick, lamina obovate to oblong, 3 - 15 × 2 - 8 cm, obtuse, basally rounded or cuneate, upper face dark green, lower face pale green, both faces velvety-pubescent; **Inf** axillary, shortly pedunculate, 2- to 16-flowered; **Ped** 2 - 5 mm; **Sep** broadly ovate, 1.5 - 3 mm; **Cl** whitish, mostly green, sometimes also purple; **Cl** tube campanulate, 2 - 3 × 2.5 - 4 mm, glabrous; **Cl** lobes ovate or oblong, 4 - 6 × 3 - 5 mm, obtuse, fleshy, curved outwards, margin revolute, outside weakly pubescent; **Cn** lobes basally 1 mm broad, 3-parted, middle appendage ascending, oblong, ventrally with 2 thick keels, tip abruptly elongate and subulate, lateral appendages small, subulate, acute, rarely 2-toothed, purple or apically olive-green and basally purple; **Fil** dull purple, linear, 1 mm; translator thinly stipitate, spoon broadly elliptic.

A large-leaved and in addition twining species and in habit somewhat reminiscent of *Fockea multiflora*. *R. procumbens* is rather close but has a very short corolla tube and is generally much more delicate.

R. zeyheri Harvey (London J. Bot. 1: 23, 1842). **T:** RSA, Eastern Cape (*Ecklon & Zeyher* s.n. [S]). – **D:** RSA (Eastern Cape). **I:** Bayer (1976).

Incl. *Raphionacme decolor* Schlechter (1916).

R tuber massively turnip-shaped; stems 5 - 13 cm tall, basally branched and woody, ± erect, terete, striate-angular, very laxly weakly pubescent; **L** sessile or very shortly petiolate, stiffly erect, lamina linear to lanceolate, 2.5 - 3.2 cm × 2 - 9 mm, basally attenuate; **Inf** axillary, 1 - 2 mm pedunculate, 5- to 10-flowered, small; **Ped** 1 - 1.5 mm; **Sep** ovate to oblong-ovate, 1.5 mm, weakly pubescent; **Cl** campanulate; **Cl** tube 1.5 - 2 mm; **Cl** lobes oblong, ± 3.5 × 1.5 mm, obtuse, outside almost glabrous to weakly pubescent; **Cn** lobes 3-parted, basally obliquely rectangular, middle appendage filiform, 3 - 4 mm, lateral appendages subulate or reduced to minute obtuse teeth, 1 mm, sparsely pubescent; **Anth** ovoid, with acute appendage; translator spatulate, 1.2 mm, spoon elliptic to circular.

Hairy corona lobes are occasionally also encountered in the closely related *R. burkei*.

RHYTIDOCAULON

U. Meve

Rhytidocaulon P. R. O. Bally (Candollea 18: 335-339, 1963). **T:** *Rhytidocaulon subscandens* P. R. O. Bally. – **Lit:** Field (1981a); Bruyns (1986b). **D:** Ethiopia, Kenya, Somalia, Oman, Saudi Arabia, Yemen. **Etym:** Gr. 'rhytidos', wrinkle, fold; and Gr. 'kaulos', stem; for the sculptured stems.

Little-branched perennial stem succulents, main stem first erect, with short decumbent to ascending **Br**, whole plant finely papillose and covered with a conspicuous wax layer and thus appearing ± grey; latex slightly milky; stems indistinctly 4- (to 6-) ribbed, cylindrical, 2 - 50 × 0.7 - 2.5 cm, brown- or olive-green, surface rough and wrinkled from irregular furrowings and tessellations; **Tu** hardly raised; **L** rudiments lanceolate, elliptic or triangular, erect, ± appressed to the stems, subsessile, rapidly caducous, basally with stipular **Gl**, midrib raised on the lower face; **Inf** appearing often already on the young main stem, laterally sunken into the stems, 1- to 3-flowered, opening in succession, numerous, often several **Inf** active at a time and scattered over the whole plant; **Ped** 0.5 - 2 mm, papillose; **Sep** triangular, 1 - 2 mm, outside papillose; **Cl** ± cream-coloured with red, red-brown or purple flecks or bands, tips of the **Cl** lobes often uniformly coloured in the same colour as the flecks, **Cl** 4 - 18 mm ⌀, rotate (tips of the lobes fused or contorted in *R. fulleri* and *R. tortum*), inside smooth or wrinkled, glabrous or papillose, outside papillose; **Cl** tube short, 2.5 - 4 mm ⌀, bowl- to cup-shaped; **Cl** lobes broadly triangular to narrowly lanceolate, 1.5 - 7 × 0.7 - 2 mm, flatly spreading or margins revolute; **Cn** biseriate, 1 - 2 × 1.2 - 3.5 mm; **Ci** basally fused into a plate-like to shortly cup-shaped structure, lobes rectangular, apically indented or deeply divided, obtuse to acutely subulate, or almost completely fused and forming a serrate rim; **Cs** 0.3 - 1 mm, broadly rectangular or linear-lanceolate, densely incumbent on the **Anth**; **Sty** head whitish to greenish, slightly convex; **Anth** ± rectangular; **Poll** yellow, pear-shaped to ovoid, attached slightly obliquely to the short caudicles, corpuscle narrowly fusiform; **Fr** fusiform, 4 - 7 cm, erect, grey-green, green or green-brown, sometimes red-spotted; **Se** pale brown, 4 - 6.5 × 2.5 - 3 mm, wing thick, undulate, cork-like, tuft of **Ha** ± 15 mm.

Rhytidocaulon is to be placed in the relationship around the genus *Caralluma* on the base of the very similar corona structures, and this is supported by the distribution range, which entirely falls within the centre of distribution of *Caralluma*. *Rhytidocaulon* differs from all other stapeliads, however, by the wrinkled rough irregularly sculptured stem surfaces and the inflorescences, which are sunken into the stems. The plants are difficult in cultivation, and permanent success is perhaps only possible by grafting.

R. ciliatum Hanácek & Ricánek (Asklepios No. 80: 19-21, ills., 2000). **T:** Yemen (*Ricánek & Hanácek* 265 [BRNM]). – **D:** Yemen.

Stems rounded 4-angled, erect to procumbent-ascending, to 15 - 30 × 1 (-1.5) cm, rigid, grey-green, surface ± rugose-tessellate, papillose; **L** rudiments triangular-lanceolate, 2.5 - 3 × 1 mm, fleshy, papillose; **Inf** 1- to 3- (to 4-) flowered; **Fl** unscented; **Ped** 2.5 - 3 mm, ± glabrous; **Sep** triangular-subulate, 1.5 mm, papillose; **Cl** rotate, 12 - 15 mm ⌀, outside grey-brown with delicate red-brown dots, ± papillose; **Cl** tube shallowly funnel-shaped, ± 0.7 mm, upper face basally yellowish, sometimes spotted purple or bright to dark purple-brown or greenish-purple like the **Cl** lobes; **Cl** lobes triangular-linear, ± 6 × 2 mm, horizontally spreading to slightly recurved, glabrous, apically and down for ½ of their length marginally ciliate, **Ha** 2 - 3.5 mm; **Cn** 2-seriate, ± 2.3 mm ⌀, fused to a shortly bowl-shaped structure, yellowish to purple; **Ci** lobes erect, ± 0.35 mm ⌀, rectangular, ± 2-bilobed into lanceolate teeth 0.6 mm long; **Cs** lobes deltoid-subquadrate, basally slightly winged, ± 1 × 1.2 mm, ± overlapping, incumbent on the **Anth**, apically subobtuse; **Poll** ± ovoid, ± 0.35 × 0.18 mm; **Fr** slenderly fusiform, erect, 5 - 6 cm × 3.5 mm, grey, glabrous.

This species is closely related with *R. piliferum* from Somalia, but differs in the non-papillose and larger corolla and the name-giving marginal hairs of the corolla lobes.

R. fulleri Lavranos & Mortimer (Nation. Cact. Succ. J. 25(1): 2-4, ills., 1970). **T:** Oman (*Lavranos & Mortimer* 7133 [PRE]). – **D:** Oman. **Fig. XLI.f**

Stems indistinctly 4-ribbed, main stem 10 - 35 × 1.5 - 2.5 cm, **Br** shorter and more slender, grey- to brown-green, surface very rough, strongly furrowed and tessellated; **Tu** raised, rounded; **L** rudiments triangular-ovate, 1.5 - 3 mm; **Inf** 1- to 3-flowered, long-lived but clustered on young **Br**; **Fl** rich in nectar, unscented; **Ped** 0.5 - 2 mm, papillose; **Sep** triangular, 1 - 2 mm, papillose; **Cl** inside greenish to cream-coloured, cross-banded with red-brown, glabrous, outside flecked with red-brown, papillose, 8 - 10 × 4 - 5 mm, cage-like from the fused lobe tips; **Cl** tube shallow, very short; **Cl** lobes acutely triangular, narrowest near the middle, 4 - 9 × 1.5 - 2 mm, erect, strongly folded back along the midrib, tips inside sometimes red; **Cn** ± 1 × 3 mm; **Ci** fused bowl-like, lobes red (-brown), broadly bifid with erect inclined triangular appendages, those in staminal position fused with the laterally neighbouring appendages to form a projecting triangular beak, beak bent downwards, canaliculate, wedged between the free bases of the **Cl** lobes; **Cs** yellow, 1 - 1.5 mm, rectangular; **Poll** (ovoid to) pear-shaped; **Fr** green-brown, banded with purple, 3 - 4 cm; **Se** ± 6 × 3 mm.

R. macrolobum Lavranos (CSJA 39(1): 3-5, ills., 1967). **T:** Yemen (*Lavranos* 4366 [K, G, PRE, P]). – **D:** Saudi Arabia, Yemen, Somalia.

The numerous and rather large flowers make *R. macrolobum* esp. attractive for greenhouse cultivation. It is very closely related to the African *R. piliferum*.

R. macrolobum ssp. **macrolobum** – **D:** Saudi Arabia, Yemen, Somalia. **I:** Collenette (1985: 71).

Stems indistinctly 4- (to 5-) ribbed, main stem 15 - 30 × 1 - 2 cm, **Br** shorter and more slender, grey- to brown-green, surface rough, finely furrowed and tessellated, rather scattered-papillose; **Tu** near the stem tips ± semiglobose, less raised on older stem parts, rounded; **L** rudiments lanceolate to narrowly triangular, 2 - 3 × 1 mm; **Inf** 1- to 2- (to 3-) flowered, massed on young **Br**; **Fl** with dull citrus-like scent; **Ped** 0.5 - 2 mm, papillose; **Sep** triangular, ± 1 × 0.5 mm, outside papillose; **Cl** rotate, 8 - 20 mm ⌀, outside pale grey-brown with fine red-brown spots, papillose; **Cl** tube basally cream-coloured or whitish, banded with purple at the transition to the lobes, shortly funnel-shaped, pentagonal, 1 × 2 - 3.5 mm, bulging to ½ the height of the **Cn** and framing the **Cn**; **Cl** lobes inside basally cream-coloured, cross-banded with red-brown, upper ½ black-purple, 5 - 9 × 1.5 - 2 mm, lanceolate to narrowly triangular, lower ½ ascending at an angle of 30°, densely papillose, outer ½ spreading horizontally, glabrous but apically with a tuft of **Ha**, **Ha** purple, fusiform, vibratile; **Cn** 2 - 2.5 mm ⌀, pentagonal, shortly fused to form a cup-shaped structure; **Ci** lobes purple, red-brown or red, erect, ± 0.5 mm, broadly triangular at the base, apically bifid into divergent subulate appendages, 0.1 - 1 mm; **Cs** purple to pale reddish, 0.5 - 2 mm, rectangular to shortly triangular, just as long as the **Anth**; **Poll** ovoid to reniform, ± 0.2 × 0.14 mm, **Fr** grey-green, erect, 5 - 7 cm × 5 - 6 mm; **Se** pale brown, 5 - 6 × 3 mm.

R. macrolobum ssp. **minimum** Meve & Collenette (Edinburgh J. Bot. 56(1): 80-83, ills., 1999). **T:** Saudi Arabia (*Collenette* 7565 [K]). – **D:** Saudi Arabia, Yemen. **I:** Collenette (1985: 72, as aff. *macrolobum*).

Vegetatively like ssp. *macrolobum*, but **Fl** unscented; **Cl** only 4.5 - 6 mm ⌀; **Cl** tube ± 1.5 mm ⌀; basal ⅗ of the **Cl** lobes white and delicately banded with pale purple to black-brown, apical ⅖ purple to black-brown, 2 - 2.5 × 1.2 - 1.4 mm, triangular-lanceolate, straight, horizontally spreading or slightly ascending, ± glabrous or slightly papillose, margins subapically each with a tuft of **Ha**, **Ha** purple, flattened, vibratile, ± 1 mm; **Cn** whitish to pale yellowish, ± 1 × 2.5 mm; **Ci** lobes ± 0.5 mm, erect, subquadrangular at the base, apically bifid into short cylindrical slightly tuberculate appendages; **Cs** sometimes pale red, 0.5 mm, triangular-subquadrangular, emarginate.

Inconspicuous plants with very small and unscented flowers and a very pale corona.

R. mccoyi Lavranos & Mies (CSJA 73(6): 299-303, ills., 2001). **T:** Yemen (*Lavranos & al.* 31332 [MO, P, UPS]). – **D:** S Yemen.

Main stem rounded 4-angled, ± erect, little branched, to 40 × 2 cm (lateral stems thinner, shorter), stiff, grey-green, surface rugose-papillose; **L** rudiments ovate-deltoid, to 5 mm, fleshy, papillate; **Inf** 1- to 3-flowered, sessile, most frequent on young lateral stems; **Ped** erect, 15 - 18 mm, ± pubescent; **Sep** deltoid-lanceolate, 1 mm, papillate; **Cl** pyramidal, cage-like, 2.5 - 4.5 mm ⌀, inside and outside pale yellowish-green, margins lime-green, inside glabrous, outside with scattered thickish white **Ha**; **Cl** tube shallow, short; **Cl** lobes narrowly triangular-lanceolate, 6 - 8 × 1.5 - 2 mm, ascending, apically fused, margins revolute, basally auriculate, prominent; **Cn** ± 1 × 1.8 - 3.5 mm, fused to form a short bowl-shaped structure, white; **Ci** deeply bilobed, erect, the filiform appendages ± 0.35 mm long, horizontally spreading, tips crossing each other at the base of the **Cs**; **Cs** ± 0.5 mm, broadly linear-obovoid, decumbent on the **Anth**, apically rounded, dorsally papillate.

R. mccoyi does not only show greatest affinity to

R. fulleri with regard to geography, but also in habit and flower structure. The long pedicels, the flower colours, and the peculiarities of the corona underline the very independent status of *R. mccoyi*.

R. paradoxum P. R. O. Bally (Candollea 18: 339, ills. (p. 336), 1963). **T:** Ethiopia, Ogaden (*Ellis* 405 [K, ZSS]). – **D:** Ethiopia, Kenya, Somalia. **I:** Bruyns (1986b).

Stems indistinctly 4- (to 5-) ribbed, main stem 2 - 11 × 1.2 - 1.5 cm, **Br** shorter, grey-olive-green, surface rough and wrinkled, furrowed and tessellated; **Tu** at the stem tips ± regularly semiglobose, less raised on older stems, rounded; **L** rudiments narrowly triangular, ± 2 × 1 mm, acute; **Inf** 1-flowered, scattered over the whole plant; **Fl** sessile, rich in nectar, scent not described; **Sep** triangular-lanceolate, ± 1.5 × 1.3 mm, outside papillose; **Cl** rotate, ± 9 mm ⌀, outside grey-reddish, somewhat papillose; **Cl** tube cream-coloured, red-dotted and -flecked, ± 1 × 5 mm, openly bowl-shaped; **Cl** lobes basally green to yellowish, flecked and dotted with red (-brown), triangular, broader than long, ± 3 × 3.5 mm, ± horizontally spreading, glabrous; **Cn** ± 1 × 4 mm, fused into a bowl-shaped structure; **Ci** lobes greenish, broadly rectangular, ± 0.3 × 2 mm, fused into a belt-like structure, only shortly incised at staminal positions, ± erect, margin serrate, inside glabrous, outside papillose; **Cs** yellowish-green, 0.3 - 1 × 0.5 - 0.8 mm, rectangular, apically straightly truncate, serrate.

A very rare and uncommon species known only from 3 collections.

R. piliferum Lavranos (CSJA 43(2): 62-64, ills., 1971). **T:** Somalia, Northern Prov. (*Lavranos* 7365 [FT]). – **D:** Somalia. **Fig. XLI.g**

Stems ± regularly 4- (to 5-) angled, erect to decumbent-ascending, to 40 × 0.5 - 1.2 cm, stiff, grey- to brown-green, surface ± strongly grooved and tessellated, very sparingly to densely papillose; **Tu** weakly prominent, rounded; **L** rudiments ovate-triangular to lanceolate, 1 - 3 × 1 mm, erectly appressed to spreading at an angle of 45°, papillose; **Inf** 1- to 2- (to 4-) flowered, massed on young lateral **Br**; **Fl** unscented; **Ped** ± 2 mm, glabrous or papillose; **Sep** triangular-subulate, 1 - 2 mm, glabrous or papillose; **Cl** rotate, 10 - 15 mm ⌀, outside pale grey-brown with ± fine red-brown spots, ± papillose; **Cl** tube shallowly funnel-shaped, ± 3 mm ⌀, white to cream and banded with purple, similar to the basal ¼ to ¾ of the **Cl** lobes, lobe tips purple; **Cl** lobes narrowly triangular-linear, 4 - 7 × 0.8 - 1.4 mm, horizontally spreading or slightly reflexed, lamina ± folded back along the midrib, papillose or glabrous, apically with purple **Ha**, these 1 - 2 mm long, fusiform, vibratile; **Cn** ± 2 mm ⌀, fused to form a shallowly bowl-shaped structure, purple or red-brown; **Ci** lobes ascending to erect, ± 0.35 mm ⌀, rectangular, ± divided into 2 short bluntly triangular appendages; **Cs** triangular-strap-shaped, ± 0.8 mm, imbricate, incumbent on the **Anth**, apically bluntish, rarely bifid; **Poll** ± ovoid, ± 0.33 × 0.25 mm, translators very short; **Fr** very narrowly fusiform, ± 5 cm, glabrous.

R. piliferum is the sister species of *R. macrolobum*.

R. richardianum Lavranos (CSJA 63(4): 167-169, ills., 1991). **T:** Somalia (*Lavranos & al.* 24895 [UPS]). – **D:** Somalia.

Stems indistinctly 4- (to 5-) ribbed, main stem 50 - 80 × 1.5 - 2 cm, lateral **Br** somewhat thinner, first erect, then ± horizontally decumbent, weakly rebranched, grey-green, with purple tinge in full sun, surface rough, wrinkled; **L** rudiments minute; **Inf** 1- to 2-flowered, predominantly near the stem tips; **Ped** ± 1.5 mm, papillose; **Sep** triangular, ± 1.2 mm, outside papillose; **Cl** yellow-green, rotate, ± 10 mm ⌀, outside papillose; **Cl** tube shortly funnel-shaped, ± 1 × 2 mm, slightly bulging around the **Cn**; **Cl** lobes ± 4.5 × 2 mm, narrowly triangular, narrowed in the middle part, then broadening into a thickened lanceolate tip, ascending at an angle of 30°, basally glabrous, upper ½ with tufts of villous **Ha**, **Ha** pale purple, very delicate, long, curved; **Cn** ± 1 × 1.5 mm, fused into a shortly bowl-shaped structure, outside with purple **Ha**-papillae; **Ci** lobes dark purple, deeply incised, the narrowly triangular appendages ± 0.4 × 0.3 mm, erect, slightly inclined, densely covered with **Ha**-papillae; **Cs** dark purple, ± 1 × 0.2 mm, narrowly triangular, apically blunt, slightly serrate; **Poll** slenderly pear-shaped, ± 0.12 × 0.05 mm.

R. richardianum is only known from a single collection and is characterized predominantly by the conspicuous tomentum of corolla and corona.

R. sheilae D. V. Field (KB 36(1): 51-54, ills., 1981). **T:** Saudi Arabia (*Collenette* 1304 [K]). – **D:** Saudi Arabia. **I:** Collenette (1985: 72). **Fig. XLI.h, XLII.a**

Stems indistinctly 4-ribbed, main stem 0.2 - 20 × 0.7 - 1.5 cm, lateral **Br** to 17 cm, olive-green, appearing often almost grey-white due to the covering by papillae and wax, surface rough, delicately grooved and tessellated; **Tu** near the stem tips ± regularly semiglobose, weakly prominent on older stems, rounded; **L** rudiments narrowly triangular, 1.5 - 3 × 0.7 - 1 mm, acute; **Inf** 1- to 2-flowered, scattered over the whole plant; **Fl** mostly pointing downwards, with little nectar, with dull citrus scent; **Ped** 1 - 2 mm, papillose; **Sep** triangular, ± 1 × 0.5 mm, outside papillose; **Cl** rotate, 8 - 12 mm ⌀, outside pale green-brown with delicate red-brown spots, papillose; **Cl** tube basally cream-coloured, banded with purple at the transition to the **Cl** lobes, shortly funnel-shaped, pentagonal, ± 1 × 2.5 mm, slightly bulging around the **Cn**; **Cl** lobes inside green, cross-banded with red-brown, apically sometimes cream-coloured, narrowly triangular, acute, 3 - 5 × 1.25 - 1.5 mm, lower ½ ascending at an angle of 30°, upper ½ bent outwards, glabrous, apically with 1 to few **Ha**, **Ha** greenish-purple, thin; **Cn** ± lentil-shaped, basally united into a shortly bowl-shaped structure, 1.25 - 2 × 0.7 - 1 mm; **Ci** lobes red-brown, ± erect, broadly triangular, ± 0.2 × 0.25 mm, apically shortly incised, appendages triangular, mostly bent inwards; **Cs** reddish, 0.5 - 1.5 mm, rectangular, apically blunt, sometimes somewhat serrate; **Poll** ovoid to pear-shaped, ± 0.11 × 0.06 mm; **Fr** grey-green, erect, ± 5 × 0.5 cm; **Se** ± 4.5 × 2.5 mm.

R. sheilae is closely related to *R. macrolobum*. Its status as separate species is warranted by differences in habit, coloration and indumentum of the corolla as well as in the corona morphology.

R. subscandens P. R. O. Bally (Candollea 18: 336-339, ills., map, 1963). **T:** Somalia (*Bally & Peck* S111 [G, K, ZSS]). – **D:** Ethiopia, Somalia.

Stems indistinctly 4- (to 5-) ribbed, main stem 2 - 11 × 1.2 - 1.5 cm, lateral **Br** shorter, grey-olive-green, surface rough and wrinkled, grooved and tessellated; **Tu** at the stem tips ± regularly semiglobose, hardly prominent on older stems, rounded; **L** rudiments narrowly triangular, ± 2 × 1 mm, acute; **Inf** 1-flowered, scattered over the whole plant; **Ped** ± 0.5 mm, papillose; **Sep** triangular-lanceolate, ± 2 × 1 mm, outside papillose; **Cl** rotate, ± 9 mm ∅, outside green, papillose; **Cl** tube whitish, broadly banded with red-brown at the transition to the **Cl** lobes, shortly cup-shaped, ± 1 × 2 mm; **Cl** lobes whitish, broadly banded with red-brown, broadly triangular, acute, 3 - 5 × 1.25 - 1.5 mm, horizontally spreading, glabrous; **Cn** ± lentil-shaped, basally shortly fused into a bowl-shaped structure, ± 2 × 3 mm; **Ci** lobes red-brown, ± erect, ± rectangular, ± 0.4 × 0.4 mm, incised for ½, appendages triangular; **Cs** red-brown, ± 0.5 mm, bluntly triangular.

The coronas of *R. subscandens* and *R. sheilae* are very similar to each other, but the smaller corolla lobes of the latter do not allow confusion.

R. tortum (N. E. Brown) M. G. Gilbert (Bradleya 8: 29, 1990). **T:** Arabia / Socotra ? (*Bent* s.n. [K]). – **D:** S Yemen. **I:** Ricánek & Hanácek (1999).
≡ *Caralluma torta* N. E. Brown (1901).

Stems indistinctly 4-ribbed, main stem 10 - 20 × 1 - 1.6 cm, lateral **Br** shorter and thinner, grey-green, surface wrinkled; **Tu** weakly prominent; **L** rudiments lanceolate, acute, 2 - 3 mm; **Inf** 1-flowered, massed on young lateral **Br**; **Ped** 1 - 3.5 mm, papillose; **Sep** triangular, 2 mm, papillose; **Cl** greenish, inside basally and outside spotted with red-brown, outside papillose, 12 - 20 × 4 mm, erect, cage-like; **Cl** tube shallow, hardly present; **Cl** lobes subulate, ± 15 × 0.7 mm, glabrous, erect, contorted column-like above the basal cage, apically somewhat clavately thickened, reflexed; **Cn** ± 1 × 3 mm; **Ci** fused into a bowl-shaped structure, lobes purple-brown, apically yellowish, broadly bifid with inclined triangular appendages, those in staminal position united with the laterally neighbouring appendages into a prominent fringe; **Cs** red-brown-purple, ± 1.5 mm, rectangular, blunt.

This species is without doubt very closely related to *R. fulleri* and *R. mccoyi* but differs by the apically column-like corolla formed by the long contorted lobes. *R. tortum* has been recollected for the first time more than 100 years after the type collection was made (Ricánek & Hanácek 1999).

RIOCREUXIA

U. Meve

Riocreuxia Decaisne (PSRV 8: 640, 1844). **T:** *Riocreuxia torulosa* Decaisne. – **Lit:** Dyer (1980); Dyer (1983). **D:** Tropical and subtropical E Africa, RSA, Lesotho, Swaziland. **Etym:** For Alfred Riocreux (1820 - 1912), French botanical artist.

The corolla in *Riocreuxia* is cage-like as in *Ceropegia*, and therefore it was classified as a section of the latter by Huber (1957). However, this character is a parallel development, and other characters point in the direction of tribe *Marsdenieae*: **R** fibrous, sometimes woody; stems annual, herbaceous or basally becoming woody; **Cn** biseriate but **Ci** sometimes strongly reduced; **Anth** erect, fleshy; **Poll** ± erect; **Sty** head fleshy.

The genus has no succulent members.

SARCORRHIZA

U. Meve

Sarcorrhiza Bullock (HIP t. 3585, in adnot., 1962). **T:** *Sarcorrhiza epiphytica* Bullock. – **D:** Tanzania, Zaïre. **Etym:** Gr. 'sarx, sarkos', flesh; and Gr. 'rhiza', root; for the root tubers.

Lianas or shrubs with white latex; **R** tubers very fleshy, ovoid-cylindrical, 5 - 10 × 3 - 5 cm, up to 5 together; stems perennial, weakly fleshy, twining or creeping, bark red, glossy; **L** decussate, 2 - 4 mm petiolate, lamina oblong-elliptic, ± 8 × 2 cm, basally cuneate to ± obtuse, apically ± acute, glabrous; **Inf** axillary, to 2 mm pedunculate, few-flowered, with triangular imbricate **Bra**; **Ped** ± 3 mm, pubescent; **Sep** free, ± 1 mm, pubescent; **Cl** small, stellate with short tube, glabrous, outside cream-coloured; **Cl** tube 1 mm; **Cl** lobes ± ascending, tortuous, basally strongly overlapping, ± ovate, ± 5 mm, inside spotted blood-red, midvein red-brown, margins greenish or yellowish; **Cn** petaloid, lobes cream-coloured, filiform from a fleshy base, originating from a common tissue bulge behind the **Fil**, erect, ± 3 mm; **Sty** head ovoid-conical, fleshy; **Anth** broadly ovoid, apically with conspicuously off-set triangular appendages, these arranged conically; translator ± 1.5 mm, spoon ovate; **Fr** in pairs, divaricate, ± 15 × 5 mm, linear-subulate.

A monotypic genus of the *Periplocoideae*, and probably closely related to *Periploca*. The flowers are rather inconspicuous. The cluster of cylindrical-ovoid root tubers is notable (similar to *Dactylorhiza*, *Orchidaceae*), as well as the at least facultatively epiphytic habit, comparable with that of the E Asian genera *Hoya* or *Dischidia*. The plants are also found creeping on rocks.

S. epiphytica Bullock (HIP t. 3585 + text, 1962). **T:** Tanzania (*Verdcourt* 1725 [K]). – **D:** Tanzania, Zaïre.

Description as for the genus.

SARCOSTEMMA

S. Liede

Sarcostemma R. Brown (Prodr., 462-463, 1810). **T:** *Euphorbia viminalis* Linné. – **Lit:** Adams & Holland (1978); Ali (1983); Forster (1992); Liede & Meve (1995); Meve & Liede (1996). **D:** Tropical and subtropical regions of the Old World. **Etym:** Gr. 'sarx, sarkos', flesh; and Gr. 'stemma', garland, wreath; for the fleshy tangling stems.
Incl. *Monostemma* Turczaninow (1848). **T:** not designated.
Incl. *Decanemopsis* Costantin & Gallaud (1906). **T:** *Decanemopsis aphylla* Costantin & Gallaud.
Incl. *Drepanostemma* Jumelle & H. Perrier (1911). **T:** *Drepanostemma luteum* Jumelle & H. Perrier.

Perennial climbing or twining stem succulents; stems strongly and often dichotomously branched, green, ≤ 6 mm ⌀, basally often corky, then ≤ 5 cm ⌀, smooth, towards the tips often ± densely hairy; **L** rudiments reduced to sessile triangular persistent **Sc** (early caducous in *S. membranaceum*); **Inf** 1 (rarely 2) per node, extra-axillary, many-flowered, pseudo-umbellate; **Cl** rotate, **Pet** basally slightly fused; **Cn** biseriate, consisting of inner free staminal parts (**Cs**) and an outer ring formed by fused staminal and interstaminal parts (**C(is)**); **Gy** sessile; **Anth** with apical appendage and conspicuously differentiated **Anth** wings, which consist of a single distal rail; **Poll** pendent, characteristically golf-club-shaped; **Sty** head with bulges at the upper end of the corpuscle, upper part umbonate to conical; **Fr** inversely clavate or oblong; **Se** testa glabrous or with 1-celled **Ha**, with an apical tuft of **Ha**.

Molecular analyses (Liede & Täuber 2002) have shown that *Sarcostemma* has its roots in the complex of stem succulent *Cynanchum* species in Madagascar, from where it has – probably along the coast lines – extended its range to West Africa and Australasia/Polynesia. During this radiation, numerous species and subspecies have arisen whose morphological differences are often only slight.

Liede & Meve (2001) have placed the Madagascan taxa of *Sarcostemma* in *Cynanchum*. The remaining taxa will also have to be placed in *Cynanchum* once a modified infraspecific concept is available. As long as the new taxonomy of the group is not complete, their coverage under the name *Sarcostemma* is to be preferred.

The following names are of unresolved application but are referred to this genus: *Asclepias acida* Roxburgh (1814) ≡ *Sarcostemma acidum* (Roxburgh) Voigt (1845); *Decanemopsis aphylla* Costantin & Gallaud (1906); *Euphorbia pendula* Boissier (s.a.); *Sarcostemma andongense* Hiern (1898); *Sarcostemma brachystigma* Wight (1834); *Sarcostemma brevistigma* Wight & Arnott (1834); *Sarcostemma intermedium* Decaisne (1844); *Sarcostemma madagascariense* Descoings (1961).

S. antsiranense Meve & Liede (KB 52(2): 491-493, ills., 1997). **T:** Madagascar, Antsiranana (*Lavranos & al.* 28772 [K, MSUN]). – **D:** N Madagascar (Antsiranana). **Fig. XLII.b**
≡ *Cynanchum antsiranense* (Meve & Liede) Liede & Meve (2001).

Stems twining to 3 m, 2.5 - 4 mm ⌀, usually regularly dichotomously branched, fresh green, glabrous, smooth; **Inf** extra-axillary, numerous, often 2 per node, 2- to 10-flowered, from horizontal 1 - 5 × 2 - 3 mm large **Br**; **Ped** 6 - 8 mm; **Fl** pointing downwards, with sweet vanilla-scent; **Pet** cream-coloured, basally flushed with pink, 4 - 4.5 × 3 - 3.5 mm, broadly ovate, horizontally spreading, basal ⅓ united; **Cn** white, ± 2.2 × 2.5 mm, outer **C(is)** ring-like, ± 0.5 mm tall; **Cs** triangular, ± 1.5 mm, shorter than the **Gy**; **Sty** head 0.7 × 0.8 mm, conical, apically incised.

S. antsiranense flowers rather easily in cultivation, and the attractive flowers are extremely pleasantly scented.

S. arabicum Bruyns & P. I. Forster (Edinburgh J. Bot. 48(3): 333-335, 1991). **T:** Saudi Arabia (*Nasher* IH137 [E]). – **D:** Saudi Arabia, Yemen. **Fig. XLII.c**

Stems to 3 m, richly branched, erect, straggling and drooping, rarely twining, grey-green, 4 - 8 mm ⌀, young hairy; **Inf** 2- to 10-flowered, terminal or extra-axillary; **Fl** sweet-scented; **Cl** pale yellow to cream-coloured, pink towards the centre; **Pet** 3 - 3.5 mm, triangular, obtuse, spreading; **Cn** white, outer **Ci** clinging to the inner **Cs**; **Cs** ± 2.5 mm, overtopping the **Gy** and incumbent above; **Fr** 5 - 10 cm, fusiform.

Closely related to *S. socotranum*, but with a hidden style head.

S. brevipedicellatum P. I. Forster (Austral. Syst. Bot. 5(1): 59-62, ills., 1992). **T:** Australia, Queensland (*Forster* 1420 [BRI, AD, CANB, CBG, DNA, K, MEL, PERTH, PRE, QRS]). – **D:** Australia (Northern Territory, Queensland). **Fig. XLIII.c**

Stems 1.5 - 3 m, twining, 1.5 - 3 mm ⌀, glabrous; **Inf** few-flowered, extra-axillary; **Ped** < 3 mm; **Cl** cream-coloured or pale yellow; **Pet** 3 - 4 mm, ovate, spreading; **Cn** white; outer **C(is)** ± 0.5 mm; inner **Cs** ± 1.5 mm, slightly shorter than the **Gy**; **Fr** 5 - 9 cm, fusiform; **Se** 6 - 9 mm.

Similar to *S. viminale*, but differing by short robust pedicels and small flowers.

S. decorsei Costantin & Gallaud (Bull. Mus. Hist. Nat. (Paris) 12: 418, 1906). **T:** Madagascar (*Decorse* s.n. [P [not traced]]). – **D:** S Madagascar (Toliara). **I:** Liede & Meve (1995).

≡ *Cynanchum decorsei* (Costantin & Gallaud) Liede & Meve (2001); **incl.** *Drepanostemma luteum* Jumelle & H. Perrier (1911).

Stems 1.5 - 3 m, ascending, twining, 1.5 - 3 mm ⌀, glabrous; **Inf** 7- to 12-flowered, extra-axillary; **Cl** yellow; **Pet** ± 3 mm; outer **C(is)** ± 0.2 mm; inner **Cs** ± 0.5 mm, shorter than the **Gy**, lingulate; **Sty** head conical; **Fr** ± 15 cm, oblong, obtuse; **Se** ± 6 mm, ovoid, papillose, winged.

A very delicate and insufficiently known species.

S. elachistemmoides Liede & Meve (Bot. J. Linn. Soc. 118: 41-43, ills., 1995). **T:** Madagascar (*Decary* 7890 [P]). – **D:** S Madagascar (Mahajanga, Toliara).

≡ *Cynanchum elachistemmoides* (Liede & Meve) Liede & Meve (2001).

Stems erect, delicately longitudinally striate, 2.5 - 4 mm ⌀, glabrous; **Inf** 4- to 8-flowered, extra-axillary; **Pet** ± 1.4 mm, spreading, colour not known with certainty; **Cn** reduced to **Sc** on the back of the **Anth**, ± 0.2 mm; **Sty** head flatly conical.

Rarely flowering and rarely collected; related to *S. decorsei*.

S. forskaolianum Schultes (Syst. Veg. 6: 117, 1820). **T** [neo]: Saudi Arabia (*Collenette* 1977 [K]). – **D:** Saudi Arabia, Yemen. **I:** Meve & Liede (1996).

Incl. *Asclepias aphylla* Forsskål (1775).

Stems to 4 m, twining, blue-green, 1.5 - 3 mm ⌀, strongly wax-covered, slightly rough, glabrescent; **Inf** 3- to 10-flowered, terminal or from 1 - 12 mm long **Br**; **Fl** sweet-scented; **Cl** cream-white to pink; **Pet** ± 3 mm, ovate-deltoid, ascending-spreading; **Cn** white; outer **C(is)** ± 1 mm; inner **Cs** ± 1.5 mm, claw-like, as tall as the **Gy**; **Sty** head flatly conical; **Fr** 5 - 7 cm, fusiform; **Se** ± 5 mm, glabrous.

S. membranaceum Liede & Meve (Bot. J. Linn. Soc. 18: 43-45, ills., 1995). **T:** Madagascar (*Liede & Conrad* 2765 [P, TAN]). – **D:** S Madagascar (Antsiranana, Mahajanga, Toliara).

≡ *Cynanchum membranaceum* (Liede & Meve) Liede & Meve (2001).

Stems 3 - 4 (-15) m, erect, twining, 4 - 6 mm ⌀ (vegetatively) or 1 - 2 mm ⌀ (flowering), glabrescent to glabrous; **Sc**-like **L** rudiments 1 × 0.8 mm, triangular, early caducous; **Inf** 10- to 15-flowered, sessile, extra-axillary, often 2 **Inf** per node; **Cl** white to very pale yellow; **Pet** 6 - 7 mm, membranous and thin-textured, spreading; **C(is)** white, ± 2 mm, ring-like; **Cs** ± 1.5 mm, shorter than the **Gy**, claw-like; **Gy** small in comparison to the **Cl**; **Sty** head conical; **Fr** ± 17 cm, oblong.

S. mulanjense Liede & Meve (Novon 2: 223-226, ills., map, 1992). **T:** Malawi (*Chapman & Chapman* 6892 [MO]). – **D:** S Malawi. **Fig. XLII.d**

Stems erect, most shortly creeping, some to 1.5 m, pale green, 3 - 5 mm ⌀, glabrous; **Inf** 5- to 7-flowered, terminal or sessile, extra-axillary; **Fl** sweet-scented; **Cl** pale cream-coloured, flushed with reddish; **Pet** 4.5 - 6 mm, ± 3× as long as broad, thin-textured, slightly undulate; **C(is)** white, ± 1.5 mm, ring-like; **Cs** ± 3 mm, shorter than the **Gy**, spoon-shaped, glabrous; **Gy** taller than broad; **Sty** head conical.

S. pearsonii N. E. Brown (BMI 1913: 301, 1913). **T:** Namibia (*Pearson* 8640 [BOL, K]). – **D:** S Namibia, RSA (Northern Cape). **I:** Liede & Meve (1989).

Stems to 50 cm, erect, shrubby, grey-green, 3 - 4 mm ⌀, young hairy, old becoming woody; **Inf** 3- to 6-flowered, terminal or extra-axillary, sessile; **Cl** bright yellow; **Pet** ± 5 mm, strongly twisted; **Ci** clinging to the **Cs**; **Cs** ± 3 mm, shorter than the **Gy**; **Poll** ellipsoid; **Sty** head conical; **Fr** 5 - 7 cm, obclavate; **Se** ovoid, winged, with papillose **Ha**.

S. resiliens B. R. Adams & R. W. Holland (CSJA 50(4): 166-167, ills., 1978). **T:** Kenya, Taita Distr. (*Adams* 147 [EA]). – **D:** Kenya, Zimbabwe. **Fig. XLII.e**

Stems to 50 cm, most shorter, some long, drooping and rooting when in contact with the ground, 3 - 5 mm ⌀, young delicately hairy; **Inf** 2- to 7-flowered, from 2 - 3 cm long **Br**; **Fl** sweet-scented; **Cl** dirty yellow-brown-purple; **Pet** ± 5 mm, ascending-spreading; **C(is)** white, ring-like; **Cs** bent inwards, as tall as the **Gy**; **Sty** head conical; **Fr** ± 7 cm, fusiform; **Se** winged.

S. socotranum Lavranos (Nation. Cact. Succ. J. 27(2): 37-38, ills., 1972). **T:** Socotra (*Smith & Lavranos* 309 [K]). – **D:** Socotra. **Fig. XLII.f**

Stems shrubby, grey-green, to 1.5 m, decumbent, 2 - 5 mm ⌀, young hairy; **Inf** 2- to 8-flowered, terminal or sessile extra-axillary; **Cl** greenish-yellow; **Pet** 3.5 mm, ascending-spreading; **Ci** clinging to the **Cs**; **Cs** ± 3 mm, slightly shorter than the **Gy**; **Sty** head conical.

Closely related to *S. arabicum*, but the style head surpasses the **Cs**-parts of the corona.

S. stocksii Hooker *fil.* (Fl. Brit. India 4: 27, 1883).

T: Pakistan (*Stocks* 509 [K]). − **D:** Arabian Peninsula, Pakistan.

≡ *Sarcostemma viminale* ssp. *stocksii* (J. D. Hooker) Ali (1983).

Stems to 1.5 m, erect, 3 - 7 mm ∅, glabrous; **Inf** many-flowered, from lateral **Br**, these thinner than the vegetative **Br** and often in chains; **Cl** greenish-white; **Pet** ± 5 mm, spreading; **C(is)** ring-like; **Cs** bent inwards, as tall as the **Gy**; **Sty** head conical; **Fr** ± 7 cm, cylindrical, woody; **Se** dark brown, ovoid to oblong, rugulose, wingless.

This species is characterized by the robust and slightly woody fruits and the dark brown rugulose and wingless seeds.

S. stoloniferum B. R. Adams & R. W. Holland (CSJA 50(3): 110-111, ills., 1978). **T:** Kenya, Rift Valley Prov. (*Adams* 138 [EA]). − **D:** Kenya, N Tanzania. **Fig. XLII.g**

Stems to 30 cm, erect (or some to 40 cm, decumbent and rooting from the nodes), dark green, flushed purple in the sun, 4 - 9 mm ∅, slowly tapering towards the tips, stiff, young hairy; **Int** very regular, usually short; **Inf** 7- to 9-flowered, terminal or at the tips of short lateral **Br**, which consist of a 2 - 7 mm long **Int**; **Fl** sweet-scented; **Cl** greenish-brown to green; **Pet** ± 5 mm; **C(is)** white, cup-shaped, ± ½ as tall as the whole **Cn**; **Cs** a little shorter than the **Gy**; **Sty** head flatly conical; **Fr** 4 - 6 cm, inversely clavate; **Se** winged, glabrous, wings with papillose **Ha**.

Easily recognized on account of the regular short internodes.

S. vanlessenii Lavranos (Nation. Cact. Succ. J. 29(2): 35, ills., 1974). **T:** Yemen (*Lavranos & van Lessen* 1832 [K, PRE]). − **D:** Yemen, Ethiopia, Kenya, Tanzania. **I:** Eggli (1994: 175). **Fig. XLII.h**

Incl. *Sarcostemma subterraneum* B. R. Adams & R. W. Holland (1978).

Stems decumbent, to 1 m, 3 - 4 mm ∅; **Inf** 4- to 8-flowered, extra-axillary, or from very short lateral **Br**; **Fl** unscented; **Cl** pale to dark purple; **Pet** 3 - 4 mm, ascending-spreading; **Cn** cream to pink; **C(is)** ring-like; **Cs** claw-like, overtopping the **Gy**; **Sty** head conical; **Fr** 3 - 4 cm, obclavate; **Se** winged.

A rather compact and freely flowering species.

S. viminale (Linné) R. Brown (Prodr., 464, 1810). **T:** [icono]: Alpinus, Pl. Aegypt., 190, t. 190. − **D:** Africa, India, SE Asia, Australia.

≡ *Euphorbia viminalis* Linné (1753); **incl.** *Asclepias aphylla* Thunberg (1794) ≡ *Cynanchum aphyllum* (Thunberg) Schlechter (s.a.) ≡ *Sarcostemma aphyllum* (Thunberg) R. Brown (1810); **incl.** *Sarcostemma tetrapterum* Turczaninow (1848) ≡ *Cynanchum tetrapterum* (Turczaninow) R. A. Dyer *ex* Bullock (1955).

Stems erect or twining, with or without a main stem; **Fl** at least slightly scented; **Cl** cream-coloured to yellow; **Pet** 4 - 6 mm; **C(is)** cream-white, ring-like, ± as tall as the **Gy**; **Cs** as broad as the **Anth**.

This extremely variable and widespread species can be divided into a series of subspecies. The names *S. aphyllum* and *Cynanchum aphyllum* in their horticultural application belong to *Cynanchum gerrardii*.

S. viminale ssp. **australe** (R. Brown) P. I. Forster (Austral. Syst. Bot. 5(1): 64, ills., 1992). **T:** Australia (*Brown* 2872 [BM]). − **D:** Australia. **Fig. XLIII.a**

≡ *Sarcostemma australe* R. Brown (1810).

Plants without main stem, forming untidy shrubs; stems blue-green; **Inf** sessile, extra-axillary; **Cl** yellow; **Fr** 8 - 12 cm, fusiform.

S. viminale ssp. **brunonianum** (Wight & Arnott) P. I. Forster (Austral. Syst. Bot. 5(1): 63, ills., 1992). **T:** sine loco (*Wight* 1557 [G-DC]). − **D:** N coast and coastal regions of Australia, China, India, Philippines, Thailand, New Caledonia.

≡ *Sarcostemma brunonianum* Wight & Arnott (1834).

Stems to 5 m, twining, with distinct main stem, glabrous; **Sc**-like **L** rudiments very short, obtuse; **Inf** sessile, extra-axillary; **Fl** sweet-scented; **Cl** yellow-green.

Closely related to ssp. *odontolepis*, but differing by the sessile inflorescences and the shorter C(is)-parts of the corona.

S. viminale ssp. **odontolepis** (Balfour *fil.*) Meve & Liede (Bot. J. Linn. Soc. 120(1): 31, ills. (p. 26), 1996). **T:** Mauritius, Rodriguez Island (*Balfour* s.n. [E]). − **D:** Ethiopia, Kenya, Mauritius (Rodriguez), Tanzania. **Fig. XLIII.d**

≡ *Sarcostemma odontolepis* Balfour *fil.* (1877).

Stems to 5 (-10) m, twining, often characteristically dichotomously branched, dark green, young glossy, glabrous; **Inf** from 2 - 30 mm long lateral **Br**, which occasionally form chains; **Ped** to 15 mm, filiform; **Cl** whitish, yellowish or greenish; **Fr** ≤ 8 mm, fusiform.

S. viminale ssp. **orangeanum** Liede & Meve (Bot. J. Linn. Soc. 112: 7-10, ills., 1993). **T:** RSA, Free State (*Liede & Meve* 579 [K, MSUN]). − **D:** Namibia, RSA (Northern Cape, Free State). **Fig. XLIII.b**

Stems erect, rarely straggling in shrubs, stoloniferous at or just below ground-level, young hairy; **Fl** somewhat smaller than in the other ssp.; **Cl** lemon-yellow; **C(is)** ± ½ as tall as the **Gy**; **Cs** claw-like, acute; **Fr** 6 - 8 cm, fusiform, slender.

S. viminale ssp. **stipitaceum** (Forsskål) Meve & Liede (Bot. J. Linn. Soc. 120(1): 32, ills. (p. 33),

1996). **T:** Yemen (*Forsskål* s.n. [C]). – **D:** Saudi Arabia, Oman, Yemen, Ethiopia, Somalia, Kenya, Tanzania. **Fig. XLIII.e**

≡ *Asclepias stipitacea* Forsskål (1775) ≡ *Sarcostemma stipitaceum* (Forsskål) Schultes (1820).

Stems to 2 m, shrubby, erect or drooping, green to blue-green, with dense hairiness present for a considerable time; **Inf** from 1 - 25 mm long lateral **Br**, which originate only from erect stems and occasionally form chains; **Cl** cream- to ivory-coloured or yellow; **Fr** 5 - 8 cm, fusiform.

S. viminale ssp. **suberosum** Meve & Liede (Bot. J. Linn. Soc. 120(1): 35-36, ills., 1996). **T:** RSA, Transvaal (*Kirk* 97 [K]). – **D:** Eritrea, Ethiopia, Kenya, Somalia, Tanzania, Zimbabwe, Malawi, Namibia, RSA; probably more widespread. **Fig. XLIII.f**

Stems to 3 m, twining, main stem distinctly corky, to 5 cm ∅; **Inf** sessile, extra-axillary, peduncle frequently persistent; **Fl** sweet-scented; **Cl** pale greenish-yellow.

S. viminale ssp. **thunbergii** (G. Don) Liede & Meve (Bot. J. Linn. Soc. 112: 10, ills., 1993). **T** [neo]: RSA, Cape Prov. (*Bayer* 68 [NBG]). – **D:** Cape Verde Islands, Namibia, RSA (Western Cape).

≡ *Sarcostemma thunbergii* G. Don (1838); **incl.** *Sarcostemma daltonii* Decaisne (1849); **incl.** *Sarcostemma nudum* C. Smith *ex* Decaisne (1849).

Stems erect to drooping, without main stem, (blue-) green, robust, strongly wax-covered; **Inf** sessile, extra-axillary; **Cl** lemon-yellow, sometimes with reddish tinge; **Fr** 4.5 - 8.5 cm, obclavate.

S. viminale ssp. **viminale** – **D:** Ethiopia, Kenya, Madagascar, Aldabra. **I:** Liede & Meve (1995).

Stems 1 - 3 m, ascending, not twining (straggling, rarely contorted screw-like), 3 - 5 mm ∅, strongly wax-covered, glabrous; **Inf** 10- to 25-flowered, pseudo-umbellate, from 4 - 6 cm long lateral **Br**; **Fl** strongly sweet-scented; **Cl** greenish-yellow; **Pet** 4 mm, spreading to incumbent; **C(is)** ± 2.5 mm; **Cs** 1.5 mm, shorter than the **Gy**; **Sty** head (flatly) conical; **Fr** ≤ 18 cm, oblong; **Se** 4 - 5 mm, oblong-ovate, smooth, winged.

SCHIZOGLOSSUM

B. Müller, P. Stegemann & F. Albers

Schizoglossum E. Meyer (Comm. Pl. Afr. Austr., 218, 1838). **T:** *Schizoglossum atropurpureum* E. Meyer. – **Lit:** Kupicha (1984); Hilliard & Burtt (1989); Bruyns (1995c). **D:** Lesotho, RSA (Eastern Cape, Free State, KwaZulu-Natal, Mpumalanga), Swaziland. **Etym:** Gr. 'schizein', to split; and Gr. 'glossa', tongue; for the often bifid corona lobes.

Erect shrubs with a **R** tuber; stems 7 - 130 cm, not or weakly branched, biseriately hairy, with latex; **L** decussate, sometimes verticillate or irregularly arranged, shortly petiolate, lamina simple, variable in shape (oblong, triangular, elliptic, cordate, ovate, linear, lanceolate), tip acute or obtuse, margins rolled backwards, upper face ± glabrous, venation softly hairy, lower face hispid, **Ha** appressed; **Inf** extra-axillary few- to many-flowered pseudo-umbels, with few to many **Inf** per stem; **Bra** filiform, inconspicuous; peduncle pubescent; **Ped** softly hairy; **Sep** triangular, outside softly appressed-hairy, inside glabrous, margins ciliate; **Cl** green, brown, yellow, partly with purple, lobes in bud contorted, overlapping to the left, elliptic, flat, weakly cucullate or margins bent outwards, glabrous or softly hairy, margins ciliate; **Ci** sometimes present, lobes minute, below the guide rail; **Cs** green, beige, white, rarely dark red, variable in shape, **Cs** lobes basally flattened, narrow, oblong, elliptic, broadly ovate, isodiametric, triangular or square, apically frequently cleft, usually with an inner appendage, often incised, sometimes longitudinally and transversally keeled; **Sty** head white or green, rarely reddish-brown, flat or scutate; **Anth** ovate, reniform or triangular, with basal appendages, laterally placed against or incumbent on the **Sty** head; **Poll** elliptic, ovate, reniform or obtriangular, attached in the middle to the translator arms, germination zone always visibly present, translators thin, slender, sometimes flattened, corpuscle with or without wing; **Fr** solitary, 4 - 7 cm, with soft appressed prickles, pubescent; **Se** dark brown, compressedly ovoid, wrinkled, apically with a tuft of **Ha**.

Closely related to *Aspidoglossum* (see there).

The following names are of unresolved application but are referred to this genus: *Schizoglossum angolense* Schlechter & Rendle (1896); *Schizoglossum capitatum* Schlechter (1895); *Schizoglossum glanvillei* Hutchinson & Dalziel (1937); *Schizoglossum graminifolium* Norman (1929); *Schizoglossum huttoniae* S. Moore (1902); *Schizoglossum pedunculatum* Schlechter (1893); *Schizoglossum tricorniculatum* K. Schumann (1893); *Schizoglossum truncatulum* K. Schumann (1895); *Schizoglossum viridulum* K. Schumann (1900); *Xysmalobium heudelotianum* Decaisne (1844) ≡ *Schizoglossum heudelotianum* (Decaisne) Roberty (1953).

S. amatolicum Hilliard (Notes Roy. Bot. Gard. Edinburgh 45(2): 182, 1989). **T:** RSA, Eastern Cape (*Hilliard & Burtt* 18876 [E]). – **D:** RSA (Eastern Cape: Amatola Mts.).

Stems 25 cm, unbranched, 2 mm ∅, biseriately hairy; **L** 2 mm petiolate, lamina 18 - 34 × 8 - 14 mm, oblong-deltoid, acute, basally ± truncate, margins and midrib of the lower face slightly hairy; **Inf** from upper nodes, with ± 6 **Fl**, lowermost peduncle 3.6 cm; **Ped** ± 1 cm; **Sep** 4 × 1.5 mm, lanceolate; **Cl** blackish-green, ± flat, incised ± to the base, lobes 9

× 5 mm, curved outwards, margins and base with few small **Ha**; **Cs** green, **Cs** lobes 2 × 1.5 mm, much shorter than the **Gy**, ± square, very fleshy, retuse, margins ± crenate, inner appendage minute, erect; **Gy** green, 4 mm; **Sty** head 2 mm ⌀; **Anth** ± 1 × 1 mm, ovoid, hiding the **Gy**; **Poll** ± 0.8 mm, compressedly oblong-triangular, subterminally attached, corpuscle ± 0.3 mm.

Closely related to *S. hamatum.*

S. atropurpureum E. Meyer (Comm. Pl. Afr. Austr., 219, 1838). **T:** RSA, Eastern Cape (*Drège* s.n. [K]). – **D:** RSA (Eastern Cape, W and S KwaZulu-Natal, Free State, Mpumalanga).

Closely related to *S. hamatum.* The subspecies are probably better recognized as separate species.

S. atropurpureum ssp. **atropurpureum** – **D:** RSA (Eastern Cape, Mpumalanga); tall grass or scrub along forest margins or near rivers, 1200 - 2040 m.
Fig. XLIII.g

Stems 30 - 130 cm, massive, rarely branched; **L** ≤ 7 mm petiolate, lamina 29 - 50 × 8 - 21 mm, oblong (-triangular); **Inf** 5 - 9, with 8 - 15 **Fl**, sometimes branched; peduncle 6 - 30 mm; **Ped** 8 - 12 mm; **Sep** 3 - 4 × 1 mm; **Cl** cucullate, outside green, striped with brown, inside dark brown, lobes 5.5 - 6 × 3.5 - 4 mm, cucullate, glabrous; **Cs** green, ± 4 mm, **Cs** lobes apically recurved, bordered by a 3-lobed groove on the inside, inner appendage ± 1 mm, erect, bi- or trifid, flanked by wing-like erect marginal appendages; **Sty** head white with brownish-purple; **Anth** 0.75 mm, triangularly ovoid, laterally placed against the **Sty** head; **Poll** 0.7 - 0.82 mm, compressedly elliptic, apically with a crest (formed by the germination crest), corpuscle 0.27 - 0.42 mm, winged, transparent; **Fr** ± 4.5 cm; **Se** 3 mm, **Ha** tuft ± 3 cm.

S. atropurpureum ssp. **tridentatum** (Schlechter) Kupicha (KB 38(4): 606, 1984). **T:** RSA, Eastern Cape (*Flanagan* 1040 [BOL, GRA, K, PRE]). – **D:** RSA (Eastern Cape); rocky grassy slopes, 450 - 1500 m.

≡ *Schizoglossum tridentatum* Schlechter (1894).

Differs from ssp. *atropurpureum*: Stems 16 - 55 cm; **L** 2 - 4 mm petiolate, lamina 21 - 36 × 6 - 22 mm, oblong or triangular; **Inf** 4 - 11 per node, with 4 - 7 **Fl**; peduncle 6 - 20 mm; **Ped** 5 - 10 mm; **Sep** 2.5 - 3.5 × 1 - 1.5 mm; **Cl** outside greenish, red-striate, inside dark red, lobes 5.5 - 6 × 3 - 4 mm; **Cs** white, ± 2 mm, inner groove simple; **Sty** head white with pale green centre; **Poll** 0.6 - 0.7 mm, corpuscle 0.22 - 0.3 mm; **Fr** 4.5 - 5.5 cm.

S. atropurpureum ssp. **virens** (E. Meyer) Kupicha (KB 38(4): 606, 1984). **T** [neo]: RSA, KwaZulu-Natal (*Gerrard* 1302 p.p. [K]). – **D:** RSA (Eastern Cape, KwaZulu-Natal); coastal scrub margins on disturbed ground, grassland on sandy or rocky soil, 30 - 900 m.

≡ *Schizoglossum virens* E. Meyer (1838); **incl.** *Schizoglossum euphorbioides* E. Meyer (1838); **incl.** *Schizoglossum oblongum* Schlechter (1894).

Differs from ssp. *atropurpureum*: Stems 16 - 70 cm; **L** 1 - 5 mm petiolate, lamina 20 - 40 × 6 - 30 mm, variable in shape (oblong (-elliptic), triangular, cordate, ovate or narrowly linear); **Inf** 5 - 13, with 5 - 15 **Fl**; peduncle 20 - 60 mm; **Ped** 10 - 15 mm; **Sep** 2.5 - 4.5 × 1 - 1.5 mm; **Cl** outside green, inside yellowish-green, purple-striate, ± flat, lobes 4.5 - 7 × 2.5 - 4 mm, sometimes slightly bulging and convex; **Cs** greenish, 2 - 3 mm, inner groove simple, inner appendage ≤ 2 mm; **Sty** head white; **Poll** 0.65 - 0.87 mm, compressedly ovoid.

S. bidens E. Meyer (Comm. Pl. Afr. Austr., 220, 1838). **T:** RSA, Eastern Cape (*Drège* s.n. [K, BM, CGE, PRE]). – **D:** E RSA, Lesotho, NW Swaziland.

S. bidens ssp. **atrorubens** (Schlechter) Kupicha (KB 38(4): 613, 1984). **T:** RSA, Eastern Cape (*Baur* 767 [K]). – **D:** RSA (Eastern Cape, border to KwaZulu-Natal and Lesotho); grassland close to rivers, 1350 - 2220 m.

≡ *Schizoglossum atrorubens* Schlechter (1894) ≡ *Lagarinthus atrorubens* (Schlechter) Bullock (1952).

Differs from ssp. *bidens*: Stems 21 - 66 cm, rarely branched; **L** 1 - 5 mm petiolate, lamina 30 - 63 × 2 - 5 mm, linear or narrowly triangular; **Inf** 4 - 10, with 4 - 13 **Fl**; peduncle 14 - 46 mm; **Ped** 3 - 14 mm; **Sep** 2.5 - 3 × 1 mm; **Cl** brown, cup-shaped, margins curved outwards, lobes 4 - 4.5 × 2 - 2.5 mm; **Cs** white with rose or purple, **Cs** lobes (1.5-) 2 (-3) mm, ± broadly ovate, apically ± strongly bifid, inner appendage sometimes weakly bifid; **Anth** 0.5 × 0.75 - 1 mm; **Poll** 0.58 - 0.8 × 0.1 - 0.15 mm, corpuscle 0.15 - 0.23 mm; **Fr** 7.3 cm; **Se** 4.5 mm.

S. bidens ssp. **bidens** – **D:** RSA (Eastern Cape, Free State), Lesotho; grassland, muddy or rocky ground, 1320 - 2600 m.

Incl. *Schizoglossum truncatum* Schlechter (1894).

Stems 9 - 22 cm, often branched; **L** verticillate or irregularly arranged, petiole 2 - 3.5 mm, lamina 34 - 60 × 2.5 - 7 mm, linear; **Inf** 3 - 7 per stem, with 5 - 8 **Fl**; peduncle 8 - 33 mm; **Ped** 3 - 8 mm; **Sep** 2 - 3 × 0.75 - 1 mm; **Cl** green or brown, lobes 3.5 - 5 × 2 - 3 mm, margins revolute, outside glabrous or weakly soft-hairy, inside glabrous or lobes basally pubescent; **Cs** white, 2 - 2.5 mm, **Cs** lobes oblong or narrowly ovate, apically bifid, inner appendage lingulate, apically bifid, curved inwards with the tip pointing outwards; **Sty** head white, conspicuously swollen above the **Anth**; **Anth** 0.5 × 0.75 - 1 mm, reniform, placed laterally against the **Sty** head; **Poll** 0.55 - 0.76 mm, sausage-shaped, slightly compressed, germination crest apical, attached in the middle, corpuscle 0.15 - 0.27 mm; **Fr** 4.5 - 5.6 cm; **Se** 3 mm, **Ha** tuft 24 mm.

S. bidens ssp. **galpinii** (Schlechter) Kupicha (KB 38(4): 617, 1984). **T:** RSA, Mpumalanga (*Galpin* 1326 [PRE, K]). — **D:** RSA (SE Mpumalanga: near Barberton); rocky grassland, 960 - 1500 m.

≡ *Schizoglossum galpinii* Schlechter (1894) ≡ *Lagarinthus galpinii* (Schlechter) Bullock (1952).

Differs from ssp. *bidens*: Stems 30 - 66 cm, unbranched; **L** 1 - 2 mm petiolate, lamina 25 - 38 × 1 - 4 mm, linear; **Inf** 5 - 8, with 3 - 10 **Fl**; peduncle 1 - 3.6 cm; **Ped** 12 - 15 mm; **Sep** 2 - 3 × 1 mm; **Cl** pale green, lobes 3.5 - 4.5 × 2 - 2.5 mm; **Cs** lobes 2 mm, inner appendage simple, apically recurved; **Anth** 0.5 × 0.75 - 1 mm; **Poll** 0.56 - 0.68 mm, corpuscle 0.23 - 0.3 mm.

S. bidens ssp. **gracile** Kupicha (KB 38(4): 612, 1984). **T:** RSA, Eastern Cape (*Hilner* 427 [GRA]). — **D:** RSA (Eastern Cape); coast, grassland, 45 - 360 m.

Incl. *Schizoglossum umbellatum* Schlechter (1894); **incl.** *Schizoglossum diversum* N. E. Brown (1907) ≡ *Lagarinthus diversus* (N. E. Brown) Bullock (1952).

Differs from ssp. *bidens*: Stems 30 - 80 cm, slender, occasionally weakly branched; **L** 0 - 1 mm petiolate, lamina 20 - 60 × 1 - 2 mm, needle-like; **Inf** 6 - 9, with 3 - 10 **Fl**; peduncle 7 - 21 mm; **Ped** 4 - 7 mm; **Sep** 2 - 2.5 × 0.5 - 1 mm; **Cl** green or yellow, lobes 2.5 - 3 × 1 - 2 mm, glabrous; **Cs** lobes 1.5 mm, inner appendage apically bifid; **Anth** 0.5 - 0.75 mm; **Poll** 0.45 - 0.49 × 0.15 - 0.25 mm, corpuscle 0.17 mm.

S. bidens ssp. **hirtum** Kupicha (KB 38(4): 615, 1984). **T:** RSA, KwaZulu-Natal (*Moll* 1329 [NU]). — **D:** RSA (border KwaZulu-Natal / Eastern Cape); grassland, 1050 - 1500 m.

Differs from ssp. *bidens*: Stems ± 30 cm, unbranched; **L** decussate, 1 - 2 mm petiolate, lamina 30 - 70 × 1.5 - 2 mm, linear; **Inf** 3 - 5, with 2 - 4 **Fl**; peduncle 15 - 28 mm; **Ped** 6 - 13 mm; **Sep** 3.5 - 4.5 × 1 mm; **Cl** brown, lobes 4.5 - 5.5 × 2.5 mm, outside glabrous, inside very densely pubescent, right-hand margins of the **Cl** lobes weakly ciliate; **Cs** lobes 2 - 2.5 mm, inner appendage apically strongly bifid; **Anth** 0.5 × 1 mm; **Poll** 0.63 - 0.73 mm, corpuscle 0.2 - 0.25 mm.

S. bidens ssp. **pachyglossum** (Schlechter) Kupicha (KB 38(4): 615, 1984). **T:** RSA, Free State (*Flanagan* 1881 [PRE, GRA, NBG, SAM]). — **D:** RSA (SW KwaZulu-Natal, Eastern Cape); grassland, often in rocky soil, 300 - 2500 m.

≡ *Schizoglossum pachyglossum* Schlechter (1894); **incl.** *Schizoglossum pachyglossum* var. *abbreviatum* N. E. Brown (1907).

Differs from ssp. *bidens*: Stems 20 - 90 cm, occasionally branched; **L** very numerous, densely arranged, petiole 2 - 3 mm, lamina 27 - 45 × 1.5 - 9 mm, linear; **Inf** 9 - 13, with ≥ 7 **Fl**; peduncle 0 - 3 cm; **Ped** 3 - 7 mm; **Sep** 2 - 3 × 0.75 - 1 mm; **Cl** yellowish-green, lobes 2.5 - 4 × 1.5 - 2.5 mm, outside glabrous or softly hairy; **Cs** lobes 1 - 1.5 mm, apically weakly bifid, inner appendage bifid; **Sty** head white; **Anth** 0.5 × 1 mm; **Poll** 0.41 - 0.58 mm, corpuscle 0.18 - 0.28 mm; **Fr** 4.7 - 6 cm.

This subspecies shows pronounced variability.

S. bidens ssp. **productum** (N. E. Brown) Kupicha (KB 38(4): 617, 1984). **T:** RSA, Transvaal (*Rehmann* 5867 [K, BM, Z]). — **D:** RSA (Northern Prov., Mpumalanga); mountainous grassland, clefts in quartzite, 1230 - 1950 m.

≡ *Schizoglossum pachyglossum* var. *productum* N. E. Brown (1907).

Differs from ssp. *bidens*: Stems (19-) 30 - 52 cm, unbranched; **L** numerous, densely arranged, petiole 1 - 3 mm, lamina 15 - 43 × 5 - 9 mm, oblong or triangular; **Inf** 5 - 9, with 5 - 9 **Fl**; peduncle 5 - 18 mm; **Ped** 7 - 14 mm; **Sep** 2.5 - 3 × 1 mm; **Cl** olive-green, venation darker or pale purple, lobes 5 - 6 × 2 - 3.5 mm; **Cs** lobes 2 - 2.5 mm, inner appendage simple, apically curved outwards; **Anth** 0.5 × 1 mm; **Poll** 0.64 - 0.75 mm, corpuscle 0.23 - 0.28 mm; **Fr** ± 6 cm.

S. cordifolium E. Meyer (Comm. Pl. Afr. Austr., 219, 1838). **T** [neo]: RSA, Eastern Cape (*MacOwan* 662 [K, BOL, GRA, Z]). — **D:** RSA (S Eastern Cape, KwaZulu-Natal, Mpumalanga); open grassland, to 1200 m.

Incl. *Schizoglossum hirsutum* Turczaninow (1848); **incl.** *Schizoglossum aemulum* Schlechter (1894); **incl.** *Schizoglossum atropurpureum* var. *lineatum* Schlechter (1894); **incl.** *Schizoglossum hollandiae* Schlechter *ex* Harvey (1894); **incl.** *Schizoglossum cordifolium* var. *centralis* N. E. Brown (1907); **incl.** *Schizoglossum divaricatum* N. E. Brown (1928).

Stems 15 - 60 cm, very rarely branched; **L** 1 - 4 mm petiolate, lamina 20 - 42 × 2 - 30 mm, (oblong or ovately) triangular, rarely elliptic or narrowly linear; **Inf** 2 - 6, with 2 - 8 **Fl**; peduncle 7 - 30 mm; **Ped** 5 - 10 mm; **Sep** 2.5 - 5 × 1 - 1.5 mm; **Cl** green, striped with brown or yellow, cup-like, lobes 5 - 7 × 3 - 4.5 mm, inside basally pubescent, margins curved outwards; **Cs** white, beige or pale green, **Cs** lobes 2 - 2.5 mm, broadly ± ovate, curved inwards, inner appendage ± 1 mm deep bifid, finger-like, parallel or divergent, curved over the **Sty** head; **Anth** 0.5 - 1 × 1 - 2 mm, reniform, incumbent on the **Sty** head; **Poll** 0.52 - 0.8 mm, reniform-ellipsoid, germination crest in the upper ½, corpuscle 0.25 - 0.45 mm; **Fr** 5.8 - 7 cm; **Se** 3.5 mm, **Ha** tuft ± 22 mm.

S. elingue N. E. Brown (BMI 1895: 149, 1895). **T:**

RSA, KwaZulu-Natal (*Evans* 358 [K]). – **D:** RSA (Eastern Cape, KwaZulu-Natal, border to Lesotho).

S. elingue ssp. **elingue** – **D:** RSA (KwaZulu-Natal, border to Lesotho); subalpine grassland, 1950 - 2400 m.

Stems 10 - 28 cm, rarely branched; **L** decussate, petiole 2 - 15 mm, lamina 37 - 47 × 7 - 15 mm, oblong (-elliptic) to oblanceolate; **Inf** 1 - 5, with 4 - 6 **Fl**; peduncle 1 - 2 cm; **Ped** 5 - 9 mm; **Sep** 4.5 - 6 × 1.5 - 2 mm, outside densely soft-hairy; **Cl** white, cucullate, lobes 8 - 10 × 3.5 - 4 mm, apical ½ of the lobes outside densely soft-hairy, inside densely pubescent, margins ciliate; **Cs** white, 4 - 5 mm, by far surpassing the **Sty** head, **Cs** lobes oblong, flat, erect, apically weakly bi- or trifid, inside with 2 inconspicuous longitudinally oriented fleshy crests; **Sty** head white; **Anth** white, ± 1 mm, square, incumbent on the **Sty** head; **Poll** 0.73 - 0.78 mm, compressedly oblong-ovoid, germination crest apical, attached subapically, corpuscle 0.38 - 0.43 mm.

S. elingue ssp. **purpureum** Kupicha (KB 38(4): 618, 1984). **T:** RSA, KwaZulu-Natal (*Killick & Vahrmeijer* 3988 [PRE, K]). – **D:** RSA (Eastern Cape, KwaZulu-Natal, border region to Lesotho); rocky grassland on steep slopes, 2300 - 2700 m.

Differs from ssp. *elingue*: Peduncle 1.5 - 6.5 cm; **Cl** purple, lobes often curved outwards; **Poll** 0.58 - 0.63 mm.

S. flavum Schlechter (J. Bot. 33: 355, 1895). **T:** RSA, KwaZulu-Natal (*Medley Wood* 5358 [K, BM, SAM]). – **D:** RSA (SE KwaZulu-Natal); stony grassland, (60-) 900 - 2250 m.

Incl. *Schizoglossum flavum* var. *lineare* N. E. Brown (1907).

Stems 7 - 27 cm, unbranched; **L** 3 - 5 mm petiolate, lamina 3 - 7 × 1 - 2 cm, oblong or ovoid-triangular or lanceolate; **Inf** 3 - 5 per node, with 5 - 9 **Fl**; peduncle 1 - 5 cm; **Ped** 5 - 9 mm, hairy; **Sep** 4 - 5.5 × 1 - 1.5 mm, outside hairy; **Cl** yellow-green, campanulate, lobes 7 - 9 × 3 - 4 mm, ascending, outside rarely weakly soft-hairy, inside glabrous, margins ciliate or glabrous; **Cs** yellow-green, **Cs** lobes 3 - 4 mm, by far surpassing the **Sty** head, narrowly elliptic to broadly sagittate, apically acute, erect, inside with median longitudinal groove; **Sty** head white; **Anth** white, ± 0.5 mm, round, incumbent on the **Sty** head; **Poll** 0.55 - 0.65 mm, compressedly pear-shaped, attached subapically, corpuscle 0.23 - 0.28 mm.

S. hamatum E. Meyer (Comm. Pl. Afr. Austr., 220, 1838). **T** [neo]: RSA, KwaZulu-Natal (*Fannin* 63 [K]). – **D:** RSA (Eastern Cape and neighbouring KwaZulu-Natal); grassland, dry slopes, 540 - 2100 m.

Incl. *Schizoglossum atropurpureum* Schlechter (1894) (*nom. illeg.*, Art. 53.1); **incl.** *Schizoglossum hamatum* var. *elegans* N. E. Brown (1907); **incl.** *Schizoglossum hamatum* var. *pallidum* N. E. Brown (1907).

Stems 25 - 60 cm, stout, sometimes weakly branching; **L** 2 - 4 mm petiolate, lamina 22 - 44 × 8 - 25 mm, (narrowly) oblong to triangular; **Inf** 3 - 7 (-15), with 4 - 10 **Fl**; peduncle 1 - 4 cm; **Ped** 8 - 10 mm, shortly hairy; **Sep** 4 - 6 × 1 - 1.5 mm, outside ± shortly hairy; **Cl** outside greenish-brown or purple, inside green, with reddish-brown or purple stripes, bowl-shaped, lobes 8 - 11 × 4 - 7 mm, outside apically occasionally tomentose, inside glabrous, margins ciliate; **Cs** green, 3.5 - 4 mm, **Cs** lobes apically curved outwards, bordered by a flat simple margin, inner appendage bifid, segments 2.5 - 3 mm, finger-like, erect, parallel, apically bent outwards, laterally flanked by vertical wing-like processes; **Anth** triangular-ovate, appendages 1 - 2 mm, laterally at the **Sty** head; **Sty** head reddish-brown or purple; **Poll** 0.7 - 0.82 mm, compressed, obtriangular, attached subterminally, corpuscle 0.4 - 0.52 mm; **Fr** ± 6 cm; **Ha** tuft of the **Se** 25 mm.

Closely related to *S. amatolicum* and *S. rubiginosum*.

S. hilliardiae Kupicha (KB 38(4): 620, 1984). **T:** RSA, KwaZulu-Natal (*Hilliard & Burtt* 11244 [E, NU]). – **D:** RSA (KwaZulu-Natal, border to Lesotho); subalpine grassland, 1800 - 2670 m.

Incl. *Schizoglossum cordifolium* J. Ross (1972) (*nom. illeg.*, Art. 53.1); **incl.** *Schizoglossum hamatum* J. Ross (1972) (*nom. illeg.*, Art. 53.1).

Stems 9 - 34 cm, branched only basally; **L** 2 - 8 (-13) mm petiolate, lamina 27 - 46 × 6 - 26 mm, (oblong-) triangular or linear; **Inf** 3 - 8 per node, with 3 - 9 **Fl**; peduncle 12 - 50 mm; **Ped** 6 - 11 mm, hairy; **Sep** 2.5 - 4 × 1 - 1.5 mm, outside hairy; **Cl** flat, outside green with brown or purple pattern, inside brownish-purple, venation white, lobes 5.5 - 7.5 × 2.5 - 3.25 mm, outside glabrous or pubescent, inside glabrous, margins weakly ciliate; **Cs** cream-coloured to pale yellow-green, **Cs** lobes 2 - 3 mm, ± square, inner appendage short, simple or deeply bifid, basal ½ united with the staminal column; **Sty** head pale green; **Anth** dull purple, margin white, 0.5 - 0.75 mm, round, ± incumbent on the **Sty** head; **Poll** compressedly pear-shaped, corpuscle 0.21 - 0.31 mm; **Fr** ± 6.5 cm.

S. ingomense N. E. Brown (FC 4(1): 602, 1907). **T:** RSA, KwaZulu-Natal (*Gerrard* 1302 p.p. [K, BM]). – **D:** RSA (KwaZulu-Natal); grassland near water, forest margins, 1200 - 1500 m.

Stems 60 - 90 cm, robust, unbranched; **L** esp. apically in ternate verticils or irregularly arranged, petiole 1 - 3 mm, lamina 29 - 34 × 4 - 15 mm, oblong-triangular; **Inf** 3 - 8 per node, with 4 - 7 **Fl**; peduncle 0 - 1 cm; **Ped** 5 - 10 mm, hairy; **Sep** 4 × 1.5 mm; **Cl** campanulate, green, venation purple, lobes 8 - 10 × 5.5 mm, margins bent outwards, outside glabrous, inside basally pubescent; **Cs** ± 5 mm, surpassing the **Sty** head, **Cs** lobes basally obtriangular, apically shortly bent horizontally outwards, comb-like, inner appendage 3 mm, narrowly oblong, erect, inside with a longitudinal groove, apically bifid; **Anth** ± 1.5 mm, triangular, ± incumbent on the **Sty** head; **Poll** 0.95 - 1.03 mm, reniform, attached subapically, corpuscle 0.4 - 0.52 mm.

S. nitidum Schlechter (BJS 20(Beiblatt 51): 18, 1895). **T:** RSA, Mpumalanga (*Schlechter* 3519 [BOL, K, Z]). – **D:** RSA (Eastern Cape, W KwaZulu-Natal, E Free State, Mpumalanga); moist grassland, swamps or near rivers, 1200 - 1920 m.

Incl. *Schizoglossum wallacei* Schlechter (1895).

Stems 7 - 30 cm, rarely branched; **L** 1 - 6 mm petiolate, lamina 23 - 85 × 4 - 31 mm, linear, elliptic, lanceolate or triangular; **Inf** 2 - 5 per node, with 3 - 7 **Fl**; peduncle (0-) 8 - 45 mm; **Ped** 3 - 12 mm, shortly hairy; **Sep** 2.5 - 3.5 × 1 mm, outside hairy; **Cl** cucullate, yellow, green, greenish-brown or white with brown or rose stripes, lobes 6 - 10 × 3 - 4.5 mm, arcuate, margins usually strongly curved outwards, glabrous; **Cs** yellow or white, lobes 1.5 - 2.5 mm, ± square, apically entire or bifid, inner appendage short, bifid, segments linguiform, several times dentate, bordered by lateral processes, these wing-like, flat, united with the staminal column; **Anth** 0.5 × 1 mm, broadly reniform, laterally ± hidden partly under the apical bulge of the **Sty** head; **Sty** head white or pink; **Poll** 0.55 - 0.8 mm, compressedly oblong-obtriangular or ovoid, attached in the middle, germination crest apical, crest-like, corpuscle 0.23 - 0.4 mm; **Fr** 5.4 - 6.5 cm.

S. quadridens N. E. Brown (BMI 1895: 252, 1895). **T:** RSA, KwaZulu-Natal (*Haygard* s.n. *in Medley Wood* 4189 [K]). – **D:** RSA (KwaZulu-Natal); only known from the type locality.

Stems 10 - 15 cm, branched only basally; **L** 2 - 4 mm petiolate, lamina 28 - 50 × 6 - 13 mm, linear, elliptic or lanceolate; **Inf** 2 - 3 per node, with ± 6 **Fl**, hardly surpassing the **L**; peduncle 11 - 32 mm; **Ped** 5 - 10 mm, hairy; **Sep** ± 4 × 1.25 mm, outside hairy, margins ciliate; **Cl** white, ± campanulate, lobes ± 6 × 3 mm, ascending, apical ½ outside softly hairy, inside densely tomentose, margins ciliate; **Cs** lobes ± 2.5 mm, ovoid-triangular, apically bifid, basally cordate, erect, inside with 2 raised longitudinal combs, apically merging into 2 short appendages, these tooth-like, curved over the **Anth**; **Anth** 1 mm, triangular, incumbent on the **Sty** head; **Poll** 0.55 - 0.57 mm, ellipsoid, attached in the middle, germination crest apical, crest-like, corpuscle 0.37 mm.

S. rubiginosum Hilliard (Notes Roy. Bot. Gard. Edinburgh 45(2): 182, 1989). **T:** RSA, KwaZulu-Natal (*Hilliard & Burtt* 16723 [E]). – **D:** RSA (KwaZulu-Natal); only known from the type locality.

R tuber ± 20 × 8 mm, carrot-like; stems 60 cm, 2.5 mm ∅, biseriately hairy, basally glabrous, apically densely leafy; **L** ± 25 - 48 × 2 (-3) mm, linear; **Inf** from apical nodes, with ± 6 **Fl**; peduncle 1 - 1.2 cm; **Ped** 7 - 10 mm; **Sep** 4 - 5 × 1.25 - 2 mm, lanceolate; **Cl** chestnut-brown-red, ± flat, incised ± to the base, lobes 10 × 6 mm, broadly elliptic, very obtuse, glabrous; **Cs** cream-coloured, lobes 1 × 1 mm, much shorter than the **Gy**, transversely rectangular, with 2 marginal lobes, which form a pouch, inner appendage bifid, segments < 1 mm, subulate, erect; **Anth** appendages 0.6 × 1 mm, transversely elliptic, just curved over the margin of the **Sty** head; **Gy** cream-coloured, 3 mm; **Sty** head 2 mm ∅; **Poll** ± 0.8 mm, compressedly oblong-triangular, attached subterminally, corpuscle ± 0.4 mm.

Closely related to *S. hamatum*.

S. saccatum Bruyns (BT 25(2): 169, 1995). **T:** Namibia (*Bruyns* 5558 [BOL, K, WIND]). – **D:** N Namibia.

Stems 7 - 20 cm, 1.5 - 2 mm ∅, basally strongly branched, ascending, hairy, **Int** 7 - 25 mm; **L** sessile, lamina 20 - 70 × 2.5 - 4 mm, narrowly linear, margins rolled outwards, upper face softly hairy, lower face glabrous, midrib softly hairy; **Inf** ± 4-flowered from all nodes except the most basal ones; **Bra** ≤ 3 mm, filiform; peduncle 7 - 20 mm, uniseriately soft-hairy; **Ped** 7 - 9 mm, softly hairy; **Sep** green, 4 - 5 × 1 mm, lanceolate, mostly curved outwards towards the apex, outside softly hairy, inside glabrous; **Cl** green, with weak purple longitudinal stripes, ± bowl-shaped, incised to the base, lobes 6 × 2.5 mm, ovate-lanceolate, ascending-spreading, margins and tips curved outwards, outside glabrous, inside densely hairy, **Ha** white, soft, at the lobe tip ≤ 1.5 mm; **Ci** lobes delicate, truncate-emarginate; **Cs** lobes fleshy, ± ovate, incurved, apical appendage broad, rounded, inner appendage short, placed on the **Anth**; **Gy** 2 × 5.5 mm; **Sty** head white, apically purple to pink; **Anth** apically slightly incised; **Poll** oblong-ovoid, attached apically on the inside.

According to Bruyns only provisionally placed in *Schizoglossum*.

S. singulare Kupicha (KB 38(4): 623, 1984). **T:** RSA, KwaZulu-Natal (*Hilliard & Burtt* 5816 [NU, NH]). – **D:** RSA (KwaZulu-Natal: Alfred Distr.).

Stems ± 9 cm, unbranched; **L** 2 - 4 mm petiolate, lamina 28 - 34 × 6 - 13 mm, oblong (-elliptic); **Inf** ± 3 per node, with 4 **Fl**; peduncle 15 - 19 mm; **Ped** 8 - 10 mm, hairy; **Sep** 3 × 1 mm, outside hairy; **Cl** bluish, venation red, ± flat, lobes ± 7 × 3 mm, sometimes weakly curved outwards, glabrous, margins weakly ciliate; **Cs** dark red, **Cs** lobes 4.5 mm, erect, basally square, apically trifid, middle appendage long-triangular, far surpassing the **Sty** head, apically curved inwards, lateral appendages tooth-like, inner appendage deeply bifid, segments linear, oriented parallel, incumbent on the **Anth**; **Sty** head white; **Anth** ± 1 mm, triangular-ovoid, incumbent on the **Sty** head; **Poll** 0.5 mm, compressedly pear-shaped, attached subapically, corpuscle 1.3 mm.

S. stenoglossum Schlechter (BJS 18(Beiblatt 45): 28, 1894). **T:** RSA, KwaZulu-Natal (*Schlechter* 3228 [PRE, BOL, GRA, K, PRE, Z]). – **D:** RSA (Eastern Cape, S KwaZulu-Natal, borders to Free State and Mpumalanga).

S. stenoglossum ssp. **flavum** (N. E. Brown) Kupicha (KB 38(4): 623, 1984). **T:** RSA, KwaZulu-Natal (*Medley Wood* 4395 [K, NH]). – **D:** RSA (KwaZulu-Natal, border to Lesotho); grassland, 1500 - 2100 m.

≡ *Schizoglossum decipiens* var. *flavum* N. E. Brown (1909); **incl.** *Schizoglossum decipiens* N. E. Brown (1907).

Differs from ssp. *stenoglossum*: Stems 18 - 48 cm; **L** 1 - 3 mm petiolate, lamina 54 - 83 × 4 - 15 mm; **Inf** ± 3, with 2 - 5 **Fl**; peduncle 10 - 27 mm; **Ped** 3 - 9 mm; **Sep** 4 - 5.5 × 1 - 1.5 mm; **Cl** greenish-yellow striped with brown, rarely brown, lobes (7.5) 9 - 10 × (3-) 3.5 - 5 mm, inside glabrous to densely pubescent, margins ciliate; **Cs** greenish-yellow, **Cs** lobes 3 mm, triangular to ovate-attenuate, incumbent over the **Anth**, apical appendage on the outside separated by a transversally oriented crest, tip sometimes bi- or trifid; **Anth** 1.5 mm; **Poll** 0.65 - 0.75 mm, rhombic, attached in the middle to subapically, germination crest apical, ± semicircular, corpuscle 0.43 - 0.45 mm.

S. stenoglossum ssp. **latifolium** Kupicha (KB 38(4): 622, 1984). **T:** RSA, KwaZulu-Natal (*Sidey* 1990 [PRE]). – **D:** RSA (borders between KwaZulu-Natal and Free State as well as Mpumalanga); mountainous grassland, 1050 - 2700 m.

Differs from ssp. *stenoglossum*: Stems 16 - 51 cm; **L** 1 - 3 (-10) mm petiolate, lamina 45 - 58 × (3-) 6 - 15 mm; **Inf** 3 - 4, with 5 - 8 **Fl**; peduncle 13 - 40 mm; **Ped** 2 - 5 mm; **Sep** 4 - 5 × 1 - 1.5 mm; **Cl** dark or chestnut-brown, rarely yellow-green with brown stripes, lobes 8 - 9 × 3 - 3.5 mm, margins ciliate, outside sometimes weakly hairy; **Cs** lobes 3 mm, surpassing the **Gy**, narrowly linear, erect or apically outcurved; **Anth** white, 1.5 mm; **Poll** 0.63 - 0.75 mm, rhombic, attached in the middle to subapically, germination crest apical, ± semicircular, corpuscle 0.35 - 0.45 mm; **Fr** ± 4.5 cm.

S. stenoglossum ssp. **stenoglossum** – **D:** RSA (Eastern Cape, S and E KwaZulu-Natal); grassland, 300 - 1500 m.

Incl. *Schizoglossum stenoglossum* var. *longipes* Schlechter (1894).

Stems 25 - 45 cm, unbranched; **L** 1 - 3 mm petiolate, lamina 40 - 70 × 2 - 9 mm, linear to lanceolate; **Inf** 2 - 5, with 3 - 6 **Fl**; peduncle 7 - 48 mm; **Ped** 3 - 8 mm, hairy; **Sep** 3 - 4.5 × 1 - 1.5 mm, outside hairy; **Cl** dark to purple-brown, cup-shaped, lobes 6 - 9 × 2.5 - 3.5 mm, outside rarely weakly pubescent, inside margins sometimes weakly ciliate; **Cs** green or yellow, dotted with purple, **Cs** lobes 2 - 3 mm, surpassing the **Sty** head, narrowly triangular, erect, inside in the middle with a longitudinal groove; **Anth** 1 - 1.5 mm, triangular-ovoid, not very closely incumbent on the **Sty** head; **Poll** 0.45 - 0.57 mm, compressedly obtriangular, attached subapically, germination crest apical, conspicuously swollen, corpuscle 0.3 - 0.35 mm; **Fr** 4.5 - 5.5 cm; **Se** 3 - 3.5 mm, **Ha** tuft 15 - 17 mm.

SCHLECHTERELLA

U. Meve

Schlechterella K. Schumann (in Engler & Prantl (eds.), Nat. Pfl.-fam. [ed. 1], Register: 462, 1899). **T:** *Pleurostelma africana* Schlechter. – **D:** E Africa; sandy ground. **Etym:** For Dr. [Friedrich Richard] Rudolf Schlechter (1872 - 1925), German traveller, collector and botanist in Berlin and specialist (amongst many other groups) for orchids.

Incl. *Pleurostelma* Schlechter (1895) (*nom. illeg.*, Art. 53.1). **T:** *Pleurostelma africana* Schlechter.

Incl. *Triodoglossum* Bullock (1962). **T:** *Raphionacme abyssinica* Chiovenda.

Erect or twining geophytic subshrubs, latex white; **R** tuber carrot-like; stems basally ± woody; **L** shortly petiolate to almost sessile, linear to obovate; **Inf** lateral, lax and delicate, cymose (appearing racemose), many-flowered; **Sep** basally shortly united; **Fl** short-lived, open only 1 day; **Cl** rotate, basally somewhat united; **Cl** tube inside keeled, basally with pouch-like outgrowths; **Cl** lobes ovate-lanceolate to broadly elliptic-obovate, delicate, glabrous; **Cn** petaloid, with ± lacerate white erect lobes originating from a basal circular bulge; **Anth** rectangular-ovate, obtuse, basally auriculate, incumbent on the conical **Sty** head; pollen in tetrads or polyads, glued together to form 2 ellipsoid **Poll** in each theca; **Fr** solitary or paired, fusiform.

This is a genus belonging to the *Periplocoideae* from among the group around *Raphionacme*. Recently it has been united with *Triodoglossum* (Venter & Verhoeven 1999). A number of characters (delicate growth, short-lived flowers, the corona fused into a ring-like structure, auriculate anthers, pollinia instead of free pollen tetrads) justify the status of a good genus for these handsome geophytes.

S. abyssinica (Chiovenda) Venter & Verhoeven (SAJB 64(6): 353, ills., 1999). **T:** Ethiopia (*Riva* 1086 [FT]). – **D:** Ethiopia, Uganda, Kenya. **I:** Asklepios 68: 28, 1996, as *Raphionacme*. **Fig. XLIV.a**

≡ *Raphionacme abyssinica* Chiovenda (1939) ≡ *Triodoglossum abyssinicum* (Chiovenda) Bullock (1962).

Stems erect or weakly twining, 10 - 30 cm × 1 - 2 mm, basally woody, hairy; **L** 1 - 4 mm petiolate, lamina 2 - 5 × 0.5 - 1.5 cm, basally cuneate-attenuate, upper face dark green, lower face usually purple-red, ± glabrous; **Inf** 10 - 30 × 1 mm pedunculate, several times branched; **Ped** 2 - 6 × 0.5 mm, pubescent; **Sep** free almost to the base, ovate-triangular or elliptic, ± obtuse, 1 - 1.2 mm, ciliate, hairy; **Cl** tube ± 1 × 1.75 mm; **Cl** lobes spreading, broadly elliptic to ovate-lanceolate, 1 - 4 × 1 - 1.5 mm, magenta-red to violet, basally sometimes yellowish; **Cn** whitish, lobes linguiform, spatulate or ovate, 0.7 - 1.5 × 0.8 mm, apically irregularly lacerate or upper ½ divided cross-like; translator 1.5 - 2 mm, rectangular-ovate, stalk very short, adhesive disc very long; **Fr** 40 - 60 × 5 mm.

S. africana (Schlechter) K. Schumann (in Engler & Prantl (eds.), Nat. Pfl.-fam. [ed. 1], Nachträge 2: 60, 1900). **T:** Tanzania (*Scott-Elliot* s.n. [BM]). – **D:** Ethiopia, Somalia, Kenya, Tanzania. **I:** Schlechter (1895b: 332).
≡ *Pleurostelma africana* Schlechter (1895) (*incorrect name*, Art. 11.4).

Stems erect or weakly twining, 2 - 100 cm × 2 mm, basally woody, ± glabrous; **L** 0.5 - 3 mm petiolate, lamina 2 - 6 × 0.2 - 1.5 cm, basally cuneate-attenuate, upper face dark green, with whitish venation, glabrous, margins ± revolute and undulate; **Inf** long-pedunculate, several times branched; **Ped** ± 1 cm, glabrous; **Sep** ovate-triangular or elliptic, glabrous; **Cl** tube ± 1 × 2 mm; **Cl** lobes spreading, broadly elliptic to obovate, 4 - 6 × 1 - 2 mm, pale greenish-cream (rarely violet), ± rolled back along the midrib, basally ± rugose, apically broadly lobate, undulate; **Cn** whitish, lobes 3 - 4 mm, basal ⅓ cylindrical, ascending, incurved, apical ⅔ connivent, filiform, twisted, basally on both sides with a triangular minute tooth; translator ± 3.5 mm, elliptic, stalk very short, adhesive disc elliptic, attached at right angles; **Poll** flat, elliptic, ± 3 × 0.5 mm.

Closely related to *S. abyssinica*, and distinguished by the large undulate corolla lobes and its colour as well as by the 3-parted corona lobes.

×STAPARESIA

B. Müller & F. Albers

×**Staparesia** G. D. Rowley (Nation. Cact. Succ. J. 37(3): 79, 1982).
Incl. ×*Tavastemon* P. V. Heath (1992).
= *Stapelia* × *Tavaresia*.

×**S. meintjesii** (R. A. Dyer *pro sp.*) G. D. Rowley (Nation. Cact. Succ. J. 37(3): 79, 1982). **T:** RSA, Northern Prov. (*Myburgh* s.n. in *Meintjes* 663 [PRE]).
≡ *Tavaresia* ×*meintjesii* R. A. Dyer (*pro sp.*) (1954) ≡ *Decabelone* ×*meintjesii* (R. A. Dyer *pro sp.*) G. D. Rowley (1958) ≡ ×*Tavastemon meintjesii* (R. A. Dyer *pro sp.*) P. V. Heath (1992).
= *S. gettliffei* × *T. barklyi*.

STAPELIA

B. Müller & F. Albers

Stapelia Linné (Spec. Pl. [ed. 1], 217, 1753). **T:** *Stapelia hirsuta* Linné [Lectotype, proposed by N. E. Brown 1909 and again by L. C. Leach, Trans. Rhodesia Sci. Assoc. 59(1): 2, 1978, and proposed for conservation by Jarvis in Taxon 41: 570, 1992, cf. l.c. 44(4): 612, 1995.]. – **Lit:** Leach (1985). **D:** Tanzania, Zambia, Angola, Namibia, Moçambique, Botswana, Zimbabwe, RSA. **Etym:** For Jan Bode van Stapel († 1636), Dutch physician (Genaust 1983).
Incl. *Stisseria* Heister *ex* Fabricius (1759) (*nom. illeg.*, Art. 52.1). **T:** *Stapelia hirsuta* Linné.
Incl. *Gonostemon* Haworth (1812). **T:** *Stapelia divaricata* Masson [Lectotype, according to L. C. Leach, Excelsa Tax. Ser. 3: 4, 1985.].

Perennial, usually basally sympodially branched small stem succulents, sometimes with rhizomatous runners; stems not numerous, erect (rarely decumbent), ± long cylindrical or slightly tapering towards the tips, unarmed, 6 - 30 × ± 0.5 - 3 cm ⌀, 4- (rarely spirally 5- to 6-) ribbed, pubescent (rarely glabrous), sides neither patterned mosaic-like nor conspicuously grooved, flat to concave; **Tu** ± conspicuously raised; **L** rudiments acute, fleshy, pubescent, erect, caducous or persistent in the dry state, basally usually with a pair of small glandular persistent stipular rudiments; **Inf** consisting of solitary or few successively (rarely simultaneously) opening **Fl**, near the stem bases or occasionally scattered along the length of the stems; peduncle short, acutely tapering, pubescent; **Fl** in their majority with extreme carrion-scent (rarely pleasantly sweetish perfumed); **Ped** ± 0.5 - 7 (-12) × ≤ 0.5 cm, pubescent; **Sep** ovate to lanceolate, long-attenuate, pubescent; **Cl** yellowish, purple or pinkish-flesh-coloured, 0.8 - 40 cm ⌀, very variable in size (even within a taxon), flat, rarely campanulate (to globose) or shallowly cup- or funnel-shaped, usually deeply dissected into lobes; annulus absent or vestigially present; **Cl** lobes usually ± ovate and attenuate, often ± strongly convex in the longitudinal direction, partly strongly curved outwards, inside ± densely transversally wrinkled, ± densely hairy; **Cn** biseriate, of the Ci+Cs type, usually shortly stipitate, stalk columnar, 5-ribbed, ± bluntly 5-angled; **Ci** 5 simple, rarely bifid, lobes, often concavely bulging in the longitudinal direction; **Cs** 5 dorsiventrally flattened lobes, densely incumbent on the **Anth**, apically ± simple or with a ± filiform or subulate appendage, dorsally in addition frequently with a raised wing, an additional appendage or basally bulging; **Anth** ± square, without apical appendage; **Poll** solitary, horizontal in each theca, ± obliquely ovate, compressed; **Fr** 10 - 12 × 1 - 2 cm, fusiform, ± pubescent (rarely completely glabrous), follicles of a pair arranged at an acute angle; **Se** brown and/or ochre-coloured, ovate, flat, margin ± smooth, one end with a tuft of simple white **Ha**.

Historically, *Stapelia* embraced numerous species which are today placed in separate genera. This is esp. true for *Orbea*, whose type species (*O. variegata*) was regrettably for a long time erroneously interpreted as the type species of *Stapelia*. The situation was only clarified by Leach (1978b).

In his revision, Leach (1985) desists from a systematic break-down of the genus (in sections etc.), as the genus cannot be divided in natural groups.

Hybrids are known with *Huernia* (= ×*Huernelia*), with *Orbea* (= ×*Orbelia*), with *Tavaresia* (= ×*Staparesia*), and with *Tromotriche* (= ×*Tromostapelia*). In addition, the hybrid *Stapelia schinzii* × *S. kwebensis* is also reported (Leach 1985: 75), and the name *Stapelia* ×*bijliae* refers to the combination *Stapelia* × *Tridentea*.

The following names are of unresolved application but are referred to this genus: *Stapelia anonymos* Poiret (s.a.); *Stapelia brevirostris* Willdenow (1814) (*nom. inval.*); *Stapelia canescens* hort. ex Haworth (1812) (*nom. inval.*); *Stapelia concinna* Masson (1797) ≡ *Gonostemon concinnus* (Masson) P. V. Heath (1992); *Stapelia congestiflora* Delile (s.a.); *Stapelia cordata* hort. ex Haworth (1812) (*nom. inval.*); *Stapelia cucullata* Graessner (s.a.) (*nom. inval.*); *Stapelia deflexa* Jacquin (1809) ≡ *Duvalia deflexa* (Jacquin) Loudon (1841) ≡ *Gonostemon deflexus* (Jacquin) P. V. Heath (1992); *Stapelia deflexa* var. *atropurpurea* Hort. ex A. Berger (1910) ≡ *Stapelia deflexa* cv. Atropurpurea (Hort. ex A. Berger) H. Jacobsen (1974) ≡ *Gonostemon deflexus* var. *atropurpureus* (A. Berger) P. V. Heath (1993); *Stapelia dentata* Forsskål (s.a.); *Stapelia emarginata* Breit. (s.a.) (*nom. inval.*); *Stapelia erivacaria* Roxburgh (s.a.) (*nom. inval.*); *Stapelia flavicomata* Haworth (1812); *Stapelia fuscopurpurea* N. E. Brown (1890) ≡ *Gonostemon fuscopurpureus* (N. E. Brown) P. V. Heath (1993); *Stapelia juvencula* Jacquin (1806) ≡ *Tridentea juvencula* (Jacquin) Sweet (1830) ≡ *Stapelia vetula* var. *juvencula* (Jacquin) A. Berger (1910); *Stapelia lunata* Sweet (s.a.); *Stapelia maccabeanae* A. C. White & B. Sloane (1933) ≡ *Gonostemon maccabeanae* (A. C. White & B. Sloane) P. V. Heath (1993); *Stapelia mastodis* St. Lager (s.a.) (*nom. inval.*); *Stapelia mermilis* Graessner (s.a.) (*nom. inval.*); *Stapelia nopenackyi* Graessner (s.a.) (*nom. inval.*); *Stapelia plantii* hort. ex Hooker fil. (1868) ≡ *Gonostemon* ×*plantii* (hort. ex Hooker) P. V. Heath (1993); *Stapelia reflexa* Haworth (1812); *Stapelia reflexa* var. *brownii* Schinz (1901); *Stapelia sororia* Masson (1797) ≡ *Gonostemon grandiflorus* var. *sororius* (Masson) P. V. Heath (1993); *Stapelia stellaris* Jacquin (1809); *Stapelia stricta* Sims (1819) ≡ *Gonostemon strictus* (Sims) Haworth (1819); *Stapelia uncinata* Jacquin fil. (s.a.); *Stapelia vittata* Sickenberger (s.a.) (*nom. inval.*).

S. acuminata Masson (Stapel. Nov., 15, t. 17, 1797). **T:** RSA, Northern Cape (*Masson* s.n. [lecto – icono: l.c. t. 17, 1797]). – **D:** RSA (Northern Cape). **Fig. XLIV.b**

≡ *Gonostemon acuminatus* (Masson) P. V. Heath (1992); **incl.** *Stapelia acuminata* var. *brevicuspis* N. E. Brown (1909) ≡ *Gonostemon acuminatus* var. *brevicuspis* (N. E. Brown) P. V. Heath (1993); **incl.** *Stapelia indocta* Nel (1948).

Stems bluish-green, flushed with purple, ± 10 × 1.5 cm, bluntly 4-ribbed, softly white-pubescent; **Tu** ± robust; **L** rudiments ± 1.5 × 1.5 mm, caducous, stipular rudiments yellow-brown, caducous with age; **Inf** scattered along the length of the stems; **Fl** buds compressedly ovoid or ovoid-pointed; **Ped** to 5 mm; **Sep** pale green, tips purple, ≤ 5 × 2 mm; **Cl** outside pale yellow-green or purple, inside pale to dark flesh-coloured, 1.7 - 4 cm ⌀, wrinkles white, cream-coloured or yellow; **Cl** lobes ± 0.4 - 1.5 cm, ovate-triangular, outside ± densely pubescent, inside occasionally weakly delicately pubescent, margins ± hairy, **Ha** white, long, simple; stalk of the **Cn** ± 2 × 3 mm; **Ci** brownish-red or purple, basally yellow, lobes ± 2 - 3 mm, rectangular, apically ± obtuse, truncate, irregularly dentate or emarginate, rarely deeply bifid; **Cs** cream-coloured or yellowish, basal portion and dorsal wings dotted with brown or purple, inner filiform appendages erectly meeting, often column-like surpassing the **Sty** head, apically ± strongly curved outwards, dorsal wing ± horizontally spreading, very variable in shape; **Anth** deeply inserted, opening of the nectar cavity exceptionally large and ± round; **Poll** orange to orange-brown, ± 0.5 - 0.6 × 0.3 - 0.4 mm, ovoid or ± D-shaped.

Closely related to *S. arenosa* and *S. kougabergensis*.

S. arenosa C. A. Lückhoff (South Afr. Gard. 25: 96, 1935). **T:** South Africa, Cape Prov. (*Beukman* s.n. in Herb. Lückhoff 224 [BOL]). – **D:** RSA (Northern Cape, Western Cape).

≡ *Gonostemon arenosus* (C. A. Lückhoff) P. V. Heath (1992); **incl.** *Stapelia beukmanii* C. A. Lückhoff (1935) (*nom. illeg.*, Art. 53.1); **incl.** *Stapelia stultitioides* C. A. Lückhoff (1937).

Stems ± squat, ± 7.5 × 1 cm, dull green, shortly hairy, sides ± deeply grooved; **Tu** rounded; **L** rudiments ± 1.5 mm, caducous, stipular rudiments yellowish, obtuse; **Inf** scattered along the length of the stems, frequently 2 **Fl** simultaneously open; **Fl** buds broadly ovoid, long-attenuate; **Ped** ± 5 mm; **Sep** ± 5 mm; **Cl** outside dull green or purple, inside black-purple, to 4 cm ⌀, basally often thickened annulus-like, united part whitish, blotched with purple; **Cl** lobes ≤ 1.5 cm, inside pubescent, margins shortly hairy, **Ha** white; stalk of the **Cn** ± 1 mm; **Ci** yellow or orange-brown, lobes ± 1.5 mm, ± square or acutely tapering, weakly concave, spreading outwards, apically truncate, emarginate or ± acute, weakly pubescent; **Cs** purple, sometimes bicoloured, lobes ± 3 mm, narrowly ovate, long-attenuate, erect, dorsal wing ± horizontally spreading, ± triangular, apically dentate or irregularly crenate; **Poll** orange-yellow to yellow-brown, germination crest transparently yellow, 0.4 × 0.3 mm, ± ovoid.

Closely related to *S. rubiginosa*. There are also similarities with *S. acuminata* and *S. rufa*.

S. arnotii N. E. Brown (HIP 20: t. 1915 + text, 1890). **T:** RSA, Eastern Cape (*Arnot s.n. in Barkly 70* [K]). – **D:** RSA (Eastern Cape).

≡ *Gonostemon grandiflorus* var. *arnotii* (N. E. Brown) P. V. Heath (1993).

Stems to 2.5 cm ⌀; **Cl** inside glossy wine-red, ± 10 cm ⌀; **Cl** lobes with ± black tips, strongly curved outwards, inside apically weakly transversally wrinkled, inside softly hairy, **Ha** pale purple, upper ½ of the lobes glabrous; **Ci** lobes ± 6 mm, linear, acute, apically curved outwards; **Cs** lobes ± 7 mm, erectly spreading, apical appendage squatly 3-angled, acute, dorsal wing broadly triangular, fused for more than ½ with the apical appendage.

This taxon is insufficiently known and is only provisionally accepted (Leach 1985: 132).

S. asterias Masson (Stapel. Nov., 14, t. 14, 1797). **T:** [lecto – icono]: l.c. t. 14, 1797. – **D:** RSA (Western Cape).

≡ *Gonostemon asterias* (Masson) P. V. Heath (1992); **incl.** *Stapelia stellaris* hort. *ex* Haworth (1812); **incl.** *Stapelia lucida* De Candolle (1813) ≡ *Stapelia asterias* var. *lucida* (De Candolle) N. E. Brown (1909) ≡ *Gonostemon asterias* var. *lucidus* (De Candolle) P. V. Heath (1993); **incl.** *Stapelia asterias* var. *gibba* N. E. Brown (1909) ≡ *Gonostemon asterias* var. *gibbus* (N. E. Brown) P. V. Heath (1993).

Stems ± long cylindrical, ± 15 × 1.5 cm, compact, green to grey-green, ± bluntly 4-ribbed; **L** rudiments ≤ 5 mm, mostly persisting in the dry state, stipular rudiments more pronounced with age; **Fl** frequently decumbent, sometimes peduncle with several consecutive **Inf**; buds broadly ovoid, ± acute; **Ped** to 6.5 cm; **Sep** green, apically purple, to 10 × 3 mm, margins often weakly hairy; **Cl** outside pale green, inside rose-red, brownish or dark purple, to 11 cm ⌀, usually glossy, wrinkles sometimes whitish or yellowish; **Cl** tube short, 5-angled, margin ± radially grooved; **Cl** lobes to 4.5 × ± 2.8 cm, inside smooth or weakly papillose, rarely also shortly hairy, **Ha** slightly crisped, tips of the lobes outside ± pubescent, margins of the lobes sometimes curved outwards, ± hairy, **Ha** white or purple, ± 6 mm, straight, robust, in between a few clavate **Ha**; **Cn** black-purple, extremely variable; **Ci** lobes ± 5 mm, oblong, apically ± 3-dentate or with a longer middle and 2 shorter lateral tips; **Cs** lobes 6 - 7 mm, erect, apical appendage ± broadly subulate, dorsal wing ± broad, free or partly or completely fused with the apical appendage; **Poll** brown, occasionally yellowish, 0.8 - 1 × 0.5 - 0.6 mm, ± D-shaped; **Fr** pale brown.

Closely related to *S. gariepensis*.

S. baylissii L. C. Leach (SAJB 3(3): 172-173, ill., 1984). **T:** RSA, Eastern Cape (*Bayliss 2786* [PRE, K, NBG]). – **D:** RSA (Eastern Cape).

≡ *Gonostemon baylissii* (L. C. Leach) P. V. Heath (1992).

Plants to ± 15 cm, compact; stems greenish, often weakly glossy, ± 1.5 cm ⌀, sides slightly concave, glabrous, sometimes slightly roughened; **Tu** with a conspicuous scar left by the **L** rudiment, this erect or slightly spreading; peduncle robust, long-attenuate, sometimes branched, persistent, contorted and button-like, with scars left by **Ped**; **Fl** buds ovoid, acute or long-attenuate, angles acute, ± winged; **Ped** sometimes roughened; **Sep** ± 6 mm, sometimes roughened; **Cl** dark purple, ± glossy, 3 - 6 cm ⌀, tube short, broadly pentagonal, margin slightly thickened, at the level of the opening with 5 centrally confluent bulges; **Cl** lobes 1.2 - 2.3 cm, long-pointedly triangular, diverging outwards, outside glabrous, midrib and lobe tips inside sometimes delicately pubescent, margins strongly curved outwards, weakly hairy, clavate **Ha** whitish; **Cn** ± sessile; **Ci** lobes 3 - 4 mm, rectangular, obtuse or truncate, apically with an obtuse median mucro or irregularly crenate; **Cs** lobes subulate or curved and filiform, dorsal wing broad, erect; **Poll** yellow, ± 0.7 × 0.4 mm, ± D-shaped.

Closely related to *S. praetermissa*, but also showing numerous similarities with *S. asterias*.

S. ×bijliae Pillans (in A. C. White & B. Sloane, Stapelieae, ed. 2, 2: 674, 1937). **T:** RSA, Eastern Cape (*van der Bijl 6* [BOL]). – **D:** RSA (Eastern Cape).

≡ ×*Gonodentea bijliae* (Pillans) P. V. Heath (1993); **incl.** *Stapelia bijlii* hort. (s.a.) (*nom. inval.*, Art. 61.1).

Insufficiently known taxon, which is in addition obviously based on a natural hybrid (*Stapelia* (*Gonostemon*) *pillansii* × *Tridentea gemmiflora*) (Leach 1985: 132-133). – [B. Müller]

S. cedrimontana Frandsen (CSJA 47(6): 260-263, ills., 1975). **T:** RSA, Western Cape (*Frandsen 59* [BOL, PRE]). – **D:** RSA (Western Cape). **Fig. XLIV.c**

Plants to 25 cm, forming compact clumps; stems ± 1 cm ⌀, bluntly 4-ribbed; **L** rudiments 2 - 2.5 mm; **Inf** sometimes scattered over the length of the stems, **Fl** pointing outwards or upwards; peduncle sometimes persistent (then contorted and button-like); **Ped** ± 2.5 (-4) cm, spreading; **Sep** ≤ 6 (-8) mm; **Cl** inside chestnut-brown or reddish-purple, 3 - 5 (-6) cm ⌀, wrinkles white or yellowish; **Cl** tube short, pentagonal; **Cl** lobes 1 - 2 (-2.4) cm, upper ½ reddish-purple, outside shortly pubescent, inside in part weakly short-pubescent, margins esp. basally strongly folded outwards, weakly hairy, clavate **Ha** whitish to pale lilac; **Cn** dark purple with irregular pale brown patterning, shortly stipitate; **Ci** lobes 3.5 - 4 mm, rectangular, slightly concave, spreading, apically weakly curved outwards, irregularly bluntly dentate or mucronate, ± truncate, obtuse, rarely emarginate; **Cs** lobes 3.5 - 4 × ± 1 mm, narrowly subulate, long-attenuate, ventrally flattened, dorsally weakly keeled, meeting each other, tips curved outwards, with a weak dorsal bulge or a small dorsal wing; **Poll** yellowish to brownish, 0.7 - 0.8 × 0.5 - 0.6 mm, ± D-shaped.

Closely related to *S. villetiae*.

S. clavicorona I. Verdoorn (FPSA 31: t. 407 + text, 1931). **T**: RSA, Northern Prov. (*Van Nouhuys* s.n. in *PRE* 9756 [PRE]). – **D**: RSA (Northern Prov.).

≡ *Gonostemon clavicoronus* (I. Verdoorn) P. V. Heath (1992).

Plants to 25 cm, forming compact clumps; stem sides deeply grooved; **Tu** wing-like compressed; **L** rudiments ≤ 2 mm, without stipular rudiments; peduncle robust; **Ped** ≤ 6 × 3.5 mm; **Sep** ± 5 × 2.5 mm; **Cl** yellow to yellowish-brown, wrinkles red-brown, 5 - 6 cm ⌀, thickened in the centre; **Cl** tube ± 2 mm, narrow, basally embracing the **Cn**; **Cl** lobes ± 1.5 - 1.8 × 1.55 cm, ± shortly pointed, outside shortly delicately pubescent, inside narrowly wrinkled, pubescent, **Ha** white, curly, margins delicately hairy, **Ha** white, short, simple, in between with delicate clavate **Ha**; **Cn** ± black, stalk basally slightly broadened; **Ci** lobes ± 2.5 × 1.25 mm, fleshy, with a deep furrow, ± erect, apically ± acute, spreading, basally with a slit-like **Nec**; **Cs** lobes with 2 appendages, apically weakly netlike-foveolate, apical appendage ± 3 × 0.8 (apically 0.5) mm, ± clavate, dorsiventrally compressed, erect-connivent, dorsal appendage ± 3.5 × 1.5 mm, ± 1 mm ⌀, clavate, compressed, erectly spreading; **Poll** brownish-yellow, ± 0.6 × 0.4 mm, ± hidden by the **Cs**.

S. divaricata Masson (Stapel. Nov., 17, t. 22, 1797). **T**: BM [alc]. – **D**: RSA (Western Cape).

≡ *Gonostemon divaricatus* (Masson) Haworth (1812); **incl.** *Stapelia pallida* H. C. Wendland (1809) ≡ *Gonostemon pallidus* (H. C. Wendland) Sweet (1827); **incl.** *Stapelia pallens* hort. ex Steudel (1841).

Stems freely branched, tapering, to 10 cm, basally ≤ 8 to apically ± 4.5 mm ⌀, 4-ribbed; **Tu** obtuse, inconspicuous with age; **L** rudiments persistent in the dry state, stipular rudiments delicate; **Inf** flowering in quick succession, often 2 **Fl** simultane-

ously open; **Ped** ≤ 2 cm; **Sep** ≤ 4.5 mm; **Cl** yellowish or purple, 4 - 5 cm ⌀, flat or curved outwards; **Cl** tube short, narrow, 5-angled, only embracing the stalk of the **Cn**, tube margin often thickened annulus-like; **Cl** lobes ± 1.5 - 1.9 cm, spreading or revolute, outside delicately pubescent, inside weakly wrinkled, densely delicately pubescent, margins curved outwards, hairy, **Ha** white, simple; **Ci** lobes ± 3 × 1.5 - 2 mm, rectangular, weakly concave, curved outwards, apically widened, obtusely mucronate, sometimes emarginate; **Cs** lobes 2 - 3 mm, acutely ovate, keeled, basally wing-like, ± erect, slightly surpassing the **Anth**; **Poll** deep orange or dark brown, 0.96 × 0.5 mm, broadly D-shaped.

S. engleriana Schlechter (BJS 38: 49, fig. 8, 1905). **T** [lecto]: RSA, Western Cape (*Pillans* 60 [BOL]). – **D**: RSA (Northern Cape, Western Cape). **I**: Leach (1984: as *Tromotriche*). **Fig. XLIV.d**

≡ *Tromotriche engleriana* (Schlechter) L. C. Leach (1982).

Stems halfways above-ground, ascending to erect, creeping halfways underground, 4-ribbed, aerial stems ± 5 - 12 × 1.5 - 2.2 cm, few-branched, bluish-grey-green, pubescent, underground stems longer and thinner, whitish; **L** rudiments minute, caducous; **Inf** 1- to few-flowered, pubescent, shortly pedunculate; **Ped** 1 - 3 cm, pubescent, usually ± curved downwards; **Sep** ovate, acute, 4 - 6 mm, pubescent; **Cl** button-like by the completely reflexed **Cl** lobes, ± 2 cm ⌀, inside with tuberculate-papillate centrifugal ridges, purple-brown, outside cream-coloured, somewhat spotted with purple, pubescent; **Cl** tube ± 2 cm ⌀, funnel-shaped, inside pentagonal; **Cl** lobes ± 2 cm; **Cn** yellowish and purplish-speckled, shortly stipitate, stalk 5-angled; **Ci** lobes erectly spreading, 3 - 4 mm, obtusely almost square, apically ± 2-toothed; **Cs** lobes bifid, inner appendages filiform, ± 8 mm, erect-connivent, apically recurved and swollen, papillose, outer appendages divaricate, linear, ± 5 mm; **Poll** D-shaped, ± 0.75 × 0.5 mm; **Fr** ± 10 × 1 cm, pubescent.

This exceptional species of *Stapelia* was often misinterpreted as a close ally of *Tromotriche revoluta* on the base of its rhizomatous growth and the completely reflexed corolla lobes.

S. erectiflora N. E. Brown (Gard. Chron., ser. 3, 6: 650, 1889). **T**: RSA, Western Cape (*Bain* s.n. in *Barkly* 80 [K]). – **D**: RSA.

≡ *Gonostemon erectiflorus* (N. E. Brown) P. V. Heath (1992).

S. erectiflora var. **erectiflora** – **D**: RSA (Northern Cape, Western Cape). **Fig. XLIV.e**

Stems shortly decumbent-erect, to 15 cm, forming clumps, ± bluntly 4-ribbed, densely white-pubescent; **Tu** small; **L** rudiments caducous; **Inf** scattered over the length of the stems, often several simultaneously on the same stem; peduncle robust; **Ped** brownish, to 12 cm, erect; **Sep** reaching the bases of the **Cl** lobes; **Cl** outside pale green, inside brownish-green or purple, transversally striped, ± 1 - 1.5 cm ∅; **Cl** tube short, narrow, pentagonal, basally embracing the **Cn**; **Cl** lobes strongly recurved and rolled towards the **Ped** (**Fl** therefore as a whole ± button-like), outside very weakly pubescent, lower ½ inside softly hairy, **Ha** white, short, appressed, directed towards the tip, margins hairy with porrect clavate **Ha**; **Ci** lobes rectangular, emarginate, ± acute, sometimes ± bifid; **Cs** lobes 3-angled-subulate or filiform, erectly outcurved; **Anth** yellow, apically strongly broadened; **Poll** yellow to brownish, ± 0.5 × 0.4 mm, broadly D-shaped.

Closely related to *S. paniculata* and *S. glanduliflora*.

S. erectiflora var. **prostratiflora** L. C. Leach (SAJB 3(3): 178, ill., 1984). **T:** RSA, Cape Prov. (*Stayner* s.n. in *Leach* 15724 [PRE, NBG]). – **D:** RSA (Northern Cape, Western Cape).

Incl. *Gonostemon erectiflorus* fa. *aberrans* P. V. Heath (1992).

Differs from var. *erectiflora*: **Fl** resting on the ground; **Cl** pale violet; **Cl** lobes less strongly reflexed; **Cs** lobes shorter, dorsal wing erect.

S. flavopurpurea Marloth (Trans. South Afr. Philos. Soc. 18: 48, t. 5: fig. 1, 1907). **T:** RSA, Northern Cape (*Marloth* 4227 [not located]). – **D:** Namibia, Botswana, RSA (Northern Cape). **Fig. XLIV.f**

≡ *Gonostemon flavopurpureus* (Marloth) P. V. Heath (1992); **incl.** *Stapelia fleckii* A. Berger & Schlechter (1909) ≡ *Stapelia flavopurpurea* var. *fleckii* (A. Berger & Schlechter) A. C. White & B. Sloane (1937) ≡ *Gonostemon flavopurpureus* var. *fleckii* (A. Berger & Schlechter) P. V. Heath (1993).

Plants to 10 cm, forming low clumps; stems ± 1 cm ∅, 4-ribbed, sides flat or slightly concave; **Tu** inconspicuous; **L** rudiments caducous, stipular rudiments pale, obtuse; **Inf** few-flowered, ± scattered along the length of the stems, **Fl** pointing outwards; peduncle robust; **Ped** 3 cm (rarely longer), erect; **Cl** outside pale green, inside yellow, brown or red, rarely green, wrinkles often contrasting in colour, ± 3 - 4 cm ∅, central depression shortly funnel-shaped; **Cl** lobes ± 1.3 - 1.6 cm, ± lanceolate, outside glabrous or weakly pubescent, inside strongly wrinkled, wrinkles often delicately roughened, around the central depression hairy, clavate **Ha** white to dark purple, lobe margins strongly recurved outwards; **Cn** whitish; **Ci** lobes ± 3 mm, rectangular, conspicuously mucronate or apically crenate-truncate, ± erect, margins curved inwards; **Cs** lobes ± 6 mm, basally ovate, apically subulate or ± filiform, erect, meeting each other and then recurved, tips ± clavate, dorsal wing ± 4 mm, laterally compressed, very narrowly triangular, ± erectly spreading from the base; **Poll** ± 0.7 × 0.3 mm, ± narrowly D-shaped, difficult to reach.

More closely related to *S. olivacea*.

S. gariepensis Pillans (South Afr. Gard. 18: 62, 1928). **T:** RSA, Northern Cape (*Pillans* 5771 [BOL]). – **D:** Namibia, RSA (Northern Cape). **Fig. XLIV.g**

≡ *Gonostemon gariepensis* (Pillans) P. V. Heath (1992).

Stems decumbent to shortly decumbent-erect, to 12.5 cm, oftten purple; stipular rudiments cream-coloured; **Fl** procumbent-spreading, buds purple, apically attenuate; **Ped** 2 - 7 cm; **Cl** outside often ± white or very pale green, ± flushed with purple, inside ± (pale) yellow with red-purple, glossy, to 10 cm ∅, centrally depressed; **Cl** lobes to 4 × 1.5 cm, red, sometimes weakly cross-banded with cream or yellow, spreading and curved outwards, tips pale yellow, margins curved outwards, midvein outside sometimes weakly pubescent, inside glabrous, central depression apically ± densely hairy, **Ha** purple, sometimes white, long, soft; **Cn** cream-coloured to yellow-brown, spotted with purple, stalk short; **Ci** lobes ± 4.5 × 1.25 mm, narrowly rectangular, ± spatulate, with a median mucro, erect-spreading, apically recurved, inside with a longitudinal groove; apical appendage of the **Cs** ≤ 9 mm, subulate, erect or ± curved, dorsal appendage ± 5.5 × 1 mm, wing-like, narrowly rectangular, porrectly spreading, free to the base; **Anth** yellowish, strongly prominent; **Poll** brownish, 0.9 - 1 × 0.55 - 0.6 mm, ± narrowly D-shaped, caudicles rather long; **Fr** erect, to 10 cm.

Closely related to *S. asterias*.

S. gettliffei Pott (Ann. Transvaal Mus. 3: 226, t. 13, 1913). **T:** RSA, Transvaal (*Gettliffe* s.n. [PRE]). – **D:** Moçambique, Zimbabwe, Botswana, RSA (Northern Prov., North-West Prov., Gauteng, Mpumalanga).

≡ *Gonostemon gettliffei* (Pott) P. V. Heath (1992).

Stems decumbent-erect, basally tapering, often limp and creeping, (greyish-) green, ± 1.2 - 1.5 cm ∅; **Tu** sometimes > 1 cm, narrowly pointed; **L** rudiments erect or slightly incurved; **Fl** decumbent, buds ovoid-acute, basally truncate; **Ped** 5 - 8 cm; **Sep** ≤ 1.5 cm; **Cl** cream-coloured, 9 - 12 cm ∅, lobes purple, wrinkles cream-coloured to yellowish, rarely uniformly cream-coloured; **Cl** lobes 3.6 - 4.8 cm, narrowly elliptic, acute, inside delicately pubescent, margins curved outwards, outside weakly pubescent, inside of the central part and margins of the lobes densely hairy, **Ha** pale purple, long, weak; **Cn** dark purple; **Ci** lobes ± 5 mm, in the middle with a deep longitudinal groove, delicately pointed, erectly spreading, apically with few robust **Ha**; apical appendage of the **Cs** 12 mm, triangular, weakly curved, erect, ventrally deeply V-shaped, dorsal wing ≤ 8 × 5 - 6 mm, ± fused with the apical appendage over the whole length; **Poll** greenish-brown, ± 1 × 0.5 mm, ± D-shaped.

Closely related to *S. hirsuta*.

S. gigantea N. E. Brown (Gard. Chron., ser. nov. 7: 684, 1877). **T:** RSA, KwaZulu-Natal (*Gerrard* 717 [K]). – **D:** Tanzania, Zambia, Moçambique, Botswana, Zimbabwe, Malawi, RSA (KwaZulu-Natal), Swaziland. **Fig. XLIV.h**

≡ *Gonostemon giganteus* (N. E. Brown) P. V. Heath (1992); **incl.** *Stapelia nobilis* N. E. Brown *ex* Hooker (1901) ≡ *Gonostemon giganteus* var. *nobilis* (Hooker) P. V. Heath (1993); **incl.** *Stapelia marlothii* N. E. Brown (1908) ≡ *Gonostemon giganteus* var. *marlothii* (N. E. Brown) P. V. Heath (1993); **incl.** *Stapelia gigantea* var. *pallida* E. Phillips (1925) ≡ *Gonostemon giganteus* nvar. *pallidus* (E. Phillips) P. V. Heath (1993); **incl.** *Stapelia youngii* N. E. Brown (1931) ≡ *Gonostemon giganteus* var. *youngii* (N. E. Brown) P. V. Heath (1993); **incl.** *Stapelia cylista* C. A. Lückhoff (1933).

Compact plants with shortly down-curved-erect stems, stems ± 15 (-25) × ± 2 cm; **Tu** inconspicuous; **L** rudiments 2 - 3 mm; **Ped** ≤ 6 cm; **Sep** ≤ 1.6 cm; **Cl** pale yellow, reddish or purple, 12.5 - 40 cm ⌀, wrinkles red-brown or purple, central part ± flat to slightly ± campanulate; **Cl** lobes 4 - 16 cm, outside delicately pubescent, inside ± weakly hairy, **Ha** whitish to purple, often ± curly, ± robust; **Cn** dark purple, rarely ± sessile; **Ci** lobes 5 - 6 × 1.5 - 2.5 mm, rectangular, sometimes ± spatulate, erectly diverging, apically crenate, with an obtuse median mucro, sometimes long-attenuate, rarely 3-toothed; **Cs** lobes 8 - 12 × 1 - 2 mm, basally broadened, sometimes hiding the **Anth**, apical appendage subulate, ± triangular, ± cylindrical, acutely tapering or dorsiventrally compressed, biconvex or ± flattened, in part keeled, dorsal wing broad, oblong, basally free or rarely united with the apical appendage, erect or diverging outwards; **Poll** brown, ± 1 × 0.7 mm, ± obliquely D-shaped; **Fr** pale yellow.

Closely related to *S. unicornis*. The flowers of *S. gigantea* are amongst the largest in the family. Commonly encountered in cultivation and worldwide introduced to many tropical countries.

S. glabricaulis N. E. Brown (HIP 20: t. 1917 + text, 1890). **T:** RSA, Eastern Cape (*Barkly* 52 [K]). – **D:** RSA (Eastern Cape).

≡ *Gonostemon glabricaulis* (N. E. Brown) P. V. Heath (1992); **incl.** *Stapelia forcipis* E. Phillips & Letty (1932) ≡ *Gonostemon glabricaulis* var. *forcipis* (E. Phillips & Letty) P. V. Heath (1993).

Stems basally frequently shortly down-curved, to 20 × 1.5 - 2 (-2.5) cm, crenately 4-ribbed, glabrous; **L** rudiments ± 2 mm, sometimes weakly pubescent; **Inf** opening in ± quick succession, sometimes 2 **Fl** simultaneously open; peduncle apically weakly pubescent; **Fl** buds ± globose, apically sometimes delicately pubescent; **Ped** ≤ 6 cm, sometimes scatteredly weakly hairy; **Sep** ± 7 mm, margins sometimes weakly hairy; **Cl** purple-brown (tips of the lobes darker), 5 - 7 (-10) cm ⌀; **Cl** lobes 1.75 - 2.5 (-3.5) cm, slightly recurved outwards, often convexely wavy near the tips, outside glabrous, ± smooth, central part and base of the lobe margins on the inside densely hairy, **Ha** pale purple or grey-white, straight, ± depressed, directed outwards, lobe margins hairy, **Ha** whitish, simple or clavate, stiff; **Cn** black-purple; **Ci** lobes sometimes obtusely rectangular, ± spatulate or irregularly 3-dentate, rarely long-pointed, spreading, curved; **Cs** lobes erect, apical appendage ± filiform, falcately curved, dorsal wing broad, rarely narrowly triangular, ± acute, for ½ or completely united with the apical appendage or free, rarely divergent; **Poll** brown, germination crest yellow, 0.9 - 1 × 0.52 - 0.57 mm, ± D-shaped.

Belonging to the complex around *S. hirsuta*, and closely related to *S. obducta*.

S. glanduliflora Masson (Stapel. Nov., 16, t. 19, 1797). **T:** [lecto – icono]: l.c. t. 19, 1797. – **D:** RSA (Western Cape). **Fig. XLV.a**

≡ *Gonostemon glanduliflorus* (Masson) P. V. Heath (1992); **incl.** *Stapelia glandulifera* Willdenow (1798); **incl.** *Stapelia hispidula* Hornemann (1813); **incl.** *Stapelia glanduliflora* var. *emarginata* N. E. Brown (1909) ≡ *Gonostemon glanduliflorus* var. *emarginatus* (N. E. Brown) P. V. Heath (1993); **incl.** *Stapelia glanduliflora* var. *haworthii* A. Berger (1910); **incl.** *Stapelia glanduliflora* var. *massonii* A. Berger (1910).

Plants to 20 cm tall, forming small clumps; **Br** bluntly 4-ribbed; stipular rudiments more conspicuous with age; **Inf** decumbent, opening in quick succession; peduncle robust, usually button-like, sometimes persistent (then woody); **Ped** rosy-brown, ≤ 5 cm, decumbent; **Sep** reaching the bases of the **Cl** lobes; **Cl** outside pale yellow-green, venation brown, inside leather-coloured or yellow, sometimes greenish, to 3.5 cm ⌀, wrinkles intermittently red-lineate; **Cl** tube short, narrow, pentagonal, embracing the **Cn**; **Cl** lobes ± 1.2 cm, sometimes ± curved outwards, outside glabrous or very weakly pubescent, inside weakly wrinkled, densely hairy (centre of the upper ⅓ ± glabrous), clavate **Ha** white, transparent; **Ci** yellow or orange, tips brownish, lobes ≤ 2 mm, ± acute; **Cs** reddish to dark purple, lobes narrowly subulate to filiform, erect, apically meeting each other, dorsally sometimes slightly keeled, basally rarely with a small dorsal bulge; **Poll** dull orange, 0.4 × 0.6 mm, ± D-shaped.

Closely related to *S. erectiflora*.

S. grandiflora Masson (Stapel. Nov., 13, t. 11, 1797). **T:** [lecto – icono]: l.c. t. 11, 1797. – **D:** RSA (Northern Cape, Western Cape, Eastern Cape, Free State). **Fig. XLIII.h**

≡ *Gonostemon grandiflorus* (Masson) P. V. Heath

(1992); **incl.** *Stapelia ambigua* Masson (1797) ≡ *Gonostemon grandiflorus* var. *ambiguus* (Masson) P. V. Heath (1993); **incl.** *Stapelia spectabilis* Haworth (1812); **incl.** *Stapelia ambigua* var. *fulva* Sweet (1830) ≡ *Gonostemon grandiflorus* var. *fulvus* (Sweet) P. V. Heath (1993); **incl.** *Stapelia grandiflora* var. *lineata* N. E. Brown (1877) ≡ *Gonostemon grandiflorus* var. *lineatus* (N. E. Brown) P. V. Heath (1993); **incl.** *Stapelia obscura* N. E. Brown (1877); **incl.** *Stapelia desmetiana* N. E. Brown (1889) ≡ *Gonostemon grandiflorus* var. *desmetianus* (N. E. Brown) P. V. Heath (1993); **incl.** *Stapelia desmetiana* var. *apicalis* N. E. Brown (1889) ≡ *Gonostemon grandiflorus* var. *apicalis* (N. E. Brown) P. V. Heath (1993); **incl.** *Stapelia flavirostris* N. E. Brown (1908); **incl.** *Stapelia desmetiana* var. *pallida* N. E. Brown (1909) ≡ *Gonostemon grandiflorus* var. *pallidus* (N. E. Brown) P. V. Heath (1993); **incl.** *Stapelia senilis* N. E. Brown (1909) ≡ *Gonostemon grandiflorus* var. *senilis* (N. E. Brown) P. V. Heath (1993); **incl.** *Stapelia desmetiana* var. *fergusoniae* R. A. Dyer (1931) ≡ *Gonostemon grandiflorus* var. *fergusoniae* (R. A. Dyer) P. V. Heath (1993) ≡ *Stapelia fergusoniae* (R. A. Dyer) P. V. Heath (1993) (*nom. inval.*, Art. 34.1).

Plants to 30 cm, forming compact clumps; stems 2 - 3 cm ∅; peduncle lump-like; **Ped** ± 25 (rarely 70) × 4 - 5 mm, ± robust, gradually tapering towards the tip; **Sep** 0.4 - 1.5 cm; **Cl** ± brownish, 8 - 22 cm ∅, flat to broadly ± funnel-shaped, narrow wrinkles purple-brown; **Cl** tube short, narrow, basally embracing the **Cn**; **Cl** lobes 2.8 - 8.8 cm, tips darker, rectangular-pointed, elliptic-acute to ovoid and long-attenuate, strongly curved outwards, outside delicately pubescent, inside ± softly hairy (lobe tips usually glabrous), **Ha** white, erect (if ± decumbent then directed to the lobe tips), margins hairy, **Ha** basally often papillose; **Cn** shortly stipitate; **Ci** lobes 7 - 9 mm, rectangular, apically occasionally slightly broadened, concave, ascending, ± erect, apically weakly outcurved, irregularly dentate, crenate or delicately pointed, rarely acute; **Cs** purple or brown, apically often yellow, lobes 6 - 11 mm, erect, apical appendage acute, narrowly or broadly triangular, rarely strongly reduced, dorsal wing broad, erect, fused for ⅔ of its length with the apical appendage; **Poll** orange or brownish, 1 - 1.25 × 0.6 - 0.75 mm, ± obliquely ovoid or D-shaped.

S. hirsuta Linné (Spec. Pl. [ed. 1], 217, 1753). **T:** BM [Herb. Clifford 77, Stapelia 2]. – **D:** RSA (Western Cape); winter-rainfall region.

≡ *Gonostemon hirsutus* (Linné) P. V. Heath (1992); **incl.** *Stapelia comata* Jacquin (1806) ≡ *Stapelia hirsuta* var. *comata* (Jacquin) N. E. Brown (1909) ≡ *Gonostemon hirsutus* var. *comatus* (Jacquin) P. V. Heath (1993); **incl.** *Stapelia depressa* Jacquin (1806) ≡ *Tridentea depressa* (Jacquin) Schultes *ex* G. Don (1838) ≡ *Stapelia patula* var. *depressa* (Jacquin) N. E. Brown (1890) ≡ *Stapelia hirsuta* var. *depressa* (Jacquin) N. E. Brown (1908) ≡ *Stapelia hirsuta* subvar. *depressa* (Jacquin) A. Berger (1910) ≡ *Gonostemon hirsutus* var. *depressus* (Jacquin) P. V. Heath (1993); **incl.** *Stapelia patula* Willdenow (1809) ≡ *Stapelia hirsuta* var. *patula* (Willdenow) N. E. Brown (1909); **incl.** *Stapelia lanifera* Haworth (1819); **incl.** *Stapelia hirsuta* var. *atra* Lindley (1823); **incl.** *Stapelia elongata* Sweet (1830); **incl.** *Stapelia lanigera* Loudon (1830); **incl.** *Stapelia patentirostris* N. E. Brown (1877); **incl.** *Stapelia unguipetala* N. E. Brown (1877) ≡ *Stapelia hirsuta* var. *unguipetala* (N. E. Brown) N. E. Brown (1908); **incl.** *Stapelia affinis* N. E. Brown (1890) ≡ *Stapelia hirsuta* var. *affinis* (N. E. Brown) N. E. Brown (1909) ≡ *Gonostemon hirsutus* var. *affinis* (N. E. Brown) P. V. Heath (1993); **incl.** *Stapelia patula* var. *longirostris* N. E. Brown (1890) ≡ *Stapelia hirsuta* var. *longirostris* (N. E. Brown) N. E. Brown (1909) ≡ *Stapelia hirsuta* subvar. *longirostris* (N. E. Brown) A. Berger (1910) ≡ *Gonostemon hirsutus* var. *longirostris* (N. E. Brown) P. V. Heath (1993); **incl.** *Stapelia villosa* N. E. Brown (1890); **incl.** *Stapelia hirsuta* var. *grata* N. E. Brown (1909) ≡ *Gonostemon hirsutus* var. *gratus* (N. E. Brown) P. V. Heath (1993); **incl.** *Stapelia hirsuta* var. *lutea* N. E. Brown (1909) ≡ *Gonostemon hirsutus* var. *luteus* (N. E. Brown) P. V. Heath (1993); **incl.** *Stapelia margarita* B. Sloane (1933) ≡ *Stapelia pulvinata* fa. *margarita* (B. Sloane) G. D. Rowley (1973).

Stems erect or down-curved-erect, creeping or forming compact clumps (to 30 × 100 cm), rarely > 12 cm, ± slender, ± 1 (-1.5) cm ∅; **L** rudiments small, sometimes persistent in the dry state; peduncle robust; **Fl** procumbent; **Ped** (2-) 6 (-11.5) cm, porrectly spreading or resting on the ground; **Sep** 0.5 - 1.3 cm; **Cl** yellowish or cream-coloured, 5 - 14 cm ∅, margins and tips of the lobes purple, central part around the **Cn** thickened; **Cl** lobes 1.6 - 5.6 cm, outside rarely ± glabrous, inside of the central region and lobe bases densely hairy, **Ha** ± slightly curly, marginal **Ha** longer, straight, porrectly spreading, base often papillose, tips of the **Cl** lobes glabrous or finely roughened (rarely **Cl** completely glabrous); **Cn** basally yellowish, apically black-purple, ± sessile; **Ci** lobes concave, ascending, apically with a median mucro or dentate, curved outwards; apical appendage of the **Cs** variable in length, (3-angled-) subulate, dorsal wing variable in length and width, variously united with the apical appendage or free, erect or widely spreading; **Poll** ± 0.9 - 1 × 0.5 - 0.6 mm, ± obliquely ovate; **Fr** yellowish-brown, brown-striped, 15 - 20 cm.

Closely related to *S. pulvinata*. Apart from *S. gigantea* and *Orbea variegata* perhaps the most frequently cultivated species of Stapeliads.

S. immelmaniae Pillans (South Afr. Gard. 18: 62, 1928). **T:** RSA, Western Cape (*Immelman* s.n. in *NBG* 581/26 [BOL]). – **D:** RSA (Western Cape).

Plants to 20 cm, forming compact clumps; stems

to 1.5 cm ⌀; **L** rudiments ± 2 mm, caducous, stipular rudiments yellowish; **Inf** with 1 to numerous **Fl**, occasionally 2 **Fl** simultaneously open, erect; peduncles scattered ± irregularly, often numerous, button-like, ± erect, persistent; **Ped** ≤ 1.5 cm, erectly spreading; **Sep** ≤ 6 mm; **Cl** pinkish-brown, to 4.5 cm ⌀, ± bulging, central part and tips of the lobes often yellowish or greenish, wrinkles ± yellowish; **Cl** lobes ≤ 1.9 cm, outside smooth, weakly pubescent, inside hairy, **Ha** apically narrower, erect (spreading), ± weak; **Cn** dark purple; **Ci** lobes ± 3 mm, rectangular, weakly concave, obtuse, mucronate, erectly spreading, rarely bidentate, appendages widely diverging; **Cs** lobes ± 2 mm, narrowly ovate, apically ± filiform, acute, 3-angled, slightly curved outwards, not meeting each other, dorsal wing or bulge variable, often crenate; **Anth** white, apically broadened, closely spaced, sometimes slightly overlapping; **Poll** brownish-orange, 0.6 × 0.45 mm, ± broadly ovoid, germination crest and translators yellow; **Fr** 10 cm.

Belonging to the *S. scitula* / *S. montana* complex and more closely related to *S. glanduliflora*.

S. kougabergensis L. C. Leach (SAJB 3(3): 176-177, ill., 1984). **T:** RSA, Eastern Cape (*Leach & Bayliss* 15659 [PRE, K, NBG]). – **D:** RSA (Eastern Cape).

≡ *Gonostemon kougabergensis* (L. C. Leach) P. V. Heath (1992).

Stems apically tapering, to 20 cm, grey-green, basally 1.2 cm ⌀; **Tu** inconspicuous; **L** rudiments persistent when dry, stipular rudiments yellowish; **Inf** often many-flowered; **Fl** buds dull purple, broadly ovoid, compressed, rarely weakly pointed; **Ped** ≤ 2 cm, robust, erectly spreading; **Sep** ≤ 5 mm; **Cl** dull purple with yellowish-purple wrinkles (or vice versa), 1.6 - 3 cm ⌀, ± flat; **Cl** tube short, bowl-like; **Cl** lobes broadly triangular, outwardly recurved and **Fl** ± globose in overall view, outside pubescent, inside glabrous, margins sometimes weakly pubescent; **Cn** dark red or ± black, robustly stipitate; **Ci** lobes ± 1.5 mm, ± square, truncate or obtuse, with a small mucro, sometimes longer, narrower, more pointed; **Cs** lobes ± 2.5 mm, robustly 3-angled-subulate, keeled, erect, basally sometimes weakly bulging; **Anth** orange-yellow; **Poll** pale orange-brown to dark brown, 0.5 × 0.4 mm, ± broadly ovoid, far distant from each other, sitting rather laxly in the thecae.

Closely allied to *S. acuminata*.

S. kwebensis N. E. Brown (FTA 4(1): 501, 1904). **T** [lecto]: Bechuanaland Protectorate, Ngamiland (*Lugard* 29 [K]). – **D:** Moçambique, Zimbabwe, Botswana, Namibia, RSA (North-West Prov., Northern Prov.). **Fig. XLVI.c**

≡ *Gonostemon kwebensis* (N. E. Brown) P. V. Heath (1992).

Stems ± tapering, to 15 cm, forming compact clumps, grey-green; **L** rudiments ≤ 7 mm, stipular rudiments pale; **Inf** irregularly scattered, ± sessile, **Fl** pointing outwards; **Ped** very short; **Cl** yellow- to black-brown, sometimes tinged greenish, with paler wrinkles, 1.5 - 3 (-4.5) cm ⌀, flat or the centre depressed cup-like, margin of the depression thickened like an annulus; **Cl** lobes ± curved outwards, outside ± pubescent, inside apically ± pubescent, margins slightly curved outwards; **Cn** shortly stipitate; **Ci** lobes transversely rectangular, sometimes semicircular, margin ± broadly acute, entire, shortly crenate or dentate, rarely emarginate; **Cs** lobes ± as long as the **Anth**, ± oblong, apically narrower, ± acute, obtuse, rarely emarginate, basally often dorsally weakly bulging; **Poll** ± 0.625 × 0.4 mm, ± obliquely ovoid.

Closely related to *S. longipedicellata, S. parvula* and *S. similis*.

S. leendertziae N. E. Brown (Ann. Transvaal Mus. 2: 168, 1910). **T:** RSA, Gauteng (*Leendertz* 2464 [K, PRE, SAM]). – **D:** RSA (Gauteng, Mpumalanga, KwaZulu-Natal), Swaziland.

≡ *Gonostemon leendertziae* (N. E. Brown) P. V. Heath (1992); **incl.** *Stapelia wilmaniae* C. A. Lückhoff (1933).

Stems shortly down-curved-erect, ± clump-forming, dull green, slightly glossy, ± slender, sides concave; **Tu** inconspicuous; **L** rudiments ± 2.5 mm, caducous, stipular rudiments small; **Fl** buds purple, ± globose or ± ovoid, abruptly long-attenuate, bases of the **Cl** lobes acutely prominent, weakly pubescent; **Ped** 2 - 3 cm; **Sep** 0.6 - 1.2 cm; **Cl** dark purple, ± 5 cm ⌀, deeply campanulate; **Cl** lobes at least 1 × 1 cm, often broader than long, ± triangular and long-attenuate, spreading and curved outwards, margin of the **Cl** tube inside densely hairy, **Ha** purple, long; margins of the **Cl** lobes in part ± weakly hairy; **Ci** brown, lobes deeply bifid, apically attenuate, erect, diverging; **Cs** dark purple, lobes ± 9 mm, apical appendage filiform, erect, slightly curved, dorsiventrally compressed, dorsally flat, basally with a dorsal wing, rectangular, thin, erect, at least as long as the apical appendage, free; **Poll** ± 0.8 × 0.6 mm, broadly ovoid; **Fr** to 15 cm.

S. longipedicellata (A. Berger) N. E. Brown (BMI 1913: 303, 1913). **T:** K. – **D:** Namibia.

≡ *Stapelia kwebensis* var. *longipedicellata* A. Berger (1910) ≡ *Gonostemon longipedicellatus* (A. Berger) P. V. Heath (1992).

Habit and stems like *S. kwebensis*, but somewhat more delicate; **L** rudiments smaller, often persisting as white **Sp**; **Inf** irregularly scattered, **Fl** pointing upwards and overtopping the stems; peduncle ± long, robust, erect; **Ped** 4 cm, erect; **Cl** 1.5 - 4.5 cm ⌀, otherwise incl. the **Cn** like *S. kwebensis*; **Poll** 0.5 - 0.7 × 0.35 - 0.55 mm.

Closely related to *S. kwebensis, S. parvula* and *S. similis*.

S. macowanii N. E. Brown (HIP 20: t. 1920 + text, 1890). **T** [lecto]: RSA, Eastern Cape (*MacOwan 909* [K]). – **D:** RSA.
≡ *Gonostemon macowanii* (N. E. Brown) P. V. Heath (1992).

S. macowanii var. **conformis** (N. E. Brown) L. C. Leach (Excelsa Tax. Ser. 3: 66-67, ill., 1985). **T** [lecto]: RSA (*Anonymus* s.n. [K]). – **D:** RSA (Eastern Cape).
≡ *Stapelia conformis* N. E. Brown (1909) ≡ *Gonostemon macowanii* var. *conformis* (N. E. Brown) P. V. Heath (1992); **incl.** *Stapelia conformis* var. *abrasa* N. E. Brown (1909) ≡ *Gonostemon macowanii* var. *abrasus* (N. E. Brown) P. V. Heath (1993).

Differs from var. *macowanii*: **Ped** ≤ 3 cm; **Sep** 1.25 cm, narrower; **Cl** to 15.5 cm ⌀, markings conspicuous, tips in part with darker margins; **Cl** lobes 4.3 - 6.2 cm, variable in shape, conspicuously hairy, clavate **Ha** variably long; **Ci** lobes longer, ± rectangular; apical appendage of the **Cs** ± slender, oblong, acutely tapering, erect, dorsal wing free to the base.

S. macowanii var. **macowanii** – **D:** RSA (Eastern Cape).
Plants to 30 cm, clump-forming; stems grey-green, ± 2 cm ⌀, squat; **Tu** sometimes wing-like; **L** rudiments 2 - 3 mm, stipular rudiments pale; **Ped** ≤ 1 cm, densely white-pubescent; **Sep** 4 - 5 mm, densely white-pubescent; **Cl** greenish-white, brownish or yellowish with darker delicate wrinkles, 5 - 6 cm ⌀, centrally slightly depressed; **Cl** lobes 1.6 - 2.1 (-2.6) × 1.25 - 1.5 (-1.9) cm, ovate-acute or ± triangular, outside densely pubescent, **Ha** brownish, inside centrally weakly pubescent, lobes rarely with few whitish **Ha**; **Ci** lobes 4 - 5 × 1.5 - 2.2 mm, ± elliptic, concave, obtuse, erectly spreading, apically with or without a small median mucro; **Cs** lobes broadly wing-like, erect, inner margin sometimes ± triangular, lobes rarely with acutely tapering apical appendage and spreading ± triangular dorsal wing; **Poll** ± 1 × 0.6 mm, ± obliquely ovoid.

S. ×meintjesii I. Verdoorn *pro sp.* (FPSA 23: t. 917 + text, 1943). **T:** RSA (*Meintjes* s.n. in *PRE* 27144 [PRE]).
≡ *Gonostemon* ×*meintjesii* (I. Verdoorn *pro sp.*) P. V. Heath (1992).
= *S. gigantea* × *S. gettliffei*.

S. montana L. C. Leach (SAJB 3(3): 175-176, ills., 1984). **T:** RSA, Western Cape (*Bayer 1602* [NBG, K, MO, PRE, SRGH]). – **D:** RSA.
≡ *Gonostemon montanus* (L. C. Leach) P. V. Heath (1992).

S. montana var. **grossa** L. C. Leach (SAJB 3(3): 176, 1984). **T:** RSA, Western Cape (*Bruyns 1436* [NBG]). – **D:** RSA (Western Cape).
≡ *Gonostemon montanus* var. *grossus* (L. C. Leach) P. V. Heath (1992).

Differs from var. *montana*: Stems more squat, to 7.5 cm; **Tu** fleshy, stipular rudiments always present; **Fl** in quicker succession, frequently 2 simultaneously open; **Ped** weakly pubescent; **Cl** larger, reddish, marginal **Ha** whitish.

S. montana var. **montana** – **D:** RSA (Western Cape).
Stems to 6 × 0.5 - 0.6 cm, glabrous or very weakly pubescent; **L** rudiments ± 2 mm, usually without stipular rudiments; peduncle lump-like; **Fl** buds ovoid, long-attenuate or ± acute, ribbed at the angles; **Ped** 1 - 2.5 cm, glabrous, erect (-spreading); **Sep** 5 mm, glabrous; **Cl** ± black (-purple), ± 3 cm ⌀, wrinkles often yellowish; **Cl** lobes ± 0.9 - 1.2 cm, often strongly recurved, twisted, inside softly hairy, **Ha** purple, along the margins ± 2 mm; **Ci** whitish, purple-spotted, lobes ± 2.5 × 1.5 mm, ± rectangular, concave; **Cs** dark purple, lobes 3.5 × 0.5 mm, narrowly ovate, apical appendage filiform, erect, ± falcately outcurved, dorsal wing ± large; **Poll** brown-orange, ± 0.75 × 0.45 mm, ± D-shaped.

Closely related to *S. scitula* and *S. immelmaniae*.

S. obducta L. C. Leach (SAJB 3(3): 169-171, ills., 1984). **T:** RSA, Eastern Cape (*Mullens 78* [PRE, NBG]). – **D:** RSA (Eastern Cape: Steytlerville Distr.). **Fig. XLV.b**
≡ *Gonostemon obductus* (L. C. Leach) P. V. Heath (1992).

Plants ± 15 cm, clump-forming; stems ± 1.5 cm ⌀, glabrous; **L** rudiments caducous; **Ped** ± 3 cm; **Sep** ± 5 × 2 mm, basally gibbose, ± glabrous, margins sometimes ± hairy; **Cl** ± dark purple-brown (tips of the lobes and margins paler), ± 4.8 cm ⌀; **Cl** lobes ± 1.4 cm, strongly revolute, outside glabrous, inside inconspicuously transversally wrinkled, densely hairy, **Ha** red-brown, long, soft, directed towards the tip, margins not or hardly inrolled; **Ci** yellow with purple-brown markings, lobes ± 5.5 × 2 mm, rectangular to spatulate, strongly diverging, apically with obtuse median mucro; **Cs** dark purple, lobes ± 6 mm, apical appendage ± oblong-triangular, converging above the **Anth**, apically slightly diverging, dorsal wings broad, triangular, erect-spreading; **Poll** dull orange, 0.75 × 0.5 mm, ± D-shaped.

Belonging to the *S. hirsuta*-complex and closely related to *S. tsomoensis* and *S. glabricaulis*.

S. olivacea N. E. Brown (Gard. Chron., ser. nov. 3: 136, 1875). **T:** K. – **D:** Namibia, RSA (Northern Cape, Western Cape, Eastern Cape).
≡ *Gonostemon olivaceus* (N. E. Brown) P. V. Heath (1992); **incl.** *Stapelia eruciformis* hort. ex N. E. Brown (1875) (*nom. inval.*, Art. 29.1).

Stems apically tapering, to 10 cm, forming compact clumps, grey-green, basally ± 1 cm ⌀, bluntly 4-ribbed, with longitudinal grooves; **L** rudiments ± 1.5 mm, caducous, stipular rudiments small; **Inf** few-flowered; **Fl** buds compressed-ovoid, obtusely 5-angled; **Fl** evil-scented; **Ped** ± 5 mm; **Sep** ± 4 × 1.5 mm; **Cl** greenish-yellow, black, sometimes olive-green, 2 - 4 cm ⌀ and more, sometimes 5-angled, campanulate, wrinkles red-brown, labyrinth-like, margin of the **Cl** tube spreading; **Cl** lobes long-attenuate-triangular, recurved, outside weakly pubescent, inside glabrous, margins ± densely hairy, **Ha** white, weak, curled; **Cn** dark purple; **Ci** ± 2.5 mm, ± narrowly triangular, weakly concave, erect-spreading, opening to the **Nec** ± circular, ± transversal; **Cs** ± 4 mm, apical appendage filiform, apically shortly subulate, connivent over the **Anth**, apically diverging, dorsal wing or appendage ± triangular, laterally compressed, spreading; **Poll** 0.5 - 0.7 × 0.4 - 0.55 mm; **Fr** pale olive-brown, to 10 cm.

Closely related to *S. pearsonii*.

S. paniculata Willdenow (Enum. Pl. Hort. Reg. Berol. Suppl., 13, 1814). **T:** B. – **D:** RSA (Western Cape).

≡ *Tridentea paniculata* (Willdenow) G. Don (1838) ≡ *Stapelia concinna* var. *paniculata* (Willdenow) N. E. Brown (1909) ≡ *Gonostemon paniculatus* (Willdenow) P. V. Heath (1992); **incl.** *Stapelia nouhuysii* E. Phillips (1929).

Stems often in large divergent clumps, 6 - 8 × 0.6 - 0.7 cm, bluntly 4-ribbed; **L** rudiments white, persistent in the dry state, stipular rudiments cream-coloured; **Inf** many-flowered, sometimes 2 **Fl** simultaneously open; peduncle clump-like; **Ped** ≤ 3.5 cm, ± erect; **Sep** ≤ 5 mm, convex; **Cl** pale lilac to purple-brown, to 3 cm ⌀, shallowly bowl-shaped; **Cl** lobes ± 9 - 11 mm, spreading, slightly curved outwards, outside delicately pubescent, inside densely hairy, **Ha** white, ± 1 mm, straight, soft, ± flatly appressed, directed towards the lobe tips, margins of the **Cl** lobes densely hairy, clavate **Ha** white, rarely brown, ≤ 3 mm, ± acute; **Cn** black-brown, apical appendage paler, basally with 5 prominent bulges, stalk short; **Ci** 2 - 2.5 × 0.75 - 1 mm, rectangular, spreading, slightly curved outwards, apically somewhat broadened, obtuse, with a short mucro, sometimes 3-dentate, rarely acutely spatulate; **Cs** lobes ± 3 × 0.75 mm, subulate, ± erect, slightly curved outwards, weakly keeled, dorsal wings raised, ± erect; **Anth** orange-yellow; **Poll** orange or brownish, 0.5 × 0.35 mm, ± broadly ovoid.

Closely related to *S. erectiflora* and *S. glanduliflora*.

S. parvula Kers (Bot. Not. 122: 173-176, 1969). **T:** Angola, Moçamedes Distr. (*Kers* 3458 [S]). – **D:** Angola.

≡ *Gonostemon parvulus* (Kers) P. V. Heath (1992).

Stems in clusters or lax clumps, 5 mm ⌀, bluntly 4-ribbed; **Tu** to ± 3 cm; **Inf** scattered over the length of the stems; peduncle sometimes persistent, elongated and contorted, with scars of the **Ped**; **Fl** decumbent on the ground; **Ped** ≤ 2.5 cm, decumbent, erect at **Fr** ripening; **Sep** ≤ 2 mm, prominent between the **Cl** lobes; **Cl** outside pale green, inside brownish, to 7.5 mm ⌀, tube very short, apically thickened annulus-like, outside delicately pubescent; margins of the **Cl** lobes and tips inside pubescent; **Cn** black, ± sessile; **Ci** lobes ± 0.5 mm, ± square, weakly furrowed, spreading, sometimes curved outwards, apically ± strongly crenate; **Cs** lobes as long as the **Anth**, ± narrowly rectangular, apically narrow, obtuse or ± acute, basally sometimes broadened and fused with the **Ci**; **Poll** ± 0.2 × 0.125 mm, ± D-shaped; **Fr** ± 2 × 0.25 cm, apically obtuse.

Closely related to *S. kwebensis, S. longipedicellata* and *S. similis*.

S. pearsonii N. E. Brown (BMI 1913: 304, 1913). **T:** Namibia (*Pearson* 8539 [BOL, PRE]). – **D:** Namibia.

≡ *Gonostemon pearsonii* (N. E. Brown) P. V. Heath (1992).

Stems slightly tapering, to 12 × basally 1.2 cm, bluntly 4-ribbed, sides ± flat or concave; **Tu** inconspicuous; **L** rudiments erect or curved towards the stems, stipular rudiments yellow or orange, sometimes absent; **Inf** frequently 1-flowered; **Ped** ≤ 4 cm, erectly spreading; **Sep** ± 3.5 × 0.5 - 0.7 mm; **Cl** outside greenish, inside dark brown, 3.5 - 5 cm ⌀; **Cl** lobes ± 1.2 - 1.75 cm, long-attenuate-triangular, longitudinally grooved, outside pubescent, inside glabrous, margins slightly curved outwards; **Cn** dark purple-brown; **Ci** lobes ≤ 4 mm, narrowly triangular, with a longitudinal groove, laying flat on the **Cl**, slightly curved upwards; apical appendage of the **Cs** ± 4 mm, filiform, upper ½ strongly outcurved, dorsal appendage ± 2.5 mm, laterally strongly compressed in a wing-like manner, broadly spreading; **Poll** 0.6 × 0.4 mm, ± obliquely ovoid.

Closely related to *S. olivacea*.

S. peglerae N. E. Brown (FC 4(1): 953, 1909). **T:** RSA, Eastern Cape (*Pegler* 760 [K]). – **D:** RSA (Eastern Cape).

≡ *Gonostemon peglerae* (N. E. Brown) P. V. Heath (1992).

Stems erect or creeping, both ends tapering, to 17.5 × 1.5 cm, obscurely 4-ribbed; **L** rudiments ≤ 1.5 mm; peduncle acutely tapering, usually glabrous; **Ped** 2 - 4 cm, usually glabrous; **Sep** ± 6 mm, ± glabrous; **Cl** glossy dark purple-brown, to 7.5 cm ⌀, outside ± glabrous, inside shortly pubescent and with weak whitish or yellowish wrinkles; **Cl** lobes to 3 cm, margins hairy, **Ha** whitish or pale lilac, ± depressed, ± curved towards the tip or incurved, intermixed with some purple clavate **Ha**; **Cn** black-purple; **Ci** lobes ± 4 mm, narrowly rectangular, ± 3-dentate, with median mucro and attenuate into an acute point; **Cs** lobes ± 8 mm, apical appendage filiform, strongly falcate-spreading, dorsal wing long, narrowly triangular, free, widely spreading; **Poll** yellow, ± 0.5 × 0.7 mm, ± D-shaped.

Erroneously identified as *S. tsomoensis* by Dyer (1941).

S. pillansii N. E. Brown (Gard. Chron., ser. 3, 35: 242, fig. 100, 1904). **T:** RSA, Western Cape (*Pillans* 38 [BOL, GRA]). – **D:** RSA.

≡ *Gonostemon pillansii* (N. E. Brown) P. V. Heath (1992).

S. pillansii var. **fontinalis** Nel (in A. C. White & B. Sloane, Stapelieae, ed. 2, 2: 576, 3: 1145, 1937). **T** [neo]: RSA, Western Cape (*Austin s.n.* in *PRE* 50789 [PRE]). – **D:** RSA (Western Cape).

≡ *Gonostemon pillansii* var. *fontinalis* (Nel) P. V. Heath (1992).

Differs from var. *pillansii*: **Cl** lemon-coloured, clavate **Ha** white.

S. pillansii var. **pillansii** – **D:** RSA (Western Cape).

Incl. *Stapelia pillansii* var. *attenuata* N. E. Brown (1909) ≡ *Gonostemon pillansii* var. *attenuatus* (N. E. Brown) P. V. Heath (1993).

Stems to 20 × 1 - 1.5 cm, in compact clumps, obscurely 4-ribbed; **L** rudiments 3 mm; **Inf** frequently 1-flowered, sometimes 2 - 3 **Fl** simultaneously open; **Ped** ≤ 5 cm; **Sep** ≤ 1 cm; **Cl** dark purple, to 20 cm ⌀, centrally ± depressed; **Cl** lobes to 8.5 cm, often caudate, tips sometimes twisted and compressed, outside weakly pubescent, inside smooth or weakly wrinkled, otherwise glabrous, margins apically strongly curved outwards, hairy, clavate **Ha** pale purple or ± colourless; **Cn** black-purple; **Ci** lobes 4 - 5 × 1.5 - 2.5 mm, rectangular, erect-spreading, apically obtuse, with a small mucro; **Cs** lobes 6 - 8 mm, apical appendage 3-angled, erectly recurved, dorsal wing broad, in part fused with the apical appendage, shorter than the apical appendage, basally often with ± broadly subulate or several tooth-like dorsal appendages, in part forming a ± toothed ring by fusion with the **Ci**; **Poll** yellow, 0.75 × 0.5 mm, ± D-shaped.

Closely related to *S. schinzii*, and with similarly stinking flowers.

S. praetermissa L. C. Leach (SAJB 3(3): 170-172, ills., 1984). **T:** RSA, Eastern Cape (*Leach & Bayliss* 15651 [PRE, K, MO, Z]). – **D:** RSA.

≡ *Gonostemon praetermissus* (L. C. Leach) P. V. Heath (1992).

S. praetermissa var. **luteola** L. C. Leach (SAJB 3(3): 172, 1984). **T:** RSA, Eastern Cape (*Stayner s.n.* in *KGW* 272/70 [NBG]). – **D:** RSA (Eastern Cape).

≡ *Gonostemon praetermissus* var. *luteolus* (L. C. Leach) P. V. Heath (1992).

Differs from var. *praetermissa*: **Cl** pale yellow, **Cl** lobes apically rose-brown; **Cn** pale yellow.

S. praetermissa var. **praetermissa** – **D:** RSA (Eastern Cape).

Stems to > 12 × ± 1 cm, in compact clumps, ± weakly glossy, slender, ± glabrous; **Tu** obscure; **L** rudiments ± 2 mm, weakly pubescent, basally sometimes without stipular rudiments; **Inf** scattered over the length of the stems, sometimes 2 - 3 per stem; peduncle robust, acutely tapering, sometimes persistent and button-like; **Fl** directed upwards, buds reddish-purple to dark chestnut-brown, narrowly ovate, angles weakly pubescent; **Ped** ≤ 6 mm; **Sep** ± 8 mm; **Cl** glossy, reddish-purple to dark red-brown, 3 - 5 cm ⌀, tube short, funnel-shaped, obtusely 5-angled; **Cl** lobes ≤ 2 cm, outside ± glabrous, inside esp. near the tube margin whitish-hairy, margins slightly curved outwards, hairy, **Ha** whitish, rarely purple, long, delicate, in part clavate, usually directed inwards, **Cl** tube and lobes apically in addition delicately pubescent; **Cn** black; **Ci** lobes ± 5.5 mm, narrowly rectangular, weakly concave, spreading, slightly curved outwards, apically obtuse or truncate, with a median mucro; **Cs** lobes ± erect, apical appendage ± long-subulate, curved, inside longitudinally strongly concave, dorsally keeled, dorsal wings small, spreading; **Poll** brownish-yellow, 0.6 - 0.65 × 0.4 - 0.45 mm, ± D-shaped.

Closely related to *S. baylissi*, but also similar to *S. asterias.*

S. pulvinata Masson (Stapel. Nov., 13, t. 13, 1797). **T:** RSA, Northern Cape (*Masson s.n.* [lecto – icono: l.c. t. 13, 1797]). – **D:** RSA (Northern Cape, Western Cape).

≡ *Gonostemon pulvinatus* (Masson) P. V. Heath (1992).

Stems down-curved-erect or creeping, irregularly branched, sometimes in large mats, ± 10 (-30) × 1 - 1.7 cm; **L** rudiments occasionally persistent as white tips, stipular rudiments yellow-brown; **Inf** 1-flowered; peduncle persistent in the dry state, weakly bracteate, sometimes with a dead bud; **Fl** frequently decumbent, **Ped** 3.5 - 10.5 (-15) cm; **Sep** ≤ 9 × 4.5 mm; **Cl** purple with cream-coloured wrinkles, 9 - 13 cm ⌀, strongly curved outwards, thickened cushion-like around the **Cn**; **Cl** lobes 2.5 - 4.3 cm, abruptly short-attenuate, outside delicately pubescent, inside with the exception of the glabrous dark purple tips densely long-hairy, **Ha** rose to purple, delicate, soft, curly, interwoven, margins delicately hairy; **Cn** purple-brown, apically frequently paler; **Ci** lobes 5 - 6 mm, narrowly spatulate, concave, erect-spreading, apically with a median mucro; **Cs** lobes ≤ 1.3 cm, irregularly 3-winged, erectly diverging, apical appendage apically ± 3-angled, dorsal wing ≤ 4 mm broad, in part or entirely fused with the apical appendage, ventral wings in pairs, narrower, acutely tapering into the apical appendage; **Poll** greenish-brown, 1 - 1.25 × 0.65 mm, ± D-shaped.

Closely related to *S. hirsuta.*

S. remota R. A. Dyer (BT 12(4): 632-633, ill., 1979). **T:** Namibia, Kaokoland (*Steenkamp* s.n. in *PRE* 57257 [PRE]). – **D:** N Namibia.

≡ *Gonostemon remotus* (R. A. Dyer) P. V. Heath (1992).

Stems to 15 × 1 - 1.5 cm, in compact clumps, 4- (rarely 6-) ribbed; **L** rudiments ± 1 mm, without stipular rudiments; peduncle bracteate, persistent; **Bra** ≤ 2 mm, pubescent, near the **Ped** base or irregularly scattered; **Ped** ≤ 1.2 cm; **Sep** green-brownish, ≤ 4 mm, basally glabrous; **Cl** outside pale green, inside dark chestnut-brown, ± 2 cm ∅; **Cl** tube yellowish, bluntly 5-angled-grooved, grooves densely papillose, papillae dark purple, clavate, forming a 5-stellate pattern, tube margin weakly irregularly pustulate, inside glabrous; **Cl** lobes strongly revolute, tips yellowish, delicately pubescent, margins densely hairy, **Ha** purple, ≤ 3 mm, simple, stiff, sometimes button-like; **Cn** dark red-brown; **Ci** basally yellow, lobes 3 mm, rectangular, concave, erect-spreading, apically somewhat broadened, irregularly 3-dentate; **Cs** lobes 4.5 mm, apical appendage obtusely 3-angled, weakly keeled, neighbouring each other in the **Fl** centre, abruptly diverging, dorsal appendage basally 0.75 mm broad, wing-like, erectly spreading, laterally compressed, apically obtuse, longer than the apical appendages; entrance to the **Nec** cavity small, narrowly elliptic; **Poll** red-brown, 0.8 × 0.5 mm, ± D-shaped.

S. rubiginosa Nel (Jahrb. Deutsche Kakt.-Ges. 1935(3): 20-21, ills., 1935). **T** [neo]: RSA, Northern Cape (*Hall* s.n. in *NBG* 422/54 [NBG]). – **D:** RSA (Northern Cape: Richtersveld).

≡ *Gonostemon rubiginosus* (Nel) P. V. Heath (1992).

Stems to 12 cm, clump-forming, purple, densely white-pubescent, bluntly 4- to 6-ribbed, sides weakly grooved; **L** rudiments ± 1.5 mm, stipular rudiments yellow or orange-brown; **Inf** scattered over the length of the stems, **Fl** solitary or fasciculate; **Ped** ≤ 6 mm; **Sep** ± 3 × 2 mm, keeled; **Cl** cream-coloured or greenish, 2 - 2.5 cm ∅, central depression 5-angled, apically thickened annulus-like; **Cl** lobes dark purple-brown, ± 5 - 7 mm, shortly pointed, spreading or recurved, with labyrinth-like wrinkles, outside weakly pubescent, inside glabrous, only the tips sometimes delicately pubescent, margins curving outwards, hairy, **Ha** white, ≤ 1 mm, simple; **Cn** brown, stalk short, ± hairy, **Ha** white, short, stiff; **Ci** lobes ± 2.5 mm, 3-angled, spreading, slightly curved outwards, deeply grooved in the middle, basally often hairy; **Cs** lobes as long as the **Anth**, ± circular or broadly elliptic, apically truncate, obtuse, emarginate or shortly obtusely 2-lobed, often irregularly toothed; **Anth** pale yellow, entrance to the **Nec** ± circular, deeply sunken; **Poll** pale yellow, ± 0.45 × 0.3 mm, ± ovate.

Closely related to *S. arenosa* and *S. rufa*.

S. rufa Masson (Stapel. Nov., 16, t. 20, 1797). **T:** [lecto – icono]: l.c. t. 20, 1797. – **D:** RSA (Northern Cape, Western Cape).

≡ *Gonostemon rufus* (Masson) P. V. Heath (1992); **incl.** *Stapelia fissirostris* Jacquin (1806) ≡ *Stapelia rufa* var. *fissirostris* (Jacquin) A. C. White & B. Sloane (1937) ≡ *Gonostemon rufus* var. *fissirostris* (Jacquin) P. V. Heath (1993); **incl.** *Stapelia rufescens* Salm-Dyck (1834); **incl.** *Stapelia rufa* var. *attenuata* N. E. Brown (1909) ≡ *Gonostemon rufus* var. *attenuatus* (N. E. Brown) P. V. Heath (1993).

Stems tapering towards both ends, rarely slenderly cylindrical, 12 × 1 - 2.5 cm, in small diverging clumps, green, flushed with purple, 4- to 6-ribbed, pubescent; **L** rudiments 2.5 × 2 mm, persistent as white tips, stipular rudiments sometimes absent; **Inf** few- to many-flowered, irregularly scattered, ± sessile; **Ped** to 6 mm; **Sep** 4 × 2.5 mm; **Cl** red-brown or purple, ± 3.6 cm ∅; **Cl** tube 1 - 1.2 × 1 - 1.2 cm, broadly campanulate; **Cl** lobes ± 1.2 cm, often caudate, spreading, outside white-pubescent, inside glabrous, apically pubescent, margins curved outwards, delicately hairy; **Cn** weakly delicately pubescent, stalk ± 2 × 3.5 mm, apically broader; **Ci** orange (-brown), lobes ± 1.5 mm, ± square, ± acute or ± truncate, horizontally spreading, irregularly dentate, sometimes upcurved; **Cs** red-brown or ± black, lobes 1.5 - 2 mm, narrowly rectangular, dorsiventrally compressed, erect, apically shortly 2-parted or elongate, slightly curved outwards, dorsal bulge small, wing-like; **Anth** pale yellow; **Poll** yellow, 0.5 × 0.25 - 0.3 mm, ± ovate.

Closely related to *S. rubiginosa*, but also with similarities to *S. arenosa*.

S. schinzii A. Berger & Schlechter (Vierteljahresschr. Naturf. Ges. Zürich 53(4): 491, 1909). **T** [lecto]: Namibia, Hereroland (*Dinter* 450 [Z, PRE [photo]]). – **D:** Angola, Botswana, Namibia.

≡ *Gonostemon schinzii* (A. Berger & Schlechter) P. V. Heath (1992).

S. schinzii var. **angolensis** Kers (Bot. Not. 122: 176-177, 1969). **T:** Angola, Moçamedes Distr. (*Kers* 3441 [S]). – **D:** S Angola, Namibia.

≡ *Gonostemon schinzii* var. *angolensis* (Kers) P. V. Heath (1992).

Differs from var. *schinzii*: Stems creeping, smaller, more slender; **Tu** more obscure; **Cl** smaller.

S. schinzii var. **bergeriana** (Dinter) L. C. Leach (Excelsa Tax. Ser. 3: 74, 1985). **T** [lecto]: Namibia, Namaland (*Dinter* 2697 [SAM]). – **D:** Namibia.

≡ *Stapelia bergeriana* Dinter (1914) ≡ *Gonostemon schinzii* var. *bergerianus* (Dinter) P. V. Heath (1992).

Differs from var. *schinzii*: **Cl** smaller, without transverse wrinkles; **Cl** lobes conspicuously short-attenuate.

S. schinzii var. **schinzii** – **D:** Namibia, Botswana.
Incl. *Stapelia johni-lavrani* Halda (1998).

Stems ± 7.5 × 1.5 - 2 cm, in small ± lax clumps, green, often purple-spotted, young delicately roughened, later more smooth; **L** rudiments ± 5 mm, ovate, long-attenuate, delicately roughened, stipular rudiments yellowish; **Inf** 1-flowered (a second bud aborting); peduncle robust, glabrous; **Ped** ≤ 6 cm, delicately roughened; **Sep** ≤ 8 mm, ± glabrous; **Cl** ± greenish- or reddish-brown, ± 10 - 22 cm ∅, with delicate wrinkles; **Cl** lobes ± 3.75 - 8.8 cm, often caudate, outside ± glabrous, inside glabrous, margins densely hairy, clavate **Ha** purple, vibratile; **Cn** dark red or ± black; **Ci** lobes oblong ± pointed, sometimes ovate and mucronate, diverging; **Cs** lobes ± 8 mm, apical appendage 3-angled, erect, dorsal wings ± as long as the apical appendage, broad, narrowly ovate-pointed or rectangular, obtuse, irregularly dentate, free; **Poll** brownish, ± 1 × 0.5 - 0.6 mm, obliquely ± D-shaped; **Fr** glabrous.

Closely related to *S. pillansii*.

S. scitula L. C. Leach (SAJB 3(3): 174-175, 1984). **T:** RSA, Western Cape (*Leach & Bayer* 15845 [PRE, B, K, NBG, SRGH]). – **D:** RSA (Western Cape).

≡ *Gonostemon scitulus* (L. C. Leach) P. V. Heath (1992).

Stems young with ± flat sides, older ± slenderly cylindrical, to 7.5 × 0.5 cm, apically branching and forming small orderly clumps, pale green, later purple; **Tu** rounded, obscure with age; **L** rudiments 1 - 1.5 mm, acutely ovate, caducous, stipular rudiments cream-coloured; peduncle ± clump-like; **Fl** buds broadly ovate, ± acute; **Ped** 0.9 - 1.5 cm; **Sep** 5 × 1.35 mm, sometimes visible between the **Cl** lobes; **Cl** outside pale green, venation darker, inside magenta or dark chestnut-brown, (1.5-) 1.9 - 2.5 cm ∅, ± strongly curved outwards; **Cl** lobes (5-) 7 - 9 × ± 5 mm, outside delicately pubescent, basis inside inconspicuously broadly wrinkled, yellow, tips greenish-yellow, inner face densely hairy, **Ha** purple, long-pointed, ± appressed; **Ci** pale brown, lobes ± 1 mm, ± square, concave, spreading; **Cs** black-purple, lobes narrowly ovate, long-attenuate, erect, slightly recurved, dorsal wings prominent, erectly spreading, broadly 3-angled; **Poll** yellowish, 0.6 × 0.4 mm, ± D-shaped.

Closely related to *S. immelmaniae*.

S. similis N. E. Brown (BMI 1911: 358, 1911). **T:** Namibia (*Pearson* 6134 [K]). – **D:** Namibia, RSA (Northern Cape). **Fig. XLV.c**

≡ *Tridentea similis* (N. E. Brown) hort. (s.a.) (*nom. inval.*, Art. 29.1) ≡ *Gonostemon similis* (N. E. Brown) P. V. Heath (1992); **incl.** *Stapelia juttae* Dinter (1914) ≡ *Gonostemon similis* var. *juttae* (Dinter) P. V. Heath (1993); **incl.** *Stapelia portae-taurinae* Dinter & A. Berger (1914) ≡ *Gonostemon similis* var. *portae-taurinae* (Dinter) P. V. Heath (1993); **incl.** *Stapelia noachabibensis* C. A. Lückhoff (1938).

Stems tapering, to 12 × basal 1 - 1.5 cm, in compact clumps, grey-green, often flushed with purple, bluntly 4-ribbed; **L** rudiments delicate, persistent as white **Sp**; **Fl** from flowering stems to ≤ 2 × 0.5 cm, opening in sequence, flatly spread on the ground; **Ped** 3 - 5.5 (-8) cm, spreading or decumbent; **Cl** yellow-brown or ± black, 1.5 - 2.2 cm ∅, flat, rarely ± campanulate, centrally with a flat cup-shaped or tubular depression, opening thickened annulus-like; **Cl** lobes shortly pointed, outside delicately pubescent, inside glabrous; **Cn** brown to black; **Ci** lobes transversally rectangular, truncate, at times broadly pointed or rounded, diverging or extremely far outcurved; **Cs** lobes oblong or ovate, acutely pointed, apically obtuse or ± acute, to as long as the **Anth**, dorsally sometimes bulging; **Poll** orange, rarely brown, ± 0.45 × 0.3 mm, ± D-shaped.

Closely related to *S. kwebensis, S. longipedicellata* and *S. parvula*.

S. surrecta N. E. Brown (FC 4(1): 970, 1909). **T** [neo]: RSA, Western Cape (*Leach & Bayer* 16070 [PRE, BOL, K, SRGH, Z]). – **D:** RSA (Northern Cape, Western Cape).

≡ *Gonostemon surrectus* (N. E. Brown) P. V. Heath (1992); **incl.** *Stapelia surrecta* var. *primosii* C. A. Lückhoff (1937) ≡ *Gonostemon surrectus* var. *primosii* (C. A. Lückhoff) P. V. Heath (1993).

Stems in small diverging clumps, green, flushed with purple, ± 1.2 cm ∅, bluntly 4-ribbed, young stems with purple axillary buds, later aborting; **L** rudiments 2.5 × 2 mm, erectly spreading, persistent as whitish points, stipular rudiments cream-coloured or brownish; **Inf** apical; peduncle sometimes persistent, spirally twisted; **Fl** erect; **Ped** ≤ 3.5 cm, usually ± erect, at times pendent; **Sep** very variable in length even within a **Fl**; **Cl** yellowish to chestnut-brown, 3 - 4 cm ∅, ± funnel-shaped; **Cl** tube paler, short, narrow, 5-angled; **Cl** lobes ± 0.9 - 1.2 cm, spreading, outside ± glabrous or delicately pubescent, inside glabrous, smooth or obscurely wrinkled; stalk of the **Cn** ± 3 mm ∅ at the base; **Ci** lobes ± 2 mm, ± rectangular, concave, fleshy, apically long-pointed on both sides, sometimes bifid, spreading, deeply longitudinally grooved; **Cs** lobes ± 3.5 mm, apical appendage subulate, recurved, dorsal wings ≥ apical appendage, variable in shape; **Poll** ± 0.5 × 0.35 - 0.4 mm, ± ovoid; **Fr** 8 × 0.8 cm.

S. tsomoensis N. E. Brown (Gard. Chron., ser. nov. 18: 168, 1882). **T:** RSA, Eastern Cape (*Barkly* 32 [K]). – **D:** RSA (Eastern Cape).

≡ *Gonostemon tsomoensis* (N. E. Brown) P. V. Heath (1992); **incl.** *Stapelia tomentosa* hort. *ex* A. C. White & B. Sloane (1933).

Stems to 15 × ± 1 cm; **Ri** 4, slightly curved, glabrous or weakly pubescent; **Tu** inconspicuous; **L** rudiments ± 2 mm, erect, weakly pubescent; peduncle weakly pubescent; **Ped** ≤ 3 cm; **Sep** 6 - 7 mm, ± pubescent; **Cl** ± yellowish, to 7 cm ⌀; **Cl** lobes purple, ≤ 2.6 cm, oblong or broadly ovate, ± long-attenuate, strongly curved outwards, with weak whitish or yellowish (rarely greenish) wrinkles, tips darker, margins not or hardly rolled inwards, venation and outside of the margins of the **Cl** lobes thinly pubescent, margins of the **Cl** tube and **Cl** lobes basally hairy, **Ha** reddish, long, ± depressed, directed towards the tip, margins of the **Cl** lobes hairy, **Ha** whitish, clavate or simple, long, robust, ± flattened; **Cn** black-purple, stalk short; **Ci** lobes 4 - 5 mm, oblong, obtuse, ± spatulate, mucronate, rarely apically attenuate; **Cs** lobes ± 8 mm, apical appendage subulate, 3-angled, falcately spreading, dorsal wings 4 - 5 mm, narrow, 3-angled, free, widely spreading; **Poll** greenish-brown, 0.8 - 0.9 × 0.5 mm, ± D-shaped.

Belonging to the *S. hirsuta*-complex and closely related to *S. glabricaulis* and *S. peglerae*.

S. unicornis C. A. Lückhoff (S.A.G. 28: 225-228, 1938). **T** [neo]: Swaziland (*Lückhoff* s.n. [BOL]). – **D:** Moçambique, RSA (KwaZulu-Natal), Swaziland.

≡ *Gonostemon unicornis* (C. A. Lückhoff) P. V. Heath (1992).

Stems creeping, decumbent-ascending, forming clumps or sometimes freely creeping, pale green, to 15 × 1 - 1.5 cm, 4-ribbed, glabrous; **L** rudiments 2.5 mm, persistent as white wrinkled tips; **Ped** ≤ 3 cm; **Sep** 5 - 9 × 2 - 2.5 mm; **Cl** ochre-yellow, 8 - 13 cm ⌀, wrinkles delicately red-brown or ± purple, fused part hypocrateriform or sometimes ± semiglobose; **Cl** lobes 2.8 - ± 4.8 cm, spreading, curved outwards, outside shortly pubescent, inside ± densely hairy, (clavate) **Ha** pale purple, long, silk-like, margins of the **Cl** lobes hairy, clavate **Ha** purple; **Cn** ± dark purple, ± sessile; **Ci** lobes 5 - 6 mm, linearly long-attenuate, concave, erectly diverging, revolute; **Cs** lobes 8 - 9 mm, erect, touching each other, dorsal wings 3 - 4 mm broad, fused with the apical appendage, ± equally long; opening to the nectar cavity transversely elliptic; **Poll** yellowish, 0.8 × 0.6 mm, ± D-shaped.

Closely related to *S. gigantea*.

S. vetula Masson (Stapel. Nov., 15, t. 16, 1797). **T:** K. – **D:** RSA (Western Cape). **Fig. XLV.d**

≡ *Tridentea vetula* (Masson) Haworth (1812) ≡ *Gonostemon vetulus* (Masson) P. V. Heath (1992); **incl.** *Tridentea simsii* Haworth (1812) ≡ *Stapelia simsii* (Haworth) Schultes (1820) ≡ *Stapelia vetula* var. *simsii* (Haworth) N. E. Brown (1909); **incl.** *Stapelia nudiflora* Pillans (1928).

Stems to 20 × 1 - 1.5 cm, clump-forming, green, flushed with purple, glabrous, 4-ribbed; **L** rudiments caducous, leaving a conspicuous whitish scar, stipular rudiments quickly caducous; peduncle very short, robust, **Ped** 1 - 2 cm; **Sep** 4 - 8 mm, weakly pubescent; **Cl** red or black-purple, 3 - 6 cm ⌀, sometimes with obscure pattern consisting of weak yellowish broadly spaced wrinkles, centre slightly depressed, frequently strongly curved outwards; **Cl** lobes 1 - 2.3 cm, outside glabrous, venation and the ± outwardly curved margins sometimes softly hairy, inside ± glabrous; **Cn** shortly stipitate; **Ci** lobes 3.5 - 6 × 1.5 - 4.5 mm, rectangular or ± spatulate, spreading, apically obtuse or truncate, with a median mucro, crenate or toothed; apical appendage of the **Cs** 6 - 8 mm, filiform, slightly recurved, dorsal wings 1 - 4.5 mm broad, erect or spreading; **Poll** 0.7 - 0.8 × 0.45 - 0.5 mm, D-shaped, apically ± acute; **Fr** glossy, glabrous.

S. villetiae C. A. Lückhoff (S.A.G. 28: 228, 1938). **T** [neo]: RSA, Northern Cape (*Stayner* s.n. [NBG]). – **D:** RSA (Northern Cape).

≡ *Gonostemon villetiae* (C. A. Lückhoff) P. V. Heath (1992).

Stems shortly decumbent-erect, squat, to 16 × ± 1.5 cm, dark green, sometimes purple, often weakly glossy, bluntly 4-ribbed, hairy; **L** rudiments slightly spreading; **Inf** 1- or few-flowered, directed upwards; **Ped** densely white-blotched, ≤ 6.5 × 0.4 cm, ± robust; **Sep** 6 - 8 mm; **Cl** outside pale green, finely stippled with purple, lobes brown, **Cl** inside dark flesh-coloured, to 6.5 cm ⌀, shallowly campanulate, wrinkles whitish or (lemon-) yellow; **Cl** tube shortly ± bowl-shaped; **Cl** lobes apically ± black, rarely greenish, 1.6 - 2.3 cm, shortly pointed, spreading, curved outwards, outside weakly pubescent, inside glabrous, apically and marginally at times delicately roughened, margins hairy, **Ha** reddish or magenta, simple; **Cn** black- or reddish-purple; **Ci** lobes 3.5 × 2 mm, rectangular-spatulate, spreading, slightly recurved, with flat longitudinal groove, apically obtuse or truncate, with short mucro or delicately crenate, placed on the **Cl**; **Cs** lobes ± 5 mm, apical appendage slightly falcate, ± 3-angled, erect, touching each other, dorsal wings ± 2 mm, ± 3-angled, spreading; opening to the nectar cavity ± elliptic, basally truncate; **Poll** glossy chestnut-brown or usually orange, ± 0.7 × 0.5 mm, ± compressed-ovate.

Closely related to *S. cedrimontana*.

STAPELIANTHUS

U. Meve

Stapelianthus Choux *ex* A. C. White & B. Sloane (Stapelieae [ed. 1], 71, 1933). **T:** *Stapeliopsis madagascariensis* Choux. – **Lit:** Morat (1994); Morat (1995). **D:** Madagascar. **Etym:** For the genus *Stapelia* (*Asclepiadaceae*); and Gr. 'anthos', flower.

Incl. *Stapeliopsis* Choux (1931) (*nom. illeg.*, Art. 53.1). **T:** *Stapeliopsis madagascariensis* Choux.

Perennials forming lax to dense groups or clumps; stems cylindrical, 2 - 30 × 0.5 - 1.2 cm, decumbent to ascending, basally branched, surface often tessellated or grooved; **Tu** in 4 - 8 **Ri** or scattered in large numbers over the whole stem; **L** rudiments obtusely triangular to subulate, caducous; **Inf** near the stem bases, few-flowered; **Fl** opening in succession; **Ped** 2 - 12 × 1 - 2 mm, glabrous; **Sep** triangular to subulate, glabrous; **Cl** reddish-brown to purple, sometimes spotted with yellow, 1 - 3.5 cm ∅, often campanulate, cylindrical or urceolate, sometimes with a glossy annulus, coarse and fleshy, variably covered with papillae or thick **Ha**; **Cl** tube round or pentagonal; **Cl** lobes (broadly) triangular, with short intermediate lobes; **Cn** biseriate, dark red or purple; **Ci** large, ± free (united for ½ only in *S. pilosus*), lobes ± erect, ovate, apically incised; **Cs** small, triangular, decumbent on and normally shorter than the **Anth**; **Sty** head small, concave; **Anth** narrowly rectangular, incumbent on the **Sty** head; **Poll** ± ovoid, ± 0.4 - 0.7 × 0.25 - 0.45 mm, caudicles apically broadened, basally connected to the elliptic-ovate corpuscle with short wing-like projections; **Fr** (narrowly) fusiform.

Characters of stem and flower morphology indicate the close relationship to *Huernia*, and an intergeneric hybrid is also known (see ×*Huernianthus*). As to flower morphology, *Tavaresia* agrees best with *Stapelianthus*.

S. arenarius Bosser & P. Morat (Adansonia, n.s., 11(2): 340-341, ills., 1971). **T:** Madagascar (*Bosser & Morat* 20392 [P]). – **D:** SE Madagascar.

Incl. *Stapelianthus calcarophilus* P. Morat (1990).

Stems 5 - 15 × 0.5 - 1 cm, decumbent-creeping, green-brown tinged with purple, surface tessellated, rough, 4-ribbed with spreading **Tu**; **L** rudiments lanceolate, reflexed, 1 - 2 mm; **Inf** 1- to 3-flowered; **Bra** 1 - 2 mm, triangular; **Ped** 5 - 10 mm; **Sep** 3 - 4 × 1.5 mm, acute; **Cl** outside grey-green spotted with purple, inside white to ivory spotted with red-brown, centre 2 - 2.5 cm ∅, rotate, delicately papillose, papillae white, short, stiff; **Cl** tube broadly funnel-shaped, ± 8 mm ∅; **Cl** lobes 10 - 12 × 4.5 - 5 mm, triangular, acute, horizontally spreading or completely reflexed (*S. calcarophilus*); **Cn** red-brown, pentagonal; **Ci** ± erect, deeply divided into 2 divergent subulate small teeth ± 2 mm long; **Cs** lobes narrowly triangular, ± 1 mm, decumbent on the **Anth**; **Poll** ± 0.7 × 0.4 mm.

The large white flowers are only rarely seen as the species is rather shy-flowering under greenhouse conditions.

Morat distinguishes *S. calcarophilus* from *S. arenarius* on the base of the strongly reflexed corolla lobes and the somewhat more delicate growth. True morphological differences are absent, however, and *S. calcarophilus* (on limestone) can at best be regarded as an ecotype of *S. arenarius* (on sand).

S. decaryi Choux (Ann. Inst. Bot.-Géol. Colon. Marseille 42(3): 13-14, t. 1-5, 1934). **T:** Madagascar (*Decary* s.n. [P]). – **D:** S Madagascar. **Fig. XLV.e**

Clump-forming, usually on rocks; stems decumbent-erect, 2.5 - 10 × 0.5 - 1.5 cm, grey-green spotted with red-brown, smooth, 6-ribbed; **Tu** flattened conical, suddenly constricted into the subulate spreading **L** rudiments, these 2 - 4 mm, persistent as dry whitish **Sp**; **Inf** 1- to 4-flowered; **Ped** 6 - 8 × 2 mm; **Sep** 4 - 7 mm, lanceolate, subulate, acute; **Cl** outside pale brown spotted with red-brown, inside cream-coloured with purple flecks, 12 - 16 mm ∅, upper ½ with elongate spine-tipped papillae; **Cl** tube ± 10 - 15 × 7 - 9 mm, cylindrical; **Cl** lobes 7 - 8 × 6 - 8 mm, triangular, acute, reflexed, intermediate lobes strongly developed; **Cn** dark purple, 5 - 6 × 3 - 4 mm, fleshy, stipitate (± 1 mm); **Ci** ± erect, 4 - 5 mm, lobes deeply divided into 2 erect or divaricate small teeth; **Cs** lobes triangular, obtuse, short; **Poll** ± 0.5 × 0.33 mm, wings of the translator ± 0.2 mm long, narrow, obtuse; **Fr** ± 10 × 0.8 cm, narrowly fusiform, acute.

S. hardyi Lavranos (Nation. Cact. Succ. J. 26(3): 67-68, ills., 1971). **T:** Madagascar (*Hardy & Jacobsen* 3569 [P, FT, K, PRE]). – **D:** SW Madagascar. **Fig. XLV.f**

Stems 2 - 7.5 × 0.6 - 0.8 cm, procumbent, creeping, greenish-red-brown, surface tessellated, uneven, 4- (to 6-) ribbed with spreading **Tu**; **L** rudiments lanceolate, reflexed, 1 - 2 mm; **Inf** 1- to 3-flowered; **Bra** ± 2 mm, lanceolate; **Ped** 4 - 10 × 1 - 2 mm; **Sep** 3 - 4 mm, triangular, acute; **Cl** outside red-brown with green to red-brown spots, inside dark purple, basally also yellow, ± 1.3 cm ∅, very coarse and fleshy, upper ½ densely hairy, **Ha** dark purple, to 3 mm, conical, coarse, stiff, rough; **Cl** tube campanulate, 5 - 10 × 9 - 15 mm; **Cl** lobes 7 - 9 × 8 - 9 mm, triangular, acute, erect to halfway reflexed; **Cn** dark purple, 5 × 5 mm, very shortly stipitate; **Ci** erect, ovate, ± 5 × 3 mm, outside concave, inside convex, apically divided into 2 triangular small teeth, centrally often with a short third small tooth; **Cs** lobes narrowly triangular, ≤ 1 mm; **Anth** slender-rectangular, incumbent on the concave **Sty** head; **Poll** ± 0.7 × 0.5 mm, wings of the translator slender, acute, ± 0.3 mm.

S. hardyi forms together with *S. madagascariensis* and *S. montagnacii* a complex, which is difficult to resolve.

S. insignis Descoings (Naturaliste Malgache 9(2): 179-182, pl. 7A, 1957). **T:** Madagascar, Toliara (*Descoings* 2751 [TAN]). – **D:** SW Madagascar.

Incl. *Stapelianthus insignis* ssp. *tongoboryensis* Rauh (1993).

Stems 3 - 10 × 0.5 - 1.2 cm, procumbent-creeping, grey-green mottled with red-brown, surface tessellated, uneven, 4- (to 5-) ribbed with broad flattened **Tu**; **L** rudiments lanceolate, reflexed, 1 - 2 mm; **Inf** 1- to 3-flowered; **Ped** ± 5 × 1.5 mm; **Sep** 5 - 6 mm, lanceolate-subulate; **Cl** outside pale green, spotted with red-brown, inside yellowish, spotted with purple, ± 2 cm ⌀; **Cl** tube doubly campanulate, lower part cup-shaped, pentagonal, ± 5 mm deep, upper part broad and shallowly urceolate, circular, sometimes pentagonal, 5 - 7 × 15 - 20 mm, apically suddenly contracted and forming a circular opening 8 - 12 mm ⌀; **Cl** lobes 2 × 3 - 5 mm, triangular, acute, ± reflexed, papillose; **Cn** dark red-brown, pentagonal, ± 6 × 5 mm, sessile or shortly stipitate; **Ci** erect to slightly divergent, ovate, ± 5 × 3 mm, outside concave, inside convex, apically divided into 2 obtuse triangular small teeth; **Cs** lobes narrowly triangular, ≤ 1 mm; **Anth** narrowly rectangular, incumbent over the concave **Sty** head; **Poll** ± 0.6 × 0.35 mm, wings of the translators slender, ± 0.35 mm long.

A very distinct but variable species. Rauh's ssp. *tongoboryensis* is not differing significantly from Descoings's type. Each individual has its own distinct shape and colour of the corolla tube.

S. keraudreniae Bosser & P. Morat (Adansonia, n.s., 11(2): 337, 1971). **T:** Madagascar, Betioky Distr. (*Bosser & Morat* 19413 [P]). – **D:** SW Madagascar. **Fig. XLV.g**

Stems 2 - 12 × 0.4 - 1 cm, procumbent-creeping, grey-green patterned with red-brown, surface tessellated, uneven, 4- (to 6-) ribbed with spreading **Tu** which are abruptly changing into **L** rudiments, these narrowly lanceolate, reflexed, 1 - 2 mm; **Inf** 1- to 3-flowered; **Bra** triangular, acute, ± 1 mm; **Ped** 3 - 6 × 1.5 mm; **Sep** 2 - 5 mm, lanceolate, acute; **Cl** outside greenish, often speckled with red-brown, 2.2 - 3.2 cm ⌀, papillose; **Cl** tube inside pale red to purple, centrally openly campanulate, pentagonal, 2 - 3 × 4 - 7 mm, then merging into a massively bulging glossy annulus of 10 - 16 mm ⌀; **Cl** lobes yellowish-green with red-brown spots, 7 - 11 × 8 - 12 mm, margins occasionally shortly hairy, intermediate lobes conspicuous; **Cn** dark purple, 3 - 4.5 × 4 mm, sessile or very shortly stipitate, **Ci** ± erect, ovate, ± 3 - 4 × 2 mm, slightly convex, apically divided for max. ½ into 2 small teeth; **Cs** lobes narrowly triangular, ± 1 mm; **Poll** ± 0.5 × 0.4 mm, wings of the translators narrowly triangular, ≤ 0.2 mm.

A conspicuous species. The flowers with the massive annulus are reminiscent of those of certain huernias, e.g. *H. zebrina*.

S. madagascariensis (Choux) Choux (Ann. Inst. Bot.-Géol. Colon. Marseille 42(3): 6-7, 1934). **T:** Madagascar (*Decary* s.n. [P ?]). – **D:** S Madagascar. **Fig. XLV.h**

≡ *Stapeliopsis madagascariensis* Choux (1931).

Stems 2 - 10 × 0.3 - 0.8 cm, procumbent-creeping, green-grey patterned with red-brown, surface tessellated, uneven, 4- (to 6-) ribbed with spreading **Tu**; **L** rudiments narrowly lanceolate, reflexed, 1 - 2 mm; **Inf** 1- to 5-flowered; **Bra** triangular, acute, 1 mm; **Ped** ± 5 × 1.5 mm; **Sep** 2 - 5 mm, triangular (-subulate), acute; **Cl** outside pale green, often dotted with red-brown, inside cream-coloured with large wine-red spots, centripetally becoming paler, 1.4 - 2.2 cm ⌀, delicate, hairy, **Ha** white, to 2 mm, cylindrical, coarse, rough, apically thickened and spine-tipped, purple; **Cl** tube campanulate, 3 - 5 × 6 - 8 mm, basally white; **Cl** lobes 6 - 9 × 5 - 6 mm; **Cn** dark purple, 3 - 3.5 × 2 - 3 mm, sessile; **Ci** erect to slightly spreading, ± 3 × 2 mm, ovate, apical ½ divided into 2 small teeth; **Cs** lobes narrowly triangular, ± 1 mm; **Poll** 0.4 - 0.5 × 0.35 mm, wings of the translators ovate, ± 0.25 mm.

See *S. montagnacii* for a comment.

S. montagnacii (Boiteau) Boiteau & Bertrand (Cactus (Paris) No. 26: 116, 1950). – **D:** SW Madagascar.

≡ *Stapelia montagnacii* Boiteau (1941).

Similar to *S. madagascariensis*, but stems to 1.2 cm ⌀; **Cl** inside (spotted) brick-red on withish background, centrally white, 18 - 25 mm ⌀, hairy, **Ha** to 3 mm; **Cl** tube campanulate, ± 5 - 6 × 8 - 11 mm ⌀; **Cl** lobes 8 - 11 × 5 - 6 mm; **Cn** ± 4 × 4 mm; **Ci** lobes ± 4 × 2.5 mm.

S. montagnacii is very closely related to or even conspecific with *S. madagascariensis*. The somewhat larger stems and flower parts, the deeper corolla tube and the predominantly red flower colour suggest that this taxon could represent the natural hybrid *S. madagascariensis* × *S. hardyi*.

S. pilosus Lavranos & D. S. Hardy (JSAB 27: 237-239, ills., 1961). **T:** Madagascar (*Decary* 4549 [P]). – **D:** S Madagascar. **Fig. XLVI.d**

Incl. *Trichocaulon decaryi* Choux (1932).

Stems 2.5 - 20 × 0.8 - 1.2 cm, procumbent, only slightly ascending, fresh green, smooth, densely covered with conical **Tu**, which pass into subulate-filiform **L** rudiments, these 2 - 4 mm, persistent as dry whitish **Sp**; **Inf** 1- to 4-flowered; **Ped** 5 - 7 × 1.5 mm; **Sep** 4 - 5 mm, lanceolate-subulate; **Cl** campanulate, 1 - 1.8 cm ⌀, both sides yellowish, spotted with red-brown, inside densely short-hairy, **Ha** cream-coloured or purple, cylindrical, rough, spine-tipped, 0.4 - 0.8 mm; **Cl** tube ± 4 × 4 mm; **Cl** lobes 4 - 7 × 4 - 7 mm, triangular, acute, reflexed, intermediate lobes hardly conspicuous; **Cn** dark red-brown, ± 4 × 3.5 mm, sessile or shortly stipitate; **Ci** ± erect with reflexed tips, basally fused into a cup-like structure, free parts of the **Ci** lobes ± deeply divided into 2 parallel small teeth; **Cs** lobes triangular, short; **Poll** ± 0.37 × 0.23 mm, wings of the translators ≤ 0.1 mm, triangular, obtuse.

The striking stems of *S. pilosus* with the numerous irregular tubercles and the long-attenuate leaf rudiments are very reminiscent of *Huernia pillansii* from RSA.

STAPELIOPSIS

B. Müller & F. Albers

Stapeliopsis Pillans (South Afr. Gard. 18: 32, 1928). **T:** *Stapeliopsis neronis* Pillans. – **Lit:** Bruyns (1981: 73-83); Bruyns (1990b). **D:** Namibia, RSA. **Etym:** Gr. '-opsis', similar to; and for the genus *Stapelia* (*Asclepiadaceae*).

Perennials; stems erect or creeping, usually with numerous subterranean runners (more slender than aerial stems and without distinct angles), richly branched, blue-green, spotted with purple, or grey-green, rarely brownish, conspicuously 4-angled, glabrous (sometimes papillose); **Tu** rounded, dome-shaped; **L** rudiments triangular, acute, hard; **Inf** 1- to 3-flowered from near the stem base, **Fl** opening successively; peduncle 2.5 cm, irregularly shaped; **Ped** 3 - 7 (-10) mm, slender, usually glabrous; **Fl** normally erect; **Sep** 2 - 4 × ± 1 mm, (ovate-) lanceolate, acute, usually glabrous; **Cl** urceolate or ± semiglobose to cylindrical, length and $\emptyset$ variable; **Cl** lobes often very small in comparison to the tube, sometimes tips united, outside normally glabrous and smooth, inside densely papillose, papillae prominent, smaller at the base of the tube and there with apical stiff **Ha**, or without papillae and then stiffly hairy with mucronate **Ha**; **Cn** of the **C(is)**+**Cs**-type, sometimes shortly stipitate; **Ci** lobes either small, truncate, sometimes bifid, just hiding the **Nec** entrance at the base of the **Cs** lobes, or fused into a tubular structure, enveloping the basal ½ of the **Gy**, free parts as small indentations or teeth at the tube mouth; **Cs** lobes laterally flattened, tips rounded, either fused laterally with the **Ci** or fused dorsally to the base of the **Ci** tube, incumbent on the **Anth**, apically ascending, usually meeting in the centre; **Anth** simple, rectangular, apically acute; **Poll** 0.25 - 0.32 × 0.2 - 0.25 mm, ovoid-elliptic, rounded or bean-shaped; **Fr** narrowly fusiform, divaricate at an acute angle; **Se** brownish, ovate, flat, margin papillate, with an apical **Ha** tuft.

Differences from *Pectinaria* are to be found in the urceolate corolla as well as in the presence of stiff hairs. The genus is divided into sections as follows:

[1] Sect. *Stapeliopsis*: Stems blue-green, spotted with purple, papillose, papillae transparent, cylindrical, rounded; **Ci** lobes forming a tube encircling the **Gy** (S Namibia, N RSA [Richtersveld]).

[2] Sect. *Cageliorona* Bruyns 1981: Stems grey-green or brownish, not papillose; **Ci** lobes not forming a tube, small, ascending, sometimes bifid, just hiding the **Nec** entrance (RSA: Western Cape, esp. near Laingsburg, Worcester and Oudtshoorn).

S. breviloba (R. A. Dyer) Bruyns (Cact. Succ. J. Gr. Brit. 43(2/3): 81-82, 1981). **T:** RSA, Western Cape (*Breda* 183 [PRE]). – **D:** RSA (Western Cape). **Fig. XLVI.e**

≡ *Pectinaria breviloba* R. A. Dyer (1954); **incl.** *Pectinaria villetii* C. A. Lückhoff (1952) (*nom. inval.*, Art. 32.1c).

[2] Stems erect, densely clustered, to 5 cm on stony ground, to 10 cm on sandy ground and then with subterranean runners; stems grey-brownish, 5 - 7 (-9) mm $\emptyset$, angles obtuse, ± prominent; **Tu** 1.5 - 2 mm, acute; **L** rudiments delicate, ovate; **Inf** 1- to 2- (rarely to 4-) flowered, often partly immersed into the ground, sometimes with underground origin; **Ped** 3 - 7 mm; **Sep** ± 2.5 mm; **Cl** outside grey-brown, transparently and densely delicately longitudinally veined, inside dark purple-red to chestnut-brown, 1 - 3 × 0.5 - 0.6 cm, tubular to ellipsoid-globose; **Cl** tube 8 - 10 mm, cylindrical, lower ½ weakly hairy with long **Ha**; **Cl** lobes yellow, 2.5 - 4 × 2.5 mm, triangular, tips united, outside glabrous, smooth, inside basally sometimes delicately tuberculate-papillose, papillae weakly prominent, densely and delicately mucronate; **Cn** shortly stipitate; **Ci** chestnut-brown, lobes 0.8 × 1 mm, forming shallow pouches, tips emarginate; **Cs** lobes 3 × 1 mm, wing-like or with a dorsal bulge, inclined; **Poll** 0.3 × 0.2 mm, bean-shaped.

S. exasperata (Bruyns) Bruyns (Cact. Succ. J. Gr. Brit. 43(2/3): 82-83, ills., 1981). **T:** RSA, Northern Cape (*Bruyns* 1345 [NBG]). – **D:** RSA (Northern Cape: Calvinia Distr.). **Fig. XLVI.a**

≡ *Pectinaria exasperata* Bruyns (1978).

[2] Stems erect, often tapering towards the tips, ± pyramidal, sometimes in dense clumps or with underground runners, dull grey-green, 1 - 4 × 0.5 - 1.5 cm; **Tu** shortly conical; **L** rudiments small, broad; **Inf** originating partly or completely underground; peduncle ≤ 1 cm; **Ped** 2 - 5 (-10) mm, elongating to 5 cm at **Fr** time; **Sep** 3 × 1 mm; **Cl** pale rose-grey, 1.8 - 2 cm; **Cl** tube basally deep purple-red, 10 - 12 × 4 - 7 mm, cylindrical-ovoid; **Cl** lobes 7 - 8 × 3 mm, lanceolate, basally with 1 mm long intermediate lobes, reclining, base of the tips weakly papillose on the inside, **Cl** tube densely papillose, papillae large, mucronate, mucro semiglobose, papillae of the tube base hairy, **Ha** ≤ 1 mm; **Cn** dark purple-red; **Ci** lobes small, bifid; **Cs** lobes 2 mm, dorsal bulge 1 mm; **Poll** 0.26 × 0.2 mm, very weakly bean-shaped, germination mouth long; **Fr** pale yellow, purple-striped, 10 - 12 cm; **Se** 4 × 2 mm.

S. neronis Pillans (South Afr. Gard. 18: 32, 1928). **T:** RSA, Northern Cape (*Pillans* 5728 [BOL]). – **D:** S Namibia, RSA (Northern Cape). **Fig. XLVI.f**

[1] Stems erect or ascending, in dense clusters to ± 30 cm $\emptyset$, usually without underground runners; stems ≤ 7 × 3.5 cm, very delicately papillose, angles blunt, hardly compressed, sides slightly convex (young concave); **L** rudiments 2 - 3 mm, broad; **Inf**

1- to 3-flowered; **Ped** 5 - 7 mm, erectly spreading, velvety; **Sep** 4 mm, delicately velvety; **Cl** outside and inside dark purple, 1.7 - 2 × ≤ 1.3 cm, fleshy; **Cl** tube 1.5 - 1.75 cm ⌀, cylindrical to globose, mouth distinctly thickened, basally with 5 longitudinal grooves; **Cl** lobes inside white or pale purple, 4 - 5 × 3.5 - 4 mm, ovate-deltoid, acute, stiff, erect, alternating with outwards-directed folds, outside papillose, papillae acutely tapering and resulting in a velvet-like appearance of the **Fl**, inside densely papillose, papillae apically hairy, **Ha** purple, ± short, simple; **Cn** sometimes with a short massive base; **Ci** purple, lobes united to form an erect tube, ≤ 1 × 1.75 mm, apically with small segments, obtuse or acute, reclining, outside ± smooth, basal ½ delicately hairy, **Ha** purple, simple, inside basally papillose; **Cs** lobes 5 mm, ascending, slightly curved, entire, laterally compressed, subulate, acute, reaching the tips of the **Ci**, united with the **Ci** for ⅓ with a wing; **Poll** pale, 0.15 × 0.1 mm, ovoid, with short translator arms, ascending, delicately furrowed.

S. pillansii (N. E. Brown) Bruyns (Cact. Succ. J. Gr. Brit. 43(2/3): 80-81, ills., 1981). **T:** RSA, Eastern Cape (*Pillans* 180 [BOL, K]). – **D:** RSA (Eastern Cape).

≡ *Pectinaria pillansii* N. E. Brown (1909).

[2] Stems decumbent, sometimes underground, only very rarely forming colonies, dark green spotted with reddish, glossy, to 15 × 1.5 - 2 cm, acutely 4-angled; **Tu** rather conical and curved downwards; **L** rudiments 3 - 4 mm; **Inf** with ≥ 3 **Fl**, originating from the lateral furrows, sometimes underground, oriented parallel to the ground or weakly bent downwards; **Ped** 2 - 3 mm; **Sep** ± 2 mm, ovate; **Cl** inside pale purple, 7 mm ⌀, ± ovoid or slightly conical, fleshy, stiff; **Cl** tube 3 mm, cup-shaped, walls > 1 mm thick; **Cl** lobes 4 - 5 × 4 - 5 mm, deltoid, tips sometimes united and **Fl** then pear-shaped, or tips spreading; **Cl** lobes basally alternating with 2 - 2.5 mm broad folds, outside smooth, glabrous, inside hairy, **Ha** sometimes apically thickened, spine-tips ± globose; **Ci** lobes ± 0.5 mm, broadly ovate, obtuse, spreading; **Cs** lobes 2.5 mm, dorsal bulge rounded, margin acute, apically erect, raised far above the **Anth**; **Poll** 0.26 × 0.16 mm, bean-shaped.

S. saxatilis (N. E. Brown) Bruyns (Cact. Succ. J. Gr. Brit. 43(2/3): 77, ills. (p. 79), 1981). **T:** RSA, Western Cape (*Pillans* 115 [BOL]). – **D:** RSA. **I:** Eggli (1994: 178).

≡ *Pectinaria saxatilis* N. E. Brown (1909).

S. saxatilis ssp. **saxatilis** – **D:** RSA (Western Cape, Eastern Cape). **Fig. XLVI.b**
Incl. *Pectinaria tulipiflora* C. A. Lückhoff (1934).

[2] Stems in large colonies, decumbent, frequently subterranean (young mostly above-ground), blue- to dull green to grey, sometimes indistinctly purple-spotted, 3.5 - 10 × 0.8 - 2.5 cm, sides flat or slightly concave, angles slightly flattened, acute; **L** rudiments ≤ 8 mm, ± broad, horizontally spreading, becoming hard-spiny; **Inf** 2- to 8-flowered, erect; **Ped** ≤ 1 cm; **Sep** tips curved outwards; **Cl** dark purple to pale rose or reddish, ≤ 1.6 × 1 cm, shortly ovoid to oblong-ellipsoid or weakly conical (then tips of the lobes free); **Cl** tube deeply ovoid to cup-shaped, sometimes conical, wall < 0.5 mm thick; **Cl** lobes 4 - 8 × 5 mm, ovate-lanceolate to acute, margins sometimes revolute, tips usually united, basally rarely alternating with delicate teeth, outside glabrous, inside mucronately hairy resulting in a frosted appearance of the **Fl**, **Ha** white, straight, rarely from small ± conical papillae; **Cn** pale yellow to dark purple; **Ci** lobes very small, 0.5 mm, broadly ovate or triangular, truncate, porrect; **Cs** dark purple-brown or yellow, lobes 2 - 3.5 mm, linear to bluntly acute, laterally flattened, dorsal bulge rounded (occasionally gladiate), comb-like; **Anth** apically with few transparent stiff **Ha**; **Poll** ± 0.25 mm ⌀, round to weakly bean-shaped.

S. saxatilis ssp. **stayneri** (M. B. Bayer) Bruyns (Cact. Succ. J. Gr. Brit. 43(2/3): 79-80, ills., 1981). **T:** RSA, Western Cape (*Stayner* s.n. [NBG]). – **D:** RSA (Western Cape).

≡ *Pectinaria stayneri* M. B. Bayer (1975).

[2] Differs from ssp. *saxatilis*: Stems 1 cm ⌀; **Cl** pale pink, whitish near the base, 6 mm ⌀, ± campanulate; **Cl** tube 4 - 6 × 4 mm, inside basally with small straight **Ha**; **Cl** lobes 4 - 6 mm; **Cn** pale yellow; **Cs** 2.8 - 3 × 0.5 mm, dorsal bulge rounded; **Poll** 0.26 × 0.18 mm, bean-shaped.

S. urniflora Lavranos (JSAB 32: 195, 1966). **T:** Namibia (*Lofty-Eaton* s.n. in *Lavranos* 2536 [PRE]). – **D:** S Namibia.

[1] Stems ascending or erect, partly subterranean, to 7 × 2 cm, delicately papillose; **Tu** obtuse, laterally compressed, **L** rudiments delicate, rapidly caducous; **Inf** 2- to 4-flowered; peduncle mostly parallel to the ground; **Bra** < 1 mm, deltoid; **Cl** outside and inside wine-red, 1 - 1.5 × ≤ 1 (basally, apically 0.5) cm; **Cl** tube ≤ 12 mm, urceolate, mouth not thickened, inside esp. near the base densely papillose, papillae apically with a long stiff simple **Ha**; **Cl** lobes 3 × 3 mm, triangular, erect, outside smooth, glossy, both sides glabrous; **Cn** sometimes with a short massive base; **Ci** 3 mm ⌀, cup-shaped, lower ½ enveloping the **Gy**, lobes bifid, segments short, deltoid, acute, erect or curved outwards; **Cs** lobes 2 mm, ascending-erect, tips rounded, not always meeting in the centre; **Poll** ± 0.25 mm ⌀, round to weakly bean-shaped.

STATHMOSTELMA

B. Müller, P. Stegemann & F. Albers

Stathmostelma K. Schumann (BJS 17: 129, 1893). **T:** *Stathmostelma gigantiflorum* K. Schumann. – **Lit:** Bullock (1953); Goyder (1998b). **D:** Tropical Africa. **Etym:** Gr. 'stathmos', plumb-line; and Gr. 'stelma', crown, garland, wreath; for the straight appendages of the inner corona segments.

Erect shrubs with **R** tubers; stems simple or branched, mostly hairy, with latex; **L** decussate, shortly petiolate, rarely sessile, lamina oblong, linear or lanceolate, apically usually acute, midrib often conspicuous, margins often rolled back, glabrous or pubescent; **Inf** extra-axillary or terminal, sometimes with **Bra**, these linear or subulate, usually hairy; peduncle usually pubescent; **Ped** glabrous or pubescent; **Sep** lanceolate, acute or attenuate, hairy, rarely glabrous, basally alternating with colleters; **Cl** white, yellow, purple, (orange-) red, often extremely large, flat, campanulate, sometimes inversely campanulate, tube only very short; **Cl** lobes oblong (-ovate), apically acute or obtuse, hairy or glabrous; **Cn** clearly stipitate; **Cs** lobes cucullate, margins curved inwards, forming a narrow pouch, margins apically each with a tooth-like appendage, sometimes with an inner appendage, this finger-like or hooked, protruding from the pouch; **Sty** head conspicuously grooved; **Anth** ovate or round, acute or obtuse, placed in the grooves of the **Sty** head; **Poll** reniform, flattened, translators large, broad, variable in shape, symmetrically hollowed, inside steeply erect, basally filiform, weak, then oblong or triangularly broadened; **Fr** solitary, erect, ± 10 cm, narrowly fusiform, apically beaked; **Se** brown, ovate or elliptic, papillate, margins inflated, wrinkled, apically with a tuft of **Ha**.

Closely related to *Asclepias* and *Gomphocarpus*, but distinguished by the conspicuously broad translators.

S. angustatum K. Schumann (BJS 17: 132, 1893). **T:** Ethiopia (*Schimper* 1589 [not located]). – **D:** Ethiopia, Uganda.

≡ *Asclepias angustata* (K. Schumann) N. E. Brown (1902); **incl.** *Gomphocarpus angustatus* Hochstetter (s.a.) (*nom. inval.*, Art. 32.1c).

S. angustatum ssp. **angustatum** – **D:** Ethiopia.

Stems 10.4 - 20.8 cm, basally weakly branched, biseriately pubescent; **L** 4 - 5 pairs, ascending, petiole 2.1 - 4.2 mm, lamina 4 - 18 × 0.1 - 0.9 cm, linear (-lanceolate), attenuate, basally narrower, margins flat or rolled back, glabrous, midrib sometimes very weakly rough-hairy; **Inf** 1 (-2), terminal; **Bra** presumably caducous; peduncle 1.5 - 8 cm, uniseriately hairy; **Ped** 1.5 - 3.5 cm, at **Fr** maturity 3.25 - 3.9 cm, pubescent; **Sep** 8.4 × 4.2 mm, ovate, glabrous; **Cl** flat, white, lobes 13 × 6.5 mm, oblong-ovate, acute; **Cs** lobes 8.4 mm, erect, oblong, apical appendages of the sides broad, triangular, lobes papillose, inside and outside below the appendages weakly pubescent; **Fr** 12 - 14 × 0.5 - 1 cm, lanceolate, apically beaked; **Se** brown, 4.2 × 3.1 mm, ovate.

S. angustatum ssp. **vomeriforme** (S. Moore) Goyder (KB 53(3): 603, 1998). **T:** Uganda (*Bagshawe* 1612 [BM]). – **D:** Uganda.

≡ *Asclepias vomeriformis* S. Moore (1908); **incl.** *Stathmostelma thomasii* Bullock (1953).

Differs from ssp. *angustatum*: **Cl** green or purple-brown; **Cl** lobes reflexed. – [U. Meve]

S. diversifolium Goyder (KB 53(3): 604-607, ills., 1998). **T:** Ethiopia (*Gilbert & Jefford* 4554 [K, ETH]). – **D:** Ethiopia, Kenya, Tanzania; open grassland.

Stems solitary, only basally branching, 6 - 30 cm, erect or ascending, pubescent; **L** 0.1 - 10 cm petiolate, lamina 4 - 10 × 0.1 - 3 cm, linear to oblong, pubescent; **Inf** extra-axillary, to 4-flowered; peduncle 0.5 - 2 (-6) cm, pubescent; **Ped** 1 - 3 cm, pubescent; **Sep** triangular, 2 - 7 × 1 - 2 mm, densely pubescent; **Cl** dull yellow-green to white or pink; **Cl** lobes recurved, 7 - 15 × 3 - 4 mm, ovate, acute; **Cn** whitish, spotted with purple; **Cs** lobes erect, concave-cucullate, 4 - 7 × 3 - 6 mm, inner apical margins drawn out into a pair of acute teeth, with an additional medianly sunken flattened 2-tipped minute tooth; **Gy** 1 - 3 mm stipitate; **Poll** reniform, 1 - 1.6 × 0.3 - 0.5 mm; **Fr** ovoid, 5 - 7 × 1.5 - 2 cm, dorsally with 1 or several rows of delicate obtuse protuberances, densely pubescent; **Se** medium brown, ovate, 4 × 2.5 mm. – [U. Meve]

S. fornicatum (N. E. Brown) Bullock (KB 8: 55, 1953). **T:** Malawi (*McClounie* 81 [K]). – **D:** Malawi, Zambia, Tanzania, Zimbabwe.

≡ *Asclepias fornicata* N. E. Brown (1906).

S. fornicatum ssp. **fornicatum** – **D:** Malawi, Zambia, Tanzania, Zimbabwe.

Stems apically curved, weakly uniseriately pubescent; **L** petiolate, lamina 8 - 20 × 0.3 - 1.3 cm, linear, acute, sometimes weakly pubescent; **Inf** with 3 - 4 **Fl**; peduncle 5 - 15 cm; **Ped** 3 - 4 cm; **Sep** 5 - 6 × 3 mm, glabrous; **Cl** flat to bowl-shaped, lobes 14 - 20 × 4 - 8 mm, oblong-elliptic, ± acute, glabrous; **Cs** lobes 7 mm, outside bulging and keeled, apical-marginal appendages oblong-falcate, overtopping the **Anth**, inside delicately papillose; **Poll** oblong, ± 1.1 × 0.35 mm; **Fr** fusiform, 11 × 0.6 cm; **Se** 5 × 3 mm, ovate.

S. fornicatum ssp. **tridentatum** Goyder (KB 53(3): 593-595, ills., 1998). **T:** Tanzania (*Brummitt & Polhill* 13667C [K, EA]). – **D:** Tanzania, between grasses in *Brachystegia* woodland.

Differs from ssp. *fornicatum*: **Cn** with a ± 2 mm long tooth-like inner appendage. – [U. Meve]

S. gigantiflorum K. Schumann (BJS 17: 129, 1893). **T:** Tanzania (*Böhm* 24 [lecto – icono: l.c. 17: t. 6: fig. A-C]). – **D:** Tanzania; seasonally inundated meadows.

≡ *Asclepias gigantiflora* (K. Schumann) N. E. Brown (1902); **incl.** *Stathmostelma gigantiflorum* Schlechter (1895) (*nom. illeg.*, Art. 53.1); **incl.** *Stathmostelma bicolor* K. Schumann (1900); **incl.** *Asclepias mihundensis* N. E. Brown (1902); **incl.** *Stathmostelma nomadacridum* Bullock (1953); **incl.** *Stathmostelma praetermissum* Bullock (1953).

Stems 30 - 100 cm, basally branching, compressed, glabrous, apically weakly pubescent; **L** obscurely petiolate, lamina 15 - 25 × 1 cm, oblong-lanceolate or linear, acute, margins and midvein pubescent; **Inf** terminal, 2-flowered; peduncle 2.5 - 18 cm; **Bra** 1.5 cm, linear, pointed, margins ciliate; **Ped** 2 - 5 cm; **Sep** 1.6 cm, ovate, acute, basally rounded, glabrous; **Cl** mauve-coloured, ≤ 6 cm ∅, flat, lobes ± 2.8 × 1 cm, lanceolate, acute, glabrous; **Cn** broadly and robustly stipitate; **Cs** lobes 1.6 cm, ± rectangular, apical-marginal appendages tooth-like, incumbent on the **Anth**; **Gy** 11 - 12 mm; **Anth** 4 - 5 mm, in the grooves of the **Sty** head; **Poll** bean-shaped, translator arms 4.5 mm, appendages tapering to a thin filiform portion; **Fr** narrowly fusiform, 10 - 16 × 0.5 - 0.8 cm; **Se** 3 × 2 mm.

S. incarnatum K. Schumann (BJS 17: 130, 1893). **T** [lecto]: Angola (*Welwitsch* 4176 [K]). – **D:** Angola; seasonally inundated grassland.

Incl. *Asclepias coccinea* N. E. Brown (1902).

Stems 10 - 35 (-40) cm, solitary, slender, basally sometimes branched, basal ½ leafy, pubescent; **L** ascending, petiole 2 mm (only in the broader **L**), lamina (4-) 6 - 8 (-9) cm × 2 - 3 (-5) mm, narrow, rarely linear-lanceolate, acute, glabrous; **Inf** 1, terminal, with 5 - 10 **Fl**; peduncle 5 - 12 cm, weakly or uniseriately pubescent; **Bra** ± 3 mm, linear; **Ped** 2 - 2.5 cm, delicate, short-hairy; **Sep** 3 - 4 mm, outside pubescent; **Cl** blood-red, flat or convexly bulging, lobes 7 - 8 × 2.5 - 3 mm, oblong, acute, outside papillose; **Cs** lobes orange-red, 3 - 5 mm, far surpassing the **Sty** head, basally ± cordate, unguiculate, apical-marginal appendages subulate, tooth-like, sometimes with an inner appendage; **Gy** 2 mm; **Anth** elliptic, obtuse; **Poll** flattened, 0.8 × 0.3 mm, translators long, appendages tapering to a filiform part.

S. katangense (De Wildeman) Goyder (KB 53(3): 612-613, 1998). **T:** Zaïre (*Verdick* s.n. [BR]). – **D:** E Zaïre; seasonally inundated grassland.

≡ *Asclepias katangensis* De Wildeman (1904); **incl.** *Asclepias dewewrei* De Wildeman (1904); **incl.** *Asclepias extensa* S. Moore (1912).

Stems solitary, 2 - 6 cm, erect, pubescent; **L** 0 - 5 mm petiolate, lamina 7 - 14 × 0.2 - 0.9 cm, narrowly linear to lanceolate, acute, laxly pubescent; **Inf** terminal, 3- to 4-flowered; peduncle 5 - 25 mm, pubescent; **Ped** 0.7 - 4 cm, pubescent; **Sep** lanceolate, 3 - 9 × 1 - 3 mm, densely pubescent; **Cl** red, rotate, inside delicately papillose; **Cl** lobes 10 - 20 × 3 - 6 mm, oblong-lanceolate, acute, folded back along the midrib; **Cn** red to orange, sessile; **Cs** lobes erect, concave-cucullate, 14 - 20 × 3 mm, inner margins of the lower ½ drawn out into a pair of falcate 3 - 4 mm long erect-connivent minute teeth, upper ½ with an erect flat appendage with an attenuate twisted tip; **Poll** oblong, flat, 0.8 × 0.3 mm. – [U. Meve]

S. odoratum K. Schumann (BJS 28: 457, 1900). **T:** Tanzania (*Goetze* 498 [B†]). – **D:** Tanzania.

≡ *Asclepias odorata* (K. Schumann) N. E. Brown (1902).

Stems 50 cm, simple, ≤ 5 mm ∅, ± robust, apically tomentose; **L** petiole 3 - 15 mm, canaliculate and pubescent, lamina 8 - 19 × 2 - 5.5 cm, oblong (-lanceolate), acute, basally broadly cuneate or rounded, very weakly pubescent and hispid; **Inf** few- to many-flowered, extra-axillary; peduncle 1 - 3.5 cm, weakly hairy; **Ped** 3 - 3.5 cm, weakly tomentose; **Sep** 7 mm, outside tomentose; **Cl** flat, orange-red, lobes 16 mm, oblong, acute, weakly papillose; **Cn** shortly stipitate; **Cs** lobes 9 mm, hood-shaped, apical-marginal appendages acute, smallish tooth-like, inside with 2 decurrent papillose lines; **Gy** 3 mm; **Anth** broad; **Fr** brown, 9.5 - 11 cm, fusiform, papillose.

Probably a synonym of *S. spectabile* (Goyder 1998b).

S. pauciflorum K. Schumann (BJS 17: 132, 1893). **T:** Moçambique (*Petas* s.n. [K [iso]]). – **D:** Kenya, Tanzania, Zaïre, Zambia, Moçambique, Malawi, Zimbabwe; sandy partly inundated grassland, to 1500 m.

≡ *Asclepias pauciflora* (K. Schumann) E. A. Bruce (1940); **incl.** *Gomphocarpus pauciflorus* Klotzsch (1861) (*nom. illeg.*, Art. 53.1); **incl.** *Stathmostelma reflexum* Britten & Rendle (1894) ≡ *Asclepias reflexa* (Britten & Rendle) N. E. Brown (1902).

Stems 30 - 100 cm, biseriately pubescent; **L** 3 - 6 pairs, petiole 1 - 6 mm, lamina 2.6 - 16 cm × 1.5 - 8.4 mm, longer than the **Int**, (lanceolate-) linear, pointed, basally acute or rounded, glabrous, rarely weakly pubescent; **Inf** 1 - 3, terminal, with 4 - 6 **Fl**; peduncle 4.5 - 28 cm, uniseriately hairy; **Bra** small or none; **Ped** 0.8 - 4 cm, pubescent; **Sep** 4.2 mm, strongly recurved; **Cl** curved outwards, red or purple, lobes 6.3 - 8.4 × 3 - 4 mm, oblong-ovate, obtuse, apically curved inwards, glabrous; **Cs** lobes 8.4 - 9.4 mm, erect, oblong-linear or linear-lanceolate, acute or obtuse, grooved, apical-marginal appendages subulate, tooth-like, incumbent on the **Anth**, sometimes with an inner appendage; **Anth** broad, ovate or round, acute or obtuse; **Poll** oblong, flat, 0.8 × 0.3 mm; **Fr** 7.5 - 8.5 × 0.8 - 1 cm, apically beaked.

S. pedunculatum (Decaisne) K. Schumann (BJS 17: 132, 1893). **T** [lecto]: Ethiopia (*Quartin-Dillon* s.n. [P]). – **D**: Ethiopia, Kenya, Uganda, Rwanda, Zaïre, Tanzania, Moçambique; seasonally flooded grassland.

≡ *Gomphocarpus pedunculatus* Decaisne (1844) ≡ *Asclepias pedunculata* (Decaisne) Dandy (1952); **incl.** *Asclepias macrantha* Hochstetter *ex* Oliver (1844) ≡ *Stathmostelma macranthum* (Hochstetter *ex* Oliver) Schlechter (1924); **incl.** *Pachycarpus corniculatus* Hochstetter (1844); **incl.** *Gomphocarpus longipes* Oliver (1875); **incl.** *Stathmostelma globuliflorum* K. Schumann (1895); **incl.** *Asclepias uvirensis* S. Moore (1908).

Stems 22 - 45 cm, simple, biseriately or completely pubescent; **L** sessile or to 5 mm petiolate, lamina 4.5 - 26 × 0.35 - 1 cm, linear-lanceolate, long-attenuate, ± pubescent; **Inf** with 3 - 6 **Fl**, extra-axillary and terminal; peduncle 7.5 - 28 cm; **Bra** 4 - 6 mm; **Ped** 2.1 - 7.1 cm, uniseriately pubescent; **Sep** 5 - 6 mm; **Cl** flat or recurved, 2.1 - 2.8 cm ∅, pink, scarlet-red or yellow, lobes oblong (-ovate), acute, outside weakly pubescent; **Cs** lobes erect, basally bulging outwards, apical-marginal appendages broad, falcate (-oblong), obtuse or acute, with an inner appendage; **Anth** broad, ± round; **Poll** ± 1 × 0.4 mm; **Fr** solitary, 7.5 - 10 × 1.2 cm, fusiform; **Se** dark brown, ovate, 5 × 3 mm.

Closely related to *S. rhacodes*.

S. propinquum (N. E. Brown) Schlechter (BJS 51: 138, 1913). **T**: Tanzania (*Smith* s.n. [K]). – **D**: Tanzania (Kilimandsharo).

≡ *Asclepias propinqua* N. E. Brown (1895).

Stems low, pubescent, **Ha** small and curved; **L** sessile, lamina 19.5 - 32.5 × 1 - 2.1 mm, linear (-filiform), pubescent or tomentose; **Inf** with 3 - 4 **Fl**, terminal; peduncle 2.6 - 4.55 cm; **Bra** ± small; **Ped** 1.63 - 2.6 cm, pubescent; **Sep** 5.25 - 6.3 mm; **Cl** yellowish-green, 2.6 cm ∅, flat, lobes 10.5 × 6.3 mm, oblong-ovate, ± obtuse, glabrous; **Cs** lobes 6.3 - 7.35 mm, far surpassing the **Sty** head, apical-marginal appendages falcate, curved upwards, inner appendage oblong, obtuse; **Anth** ovate.

A doubtful taxon and presumably synonymous with *S. welwitschii* (Goyder 1998b).

S. rhacodes K. Schumann (BJS 17: 131, t. 6: D-F, 1893). **T:** [lecto – icono]: l.c. t. 6D-F. – **D:** Kenya; swampy grassland, 900 - 2250 m.

≡ *Asclepias rhacodes* (K. Schumann) N. E. Brown (1902); **incl.** *Stathmostelma pedunculatum* Schlechter (1895) (*nom. illeg.*, Art. 53.1).

Stems 10 - 50 cm, simple, sometimes branched in the **Inf** region, terete, basally conspicuously compressed, weakly tomentose; **L** shortly petiolate, lamina 5 - 15 × 0.1 - 1 cm, narrowly linear, pointed, basally narrowly obtuse, margins undulate, glabrous, midrib hispid; **Inf** with 2 - 4 **Fl**, extra-axillary and terminal; peduncle 0.6 - 15 cm, uniseriately hairy; **Bra** 3 mm, filiform; **Ped** 2 - 3.5 cm, pubescent; **Sep** 4 × 2 mm, lanceolate, acute, outside pubescent or glabrous; **Cl** outside greenish, red and yellow, inside scarlet-red, inversely bowl-shaped, lobes ± 12 × 4.5 mm, oblong-lanceolate, acute, glabrous; **Cn** 2 mm stipitate; **Cs** basally reddish-yellow, appendages yellow, lobes ≤ 9 mm, fleshy, erect, outside convexely bulging, keeled, apical-marginal appendages erect, falcate-ovate, with an inner appendage; **Gy** 3.5 - 4 mm; **Sty** head strongly fissured; **Anth** round, apically obtuse; **Poll** oblong-reniform, ± 0.8 × 0.3 mm, translators tapering to a thin thread; **Fr** fusiform, ± 11 × 0.5 cm; **Se** 8 × 4 mm.

Closely related to *S. pedunculatum*.

S. spectabile (N. E. Brown) Schlechter (BJS 51: 138, 1913). **T** [lecto]: Malawi (*Buchanan* 553 [K]). – **D**: Tanzania, Zambia, Zaïre, Zimbabwe, Malawi.

≡ *Asclepias spectabilis* N. E. Brown (1895); **incl.** *Stathmostelma pachycladum* K. Schumann (1900) ≡ *Asclepias pachyclada* (K. Schumann) N. E. Brown (1902); **incl.** *Stathmostelma macropetalum* Schlechter & K. Schumann (1903) ≡ *Asclepias macropetala* (Schlechter & K. Schumann) N. E. Brown (1904).

S. spectabile ssp. **frommii** (Schlechter) Goyder (KB 53(3): 585, 1998). **T**: Malawi (*Fromm* 89 [ST]). – **D**: Malawi.

≡ *Stathmostelma frommii* Schlechter (1913).

Differs from ssp. *spectabile*: **Bra** ovate, 4 mm broad; **Cl** lobes spreading, yellow to red; **Fr** 8 × 3 cm. – [U. Meve]

S. spectabile ssp. **spectabile** – **D**: Tanzania, Zambia, Zaïre, Zimbabwe, Malawi.

Stems 1 - 1.3 m, robust, pubescent; **L** ascending, petiole 2.1 - 12.6 mm, lamina 9.1 - 18.2 × 2.6 - 5.2 cm, lanceolate, acute, basally ± rounded, pubescent; **Inf** with 5 - 6 **Fl**, extra-axillary and terminal; peduncle 2 - 10 cm; **Bra** 4.2 - 5.25 mm, subulate; **Ped** 2 - 5 cm, uniseriately pubescent; **Sep** 4 - 16 × 1 - 2 mm; **Cl** carmine to scarlet-red, 3.25 cm ∅, flat to campanulate, lobes 15 - 20 × 6 - 10 mm, oblong, obtuse, apically ± recurved, outside weakly tomentose towards the tips, lobe margins inside pubescent; **Cs** yellow, **Cs** lobes ± 10 mm, erect, oblong-ovate, apical-marginal appendages falcate, acute or linear-oblong, truncate, basally on both sides with a gibbosity, inner appendage subulate or oblong; **Anth** round, obtuse; **Poll** oblong, flat, ± 1.5 × 0.4 mm; **Fr** 10 - 14 × 1.5 - 2 cm, fusiform, oblong, erect; **Se** ovate with papillate margin, 5 × 3 mm.

S. welwitschii Britten & Rendle (Trans. Linn. Soc. London, Bot. 4: 28, 1894). **T**: Angola (*Welwitsch*

4168 [BM, K]). – **D:** Angola, Sudan, Uganda, Zaïre, Tanzania, Zambia.

≡ *Asclepias welwitschii* (Britten & Rendle) N. E. Brown (1902); **incl.** *Stathmostelma laurentianum* Dewèvre (1895) ≡ *Asclepias laurentiana* (Dewèvre) N. E. Brown (1902); **incl.** *Stathmostelma chironioides* K. Schumann *ex* De Wildeman & T. Durand (1906).

S. welwitschii var. **bagshawei** (S. Moore) Goyder (KB 53(3): 599, 1998). **T:** Sudan (*Bagshawe* 1640 [BM, K]). – **D:** Sudan, Uganda, Zaïre, Tanzania.

≡ *Asclepias bagshawei* S. Moore (1908).

Differs from var. *welwitschii*: **Cn** lobes with a short rounded inner appendage. – [U. Meve]

S. welwitschii var. **welwitschii** – **D:** Angola, Zaïre, S Sudan, Uganda, N and SW Tanzania, N Zambia; in seasonally inundated grassland.

Stems 20 - 120 cm, terete, branched or simple, biseriately or only the apical stem parts pubescent, sometimes glabrous; **L** petiole 2.1 × 4.2 mm, lamina blue-green, 7.5 - 17.5 cm × 2 - 8.4 mm, linear-lanceolate, apically pointed, basally cuneate or abruptly narrowed, upper face, margins and midrib of the lower face hispidly pubescent; **Inf** with 2 - 6 **Fl**, extra-axillary or terminal; peduncle 1.2 - 7.5 cm; **Bra** 3 - 4 mm, subulate; **Ped** 1.5 - 3.25 cm, uniseriately pubescent; **Sep** 6.3 × 2.1 mm, weakly hairy; **Cl** intensely orange, flat or convex, lobes 10.5 × 4.1 mm, ovate, acute, glabrous; **Cs** lobes 5.25 mm, far surpassing the **Sty** head, basally ± cordate, apical-marginal appendages surpassing the **Anth**, with an acute inner appendage; **Poll** 1 × 0.3 mm, translators oblong, curved at the end, abruptly thin and filiform, then broadened again in the triangular **Poll**-bearing part; **Fr** 5 - 10 × 0.5 - 1 cm; **Se** ovate with papillate margin, 5 × 3 mm.

S. wildemanianum T. Durand (Syll. Fl. Congol., 361, 1909). **T:** Zaïre (*Verdick* 1899 [BR]). – **D:** Zaïre.

Incl. *Asclepias verdickii* De Wildeman (1906); **incl.** *Stathmostelma verdickii* De Wildeman (1906) (*nom. inval.*, Art. 34.1b).

Stems glabrous or weakly hairy; **L** sessile, 14 cm × 7 - 8 mm, lanceolate, apically acute, basally rounded, midrib on the lower face weakly hairy, upper face glabrous, margins hispid; **Inf** terminal, 5-flowered; **Ped** 8 - 9 cm, slender, weakly hairy; **Sep** 50 - 60 × 2.5 mm, glabrous, margins ciliate; **Cl** convex, colour unknown, lobes 18 × 9 mm, glabrous, inside sometimes tomentose; **Cs** lobes 1 cm, apical-marginal appendages surpassing the **Sty** head; **Poll** ± 1.3 × 0.25 mm.

STENOSTELMA

B. Müller, P. Stegemann & F. Albers

Stenostelma Schlechter (BJS 18(Beiblatt 45): 6, 1894). **T:** *Stenostelma capense* Schlechter. – **D:** S Zimbabwe, RSA (Northern Prov., Gauteng, Mpumalanga, Eastern Cape, KwaZulu-Natal, Free State). **Etym:** Gr. 'stenos', narrow, slender; and Gr. 'stelma', crown, garland, wreath; application obscure.

Incl. *Krebsia* Harvey (1868) (*nom. illeg.*, Art. 53.1). **T:** *Lagarinthus corniculatus* E. Meyer.

Erect or decumbent shrubs, basally branched, with **R** tuber; stems hairy or glabrous; **L** opposite, shortly petiolate, erect, lamina narrowly linear, margins often rolled back; **Inf** extra-axillary, laterally and terminally, globose, pedunculate; **Bra** subulate or filiform; **Ped** short, pubescent; **Sep** lanceolate, pointed, ± strongly softly hairy; **Cl** (inversely) campanulate, lobes lanceolate, triangular to elliptic, pointed, sometimes curved back only apically; **Ci** only partly formed, lobes minute, below the guide rail; **Cs** lobes erect, basally bulging outwards, apically bulging inwards, completely covering the **Sty** head, basally with small lateral wings, fleshy; **Sty** head conical; **Anth** ± square, membranous, triangular or rounded, erect, incumbent on and covering the **Sty** head; **Poll** ovate, partly curved inwards, compressed, apically obtuse without defined germination crest, translators long, flattened, conspicuously geniculate and variously thickened, broadly attached at the corpuscle, pollinia attached apically at the translator arms, corpuscle ovoid; **Fr** commonly solitary, erect, basally slightly thickened, apically beaked, glabrous; **Se** unknown.

Closely related to *Gomphocarpus*. *Stenostelma* is distinguished by the conical shape of the style head, which is covered by conspicuous connective appendages. Additional differences from *Gomphocarpus* are the presence of root tubers and the lobes of the staminal corona, which are cucullate or hood-shaped in *Gomphocarpus*, but form small hollows in *Stenostelma*.

The following names are of unresolved application but are referred to this genus: *Lagarinthus eustegioides* E. Meyer (1837) ≡ *Schizoglossum eustegioides* (E. Meyer) Druce (1917); *Schizoglossum crassipes* S. Moore (1902); *Schizoglossum orbiculare* Schlechter (1895); *Schizoglossum periglossoides* Schlechter (1895); *Schizoglossum umbelliferum* Schlechter (1895).

S. capense Schlechter (BJS 18(Beiblatt 45): 6, 1894). **T:** RSA, Northern Cape (*Schlechter* 1693 [B ?]). – **D:** Botswana, RSA (Northern Cape, Northern Prov.). **Fig. XLVI.g**

≡ *Schizoglossum capense* (Schlechter) H. Huber (1967); **incl.** *Gomphocarpus stenoglossus* Schlechter (1894) ≡ *Krebsia stenoglossa* (Schlechter)

Schlechter (1894); **incl.** *Schizoglossum aciculare* N. E. Brown (1902).

Erect, 16 - 20 cm, basally branched; **R** tuber pear-shaped; stems greenish-red, compressed, softly uniseriately hairy (the side with the **Ha** changes after each node), sometimes glabrous; **L** longer than the **Int**, lamina ≤ 40 × 5 mm, margins curved outwards, upper face shortly soft-hairy, lower face ± glabrous, midrib hairy; **Inf** 5 - 6, with 8 - 12 **Fl**, ± globose; peduncle ≤ 1 cm, erect, pubescent; **Ped** ± 3 mm, **Ha** white, short; **Sep** 1.5 × ± 1 mm; **Cl** calyx-like, outside greenish, lobes apically reddish, inside cream-greenish, 3 × 2 mm, narrowly triangular, fleshy, erect, margins inwards ± undulate, tips recurved and thus falcate, papillose; **Ci** lobes minute, tooth-like, erect; **Cs** lobes white, 8 mm, basally stalk-like, abruptly triangularly broadened, apically long-attenuate, narrow, erect, far elongated over the conical **Sty** head, tips sometimes keeled on the outside; **Anth** triangular, emarginate, appendages marginally touching the **Sty** head and ascending to meet in the centre; **Poll** curved inwards, translators broad.

S. carinatum (Schlechter) Bullock (KB 11: 521, 1956). **T:** RSA, Eastern Cape (*Tyson* 1439 [not located]). – **D:** RSA (Eastern Cape); moist grassland.

≡ *Krebsia carinata* Schlechter (1895) ≡ *Gomphocarpus carinatus* (Schlechter) Schlechter (1904) ≡ *Xysmalobium carinatum* (Schlechter) N. E. Brown (1907).

Stems 20 - 35 cm, thin, biseriately pubescent; **L** numerous, 2 - 9 × 0.2 - 0.5 cm, glabrous or very delicately hairy; **Inf** with 7 - 10 **Fl**, extra-axillary; **Ped** 5 mm, filiform; **Sep** ± 2 mm, acute, erect or somewhat diverging; **Cl** lobes 4 × ≤ 2 mm, elliptic, apically thickened, margins curved outwards, glabrous, tips inside velvety; **Cs** lobes falcate, obtuse; **Anth** triangular, incumbent on the convex **Sty** head, basally with triangular wings; **Poll** compressed, apically obtuse, translators filiform, basally thickened.

The placement in this genus is uncertain.

S. corniculatum (E. Meyer) Bullock (KB 7: 417, 1952). **T:** RSA, Eastern Cape (*Drège* s.n. [not located]). – **D:** RSA (North-West Prov., Northern Cape, Mpumalanga, KwaZulu-Natal, Eastern Cape).

≡ *Lagarinthus corniculatus* E. Meyer (1837) ≡ *Gomphocarpus corniculatus* (E. Meyer) Dietrich (1840) ≡ *Krebsia corniculata* (E. Meyer) Schlechter (1895) ≡ *Schizoglossum corniculatum* (E. Meyer) R. A. Dyer (1971).

Very similar to *S. capense*, but **Ped** shorter; **Cl** whitish, glabrous; **Cs** lobes to 10 mm. – [U. Meve]

S. eminens (Harvey) Bullock (KB 8: 342, 1953). **T:** RSA, KwaZulu-Natal (*Gerrard & M'Ken* 1291 [not located]). – **D:** Zimbabwe, RSA (KwaZulu-Natal, Northern Prov., Mpumalanga); moist loamy ground.

≡ *Gomphocarpus eminens* Harvey (1863) ≡ *Asclepias eminens* (Harvey) Schlechter (1896).

Stems prostrate or ascending, branched, biseriately hairy; **L** conspicuously petiolate, lamina much shorter than the **Int**, hastate-linear, margins curved outwards, basally obtuse, apically pointed, glabrous; **Inf** terminal, with ≥ 5 **Fl**; **Bra** small, acute; **Sep** as long as the **Cl** lobes; **Cl** flat, somewhat recurved, greenish, lobes elliptic, margins curved outwards; **Cn** stipitate; sides of the **Cs** lobes folded inwards, apically rounded, outside keeled, keel apically elongated in a spine-like tip; **Anth** ± circular, apically curved inwards; translators of the **Poll** conspicuously geniculate.

STOMATOSTEMMA

U. Meve

Stomatostemma N. E. Brown (FTA 4(1:2): 252, 1902). **T:** *Cryptolepis monteiroae* Oliver. – **Lit:** Venter & Verhoeven (1993). **D:** SE Africa. **Etym:** Gr. 'stoma', mouth; and Gr. 'stemma', garland, wreath; for the coronal outgrowths at the mouth of the corolla tube.

Lianas or shrubs with white latex, partly with **R** tubers; stems perennial, woody; **Inf** terminal or axillary; **Sep** free; **Cl** campanulate; **Cn** petaloid, clavate, originating in the sinuses between the **Cl** lobes and alternating with the **Cl** lobes; **Anth** lanceolate, acute, whitish, arranged pyramidally; pollen in tetrads.

The genus, belonging to the *Periplocoideae*, numbers only 2 species. The type species *S. monteiroae* is a root-succulent liana differing in all respects except similarities in corona morphology from the shrubby and non-succulent *S. pendulina*. The placement of the 2 species in a single genus (Venter & Verhoeven 1993) appears improbable.

S. monteiroae (Oliver) N. E. Brown (FTA 4(1:2): 253, 1902). **T:** Moçambique (*Monteiro* s.n. [K, P]). – **D:** Moçambique, Botswana, Zambia, Zimbabwe, RSA (KwaZulu-Natal, Mpumalanga, North-West Prov.), Swaziland. **I:** Venter & Verhoeven (1993). **Fig. XLVII.a**

≡ *Cryptolepis monteiroae* Oliver (1887).

R tuber globose or irregularly shaped, ± 5 - 20 cm ⌀, sometimes several; stems to 10 m, twining, becoming woody, bark brown, glabrous, somewhat tuberculate; **L** decussate but arranged ± bifarious, 0.1 - 1 (-3) cm petiolate, lamina ovate, elliptic or obovate, 3 - 10 × 0.5 - 2 (-3) cm, basally ± cuneate, apically ± acute, hardly coriaceous, glabrous; **Inf** 5 - 25 mm pedunculate, few- to many-flowered; **Ped** 4 - 13 mm, glabrous, glossy; **Sep** 2 - 3 mm; **Cl** horizontally oriented, campanulate, large, glabrous, outside cream-coloured; **Cl** tube 5 - 6 mm, inside green and purple; **Cl** lobes ascending, tips revolute and slightly obliquely twisted, ± ovate, 12 - 16 × 5 - 6 mm, cream-coloured, basally tinged purple, margins revolute; **Cs** lobes clavate-ovoid, 2 - 3 mm, erect, glossy; **Anth** greenish-white, translator 2 mm, spoon rhomboid, stalk very short; **Fr** in pairs, thickly fusiform, grooved or slightly winged, 6 - 9 × 1.5 - 2.5 cm.

An attractive large-flowered and rather weakly twining liana, which flowers in cultivation when only some 10 cm tall.

TAVARESIA

U. Meve

Tavaresia Welwitsch (Bol. Ann. Cons. Ultramar. No. 7: 79, 1854). **T:** *Tavaresia angolensis* Welwitsch. – **Lit:** White & Sloane (1937). **D:** Angola, Botswana, Namibia, Zimbabwe, RSA. **Etym:** For José Tavares de Macedo (fl. ± 1850), superior official in the Portuguese Ministry of Marine and the Colonies, and amateur botanist.

Incl. *Decabelone* Decaisne (1871). **T:** *Decabelone elegans* Decaisne.

Compact basally branching perennial stem succulents, few- to many-stemmed, forming clusters, 5 - 100 cm ⌀, spiny; stems ± erect, 5- to 14-ribbed, cylindrical, 2 - 30 × 1.5 - 2.5 cm, green; **Ri** prominent; **Tu** conical, with a 3-partite **Sp** derived from the **L** rudiments; **Inf** basal, ± sessile, few-flowered, only 1 **Fl** open at a time, carrion-scented; **Ped** 5 - 15 × 1.5 - 2.5 mm; **Sep** lanceolate-subulate, acute, 5 - 15 × ± 2 mm, glabrous; **Cl** yellowish with red-brown spots and flecks, deeply tubular to funnel-shaped, 3 - 15 × 1.5 - 3.5 cm, outside ± glabrous, inside densely beset with **Ha** (-papillae), these ± cylindrical or clavate, 0.02 - 0.2 mm, longer towards the **Fl** centre, apically mostly with an acute **Bri**; **Cl** tube 3 - 14 × 2 - 5 cm; **Cl** lobes triangular, acute, 1 - 1.5 × 1 - 2 cm, ± spreading horizontally; **Cn** biseriate, shortly stipitate, 12 - 18 × 4 - 7 mm, basally whitish, inside towards the tips more and more striped brown-orange / red-brown; **Ci** fused to form a (bowl-shaped to) cup-shaped structure, basally somewhat saccate; **Ci** lobes ± erect, bifid, all 10 appendages ± 8 - 12 mm, drawn out filiform and crowned by a drop-shaped purple mobile finely papillose structure; **Cs** 1 - 2 mm, linear, incumbent on the **Anth**; **Anth** ± triangular (to spoon-shaped), densely incumbent on the flat and concave **Sty** head; **Poll** obovoid, ± 0.4 × 0.2 mm, caudicles ± 0.12 mm, ± erect, corpuscle ± 0.15 mm, basally with narrow wing-like broadenings; **Fr** coarse, fusiform, divergent at an angle of 30°, glabrous; **Se** pale brown, ± 4 - 6 × 3 - 4 mm, margin thick, undulate, cork-like.

The genus shares the characteristic indumentum of the corolla as well as the morphology of the papillae with *Huernia*, and the 2 are probably closely related. *Tavaresia* is notable not only for the spiny stems and the large funnel-shaped flowers, but *T. barklyi* also has one of the largest distribution ranges of all stapeliads.

The flowers of the 2 species are almost identical (leaving apart the dubious *T. thompsoniorum*), and the following descriptions are limited to the diagnostic vegetative characters.

For a natural hybrid with *Stapelia* see ×*Staparesia*.

T. angolensis Welwitsch (Bol. Ann. Cons. Ultramar. no. 7: 79, 1854). **T:** Angola, Luanda Distr. (*Welwitsch* 4262 [BM, G, K]). – **D:** Angola, Namibia. **Fig. XLVII.b**

Incl. *Huernia tavaresii* Welwitsch (1856); **incl.** *Decabelone elegans* Decaisne (1871); **incl.** *Stapelia digitaliflora* Pfersdorf *ex* Decaisne (1871) (*nom. inval.*, Art. 34.1c); **incl.** *Decabelone sieberi* Pfersdorf *ex* Hooker *fil.* (1874) (*nom. inval.*, Art. 34.1c).

Stems 5- to 6-ribbed, the 3-parted **Sp** ± horizontally spreading, lateral **Sp** somewhat ascending, conspicuously shorter than the central **Sp**.

T. barklyi (Dyer) N. E. Brown (FTA 4(1): 494, 1903). – **D:** Botswana, Namibia, Zimbabwe, RSA (widespread). **Fig. XLVII.c**

≡ *Decabelone barklyi* Dyer (1875); **incl.** *Euphorbia antunesii* Pax (1904); **incl.** *Decabelone grandiflora* Dinter (1909) ≡ *Tavaresia grandiflora* (Dinter) A. Berger (1910); **incl.** *Hoodia senilis* H. Jacobsen (1933); **incl.** *Hoodia genilis* Jacobsen *ex* Bruyns (1993) (*nom. inval.*, Art. 61.1).

Stems 5- to 14-ribbed, with 3-parted very acute **Sp**, central **Sp** horizontal to erect, lateral **Sp** shorter, oriented downwards and ± parallel to the stem surface.

T. thompsoniorum van Jaarsveld & Nagel (Asklepios No. 76: 9-10, ills., 1999). **T:** Angola (*Bell* s.n. [NBI]). – **D:** S Angola.

Stems to 12-ribbed, spination like *T. angolensis*; **Cl** tube bowl-shaped, ± 15 mm, **Cs** lobes spreading star-like, similar to those of *Stapelia*.

The taxon was described on the base of a solitary collection from the region where the ranges of *T. angolensis*, *T. barklyi* and *Stapelia similis* overlap, and presumably represents the hybrid between the 2 last-named taxa.

TRACHYCALYMMA

U. Meve

Trachycalymma (K. Schumann) Bullock (KB 8: 348, 1953). **T:** *Gomphocarpus cristatus* Decaisne. – **D:** Tropical Africa. **Etym:** Gr. 'trachys', rough; and Gr. 'calymma', covering; perhaps for the ± coarse hairiness of the plants.

≡ *Gomphocarpus* Subsect. *Trachycalymma* K. Schumann (1895).

Medium-tall perennials with annual herbaceous stems and white latex; **R** tuberous and/or with fusiform fleshy lateral **R**; stems erect, usually solitary and unbranched, densely pubescent; **L** sessile or petiolate, lamina linear or lanceolate, oblong or ovate, mostly with recurved margins, ± pubescent; **Inf** extra-axillary, pedunculate; **Fl** petiolate, ± nodding; **Sep** ± lanceolate, acute, hairy; **Cl** rotate or campanulate; **Cl** lobes occasionally recurved, tips usually curved inwards, outside pubescent, inside glabrous; **Cn** 1-seriate; **Cs** lobes various in shape, always very fleshy and fused with the base of the **Gy**, weakly to strongly cucullate, ± pouch-like, cavity mostly papillose, inside with or without a flattened small tooth; **Anth** appendages membranous, roundish to rectangular, incumbent on the **Sty** head; **Poll** flattened obovate, reniform oder drop-shaped, caudiculae flattened, geniculate; **Fr** usually solitary, erect, fusiform to ovoid, smooth or softly spiny, pubescent, **Ped** contorted or not; **Se** ovate.

Goyder (2001) recognizes 10 species of these geophytic *Asclepias* relatives from tropical African grasslands and deciduous forests in this genus. Apart from the erect solitary stems, *Trachycalymma* is characterized by its gynostegium. The fleshy corona lobes are usually bearded inside and dorsally rounded, not folded or compressed, and are basally adnate to the gynostegium or, respectively, the guide rails. As species of this genus are hardly cultivated, no individual taxa are covered here, though *T. cristatum* is figured (**Fig. XLVI.h**).

TRIDENTEA

B. Müller & F. Albers

Tridentea Haworth (Synops. Pl. Succ., 34, 1812). **T**: *Stapelia gemmiflora* Masson [Lectotype, designated by L. C. Leach, Trans. Rhodesia Sci. Assoc. 59(1): 2, 1978.]. – **Lit**: Leach (1980); Bruyns (1995b). **D**: Namibia, Botswana, RSA. **Etym**: Lat. 'tri-', three-; and Lat. 'dens, dentis', tooth; referring to the frequently 3-toothed intrastaminal corona segments.

Small perennial stem succulents; stems procumbent-erect, basally freely branched (rarely from further up), clustered and forming compact clumps, rarely weakly rhizomatous, (2-) 5 - 15 (-20) × (0.7-) 1 - 2 (-2.5) cm, softly fleshy, obtusely 4-angled, sides flat or slightly convex, conspicuously longitudinally grooved, glabrous (rarely young weakly roughened to delicately hairy), glossy, edible (taste lettuce-like); **Tu** large, obtuse, rectangular, acutely tapering, those of young stems with 1.5 - 10 mm long **L** rudiments, these subulate, horizontally spreading, glabrous, rapidly caducous, leaving a conspicuous slightly raised whitish scar, basally (sometimes also scattered along the margins of the **L** rudiments towards the tip) with clusters of 1 to several stipular **Ha** (incumbent on the **L** rudiment surface); **Inf** mostly few-flowered, from the sides of the basal ½ of the stems (sometimes scattered over the stem length), successively opening; peduncle (when present) gradually elongating, glabrous; **Ped** 25 - 80 × ± 2 mm, spreading to decurved, glabrous; **Sep** mostly ovate, acute, glabrous or with few marginal **Ha**; **Cl** outside pale green, inside pale with darker patterning, 2 - 10 cm ⌀, mostly flat, rarely campanulate, incised for max. ½, central hollowing shallowly funnel-shaped, frequently thickened annulus-like at the level of the **Ci** lobes; **Cl** lobes usually ± broadly triangular with ovate base, horizontally spreading to ascending, margins sometimes abruptly folded upwards, outside glabrous and smooth, inside glabrous, smooth to slightly wrinkled or papillose, papillae columnar, usually basally confluent, frequently apically with a **Ha** or **Bri**, lobe margins usually hairy; **Cn** of 2 series, of the C(is) + Cs type, shortly stipitate, foot pentagonal, stubby, glabrous; **Ci** lobes ascending to erect, flattened, simple to bifid or 3-parted, only basally connected with the **Cs**; **Cs** lobes incumbent on the **Anth**, mostly much longer than the **Anth**, basally flattened, apically frequently cylindrical, in the middle raised column-like over the **Sty** head, then diverging, in the lower ½ with a dorsal appendage, these ± deltoid, flattened, fin-like; entry to the nectar cavity large, ± round or broadly transversely elliptic; **Anth** ± square, incumbent on the **Sty** head; **Poll** D-shaped, ± horizontally in each theca, caudicles short, corpuscle ± as long as broad, winged; **Sty** head not surpassing the **Anth**, truncate-compressed; **Fr** purple-spotted, cylindrical-fusiform, glabrous, smooth, diverging at an angle of 30° - 60°; **Se** flat, ovate, margins mostly ± flat, rarely thickened.

While Leach (1980) supposes a closer relation to *Stapelia*, Bruyns (1995b) postulates a close relationship to *Hoodia* and *Lavrania*. *Orbea* is similar as to stem morphology. The genus is divided as follows:

[1] Sect. *Tridentea*: **Tu** of the stems not conspicuously thickened; **L** rudiments ≤ 1 (-1.5) cm; **Cl** unicoloured or inconspicuously blotched, inside rough and roughly papillose; **Ci** lobes deeply 3-parted; **Cs** lobes with a dorsal appendage; **Poll** often much longer than broad.

[2] Sect. *Parvipunctia* L. C. Leach 1980: **Tu** of the stems conspicuously thickened, stems often inconspicuously divided into oblong sections; **L** rudiments ≤ 5 mm; **Cl** blotched, sometimes with dark margins, inside papillose to ± smooth; **Ci** lobes simple, bifid or 3-parted; **Cs** lobes often without a dorsal appendage; **Poll** broadly D-shaped.

T. dwequensis (C. A. Lückhoff) L. C. Leach (Excelsa Tax. Ser. 2: 35-37, ills., 1980). **T** [neo]: RSA,

Northern Cape (*Hardy* 738 [PRE]). – **D:** S and SW Namibia, RSA (Northern Cape). **I:** Bruyns (1995b: 197, 199). **Fig. XLVII.d**

≡ *Stapelia dwequensis* C. A. Lückhoff (1937).

[2] Stems rarely branched towards the tips, in cultivation with weak rhizomatose tendency, grey-green, ± robust, to 7.5 × basally 1.5 cm; **L** rudiments ≤ 2.5 mm, without stipular rudiments; peduncle short; **Ped** 3 cm; **Sep** ± 4 × 1.5 mm; **Cl** inside pale yellow or cream-coloured, dotted to blotched with purple, ± 3.5 cm ⌀, campanulate; **Cl** tube ± 1.4 cm, completely embracing the **Cn**, margin with 5 grooves that start at the base of the **Cl** lobes, without annulus-like thickening; **Cl** lobes ± 1.1 × 1.2 cm, slightly curved outwards, inside densely wrinkled-papillose (esp. towards the tips, confluent towards the tube), papilla-**Ha** delicate, rather short, margins of the **Cl** lobes basally sometimes with few small clavate **Ha**; **Cn** black; **Ci** lobes 6 - 6.5 (-8) mm, linear-lanceolate, lower parts of the margins broadened wing-like, folded upwards; **Cs** lobes ≤ 1 cm, basally 1.25 mm broad, slender towards the tips, tips transparent, obtuse or clavate, without dorsal appendage; **Poll** ± 0.6 × 0.4 mm.

Closely related to *T. jucunda*.

T. gemmiflora (Masson) Haworth (Synops. Pl. Succ., 34, 1812). **T:** [lecto – icono]: Masson, Stapel. Nov. t. 15, 1797. – **D:** RSA (Mpumalanga?, Free State, Northern Cape, Eastern Cape). **I:** Bruyns (1995b: 193, 195).

≡ *Stapelia gemmiflora* Masson (1797); **incl.** *Stapelia hircosa* Jacquin (1806) ≡ *Stapelia gemmiflora* var. *hircosa* (Jacquin) N. E. Brown (1909) ≡ *Tridentea gemmiflora* var. *hircosa* (Jacquin) P. V. Heath (1993); **incl.** *Tridentea stygia* Haworth (1812) ≡ *Stapelia stygia* (Haworth) Schultes (1820); **incl.** *Stapelia hircola* Poiret (1817); **incl.** *Tridentea moschata* hort. ex Haworth (1819) (*nom. inval.*, Art. 32.1c) ≡ *Stapelia moschata* (hort ex Haworth) Schultes (1820) (*nom. nud.*); **incl.** *Stapelia hircosa* var. *densa* N. E. Brown (1890) ≡ *Stapelia gemmiflora* var. *densa* (N. E. Brown) N. E. Brown (1909) ≡ *Tridentea gemmiflora* var. *densa* (N. E. Brown) P. V. Heath (1993).

[1] Stems in large clumps, pale green to greenish-grey, ± 7 × ± 1.5 cm, stubby; **Tu** porrect; **L** rudiments ≤ 1 cm, stipular rudiments ≤ 2.5 mm; **Fl** of each **Inf** usually opening in quick succession; peduncle 1 - 2 (-4) cm, robust, long-tappering, sometimes sessile; **Ped** < 5 cm, ascending or spreading; **Sep** ± 5 - 6 × 2.5 - 3 mm, sometimes with few short stout colourless **Ha**; **Cl** outside dotted with purple around the **Sep**, lobes apically brownish, inside pale yellow, irregularly spotted with purple, 4.5 - 10 cm ⌀, **Cn** sunken for ¾; **Cl** lobes broadly dark-margined, 3.5 × ± 3 cm, margins crenate, abruptly folded upwards and inwards, basally transitional into intermediate lobes, inside grooved, ± strongly wrinkled labyrinth-like (± concentrically around the tube), densely papillose, papillae apically ± without **Bri**, margins of the **Cl** lobes ± densely hairy, **Ha** white to dark purple, ± 3 mm, ± clavate or simple, vibratile, weak, straight or slightly undulate, often mixed; **Cn** yellow, with strong dark purple pattern; **Ci** lobes ± 8 × 2.5 mm, basally ± square, longitudinally grooved, deeply 3-parted, middle appendage ± 5 mm, ± narrowly triangular, apically acutely tapering, frequently variably toothed, often also asymmetrically bifid, lateral appendages ± 2.5 mm, subulate, widely diverging; **Cs** lobes ± 6 mm, dorsal appendage ± 2 mm, often wing-like, laterally flattened, erectly spreading, apically truncate or irrgularly toothed; **Anth** touching each other apically-marginally, wings rather long; **Poll** orange-brown, ± 0.65 - 0.7 × 0.35 - 0.4 mm, germination crest ± 0.35 mm.

Closely related to *T. pachyrrhiza*.

T. jucunda (N. E. Brown) L. C. Leach (Trans. Rhodesia Sci. Assoc. 59(1): 3, 1978). **T** [lecto]: RSA, Northern Cape (*Pillans* 644 [BOL]). – **D:** Namibia, RSA (Northern Cape, Western Cape, Free State). **I:** Meve (1988); Bruyns (1995b: 196). **Fig. XLVII.e**

≡ *Stapelia jucunda* N. E. Brown (1909); **incl.** *Tridentea jucunda* var. *jucunda*; **incl.** *Stapelia jucunda* var. *deficiens* N. E. Brown (1909) ≡ *Tridentea jucunda* var. *deficiens* (N. E. Brown) P. V. Heath (1993); **incl.** *Stapelia cincta* Marloth (1913) ≡ *Tridentea jucunda* var. *cincta* (Marloth) L. C. Leach (1980); **incl.** *Stapelia dinteri* A. Berger (1914) ≡ *Tridentea jucunda* var. *dinteri* (A. Berger) L. C. Leach (1980); **incl.** *Stapelia ausana* Dinter & A. Berger (1928); **incl.** *Stapelia dinteri* var. *capensis* C. A. Lückhoff (1938) (*nom. inval.*, Art. 36.1); **incl.** *Stapelia dinteri* var. *pseudocapensis* C. A. Lückhoff (1938) (*nom. inval.*, Art. 36.1).

[2] Stems rarely branched towards the tips, grey-green, rather stout, to 7.5 × basally 1.5 cm; **L** rudiments ± 5 mm; peduncle erect or spreading; **Ped** ≤ 3 cm; **Sep** ± 2.5 × 1.75 mm, or ± 4 × 2 mm, then narrowly triangular; **Cl** outside sometimes spotted with dark purple, inside cream-coloured, ± densely spotted with purple-brown, 2 - 3.5 cm ⌀, lobes with broad purple-brown margins, sometimes curved outwards, intermediate lobes obtuse, inside ± smooth, rarely weakly papillose, margins of the lobes ciliate, with vibratile clavate purple **Ha**; **Ci** lobes ± 2.5 - 3 × 1 mm, simple, ± narrowly triangular, weakly longitudinally grooved, rarely ± rectangular or ± deeply bifid; **Cs** lobes 4 - 5 mm, apically long-attenuate, filiform, sometimes with weak dorsal bulge; **Poll** ± 0.5 - 0.6 × 0.3 - 0.4 mm.

Closely related to *T. dwequensis*.

T. marientalensis (Nel) L. C. Leach (Trans. Rhodesia Sci. Assoc. 59(1): 3, 1978). **T** [neo]: Namibia (*Leistner* 1819 [PRE, KMG]). – **D:** Namibia, RSA. **I:** Bruyns (1995b: 193-195).

≡ *Stapelia marientalensis* Nel (1935).

T. marientalensis ssp. **albipilosa** (Giess) L. C. Leach (Excelsa Tax. Ser. 2: 22, ill., 1980). **T**: Namibia (*Giess* 10262 [WIND, M, PRE]). – **D**: S Namibia (Karasberge). **Fig. XLVII.f**

≡ *Stapelia albipilosa* Giess (1974) ≡ *Tridentea marientalensis* var. *albipilosa* (Giess) P. V. Heath (1993).

[1] Differs from ssp. *marientalensis*: **Cl** yellowish to cream-coloured, campanulate to bowl-shaped, ± the upper ½ of the lobes reddish to blackish; **Cl** tube ± asymmetrically cup-shaped with the side towards the ground longer, inside densely hairy, **Ha** white, long.

Specimens from SE of the Karasberge are ± intermediate between the two subspecies, and the complex is perhaps better treated as a single variable species. (Bruyns 1995b).

T. marientalensis ssp. **marientalensis** – **D**: S and SE Namibia, Botswana, RSA (Northern Cape).

Incl. *Stapelia auobensis* Nel (1937) ≡ *Tridentea marientalensis* var. *auobensis* (Nel) P. V. Heath (1993).

[1] Stems apically tapering to 15 cm; **L** rudiments ± 1.2 cm; **Inf** sometimes 1-flowered, near the stem base; peduncle ≤ 10 cm, long-attenuate, **Fl** frequently decumbent, with horse-like scent; **Ped** usually > 7 (10 - 13) cm, sometimes delicately roughened; **Sep** 6 - 7 × ± 2 mm; **Cl** inside centrally weakly yellowish, otherwise reddish to blackish, ± 5 - 7.5 cm ∅, lobes 1.8 - 2.5 × ± 1.25 cm, frequently widely recurved, inside ± densely papillose, papillae apically with a **Ha** or long **Bri**, white, slightly clavate, **Bri** ± 5× as long as the papillae, margins of the **Cl** lobes hairy, clavate **Ha** white, pink, purple (sometimes colours mixed), very long, vibratile; **Ci** lobes ± 6 × 1.5 mm, deeply 3-parted (sometimes bifid), middle appendage ± 5 mm, usually ± truncate, apically irregularly bluntly dentate, lateral appendages ± 3.5 mm, acutely pointed, ascending, weakly longitudinally canaliculate, basally ± square, margins basally incurved; **Cs** lobes ± 3.5 mm, apically ± falcate, dorsal appendage 1 × 0.6 mm, laterally flattened, wing-like, erectly spreading, acutely tapering, tip irregularly dentate; **Poll** ± 0.6 × 0.3 mm.

T. pachyrrhiza (Dinter) L. C. Leach (Excelsa Tax. Ser. 2: 14, ills. (p. 15), 1980). **T** [neo]: Namibia (*Dinter* 6491 [SAM, BOL, G, GRA, KMG, M, PRE]). – **D**: S Namibia (Namib Desert); RSA (Northern Cape: Richtersveld; North-West Prov.). **I**: Bruyns (1995b: 193).

≡ *Stapelia pachyrrhiza* Dinter (1923); **incl.** *Stapelia umbonata* Pillans (1928).

[1] Plants ± 50 cm ∅; stems stubby, ± 1.5 cm ∅; **Tu** raised, obtuse; **L** rudiments < 2 mm; **Fl** of each **Inf** opening in quick succession (sometimes 2 or 3 simultaneously open); peduncle short, robust, pointed; **Ped** usually < 5 × ± 0.5 cm; **Sep** ± 7 × 4 mm, ovate-lanceolate; **Cl** outside brown-red, inside yellow, densely blotched with reddish to ± black, ≤ 7.5 cm ∅, convexly bulging, without annulus-like thickening; **Cl** lobes ± 2.5 × 2.5 cm, curved outwards, margins ≤ 1.5 mm abruptly folded upwards, fleshy, crenate, curly, inside densely wrinkled-papillose, papillae ± without apical **Bri**, margins of the **Cl** lobes densely hairy, clavate **Ha** dark purple, ± 3 mm, vibratile; **Ci** lobes 4 mm, 3-parted, middle appendage ± 2 mm, margins thickened, strongly incurved and forming a groove, lateral appendages ± 0.5 mm, basal part ± square, slightly concave; **Cs** lobes ≤ 7 mm, dorsal appendage ± 2 mm, laterally flattened, wing-like, erectly spreading, apically narrower; **Poll** 0.5 - 0.55 × 0.3 mm; **Se** 9 × 6 mm (extremely large).

Closely related to *T. gemmiflora*.

T. parvipuncta (N. E. Brown) L. C. Leach (Trans. Rhodesia Sci. Assoc. 59(1): 3, 1978). **T**: RSA, Northern Cape (*Barkly* s.n. [K]). – **D**: RSA. **I**: Bruyns (1995b: 194, 197).

≡ *Stapelia parvipuncta* N. E. Brown (1890); **incl.** *Stapelia parvipunctata* K. Schumann (s.a.) (*nom. inval.*, Art. 32.1c).

T. parvipuncta ssp. **parvipuncta** – **D**: RSA (Northern Cape, Western Cape).

Incl. *Stapelia parvipuncta* var. *parvipuncta*; **incl.** *Tridentea parvipuncta* var. *parvipuncta*.

[2] Stems rarely branched towards the tips, grey-green, rather robust, to 7.5 × basally 1.5 cm; **L** rudiments ± 3 (-5) mm; peduncle spreading; **Fl** frequently decumbent; **Ped** ± 6 cm; **Sep** ± 3 × 1.25 mm, ± narrowly triangular; **Cl** outside dotted with purple-brown, inside pale cream-coloured to greenish, with raised purple-brown dots, central depression embracing the **Cn**; **Cl** lobes sometimes with narrow purple-brown margin, ≤ 1 × ± 1.2 cm, slightly bulging outwards, intermediate lobes small, tooth-like, inside weakly and irregularly wrinkled, not papillose, margins of the **Cl** lobes hairy, **Ha** purple, simple or ± clavate; **Ci** lobes 2 - 3 mm, usually deeply bifid, appendages subulate, widely diverging and curved outwards; **Cs** lobes ± pointed, usually not surpassing the **Anth**, sometimes meeting in the centre, at times dorsally weakly bulging; **Poll** ± 0.5 × 0.3 mm.

T. parvipuncta ssp. **truncata** (C. A. Lückhoff) Bruyns (SAJB 61(4): 194, fig. 18 (p. 197), 1995). **T**: RSA, Northern Cape (*Hall* s.n. in *NBG* 229/56 [NBG]). – **D**: RSA (Northern Cape, Western Cape).

≡ *Stapelia parvipuncta* var. *truncata* C. A. Lückhoff (1937) ≡ *Tridentea parvipuncta* var. *truncata* (C. A. Lückhoff) L. C. Leach (1980); **incl.** *Tridentea pusilla* R. Frandsen (1993).

[2] Differs from ssp. *parvipuncta*: Margins of the **Cl** lobes glabrous; **Ci** lobes rectangular, ± truncate.

T. peculiaris (C. A. Lückhoff) L. C. Leach (Trans. Rhodesia Sci. Assoc. 59(1): 4, 1978). **T** [neo]: RSA, Western Cape (*Lückhoff* 280 [BOL]). – **D:** RSA (Western Cape: Little Namaqualand, Knersvlakte). **I:** Bruyns (1995b: 196-197, 199).

≡ *Stapelia peculiaris* C. A. Lückhoff (1938).

[2] Stems tapering towards the tips, rhizomatous, grey-green, basally 2 cm ⌀; **L** rudiments ≤ 3 mm; **Inf** sometimes 1-flowered; peduncle short; **Ped** ≤ 2.5 cm; **Sep** ± 3 mm; **Cl** reddish-brown, sometimes spotted with greenish, brown or dull purple, 3 - 4 cm ⌀, with ± thick and usually paler annulus around the **Cn**; **Cl** lobes ± 1.25 × 1.4 cm, margins in part basally folded inwards, inside densely papillose, papillae in part or completely confluent, apically free, apical **Bri** dark, small (esp. at the lobe tips and in the **Fl** centre), margins of the **Cl** lobes hairy, clavate **Ha** purple, ± 1.5 mm, vibratile; **Cn** flatly bowl-shaped; **Ci** lobes ± 2 × 2 mm, conspicuously 3-parted, margins of the middle appendage thickened, slightly incurved and forming a longitudinal groove, apically obtuse or irregularly crenate, lateral appendages ± 1 mm, ± laterally diverging; **Cs** lobes (narrowly) ovate-acute, not surpassing the **Anth**, with ± weak dorsal bulge; **Anth** apically obtusely broadened, not overlapping; **Poll** ± 0.6 × 0.4 mm.

T. virescens (N. E. Brown) L. C. Leach (Trans. Rhodesia Sci. Assoc. 59(1): 3, 1978). – **D:** S Namibia, N RSA (Northern Cape, Western Cape). **I:** Eggli (1994: 182); Bruyns (1995b: 194, 196). **Fig. XLVII.g**

≡ *Stapelia virescens* N. E. Brown (1890); **incl.** *Stapelia aurea* Dinter (1928).

[1] Stems with rounded **Ri**; **L** rudiments ≤ 1 cm, without stipular rudiments; **Inf** usually many-flowered, flowering several times; peduncle (when present) ≤ 4.5 cm, robust, erect; **Fl** strongly excrement-scented; **Ped** ≤ 8 cm; **Sep** 2.5 - 4 × ± 1.25 mm, sometimes weakly ciliate, **Ha** weakly clavate; **Cl** uniformly pale greenish-yellow or yellow, 2.5 - 3 cm ⌀, deeply incised, **Cn** sunken for ½; **Cl** lobes ± 0.8 - 1.2 × 0.7 - 0.8 cm, broadly lanceolate, margins outcurved, apically obtuse, inside densely papillose and wrinkled, papillae conspicuously raised, only apically free, apical **Bri** minute, margins of the **Cl** lobes very rarely hairy, **Ha** white, simple, sometimes very weak; **Cn** very shallowly bowl-shaped; **Ci** lobes ± 3 mm, deeply 3-parted, middle appendage much larger and longer than the widely divergent lateral appendages, apically entire or ± toothed, inside concave; **Cs** lobes ± 3.5 mm, filiform, dorsal appendage ± 2 mm, laterally flattened, wing-like, ± narrowly triangular; nectar cavity ± obtusely triangular; **Poll** ± 0.53 × 0.33 mm.

×TROMOSTAPELIA

B. Müller & F. Albers

×Tromostapelia P. V. Heath (Calyx 1(1): 21, 1992).
Incl. ×*Gonotriche* P. V. Heath (1992).
= *Tromotriche* × *Stapelia*.

The following names are of unresolved application but are referred to this genus: *Stapelia* ×*bella* A. Berger (*pro sp.*) (1902) ≡ ×*Gonotriche bella* (A. Berger) P. V. Heath (1992); *Stapelia* ×*cupularis* N. E. Brown (*pro sp.*) (1897) ≡ ×*Tromostapelia mutabilis* var. *cupularis* (N. E. Brown) P. V. Heath (1992); *Stapelia* ×*maculosa* Donn (*pro sp.*) (1804) ≡ *Orbea* ×*maculosa* (Donn *pro sp.*) Haworth (1812) ≡ ×*Gonostapelia maculosa* (Donn) P. V. Heath (1992).

TROMOTRICHE

B. Müller & F. Albers

Tromotriche Haworth (Synops. Pl. Succ., 36, 1812). **T:** *Stapelia revoluta* Masson [Lectotype, designated by L. C. Leach, J. South Afr. Bot. 48: 425-426, 1982.]. – **Lit:** Leach (1984); Bruyns (1995b). **D:** S Namibia, NW and SW RSA. **Etym:** Gr. 'tromos', trembling; and Gr. 'thrix, trichos', hair; referring to the vibratile corolla hairs of some taxa.

Incl. *Caruncularia* Haworth (1812). **T:** *Stapelia pedunculata* Masson [Lectotype, designated by L. C. Leach, Excelsa Tax. Ser. 2: 4, 1980].

Stems decumbent-erect to ascending, ± clustered and basally branched, sometimes in compact tufts, sometimes rhizomatous with distant aerial **Br**, rarely stems pendent or creeping, ± 5 - 300 × 0.6 - 2.5 cm, obtusely 4- (to 6-) ribbed; **Tu** inconspicuous and stems therefore sometimes with ± mosaic-like surface, ± cylindrical, old **Tu** apically mostly with a yellow-brown cork layer, mostly without **L** and stipular rudiments, or **L** rudiments when present small, deltoid, with lateral stipular rudiments; **Inf** 1- to few-flowered with successive development, mostly scattered along the stems from the lateral furrows; peduncle short, stout, tapered, ± lump-like, elongating with time; **Ped** (0.6-) 1 - 19 cm, spreading to ascending, glabrous; **Sep** 5, acute, thickly fleshy, glabrous; **Cl** (1.2-) 1.5 - 7 cm ⌀, inverted saucer-shaped, flat or tubular-campanulate, deeply (rarely shallowly) incised, ± 5-angular, thickened near the tube base, mostly ascending into an inwards-directed annulus; **Cl** lobes ± ovate, acute, mostly widely spreading, sometimes with intermediate lobes as in *Huernia*, inside frequently rugose, mostly ± densely papillose, margins often with clavate **Ha**; **Cn** biseriate, glabrous, ± conspicuously stipitate, foot 5-angled; **Ci** lobes ascending or horizontally spreading, fused basally or laterally with the **Cs** lobes to form a cup-shaped structure; **Cs** lobes incumbent on the **Anth**, commonly much overtopping the **Anth**, basally flattened, apical appendage frequently cylindrical or clavate, ascending above the **Sty** head, tip often ± papillate, lower ½ often with a dorsal appendage, this often laterally compressed, deltoid to clavate or horn- or wing-like, mostly shorter than the apical appendage; **Anth** short, ± rectangular, without apical appendage, incumbent on the **Sty** head, partly thickened, laterally broadened towards the tip, apically meeting one another; entry to the nectar cavity mostly small, narrowly oblong-elliptic; **Poll** ± 0.3 - 0.5 (-0.7) × 0.15 - 0.5 mm, ± round to ellipsoid or D-shaped, corpuscle ± 2× as long as broad, translators ± long, slender, apically ± acute, wings short, broad; **Fr** ± 6 - 10 cm, fusiform, paired, glabrous, smooth, widely divergent; **Se** > 2× as long as broad, strongly convex, ± narrowly ovate, margins strongly thickened, bent inwards.

Bruyns (1995b) postulates a close relationship to *Tridentea* and *Stapelia*. The genus can be divided as follows:

[1] Sect. *Tromotriche*: Stems pendent, creeping or rhizomatous; **L** rudiments (when present) delicate, deltoid, with stipular rudiments; **Cl** inverted saucer-shaped, tube flatly bowl-shaped, or tubular-campanulate; **Ci** lobes longer than the **Cs** lobes, simple or bifid, tips strongly thickened; apical appendage of the **Cs** lobes cylindrical, tips often thickened, mostly with a dorsal appendage; **Anth** not broadened and not meeting one another.

[2] Sect. *Caruncularia* (Haworth) Bruyns 1995: Stems ± clustered, erect (sometimes creeping), without **L** and stipular rudiments; **Cl** flat to doubly campanulate; **Ci** lobes much shorter than the **Cs** lobes, simple, spreading or laterally fused with the **Cs** to form a cup, margins esp. near the tips frequently crenate, tips not thickened; **Cs** lobes short, simple or with an apical appendage, this cylindrical, elongating over the **Sty** head, tips clavate, mostly with a dorsal appendage; **Anth** laterally broadened, tips meeting one another, sometimes imbricate.

Natural hybrids have been reported with species of the genus *Stapelia* (= ×*Tromostapelia*) as well as the combination *Tromotriche ruschiana* × *Hoodia officinalis* ssp. *delaetiana*.

T. aperta (Masson) Bruyns (SAJB 61: 208, 1995). **T:** RSA, Northern Cape (*Masson* s.n. [lecto − icono: Masson, Stapel. Nov. t. 37, 1797]). − **D:** S Namibia, RSA (NW Northern Cape). **I:** Bruyns (1995b: 205, 208). **Fig. XLVII.h**

≡ *Stapelia aperta* Masson (1797) ≡ *Orbea aperta* (Masson) Sweet (1827) ≡ *Caruncularia aperta* (Masson) Sweet (1830) ≡ *Caralluma aperta* (Masson) N. E. Brown (1890) ≡ *Tridentea aperta* (Masson) L. C. Leach (1978).

[2] Stems spreading to ascending, sometimes horizontal, conical, 5 - 7.5 × (basally) 1 cm, young parts commonly ± distinctly finely roughened; **Inf** 1-flowered near the stem base, opening in long intervals; **Ped** ≤ 14 cm, slender, normally rather weak and ± creeping; **Sep** ≤ ± 5 mm, sometimes finely roughened; **Cl** outside pale green, dotted chestnut-brown, inside normally brown or green (often red around the **Cn**, sometimes white or pink), to ± 4 cm ∅, ± conical; **Cl** tube small, cup-shaped, completely enclosing the **Cn**, wrinkles on the tube margin and basal ½ of **Cl** lobes commonly whitish, apically darker; **Cl** lobes (often widely) spreading, inside finely pubescent, outside sometimes indistinctly roughened; **Ci** lobes with deep pouches, laterally ± strongly fused with the dorsal sides of the **Cs** to form a cup-like structure, inside wrinkled, apically papillose, strongly crenate, margins dentate, sometimes apically with a dentate appendage bent outwards; **Cs** lobes very variable in length, simple, apically clavate or clavate-spatulate, erectly connivent, apically papillate, dorsally with a broad irregularly papillose crest, rarely with a dorsal bulge, sometimes as appendage; entry to the nectar cavity often large, broadly transversely elliptic; **Poll** ± 0.3 - 0.4 mm, ± broadly D-shaped.

Closely related to *T. umdausensis*.

T. baylissii (L. C. Leach) Bruyns (SAJB 61(4): 206, fig. 28 (p. 203), 1995). **T:** RSA, Eastern Cape (*Leach & Bayliss* 13617 [PRE, K, NBG, SRGH]). − **D:** RSA (Eastern Cape). **Fig. XLVIII.a**

≡ *Stapelianthus baylissii* L. C. Leach (1968) ≡ *Tridentea baylissii* (L. C. Leach) L. C. Leach (1978); **incl.** *Tridentea baylissii* var. *baylissii*; **incl.** *Tridentea baylissii* var. *ciliata* L. C. Leach (1980).

[1] Stems pendent from rocks, to 3 m, in cultivation rhizomatous, rarely branching over the whole length, ≤ 1.5 cm ∅, very rarely 6-ribbed, sides ± flat, furrows insignificant with age; **Inf** 1-flowered; peduncle 2.5 - 3 mm; **Ped** ≤ 1.3 cm; **Sep** 2.5 - 3 mm; **Cl** outside purple-rose, inside red-purple, campanulate, 1.2 - 1.5 × 1.2 - 1.3 cm; **Cl** lobes apically darker, ± 5 × 7.5 mm, ascending to widely spreading, outside sometimes indistinctly roughened; **Cl** tube inside ± strongly transversely wrinkled, lobes reticulately wrinkled, mouth of tube sometimes papillose, margins of the **Cl** lobes basally sometimes hairy; **Cn** dark purple, ≤ 7.5 mm; **Ci** lobes large and broad, conspicuously spreading-ascending, inside doubly keeled, apically 3-dentate, middle tooth smaller or larger than the lateral teeth; **Cs** lobes ± 5.5 mm, apically cylindrical, slightly thickened and papillate, ascending over the **Sty** head, then divergent-spreading, dorsal appendage to ± 3 mm, laterally strongly flattened, horn- or wing-like; **Poll** ± 0.5 mm, narrowly ellipsoid (to triangular); **Fr** ≤ 6 cm.

Closely related to *T. choanantha*.

T. choanantha (Lavranos & Hall) Bruyns (SAJB 61(4): 204, fig. 27 (p. 202), 1995). **T**: RSA, Western Cape (*Hall* 2579 [BOL]). – **D:** RSA (Western Cape: Little Karoo). **I:** FPA 37: t. 1459, 1965, as *Stapelianthus*. **Fig. XLVIII.b**

≡ *Stapelia choanantha* Lavranos & Hall (1964) ≡ *Stapelianthus choananthus* (Lavranos & Hall) R. A. Dyer (1965) ≡ *Tridentea choanantha* (Lavranos & Hall) L. C. Leach (1978).

[1] Stems often pendent from rocks or rarely in part underground between gravel, poorly branched, to 3 m, ≤ 1.5 cm ∅; **Inf** 1-flowered, mostly near the stem base; **Ped** ± 1 cm; **Sep** ± 3.5 mm, sometimes very slightly roughened like the **Ped**; **Cl** outside basally rose-coloured, towards the tips and inside red-purple, tubular-campanulate, tube ± 1.5 × 1.1 cm; **Cl** lobes ± 6 × 7 mm, erect to widely spreading, sometimes slightly curved, inside ± smooth, velvet-like glossy; **Cn** red-brown; **Ci** lobes ± 2 - 3 mm, distinctly spreading-ascending, inside longitudinally canaliculate, apically 3-dentate, middle tooth smaller and shorter than the lateral teeth; **Cs** lobes ± 5 mm, apically cylindrical and filiform, erect, apically indistinctly wrinkled, sometimes closely meeting and forming a slender column above the **Sty** head, dorsal appendage laterally flattened, horn-like, ± erect; **Poll** 0.45 - 0.5 × 0.15 - 0.2 mm, ± erect, somewhat covered by the **Ci**; **Fr** to ± 10 cm.

Closely related to *T. baylissii*.

T. herrei (Nel) Bruyns (SAJB 61(4): 208, fig. 34 (p. 207), 1995). **T** [neo]: RSA, Northern Cape (*Herre* s.n. [STE 1571]). – **D:** RSA (Northern Cape).

≡ *Stapelia herrei* Nel (1933) ≡ *Tridentea herrei* (Nel) L. C. Leach (1978); **incl.** *Stapelia tigrina* Nel (1933) (*nom. illeg.*, Art. 52.1); **incl.** *Stapelia neliana* A. C. White & B. Sloane (1937) ≡ *Tridentea herrei* var. *neliana* (A. C. White & B. Sloane) P. V. Heath (1993).

[2] Stems compact, apically tapering and rarely branched, basally shortly decumbent-erect, pale grey-green, 7 - 12 × (basally) ≤ 1.6 cm, very bluntly ribbed, rarely distinctly 6-ribbed, sometimes finely roughened; **Inf** 1- to few-flowered, normally from young stem parts (sometimes along the whole stem length); peduncle persistent, often lacking; **Ped** to ± 3 (-4.5) cm, erect or ascending, often finely roughened; **Sep** ± 4 × 1.25 mm, ± narrowly triangular; **Cl** outside pale green with darker venation, lobes brownish, inside brownish, purple, green or red-brown, shallowly doubly campanulate, wrinkles white or cream-coloured; **Cl** tube basally often red or cream-coloured, to ± 2.5 cm ∅, short, bowl-shaped or basally urceolate, closely embracing the **Cn**; **Cl** lobes ± 2.5 × 1.5 cm, convex, margins strongly bulging outwards, sometimes with intermediate lobes like *Huernia*, outside sometimes finely roughened; **Ci** orange-brown, lobes ± 2.5 × 2 mm, oblong, obtuse, ascending, apically canaliculate, tips spreading, crenate; **Cs** glossy purple-black, rarely paler, lobes ± 4.5 mm, apically cylindrical, very strongly thickened (inflated), ± 1.75 mm ∅, tips smooth, meeting one another and erect, then divergent, dorsal appendage laterally flattened, horn-like, apically thickened; **Anth** lemon-yellow; **Poll** orange, ± 0.6 - 0.7 × 0.4 - 0.5 mm, germination mouth conspicuously raised but short.

T. longii (C. A. Lückhoff) Bruyns (SAJB 61(4): 204, fig. 26 (p. 201), 1995). **T**: RSA, Eastern Cape (*Leach & Bayliss* 15680 [SRGH, K, PRE]). – **D:** RSA (Eastern Cape). **Fig. XLVIII.c**

≡ *Stapelia longii* C. A. Lückhoff (1935) ≡ *Tridentea longii* (C. A. Lückhoff) L. C. Leach (1978).

[1] Stems pendent from rocks, to 30 × 0.6 - 0.7 cm, young finely roughened; **Tu** young with an acute tooth to ± 1 mm, spreading, sometimes contorted; **Inf** 1-flowered; peduncle short, from the **Tu** side (not from the furrows); **Bra** 1.5 mm, acute, canaliculate; **Ped** ± 1 - 4 × 0.15 cm, sometimes slightly roughened; **Sep** ± 3 mm; **Cl** outside dull green, inside yellowish- to reddish-brown (centrally paler), 2 - 2.5 cm ∅, occasionally shallowly bowl-shaped, central sunken area ± 0.75 mm; **Cl** lobes 7 - 8 × 7 - 8 mm, like the margins slightly revolute, intermediate lobes strongly reduced, sometimes more prominent, outside sometimes slightly finely roughened, inside ± smooth or very indistinctly concentrically wrinkled, basal ½ of the margins of the lobes hairy, clavate **Ha** dark purple, ± 2 mm, vibratile; **Ci** lobes ± 2.5 mm, spreading-ascending, oblong-rectangular, basally with thickened lateral ribs, apically divided with obtuse papillate globose tips; **Cs** lobes ± 2 mm, cylindrical, apically strongly thickened, papillate, meeting one another and erect, dorsal appendage laterally flattened, horn-like, apically thickened, spreading outwards; **Anth** obtusely emarginate, dorsally winged; **Poll** ± 0.5 × 0.35 mm, oblong-ellipsoid, ascending.

T. longipes (C. A. Lückhoff) Bruyns (SAJB 61(4): 208, fig. 33: B-D,F (p. 206), 1995). **T**: Namibia (*Rusch* s.n. [BOL 31684]). – **D:** S Namibia, RSA (Northern Cape). **Fig. XLVIII.d**

≡ *Stapelia longipes* C. A. Lückhoff (1934) ≡ *Tridentea longipes* (C. A. Lückhoff) L. C. Leach (1980); **incl.** *Stapelia longipes* var. *namaquensis* C. A. Lückhoff (1937) ≡ *Tridentea longipes* var. *namaquensis* (C. A. Lückhoff) P. V. Heath (1993).

[2] Forming clusters to 40 cm ∅; stems apically tapering, spreading to ascending (rarely horizontal), rooting when touching the ground, ± 5 - 12 × (basally) ≤ 2 cm, ± finely roughened; **L** rudiments (when present) very delicate; **Inf** several-flowered, **Fl** opening in succession; peduncle very short; **Ped** 3 - 19 cm, first erect, then elongating and decumbent, apically often ± bumpy; **Sep** 3 - 4 × 1.5 mm, narrowly triangular; **Cl** outside pale green, irregularly spotted with purple, inside purple, with broad

whitish wrinkles, ± 3 - 7 cm ⌀, ± flat, deeply incised; **Cl** tube ± 2 mm, shallowly funnel-shaped, closely embracing the **Cn**; **Cl** lobes apically red or (dark) brown, margins somewhat revolute, outside sometimes finely roughened, tube inside basally smooth, margins of the **Cl** lobes basally densely hairy in tufts, **Ha** ± 2.5 mm, fusiform or slender-clavate, usually acute; **Ci** (pale) brown, lobes small, ascending, apically shortly spreading, emarginate or sometimes ± bifid, crenate or sometimes irregularly bumpy, inside longitudinally canaliculate; **Cs** glossy black, red-brown or yellow-brown, dorsal appendage sometimes differently coloured, lobes apically cylindrical (spatulate), strongly thickened, papillate, erect and meeting one another, dorsal appendage laterally flattened, horn-like, apically thickened and papillate; **Poll** 0.55 - 0.6 × 0.45 mm, germination mouth ± 0.3 mm, ± erect, translator ≤ 0.4 mm.

Closely related to *T. pedunculata* and *T. ruschiana*.

T. pedunculata (Masson) Haworth (SAJB 61: 208, 1995). **T:** RSA, Western Cape (*Masson* s.n. [lecto – icono: Masson, Stapel. Nov. t. 21, 1797]). – **D:** RSA (Northern Cape, Western Cape). **I:** Bruyns (1995b: 206, 208). **Fig. XLVIII.e**

≡ *Stapelia pedunculata* Masson (1797) ≡ *Caruncularia pedunculata* (Masson) Haworth (1812) ≡ *Duvalia pedunculata* (Masson) Loudon (1841) ≡ *Tridentea pedunculata* (Masson) L. C. Leach (1978); **incl.** *Caruncularia jacquinii* Sweet (1830); **incl.** *Caruncularia massonii* Sweet (1830); **incl.** *Caruncularia penduliflora* Sweet (1830) ≡ *Stapelia penduliflora* (Sweet) Steudel (1840); **incl.** *Caruncularia simsii* Sweet (1830); **incl.** *Stapelia laevis* Decaisne (1844).

[2] Stems apically slightly tapering, spreading to ascending (rarely horizontal), laxly branched, at times forming compact clusters, pale green (later grey-green), ± 5 - 12 × (basally) ± 1.5 cm, very bluntly ribbed; **L** rudiments (when present) very delicate; **Inf** often 1-flowered, additional buds caducous, over a long period of time; peduncle very short; **Ped** 4.5 - 19 cm, first erect, then elongating and often creeping; **Sep** brownish, ± 4 × 1.5 mm, ± narrowly triangular, sometimes like the **Ped** finely roughened; **Cl** outside pale green, venation darker, inside dark purple-brown, ± densely spotted or blotched with white or dark purple, ± 7 cm ⌀, ± flat, deeply incised; **Cl** tube ± 2 mm, shallowly funnel-shaped, basally embracing the **Cn**, margin of the tube and **Cl** lobes basally whitish (ash-coloured), sometimes with faint rosy or yellowish hue, usually dotted with red-brown, upper ⅔ of the **Cl** lobes olive-green, greenish-yellow, yellow-brown or (purple-) brown, to ± 3 × 1.6 cm, margins strongly revolute and thus lobes appearing conical, margins basally densely hairy in tufts, clavate **Ha** dark purple, ≤ 2.5 mm, vibratile; **Ci** purple-black, lobes ± 2 mm, ± rectangular, ascending, inside longitudinally canaliculate, apically obtuse, retuse or ± acute, often crenate; **Cs** glossy black or often brown (or bicoloured), lobes apically cylindrical (clavate to spatulate), strongly thickened, papillate, meeting one another and erect, dorsal appendage laterally flattened, horn-like, apically thickened; **Poll** ± 0.6 × 0.4 mm, ± ovoid, germination mouth ± 0.3 mm.

Closely related to *T. longipes* and *T. ruschiana*. *T. pedunculata* in the sense of Dyer (1962) is *T. longipes* according to Bruyns (1995b).

T. revoluta (Masson) Haworth (Synops. Pl. Succ., 36, 1812). **T:** [lecto – icono]: Masson, Stapel. Nov. t. 10, 1796. – **D:** RSA (Western Cape). **I:** Bruyns (1995b: 200, 203). **Fig. XLVIII.f**

≡ *Stapelia revoluta* Masson (1796); **incl.** *Stapelia fuscata* Jacquin (1806) ≡ *Tromotriche fuscata* (Jacquin) Haworth (1819) ≡ *Stapelia revoluta* var. *fuscata* (Jacquin) N. E. Brown (1909) ≡ *Tromotriche revoluta* var. *fuscata* (Jacquin) P. V. Heath (1993); **incl.** *Stapelia glauca* Jacquin (1809) ≡ *Tromotriche glauca* (Jacquin) Haworth (1812); **incl.** *Stapelia potensa* Hornemann (1819); **incl.** *Stapelia protensa* Hornemann (1819); **incl.** *Stapelia tigrida* Decaisne (1844) ≡ *Stapelia revoluta* var. *tigrida* (Decaisne) N. E. Brown (1909) ≡ *Tromotriche revoluta* var. *tigrida* (Decaisne) P. V. Heath (1993); **incl.** *Stapelia revoluta* var. *glaucescens* Rüst *ex* A. Berger (1910); **incl.** *Stapelia glaucescens* Rüst *ex* A. Berger (1910) (*nom. inval.*, Art. 34.1c).

[1] Stems few, erect, sometimes scattered over large areas, rarely in clumps, rhizomatous stolons slender, ± cylindrical, horizontal; stems 15 - 35 cm, sometimes poorly segmented; **Tu** small, prominent and stems conspicuously 4-angled; **L** rudiments minute, ± conical, stipular rudiments conical, persistent, corky with age, yellow-brown; **Inf** 1- to 2-flowered, sessile; **Ped** ≤ 1.5 cm; **Sep** ≤ 8 mm; **Cl** inside yellow to red-brown (centrally paler), very variable in size, sometimes spreading to 6.8 cm ⌀, incised to ± ½ and lobes completely revolute; **Cl** tube wide, embracing the **Cn** partly (rarely completely), annulus strongly thickened, inside smooth, base of the tube papillose, margins of the **Cl** lobes hairy, clavate **Ha** purple, vibratile; **Cn** sometimes ± sessile; **Ci** brown, lobes ± rectangular, ascending or horizontally spreading, bluntly bifid or usually 3-toothed, sometimes ± retuse; **Cs** yellow, speckled with purple, lobes apically cylindrical, tips thickened and papillate, meeting one another and apically reclining, dorsal appendage ≤ 4 mm, horn-like, laterally flattened, narrowly triangular, obtuse, spreading, mostly reduced to a ± small bulge; **Poll** 0.8 × 0.5 mm, broadly D-shaped.

Closely related to *T. thudichumii*.

T. ruschiana (Dinter) Bruyns (SAJB 61(4): 208, 1995). **T** [neo]: Namibia (*Plowes* 4988 [PRE]). – **D:** S Namibia.

≡ *Stapelia ruschiana* Dinter (1923) ≡ *Tridentea ruschiana* (Dinter) L. C. Leach (1978).

[2] Forming clumps to 20 cm ∅; stems apically tapering, sometimes laxly branched, basally 2 cm ∅, young sometimes finely roughened; **Inf** usually rather close to the stem base; peduncle ≤ 4 × 0.6 cm, sometimes persistent and repeatedly flowering, morphologically variable; **Ped** irregularly longitudinally ribbed; **Sep** ± 5 × 1.5 mm, margins frequently with few short obtuse transparent **Ha**; **Cl** 4.5 cm ∅, campanulate with spreading or sometimes revolute apically darker lobes; **Cl** tube ± 8 - 10 × 10 - 12 mm, cream- to sand-coloured, spotted with purple, elongately bowl-shaped; **Cl** lobes ≤ 2.5 × 1 cm, red-brown with greenish-purple, inside partly with 20 radial furrows, sometimes very delicately pubescent, **Ha** ± 0.25 mm, clavate or simple, erect, stout, margins of the lobes revolute, basally densely hairy, clavate **Ha** dark purple-red, ≤ 1.5 mm, sometimes fusiform; **Cn** ± 8 mm; **Ci** yellow (-brown), lobes ± 3 × 1.5 mm, ascending, inside bulging and concave, apically obtuse, variably crenate or bluntly toothed; **Cs** glossy purple-black, lobes ± 5 mm, apically cylindrical, strongly thickened, papillate, meeting one another and erect, then divergent, dorsal appendage laterally flattened, horn-like, apically thickened, basally ± triangular; **Poll** ± 0.6 × 0.45 mm, ± broadly elliptic, germination mouth 0.3 mm, distinctly prominent.

Closely related to *T. longipes* and *T. pedunculata*.

T. thudichumii (Pillans) L. C. Leach (JSAB 48(3): 426, 1982). **T:** RSA, Northern Cape (*Thudichum s.n.* in *KG* 583/54 [BOL 26740]). – **D:** RSA (Northern Cape: W Great Karoo). **I:** Bruyns (1995b: 201, 204). **Fig. XLVIII.g**

≡ *Stapelia thudichumii* Pillans (1959).

[1] Stems solitary or few together, scattered, rhizomatous stolons slender, ± cylindrical, horizontal; stems ± 10 × ≤ 1.5 cm, minutely papillose; **L** rudiments minute, conical, stipular rudiments conical, persistent, corky with age, yellow-brown; **Inf** 1-flowered, sessile; **Ped** ≤ 1 cm, sometimes like the **Sep** finely roughened; **Cl** inside greenish-purple, ≤ 3 cm ∅ (when spread out), incised for ½; **Cl** tube shallowly bowl-shaped, inside finely pubescent, annulus conspicuous; **Cl** lobes strongly revolute, sometimes broader than long, margins ± transparent, flat, hairy, clavate **Ha** dark purple, vibratile; **Ci** dark brown, lobes ≤ 2.5 mm, ± rectangular, slightly ascending, with slight longitudinal furrow, apically bifid; **Cs** pale brown, lobes apically cylindrical, meeting and crossing each other in the centre, or sometimes shortly ascending, tips thickened and papillate, dorsal appendage laterally flattened, horn-like; **Poll** brown, D-shaped, germination mouth yellow, ± 0.5 × 0.375 mm.

Closely related to *T. revoluta*.

T. umdausensis (Nel) Bruyns (SAJB 61(4): 207, fig. 30 (p. 204), 1995). **T:** RSA, Northern Cape (*Herre s.n.* in *Stellenbosch* 7068 [STE]). – **D:** S Namibia, NW RSA (Northern Cape).

≡ *Caralluma umdausensis* Nel (1935) ≡ *Tridentea umdausensis* (Nel) L. C. Leach (1978).

[2] Stems apically slightly tapering, basally shortly decumbent, otherwise erect, ≤ 9 × ± 1 cm; **Inf** frequently 1-flowered; peduncle very short; **Ped** usually rather weak and thus **Fl** pendent, sometimes slightly roughened; **Sep** ± 3 × 1.5 mm; **Cl** ± red, mostly radially white-lineate, campanulate with spreading ± revolute lobes; **Cl** tube embracing the **Cn** completely; **Cl** lobes basally mostly whitish or yellowish, apically greenish-yellow or brown, ≤ 1 × 0.5 cm, papillae dark purple, pink or white (sometimes colours mixed), outside sometimes very finely roughened; **Cn** purple-black to pale brown, margin variably crenate-dentate; **Ci** lobes laterally ± strongly united with the dorsal side of the **Cs** lobes; **Cs** short, ± ovate, apically obtuse, rarely acute, not overtopping the **Anth**, with a dorsal appendage or bulge; **Anth** apically emarginate, partly imbricate and overlapping; nectar cavity large, shallow; **Poll** orange-yellow, ± 0.5 × 0.4 mm, ± broadly D-shaped, germination mouth ± 0.3 mm, conspicuously prominent.

Closely related to *T. aperta*.

WHITE-SLOANEA

U. Meve

White-sloanea Chiovenda (Malpighia 34: 541, 1937). **T:** *Caralluma crassa* N. E. Brown. – **Lit:** Bally & al. (1975). **D:** Somalia. **Etym:** For Alain C. White (1880 - 1951) and Boyd L. Sloane (1886 - 1955), US-American authors of important books on euphorbias and asclepiads.

Incl. *Drakebrockmania* A. C. White & B. Sloane (1937) (*nom. illeg.*, Art. 53.1). **T:** *Caralluma crassa* N. E. Brown.

Compact 1-stemmed stem succulents; stem ascending to erect, 4-ribbed, rectangular, 3 - 13 × 4.5 - 5.5 cm, grey-green to pale brown, with conspicuous wax covering; **Tu** small, hardly porrect, laterally compressed, without **L** rudiment; **Inf** basal, peduncle short, with numerous **Bra**, many-flowered but only 1 **Fl** open at a time; **Ped** 5 - 15 × 1.5 - 2 mm, horizontally spreading or somewhat ascending; **Sep** lanceolate, acute, 2 - 3 mm, glabrous; **Cl** outside grey, delicately spotted with reddish, inside cream-coloured, dotted and spotted with red, deeply funnel-shaped, ± 2 × 2 cm, outside glabrous, inside densely covered with **Ha**-papilla, these semiglobose to conical, apically sometimes with a **Bri**, **Bri** very acute, short and straight to long and geniculate; **Cl** tube 10 - 12 × 7 - 9 mm; **Cl** lobes triangular, acute, 12 - 14 × 5 - 8 mm, ± spreading at an angle of 45°, margins basally with purple clavate **Ha** 1.5 - 2 mm long; **Cn** biseriate, shortly stipitate, deeply divided into 10 free lobes, ± 6 - 7 × 6 mm, basally whitish, inside towards the tips more and more striped with brown-orange and red-brown; **Ci** lobes red-brown, inserted at the **Gy** base, ascending-erect, apically somewhat inflexed, 5.5 - 6.5 mm, flattened and inner face somewhat canaliculate, bifid to the middle, appendages 2.5 - 3.5 mm, subulate; **Cs** lobes pale red with red-brown stripes and blotches, 3 - 4 mm, inserted high up on the **Anth** back, erect, broadened from the narrow base in an elongate-trapezoid shape, lobes united to form a funnel-shaped structure, margins slightly revolute, apically truncate but deep and irregularly toothed; guide rails short, vertical and inserted high up on the elongate **Gy**; **Anth** ± ovoid, somewhat pointed, closely incumbent on the slightly convex **Sty** head; **Poll** ± cylindrical, ± 0.55 × 0.25 mm, germination mouths reaching almost to the base of the **Poll** along the inner margins, ± 0.3 mm, caudicles ± 0.1 mm, ± erect, with apical disc-like broadening, corpuscle broadly obovate, compact, ± 0.2 × 0.15 mm, basally with narrow tringular wing-like extensions; **Fr** coarse, shortly fusiform; **Se** pale brown, ± 5 × 3.5 mm, margin flat.

This monotypic genus is – despite the similarity to various 4-ribbed species of *Caralluma* – closely related to *Duvaliandra*, with which it shares the compact leafless stems and the winged pollinaria. Such winged pollinaria are also characteristic for *Duvalia* and *Huernia*, and this is the place where the evolutionary lines probably evenutally lead to.

W. crassa is one of the rarest stapeliads and has been collected only 2 or 3 times. A hybrid with *Huerniopsis* has been obtained in cultivation (see ×*Whitesloaniopsis*).

W. crassa (N. E. Brown) Chiovenda (Malpighia 34: 541, 1937). **T:** Somalia (*Drake-Brockman* 1132 [K [lecto – icono]]). – **D:** Somalia; very local and exceedingly rare. **I:** Aloe 23: 11, 1986. **Fig. XLVIII.h**

≡ *Caralluma crassa* N. E. Brown (1935) ≡ *Drakebrockmania crassa* (N. E. Brown) A. C. White & B. Sloane (1937).

Description as for the genus.

×WHITESLOANIOPSIS

B. Müller & F. Albers

×**Whitesloaniopsis** Barad (CSJA 67(4): 255, 1995).

= *White-sloanea* × *Huerniopsis.*

×**W. 'Gray Ghost'** Barad (CSJA 67(4): 254-255, ill., 1995).

= *W. crassa* × *H. decipiens.*

XYSMALOBIUM

U. Meve

Xysmalobium R. Brown (Mem. Wern. Nat. Hist. Soc. 1810: 38, 1810). **T:** *Asclepias undulata* Linné. – **D:** Tropical Africa. **Etym:** Gr. 'xysma', cleft or scraped place; and Gr. 'lobion', small lobe; for the often cleft corona lobes.

Incl. *Xyomalobium* Weale (1873) (*nom. inval.*, Art. 61.1).

Geophytic perennials, above-ground parts deciduous, with white latex; **R** tuberous or with subulate fleshy lateral **R**; stems erect, ± unbranched, basally often becoming woody; **L** decussate, ± sessile; **Inf** terminal or lateral near the stem tips, sessile or pedunculate, pseudo-umbellate; **Cl** stellate or campanulate; **Cn** uniseriate; **Cs** lobes variably shaped, always very fleshy, apically sometimes once or doubly incised, insides with or without keels or teeth; **Gy** sessile; **Anth** appendages membranous, ± ovate; **Poll** flat, ovoid to drop-shaped, pendent; **Fr** mostly solitary, usually erect, fusiform to ovoid, smooth or softly prickly.

The genus belongs to the *Asclepiadinae* of tropical grasslands, esp. in E Africa. Storage roots are present in most taxa. 18 of the ± 45 species occur in RSA. *Xysmalobium* is closely related to *Pachycarpus* and *Gomphocarpus* as well as *Asclepias*.

References

Adams, B. R. & Holland, R. W. K. (1978) The genus *Sarcostemma* in East Africa. Cact. Succ. J. (US) 50(3): 107-111, (4): 166-169, key, ills.

Agnew, A. D. Q. & Agnew, S. (1994) Upland Kenya Wild Flowers. Nairobi (Kenya): East African Natural History Society. Ed. 2; 374 pp., keys, ills.

Ali, S. I. (1983) *Asclepiadaceae*. In: Nasir, E. & Ali, S. I. (eds.), Flora of Pakistan, No. 150. Karachi (Pakistan): Department of Botany, University of Karachi. 65 pp., ills., keys, map.

Alp, P. R. (1998) A new record for *Brachystelma glenense* R. A. Dyer from the Thabazimbi Area. Asklepios No. 74: 19-22.

Ansari, M. Y. (1971) *Ceropegia media* (Huber) Ansari stat. nov. from Western Ghats (Maharashtra). Bull. Bot. Surv. India 11(1-2): 199-201, ills.

Ansari, M. Y. (1984) *Asclepiadaceae*: Genus *Ceropegia*. Fascicles of Flora of India, no. 16. Howrah (India): Botanical Survey of India. 34 pp., ills., key.

Archer, P. G. (1992) Kenya *Ceropegia* scrapbook. Notes and records of some Kenya *Ceropegia*. Hobart (AUS): Artemis Publishing Consultants. 183 pp., ills., maps.

Arekal, G. D. & Ramakrishna, T. M. (1981) A new species of *Brachystelma* (*Asclepiadaceae*) from India. Curr. Sci. 50(3): 145-146.

Bakhuizen van den Brink, R. C. & Backer, C. A. (1950) *Asclepiadaceae*. In: Notes on the Flora of Java. VI. Blumea 6(2): 368-382.

Bally, P. R. O. (1959) *Lithocaulon* Bally (*Asclepiadaceae*), a new genus from Somaliland. Candollea 17: 53-59, ills., maps.

Bally, P. R. O. (1965) Miscellaneous notes on the flora of Tropical East Africa, including descriptions of new taxa, 23-28. Candollea 20: 13-41, ills., maps.

Bally, P. R. O. & al. (1975) A monograph of the genera *Pseudolithos* and *Whitesloanea*. Nation. Cact. Succ. J. 30(2): 31-36, (4): 88-93, ills.

Bayer, M. B. (1976) Notes on species of the genera *Fockea* and *Raphionacme*. Excelsa 6: 61, 87-91.

Bayer, M. B. (1984) Duvalias. Veld. Fl. (1975+) 70: 11-14, ills.

Beddome, R. H. (1874) Icones Plantarum Indiae Orientalis. Vol. 1. Madras (India): Gantz.

Berhaut, J. (1971) Flore illustrée du Sénégal. Tome 1. Dakar (Sénégal): Gouvernement du Sénégal. 626 pp., ills.

Boele, C. (1989) Notes on the genus *Brachystelma* Sims – Part 1. Asklepios No. 47: 3-7, ills.

Bridson, G. D. R. & Smith, E. R. (1991) Botanico-Periodicum-Huntianum / Supplementum. Pittsburgh (US: PA): Hunt Institute for Botanical Documentation, Carnegie Mellon University. 1068 pp.

Brodie, S. (1998) *Huernia macrocarpa* and *Huernia kennedyana*. Curtis's Bot. Mag., ser. nov., 15(1): 2-11, ills.

Brown, N. E. (1903) *Asclepiadaceae*. In: Thiselton-Dyer, W. T. (ed.): Flora of Tropical Africa, 4(1): 231-502, keys. London (GB): Lovell Reeve & Co. Ltd.

Brown, R. (1810) On the *Asclepiadeae*, a natural order of plants separated from the *Apocinae* of Jussieu. Mem. Wern. Nat. Hist. Soc. 1: 15-58.

Bruce, E. A. (1948) Tropical African plants. XX. *Asclepiadaceae*. Kew Bull. 1948: 462-464.

Brummitt, R. K. & Powell, C. E. (eds.) (1992) Authors of plant names. A list of authors of scientific names of plants, with recommended standard forms of their names, including abbreviations. Richmond (GB): The Board of Trustees of The Royal Botanic Gardens, Kew. 732 pp.

Bruyns, P. V. (1980) Ceropegias of the Pretoria area. Asclepiadaceae No. 21: 2-20, ills.

Bruyns, P. V. (1981) A review of *Pectinaria* Haw., *Stapeliopsis* Pillans and a new genus, *Ophionella* (*Asclepiadaceae*). Cact. Succ. J. Gr. Brit. 43(2-3): 61-83, ills., keys.

Bruyns, P. V. (1982a) A few Southern African Brachystelmas. Asklepios No. 25: 107-113, ills.

Bruyns, P. V. (1982b) *Raphionacme* in the Pretoria District. Cact. Succ. J. (US) 54(2): 56-63, ills.

Bruyns, P. V. (1982c) Notes on two little-known stapeliads from Southern Africa. Excelsa 10: 103-111, ills.

Bruyns, P. V. (1983) Resurrection of *Quaqua* N. E. Brown (*Asclepiadaceae* – *Stapelieae*) with a critical review of the species. Bradleya 1: 33-78, ills., key, maps.

Bruyns, P. V. (1984) *Ceropegia*, *Brachystelma* and *Tenaris* in South West Africa. Dinteria 17: 3-80, ills., key.

Bruyns, P. V. (1985) Notes on Ceropegias of the Cape Province. Bradleya 3: 1-47, ills., key.

Bruyns, P. V. (1986a) The genus *Ceropegia* on the Canary Islands (*Asclepiadaceae* – *Ceropegieae*). A morphological and taxonomic account. Beitr. Biol. Pfl. 60(3): 427-458, ills.

Bruyns, P. V. (1986b) Miscellaneous notes on *Ceropegieae* (*Asclepiadaceae*). Bradleya 4: 29-38, ills.

Bruyns, P. V. (1987a) Miscellaneous notes on *Stapelieae* (*Asclepiadaceae*). Bradleya 5: 77-90, ills., maps.

References

Bruyns, P. V. (1987b) The genus *Caralluma* R. Brown (*Asclepiadaceae*) in Israel. Israel J. Bot. 36(2): 73-86, ills., maps.

Bruyns, P. V. (1988a) A revision of the genus *Echidnopsis* Hook. f. (*Asclepiadaceae*). Bradleya 6: 1-48, ills., maps, keys.

Bruyns, P. V. (1988b) Studies in the Flora of Arabia XXIV: The genus *Ceropegia* in Arabia. Notes Roy. Bot. Gard. Edinburgh 45(2): 287-326, maps, ills., key.

Bruyns, P. V. (1989) Miscellaneous notes on *Stapelieae* (*Asclepiadaceae*). Bradleya 7: 63-68, ills., maps.

Bruyns, P. V. (1990a) Miscellaneous notes on *Stapelieae* (*Asclepiadaceae*). 9. A review of *Pseudolithos*. 10. A note on *Caralluma dodsoniana* Lavranos. Bradleya 8: 33-38, ills.

Bruyns, P. V. (1990b) *Stapeliopsis* figures. Asklepios No. 51: 48-51, ills.

Bruyns, P. V. (1992) Miscellaneous notes on *Stapelieae* (*Asclepiadaceae*). 11. A note on *Caralluma adscendens* and its synonyms *C. subulata* and *C. dalzielii*. Bradleya 10: 95-96, ills.

Bruyns, P. V. (1993) A revision of *Hoodia* and *Lavrania* (*Asclepiadaceae* – *Stapelieae*). Bot. Jahrb. Syst. 115(2): 145-270, ills., maps, keys.

Bruyns, P. V. (1995a) A note on *Raphionacme namibiana* (*Asclepiadaceae* – *Periploceae*). Aloe 31(3/4): 63-67, ills., map.

Bruyns, P. V. (1995b) A re-assessment of the genera *Tridentea* Haw. and *Tromotriche* Haw. South Afr. J. Bot. 61(4): 180-208, ills., maps, keys.

Bruyns, P. V. (1995c) New records and new species of *Asclepiadaceae* from Namibia. Bothalia 25(2): 155-172, ills., maps.

Bruyns, P. V. (1999a) The relationships of *Huerniopsis* and *Piaranthus* (*Apocynaceae* – *Asclepiadoideae*). Syst. Bot. 24(3): 379-397, ills., SEM-ills., map, diags.

Bruyns, P. V. (1999b) The systematic position of *Quaqua* (*Apocynaceae* – *Asclepiadoideae*) with a critical revision of the species. Bot. Jahrb. Syst. 121(3): 311-402, ills., key, maps.

Bruyns, P. V. (1999c) A systematic assessment of *Ophionella* (*Apocynaceae*; *Asclepiadoideae*; *Stapelieae*). Bot. J. Linn. Soc. 131(4): 383-398, ills., map, key.

Bruyns, P. V. (1999d) A systematic analysis with notes on the taxonomy of *Notechidnopsis* (*Apocynaceae*: *Asclepiadoideae*). Kew Bull. 54(2): 327-345, ills., map, key.

Bruyns, P. V. (2001) New combinations in the genus *Orbea*. Aloe 37(4): 72-76, ills.

Bruyns, P. V. & Forster, P. I. (1991) Recircumscription of the *Stapelieae* (*Asclepiadaceae*). Taxon 40(3): 381-391, ills.

Bruyns, P. V. & Jonkers, H. A. (1994) The genus *Caralluma* R. Br. (*Asclepiadaceae*) in Oman. Bradleya 11: 51-69, ills., maps, key.

Bruyns, P. V. & Meve, U. (1995) The generic position of *Caralluma dodsoniana*. Edinburgh J. Bot. 52(2): 195-203, ills.

Bullock, A. A. (1952) Notes on African *Asclepiadaceae* 1. Kew Bull. 7: 405-426.

Bullock, A. A. (1953) Notes on African *Asclepiadaceae*. II. Kew Bull. 8: 51-67.

Bullock, A. A. (1963) *Asclepiadaceae*. In: Hutchinson, J. & Dalziel, J. M. (eds.): Flora of West Tropical Africa, 2: 85-103. London (GB): Crown Agents for Oversea Governments and Administrations.

Choux, P. (1914) Études biologiques sur les Asclépiadacées de Madagascar. Ann. Inst. Bot.-Géol. Colon. Marseille, sér. 3, 2: 211-459, ills., 50 tt.

Choux, P. (1923) Nouvelles études biologiques sur les Asclépiadacées de Madagascar. Ann. Inst. Bot.-Géol. Colon. Marseille, sér. 4, 1(2): 1-51, ills.

Clark, P. (1996) *Hoya* cultivar names found. Asklepios No. 67: 33-39.

Collenette, S. (1985) An illustrated guide to the flowers of Saudi Arabia. Buckhurst Hill (GB): Scorpion Publishing Ltd. 514 pp., ill., maps.

Costantin, J. N. (1912) Asclepiadacées. In: Lecomte, H. (ed.): Flore générale de l'Indochine, 4(1): 1-154. Paris (F): Masson & Cie / Muséum National d'Histoire Naturelle.

Court, G. D. (1987) *Fockea* Endl. – an African genus. Asklepios No. 40: 69-74, ills.

Craib, C. (1994) Field studies and cultivation of some *Brachystelma* species in the Eastern Cape, the Eastern Free State (including Qua Qua) and the Transvaal. Excelsa 16: 11-23, ills.

Cribb, P. J. & Leedal, G. P. (1982) The mountain flowers of Southern Tanzania. Rotterdam (NL): A. A. Balkema. 244 pp., ills., maps.

Dahlgren, R. M. T. & al. (1985) The families of the Monocotyledons. Structure, evolution, and taxonomy. Berlin, Heidelberg (D) etc.: Springer-Verlag. xi + 520 pp., ills., keys.

Decaisne, J. (1844) *Asclepiadeae*. In: De Candolle, A. P. (ed.): Prodromus systematis naturalis regni vegetabilis; 8: 490-655. Paris (F): Treuttel & Würtz.

Deflers, A. (1896) Descriptions de quelques plantes nouvelles ou peu connues de l'Arabie méridionale. Decas II / Decas III. Bull. Soc. Bot. France 34: 104-119.

Descoings, B. (1961) Notes taxinomiques et descriptives sur quelques Asclépiadées Cynanchées (Asclépiadacées) aphylles de Madagascar. Adansonia, n.s., 2(1): 299-341, ills.

Dyer, R. A. (1941) *Ceropegia stapeliiformis*. *Stapelia woodii* var. *westii*. *Stapelia tsomoensis*. Flow. Pl. South Afr. 21: tt. 809, 811, 840 + text.

Dyer, R. A. (1962) *Stapelia pedunculata* Masson. Flow. Pl. Afr. 35(3): t. 1388 + text.

Dyer, R. A. (1980) *Asclepiadaceae* (*Brachystelma, Ceropegia, Riocreuxia*). In: Leistner, O. A. (ed.): Flora of Southern Africa, Vol. 27(4). Pretoria (RSA): Department of Agricultural Technical Services. 91 pp., ills., keys.

Dyer, R. A. (1983) *Ceropegia, Brachystelma* and *Riocreuxia* in Southern Africa. Rotterdam (NL): A. A. Balkema. 242 pp., ills., maps, keys.

Eggli, U. (1985) A bibliography of succulent plant periodicals. Bradleya 3: 103-119.

Eggli, U. (1994) Sukkulenten. Stuttgart (D): Ulmer Verlag. 336 pp., ills., maps, keys.

Eggli, U. (1998a) Bibliography of succulent plant periodicals. Bibliografie casopisu o sukulentních rostlinách. Friciana 60: 139 pp.

Endress, M. E & Bruyns, P. V. (2000) A revised classification of the *Apocynaceae* s.l. Bot. Rev. (Lancaster) 66(1): 1-56.

Fabian, A. & Germishuizen, G. (1982) Transvaal wild flowers. Johannesburg (RSA): Macmillan South Africa Ltd. 292 pp., ills.

Field, D. V. (1981a) A new species of *Rhytidocaulon* (*Asclepiadaceae*) from Saudi Arabia. Kew Bull. 36(1): 51-54, ills., key.

Field, D. V. (1981b) *Ceropegia distincta* (*Asclepiadaceae*) and some related species in tropical East and southern Africa. Kew Bull. 36(3): 441-450, key.

Field, D. V. (1982) Two new species of *Ceropegia* (*Asclepiadaceae*) and a reconsideration of *C. subaphylla*. Kew Bull. 37(2): 305-313, ills.

Forster, P. I. (1985) (790) Proposal to conserve 6870 *Brachystelma* against *Microstemma* (*Asclepiadaceae*). Taxon 34(2): 318-319.

Forster, P. I. (1986) *Asclepiadaceae*. The nomenclature of several *Brachystelma* species from Southern Africa. Bothalia 16(2): 227-228, ills.

Forster, P. I. (1990) *Brachystelma glabriflorum* (F. Muell.) Schltr. (*Asclepiadaceae*). Asklepios No. 49: 77-84, ills. map.

Forster, P. I. (1991) Variation in *Hoya australis* R. Br. *ex* Traill (*Asclepiadaceae*). Austrobaileya 3(3): 503-521, ills., maps.

Forster, P. I. (1992) A taxonomic revision of *Sarcostemma* R. Br. subgenus *Sarcostemma* (*Asclepiadaceae: Asclepiadeae*) in Australia. Austral. Syst. Bot. 5: 53-70.

Forster, P. I. (1995) Circumscription of *Marsdenia* (*Asclepiadaceae* – *Marsdenieae*), with a revision of the genus in Australia and Papuasia. Austral. Syst. Bot. 8: 703-933, ills., maps, keys.

Forster, P. I. & Liddle, D. J. (1988) Studies on the Australasian *Asclepiadaceae*, IV. *Dischidia* R. Br. in Australia. Austrobaileya 2(5): 507-514, ills., maps, key.

Forster, P. I. & Liddle, D. J. (1990) *Hoya* R. Br. (*Asclepiadaceae*) in Australia – an alternative classification. Austrobaileya 3(2): 217-234, key, ills., maps.

Forster, P. I. & Liddle, D. J. (1992) Taxonomic studies on the genus *Hoya* R. Br. (*Asclepiadaceae*) in Papuasia, 1-5. Austrobaileya 3(4): 627-641, ills., maps.

Forster, P. I. & al. (1995) Taxonomic studies on the genus *Hoya* R. Br. (*Asclepiadaceae: Marsdenieae*) in Papuasia, 7. Austrobaileya 4(3): 401-406, ills., key.

Forster, P. I. & al. (1998) Diversity in the genus *Hoya* (*Asclepiadaceae* – *Marsdenieae*). Aloe 35(2): 44-48, ills.

Geesink, R. & al. (1981) Thonner's analytical key to the families of flowering plants. Den Haag (NL): Leiden University Press. 231 pp.

Gilbert, M. G. (1977) *Caralluma* Sect. *Caralluma* in Ethiopia. Nation. Cact. Succ. J. 32(2): 26-31, ills., map, key.

Gilbert, M. G. (1978) The 'Ango Group' of *Caralluma* in Ethiopia. Cact. Succ. J. Gr. Brit. 40(2): 39-50, ills., key.

Gilbert, M. G. (1990) A review of *Caralluma* R. Br. and its segregates. Bradleya 8: 1-32, ills., key.

Gilbert, M. G. & al. (1995) Notes on the *Asclepiadaceae* of China. Novon 5(1): 1-16.

Goyder, D. J. (1995) Notes on the African genus *Glossostelma* (*Asclepiadaceae*). Kew Bull. 50(3): 527-555, ills., maps, key.

Goyder, D. J. (1998a) A revision of *Pachycarpus* E. Mey. (*Asclepiadaceae: Asclepiadeae*) in tropical Africa with notes on the genus in southern Africa. Kew Bull. 53(2): 335-374, key.

Goyder, D. J. (1998b) A revision of the African genus *Stathmostelma* K. Schum. (*Apocynaceae*: *Asclepiadeae*). Kew Bull. 53(3): 577-616, ills., key.

Goyder, D. J. (2001) A revision of the tropical African genus *Trachycalymma* (K. Schum.) Bullock (*Apocynaceae: Asclepiadoideae*). Kew Bull. 56(1): 129-161, key.

Goyder, D. J. & Kent, D. H. (1994) *Micholitzia obcordata* N. E. Br. (*Asclepiadaceae* – *Marsdenieae*) reinstated. Asklepios No. 62: 13-19, ills., map, 1 t.

Gravely, F. H. & Mayuranathan, P. V. (1931) The Indian species of the genus *Caralluma*. Bull. Madras Gov. Mus. 4(1): 1-28, key, ills.

Green, T. (1994) A broom: *Absolmsia spartiodes* [sic] Kuntze. Fraterna 1993(4): 3-4, ills.

Gunn, M. & Codd, L. E. (1981) Botanical exploration of Southern Africa. Cape Town (RSA): A. A. Balkema for the Botanical Research Institute. 400 pp., ills., maps.

Hara, H. (1971) *Asclepiadaceae*. In: Hara, H. (ed.): The flora of the eastern Himalaya, second report; pp. 108-110. Tokyo (J): University of Tokyo Press.

Hart, A. (1988) Some Nepali Ceropegias. Asklepios No. 43: 71-74, ills.

References

Hegnauer, R. (1989) *Asclepiadaceae*. In: Chemotaxonomie der Pflanzen, 8: 84-96. Basel (CH): Birkhäuser.

Hiern, W. P. (1898) Catalogue of the African plants collected by Dr. Friedrich Welwitsch in 1853 - 1861. Vol. 1, part 2. London (GB): British Museum.

Hilliard, O. M. & Burtt, B. L. (1987) The botany of the Southern Natal Drakensberg. Ann. Kirstenbosch Bot. Gard. 15: 253 pp.

Hilliard, O. M. & Burtt, B. L. (1989) Notes on some plants of southern Africa, chiefly from Natal: XV. Notes Roy. Bot. Gard. Edinburgh 45(2): 179-223.

Hooker, J. D. (1885) *Asclepiadaceae*. In: The flora of British India; 4: 1-78. London (GB): Reeve & Co.

Huber, H. (1957) Revision der Gattung *Ceropegia*. Mem. Soc. Brot. 12: 203 pp., 16 pl., maps, keys.

Huber, H. (1967) *Asclepiadaceae*. In: Merxmüller, H. (ed.): Prodromus einer Flora von Südwestafrika. 114. Lehre (D): J. Cramer. 71 pp., key.

Hutchinson, J. (1921) *Asclepiadaceae*. Bull. Misc. Inform. [Kew] 1921: 386-389.

Jacobsen, H. (1954a) Handbuch der sukkulenten Pflanzen. Band 1. *Abromeitiella* bis *Euphorbia*. Jena (DDR): VEB Gustav Fischer Verlag. 614 pp., ills.

Jacobsen, H. (1954b) Handbuch der sukkulenten Pflanzen. Band 2. *Fockea* bis *Zygophyllum*. Jena (DDR): VEB Gustav Fischer Verlag. pp. 615-1124, ills.

Jacobsen, H. (1955) Handbuch der sukkulenten Pflanzen. Mesembryanthemaceae. Band 3. Jena (DDR): VEB Gustav Fischer Verlag. 1126-1716 pp., ill., maps.

Jacobsen, H. (1970) Das Sukkulentenlexikon. Kurze Beschreibung, Herkunftsangaben und Synonymie der sukkulenten Pflanzen mit Ausnahme der *Cactaceae*. Stuttgart (D): Gustav Fischer Verlag. 589 pp., ill.

Jumelle, H. & Perrier, H. (1908) Notes biologiques sur la végétation du nord-ouest de Madagascar: Les Asclépiadacées. Ann. Inst. Bot.-Géol. Colon. Marseille, sér. 2, 6: 131-239, map, ills., t. 1-5.

Jumelle, H. & Perrier, H. (1911) Les Asclépiadacées aphylles dans l'ouest de Madagascar. Rev. Gén. Bot. 23: 248-269.

Kerr, A. F. G. & Craib, W. G. (1951) *Asclepiadaceae*. In: Craib, W. G. & Kerr, A. F. G. (eds.): Florae Siamensis Enumeratio; 3: 1-51. Bangkok (Thailand): Siam Society.

Koorders, S. H. (1919) Notiz über *Hoya imbricata* und *Hoya pseudomaxima*. Philipp. J. Sci. 15: 263-267.

Kunze, H. & al. (1994) *Cibirhiza albersiana*, a new species of *Asclepiadaceae*, and establishment of the tribe *Fockeeae*. Taxon 43(3): 367-376, ills., map.

Kupicha, F. K. (1984) Studies on African *Asclepiadaceae*. Kew Bull. 38(4): 599-673, ills., keys.

Land, M. (1992) Two stapelia pairs: Notes on *Huerniopsis* and *Orbeanthus*. Asklepios No. 57: 16-21, 1 t., ills.

Lawrence, G. H. M. & al. (eds.) (1968) Botanico-Periodicum-Huntianum. Pittsburgh (USA: PA): Hunt Botanical Library. 1063 pp.

Leach, L. C. (1969) *Stapelieae* from South Tropical Africa, Part V. Bothalia 10(1): 45-54, ills.

Leach, L. C. (1975) The lectotype species of *Stapelia* L. and the reinstatement of *Orbea* Haw. (*Asclepiadaceae*). Kirkia 10(1): 287-291.

Leach, L. C. (1978a) A contribution towards a new classification of *Stapelieae* (*Asclepiadaceae*) with a preliminary report of *Orbea* Haw. and descriptions of three new genera. Excelsa Tax. Ser. 1: 1-75.

Leach, L. C. (1978b) On the classification of *Stapelieae* and the reinstatement of *Tridentea* Haw. (*Asclepiadaceae*). Trans. Rhodesia Sci. Assoc. 59(1): 1-5.

Leach, L. C. (1980) A review of *Tridentea* Haw. (*Asclepiadaceae*). Excelsa Tax. Ser. 2: 68 pp., key, ills., maps.

Leach, L. C. (1984) A revision of *Tromotriche* Haw. (*Asclepiadaceae*). J. South Afr. Bot. 50(4): 549-562, ills., map, key.

Leach, L. C. (1985) A revision of *Stapelia* L. (*Asclepiadaceae*). Excelsa Tax. Ser. 3: 157 pp., ills., maps, keys.

Leach, L. C. (1988) A revision of *Huernia* R. Br. (*Asclepiadaceae*). Excelsa Tax. Ser. 4: v + 197 pp., ills., maps, key.

Lebrun, J.-P. & al. (1994) *Brachystelma letestui* Pellegrin, une rare *Asclepiadaceae* d'Afrique équatoriale. Candollea 49(1): 183-186, ills.

Li, P. T. (1992) *Antiostelma* (*Asclepiadaceae*), a new genus from China. Novon 2: 218-219.

Liede, S. (1993a) A taxonomic revision of the genus *Cynanchum* L. (*Asclepiadaceae*) in southern Africa. Bot. Jahrb. Syst. 114(4): 503-550, ills., map, key.

Liede, S. (1993b) New species and some important name changes in Malagasy leafy *Cynanchum* (*Asclepiadaceae*). Bull. Mus. Nation. Hist. Nat., Sect. B, Adansonia 14(3-4): 429-453, ills.

Liede, S. (1996a) A revision of the genus *Cynanchum* (*Asclepiadaceae*) on the African mainland. Ann. Missouri Bot. Gard. 83: 283-345.

Liede, S. (1996b) The *Cynanchinae* (*Asclepiadaceae*) in Madagascar: More new leafy and leafless species and subspecies. Bull. Mus. Nation. Hist. Nat., Sect. B, Adansonia 18(1-2): 103-135, ills., maps.

Liede, S. (1997) Subtribes and genera of the tribe *Asclepiadeae* (*Apocynaceae, Asclepiadoideae*) – a synopsis. Taxon 46(2): 233-247.

Liede, S. & Albers, F. (1994) Tribal disposition of genera in the *Asclepiadaceae*. Taxon 43(2): 201-231.

Liede, S. & Meve, U. (1989) *Sarcostemma pearsonii* N. E. Br.: A neglected species from Southern Africa. Bradleya 7: 69-72, ills., map.

Liede, S. & Meve, U. (1995) The genus *Sarcostemma* R. Br. (*Asclepiadaceae*) in Madagascar. Bot. J. Linn. Soc. 118: 37-51, ills., key.

Liede, S. & Meve, U. (1996) The circumscription of *Karimbolea* Descoings (*Asclepiadaceae*). Brittonia 48(4): 501-507, ills., key.

Liede, S. & Meve, U. (2001) New combinations and new names in Malagasy *Asclepiadoideae* (*Apocynaceae*). Adansonia, sér. 3, 23(2): 347-351.

Liede, S. & Täuber, A. (2002) Circumscription of the genus *Cynanchum* (*Apocynaceae – Asclepiadoideae*). Syst. Bot. [in press].

Liede, S. & al. (1993) On the position of the genus *Karimbolea* (*Asclepiadaceae*). Amer. J. Bot. 80(2): 215-221, ills.

Lu, F.-Y. & Kao, M.-T. (1978) *Asclepiadaceae*. In: Li, H.-L. & al. (eds.): Flora of Taiwan, 4: 222-246, ills., keys. Taipeh (Taiwan): Epoch Publishing.

Lückhoff, C. A. (1952) The *Stapelieae* of southern Africa. Amsterdam (NL): A. A. Balkema. 283 pp., ills.

Mabberley, D. J. (1987) The plant-book. A portable dictionary of the higher plants. Cambridge (GB): Cambridge University Press. 706 pp.

Malaisse, F. (1984) Recherches sur les *Asclepiadaceae* du Shaba (Zaïre). I. Nouvelles observations sur le genre *Ceropegia* L. Bull. Jard. Bot. Belg. 54(1-2): 213-234, keys, ills.

Malaisse, F. & Schaijes, M. (1993) Notes on the Ceropegias of South East Zaïre. Asklepios No. 58: 21-30, ills., 2 col. pl.

Marloth, R. (1913-1932) The flora of South Africa with synoptical tables of the genera of the higher plants. Vols. I-IV. Cape Town (RSA): Darter Bros. & Co. 1093 pp., ills., 215 pl.

Matthew, K. (ed.) (1983) The Flora of the Tamilnadu Carnatic. Volume 3. Tiruchirapalli (India): Rapinat Herbarium, St. Joseph's College. 3 parts, 2152 pp., keys.

Maxwell, J. F. (1991) Botanical notes on the vascular flora of Chiang May Province, Thailand. 2. Nat. Hist. Bull. Siam Soc. 39: 71-83.

Meve, U. (1988) The differing status of varieties in the *Stapelieae*. Asklepios No. 44: 2-7, ills.

Meve, U. (1993) A new Zimbabwean locality for the variable *Brachystelma gracile* E. A. Bruce. Asklepios No. 59: 22-25, ills., map.

Meve, U. (1994) The genus *Piaranthus* R. Br. (*Asclepiadaceae*). Bradleya 12: 57-102, ills., maps, diags., key.

Meve, U. (1995) Cytological and morphological differentiation in *Caralluma burchardii* (*Asclepiadaceae*). Nordic J. Bot. 15(5): 459-467, ills., map.

Meve, U. (1997) The genus *Duvalia* (*Stapelieae*). Stem-succulents between the Cape and Arabia. Wien (A) / New York (US: NY): Springer. x + 132 pp., ills., keys.

Meve, U. (1999) Bemerkungen zur Stapelieen-Flora von Tansania. Kakt. and. Sukk. 50(9): 219-224, ills.

Meve, U. (2000a) A new species of *Raphionacme* in Tanzania. Bradleya 18: 55-58, ills.

Meve, U. (2000b) Die Gattungszugehörigkeit von *Stapelia miscella* N. E. Brown. Kakt. and. Sukk. 51(7): 183-188, ills.

Meve, U. (2001) *Hoya heuschkeliana*, a neo-endemic of Mt. Balusan (Philippines, Luzon), and remarks on the urceolate flower type in *Asclepiadoideae*. Asklepios No. 82: 7-10, ills.

Meve, U. & Albers, F. (1990) The species concept in *Duvalia* (*Asclepiadaceae*) – a preliminary revision of the genus. Mitt. Inst. Allg. Bot. Hamburg 23b: 595-604, ills., map.

Meve, U. & Liede, S. (1994) A conspectus of *Ceropegia* L. (*Asclepiadaceae*) in Madagascar, and the establishment of *C.* Sect. *Dimorpha*. Phyton (Horn) 34(1): 131-141, ills.

Meve, U. & Liede, S. (1996) *Sarcostemma* R. Br. (*Asclepiadaceae*) in East Africa and Arabia. Bot. J. Linn. Soc. 120(1): 21-38, ills., maps.

Meve, U. & Liede, S. (1998) A surprising discovery in the stem-succulent *Cynanchum* complex in Madagascar. Bradleya 16: 9-13, ills.

Meve, U. & Liede, S. (2001a) Inclusion of *Tenaris* and *Macropetalum* in *Brachystelma* (*Apocynaceae – Asclepiadoideae – Ceropegieae*) inferred from non-coding nuclear and chloroplast DNA sequences. Pl. Syst. Evol. 228(1-2): 89-105, ills.

Meve, U. & Liede, S. (2001b) Reconsideration of the status of *Lavrania, Larryleachia* and *Notechidnopsis* (*Asclepiadoideae – Ceropegieae*). South Afr. J. Bot. 67(2): 161-168, ills., SEM-ills.

Meve, U. & Porembski, S. (1993) *Brachystelma* Sims (*Asclepiadaceae*) in West Tropical Africa. Bot. Jahrb. Syst. 115(3): 315-324, ills., chrom. nos.

Meve, U. & Wolf, F. (2001) *Echidnopsis bentii* N. E. Brown (*Ceropegieae*) auf Sokotra gefunden. Kakt. and. Sukk. 52(5): 113-119, ills.

Miller, A. G. & Morris, M. (1988) Plants of Dhofar. The southern region of Oman. Traditional, economic and medicinal uses. Sultanate of Oman: Government of Oman. 388 pp., ills., maps.

References

Morat, P. (1994) *Stapelianthus* Choux *ex* A. C. White et B. Sloane, genre endémique malgache de *Stapelieae* (*Asclepiadaceae*). Succulentes 17(4): 18-28, ills., key.

Morat, P. (1995) *Stapelianthus* Choux *ex* A. C. White et B. Sloane, genre endémique malgache de *Stapelieae* (*Asclepiadaceae*). Suite. Succulentes 18(1): 30-32, maps.

Newton, L. E. (1974) Succulents in the forest region of West Tropical Africa. Cact. Succ. J. (US) 46: 55-59, ills.

Newton, L. E. (1996) The genus *Brachystelma* (*Asclepiadaceae*) in East Africa, with a new record and description of a new species. Bradleya 14: 94-98, ills.

Ngwenya, M. A. & al. (1995) *Brachystelma natalense*: Rediscovered and redescribed. Aloe 32(2): 44-45, ills.

Nicholas, A. & Goyder, D. J. (1992) *Aspidonepsis* (*Asclepiadaceae*), a new southern African genus. Bothalia 22(1): 23-35, ills., maps, key.

Nicholas, A. & Goyder, D. J. (1993) *Asclepiadaceae*. Bothalia 23(2): 236-237.

Olmstead, R. G. & al. (1993) A parsimony analysis of the *Asteridae* sensu lato based on rbcL sequences. Ann. Missouri Bot. Gard. 80: 700-722.

Omlor, R. (1998) Generische Revision der *Marsdenieae* (*Asclepiadaceae*). Aachen (D): Shaker Verlag. 257 pp.

Onderstall, J. (1996) Wild flower guide Mpumalanga and Northern Province. Nelspruit (RSA): Dynamic. 250 pp., ills.

Peckover, R. G. (1993a) The Ceropegias of the Makatini Flats. Aloe 30(1): 20-22, ills.

Peckover, R. G. (1993b) The unusual odd man out – *Macropetalum burchellii*. Aloe 30(1): 59-61, ills.

Peckover, R. G. (1994) An overlooked miniature *Brachystelma* species, *B. angustum* Peckover sp. nov. (*Asclepiadaceae*) from near Carolina, eastern Transvaal. Aloe 31: 59-61, ills.

Peckover, R. G. (1996) The inclusion of the genera *Tenaris* E. Mey. and *Macropetalum* Burch. *ex* Decne. into *Brachystelma* R. Br. Aloe 33(2-3): 41-43, ills.

Percy-Lancaster, A. (1989) *Brachystelma* in Zimbabwe. Excelsa 13: 63-77, ills., maps.

Pham-Hoang, H. (1972) *Asclepiadaceae*. In: An illustrated Flora of Vietnam, 2: 176-202, keys, ills. Hanoi (Vietnam).

Plowes, D. C. H. (1990) On the identity of *Caralluma cicatricosa* and *C. quadrangula*. Asklepios No. 50: 3-20, ills., map.

Plowes, D. C. H. (1992) A preliminary reassessment of the genera *Hoodia* and *Trichocaulon* (*Stapelieae: Asclepiadaceae*). Asklepios No. 56: 5-15, ills.

Plowes, D. C. H. (1993) A new account of *Echidnopsis* Hook. f. (*Asclepiadaceae: Stapelieae*). Haseltonia 1: 65-85, ills.

Plowes, D. C. H. (1994a) The taxonomy of *Quaqua* N. E. Brown. Excelsa 16: 83-102, ills.

Plowes, D. C. H. (1994b) The taxonomy of the genera *Pachycymbium* Leach and *Angolluma* Munster (*Stapelieae: Asclepiadaceae*). Excelsa 16: 103-123, ills.

Plowes, D. C. H. (1995) A reclassification of *Caralluma* R. Brown (*Stapelieae: Asclepiadaceae*). Haseltonia 3: 49-70, ills., key.

Plowes, D. C. H. (1997) *Larryleachia* and *Hoodia* (*Stapelieae: Asclepiadaceae*): Some new nomenclatural proposals. Excelsa 17: 3-28, 68, ills.

Plowes, D. C. H. & Drummond, R. B. (1976) Wild flowers of Rhodesia. A guide to some of the common wild flowers of Rhodesia. Salisbury (Rhodesia): Longman Rhodesia. 170 pp., ill.

Radcliffe-Smith, R. (1967) *Riocreuxia* Decne., an Asclepiadaceous genus new to Nepal. Kew Bull. 21: 295-298.

Rauh, W. (1995a) Succulent and xerophytic plants of Madagascar. Vol. 1. Mill Valley (US: CA): Strawberry Press. 343 pp., ills., maps.

Rauh, W. (1998) Succulent and xerophytic plants of Madagascar. Vol. 2. Mill Valley (US: CA): Strawberry Press. 385 pp., ills.

Ricánek, M. & Hanácek, P. (1999) *Rhytidocaulon tortum* (N. E. Br.) M. Gilbert refound after 100 years. Cact. Succ. J. (US) 71(2): 81-85, ills.

Ricánek, M. & Hanácek, P. (2001) Two Stapeliad species from Iran. Asklepios No. 82: 21-26, ills.

Rintz, R. E. (1978) The peninsular Malaysian species of *Hoya* (*Asclepiadaceae*). Malayan Nat. J. 30(3/4): 467-522, ills., key.

Rintz, R. E. (1980) The peninsular Malayan species of *Dischidia* (*Asclepiadaceae*). Blumea 26: 81-126, ills., key.

Rothe, W. (1915) Über die Gattung *Marsdenia* und die Stammpflanze der Condurangarinde. Bot. Jahrb. Syst. 52: 354-434.

Rowley, G. D. (1980) Name that succulent. Keys to the families and genera of succulent plants in cultivation. Cheltenham (GB): Stanley Thornes (Publishers) Ltd. 268 pp., ill., keys.

Rowley, G. D. (1987) Caudiciform and pachycaul succulents. Pachycauls, bottle-, barrel- and elephant-trees and their kin: A collector's miscellany. Mill Valley (US: CA): Strawberry Press. 282 pp., ills.

Schlechter, R. (1895a) Beiträge zur Kenntnis südafrikanischer Asclepiadeen. Bot. Jahrb. Syst. 20(Beibl. 51): 1-56.

Schlechter, R. (1895b) *Asclepiadaceae* Elliotianae. J. Bot. 33: 300-307, 332-339.

Schumann, K. (1895) *Asclepiadaceae*. In: Engler, A. & Prantl, K. (eds.): Die natürlichen Pflanzenfamilien. IV(2): 189-306. Leipzig (D): Verlag Wilhelm Engelmann.

Sennblad, B. & Bremer, B. (1996) The familial and subfamilial relationships of *Apocynaceae* and *Asclepiadaceae* evaluated with rbcL data. Pl. Syst. Evol. 202(3): 153-175.

Smith, D. M. N. (1988) A revision of the genus *Pachycarpus* in southern Africa. South Afr. J. Bot. 54(5): 399-439, key, ills.

Stafleu, F. A. & Cowan, R. S. (1976-1988) Taxonomic literature. Utrecht (NL): Bohn, Scheltema & Holkema, etc. 2. Ed.; 7 vols.

Stafleu, F. A. & Mennega, E. A. (1992-2000) Taxonomic literature. Königstein (D): Koeltz Scientific Books. Supplements to Ed. 2; 6 vols.

Stearn, W. T. (1992) Botanical Latin. Newton Abbot (GB): David & Charles Publishers. Ed. 4; 560 pp.

Steenis, C. G. van (1972) The Mountain Flora of Java. Leiden (NL): E. J. Brill. 98 pp., 57 pl.

Stevens, W. D. (1976) *Asclepiadaceae*. In: Saldanha, C. J. & Nicolson, D. H. (eds.): Flora of Hassan District, Karnataka, India; pp. 437-455. New Delhi (India): Amerind.

Stevens, W. D. (1988) A synopsis of *Matelea* subg. *Dictyanthus* (*Apocynaceae – Asclepiadoideae*). Ann. Missouri Bot. Gard. 75: 1533-1564, ills., key.

Stevens, W. D. (2000) New and interesting Milkweeds (*Apocynaceae, Asclepiadoideae*). Novon 10(3): 242-256.

Swarupanandan, K. & al. (1996) The subfamilial and tribal classification of the family *Asclepiadaceae*. Bot. J. Linn. Soc. 120: 327-369.

Talbot, W. A. (1909-1911) Forest flora of the Bombay Presidency and Sind. Poona (India): Printed by the Government. 2 vols.

Tsiang, Y. (1939) Notes on Asiatic *Apocynales*, IV. Sunyatsenia 4(1-2): 88-94.

Tsiang, Y. & Li, P. T. (1977) *Asclepiadaceae*. In: Tsiang, Y. & Li, P. T. (eds.): Flora Reipublica Popularis Sinica; 63: 249-575, ills., key. Beijing (China): Science Press.

Venter, F. (1993) Rare *Ceropegia* re-discovered. Veld. Fl. (1975+) 79(4): 120, ills.

Venter, H. J. T. & Verhoeven, R. L. (1993) A taxonomic account of *Stomatostemma* (*Periplocaceae*). South Afr. J. Bot. 59(1): 50-56, key, ills.

Venter, H. J. T. & Verhoeven, R. L. (1999) A taxonomic revision of *Schlechterella* (*Periplocoideae, Apocynaceae*). South Afr. J. Bot. 64(6): 350-355, ills., key, map.

Venter, H. J. T. & Verhoeven, R. L. (2001) Diversity and relationships within the *Periplocoideae* (*Apocynaceae*). Ann. Missouri Bot. Gard. 88(4): 550-568, diag., key.

Venter, H. J. T. & al. (1990) The genus *Petopentia* (*Periplocaceae*). South Afr. J. Bot. 56(3): 393-398, ills.

Verhoeven, R. L. & Venter, H. J. T. (1997) The translator of *Raphionacme* (*Periplocaceae*). South Afr. J. Bot. 63(1): 46-54, ills.

Walker, C. C. (1982) *Brachystelma* – an introduction and checklist. Asklepios No. 25: 92-106, ills.

Werdermann, E. (1939) Revision der ostafrikanischen Arten der Gattung *Ceropegia*. Bot. Jahrb. Syst. 70: 189-232.

White, A. C. & Sloane, B. L. (1937) The *Stapelieae*. Pasadena (US: CA): Abbey San Encino Press. 3 vols.; 1207 pp., ills., maps, keys.

White, C. T. (1928) Plants collected in the Mandated Territory of New-Guinea by C. E. Lane-Poole. *Asclepiadaceae*. Proc. Roy. Soc. Queensland 39(6): 69.

Woodson, R. E. (1954) The North American species of *Asclepias* L. Ann. Missouri Bot. Gard. 41(1): 1-211, ills., maps, keys.

Yadav, S. R. & al. (1993) A new species of *Brachystelma* (*Asclepiadaceae*) from India. Kew Bull. 48(1): 59-61, ills.

Zizka, G. (1990) Pflanzen und Ameisen. Palmengarten Sonderheft No. 15: 125 pp., ills.

Zomlefer, W. B. (1994) Guide to flowering plant families. Chapel Hill (US: NC) / London (GB): University of North Carolina Press. 430 pp., ills.

Taxonomic Cross-Reference Index

Absolmsia : 9
– **spartioides**: 9
Acanthostemma → Hoya: 147
Acerates → Asclepias: 9
Adenium namaquarium → Hoodia currorii ssp. currorii: 143
Anantherix → Asclepias: 9
Angolluma → Orbea: 187
– abayensis → Orbea abayensis: 188
– araysiana → Orbea araysiana: 188
– baldratii → Orbea baldratii: 189
– chrysostephana → Orbea chrysostephana: 190
– circes → Orbea dummeri: 192
– commutata → Orbea commutata: 190
– – ssp. commutata → Orbea commutata: 190
– – – sheilae → Orbea commutata: 190
– decaisneana → Orbea decaisneana: 191
– deflersiana → Orbea deflersiana: 191
– denboefii → Orbea denboefii: 191
– distincta → Orbea distincta: 192
– dummeri → Orbea dummeri: 192
– eremastrum → Orbea wissmannii var. eremastrum: 205
– foetida → Orbea sprengeri ssp. foetida: 200
– gemugofana → Orbea gemugofana: 192
– gilbertii → Orbea abayensis: 188
– hesperidum → Orbea decaisneana: 191
– huernioides → Orbea huernioides: 193
– kochii → Orbea sacculata: 199
– laikipiensis → Orbea laikipiensis: 194
– laticorona → Orbea laticorona: 195
– lenewtonii → Orbea subterranea: 201
– lugardii → Orbea lugardii: 195
– luntii → Orbea luntii: 195
– miscella → Orbea miscella: 197
– ogadensis → Orbea sprengeri ssp. ogadensis: 201
– rogersii → Orbea rogersii: 199
– sacculata → Orbea sacculata: 199
– schweinfurthii → Orbea schweinfurthii: 199
– semitubiflora → Orbea semitubiflora: 200
– sprengeri → Orbea sprengeri: 200
– subterranea → Orbea subterranea: 201
– tubiformis → Orbea tubiformis: 201
– ubomboensis → Orbea ubomboensis: 202
– venenosa → Orbea decaisneana: 191
– vibratilis → Orbea vibratilis: 204
– wilsonii → Orbea wilsonii: 204
– wissmannii → Orbea wissmannii: 205
Anomalluma → Pseudolithos: 212
– dodsoniana → Pseudolithos dodsonianus: 212
Anthonotis → Asclepias: 9
Antiostelma → Micholitzia: 183
– lantsangense → Micholitzia obcordata: 183
– manipurense → Micholitzia obcordata: 183
Apegia → Ceropegia: 63

Apoxanthera → Raphionacme: 220
– pubescens → Raphionacme hirsuta: 224
Apteranthes → Caralluma: 46
– burchardii → Caralluma burchardii: 50
– cylindrica → Echidnopsis cereiformis: 131
– europaea → Caralluma europaea: 53
– – ssp. gussoneana → Caralluma europaea: 53
– – – maroccana → Caralluma europaea: 53
– – var. affinis → Caralluma europaea: 53
– – – albotigrina → Caralluma europaea: 53
– – – barrueliana → Caralluma europaea: 53
– – – confusa → Caralluma europaea: 53
– – – decipiens → Caralluma europaea: 53
– – – gattefossei → Caralluma europaea: 53
– – – judaica → Caralluma europaea var. judaica: 54
– – – marmaricensis → Caralluma europaea: 53
– – – micrantha → Caralluma europaea: 53
– – – schmuckiana → Caralluma europaea: 53
– – – simonis → Caralluma europaea: 53
– – – tristis → Caralluma europaea: 53
– gussoneana → Caralluma europaea: 53
– joannis → Caralluma joannis: 56
– negevensis → Caralluma europaea var. judaica: 54
– tessellata → Echidnopsis cereiformis: 131
Aristolochia blinii → Ceropegia mairei: 89
– mairei → Ceropegia mairei: 89
– viridiflora → Ceropegia mairei: 89
– – var. occlusa → Ceropegia mairei: 89
Asclepiadaceae : 5
Asclepias : 9
– acida → Sarcostemma sp.: 233
– angustata → Stathmostelma angustatum: 260
– aphylla → Cynanchum luteifluens var. luteifluens: 113
– – → Sarcostemma forskaolianum: 234
– – → Sarcostemma viminale: 235
– bagshawei → Stathmostelma welwitschii var. bagshawei: 263
– cabrae → Glossostelma cabrae: 140
– carnosa → Hoya carnosa: 149
– carsonii → Glossostelma carsonii: 140
– coccinea → Stathmostelma incarnatum: 261
– cognata → Aspidonepsis cognata: 18
– congolensis → Glossostelma lisianthoides: 141
– dewewrei → Stathmostelma katangense: 261
– diploglossa → Aspidonepsis diploglossa: 19
– dissoluta → Glossostelma lisianthoides: 141
– eminens → Stenostelma eminens: 264
– erecta → Glossostelma erectum: 140
– extensa → Stathmostelma katangense: 261
– flava → Aspidonepsis flava: 19
– fornicata → Stathmostelma fornicatum: 260
– gigantiflora → Stathmostelma gigantiflorum: 261

[Asclepias]
- katangensis → Stathmostelma katangense: 261
- laurentiana → Stathmostelma welwitschii: 263
- lisianthoides → Glossostelma lisianthoides: 141
- macrantha → Stathmostelma pedunculatum: 262
- macropetala → Stathmostelma spectabile: 262
- mihundensis → Stathmostelma gigantiflorum: 261
- nemorensis → Glossostelma lisianthoides: 141
- nuttii → A. sp.: 10
- odorata → Stathmostelma odoratum: 261
- pachyclada → Stathmostelma spectabile: 262
- parasitica → Hoya verticillata: 158
- pauciflora → Stathmostelma pauciflorum: 261
- pedunculata → Stathmostelma pedunculatum: 262
- propinqua → Stathmostelma propinquum: 262
- reenensis → Aspidonepsis reenensis: 19
- reflexa → Stathmostelma pauciflorum: 261
- rhacodes → Stathmostelma rhacodes: 262
- schizoglossoides → Aspidonepsis diploglossa: 19
- spectabilis → Stathmostelma spectabile: 262
- stipitacea → Sarcostemma viminale ssp. stipitaceum: 236
- **subulata**: 10
- uvirensis → Stathmostelma pedunculatum: 262
- verdickii → Stathmostelma wildemanianum: 263
- vomeriformis → Stathmostelma angustatum ssp. vomeriforme: 260
- welwitschii → Stathmostelma welwitschii: 263
- xysmalobioides → Glossostelma xysmalobioides: 142
Asclepiodella → Asclepias: 9
Asclepiodora → Asclepias: 9
Aspidoglossum : 10
- **angustissimum**: 11
- **araneiferum**: 11
- **biflorum**: 11
- **breve**: 11
- **carinatum**: 12
- **connatum**: 12
- **crebrum**: 12
- **delagoense**: 12
- **demissum**: 12
- **difficile**: 12
- **dissimile**: 13
- **elliotii**: 13
- **erubescens**: 13
- **eylesii**: 13
- **fasciculare**: 13
- **flanaganii**: 14
- **glabellum**: 14
- **glabrescens**: 14
- **glanduliferum**: 14
- **gracile**: 14
- **grandiflorum**: 15
- **heterophyllum**: 15

[Aspidoglossum]
- **hirundo**: 15
- **interruptum**: 15
- kulsii → A. masaicum: 16
- **lamellatum**: 15
- **lanatum**: 16
- **masaicum**: 16
- **nyasae**: 16
- **ovalifolium**: 16
- **restioides**: 17
- **rhodesicum**: 17
- **uncinatum**: 17
- **validum**: 17
- **virgatum**: 17
- whytei → A. angustissimum: 11
- **woodii**: 18
- **xanthosphaerum**: 18
Aspidonepsis : 18
- **cognata**: 18
- **diploglossa**: 19
- **flava**: 19
- **reenensis**: 19
- **shebae**: 19
Astrostemma → Absolmsia: 9
- spartioides → Absolmsia spartioides: 9
Aulostephanus → Brachystelma: 20
- natalensis → Brachystelma natalense: 37
Australluma → Caralluma: 46
- peschii → Caralluma peschii: 58
Ballyanthus → Orbea: 187
- prognathus → Orbea prognatha: 198
Baxtera → Marsdenia: 178
Baynesia : 20
- **lophophora**: 20
Bidaria leptophylla → Marsdenia viridiflora: 180
Biventraria → Asclepias: 9
Blepharanthera → Brachystelma: 20
- dinteri → Brachystelma blepharanthera: 22
- edulis → Brachystelma blepharanthera: 22
Borealluma → Caralluma: 46
- munbyana → Caralluma munbyana: 57
- plicatiloba → Caralluma tuberculata: 61
- staintonii → Caralluma staintonii: 61
- tuberculata → Caralluma tuberculata: 61
Boucerosia → Caralluma: 46
- aaronis → Caralluma europaea var. judaica: 54
- acutangula → Caralluma acutangula: 47
- adenensis → Caralluma adenensis: 48
- aucheriana → Caralluma sp.: 47
- awdeliana → Caralluma awdeliana: 49
- campanulata → Caralluma umbellata: 62
- cicatricosa → Caralluma cicatricosa: 51
- crenulata → Caralluma crenulata: 51
- cylindrica → Echidnopsis cereiformis: 131
- decaisneana → Orbea decaisneana: 191
- diffusa → Caralluma diffusa: 52
- edulis → Caralluma edulis: 52
- europaea → Caralluma europaea: 53
- forskaolii → Caralluma quadrangula: 59

[Boucerosia]
- gussoneana → Caralluma europaea: 53
- hispanica → Caralluma munbyana: 57
- hutchinia → Caralluma indica: 56
- incarnata → Quaqua incarnata: 217
- indica → Caralluma indica: 56
- lasiantha → Caralluma umbellata: 62
- mammillaris → Quaqua mammillaris: 217
- maroccana → Caralluma europaea: 53
- munbyana → Caralluma munbyana: 57
- – var. hispanica → Caralluma munbyana: 57
- nilagiriana → Caralluma crenulata: 51
- pauciflora → Caralluma pauciflora: 57
- penicillata → Caralluma penicillata: 58
- procumbens → Caralluma procumbens: 59
- quadrangula → Caralluma quadrangula: 59
- russelliana → Caralluma acutangula: 47
- sinaica → Caralluma sinaica: 59
- socotrana → Caralluma socotrana: 60
- stocksiana → Caralluma edulis: 52
- tombuctuensis → Caralluma acutangula: 47
- truncato-coronata → Caralluma crenulata: 51
- umbellata → Caralluma umbellata: 62

Brachystelma : 20
- **albipilosum**: 21
- **alpinum**: 21
- angustum → B. nanum: 37
- **arenarium**: 21
- **arnotii**: 21
- asmarense → B. lineare: 34
- atacorense → B. togoense: 45
- **attenuatum**: 21
- **australe**: 21
- bagshawii → B. johnstonii: 32
- **barberae**: 22
- beddomei → B. brevitubulatum: 23
- **bikitaense**: 22
- bingeri → Raphionacme splendens: 227
- **blepharanthera**: 22
- bolusii → B. circinatum: 26
- bourneae → B. maculatum: 34
- **bracteolatum**: 22
- **brevipedicellatum**: 23
- **brevitubulatum**: 23
- **brownianum**: 23
- **bruceae**: 23
- – – ssp. hirsutum → B. bruceae: 23
- **buchananii**: 23
- **burchellii**: 24
- – – var. **burchellii**: 24
- – – – **grandiflorum**: 24
- **caffrum**: 24
- – – → B. meyerianum: 35
- **campanulatum**: 24
- **canum**: 25
- **cathcartense**: 25
- caudatum → B. tuberosum: 45
- **chloranthum**: 25
- **chlorozonum**: 25

[Brachystelma]
- **christianeae**: 25
- ciliatum → B. laevigatum: 33
- cinereum → B. circinatum: 26
- **circinatum**: 26
- **coddii**: 26
- **codonanthum**: 26
- comaru → Fockea comaru: 138
- commixtum → B. circinatum: 26
- **comptum**: 26
- constrictum → B. lineare: 34
- crispum → B. tuberosum: 45
- – – → Fockea capensis: 138
- **cupulatum**: 27
- **decipiens**: 27
- **delicatum**: 27
- **dimorphum**: 27
- – – ssp. gratum → B. dimorphum: 27
- **dinteri**: 27
- – – → B. blepharanthera: 22
- **discoideum**: 28
- distinctum → B. elongatum: 28
- **duplicatum**: 28
- dyeri → B. nanum: 37
- **edule**: 28
- **elegantulum**: 28
- elenaduense → B. edule: 28
- ellipticum → B. lineare: 34
- **elongatum**: 28
- erianthum → B. oianthum: 38
- **exile**: 29
- **festucifolium**: 29
- **filifolium**: 29
- filiforme → B. circinatum: 26
- flavidum → B. pygmaeum ssp. flavidum: 40
- **floribundum**: 29
- – – → B. duplicatum: 28
- **foetidum**: 29
- **franksiae**: 30
- **furcatum**: 30
- galpinii → B. circinatum: 26
- gemmeum → B. elongatum: 28
- **gerrardii**: 30
- **glabriflorum**: 30
- **glabrum**: 30
- **glenense**: 31
- **gracile**: 31
- **gracillimum**: 31
- grossartii → B. arnotii: 21
- **gymnopodum**: 31
- hirsutum → Raphionacme hirsuta: 224
- **hirtellum**: 31
- **huttonii**: 31
- inandense → B. natalense: 37
- **incanum**: 32
- inconspicuum → B. tenue: 44
- **johnstonii**: 32
- **keniense**: 32
- kerrii → B. glabriflorum: 30

[Brachystelma]
- **kerzneri**: 32
- kituloense → B. coddii: 26
- **kolarensis**: 33
- **laevigatum**: 33
- **lancasteri**: 33
- lanceolatum → B. johnstonii: 32
- **lankanum**: 33
- **letestui**: 33
- **lineare**: 34
- linearifolium → B. plocamoides: 39
- **longifolium**: 34
- luteum → B. huttonii: 32
- **macropetalum**: 34
- macrorrhizum → Fockea edulis: 138
- **maculatum**: 34
- **mafekingense**: 34
- magicum → B. buchananii: 23
- malwanense → B. kolarensis: 33
- **maritae**: 35
- medusanthemum → B. johnstonii: 32
- **megasepalum**: 35
- merrillii → B. glabriflorum: 30
- **meyerianum**: 35
- **micranthum**: 35
- microstemma → B. glabriflorum: 30
- **minimum**: 35
- **minor**: 36
- **modestum**: 36
- **molaventi**: 36
- **montanum**: 36
- **mortonii**: 36
- **nanum**: 36
- naorojii → B. kolarensis: 33
- **natalense**: 37
- nauseosum → B. buchananii: 24
- **nepalense**: 37
- **ngomense**: 37
- nigrum → B. gerrardii: 30
- **occidentale**: 37
- **oianthum**: 38
- **omissum**: 38
- ovatum → B. circinatum: 26
- **pachypodium**: 38
- pallidum → B. circinatum: 26
- papuanum → B. glabriflorum: 30
- **parviflorum**: 38
- – – → B. mortonii: 36
- **parvulum**: 38
- **pauciflorum**: 38
- **pellacibellum**: 39
- **perditum**: 39
- **petraeum**: 39
- phyteumoides → B. lineare: 34
- pilosum → B. hirtellum: 31
- **plocamoides**: 39
- **praelongum**: 39
- – – ssp. **praelongum**: 40
- – – – **thunbergii**: 40

[Brachystelma]
- **prostratum**: 40
- **pulchellum**: 40
- **punctatum**: 40
- **pygmaeum**: 40
- – – ssp. **flavidum**: 40
- – – – **pygmaeum**: 40
- – – var. breviflorum → B. pygmaeum ssp. pygmaeum: 40
- **ramosissimum**: 41
- rangacharii → B. maculatum: 34
- **recurvatum**: 41
- rehmannii → B. foetidum: 29
- **remotum**: 41
- **richardsii**: 41
- ringens → B. brevipedicellatum: 23
- **rubellum**: 41
- **sandersonii**: 42
- **schinzii**: 42
- **schizoglossoides**: 42
- **schoenlandianum**: 42
- **schultzei**: 42
- **setosum**: 43
- shirense → B. buchananii: 23
- **simplex**: 43
- – – ssp. banforae → B. simplex: 43
- sinuatum → Fockea sinuata: 139
- spathulatum → B. tuberosum: 45
- **stellatum**: 43
- **stenophyllum**: 43
- subaphyllum → Ceropegia sp.: 63
- **swarupa**: 43
- **swazicum**: 44
- **tabularium**: 44
- **tavalla**: 44
- **tenellum**: 44
- **tenue**: 44
- thunbergii → B. praelongum ssp. thunbergii: 40
- **togoense**: 45
- **tuberosum**: 45
- – – → B. decipiens: 27
- undulatum → B. circinatum: 26
- **vahrmeijeri**: 45
- **villosum**: 45
- viridiflorum → Raphionacme burkei: 222
- **volubile**: 45
- zeyheri → B. circinatum: 26

Brachystelmaria → Brachystelma: 20
- longifolia → Brachystelma longifolium: 34
- macropetala → Brachystelma macropetalum: 34
- natalensis → Brachystelma sandersonii: 42
- occidentalis → Brachystelma occidentale: 37
- ramosissima → Brachystelma ramosissimum: 41

Caralluma : 46
- aaronis → C. europaea var. judaica: 54
- abayensis → Orbea abayensis: 188
- **acutangula**: 47
- acutiloba → Quaqua acutiloba: 214
- **adenensis**: 47

[Caralluma]
- **adscendens**: 48
- – var. **adscendens**: 48
- – – **attenuata**: 48
- – – **carinata**: 48
- – – **fimbriata**: 48
- – – **geniculata**: 48
- – – **gracilis**: 49
- affinis → C. europaea: 53
- albocastanea → Orbea albocastanea: 188
- anemoniflora → Duvalia anemoniflora: 124
- ango → C. sp.: 47
- aperta → Tromotriche aperta: 270
- **arabica**: 49
- **arachnoidea**: 49
- – var. **arachnoidea**: 49
- – – **breviloba**: 49
- arenicola → Quaqua arenicola: 214
- arida → Quaqua arida: 215
- armata → Quaqua armata: 215
- atrosanguinea → Huerniopsis atrosanguinea: 175
- attenuata → C. adscendens var. attenuata: 48
- aucheriana → C. sp.: 47
- aurea → Quaqua aurea: 215
- ausana → Quaqua hottentotorum: 216
- australis → Orbea melanantha: 197
- **awdeliana**: 49
- baldratii → Orbea baldratii: 189
- **baradii**: 49
- **bhupinderiana**: 50
- bredae → Orbea miscella: 198
- – var. thomallae → Orbea miscella: 198
- brownii → Orbea lutea ssp. vaga: 196
- **burchardii**: 50
- – fa. grandiflora → C. burchardii ssp. maura: 50
- – – sordida → C. burchardii ssp. maura: 50
- – – viridis → C. burchardii ssp. maura: 50
- – ssp. **burchardii**: 50
- – – **maura**: 50
- – var. maura → C. burchardii ssp. maura: 50
- – – purpurascens → C. burchardii ssp. burchardii: 50
- – – sventenii → C. burchardii ssp. burchardii: 50
- campanulata → C. umbellata: 62
- carnosa → Orbea carnosa: 189
- caudata → Orbea caudata: 189
- – ssp. caudata → Orbea caudata ssp. caudata: 189
- – – rhodesiaca → Orbea caudata ssp. rhodesiaca: 190
- – var. chibensis → Orbea caudata ssp. rhodesiaca: 190
- – – fusca → Orbea caudata ssp. caudata: 189
- – – milleri → Orbea caudata ssp. rhodesiaca: 190
- – – stevensonii → Orbea caudata ssp. rhodesiaca: 190

[Caralluma]
- chibensis → Orbea caudata ssp. rhodesiaca: 190
- chlorantha → C. sp.: 47
- chrysostephana → Orbea chrysostephana: 190
- **cicatricosa**: 50
- cincta → Quaqua cincta: 215
- circes → Orbea dummeri: 192
- codonoides → C. speciosa: 61
- commutata → Orbea commutata: 190
- – ssp. eu-commutata → Orbea commutata: 190
- – – hesperidum → Orbea decaisneana: 191
- – var. hesperidum → Orbea decaisneana: 191
- compta → Piaranthus comptus: 210
- confusa → C. europaea: 53
- **congestiflora**: 51
- corrugata → C. socotrana: 60
- crassa → White-sloanea crassa: 274
- **crenulata**: 51
- cucullata → C. sp.: 47
- dalzielii → C. subulata: 61
- decaisneana → Orbea decaisneana: 191
- – ssp. hesperidum → Orbea decaisneana: 191
- decora → Piaranthus decorus: 210
- deflersiana → Orbea deflersiana: 191
- denboefii → Orbea denboefii: 191
- dependens → Quaqua dependens: 216
- **dicapuae**: 51
- – ssp. dicapuae → C. dicapuae: 51
- – – seticorona → C. dicapuae: 51
- – – turneri → C. turneri: 62
- – – ukambensis → C. turneri ssp. ukambensis: 62
- **diffusa**: 52
- dioscoridis → Duvaliandra dioscoridis: 129
- distincta → Orbea distincta: 192
- dodsoniana → Pseudolithos dodsonianus: 212
- **dolichocarpa**: 52
- dummeri → Orbea dummeri: 192
- – fa. colorata → Orbea dummeri: 192
- **edithae**: 52
- **edulis**: 52
- – – → C. edulis: 52
- **edwardsiae**: 53
- elata → C. priogonium: 58
- eremastrum → Orbea wissmannii var. eremastrum: 205
- ericeta → Quaqua gracilis: 216
- **europaea**: 53
- – fa. parviflora → C. europaea: 53
- – ssp. gussoneana → C. europaea: 53
- – – maroccana → C. europaea: 53
- – var. affinis → C. europaea: 53
- – – albotigrina → C. europaea: 53
- – – barrueliana → C. europaea: 53
- – – confusa → C. europaea: 53
- – – decipiens → C. europaea: 53
- – – **europaea**: 53
- – – gattefossei → C. europaea: 53
- – – **judaica**: 53

[Caralluma europaea var.]
– – – marmaricensis → C. europaea: 53
– – – maroccana → C. europaea: 53
– – – micrantha → C. europaea: 53
– – – schmuckiana → C. europaea: 53
– – – simonis → C. europaea: 53
– – – tristis → C. europaea: 53
– fimbriata → C. adscendens var. fimbriata: 48
– **flava**: 54
– **flavovirens**: 54
– **foetida**: 54
– fosteri → Orbea carnosa: 189
– foulcheri-delboscii → C. hexagona: 55
– – var. greenbergiana → C. hexagona: 55
– framesii → Quaqua framesii: 216
– **frerei**: 54
– **furta**: 55
– geminata → Piaranthus geminatus: 211
– gemugofana → Orbea gemugofana: 192
– gerstneri → Orbea gerstneri: 192
– – ssp. elongata → Orbea gerstneri ssp. elongata: 193
– – – gerstneri → Orbea gerstneri ssp. gerstneri: 193
– gossweileri → Orbea huillensis: 194
– **gracilipes**: 55
– – ssp. arachnoidea → C. arachnoidea: 49
– – – breviloba → C. arachnoidea var. breviloba: 49
– – – edwardsiae → C. edwardsiae: 53
– gracilis → Quaqua gracilis: 216
– grandidens → Orbea maculata ssp. maculata: 197
– grivana → Huerniopsis decipiens: 175
– hahnii → Orbea lutea ssp. vaga: 196
– hesperidum → Orbea decaisneana: 191
– **hexagona**: 55
– – var. septentrionalis → C. hexagona: 55
– hirtiflora → C. acutangula: 47
– hottentotorum → Quaqua hottentotorum: 216
– – var. major → Quaqua hottentotorum: 216
– – – minor → Quaqua hottentotorum: 216
– – – tubata → Quaqua hottentotorum: 216
– huernioides → Orbea huernioides: 193
– huillensis → Orbea huillensis: 194
– incarnata → Quaqua incarnata: 217
– – var. alba → Quaqua incarnata: 217
– **indica**: 56
– intermedia → Quaqua sp.: 214
– inversa → Quaqua inversa: 217
– israelitica → C. europaea var. judaica: 54
– **joannis**: 56
– kalaharica → Orbea knobelii: 194
– kalmbacheriana → C. adenensis: 48
– keithii → Orbea carnosa: 189
– knobelii → Orbea knobelii: 194
– – var. langii → Orbea knobelii: 194
– kochii → Orbea sacculata: 199
– langii → Orbea knobelii: 194

[Caralluma]
– lasiantha → C. umbellata: 62
– lateritia → Orbea lutea ssp. lutea: 196
– – var. stevensonii → Orbea lutea ssp. lutea: 196
– **lavrani**: 56
– leendertziae → Orbea melanantha: 197
– linearis → Quaqua linearis: 217
– longecornuta → Orbea caudata ssp. caudata: 189
– longicuspis → Orbea lugardii: 195
– longidens → C. edulis: 52
– **longiflora**: 56
– longipes → Pectinaria longipes: 208
– – var. villettii → Pectinaria longipes: 208
– lugardii → Orbea lugardii: 195
– luntii → Orbea luntii: 195
– lutea → Orbea lutea: 196
– – ssp. lutea → Orbea lutea ssp. lutea: 196
– – – vaga → Orbea lutea ssp. vaga: 196
– – var. lateritia → Orbea lutea ssp. lutea: 196
– – – vansonii → Orbea lutea ssp. lutea: 196
– maculata → Orbea maculata: 197
– – var. brevidens → Orbea maculata ssp. rangeana: 197
– mammillaris → Quaqua mammillaris: 217
– maris-mortui → C. sinaica: 59
– marlothii → Quaqua arida: 215
– – var. viridis → Quaqua arida: 215
– maroccana → C. europaea: 53
– maughanii → Pectinaria maughanii: 208
– meintjesiana → Orbea wissmannii var. wissmannii: 205
– melanantha → Orbea melanantha: 197
– – fa. rubiginosa → Orbea melanantha: 197
– – var. sousae → Orbea melanantha: 197
– **mireillae**: 57
– mogadoxensis → C. priogonium: 58
– **moniliformis**: 57
– montana → Echidnopsis montana: 133
– mouretii → C. edulis: 52
– multiflora → Quaqua multiflora: 218
– **munbyana**: 57
– nebrownii → Orbea lutea ssp. vaga: 196
– – var. discolor → Orbea lutea ssp. vaga: 196
– – – pseudonebrownii → Orbea lutea ssp. vaga: 196
– negevensis → C. europaea var. judaica: 54
– nilagiriana → C. crenulata: 51
– ortholoba → Quaqua acutiloba: 214
– oxydonta → C. speciosa: 61
– parviflora → Quaqua parviflora: 218
– **pauciflora**: 57
– **peckii**: 58
– **penicillata**: 58
– – var. robusta → C. penicillata: 58
– **peschii**: 58
– petraea → C. awdeliana: 49
– piaranthoides → Orbea schweinfurthii: 199
– pillansii → Quaqua pillansii: 219
– plicatiloba → C. tuberculata: 61

[Caralluma]
- plurifasciculata → C. congestiflora: 51
- praegracilis → Orbea caudata ssp. caudata: 189
- **priogonium**: 58
- **procumbens**: 59
- pruinosa → Quaqua pruinosa: 219
- – var. nigra → Quaqua pruinosa: 219
- pseudonebrownii → Orbea lutea ssp. vaga: 196
- punctata → Piaranthus punctatus: 211
- **quadrangula**: 59
- ramosa → Quaqua ramosa: 220
- rangeana → Orbea maculata ssp. rangeana: 197
- rangei → Orbea maculata ssp. rangeana: 197
- rauhii → C. adenensis: 48
- reflexa → Quaqua dependens: 216
- retrospiciens → C. acutangula: 47
- – ssp. tombuctuensis → C. acutangula: 47
- – var. acutangula → C. acutangula: 47
- – – glabra → C. acutangula: 47
- – – hirtiflora → C. acutangula: 47
- – – laxiflora → C. acutangula: 47
- – – tombuctuensis → C. acutangula: 47
- rivae → C. socotrana: 60
- robusta → C. penicillata: 58
- rogersii → Orbea rogersii: 199
- rosengrenii → C. socotrana: 60
- rubiginosa → Orbea melanantha: 197
- russelliana → C. acutangula: 47
- sacculata → Orbea sacculata: 199
- **sarkariae**: 59
- schweickerdtii → Orbea carnosa: 189
- schweinfurthii → Orbea schweinfurthii: 199
- scutellata → Echidnopsis scutellata: 134
- serpentina → Notechidnopsis tessellata: 186
- serrulata → Piaranthus decorus ssp. decorus: 210
- shadhbana → C. hexagona: 55
- – var. barhana → C. hexagona: 55
- simonis → C. europaea: 53
- simulans → Quaqua arida: 215
- **sinaica**: 59
- – var. baradii → C. sinaica: 59
- – – sinaica → C. sinaica: 59
- **socotrana**: 60
- **solenophora**: 60
- **somalica**: 60
- **speciosa**: 60
- sprengeri → Orbea sprengeri: 200
- – ssp. foetida → Orbea sprengeri ssp. foetida: 200
- – – laticorona → Orbea laticorona: 195
- – – ogadensis → Orbea sprengeri ssp. ogadensis: 201
- **staintonii**: 61
- **stalagmifera**: 61
- stapeliiformis → C. sp.: 47
- subterranea → Orbea subterranea: 201
- **subulata**: 61
- swanepoelii → Quaqua swanepoelii: 220
- tessellata → Notechidnopsis tessellata: 186

[Caralluma]
- tombuctuensis → C. acutangula: 47
- torta → Rhytidocaulon tortum: 232
- truncato-coronata → C. crenulata: 51
- truncatocorona → C. crenulata: 51
- tsumebensis → Orbea valida ssp. valida: 202
- **tuberculata**: 61
- tubiformis → Orbea tubiformis: 201
- **turneri**: 62
- – – ssp. **turneri**: 62
- – – – **ukambensis**: 62
- ubomboensis → Orbea ubomboensis: 202
- **umbellata**: 62
- umdausensis → Tromotriche umdausensis: 273
- **vaduliae**: 62
- vaga → Orbea lutea ssp. vaga: 196
- valida → Orbea valida: 202
- vansonii → Orbea lutea ssp. lutea: 196
- venenosa → Orbea decaisneana: 191
- vibratilis → Orbea vibratilis: 204
- villetii → Quaqua inversa: 217
- virescens → Quaqua parviflora: 218
- vittata → C. edulis: 52
- wilfriedii → Quaqua acutiloba: 214
- wilsonii → Orbea wilsonii: 204
- winkleri → Quaqua mammillaris: 217
- winkleriana → Quaqua mammillaris: 217
- wissmannii → Orbea wissmannii: 205
Carapelia tarantuloides → Orbelia sp.: 207
Caruncularia → Tromotriche: 269
- aperta → Tromotriche aperta: 270
- jacquinii → Tromotriche pedunculata: 272
- massonii → Tromotriche pedunculata: 272
- pedunculata → Tromotriche pedunculata: 272
- penduliflora → Tromotriche pedunculata: 272
- simsii → Tromotriche pedunculata: 272
Cathetostemma → Hoya: 147
Caudanthera → Caralluma: 46
- mireillae → Caralluma mireillae: 57
- sinaica → Caralluma sinaica: 59
Centrostemma → Hoya: 147
Ceropegia : 63
- aberrans → C. stenoloba: 102
- abinsica → C. campanulata: 70
- **abyssinica**: 63
- – – → C. loranthiflora: 88
- – – var. songeensis → C. abyssinica: 63
- acacietorum → C. pachystelma: 93
- **achtenii**: 64
- – – ssp. adolfii → C. achtenii: 64
- – – – togoensis → C. achtenii: 64
- acuminata → C. bulbosa: 70
- adolfii → C. achtenii: 64
- – – var. gracillima → C. achtenii: 64
- adrienneae → C. bosseri: 69
- **affinis**: 64
- **africana**: 65
- – – ssp. **africana**: 65
- – – – **barklyi**: 65

[Ceropegia africana ssp.]
– – – fortuita → C. fortuita: 79
– **ahmarensis**: 65
– albertina → C. aristolochioides ssp. aristolochioides: 68
– albiflora → C. ensifolia: 77
– **albisepta**: 65
– – var. bruceana → C. albisepta: 65
– – – robynsiana → C. albisepta: 65
– – – truncata → C. albisepta: 65
– – – viridis → C. albisepta: 65
– **ambovombensis**: 66
– **ampliata**: 66
– – ssp. insulicola → C. ampliata: 66
– – – madagascariensis → C. ampliata: 66
– – var. oxyloba → C. ampliata: 66
– angiensis → C. meyeri-johannis: 90
– angusta → C. affinis: 64
– angustifolia → C. longifolia: 87
– – → C. stenantha: 102
– angustilimba → C. trichantha: 104
– **antennifera**: 66
– aphylla → C. dichotoma ssp. dichotoma: 75
– – → C. sp.: 63
– apiculata → C. lugardiae: 88
– **arabica**: 66
– – var. **abbreviata**: 67
– – – **arabica**: 67
– – – **powysii**: 67
– – – **superba**: 67
– **arenaria**: 67
– **aridicola**: 67
– **aristolochioides**: 68
– – ssp. albertina → C. aristolochioides ssp. aristolochioides: 68
– – – **aristolochioides**: 68
– – – **deflersiana**: 68
– – var. wittei → C. aristolochioides ssp. aristolochioides: 68
– **armandii**: 68
– – var. petignatii → C. petignatii: 94
– **arnottiana**: 68
– assimilis → C. cancellata: 71
– atacorensis → C. affinis: 64
– **attenuata**: 69
– balfouriana → C. mairei: 89
– **ballyana**: 69
– barbata → C. filiformis: 77
– barbertonensis → C. linearis ssp. woodii: 87
– barbigera → C. arabica var. powysii: 67
– barkleyi → C. africana ssp. barklyi: 65
– – var. tugelensis → C. africana ssp. barklyi: 65
– beccariana → C. aristolochioides ssp. aristolochioides: 68
– bequaertii → C. abyssinica: 63
– biddumana → C. affinis: 64
– biflora → C. candelabrum: 71
– blatteri → C. odorata: 93
– boerhaaviifolia → C. pachystelma: 93

[Ceropegia boerhaaviifolia]
– – → C. sp.: 63
– **bonafouxii**: 69
– – var. linearifolia → C. abyssinica: 63
– borii → C. longifolia: 87
– **bosseri**: 69
– – var. razafindratsirana → C. bosseri: 69
– botrys → C. subaphylla: 103
– boussingaultifolia → C. nilotica: 92
– **bowkeri**: 70
– – ssp. sororia → C. bowkeri: 70
– brachyceras → C. crassifolia var. crassifolia: 73
– brevicollis → C. decaisneana: 74
– breviloba → C. madagascariensis: 88
– brevirostris → C. distincta: 76
– brevitubulata → Brachystelma brevitubulatum: 23
– brosima → C. bulbosa: 70
– brownii → C. denticulata: 75
– **bulbosa**: 70
– – var. esculenta → C. bulbosa: 70
– – – lushii → C. bulbosa: 70
– butaguensis → C. affinis: 64
– caffrorum → C. linearis ssp. linearis: 86
– – var. dubia → C. linearis ssp. linearis: 86
– calcarata → C. meyeri-johannis: 90
– **campanulata**: 70
– – var. abinsica → C. campanulata: 70
– – – porphyrotricha → C. porphyrotricha: 95
– – – pulchella → C. insignis: 83
– **cancellata**: 71
– candelabriformis → C. candelabrum: 71
– **candelabrum**: 71
– – → C. sp.: 63
– – ssp. tuberosa → C. candelabrum: 71
– – var. biflora → C. candelabrum: 71
– – – tuberosa → C. candelabrum: 71
– **carnosa**: 71
– **cataphyllaris**: 72
– ceratophora → C. dichotoma ssp. krainzii: 76
– chipiaensis → C. umbraticola: 105
– chortophylla → C. stenoloba: 102
– chrysantha → C. dichotoma ssp. krainzii: 76
– **ciliata**: 72
– – ssp. ensifolia → C. ensifolia: 77
– **cimiciodora**: 72
– **claviloba**: 72
– connivens → C. fimbriata ssp. connivens: 78
– – fa. angustata → C. fimbriata ssp. connivens: 78
– **conrathii**: 72
– constricta → C. nilotica: 92
– contorta → C. saxatilis: 99
– **convolvuloides**: 73
– copleyae → C. crassifolia var. copleyae: 73
– corallicorona → C. linearis ssp. woodii: 87
– cordata → C. sp.: 63
– cordiloba → C. papillata: 94
– craibii → C. antennifera: 66

[Ceropegia]
- **crassifolia**: 73
- – var. **copleyae**: 73
- – – **crassifolia**: 73
- crassula → C. aristolochioides ssp. aristolochioides: 68
- criniticaulis → C. meyeri-johannis: 90
- crispata → C. crassifolia var. crassifolia: 73
- **cufodontii**: 73
- **cycniflora**: 74
- cynanchoides → C. affinis: 64
- cyrtoidea → C. distincta: 76
- dalzielii → C. campanulata: 70
- **damannii**: 74
- debilis → C. linearis ssp. debilis: 86
- **decaisneana**: 74
- – var. brevicollis → C. decaisneana: 74
- decaryi → C. albisepta: 65
- **decidua**: 74
- – ssp. **decidua**: 74
- – – **pretoriensis**: 74
- decumbens → C. nilotica: 92
- deflersii → C. sp.: 63
- **deightonii**: 75
- – ssp. conjuncta → C. deightonii: 75
- – – tisserantii → C. deightonii: 75
- dentata → C. paricyma: 94
- **denticulata**: 75
- – var. brownii → C. denticulata: 75
- devecchii → C. variegata: 105
- – var. adelaidae → C. variegata: 105
- dewevrei → C. volubilis: 106
- **dichotoma**: 75
- – ssp. **dichotoma**: 75
- – – fusca → C. fusca: 80
- – – **krainzii**: 76
- dichroantha → C. filipendula: 78
- **dimorpha**: 76
- **dinteri**: 76
- discreta → C. candelabrum: 71
- **distincta**: 76
- – fa. pubescens → C. lugardiae: 88
- – ssp. haygarthii → C. haygarthii: 81
- – – lugardiae → C. lugardiae: 88
- – – verruculosa → C. lugardiae: 88
- **dolichophylla**: 77
- – var. brachyloba → C. dolichophylla: 77
- – – purpureobarbata → C. dolichophylla: 77
- driophila → Ceropegia trichantha: 104
- dubia → C. meyeri-johannis: 90
- ellenbeckii → C. convolvuloides: 73
- elliotii → C. candelabrum: 71
- **ensifolia**: 77
- erecta → C. wallichii: 106
- esculenta → C. bulbosa: 70
- estelleana → C. fimbriata ssp. fimbriata: 78
- euryacme → C. linearis ssp. woodii: 87
- **evansii**: 77
- – var. media → C. media: 89

[Ceropegia]
- evelynae → C. albisepta: 65
- **exigua**: 77
- **fantastica**: 77
- farrokhii → C. sp.: 63
- filicalyx → C. abyssinica: 63
- **filiformis**: 77
- – – → C. cataphyllaris: 72
- **filipendula**: 78
- **fimbriata**: 78
- – ssp. **connivens**: 78
- – – **fimbriata**: 78
- – – **geniculata**: 79
- **fimbriifera**: 79
- **floribunda**: 79
- **foliosa**: 79
- **fortuita**: 79
- **fusca**: 80
- **galeata**: 80
- galpinii → C. rendallii: 96
- gemmifera → C. nilotica: 92
- geniculata → C. fimbriata ssp. geniculata: 79
- **gikyi**: 80
- **gilgiana**: 80
- gilletii → C. abyssinica: 63
- glabripedicellata → C. affinis: 64
- gossweileri → C. nilotica: 92
- gourmacea → C. affinis: 64
- gracilis → C. decaisneana: 74
- grandis → C. nilotica: 92
- gymnopoda → Brachystelma gymnopodum: 31
- hastata → C. linearis ssp. woodii: 87
- **haygarthii**: 81
- helicoidea → C. albisepta: 65
- helicoides → C. ballyana: 69
- hepburnii → C. campanulata: 70
- **hermannii**: 81
- hians → C. dichotoma ssp. dichotoma: 75
- – var. striata → C. dichotoma ssp. dichotoma: 75
- **hirsuta**: 81
- – – → C. abyssinica: 63
- – – → C. ciliata: 72
- – var. jacquemontiana → C. hirsuta: 81
- – – ophiocephala → C. hirsuta: 81
- – – stenophylla → C. hirsuta: 81
- – – vincifolia → C. vincifolia: 106
- hispida → C. hirsuta: 81
- hispidipes → C. abyssinica: 63
- hochstetteri → C. affinis: 64
- **hofstaetteri**: 81
- **hookeri**: 82
- – var. mollis → C. hookeri: 82
- huberi → C. santapaui: 98
- **humbertii**: 82
- **illegitima**: 82
- **imbricata**: 83
- infausta → C. stenantha: 102
- **inflata**: 83

[Ceropegia]
- infundibuliformis → C. filiformis: 77
- **inornata**: 83
- **insignis**: 83
- intracolor → C. imbricata: 83
- jacquemontiana → C. hirsuta: 81
- **jainii**: 83
- **johnsonii**: 84
- jucunda → C. trichantha: 104
- **juncea**: 84
- kamerunensis → C. affinis: 64
- kassneri → C. purpurascens: 96
- **keniensis**: 84
- kerstingii → C. campanulata: 70
- **kituloensis**: 84
- **konasita**: 84
- krainziana → C. dichotoma ssp. krainzii: 76
- krainzii → C. dichotoma ssp. krainzii: 76
- kroboensis → C. nigra: 92
- **kundelunguensis**: 85
- kwebensis → C. purpurascens: 96
- lanceolata → C. longifolia: 87
- **lawii**: 85
- **ledermannii**: 85
- leptocarpa → C. linearis ssp. woodii: 87
- **leroyi**: 86
- leucotaenia → C. abyssinica: 63
- **lindenii**: 86
- **linearis**: 86
- – ssp. **debilis**: 86
- – – – **linearis**: 86
- – – – **tenuis**: 86
- – – – **woodii**: 87
- **linophylla**: 87
- longiflora → C. candelabrum: 71
- **longifolia**: 87
- – ssp. sinensis → C. dolichophylla: 77
- – var. exigua → C. exigua: 77
- **loranthiflora**: 87
- loureiroi → C. sp.: 63
- lucida → Ceropegia trichantha: 104
- **lugardiae**: 88
- lujai → C. johnsonii: 84
- lushii → C. bulbosa: 70
- maasaiorum → C. aristolochioides ssp. aristolochioides: 68
- maccannii → C. lawii: 85
- **macrantha**: 88
- – var. thorelii → C. macrantha: 88
- **madagascariensis**: 88
- mafekingensis → Brachystelma mafekingense: 34
- **mahabalei**: 89
- **mairei**: 89
- – var. tenella → C. mairei: 89
- **maiuscula**: 89
- mansouriana → C. subaphylla: 103
- mazoensis → C. stenantha: 102
- **media**: 89

[Ceropegia]
- medoensis → C. filipendula: 78
- melanops → C. nigra: 92
- **meleagris**: 89
- **mendesii**: 90
- **meyeri**: 90
- **meyeri-johannis**: 90
- – – var. angiensis → C. meyeri-johannis: 90
- – – – verdickii → C. meyeri-johannis: 90
- mirabilis → C. filipendula: 78
- monteiroae → C. sandersonii: 98
- mozambicensis → C. nilotica: 92
- – – var. ulugurensis → C. nilotica: 92
- mucronata → C. candelabrum: 71
- **muliensis**: 91
- **multiflora**: 91
- – – fa. puberula → C. multiflora ssp. tentaculata: 91
- – – – pubescens → C. multiflora ssp. multiflora: 91
- – – ssp. **multiflora**: 91
- – – – **tentaculata**: 91
- – – var. latifolia → C. multiflora ssp. multiflora: 91
- munronii → C. spiralis: 101
- mutabilis → C. paricyma: 94
- **muzingana**: 91
- **nana**: 91
- **nigra**: 92
- **nilotica**: 92
- – – var. plicata → C. nilotica: 92
- – – – simplex → C. denticulata: 75
- **noorjahaniae**: 92
- nuda → C. subaphylla: 103
- obtusa → C. sp.: 63
- **occidentalis**: 93
- **occulta**: 93
- **oculata**: 93
- – – var. subhirsuta → C. vincifolia: 106
- **odorata**: 93
- ophiocephala → C. hirsuta: 81
- **pachystelma**: 93
- – – ssp. undulata → C. pachystelma: 93
- panchganiensis → C. lawii: 85
- **papillata**: 94
- – – var. cordiloba → C. papillata: 94
- **paricyma**: 94
- patersoniae → C. zeyheri: 107
- patriciae → Brachystelma mafekingense: 34
- pedunculata → C. affinis: 64
- perrieri → C. madagascariensis: 88
- perrottetii → C. aristolochioides ssp. aristolochioides: 68
- peteri → C. filipendula: 78
- **petignatii**: 94
- petiolata → C. madagascariensis: 88
- plicata → C. nilotica: 92
- **poluniniana**: 95
- polyantha → C. vincifolia: 106

[Ceropegia]
- **porphyrotricha**: 95
- powysii → C. arabica var. powysii: 67
- **praetermissa**: 95
- prainii → C. jainii: 83
- profundorum → C. dolichophylla: 77
- pseudodimorpha → C. dimorpha: 76
- **pubescens**: 95
- – – → C. meyeri: 90
- pumila → Brachystelma gymnopodum: 31
- **purpurascens**: 96
- – – ssp. thysanotos → C. purpurascens: 96
- **pusilla**: 96
- pygmaea → Brachystelma gymnopodum: 31
- – – var. pumila → Brachystelma gymnopodum: 31
- quarrei → C. stenantha: 102
- racemosa → C. affinis: 64
- – – ssp. glabra → C. affinis: 64
- – – – secamonoides → C. affinis: 64
- – – – setifera → C. affinis: 64
- **radicans**: 96
- – – ssp. **radicans**: 96
- – – – **smithii**: 96
- – – var. smithii → C. radicans ssp. smithii: 96
- raizadiana → C. macrantha: 88
- rara → C. vanderystii: 105
- razafindratsirana → C. bosseri: 69
- **rendallii**: 96
- renzii → C. filipendula: 78
- **ringens**: 97
- – – → C. convolvuloides: 73
- **ringoetii**: 97
- **robivelonae**: 97
- robynsiana → C. albisepta: 65
- rollae → C. lawii: 85
- rostrata → C. umbraticola: 105
- **rudatisii**: 97
- **rupicola**: 97
- – – var. **stictantha**: 98
- ruspoliana → C. affinis: 64
- sahyadrica → C. lawii: 85
- **salicifolia**: 98
- **sandersonii**: 98
- **santapaui**: 98
- **saxatilis**: 99
- – – → C. bonafouxii: 69
- **scabra**: 99
- scabriflora → C. tomentosa: 104
- scandens → C. volubilis: 106
- schaijesiorum → C. umbraticola: 105
- schinziana → C. pachystelma: 93
- schlechteriana → C. ringoetii: 97
- schliebenii → C. stenoloba: 102
- schoenlandii → C. linearis ssp. woodii: 87
- secamonoides → C. affinis: 64
- senegalensis → C. linophylla: 87
- **sepium**: 99
- serpentina → C. stapeliiformis ssp. serpentina: 101

[Ceropegia]
- seticorona → C. aristolochioides ssp. aristolochioides: 68
- – – var. dilatiloba → C. aristolochioides ssp. aristolochioides: 68
- setifera → C. affinis: 64
- – – var. natalensis → C. affinis: 64
- **simoneae**: 99
- **sinoerecta**: 100
- sinuata → C. ringens: 97
- – – → Ceropegia ringens: 97
- smithii → C. radicans ssp. smithii: 96
- **sobolifera**: 100
- – – var. **nephroloba**: 100
- – – – **sobolifera**: 100
- **somalensis**: 100
- – – fa. erostrata → C. somalensis: 100
- **sootepensis**: 100
- sororia → C. bowkeri: 70
- **speciosa**: 101
- **spiralis**: 101
- squamulata → Echidnopsis squamulata: 135
- **stapeliiformis**: 101
- – – ssp. **serpentina**: 101
- – – – **stapeliiformis**: 101
- – – var. serpentina → C. stapeliiformis ssp. serpentina: 101
- **stenantha**: 101
- – – var. parviflora → C. stenantha: 102
- **stenoloba**: 102
- – – var. australis → C. stenoloba: 102
- – – – moyalensis → C. stenoloba: 102
- **stenophylla**: 102
- **stentiae**: 102
- steudneri → C. abyssinica: 63
- steudneriana → C. abyssinica: 63
- stocksii → C. vincifolia: 106
- **striata**: 103
- **subaphylla**: 103
- suberosa → Fockea edulis: 138
- subtruncata → C. sobolifera var. sobolifera: 100
- succulenta → C. albisepta: 65
- superba → C. arabica var. superba: 67
- **swaziorum**: 103
- tamalensis → C. campanulata: 70
- **taprobanica**: 103
- tentaculata → C. multiflora ssp. tentaculata: 91
- – – var. puberula → C. multiflora ssp. tentaculata: 91
- tenuis → C. linearis ssp. tenuis: 86
- tenuissima → C. stenantha: 102
- thorelii → C. macrantha: 88
- thorncroftii → C. crassifolia var. crassifolia: 73
- thysanotos → C. purpurascens: 96
- **tihamana**: 104
- **tomentosa**: 104
- **trichantha**: 104
- triebneri → C. ampliata: 66
- tristis → C. haygarthii: 81

293

[Ceropegia]
- tsaiana → C. pubescens: 95
- tuberculata → C. crassifolia var. crassifolia: 73
- tuberosa → C. bulbosa: 70
- – – → C. candelabrum: 71
- tubulifera → C. variegata: 105
- **turricula**: 104
- **umbraticola**: 105
- undulata → C. pachystelma: 93
- **vanderystii**: 105
- **variegata**: 105
- – – var. adelaidae → C. variegata: 105
- – – – cornigera → C. variegata: 105
- ventricosa → C. inflata: 83
- verdickii → C. meyeri-johannis: 90
- verrucosa → C. albisepta: 65
- verruculosa → C. lugardiae: 88
- **verticillata**: 106
- vignaldiana → C. bulbosa: 70
- **vincifolia**: 106
- viridis → C. albisepta: 65
- – – var. truncata → C. albisepta: 65
- **volubilis**: 106
- – – var. crassicaulis → C. aristolochioides ssp. aristolochioides: 68
- **wallichii**: 106
- wellmanii → C. umbraticola: 105
- woodii → C. linearis ssp. woodii: 87
- **yemenensis**: 106
- **yorubana**: 107
- **zeyheri**: 107

Chlorochlamys → Marsdenia: 179
Chlorocyathus → Raphionacme: 220
- monteiroae → Raphionacme monteiroae: 226
- welwitschii → Raphionacme welwitschii: 228
Chymocormus → Fockea: 137
- edulis → Fockea edulis: 138
Cibirhiza : 107
- **albersiana**: 107
- **dhofarensis**: 107
Cinclia → Ceropegia: 63
Collyris → Dischidia: 118
- major → Dischidia major: 121
- – – → Hoya imbricata: 152
- minor → Dischidia nummularia: 122
Conchophyllum → Dischidia: 118
- angulatum → Dischidia astephana: 119
- imbricatum → Dischidia imbricata: 121
- maximum → Hoya imbricata: 152
- merrillii → Dischidiopsis parasitica: 124
Craterostemma → Brachystelma: 20
- schinzii → Brachystelma schinzii: 42
Crenulluma → Caralluma: 46
- adenensis → Caralluma adenensis: 48
- arabica → Caralluma arabica: 49
- aucheriana → Caralluma sp.: 47
- awdeliana → Caralluma awdeliana: 49
- dolichocarpa → Caralluma dolichocarpa: 52
- flava → Caralluma flava: 54

[Crenulluma]
- kalmbacheriana → Caralluma adenensis: 48
- lavrani → Caralluma lavrani: 56
- petraea → Caralluma awdeliana: 49
- rauhii → Caralluma adenensis: 48
Cryptolepis monteiroae → Stomatostemma monteiroae: 264
Cryptolluma → Caralluma: 46
- edulis → Caralluma edulis: 52
Cyathella → Cynanchum: 108
Cylindrilluma → Caralluma: 46
- solenophora → Caralluma solenophora: 60
Cynanchum : 108
- **aculeatum**: 108
- aequilongum → C. implicatum: 112
- ambositrense → C. mahafalense: 114
- **ambovombense**: 109
- **ampanihense**: 109
- **andringitrense**: 109
- **angavokeliense**: 109
- **ansamalense**: 109
- antandroy → C. menarandrense: 115
- antsiranense → Sarcostemma antsiranense: 233
- aphyllum → Sarcostemma viminale: 235
- **appendiculatopsis**: 109
- **arenarium**: 109
- bekinolense → C. gerrardii ssp. bekinolense: 111
- **bisinuatum**: 110
- bojerianum → C. luteifluens var. luteifluens: 113
- carnosum → Hoya carnosa: 149
- **compactum**: 110
- – – var. **compactum**: 110
- – – – **imerinense**: 110
- **crassipedicellatum**: 110
- crispum → Brachystelma praelongum ssp. thunbergii: 40
- – – → Fockea capensis: 138
- **cucullatum**: 110
- decaisneanum → C. luteifluens: 113
- decorsei → Sarcostemma decorsei: 234
- **descoingsii**: 110
- edule → C. gerrardii ssp. gerrardii: 111
- elachistemmoides → Sarcostemma elachistemmoides: 234
- **fimbriatum**: 111
- **floriferum**: 111
- **folotsioides**: 111
- **gerrardii**: 111
- – – ssp. **bekinolense**: 111
- – – – **gerrardii**: 111
- **grandidieri**: 111
- **hardyi**: 112
- helicoideum → C. madagascariense: 114
- **humbert-capuronii**: 112
- humbertii → C. ampanihense: 109
- **implicatum**: 112
- **insigne**: 112
- **juliani-marnieri**: 112

[Cynanchum]
- **jumellei**: 112
- **junciforme**: 112
- **lecomtei**: 113
- **lenewtonii**: 113
- **lineare**: 113
- – ssp. **keraudreniae**: 113
- – – **lineare**: 113
- **luteifluens**: 113
- – var. **longicoronae**: 113
- – – **luteifluens**: 113
- **macranthum**: 114
- **macrolobum**: 114
- **madagascariense**: 114
- madecassum → C. arenarium: 110
- **mahafalense**: 114
- **mariense**: 114
- **marnierianum**: 114
- membranaceum → Sarcostemma membranaceum: 234
- **menarandrense**: 115
- **messeri**: 115
- **mevei**: 115
- **moramangense**: 115
- **napiferum**: 115
- **nematostemma**: 115
- nodosum → C. arenarium: 109
- omissum → Fockea angustifolia: 138
- **orangeanum**: 116
- **pachycladon**: 116
- **papillatum**: 116
- **perrieri**: 116
- **petignatii**: 116
- **phillipsonianum**: 116
- **praecox**: 117
- **pycnoneuroides**: 117
- pygmaeum → C. praecox: 117
- **radiatum**: 117
- **rauhianum**: 117
- **rossii**: 117
- rusillonii → C. junciforme: 112
- sarcostemmatoides → C. gerrardii ssp. gerrardii: 111
- **sessiliflorum**: 117
- **sigridiae**: 118
- **subtile**: 118
- tetrapterum → Sarcostemma viminale: 235
- **toliari**: 118
- tuberosum → C. sp.: 108
- **verrucosum**: 118

Cynoctonum → Cynanchum: 108
Cyrtoceras → Hoya: 147
Cystidianthus → Hoya: 147
Dalzielia → Marsdenia: 179
Decabelone → Tavaresia: 265
- barklyi → Tavaresia barklyi: 265
- elegans → Tavaresia angolensis: 265
- grandiflora → Tavaresia barklyi: 265
- meintjesii → Staparesia meintjesii: 242

[Decabelone]
- sieberi → Tavaresia angolensis: 265

Decaceras → Brachystelma: 20
- arnotii → Brachystelma arnotii: 21
- huttonii → Brachystelma huttonii: 32

Decanema → Cynanchum: 108
- bojerianum → Cynanchum luteifluens var. luteifluens: 113
- grandiflorum → Cynanchum grandidieri: 111
- luteifluens → Cynanchum luteifluens: 113

Decanemopsis → Sarcostemma: 233
- aphylla → Sarcostemma sp.: 233

Dernia : 118
- – **Surprise** : 118

Desmidorchis → Caralluma: 46
- acutangula → Caralluma acutangula: 47
- aucheriana → Caralluma sp.: 47
- crenulata → Caralluma crenulata: 51
- decaisneana → Orbea decaisneana: 191
- diffusa → Caralluma diffusa: 52
- edithae → Caralluma edithae: 52
- europaea → Caralluma europaea: 53
- foetida → Caralluma foetida: 54
- indica → Caralluma indica: 56
- pauciflora → Caralluma pauciflora: 57
- penicillata → Caralluma penicillata: 58
- quadrangula → Caralluma quadrangula: 59
- retrospiciens → Caralluma acutangula: 47
- somalica → Caralluma somalica: 60
- speciosa → Caralluma speciosa: 61
- stocksiana → Caralluma edulis: 52
- umbellata → Caralluma umbellata: 62

Dichaelia → Brachystelma: 20
- brachylepis → Brachystelma circinatum: 26
- breviflora → Brachystelma pygmaeum ssp. pygmaeum: 40
- cinerea → Brachystelma circinatum: 26
- circinata → Brachystelma circinatum: 26
- elongata → Brachystelma elongatum: 28
- filiformis → Brachystelma circinatum: 26
- forcipata → Brachystelma circinatum: 26
- galpinii → Brachystelma circinatum: 26
- gracillima → Brachystelma gracillimum: 31
- macra → Brachystelma circinatum: 26
- microphylla → Brachystelma circinatum: 26
- natalensis → Brachystelma sandersonii: 42
- ovata → Brachystelma circinatum: 26
- pallida → Brachystelma circinatum: 26
- pygmaea → Brachystelma pygmaeum: 40
- undulata → Brachystelma circinatum: 26
- villosa → Brachystelma villosum: 45
- zeyheri → Brachystelma circinatum: 26

Dictyanthus → Matelea: 181
- – → Matelea: 181
- campanulatus → Matelea pavonii: 182
- macvaughianus → Matelea macvaughiana: 182
- parviflorus → Matelea hemsleyana: 182
- pavonii → Matelea pavonii: 182
- prostratus → Matelea hemsleyana: 182

[Dictyanthus]
– reticulatus → Matelea dictyantha: 181
– sepicola → Matelea sepicola: 182
– stapeliiflorus → Matelea pavonii: 182
– tigrinus → Matelea standleyana: 183
– tuberosus → Matelea tuberosa: 183
Diplocyatha → Orbea: 187
– ciliata → Orbea ciliata: 190
Dischidia : 118
– actephila → D. nummularia: 122
– **acutifolia**: 119
– – ssp. **acutifolia**: 119
– – – **klossii**: 119
– aemula → D. nummularia: 122
– **albida**: 119
– **astephana**: 119
– bauerlenii → D. major: 121
– beiningiana → D. nummularia: 122
– **bengalensis**: 120
– brachystele → D. acutifolia ssp. acutifolia: 119
– brunoniana → D. hirsuta: 120
– clavata → D. major: 121
– coccinea → D. cochleata: 120
– **cochleata**: 120
– copelandii → D. nummularia: 122
– cuneifolia → D. bengalensis: 120
– decipiens → D. nummularia: 122
– depressa → D. imbricata: 121
– dirhiza → D. nummularia: 122
– **dolichantha**: 120
– euryloma → D. hirsuta: 120
– fasciculata → D. hirsuta: 120
– **formosana**: 120
– **fruticulosa**: 120
– fultonii → D. albida: 119
– gaudichaudii → D. nummularia: 122
– glabra → D. nummularia: 122
– glaucescens → Dischidia nummularia: 122
– **hirsuta**: 120
– horsfieldiana → D. rhombifolia: 123
– hoyoides → D. acutifolia ssp. acutifolia: 119
– **imbricata**: 121
– immortalis → D. nummularia: 122
– joloensis → D. punctata: 122
– kawengica → D. albida: 119
– klossii → D. acutifolia ssp. klossii: 119
– kutchinensis → D. albida: 119
– lagenifera → D. albida: 119
– **lanceolata**: 121
– – → D. vidalii: 123
– lancifolia → D. acutifolia ssp. klossii: 119
– **latifolia**: 121
– **livida**: 121
– loeseneriana → D. bengalensis: 120
– **major**: 121
– maxima → Hoya imbricata: 152
– merguiensis → D. major: 121
– merrillii → Dischidiopsis parasitica: 124
– microphylla → D. nummularia: 122

[Dischidia]
– minor → D. nummularia: 122
– **nummularia**: 122
– – var. gaudichaudii → D. nummularia: 122
– – – glabra → D. nummularia: 122
– – – gracilis → D. nummularia: 122
– – – rhombifolia → D. rhombifolia: 123
– obcordata → Micholitzia obcordata: 183
– orbicularis → D. nummularia: 122
– **ovata**: 122
– parasitica → Dischidiopsis parasitica: 124
– pectenoides → D. vidalii: 123
– pedunculata → D. acutifolia ssp. acutifolia: 119
– picta → D. ovata: 122
– pubicaulis → Dischidia nummularia: 122
– pubiflora → D. major: 121
– pulchella → D. hirsuta: 121
– **punctata**: 122
– punctatoides → D. punctata: 122
– rafflesiana → D. major: 121
– **rhombifolia**: 122
– ridleyana → D. nummularia: 122
– **ruscifolia**: 123
– **sagittata**: 123
– schumanniana → D. nummularia: 122
– semperflorens → D. albida: 119
– sepikana → D. nummularia: 122
– **singularis**: 123
– spatulata → D. bengalensis: 120
– subpeltigera → D. hirsuta: 121
– timorensis → D. major: 121
– **truncata**: 123
– – var. celebica → D. truncata: 123
– tubiflora → D. dolichantha: 120
– verruculosa → D. hirsuta: 121
– **vidalii**: 123
– viridiflora → D. punctata: 122
– zollingeri → D. acutifolia ssp. acutifolia: 119
Dischidiopsis : 123
– **parasitica**: 124
– philippinensis → D. parasitica: 124
Drakebrockmania → White-sloanea: 273
– crassa → White-sloanea crassa: 274
Dregea → Marsdenia: 178
Drepanostemma → Sarcostemma: 233
– luteum → Sarcostemma decorsei: 234
Duvalia : 124
– andreaeana → Huernia andreaeana: 159
– **anemoniflora**: 124
– **angustiloba**: 124
– **caespitosa**: 125
– – var. **caespitosa**: 125
– – – **compacta**: 125
– compacta → D. caespitosa var. compacta: 125
– concolor → D. caespitosa var. caespitosa: 125
– **corderoyi**: 125
– deflexa → Stapelia sp.: 243
– dentata → D. polita: 127
– **eilensis**: 125

[Duvalia]
- **elegans**: 126
- – fa. magnicorona → D. elegans: 126
- – var. namaquana → D. pubescens: 127
- – – seminuda → D. elegans: 126
- emiliana → D. caespitosa var. caespitosa: 125
- **galgallensis**: 126
- glomerata → D. caespitosa var. caespitosa: 125
- **gracilis**: 126
- hirtella → D. caespitosa var. caespitosa: 125
- – var. minor → D. caespitosa var. caespitosa: 125
- – – obscura → D. caespitosa var. caespitosa: 125
- **immaculata**: 126
- jacquiniana → D. elegans: 126
- jacquinii → D. elegans: 126
- laevigata → D. caespitosa var. caespitosa: 125
- **maculata**: 126
- – var. immaculata → D. immaculata: 126
- marlothii → D. caespitosa var. caespitosa: 125
- mastodes → D. caespitosa var. compacta: 125
- minuta → D. maculata: 126
- **modesta**: 127
- **parviflora**: 127
- pedunculata → Tromotriche pedunculata: 272
- **pillansii**: 127
- – var. albanica → D. pillansii: 127
- **polita**: 127
- – fa. intermedia → D. polita: 127
- – var. transvaalensis → D. polita: 127
- procumbens → Huernia procumbens: 170
- propinqua → D. caespitosa var. caespitosa: 125
- **pubescens**: 127
- – var. major → D. pubescens: 128
- – radiata → D. caespitosa var. caespitosa: 125
- – var. hirtella → D. caespitosa var. caespitosa: 125
- – – minor → D. caespitosa var. caespitosa: 125
- – – obscura → D. caespitosa var. caespitosa: 125
- reclinata → D. caespitosa var. caespitosa: 125
- – var. angulata → D. caespitosa var. caespitosa: 125
- – – bifida → D. caespitosa var. caespitosa: 125
- replicata → D. caespitosa var. caespitosa: 125
- serrulata → Piaranthus decorus ssp. decorus: 210
- somalensis → D. sulcata ssp. somalensis: 128
- **sulcata**: 128
- – ssp. **seminuda**: 128
- – – **somalensis**: 128
- – – **sulcata**: 128
- – var. seminuda → D. sulcata ssp. seminuda: 128
- tanganyikensis → Huernia tanganyikensis: 172
- transvaalensis → D. polita: 127
- – var. parviflora → D. polita: 127
- tuberculata → D. caespitosa var. caespitosa: 125
- **velutina**: 128

[Duvalia]
- **vestita**: 129

Duvaliandra: 129
- **dioscoridis**: 129

Duvaliaranthus: 129
- **albostriatus**: 129

Echidnopsis: 129
- adamsonii → E. radians: 134
- **angustiloba**: 130
- **archeri**: 130
- **ballyi**: 130
- bavazzanoi → E. sharpei ssp. ciliata: 135
- **bentii**: 130
- **bihendulensis**: 131
- **cereiformis**: 131
- – var. brunnea → E. cereiformis: 131
- – – obscura → E. cereiformis: 131
- **chrysantha**: 131
- – ssp. **chrysantha**: 131
- – – **filipes**: 131
- – var. filipes → E. chrysantha ssp. filipes: 131
- ciliata → E. sharpei ssp. ciliata: 135
- columnaris → Notechidnopsis columnaris: 185
- cylindrica → E. cereiformis: 131
- **dammanniana**: 131
- **ericiflora**: 132
- ethiopica → E. mariae: 133
- flavicorona → E. planiflora: 133
- framesii → Notechidnopsis tessellata: 186
- **globosa**: 132
- golathii → Caralluma penicillata: 58
- hirsuta → E. planiflora: 133
- **insularis**: 132
- **jacksonii**: 132
- lavraniana → E. sharpei ssp. ciliata: 135
- **leachii**: 132
- **malum**: 132
- **mariae**: 132
- **mijerteina**: 133
- – var. marchandii → E. mijerteina: 133
- **milleri**: 133
- modesta → E. radians: 134
- **montana**: 133
- multangula → E. sp.: 130
- **nubica**: 133
- **planiflora**: 133
- quadrangula → Caralluma quadrangula: 59
- **radians**: 134
- **repens**: 134
- **rubrolutea**: 134
- **scutellata**: 134
- – ssp. australis → E. mariae: 133
- – – **dhofarensis**: 134
- – – planiflora → E. planiflora: 133
- – – **scutellata**: 134
- **seibanica**: 134
- serpentina → Notechidnopsis tessellata: 186
- **sharpei**: 135
- – ssp. **ciliata**: 135

[Echidnopsis sharpei ssp.]
− − − repens → E. repens: 134
− − − **sharpei**: 135
− similis → E. archeri: 130
− **socotrana**: 135
− somalensis → E. dammanniana: 131
− **squamulata**: 135
− stellata → E. virchowii var. stellata: 136
− tessellata → E. cereiformis: 131
− urceolaris → E. urceolata: 135
− **urceolata**: 135
− **virchowii**: 136
− − var. **stellata**: 136
− − − **virchowii**: 136
− **watsonii**: 136
− **yemenensis**: 136
Edithcolea : 136
− **grandis**: 137
− − var. baylissiana → E. grandis: 137
− sordida → E. grandis: 137
Eriopetalum → Brachystelma: 20
− attenuatum → Brachystelma attenuatum: 21
− laevigatum → Brachystelma laevigatum: 33
− parviflorum → Brachystelma parviflorum: 38
Eriostemma → Hoya: 147
Euphorbia antunesii → Tavaresia barklyi: 265
− pendula → Sarcostemma sp.: 233
− viminalis → Sarcostemma viminale: 235
Fanninia : 137
− **caloglossa**: 137
Flanagania → Cynanchum: 108
− orangeana → Cynanchum orangeanum: 116
Fockea : 137
− **angustifolia**: 137
− − var. volkii → F. angustifolia: 138
− **capensis**: 138
− **comaru**: 138
− crispa → F. capensis: 138
− cylindrica → F. edulis: 138
− dammarana → F. angustifolia: 138
− **edulis**: 138
− − var. capensis → F. capensis: 138
− glabra → F. edulis: 138
− gracilis → F. comaru: 138
− lugardii → F. angustifolia: 137
− mildbraedii → F. angustifolia: 138
− monroi → F. angustifolia: 138
− **multiflora**: 138
− natalensis → Petopentia natalensis: 209
− schinzii → F. multiflora: 138
− sessiliflora → F. angustifolia: 137
− **sinuata**: 139
− tugelensis → F. angustifolia: 138
− − → Petopentia natalensis: 209
− undulata → F. sinuata: 139
Folotsia → Cynanchum: 108
− aculeata → Cynanchum aculeatum: 108
− ambovombensis → Cynanchum ambovombense: 109

[Folotsia]
− floribunda → Cynanchum floriferum: 111
− grandiflora → Cynanchum grandidieri: 111
− humbertii → Cynanchum humbert-capuronii: 112
− madagascariensis → Cynanchum toliari: 118
− sarcostemmoides → Cynanchum grandidieri: 111
Frerea → Caralluma: 46
− indica → Caralluma frerei: 54
Glossostelma : 139
− **angolense**: 139
− **brevilobum**: 139
− **cabrae**: 140
− **carsonii**: 140
− **ceciliae**: 140
− **erectum**: 140
− **lisianthoides**: 141
− **mbisiense**: 141
− **nyikense**: 141
− **rusapense**: 141
− **spathulatum**: 141
− **xysmalobioides**: 142
Gomphocarpus → Asclepias: 10
− angustatus → Stathmostelma angustatum: 260
− carinatus → Stenostelma carinatum: 264
− chironioides → Glossostelma lisianthoides: 141
− chlorojodinus → Glossostelma carsonii: 140
− corniculatus → Stenostelma corniculatum: 264
− diploglossus → Aspidonepsis diploglossa: 19
− eminens → Stenostelma eminens: 264
− laevigatus → Brachystelma laevigatum: 33
− lisianthoides → Glossostelma lisianthoides: 141
− longipes → Stathmostelma pedunculatum: 262
− pauciflorus → Stathmostelma pauciflorum: 261
− pedunculatus → Stathmostelma pedunculatum: 262
− spathulatus → Glossostelma spathulatum: 141
− stenoglossus → Stenostelma capense: 263
Gonodentea bijliae → Stapelia bijliae: 244
Gonolobus cyclophyllus → Matelea cyclophylla: 181
Gonostapelia barklyi → Orbelia barklyi: 207
− maculosa → Tromostapelia sp.: 269
Gonostemon → Stapelia: 242
− acuminatus → Stapelia acuminata: 243
− − var. brevicuspis → Stapelia acuminata: 243
− arenosus → Stapelia arenosa: 243
− asterias → Stapelia asterias: 244
− − var. gibbus → Stapelia asterias: 244
− − − lucidus → Stapelia asterias: 244
− baylissii → Stapelia baylissii: 244
− clavicoronus → Stapelia clavicorona: 245
− concinnus → Stapelia sp.: 243
− deflexus → Stapelia sp.: 243
− − var. atropurpureus → Stapelia sp.: 243
− divaricatus → Stapelia divaricata: 245
− erectiflorus → Stapelia erectiflora: 245
− − fa. aberrans → Stapelia erectiflora var. prostratiflora: 246

[Gonostemon]
- flavopurpureus → Stapelia flavopurpurea: 246
- – var. fleckii → Stapelia flavopurpurea: 246
- fuscopurpureus → Stapelia sp.: 243
- gariepensis → Stapelia gariepensis: 246
- gettliffei → Stapelia gettliffei: 246
- giganteus → Stapelia gigantea: 247
- – nvar. pallidus → Stapelia gigantea: 247
- – var. marlothii → Stapelia gigantea: 247
- – – – nobilis → Stapelia gigantea: 247
- – – – youngii → Stapelia gigantea: 247
- glabricaulis → Stapelia glabricaulis: 247
- – var. forcipis → Stapelia glabricaulis: 247
- glanduliflorus → Stapelia glanduliflora: 247
- – var. emarginatus → Stapelia glanduliflora: 247
- gordonii → Hoodia gordonii: 143
- grandiflorus → Stapelia grandiflora: 247
- – var. ambiguus → Stapelia grandiflora: 248
- – – – apicalis → Stapelia grandiflora: 248
- – – – arnotii → Stapelia arnotii: 244
- – – – desmetianus → Stapelia grandiflora: 248
- – – – fergusoniae → Stapelia grandiflora: 248
- – – – fulvus → Stapelia grandiflora: 248
- – – – lineatus → Stapelia grandiflora: 248
- – – – pallidus → Stapelia grandiflora: 248
- – – – senilis → Stapelia grandiflora: 248
- – – – sororius → Stapelia sp.: 243
- hirsutus → Stapelia hirsuta: 248
- – var. affinis → Stapelia hirsuta: 248
- – – – comatus → Stapelia hirsuta: 248
- – – – depressus → Stapelia hirsuta: 248
- – – – gratus → Stapelia hirsuta: 248
- – – – longirostris → Stapelia hirsuta: 248
- – – – luteus → Stapelia hirsuta: 248
- kougabergensis → Stapelia kougabergensis: 249
- kwebensis → Stapelia kwebensis: 249
- leendertziae → Stapelia leendertziae: 249
- longipedicellatus → Stapelia longipedicellata: 249
- maccabeanae → Stapelia sp.: 243
- macowanii → Stapelia macowanii: 250
- – var. abrasus → Stapelia macowanii var. conformis: 250
- – – – conformis → Stapelia macowanii var. conformis: 250
- meintjesii → Stapelia meintjesii: 250
- montanus → Stapelia montana: 250
- – var. grossus → Stapelia montana var. grossa: 250
- obductus → Stapelia obducta: 250
- olivaceus → Stapelia olivacea: 250
- pallidus → Stapelia divaricata: 245
- paniculatus → Stapelia paniculata: 251
- parvulus → Stapelia parvula: 251
- pearsonii → Stapelia pearsonii: 251
- peglerae → Stapelia peglerae: 251
- pillansii → Stapelia pillansii: 252
- – var. attenuatus → Stapelia pillansii var. pillansii: 252

[Gonostemon pillansii var.]
- – – fontinalis → Stapelia pillansii var. fontinalis: 252
- plantii → Stapelia sp.: 243
- praetermissus → Stapelia praetermissa: 252
- – var. luteolus → Stapelia praetermissa var. luteola: 252
- pulvinatus → Stapelia pulvinata: 252
- remotus → Stapelia remota: 253
- rubiginosus → Stapelia rubiginosa: 253
- rufus → Stapelia rufa: 253
- – var. attenuatus → Stapelia rufa: 253
- – – – fissirostris → Stapelia rufa: 253
- schinzii → Stapelia schinzii: 253
- – var. angolensis → Stapelia schinzii var. angolensis: 253
- – – – bergerianus → Stapelia schinzii var. bergeriana: 253
- scitulus → Stapelia scitula: 254
- similis → Stapelia similis: 254
- – var. juttae → Stapelia similis: 254
- – – – portae-taurinae → Stapelia similis: 254
- strictus → Stapelia sp.: 243
- surrectus → Stapelia surrecta: 254
- – var. primosii → Stapelia surrecta: 254
- tsomoensis → Stapelia tsomoensis: 254
- unicornis → Stapelia unicornis: 255
- vetulus → Stapelia vetula: 255
- villetiae → Stapelia villetiae: 255
Gonotriche → Tromostapelia: 269
- bella → Tromostapelia sp.: 269
Gymnema recurvifolium → Hoya australis ssp. tenuipes: 148
Harrisonia → Marsdenia: 178
Hoodia : 142
- albispina → H. gordonii: 144
- **alstonii** : 142
- annulata → H. pilifera ssp. annulata: 145
- bainii → H. gordonii: 144
- – var. juttae → H. juttae: 144
- barklyi → H. gordonii: 144
- burkei → H. gordonii: 144
- cactiformis → Larryleachia cactiformis: 176
- colei → H. pilifera ssp. pillansii: 146
- **currorii** : 143
- – ssp. **currorii** : 143
- – – – **lugardii** : 143
- – var. minor → H. currorii ssp. currorii: 143
- delaetiana → H. officinalis ssp. delaetiana: 144
- dinteri → H. gordonii: 144
- – – → Larryleachia marlothii: 177
- **dregei** : 143
- felina → Larryleachia cactiformis: 176
- **flava** : 143
- foetida → H. triebneri: 146
- genilis → Tavaresia barklyi: 265
- gibbosa → H. currorii ssp. currorii: 143
- **gordonii** : 143
- grandis → H. pilifera ssp. pillansii: 146

[Hoodia]
- haagnerae → Lavrania haagnerae: 178
- husabensis → H. gordonii: 144
- **juttae**: 144
- langii → H. gordonii: 144
- **longispina**: 144
- lugardii → H. currorii ssp. lugardii: 143
- macrantha → H. currorii ssp. currorii: 143
- marlothii → Larryleachia marlothii: 177
- meloformis → Larryleachia picta: 177
- montana → H. currorii ssp. currorii: 143
- **mossamedensis**: 144
- **officinalis**: 144
- – ssp. **delaetiana**: 144
- – – – **officinalis**: 145
- **parviflora**: 145
- **pedicellata**: 145
- perlata → Larryleachia perlata: 177
- picta → Larryleachia picta: 177
- **pilifera**: 145
- – ssp. **annulata**: 145
- – – – **pilifera**: 145
- – – – **pillansii**: 146
- pillansii → H. gordonii: 144
- rosea → H. gordonii: 144
- **ruschii**: 146
- rustica → H. officinalis ssp. officinalis: 145
- senilis → Tavaresia barklyi: 265
- similis → Larryleachia cactiformis: 176
- – – → H. sp.: 142
- sociarum → Larryleachia sp.: 176
- tirasmontana → Larryleachia tirasmontana: 177
- **triebneri**: 146
- whitesloaneana → H. gordonii: 144

Hoodialluma : 146
Hoodiopsis → Hoodialluma: 146
Hoya : 146
- acuta → H. verticillata: 158
- – var. citrina → H. verticillata: 158
- alata → H. anulata: 147
- albens → H. verticillata: 158
- amoena → H. verticillata: 158
- angustifolia → H. pottsii: 156
- – – → H. tsangii: 158
- **anulata**: 147
- **archboldiana**: 147
- **australis**: 148
- – ssp. **australis**: 148
- – – – **oramicola**: 148
- – – – **rupicola**: 148
- – – – **sanae**: 148
- – – – **tenuipes**: 148
- barrackii → H. australis ssp. tenuipes: 148
- bella → H. lanceolata ssp. bella: 153
- **benguetensis**: 148
- bicarinata → H. australis ssp. tenuipes: 148
- **bilobata**: 148
- **bordenii**: 149
- browniana → H. macrophylla: 154

[Hoya]
- **calycina**: 149
- – ssp. **calycina**: 149
- – – – **glabrifolia**: 149
- **carnosa**: 149
- – – → H. meliflua: 154
- – – var. compacta → H. carnosa: 149
- – – – gushanica → H. carnosa: 149
- – – – japonica → H. carnosa: 149
- – – – marmorata → H. carnosa: 149
- – – – variegata → H. carnosa: 149
- **caudata**: 149
- chinensis → H. carnosa: 149
- **cinnamomifolia**: 150
- – – var. purpureofusca → H. purpureofusca: 156
- citrina → H. verticillata: 158
- clandestina → H. macrophylla: 154
- **crassicaulis**: 150
- crassifolia → H. carnosa: 149
- – – → H. caudata: 149
- crassipes → H. diversifolia: 150
- dalrympleana → H. australis ssp. australis: 148
- **diptera**: 150
- **diversifolia**: 150
- – – → H. meliflua: 154
- dodecatheiflora → H. inconspicua: 152
- **eitapensis**: 150
- **elliptica**: 150
- **engleriana**: 151
- **erythrostemma**: 151
- esculenta → H. diversifolia: 150
- **finlaysonii**: 151
- flagellata → H. caudata: 149
- **fuscomarginata**: 151
- globifera → H. verticillata: 158
- **globulosa**: 151
- hellwigiana → H. nicholsoniae: 155
- hellwigii → H. nicholsoniae: 155
- **heuschkeliana**: 151
- hookeriana → H. verticillata: 158
- **hypolasia**: 152
- **imbricata**: 152
- – – fa. typica → H. imbricata: 152
- – – var. basi-subcordata → H. imbricata: 152
- **inconspicua**: 152
- intermedia → H. carnosa: 149
- **kerrii**: 152
- keysii → H. australis ssp. australis: 148
- lactea → H. australis ssp. tenuipes: 148
- **lacunosa**: 153
- – – var. pallidiflora → H. lacunosa: 153
- **lanceolata**: 153
- – – ssp. **bella**: 153
- – – – **lanceolata**: 153
- lantsangensis → Micholitzia obcordata: 183
- **latifolia**: 153
- – – ssp. kinabaluensis → H. latifolia: 153
- **limoniaca**: 153
- **linearis**: 153

[Hoya linearis]
– – var. nepalensis → H. linearis: 153
– – – sikkimensis → H. linearis: 153
– litoralis → H. inconspicua: 152
– **longifolia**: 154
– loyceandrewsiana → H. latifolia: 153
– luzonica → H. meliflua: 155
– **macgillivrayi**: 154
– **macrophylla**: 154
– – → H. latifolia: 153
– **magnifica**: 154
– manipurensis → Micholitzia obcordata: 183
– maxima → H. imbricata: 152
– **meliflua**: 154
– **meredithii**: 155
– motoskei → H. carnosa: 149
– **nicholsoniae**: 155
– **nummularioides**: 155
– **obovata**: 155
– – var. kerrii → H. kerrii: 152
– obscurinerva → H. pottsii: 156
– oligotricha → H. australis ssp. australis: 148
– – ssp. tenuipes → H. australis ssp. tenuipes: 148
– orbiculata → H. diversifolia: 150
– ovalifolia → H. revoluta: 157
– **pachyclada**: 155
– pallida → H. verticillata: 158
– papillantha → H. australis ssp. tenuipes: 148
– parasitica → H. meliflua: 154
– – → H. verticillata: 158
– – var. citrina → H. verticillata: 158
– – – geoffrayi → H. verticillata: 158
– – – hendersonii → H. verticillata: 158
– – – spirei → H. verticillata: 158
– **parviflora**: 156
– **parvifolia**: 156
– **pauciflora**: 156
– paxtonii → H. lanceolata ssp. bella: 153
– picta → H. carnosa: 149
– pilosa → H. australis ssp. australis: 148
– polystachya → H. latifolia: 153
– poolei → H. anulata: 147
– **pottsii**: 156
– – var. angustifolia → H. pottsii: 156
– pseudolittoralis → H. anulata: 147
– pseudomaxima → H. imbricata: 152
– pubescens → H. australis ssp. australis: 148
– **purpureofusca**: 156
– **retusa**: 157
– **revoluta**: 157
– ridleyi → H. verticillata: 158
– rotundifolia → H. carnosa: 149
– rupicola → H. australis ssp. rupicola: 148
– sanae → H. australis ssp. sanae: 148
– schlechteriana → H. anulata: 147
– **serpens**: 157
– **shepherdii**: 157
– **siamica**: 157
– sikkimensis → H. lanceolata ssp. lanceolata: 153

[Hoya]
– sogerensis → H. nicholsoniae: 155
– spartioides → Absolmsia spartioides: 9
– suaveolens → H. lacunosa: 153
– **thailandica**: 157
– **thomsonii**: 158
– trinervis → H. pottsii: 156
– **tsangii**: 158
– variegata → H. carnosa: 149
– **verticillata**: 158
– – var. citrina → H. verticillata: 158
– – – hendersonii → H. verticillata: 158
– **vitellina**: 158
– wightiana → H. pauciflora: 156
– yunnanensis → Micholitzia obcordata: 183
– zollingeriana → H. diversifolia: 150
Huernelia : 158
Huernia : 159
– albomaculata → H. sp.: 159
– **andreaeana**: 159
– appendiculata → H. hystrix: 164
– arabica → H. macrocarpa: 167
– **archeri**: 159
– **aspera**: 159
– **barbata**: 160
– – var. crispa → H. sp.: 159
– – – griquensis → H. barbata: 160
– – – tubata → H. barbata: 160
– **bayeri**: 160
– bicampanulata → H. kirkii: 165
– blackbeardiae → H. zebrina ssp. magniflora: 174
– **boleana**: 160
– **brevirostris**: 160
– – ssp. **baviaana**: 160
– – – **brevirostris**: 160
– – – **intermedia**: 161
– – var. histrionica → H. brevirostris ssp. intermedia: 161
– – – immaculata → H. brevirostris ssp. intermedia: 161
– – – intermedia → H. brevirostris ssp. intermedia: 161
– – – longula → H. brevirostris ssp. intermedia: 161
– – – pallida → H. brevirostris ssp. intermedia: 161
– – – parvipuncta → H. brevirostris ssp. intermedia: 161
– – – scabra → H. brevirostris ssp. intermedia: 161
– **campanulata**: 161
– – var. denticoronata → H. campanulata: 161
– **clavigera**: 161
– – var. maritima → H. clavigera: 161
– **concinna**: 161
– confusa → H. insigniflora: 164
– crispa → H. sp.: 159
– decemdentata → H. clavigera: 161
– **distincta**: 162

[Huernia]
- duodecimfida → H. barbata: 160
- **echidnopsioides**: 162
- **erectiloba**: 162
- **erinacea**: 162
- **formosa**: 162
- **guttata**: 162
- – ssp. **calitzdorpensis**: 163
- – – **guttata**: 163
- **hadhramautica**: 163
- **hallii**: 163
- herrei → H. namaquensis: 167
- – var. immaculata → H. namaquensis: 167
- **hislopii**: 163
- – ssp. **hislopii**: 163
- – – **robusta**: 163
- **humilis**: 163
- **hystrix**: 164
- – var. appendiculata → H. hystrix: 164
- – – **hystrix**: 164
- – – **parvula**: 164
- ingeae → H. clavigera: 161
- inornata → H. thuretii: 172
- **insigniflora**: 164
- **keniensis**: 164
- – var. **globosa**: 164
- – – **grandiflora**: 164
- – – **keniensis**: 164
- – – **molonyae**: 165
- – – **nairobiensis**: 165
- – – quintitia → H. keniensis var. keniensis: 164
- **kennedyana**: 165
- **kirkii**: 165
- **laevis**: 165
- **lavrani**: 165
- **leachii**: 166
- **lenewtonii**: 166
- lentiginosa → H. guttata: 162
- **levyi**: 166
- **lodarensis**: 166
- **loeseneriana**: 166
- **longii**: 166
- – ssp. echidnopsioides → H. echidnopsioides: 162
- **longituba**: 167
- – ssp. **cashelensis**: 167
- – – **longituba**: 167
- **macrocarpa**: 167
- – – fa. schimperi → H. macrocarpa: 167
- – – ssp. concinna → H. concinna: 161
- – – var. arabica → H. macrocarpa: 167
- – – – cerasina → H. macrocarpa: 167
- – – – flavicoronata → H. macrocarpa: 167
- – – – penzigii → H. macrocarpa: 167
- – – – schweinfurthii → H. macrocarpa: 167
- **marnieriana**: 167
- montana → H. volkartii: 173
- **namaquensis**: 167
- – ssp. hallii → H. hallii: 163

[Huernia]
- **nigeriana**: 168
- **nouhuysii**: 168
- **occulta**: 168
- ocellata → H. guttata: 162
- **oculata**: 168
- owamboensis → H. namaquensis: 167
- **pendula**: 169
- penzigii → H. macrocarpa: 167
- – – var. arabica → H. macrocarpa: 167
- – – – schimperi → H. macrocarpa: 167
- – – – schweinfurthii → H. macrocarpa: 167
- **piersii**: 169
- **pillansii**: 169
- – ssp. echidnopsioides → H. echidnopsioides: 162
- – – – pillansii → H. pillansii: 169
- **plowesii**: 169
- **praestans**: 169
- primulina → H. thuretii var. primulina: 172
- – var. rugosa → H. thuretii var. primulina: 172
- **procumbens**: 170
- **quinta**: 170
- – var. **blyderiverensis**: 170
- – – **quinta**: 170
- **recondita**: 170
- repens → H. volkartii var. repens: 173
- **reticulata**: 170
- rogersii → H. oculata: 168
- **rosea**: 171
- **rubra**: 171
- **saudi-arabica**: 171
- scabra → H. brevirostris ssp. intermedia: 161
- – var. immaculata → H. brevirostris ssp. intermedia: 161
- – – – longula → H. brevirostris ssp. intermedia: 161
- – – – pallida → H. brevirostris ssp. intermedia: 161
- – – – quinta → H. quinta: 170
- **schneideriana**: 171
- **similis**: 171
- simplex → H. humilis: 163
- **somalica**: 171
- sprengeri → Orbea sprengeri: 200
- **stapelioides**: 172
- striata → H. thuretii: 172
- **tanganyikensis**: 172
- tavaresii → Tavaresia angolensis: 265
- **thudichumii**: 172
- **thuretii**: 172
- – var. **primulina**: 172
- – – – rugosa → H. thuretii var. primulina: 172
- – – – **thuretii**: 172
- transmutata → H. sp.: 159
- **transvaalensis**: 173
- tubata → H. barbata: 160
- **urceolata**: 173
- venusta → H. sp.: 159

[Huernia]
- **verekeri**: 173
- – var. **angolensis**: 173
- – – **pauciflora**: 173
- – – stevensonii → H. verekeri: 173
- – – **verekeri**: 173
- – vogtsii → H. stapelioides: 172
- **volkartii**: 173
- – var. nigeriana → H. nigeriana: 168
- – – **repens**: 173
- – – **volkartii**: 173
- **whitesloaneana**: 174
- **witzenbergensis**: 174
- **zebrina**: 174
- – ssp. **magniflora**: 174
- – – **zebrina**: 174
- – var. magniflora → H. zebrina ssp. magniflora: 174
- – – zebrina → H. zebrina ssp. zebrina: 174

Huernianthus : 174
- – **Alexis** : 174

Huerniopsis : 174
- **atrosanguinea**: 175
- – var. gibbosa → H. atrosanguinea: 175
- **decipiens**: 175
- – gibbosa → H. atrosanguinea: 175
- – papillata → H. atrosanguinea: 175

Hutchinia → Caralluma: 46
- indica → Caralluma indica: 56

Ischnolepis : 175
- **graminifolia**: 176
- natalensis → Petopentia natalensis: 209
- tuberosa → I. graminifolia: 176

Jasminanthes → Marsdenia: 179

Karimbolea → Cynanchum: 108
- macrantha → Cynanchum macranthum: 114
- mariensis → Cynanchum mariense: 114
- verrucosa → Cynanchum verrucosum: 118

Kinepetalum → Brachystelma: 20
- schultzei → Brachystelma schultzei: 42

Krebsia → Stenostelma: 263
- carinata → Stenostelma carinatum: 264
- corniculata → Stenostelma corniculatum: 264
- stenoglossa → Stenostelma capense: 263

Lagarinthus abyssinicus → Aspidoglossum interruptum: 15
- atrorubens → Schizoglossum bidens ssp. atrorubens: 237
- corniculatus → Stenostelma corniculatum: 264
- diversus → Schizoglossum bidens ssp. gracile: 238
- eustegioides → Stenostelma sp.: 263
- exilis → Aspidoglossum gracile: 14
- galpinii → Schizoglossum bidens ssp. galpinii: 238
- gracilis → Aspidoglossum gracile: 14
- interruptus → Aspidoglossum interruptum: 15
- virgatus → Aspidoglossum virgatum: 17

Larryleachia : 176

[Larryleachia]
- **cactiformis**: 176
- dinteri → L. marlothii: 177
- felina → L. cactiformis: 176
- **marlothii**: 177
- meloformis → L. picta: 177
- **perlata**: 177
- **picta**: 177
- similis → L. cactiformis: 176
- sociarum → L. sp.: 176
- **tirasmontana**: 177

Lasiostelma → Brachystelma: 20
- benthamii → Brachystelma sandersonii: 42
- longifolium → Brachystelma longifolium: 34
- macropetalum → Brachystelma macropetalum: 34
- nanum → Brachystelma nanum: 37
- occidentale → Brachystelma occidentale: 37
- ramosissimum → Brachystelma ramosissimum: 41
- sandersonii → Brachystelma sandersonii: 42
- somalense → Caralluma sp.: 47

Lavrania : 178
- cactiformis → Larryleachia cactiformis: 176
- **haagnerae**: 178
- marlothii → Larryleachia marlothii: 177
- perlata → Larryleachia perlata: 177
- picta → Larryleachia picta: 177
- – ssp. parvipunctata → Larryleachia tirasmontana: 178
- – – picta → Larryleachia picta: 177

Leachia → Larryleachia: 176
- cactiformis → Larryleachia cactiformis: 176
- dinteri → Larryleachia marlothii: 177
- felina → Larryleachia cactiformis: 176
- marlothii → Larryleachia marlothii: 177
- meloformis → Larryleachia picta: 177
- perlata → Larryleachia perlata: 177
- picta → Larryleachia picta: 177
- similis → Larryleachia cactiformis: 176
- sociarum → Larryleachia sp.: 176
- tirasmontana → Larryleachia tirasmontana: 177

Leachiella → Larryleachia: 176
- cactiformis → Larryleachia cactiformis: 176
- dinteri → Larryleachia marlothii: 177
- felina → Larryleachia cactiformis: 176
- marlothii → Larryleachia marlothii: 177
- meloformis → Larryleachia picta: 177
- perlata → Larryleachia perlata: 177
- picta → Larryleachia picta: 177
- similis → Larryleachia cactiformis: 176
- sociarum → Larryleachia sp.: 176
- tirasmontana → Larryleachia tirasmontana: 177

Leichardtia → Marsdenia: 178
- australis → Marsdenia australis: 179

Leptostemma → Dischidia: 118
- fasciculatum → Dischidia hirsuta: 120
- hirsutum → Dischidia hirsuta: 120
- lanceolatum → Dischidia lanceolata: 121

[Leptostemma]
– latifolium → Dischidia latifolia: 121
– pictum → Dischidia ovata: 122
– punctatum → Dischidia punctata: 122
– sagitattum → Dischidia sagittata: 123
– truncatum → Dischidia rhombifolia: 123
– – → Dischidia truncata: 123
Lithocaulon → Pseudolithos: 212
– cubiforme → Pseudolithos cubiformis: 212
– sphaericum → Pseudolithos migiurtinus: 213
Loniceroides → Marsdenia: 179
Macropetalum → Brachystelma: 20
– burchellii → Brachystelma burchellii: 24
– – var. burchellii → Brachystelma burchellii var. burchellii: 24
– – – grandiflorum → Brachystelma burchellii var. grandiflorum: 24
– filifolium → Brachystelma filifolium: 29
Madangia : 178
– **inflata**: 178
Mahafalia → Cynanchum: 108
– nodosa → Cynanchum arenarium: 109
Marsdenia : 178
– **australis**: 179
– **brevis**: 179
– **coronata**: 179
– **gillespieae**: 179
– **guanchezii**: 180
– leptophylla → M. viridiflora: 180
– **megalantha**: 180
– **microlepis**: 180
– parasitica → Dischidiopsis parasitica: 124
– **pumila**: 180
– **rara**: 180
– **viridiflora**: 180
– – ssp. tropica → M. viridiflora: 180
Matelea : 181
– **cyclophylla**: 181
– **decumbens**: 181
– **dictyantha**: 181
– diffusa → M. hemsleyana: 182
– **hemsleyana**: 182
– **macvaughiana**: 182
– **pavonii**: 182
– **sepicola**: 182
– **standleyana**: 182
– stapeliiflora → M. pavonii: 182
– **tuberosa**: 183
Micholitzia : 183
– **obcordata**: 183
Micraster → Brachystelma: 20
– pulchellus → Brachystelma pulchellum: 40
Microstemma → Brachystelma: 20
– glabriflorum → Brachystelma glabriflorum: 30
– tuberosum → Brachystelma glabriflorum: 30
Miraglossum : 183
– **anomalum**: 184
– **davyi**: 184
– **laeve**: 184

[Miraglossum]
– **pilosum**: 184
– **pulchellum**: 184
– **superbum**: 185
– **verticillare**: 185
Monolluma → Caralluma: 46
– cicatricosa → Caralluma cicatricosa: 51
– quadrangula → Caralluma quadrangula: 59
Monostemma → Sarcostemma: 233
Monothylaceum → Hoodia: 142
– gordonii → Hoodia gordonii: 143
Nematostemma → Cynanchum: 108
– perrieri → Cynanchum nematostemma: 116
Niota → Ceropegia: 63
Notechidnopsis : 185
– **columnaris**: 185
– **tessellata**: 186
Obesia → Piaranthus: 209
– arida → Quaqua arida: 215
– decora → Piaranthus decorus: 210
– geminata → Piaranthus geminatus: 211
– punctata → Piaranthus punctatus: 211
– serrulata → Piaranthus decorus ssp. decorus: 210
Odontostelma : 186
– **minus**: 186
– **welwitschii**: 186
Oligoron → Asclepias: 9
Ophionella : 186
– **arcuata**: 186
– – ssp. **arcuata**: 187
– – – **mirkinii**: 187
– – var. mirkinii → O. arcuata ssp. mirkinii: 187
– **willowmorensis**: 187
Orbea : 187
– **abayensis**: 188
– **albocastanea**: 188
– anguinea → O. variegata: 203
– aperta → Tromotriche aperta: 270
– **araysiana**: 188
– arida → Quaqua arida: 215
– **baldratii**: 189
– – ssp. **baldratii**: 189
– – – **somalensis**: 189
– bisulca → O. variegata: 203
– bufonia → O. variegata: 203
– **carnosa**: 189
– **caudata**: 189
– – ssp. **caudata**: 189
– – – **rhodesiaca**: 190
– **chrysostephana**: 190
– **ciliata**: 190
– clypeata → O. variegata: 203
– **commutata**: 190
– conjuncta → Orbeanthus conjunctus: 206
– conspurcata → O. variegata: 203
– **cooperi**: 191
– curtisii → O. variegata: 203
– **decaisneana**: 191
– **deflersiana**: 191

[Orbea]
- **denboefii**: 191
- **distincta**: 192
- **dummeri**: 192
- – ssp. circes → O. dummeri: 192
- **gemugofana**: 192
- **gerstneri**: 192
- – ssp. **elongata**: 192
- – – – **gerstneri**: 193
- **halipedicola**: 193
- – ssp. **halipedicola**: 193
- – – – **septentrionalis**: 193
- hardyi → Orbeanthus hardyi: 206
- **huernioides**: 193
- **huillensis**: 194
- – ssp. **flava**: 194
- – – – **huillensis**: 194
- inodora → O. variegata: 203
- irrorata → O. verrucosa var. fucosa: 204
- **knobelii**: 194
- **laikipiensis**: 194
- **laticorona**: 195
- lepida → O. variegata: 203
- **longidens**: 195
- **lugardii**: 195
- **luntii**: 195
- **lutea**: 196
- – ssp. **lutea**: 196
- – – – **vaga**: 196
- **macloughlinii**: 196
- **maculata**: 196
- – ssp. **kaokoensis**: 197
- – – – **maculata**: 197
- – – – **rangeana**: 197
- maculosa → Tromostapelia sp.: 269
- marginata → O. variegata: 203
- marmorata → O. variegata: 203
- **melanantha**: 197
- **miscella**: 197
- mixta → O. variegata: 203
- **namaquensis**: 198
- normalis → O. variegata: 203
- orbicularis → O. variegata: 203
- **paradoxa**: 198
- picta → O. variegata: 203
- planiflora → O. variegata: 203
- – var. marginata → O. variegata: 203
- **prognatha**: 198
- **pulchella**: 199
- quinquenervia → O. variegata: 203
- rangeana → O. maculata ssp. rangeana: 197
- retusa → O. variegata: 203
- **rogersii**: 199
- rugosa → O. variegata: 203
- **sacculata**: 199
- **schweinfurthii**: 199
- **semitubiflora**: 200
- **semota**: 200
- **speciosa**: 200

[Orbea]
- **sprengeri**: 200
- – – ssp. **foetida**: 200
- – – – **ogadensis**: 201
- – – – **sprengeri**: 201
- **subterranea**: 201
- **tapscottii**: 201
- trisulca → O. variegata: 203
- **tubiformis**: 201
- **ubomboensis**: 202
- **umbracula**: 202
- **valida**: 202
- – ssp. **occidentalis**: 202
- – – – **valida**: 202
- **variegata**: 203
- – var. clypeata → O. variegata: 203
- **verrucosa**: 204
- – var. **fucosa**: 204
- – – – **verrucosa**: 204
- **vibratilis**: 204
- wendlandiana → O. verrucosa var. verrucosa: 204
- **wilsonii**: 204
- **wissmannii**: 205
- – var. **eremastrum**: 205
- – – – **parviloba**: 205
- – – – **wissmannii**: 205
- woodfordiana → O. variegata: 203
- **woodii**: 205

Orbeanthus : 206
- **conjunctus**: 206
- **hardyi**: 206

Orbelia : 207
- **barklyi**: 207

Orbeopelia → Orbelia: 207
Orbeopsis → Orbea: 187
- albocastanea → Orbea albocastanea: 188
- caudata → Orbea caudata: 189
- – ssp. caudata → Orbea caudata ssp. caudata: 189
- – – – rhodesiaca → Orbea caudata ssp. rhodesiaca: 190
- gerstneri → Orbea gerstneri: 192
- – ssp. elongata → Orbea gerstneri ssp. elongata: 193
- – – – gerstneri → Orbea gerstneri ssp. gerstneri: 193
- gossweileri → Orbea huillensis: 194
- huillensis → Orbea huillensis: 194
- knobelii → Orbea knobelii: 194
- lutea → Orbea lutea: 196
- – ssp. lutea → Orbea lutea ssp. lutea: 196
- – – – vaga → Orbea lutea ssp. vaga: 196
- melanantha → Orbea melanantha: 197
- tsumebensis → Orbea valida ssp. valida: 202
- valida → Orbea valida: 202

Orbeostemon → Orbelia: 207
- tarantuloides → Orbelia sp.: 207
Otanema → Asclepias: 9

[]
Otaria → Asclepias: 9
Otostemma → Hoya: 147
− lacunosum → Hoya lacunosa: 153
Oxypteryx → Asclepias: 9
Pachycarpus : 207
− corniculatus → Stathmostelma pedunculatum: 262
Pachycymbium → Orbea: 187
− abayense → Orbea abayensis: 188
− araysianum → Orbea araysiana: 188
− baldratii → Orbea baldratii: 189
− − ssp. baldratii → Orbea baldratii ssp. baldratii: 189
− − − subterraneum → Orbea subterranea: 201
− carnosum → Orbea carnosa: 189
− chrysostephanum → Orbea chrysostephana: 190
− circes → Orbea dummeri: 192
− commutatum → Orbea commutata: 190
− decaisneanum → Orbea decaisneana: 191
− deflersianum → Orbea deflersiana: 191
− denboefii → Orbea denboefii: 191
− distinctum → Orbea distincta: 192
− dummeri → Orbea dummeri: 192
− eremastrum → Orbea wissmannii var. eremastrum: 205
− gemugofanum → Orbea gemugofana: 192
− huernioides → Orbea huernioides: 193
− keithii → Orbea carnosa: 189
− kochii → Orbea sacculata: 199
− laikipiense → Orbea laikipiensis: 194
− lancasteri → Orbea carnosa: 189
− laticoronum → Orbea laticorona: 195
− lugardii → Orbea lugardii: 195
− luntii → Orbea luntii: 195
− meintjesianum → Orbea wissmannii var. wissmannii: 205
− miscellum → Orbea miscella: 197
− rogersii → Orbea rogersii: 199
− sacculatum → Orbea sacculata: 199
− schweinfurthii → Orbea schweinfurthii: 199
− sprengeri → Orbea sprengeri: 200
− − ssp. foetidum → Orbea sprengeri ssp. foetida: 200
− − − ogadense → Orbea sprengeri ssp. ogadensis: 201
− tubiforme → Orbea tubiformis: 201
− ubomboense → Orbea ubomboensis: 202
− vibratile → Orbea vibratilis: 204
− wilsonii → Orbea wilsonii: 204
− wissmannii → Orbea wissmannii: 205
Panninia → Fanninia: 137
Pectinaria : 207
− arcuata → Ophionella arcuata: 186
− **articulata**: 207
− − ssp. **articulata**: 207
− − − **asperiflora**: 208
− − − **borealis**: 208
− − − **namaquensis**: 208

[Pectinaria articulata]
− − var. namaquensis → P. articulata ssp. namaquensis: 208
− asperiflora → P. articulata ssp. asperiflora: 208
− breviloba → Stapeliopsis breviloba: 258
− exasperata → Stapeliopsis exasperata: 258
− **longipes**: 208
− mammillaris → Quaqua mammillaris: 217
− **maughanii**: 208
− mirkinii → Ophionella arcuata ssp. mirkinii: 187
− pillansii → Stapeliopsis pillansii: 259
− saxatilis → Stapeliopsis saxatilis: 259
− stayneri → Stapeliopsis saxatilis ssp. stayneri: 259
− tulipiflora → Stapeliopsis saxatilis ssp. saxatilis: 259
− villetii → Stapeliopsis breviloba: 258
Pentagonanthus caeruleus → Raphionacme caerulea: 222
− grandiflorus → Raphionacme grandiflora: 223
− − ssp. glabrescens → Raphionacme grandiflora: 223
Pentopetia graminifolia → Ischnolepis graminifolia: 176
− natalensis → Petopentia natalensis: 209
Pergularia edulis → Fockea edulis: 138
Petopentia : 208
− **natalensis**: 209
Physostelma → Hoya: 147
Piaranthus : 209
− aridus → Quaqua arida: 215
− atrosanguineus → Huerniopsis atrosanguinea: 175
− **barrydalensis**: 209
− **comptus**: 210
− − var. ciliatus → P. comptus: 210
− cornutus → P. decorus ssp. cornutus: 210
− − var. grandis → P. decorus ssp. cornutus: 210
− decipiens → Huerniopsis decipiens: 175
− **decorus**: 210
− − ssp. **cornutus**: 210
− − − **decorus**: 210
− disparilis → P. geminatus var. geminatus: 211
− − var. immaculatus → P. geminatus var. geminatus: 211
− fascicularis → Echidnopsis cereiformis: 131
− fasciculatus → Duvalia caespitosa: 125
− foetidus → P. geminatus var. foetidus: 211
− − var. diversus → P. geminatus var. foetidus: 211
− − − multipunctatus → P. geminatus var. foetidus: 211
− − − pallidus → P. geminatus var. foetidus: 211
− − − purpureus → P. geminatus var. foetidus: 211
− **framesii**: 210
− **geminatus**: 211
− − var. **foetidus**: 211
− − − **geminatus**: 211
− globosus → P. geminatus var. geminatus: 211

[Piaranthus]
- grivanus → Huerniopsis decipiens: 175
- incarnatus → Quaqua incarnata: 217
- – var. albus → Quaqua incarnata: 217
- mammillaris → Quaqua mammillaris: 217
- mennellii → P. decorus ssp. cornutus: 210
- nebrownii → P. decorus ssp. cornutus: 210
- pallidus → P. decorus ssp. cornutus: 210
- parviflorus → Quaqua parviflora: 218
- **parvulus**: 211
- piliferus → Hoodia pilifera: 145
- pillansii → P. geminatus var. geminatus: 211
- – var. fuscatus → P. geminatus var. geminatus: 211
- – – inconstans → P. geminatus var. geminatus: 211
- pulcher → P. decorus ssp. cornutus: 210
- – var. nebrownii → P. decorus ssp. cornutus: 210
- pullus → Quaqua mammillaris: 217
- **punctatus**: 211
- ramosus → Quaqua ramosa: 220
- rorifluus → Orbea verrucosa var. verrucosa: 204
- ruschii → P. decorus ssp. cornutus: 210
- serrulatus → P. decorus ssp. decorus: 210
- streyianus → Orbea maculata ssp. rangeana: 197
Platykeleba → Cynanchum: 108
- insignis → Cynanchum insigne: 112
Pleurostelma → Schlechterella: 241
- africana → Schlechterella africana: 242
Plocostemma → Hoya: 147
Podanthes → Orbea: 187
- ciliata → Orbea ciliata: 190
- geminata → Piaranthus geminatus: 211
- incarnata → Quaqua incarnata: 217
- irrorata → Orbea verrucosa var. fucosa: 204
- lepida → Orbea variegata: 203
- pulchellus → Orbea pulchella: 199
- pulchra → Orbea verrucosa var. verrucosa: 204
- – var. major → Orbea verrucosa var. verrucosa: 204
- – – verrucosa → Orbea verrucosa var. verrucosa: 204
- roriflua → Orbea verrucosa var. verrucosa: 204
- verrucosa → Orbea verrucosa: 204
Podostemma → Asclepias: 9
Podostigma → Asclepias: 9
Polyotus → Asclepias: 9
Prosopostelma aculeatum → Cynanchum aculeatum: 108
- grandiflorum → Cynanchum floriferum: 111
- madagascariense → Cynanchum toliari: 118
Pseudolithos : 212
- **caput-viperae**: 212
- **cubiformis**: 212
- – var. viridiflorus → P. cubiformis: 212
- **dodsonianus**: 212
- **horwoodii**: 213
- **mccoyi**: 213

[Pseudolithos]
- **migiurtinus**: 213
- sphaericus → P. migiurtinus: 213
Pseudomarsdenia → Marsdenia: 179
Pseudopectinaria → Echidnopsis: 129
- malum → Echidnopsis malum: 132
Pterophora → Marsdenia: 178
Pterostelma → Hoya: 147
Pterygocarpus → Marsdenia: 178
Pycnoneurum → Cynanchum: 108
- junciforme → Cynanchum junciforme: 112
- sessiliflorum → Cynanchum sessiliflorum: 117
Quaqua : 213
- **acutiloba**: 214
- **albersii**: 214
- **arenicola**: 214
- – ssp. pilifera → Q. pilifera: 219
- **arida**: 214
- **armata**: 215
- – ssp. arenicola → Q. arenicola: 214
- – – maritima → Q. maritima: 218
- – – pilifera → Q. pilifera: 219
- **aurea**: 215
- **bayeriana**: 215
- **cincta**: 215
- **confusa**: 216
- **dependens**: 216
- **framesii**: 216
- **gracilis**: 216
- **hottentotorum**: 216
- **incarnata**: 217
- – ssp. aurea → Q. aurea: 215
- – – hottentotorum → Q. hottentotorum: 216
- – – tentaculata → Q. tentaculata: 220
- – var. tentaculata → Q. tentaculata: 220
- **inversa**: 217
- – var. cincta → Q. cincta: 215
- **linearis**: 217
- **mammillaris**: 217
- **maritima**: 218
- marlothii → Q. arida: 215
- **multiflora**: 218
- **pallens**: 218
- **parviflora**: 218
- – ssp. bayeriana → Q. bayeriana: 215
- – – confusa → Q. confusa: 216
- – – dependens → Q. dependens: 216
- – – gracilis → Q. gracilis: 216
- – – pulchra → Q. pulchra: 219
- – – swanepoelii → Q. swanepoelii: 220
- **pilifera**: 218
- **pillansii**: 219
- **pruinosa**: 219
- **pulchra**: 219
- **radiata**: 219
- **ramosa**: 219
- **swanepoelii**: 220
- **tentaculata**: 220
Raphiacme → Raphionacme: 220

[Raphiacme]
- macrostemon → Raphionacme splendens: 227
Raphionacme : 220
- abyssinica → Schlechterella abyssinica: 241
- **angolensis**: 221
- **arabica**: 221
- bagshawei → R. brownii: 221
- baguirmiensis → R. sp.: 221
- bingeri → R. splendens: 227
- **borenensis**: 221
- **brownii**: 221
- – var. longifolia → R. brownii: 221
- **burkei**: 222
- **caerulea**: 222
- chimanimaniana → R. splendens: 227
- daronii → R. splendens: 227
- decolor → R. zeyheri: 229
- denticulata → R. welwitschii: 228
- dinteri → R. burkei: 222
- divaricata → R. hirsuta: 224
- – var. glabra → R. hirsuta: 224
- **dyeri**: 222
- elata → R. globosa: 223
- **elsana**: 222
- engleriana → R. sp.: 221
- ernstiana → R. longituba: 225
- excisa → R. splendens: 227
- **flanaganii**: 223
- **galpinii**: 223
- **globosa**: 223
- gossweileri → R. splendens: 227
- **grandiflora**: 223
- – ssp. glabrescens → R. grandiflora: 223
- **haeneliae**: 224
- **hirsuta**: 224
- **inconspicua**: 224
- jurensis → R. brownii: 221
- **keayi**: 225
- kubangensis → R. angolensis: 221
- **lanceolata**: 225
- – var. latifolia → R. lanceolata: 225
- **loandae**: 225
- **lobulata**: 225
- **longifolia**: 225
- **longituba**: 225
- lucens → R. globosa: 223
- macrorrhiza → R. galpinii: 223
- **madiensis**: 226
- **michelii**: 226
- **monteiroae**: 226
- moyalica → R. borenensis: 221
- **namibiana**: 226
- obovata → R. hirsuta: 224
- pachyodon → R. burkei: 222
- **palustris**: 227
- **procumbens**: 227
- pubescens → R. hirsuta: 224
- **pulchella**: 227
- purpurea → R. hirsuta: 224

[Raphionacme]
- scandens → R. flanaganii: 223
- seineri → R. sp.: 221
- **splendens**: 227
- sudanica → R. sp.: 221
- **sylvicola**: 227
- **utilis**: 228
- **velutina**: 228
- verdickii → R. welwitschii: 228
- **vignei**: 228
- virgultorum → R. galpinii: 223
- volubilis → Raphionacme: 220
- **welwitschii**: 228
- wilczekiana → R. brownii: 221
- **zeyheri**: 229
Rhaphionacme → Raphionacme: 220
Rhinolobium → Aspidoglossum: 10
- lineare → Aspidoglossum gracile: 14
- tenue → Aspidoglossum gracile: 14
Rhytidocaulon : 229
- **ciliatum**: 229
- **fulleri**: 230
- **macrolobum**: 230
- – ssp. **macrolobum**: 230
- – – **minimum**: 230
- **mccoyi**: 230
- **paradoxum**: 231
- **piliferum**: 231
- **richardianum**: 231
- **sheilae**: 231
- **subscandens**: 232
- **tortum**: 232
Riocreuxia : 232
- longiflora → Ceropegia stenantha: 101
- nepalensis → Brachystelma nepalense: 37
Ruehssia → Marsdenia: 179
Rytidoloma reticulatum → Matelea dictyantha: 181
Sanguilluma → Caralluma: 46
- socotrana → Caralluma socotrana: 60
Sarcocodon → Caralluma: 46
- speciosa → Caralluma speciosa: 61
Sarcocyphula → Cynanchum: 108
- gerrardii → Cynanchum gerrardii: 111
Sarcophagophilus → Quaqua: 213
- armatus → Quaqua armata: 215
- winkleri → Quaqua mammillaris: 217
- winklerianus → Quaqua mammillaris: 217
Sarcorrhiza : 232
- **epiphytica**: 233
Sarcostemma : 233
- acidum → S. sp.: 233
- andongense → S. sp.: 233
- **antsiranense**: 233
- aphyllum → S. viminale: 235
- **arabicum**: 233
- australe → S. viminale ssp. australe: 235
- brachystigma → S. sp.: 233
- **brevipedicellatum**: 233

[Sarcostemma]
- brevistigma → S. sp.: 233
- brunonianum → S. viminale ssp. brunonianum: 235
- daltonii → S. viminale ssp. thunbergii: 236
- **decorsei**: 234
- **elachistemmoides**: 234
- **forskaolianum**: 234
- implicatum → Cynanchum implicatum: 112
- insigne → Cynanchum insigne: 112
- intermedium → S. sp.: 233
- madagascariense → S. sp.: 233
- mauritianum → Cynanchum luteifluens var. luteifluens: 113
- **membranaceum**: 234
- **mulanjense**: 234
- nudum → S. viminale ssp. thunbergii: 236
- odontolepis → S. viminale ssp. odontolepis: 235
- **pearsonii**: 234
- **resiliens**: 234
- **socotranum**: 234
- stipitaceum → S. viminale ssp. stipitaceum: 236
- **stocksii**: 234
- **stoloniferum**: 235
- subterraneum → S. vanlessenii: 235
- tetrapterum → S. viminale: 235
- thunbergii → S. viminale ssp. thunbergii: 236
- **vanlessenii**: 235
- **viminale**: 235
- – ssp. **australe**: 235
- – – **brunonianum**: 235
- – – **odontolepis**: 235
- – – **orangeanum**: 235
- – – **stipitaceum**: 235
- – – stocksii → S. stocksii: 235
- – – **suberosum**: 236
- – – **thunbergii**: 236
- – – **viminale**: 236
Saurolluma → Caralluma: 46
- furta → Caralluma furta: 55
Schizoglossum: 236
- aciculare → Stenostelma capense: 264
- addoense → Aspidoglossum gracile: 14
- aemulum → S. cordifolium: 238
- alpestre → Aspidoglossum sp.: 11
- altissimum → Aspidoglossum interruptum: 15
- altum → Aspidoglossum masaicum: 16
- **amatolicum**: 236
- angolense → S. sp.: 236
- angustissimum → Aspidoglossum angustissimum: 11
- anomalum → Miraglossum anomalum: 184
- araneiferum → Aspidoglossum araneiferum: 11
- aschersonianum → Aspidoglossum sp.: 11
- – var. longipes → Aspidoglossum sp.: 11
- – – pygmaeum → Aspidoglossum sp.: 11
- – – radiatum → Aspidoglossum sp.: 11
- **atropurpureum**: 237
- – – → S. hamatum: 239

[Schizoglossum atropurpureum]
- – ssp. **atropurpureum**: 237
- – – **tridentatum**: 237
- – – **virens**: 237
- – var. lineatum → S. cordifolium: 238
- atrorubens → S. bidens ssp. atrorubens: 237
- auriculatum → Aspidoglossum glanduliferum: 14
- barbatum → Aspidoglossum glabrescens: 14
- – – → Aspidoglossum sp.: 11
- barberae → Aspidoglossum interruptum: 15
- baumii → Aspidoglossum masaicum: 16
- biauriculatum → Aspidoglossum delagoense: 12
- **bidens**: 237
- – ssp. **atrorubens**: 237
- – – **bidens**: 237
- – – **galpinii**: 238
- – – **gracile**: 238
- – – **hirtum**: 238
- – – **pachyglossum**: 238
- – – **productum**: 238
- biflorum → Aspidoglossum biflorum: 11
- – var. concinnum → Aspidoglossum biflorum: 11
- – – integrum → Aspidoglossum biflorum: 11
- bilamellatum → Aspidoglossum lamellatum: 16
- – var. cordylogynoides → Aspidoglossum lamellatum: 16
- bolusii → Aspidoglossum gracile: 14
- bowkerae → Aspidoglossum gracile: 14
- buchananii → Aspidoglossum gracile: 14
- burchellii → Aspidoglossum gracile: 14
- cabrae → Pachycarpus sp.: 207
- capense → Stenostelma capense: 263
- capitatum → S. sp.: 236
- carinatum → Aspidoglossum carinatum: 12
- carsonii → Glossostelma carsonii: 140
- chlorojodinum → Glossostelma carsonii: 140
- ciliatum → Aspidoglossum fasciculare: 13
- commixtum → Aspidoglossum glanduliferum: 14
- connatum → Aspidoglossum connatum: 12
- conrathii → Aspidoglossum biflorum: 11
- consimile → Aspidoglossum heterophyllum: 15
- contracurvum → Aspidoglossum ovalifolium: 16
- **cordifolium**: 238
- – – → S. hilliardiae: 239
- – var. centralis → S. cordifolium: 238
- corniculatum → Stenostelma corniculatum: 264
- crassipes → Stenostelma sp.: 263
- davyi → Miraglossum davyi: 184
- debile → Aspidoglossum elliotii: 13
- decipiens → S. stenoglossum ssp. flavum: 241
- – var. flavum → S. stenoglossum ssp. flavum: 241
- delagoense → Aspidoglossum delagoense: 12
- dissimile → Aspidoglossum dissimile: 13
- – var. pubiflorum → Aspidoglossum dissimile: 13

[Schizoglossum]
- divaricatum → S. cordifolium: 238
- diversum → S. bidens ssp. gracile: 238
- dregei → Aspidoglossum gracile: 14
- elatum → Aspidoglossum angustissimum: 11
- **elingue**: 238
- – ssp. **elingue**: 239
- – – – **purpureum**: 239
- elliotii → Aspidoglossum elliotii: 13
- erubescens → Aspidoglossum erubescens: 13
- euphorbioides → S. atropurpureum ssp. virens: 237
- eustegioides → Stenostelma sp.: 263
- excisum → Aspidoglossum biflorum: 11
- exile → Aspidoglossum gracile: 14
- eylesii → Aspidoglossum eylesii: 13
- fasciculare → Aspidoglossum fasciculare: 13
- filifolium → Aspidoglossum gracile: 14
- filiforme → Aspidoglossum sp.: 11
- filipes → Aspidoglossum gracile: 14
- flanaganii → Aspidoglossum flanaganii: 14
- **flavum**: 239
- – var. lineare → S. flavum: 239
- fusco-purpureum → Aspidoglossum masaicum: 16
- galpinii → S. bidens ssp. galpinii: 238
- garcianum → Aspidoglossum sp.: 11
- garuanum → Aspidoglossum interruptum: 15
- glabrescens → Aspidoglossum glabrescens: 14
- – var. longirostre → Aspidoglossum glabrescens: 14
- glanduliferum → Aspidoglossum glanduliferum: 14
- glanvillei → S. sp.: 236
- gossweileri → Pachycarpus sp.: 207
- gracile → Aspidoglossum interruptum: 15
- graminifolium → S. sp.: 236
- grandiflorum → Aspidoglossum grandiflorum: 15
- guthriei → Aspidoglossum gracile: 14
- gwelense → Aspidoglossum biflorum: 11
- **hamatum**: 239
- – – → S. hilliardiae: 239
- – – var. elegans → S. hamatum: 239
- – – – pallidum → S. hamatum: 239
- harveyi → Aspidoglossum heterophyllum: 15
- heterophyllum → Aspidoglossum heterophyllum: 15
- – – var. majus → Aspidoglossum heterophyllum: 15
- – – – schinzianum → Aspidoglossum heterophyllum: 15
- heudelotianum → S. sp.: 236
- **hilliardiae**: 239
- hirsutum → S. cordifolium: 238
- hirtiflorum → Aspidoglossum glabrescens: 14
- hollandiae → S. cordifolium: 238
- huttoniae → S. sp.: 236
- **ingomense**: 239

[Schizoglossum]
- interruptum → Aspidoglossum interruptum: 15
- kamerunense → Aspidoglossum interruptum: 15
- kassneri → Glossostelma carsonii: 140
- lamellatum → Aspidoglossum lamellatum: 16
- lanatum → Aspidoglossum lanatum: 16
- lasiopetalum → Aspidoglossum interruptum: 15
- ledermannii → Aspidoglossum angustissimum: 11
- leptoglossum → Aspidoglossum nyasae: 16
- linifolium → Aspidoglossum sp.: 11
- – – var. centrirostratum → Aspidoglossum sp.: 11
- lividiflorum → Glossostelma sp.: 139
- longirostre → Aspidoglossum glabrescens: 14
- loreum → Aspidoglossum glabrescens: 14
- lunatum → Aspidoglossum gracile: 14
- macowanii → Aspidoglossum grandiflorum: 15
- – – var. tugelense → Aspidoglossum grandiflorum: 15
- macroglossum → Glossostelma sp.: 139
- masaicum → Aspidoglossum masaicum: 16
- montanum → Aspidoglossum sp.: 11
- monticola → Aspidoglossum gracile: 14
- morumbenense → Aspidoglossum interruptum: 15
- multifolium → Aspidoglossum nyasae: 16
- **nitidum**: 240
- nyasae → Aspidoglossum nyasae: 16
- oblongum → S. atropurpureum ssp. virens: 237
- orbiculare → Stenostelma sp.: 263
- ovalifolium → Aspidoglossum ovalifolium: 16
- pachyglossum → S. bidens ssp. pachyglossum: 238
- – – var. abbreviatum → S. bidens ssp. pachyglossum: 238
- – – – productum → S. bidens ssp. productum: 238
- pallidum → Aspidoglossum restioides: 17
- parcum → Aspidoglossum gracile: 14
- parile → Aspidoglossum glanduliferum: 14
- parvulum → Aspidoglossum gracile: 14
- – – var. sessile → Aspidoglossum gracile: 14
- pedunculatum → S. sp.: 236
- peglerae → Aspidoglossum sp.: 11
- pentheri → Aspidoglossum carinatum: 12
- periglossoides → Stenostelma sp.: 263
- pilosum → Miraglossum pilosum: 184
- polynema → Aspidoglossum araneiferum: 11
- propinquum → Aspidoglossum lamellatum: 16
- pulchellum → Miraglossum pulchellum: 184
- pumilum → Aspidoglossum ovalifolium: 16
- pygmaeum → Aspidoglossum sp.: 11
- **quadridens**: 240
- randii → Aspidoglossum restioides: 17
- restioides → Aspidoglossum restioides: 17
- rhodesicum → Aspidoglossum rhodesicum: 17
- robustum → Aspidoglossum ovalifolium: 16
- – – var. inandense → Aspidoglossum ovalifolium: 16
- – – – pubiflorum → Aspidoglossum ovalifolium: 16

[Schizoglossum]
- **rubiginosum**: 240
- **saccatum**: 240
- schinzianum → Aspidoglossum heterophyllum: 15
- schlechteri → Aspidoglossum glabrescens: 14
- semlikense → Aspidoglossum masaicum: 16
- shirense → Aspidoglossum biflorum: 11
- **singulare**: 240
- spathulatum → Glossostelma spathulatum: 141
- **stenoglossum**: 241
- – ssp. **flavum**: 241
- – – **latifolium**: 241
- – – **stenoglossum**: 241
- – var. longipes → S. stenoglossum ssp. stenoglossum: 241
- striatum → Aspidoglossum ovalifolium: 16
- strictissimum → Aspidoglossum erubescens: 13
- strictum → Aspidoglossum biflorum: 11
- tenellum → Aspidoglossum sp.: 11
- tenue → Aspidoglossum gracile: 14
- tenuissimum → Aspidoglossum glabrescens: 14
- theileri → Pachycarpus sp.: 207
- togoense → Aspidoglossum interruptum: 15
- tricorniculatum → S. sp.: 236
- tricuspidatum → Aspidoglossum carinatum: 12
- tridens → Aspidoglossum glabrescens: 14
- tridentatum → S. atropurpureum ssp. tridentatum: 237
- truncatulum → S. sp.: 236
- truncatum → S. bidens ssp. bidens: 237
- tubulosum → Aspidoglossum biflorum: 11
- umbellatum → S. bidens ssp. gracile: 238
- umbelliferum → Stenostelma sp.: 263
- uncinatum → Aspidoglossum uncinatum: 17
- unicum → Aspidoglossum glabrescens: 14
- venustum → Aspidoglossum biflorum: 11
- – var. concinnum → Aspidoglossum biflorum: 11
- – – gwelense → Aspidoglossum biflorum: 11
- verticillare → Miraglossum verticillare: 185
- villosum → Aspidoglossum heterophyllum: 15
- violaceum → Glossostelma sp.: 139
- virens → S. atropurpureum ssp. virens: 237
- virgatum → Aspidoglossum virgatum: 17
- viridulum → S. sp.: 236
- vulcanorum → Aspidoglossum connatum: 12
- wallacei → S. nitidum: 240
- welwitschii → Odontostelma welwitschii: 186
- whytei → Aspidoglossum angustissimum: 11
- woodii → Aspidoglossum woodii: 18
- zernyi → Aspidoglossum angustissimum: 11

Schizonotus → Asclepias: 9

Schlechterella: 241
- **abyssinica**: 241
- **africana**: 242

Schollia → Hoya: 147
- carnosa → Hoya carnosa: 149
- chinensis → Hoya carnosa: 149

[Schollia]
- crassifolia → Hoya carnosa: 149

Scytanthus → Hoodia: 142
- burkei → Hoodia currorii ssp. currorii: 143
- currorii → Hoodia currorii: 143
- gordonii → Hoodia gordonii: 144

Siphonostelma → Brachystelma: 20
- stenophyllum → Brachystelma stenophyllum: 43

Sisyranthus schizoglossoides → Brachystelma schizoglossoides: 42

Solanoa → Asclepias: 9
Solanoana → Asclepias: 9
Somalluma → Caralluma: 46
- baradii → Caralluma baradii: 50

Spathidolepis → Dischidia: 118
Spathulopetalum → Caralluma: 46
- arachnoideum → Caralluma arachnoidea: 49
- congestiflorum → Caralluma congestiflora: 51
- dicapuae → Caralluma dicapuae: 51
- edwardsiae → Caralluma edwardsiae: 53
- gracilipes → Caralluma gracilipes: 55
- longiflorum → Caralluma longiflora: 56
- mogadoxense → Caralluma priogonium: 58
- moniliforme → Caralluma moniliformis: 57
- peckii → Caralluma peckii: 58
- priogonium → Caralluma priogonium: 58
- turneri → Caralluma turneri: 62
- vaduliae → Caralluma vaduliae: 62

Sperlingia → Hoya: 147
- verticillata → Hoya verticillata: 158

Spiralluma → Caralluma: 46
- longidens → Caralluma edulis: 52
- mouretii → Caralluma edulis: 52

Staparesia: 242
- **meintjesii**: 242

Stapelia: 242
- **acuminata**: 243
- – var. brevicuspis → S. acuminata: 243
- adscendens → Caralluma adscendens: 48
- affinis → S. hirsuta: 248
- albipilosa → Tridentea marientalensis ssp. albipilosa: 268
- albocastanea → Orbea albocastanea: 188
- ambigua → S. grandiflora: 248
- – var. fulva → S. grandiflora: 248
- anemoniflora → Duvalia anemoniflora: 124
- ango → Caralluma sp.: 47
- anguinea → Orbea variegata: 203
- anonymos → S. sp.: 243
- aperta → Tromotriche aperta: 270
- **arenosa**: 243
- arida → Quaqua arida: 215
- **arnotii**: 244
- articulata → Pectinaria articulata: 207
- **asterias**: 244
- – var. gibba → S. asterias: 244
- – – lucida → S. asterias: 244
- atrata → Orbea variegata: 203
- atropurpurea → Orbea variegata: 203

[Stapelia]
- atrosanguinea → Huerniopsis atrosanguinea: 175
- auobensis → Tridentea marientalensis ssp. marientalensis: 268
- aurea → Tridentea virescens: 269
- ausana → Tridentea jucunda: 267
- barbata → Duvalia caespitosa var. caespitosa: 125
- – – → Huernia barbata: 160
- barklyi → Orbelia barklyi: 207
- **baylissii**: 244
- beffoniana → Orbea variegata: 203
- bella → Tromostapelia sp.: 269
- bergeriana → S. schinzii var. bergeriana: 253
- beukmanii → S. arenosa: 243
- bidentata → Orbea variegata: 203
- bifolia → Orbea variegata: 203
- **bijliae**: 244
- bijlii → S. bijliae: 244
- bisulca → Orbea variegata: 203
- brevirostris → S. sp.: 243
- buffoniana → Orbea variegata: 203
- buffonis → Orbea variegata: 203
- bufonia → Orbea variegata: 203
- bufoniana → Orbea variegata: 203
- cactiformis → Larryleachia cactiformis: 176
- caespitosa → Duvalia caespitosa: 125
- – – var. hirtella → Duvalia caespitosa var. caespitosa: 125
- callamulia → Caralluma umbellata: 62
- campanulata → Huernia campanulata: 161
- – – → Matelea pavonii: 182
- canescens → S. sp.: 243
- caroli-schmidtii → Orbea albocastanea: 188
- caudata → Brachystelma tuberosum: 45
- **cedrimontana**: 244
- chinensis → Hoya carnosa: 149
- choanantha → Tromotriche choanantha: 271
- chrysostephana → Orbea chrysostephana: 190
- ciliata → Orbea ciliata: 190
- ciliolata → Orbea namaquensis: 198
- – – → Orbea variegata: 203
- cincta → Tridentea jucunda: 267
- clavata → Larryleachia sp.: 176
- **clavicorona**: 245
- clavigera → Huernia clavigera: 161
- clypeata → Orbea variegata: 203
- comata → S. hirsuta: 248
- compacta → Duvalia caespitosa var. compacta: 125
- concinna → S. sp.: 243
- – – var. paniculata → S. paniculata: 251
- concolor → Duvalia caespitosa var. caespitosa: 125
- conformis → S. macowanii var. conformis: 250
- – – var. abrasa → S. macowanii var. conformis: 250
- congestiflora → S. sp.: 243

[Stapelia]
- conjuncta → Orbea paradoxa: 198
- conspurcata → Orbea variegata: 203
- cooperi → Orbea cooperi: 191
- cordata → S. sp.: 243
- corderoyi → Duvalia corderoyi: 125
- crassa → Huernia reticulata: 170
- crispata → Huernia barbata: 160
- cucullata → S. sp.: 243
- cupularis → Tromostapelia sp.: 269
- curtisii → Orbea variegata: 203
- cylindrica → Echidnopsis cereiformis: 131
- cylista → S. gigantea: 247
- cymosa → Duvalia caespitosa var. caespitosa: 125
- decaisneana → Orbea decaisneana: 191
- decora → Piaranthus decorus: 210
- deflexa → S. sp.: 243
- – – cv. Atropurpurea → S. sp.: 243
- – – var. atropurpurea → S. sp.: 243
- dentata → S. sp.: 243
- depressa → S. hirsuta: 248
- desmetiana → S. grandiflora: 248
- – – var. apicalis → S. grandiflora: 248
- – – – fergusoniae → S. grandiflora: 248
- – – – pallida → S. grandiflora: 248
- desmidorchis → Caralluma acutangula: 47
- digitaliflora → Tavaresia angolensis: 265
- dinteri → Tridentea jucunda: 267
- – – var. capensis → Tridentea jucunda: 267
- – – – pseudocapensis → Tridentea jucunda: 267
- discoidea → Orbea semota: 200
- **divaricata**: 245
- dummeri → Orbea dummeri: 192
- duodecimfida → Huernia barbata: 160
- dwequensis → Tridentea dwequensis: 267
- echinata → Duvalia polita: 127
- elegans → Duvalia elegans: 126
- elongata → S. hirsuta: 248
- emarginata → S. sp.: 243
- **engleriana**: 245
- **erectiflora**: 245
- – – var. **erectiflora**: 245
- – – – **prostratiflora**: 246
- erivacaria → S. sp.: 243
- eruciformis → S. olivacea: 250
- europaea → Caralluma europaea: 53
- fasciculata → Duvalia caespitosa: 125
- fergusoniae → S. grandiflora: 248
- fissirostris → S. rufa: 253
- flavicomata → S. sp.: 243
- flavirostris → S. grandiflora: 248
- **flavopurpurea**: 246
- – – var. fleckii → S. flavopurpurea: 246
- fleckii → S. flavopurpurea: 246
- forcipis → S. glabricaulis: 247
- fucosa → Orbea verrucosa var. fucosa: 204
- furcata → Orbea melanantha: 197
- fuscata → Tromotriche revoluta: 272

[Stapelia]
- fuscopurpurea → S. sp.: 243
- **gariepensis**: 246
- geminata → Piaranthus geminatus: 211
- gemmiflora → Tridentea gemmiflora: 267
- – var. densa → Tridentea gemmiflora: 267
- – – hircosa → Tridentea gemmiflora: 267
- **gettliffei**: 246
- **gigantea**: 247
- – var. pallida → S. gigantea: 247
- **glabricaulis**: 247
- glandulifera → S. glanduliflora: 247
- **glanduliflora**: 247
- – var. emarginata → S. glanduliflora: 247
- – – haworthii → S. glanduliflora: 247
- – – massonii → S. glanduliflora: 247
- glauca → Tromotriche revoluta: 272
- glaucescens → Tromotriche revoluta: 272
- glomerata → Duvalia caespitosa var. caespitosa: 125
- gordonii → Hoodia gordonii: 143
- grandidens → Orbea maculata ssp. maculata: 197
- **grandiflora**: 247
- – var. lineata → S. grandiflora: 248
- gussoneana → Caralluma europaea: 53
- guttata → Huernia guttata: 162
- herrei → Tromotriche herrei: 271
- hircola → Tridentea gemmiflora: 267
- hircosa → Tridentea gemmiflora: 267
- – var. densa → Tridentea gemmiflora: 267
- **hirsuta**: 248
- – – subvar. depressa → S. hirsuta: 248
- – – – longirostris → S. hirsuta: 248
- – – var. affinis → S. hirsuta: 248
- – – – atra → S. hirsuta: 248
- – – – comata → S. hirsuta: 248
- – – – depressa → S. hirsuta: 248
- – – – grata → S. hirsuta: 248
- – – – longirostris → S. hirsuta: 248
- – – – lutea → S. hirsuta: 248
- – – – patula → S. hirsuta: 248
- – – – unguipetala → S. hirsuta: 248
- hirtella → Duvalia caespitosa var. caespitosa: 125
- hispida → Orbea variegata: 203
- hispidula → S. glanduliflora: 247
- horizontalis → Orbea variegata: 203
- humilis → Huernia humilis: 163
- hystrix → Huernia hystrix: 164
- **immelmaniae**: 248
- incarnata → Quaqua incarnata: 217
- indocta → S. acuminata: 243
- inodora → Orbea variegata: 203
- intermedia → Quaqua sp.: 214
- irrorata → Orbea verrucosa: 204
- – – → Orbea verrucosa var. fucosa: 204
- italica → Caralluma europaea: 53
- jacquiniana → Duvalia elegans: 126

[Stapelia]
- jacquinii → Duvalia caespitosa var. caespitosa: 125
- johni-lavrani → S. schinzii var. schinzii: 254
- jucunda → Tridentea jucunda: 267
- – – var. deficiens → Tridentea jucunda: 267
- juttae → S. similis: 254
- juvencula → S. sp.: 243
- kagerensis → Orbea semota: 200
- knobelii → Orbea knobelii: 194
- **kougabergensis**: 249
- **kwebensis**: 249
- – – var. longipedicellata → S. longipedicellata: 249
- laevigata → Duvalia caespitosa var. caespitosa: 125
- laevis → Tromotriche pedunculata: 272
- lampadosa → Caralluma europaea: 53
- lanifera → S. hirsuta: 248
- lanigera → S. hirsuta: 248
- **leendertziae**: 249
- lentiginosa → Huernia guttata: 162
- lepida → Orbea variegata: 203
- limosa → Orbea variegata: 203
- longidens → Orbea longidens: 195
- longii → Tromotriche longii: 271
- **longipedicellata**: 249
- longipes → Tromotriche longipes: 271
- – – var. namaquensis → Tromotriche longipes: 271
- lucida → S. asterias: 244
- lunata → S. sp.: 243
- maccabeanae → S. sp.: 243
- macloughlinii → Orbea macloughlinii: 196
- **macowanii**: 250
- – – var. **conformis**: 250
- – – – **macowanii**: 250
- macrocarpa → Huernia macrocarpa: 167
- maculosa → Tromostapelia sp.: 269
- mammillaris → Quaqua mammillaris: 217
- margarita → S. hirsuta: 248
- marginata → Orbea variegata: 203
- marientalensis → Tridentea marientalensis: 267
- marlothii → S. gigantea: 247
- marmorata → Orbea variegata: 203
- mastodes → Duvalia caespitosa var. compacta: 125
- mastodis → S. sp.: 243
- **meintjesii**: 250
- melanantha → Orbea melanantha: 197
- meliflua → Hoya meliflua: 154
- mermilis → S. sp.: 243
- miscella → Orbea miscella: 197
- mixta → Orbea variegata: 203
- molonyae → Orbea semota: 200
- monstrosa → Orbea variegata: 203
- montagnacii → Stapelianthus montagnacii: 257
- **montana**: 250
- – – var. **grossa**: 250

[Stapelia montana var.]
– – – **montana**: 250
– moschata → Tridentea gemmiflora: 267
– multangula → Echidnopsis sp.: 130
– namaquensis → Orbea namaquensis: 198
– – var. bidens → Orbea namaquensis: 198
– – – ciliolata → Orbea namaquensis: 198
– – – minor → Orbea namaquensis: 198
– – – tridentata → Orbea namaquensis: 198
– neliana → Tromotriche herrei: 271
– noachabibensis → S. similis: 254
– nobilis → S. gigantea: 247
– nopenackyi → S. sp.: 243
– normalis → Orbea variegata: 203
– nouhuysii → S. paniculata: 251
– nudiflora → S. vetula: 255
– **obducta**: 250
– obliqua → Orbea variegata: 203
– obscura → S. grandiflora: 248
– ocellata → Huernia guttata: 162
– **olivacea**: 250
– ophiuncula → Orbea variegata: 203
– orbicularis → Orbea variegata: 203
– orbiculata → Orbea variegata: 203
– pachyrrhiza → Tridentea pachyrrhiza: 268
– pallens → S. divaricata: 245
– pallida → S. divaricata: 245
– **paniculata**: 251
– paradoxa → Orbea paradoxa: 198
– parviflora → Quaqua parviflora: 218
– parvipuncta → Tridentea parvipuncta: 268
– – var. parvipuncta → Tridentea parvipuncta ssp. parvipuncta: 268
– – – truncata → Tridentea parvipuncta ssp. truncata: 268
– parvipunctata → Tridentea parvipuncta: 268
– **parvula**: 251
– patentirostris → S. hirsuta: 248
– patula → S. hirsuta: 248
– – var. depressa → S. hirsuta: 248
– – – longirostris → S. hirsuta: 248
– **pearsonii**: 251
– peculiaris → Tridentea peculiaris: 269
– pedunculata → Tromotriche pedunculata: 272
– **peglerae**: 251
– penduliflora → Tromotriche pedunculata: 272
– picta → Orbea variegata: 203
– pilifera → Hoodia pilifera: 145
– **pillansii**: 252
– – var. attenuata → S. pillansii var. pillansii: 252
– – – **fontinalis**: 252
– – – **pillansii**: 252
– planiflora → Orbea variegata: 203
– – var. marginata → Orbea variegata: 203
– plantii → S. sp.: 243
– polita → Duvalia polita: 127
– portae-taurinae → S. similis: 254
– potensa → Tromotriche revoluta: 272
– **praetermissa**: 252

[Stapelia praetermissa]
– – var. **luteola**: 252
– – – **praetermissa**: 252
– prognatha → Orbea prognatha: 198
– protensa → Tromotriche revoluta: 272
– pruinosa → Quaqua pruinosa: 219
– pulchella → Orbea pulchella: 199
– pulchra → Orbea verrucosa var. verrucosa: 204
– pulla → Quaqua mammillaris: 217
– **pulvinata**: 252
– – fa. margarita → S. hirsuta: 248
– punctata → Piaranthus punctatus: 211
– quadrangula → Caralluma europaea: 53
– – → Caralluma quadrangula: 59
– – var. ramosa → Caralluma quadrangula: 59
– quinquenervis → Orbea variegata: 203
– radiata → Duvalia caespitosa var. caespitosa: 125
– – → Duvalia elegans: 126
– ramosa → Quaqua ramosa: 220
– rangeana → Orbea maculata ssp. rangeana: 197
– reclinata → Duvalia caespitosa var. caespitosa: 125
– reflexa → S. sp.: 243
– – var. brownii → S. sp.: 243
– **remota**: 253
– replicata → Duvalia caespitosa var. caespitosa: 125
– reticulata → Huernia reticulata: 170
– – var. deformis → Huernia reticulata: 170
– retusa → Orbea variegata: 203
– revoluta → Tromotriche revoluta: 272
– – var. fuscata → Tromotriche revoluta: 272
– – – glaucescens → Tromotriche revoluta: 272
– – – tigrida → Tromotriche revoluta: 272
– rogersii → Orbea rogersii: 199
– roriflua → Orbea verrucosa var. verrucosa: 204
– **rubiginosa**: 253
– **rufa**: 253
– – var. attenuata → S. rufa: 253
– – – fissirostris → S. rufa: 253
– rufescens → S. rufa: 253
– rugosa → Orbea variegata: 203
– – → Orbea verrucosa: 204
– ruschiana → Tromotriche ruschiana: 273
– sarmentosa → Ceropegia variegata: 105
– **schinzii**: 253
– – var. **angolensis**: 253
– – – **bergeriana**: 253
– – – **schinzii**: 254
– **scitula**: 254
– scutellata → Orbea variegata: 203
– scylla → Orbea variegata: 203
– semota → Orbea semota: 200
– senilis → S. grandiflora: 248
– serrulata → Piaranthus decorus ssp. decorus: 210
– **similis**: 254
– simsii → S. vetula: 255
– sororia → S. sp.: 243

[Stapelia]
- spectabilis → S. grandiflora: 248
- stellaris → S. asterias: 244
- — — → S. sp.: 243
- stricta → S. sp.: 243
- stultitioides → S. arenosa: 243
- stygia → Tridentea gemmiflora: 267
- subulata → Caralluma subulata: 61
- **surrecta**: 254
- — var. primosii → S. surrecta: 254
- tapscottii → Orbea tapscottii: 201
- tarantuloides → Orbelia sp.: 207
- thudichumii → Tromotriche thudichumii: 273
- thuretii → Huernia thuretii: 172
- tigrida → Tromotriche revoluta: 272
- tigrina → Tromotriche herrei: 271
- tomentosa → S. tsomoensis: 254
- tridentata → Orbea namaquensis: 198
- trisulca → Orbea variegata: 203
- **tsomoensis**: 254
- — — → S. peglerae: 252
- tubata → Huernia barbata: 160
- tuberculata → Duvalia caespitosa var. caespitosa: 125
- tuberosa → Brachystelma tuberosum: 45
- tubulosa → Huernia barbata: 160
- umbellata → Caralluma umbellata: 62
- umbonata → Tridentea pachyrrhiza: 268
- umbracula → Orbea umbracula: 202
- uncinata → S. sp.: 243
- unguipetala → S. hirsuta: 248
- **unicornis**: 255
- vaga → Orbea lutea ssp. vaga: 196
- variegata → Ceropegia variegata: 105
- — — → Orbea variegata: 203
- — var. asparagensis → Orbea variegata: 203
- — — — atrata → Orbea variegata: 203
- — — — atropurpurea → Orbea variegata: 203
- — — — brevicornis → Orbea variegata: 203
- — — — bufonia → Orbea variegata: 203
- — — — clypeata → Orbea variegata: 203
- — — — conspurcata → Orbea variegata: 203
- — — — horizontalis → Orbea variegata: 203
- — — — laeta → Orbea variegata: 203
- — — — marginata → Orbea variegata: 203
- — — — marmorata → Orbea variegata: 203
- — — — mixta → Orbea variegata: 203
- — — — normalis → Orbea variegata: 203
- — — — pallida → Orbea variegata: 203
- — — — picta → Orbea variegata: 203
- — — — planiflora → Orbea variegata: 203
- — — — prometheus → Orbea variegata: 203
- — — — retusa → Orbea variegata: 203
- — — — rugosa → Orbea variegata: 203
- — — — trisulca → Orbea variegata: 203
- venusta → Huernia guttata: 163
- — — → Huernia sp.: 159
- — var. minor → Huernia guttata: 162
- verrucosa → Orbea verrucosa: 204

[Stapelia verrucosa]
- — var. conspicua → Orbea verrucosa var. verrucosa: 204
- — — — fucosa → Orbea verrucosa var. fucosa: 204
- — — — pallescens → Orbea verrucosa var. verrucosa: 204
- — — — pulchra → Orbea verrucosa var. verrucosa: 204
- — — — punctifera → Orbea verrucosa var. verrucosa: 204
- — — — robusta → Orbea verrucosa var. verrucosa: 204
- — — — roriflua → Orbea verrucosa var. verrucosa: 204
- — — — verrucosa → Orbea verrucosa var. verrucosa: 204
- **vetula**: 255
- — var. juvencula → S. sp.: 243
- — — — simsii → S. vetula: 255
- **villetiae**: 255
- villosa → S. hirsuta: 248
- virescens → Tridentea virescens: 269
- vittata → S. sp.: 243
- wendlandiana → Orbea verrucosa var. verrucosa: 204
- wilmaniae → S. leendertziae: 249
- woodfordiana → Orbea variegata: 203
- woodii → Orbea woodii: 205
- — var. westii → Orbea woodii: 205
- youngii → S. gigantea: 247

Stapelianthus : 255
- **arenarius**: 256
- baylissii → Tromotriche baylissii: 270
- calcarophilus → S. arenarius: 256
- choananthus → Tromotriche choanantha: 271
- **decaryi**: 256
- **hardyi**: 256
- **insignis**: 256
- — ssp. tongoboryensis → S. insignis: 256
- **keraudreniae**: 257
- **madagascariensis**: 257
- **montagnacii**: 257
- **pilosus**: 257

Stapeliopsis → Orbea: 187
- — — → Stapelianthus: 255
- — —: 258
- ballyi → Echidnopsis ballyi: 130
- **breviloba**: 258
- cooperi → Orbea cooperi: 191
- **exasperata**: 258
- madagascariensis → Stapelianthus madagascariensis: 257
- **neronis**: 258
- **pillansii**: 259
- **saxatilis**: 259
- — — ssp. **saxatilis**: 259
- — — — **stayneri**: 259
- **urniflora**: 259

Stathmostelma : 260

[Stathmostelma]
- **angustatum**: 260
- – ssp. **angustatum**: 260
- – – **vomeriforme**: 260
- bicolor → S. gigantiflorum: 261
- chironioides → S. welwitschii: 263
- **diversifolium**: 260
- **fornicatum**: 260
- – ssp. **fornicatum**: 260
- – – **tridentatum**: 260
- frommii → S. spectabile ssp. frommii: 262
- **gigantiflorum**: 261
- – – → S. gigantiflorum: 261
- globuliflorum → S. pedunculatum: 262
- **incarnatum**: 261
- **katangense**: 261
- laurentianum → S. welwitschii: 263
- macranthum → S. pedunculatum: 262
- macropetalum → S. spectabile: 262
- nomadacridum → S. gigantiflorum: 261
- nuttii → Asclepias sp.: 10
- **odoratum**: 261
- pachycladum → S. spectabile: 262
- **pauciflorum**: 261
- **pedunculatum**: 262
- – – → S. rhacodes: 262
- praetermissum → S. gigantiflorum: 261
- **propinquum**: 262
- reflexum → S. pauciflorum: 261
- **rhacodes**: 262
- **spectabile**: 262
- – ssp. **frommii**: 262
- – – **spectabile**: 262
- thomasii → S. angustatum ssp. vomeriforme: 260
- verdickii → S. wildemanianum: 263
- **welwitschii**: 262
- – var. **bagshawei**: 263
- – – **welwitschii**: 263
- **wildemanianum**: 263

Stenostelma : 263
- **capense**: 263
- **carinatum**: 264
- **corniculatum**: 264
- **eminens**: 264

Stephanotella → Marsdenia: 179
Stephanotis → Marsdenia: 178
Stisseria → Stapelia: 242
Stomatostemma : 264
- **monteiroae**: 264

Stultitia → Orbea: 187
- araysiana → Orbea araysiana: 188
- conjuncta → Orbeanthus conjunctus: 206
- cooperi → Orbea cooperi: 191
- hardyi → Orbeanthus hardyi: 206
- miscella → Orbea miscella: 197
- paradoxa → Orbea paradoxa: 198
- tapscottii → Orbea tapscottii: 201
- umbracula → Orbea umbracula: 202

[]
Stylandra → Asclepias: 9
Sulcolluma → Caralluma: 46
- foulcheri-delboscii → Caralluma hexagona: 55
- – var. greenbergiana → Caralluma hexagona: 55
- hexagona → Caralluma hexagona: 55
- – var. septentrionalis → Caralluma hexagona: 55
- shadhbana → Caralluma hexagona: 55
- – var. barhana → Caralluma hexagona: 55
Sussuela esculenta → Hoya diversifolia: 150
Systrepha → Ceropegia: 63
- filiforme → Ceropegia filiformis: 77
- multiflorum → Ceropegia multiflora: 91
Tacazzea natalensis → Petopentia natalensis: 209
Tapeinostelma → Brachystelma: 20
- caffrum → Brachystelma caffrum: 24
Tavaresia : 265
- **angolensis**: 265
- **barklyi**: 265
- grandiflora → T. barklyi: 265
- meintjesii → Staparesia meintjesii: 242
- **thompsoniorum**: 265
Tavastemon → Staparesia: 242
- meintjesii → Staparesia meintjesii: 242
Tenaris → Brachystelma: 20
- bikitaensis → Brachystelma bikitaense: 22
- browniana → Brachystelma brownianum: 23
- chlorantha → Brachystelma chloranthum: 25
- christianeae → Brachystelma christianeae: 25
- filifolia → Brachystelma filifolium: 29
- rostrata → Brachystelma rubellum: 41
- rubella → Brachystelma rubellum: 41
- schultzei → Brachystelma schultzei: 42
- simulans → Brachystelma rubellum: 41
- somalensis → Caralluma sp.: 47
- subaphylla → Ceropegia sp.: 63
- volkensii → Brachystelma rubellum: 41
Tetragonocarpus → Marsdenia: 179
Trachycalymma : 265
Traunia → Marsdenia: 179
Trichocaulon → Hoodia: 142
- alstonii → Hoodia alstonii: 142
- annulatum → Hoodia pilifera ssp. annulata: 145
- cactiforme → Larryleachia cactiformis: 176
- cinereum → Larryleachia perlata: 177
- clavatum → Larryleachia sp.: 176
- columnare → Notechidnopsis columnaris: 185
- decaryi → Stapelianthus pilosus: 257
- delaetianum → Hoodia officinalis ssp. delaetiana: 144
- dinteri → Larryleachia marlothii: 177
- engleri → Larryleachia picta: 177
- felinum → Larryleachia cactiformis: 176
- flavum → Hoodia flava: 143
- grande → Hoodia pilifera ssp. pillansii: 146
- halenbergense → Hoodia alstonii: 142
- karasmontanum → Hoodia sp.: 142

[Trichocaulon]
- keetmanshoopense → Larryleachia marlothii: 177
- kubusense → Larryleachia perlata: 177
- marlothii → Larryleachia marlothii: 177
- meloforme → Larryleachia picta: 177
- mossamedense → Hoodia mossamedensis: 144
- officinale → Hoodia officinalis: 144
- pedicellatum → Hoodia pedicellata: 145
- perlatum → Larryleachia perlata: 177
- pictum → Larryleachia picta: 177
- piliferum → Hoodia pilifera: 145
- pillansii → Hoodia pilifera ssp. pillansii: 146
- – var. major → Hoodia pilifera ssp. pillansii: 146
- pubiflorum → Hoodia officinalis ssp. officinalis: 145
- rusticum → Hoodia officinalis ssp. officinalis: 145
- simile → Larryleachia cactiformis: 176
- sinus-luederitzii → Larryleachia marlothii: 177
- sociarum → Larryleachia sp.: 176
- somaliense → Echidnopsis planiflora: 133
- triebneri → Hoodia triebneri: 146
- truncatum → Larryleachia perlata: 177

Tridentea : 266
- aperta → Tromotriche aperta: 270
- baylissii → Tromotriche baylissii: 270
- – var. baylissii → Tromotriche baylissii: 270
- – – ciliata → Tromotriche baylissii: 270
- choanantha → Tromotriche choanantha: 271
- depressa → Stapelia hirsuta: 248
- **dwequensis**: 266
- **gemmiflora**: 267
- – var. densa → T. gemmiflora: 267
- – – hircosa → T. gemmiflora: 267
- herrei → Tromotriche herrei: 271
- – var. neliana → Tromotriche herrei: 271
- **jucunda**: 267
- – var. cincta → T. jucunda: 267
- – – deficiens → T. jucunda: 267
- – – dinteri → T. jucunda: 267
- – – jucunda → T. jucunda: 267
- juvencula → Stapelia sp.: 243
- longii → Tromotriche longii: 271
- longipes → Tromotriche longipes: 271
- – var. namaquensis → Tromotriche longipes: 271
- **marientalensis**: 267
- – ssp. **albipilosa**: 268
- – – **marientalensis**: 268
- – var. albipilosa → T. marientalensis ssp. albipilosa: 268
- – – auobensis → T. marientalensis ssp. marientalensis: 268
- moschata → T. gemmiflora: 267
- **pachyrrhiza**: 268
- paniculata → Stapelia paniculata: 251
- **parvipuncta**: 268

[Tridentea parvipuncta]
- – ssp. **parvipuncta**: 268
- – – **truncata**: 268
- – var. parvipuncta → T. parvipuncta ssp. parvipuncta: 268
- – – truncata → T. parvipuncta ssp. truncata: 268
- **peculiaris**: 269
- pedunculata → Tromotriche pedunculata: 272
- pusilla → T. parvipuncta ssp. truncata: 268
- rugosa → Orbea variegata: 203
- ruschiana → Tromotriche ruschiana: 273
- similis → Stapelia similis: 254
- simsii → Stapelia vetula: 255
- stygia → T. gemmiflora: 267
- umdausensis → Tromotriche umdausensis: 273
- vetula → Stapelia vetula: 255
- **virescens**: 269

Triodoglossum → Schlechterella: 241
- abyssinicum → Schlechterella abyssinica: 241
Triplosperma → Dischidia: 118
Tromostapelia : 269
- mutabilis var. cupularis → T. sp.: 269
Tromotriche : 269
- **aperta**: 270
- **baylissii**: 270
- **choanantha**: 271
- ciliata → Orbea ciliata: 190
- engleriana → Stapelia engleriana: 245
- fuscata → T. revoluta: 272
- glauca → T. revoluta: 272
- **herrei**: 271
- **longii**: 271
- **longipes**: 271
- **pedunculata**: 272
- pruinosa → Quaqua pruinosa: 219
- **revoluta**: 272
- – var. fuscata → T. revoluta: 272
- – – tigrida → T. revoluta: 272
- **ruschiana**: 272
- **thudichumii**: 273
- **umdausensis**: 273

Tympananthe suberosa → Matelea pavonii: 182
Verlotia → Marsdenia: 179
Vincetoxicum cyclophyllum → Matelea cyclophylla: 181
- madagascariense → Cynanchum madagascariense: 114
Virchowia → Echidnopsis: 129
- africana → Echidnopsis virchowii: 136
Voharanga → Cynanchum: 108
- madagascariensis → Cynanchum arenarium: 109
Vohemaria → Cynanchum: 108
- implicata → Cynanchum implicatum: 112
- messeri → Cynanchum messeri: 115
White-sloanea : 273
- **crassa**: 274
- migiurtina → Pseudolithos migiurtinus: 213
Whitesloaniopsis : 274
- – **Gray Ghost** : 274

[]
Xyomalobium → Xysmalobium: 274
Xysmalobium : 274
– bellum → Glossostelma spathulatum: 141
– carinatum → Stenostelma carinatum: 264
– carsonii → Glossostelma carsonii: 140
– ceciliae → Glossostelma ceciliae: 140
– dissolutum → Glossostelma lisianthoides: 141
– fritillarioides → Glossostelma lisianthoides: 141
– grande → Glossostelma angolense: 139
– heudelotianum → Schizoglossum sp.: 236
– spathulatum → Glossostelma spathulatum: 141
– speciosum → Glossostelma cabrae: 140
Zaczatea → Raphionacme: 220
– angolensis → Raphionacme angolensis: 221
Zucchellia → Raphionacme: 220
– angolensis → Raphionacme angolensis: 221

Colour Plates

Absolmsia, Aspidoglossum, Brachystelma

a *Absolmsia spartioides*

b *Aspidoglossum nyasae*

c *Brachystelma buchananii*

d *Brachystelma burchellii*

e *Brachystelma caffrum*

f *Brachystelma christianeae*

g *Brachystelma coddii*

a *Asclepias subulata*

b *Brachystelma huttonii*

c *Brachystelma meyerianum*

d *Brachystelma mortonii*

e *Brachystelma pulchellum*

f *Brachystelma nanum* ('*B. angustum*')

g *Brachystelma nanum*

Brachystelma

a *Brachystelma gracile*
b *Brachystelma lineare*
c *Brachystelma macropetalum*
d *Brachystelma megasepalum*
e *Brachystelma occidentale*
f *Brachystelma simplex*
g *Brachystelma rubellum*
h *Brachystelma stellatum*

a *Brachystelma tenue*

b *Brachystelma tuberosum*

c *Caralluma acutangula*

d *Caralluma adscendens* var. *adscendens*

e *Caralluma acutangula*

f *Caralluma adscendens* var. *geniculata*

g *Caralluma arachnoidea* var. *arachnoidea*

Caralluma

a *Caralluma cicatricosa*
b *Caralluma baradii*
c *Caralluma burchardii* ssp. *burchardii*
d *Caralluma crenulata*
e *Caralluma dicapuae*
f *Caralluma europaea* ('*C. maroccana*')
g *Caralluma frerei*

a *Caralluma diffusa* b *Caralluma edulis* c *Caralluma flava* d *Caralluma furta* e *Caralluma peschii* f *Caralluma sinaica* g *Caralluma lavrani* h *Caralluma munbyana*

Caralluma, Ceropegia

a *Caralluma tuberculata*

b *Caralluma peckii*

c *Caralluma speciosa*

d *Ceropegia affinis*

e *Caralluma turneri* ssp. *turneri*

f *Ceropegia crassifolia* var. *crassifolia*

a *Ceropegia albisepta*
b *Ceropegia ampliata*
c *Ceropegia arabica* var. *arabica*
d *Ceropegia arabica* var. *powysii* ('*C. barbigera*')
e *Ceropegia aristolochioides* ssp. *aristolochioides*
f *Ceropegia aristolochioides* ssp. *deflersiana*
g *Ceropegia carnosa*
h *Ceropegia dichotoma* ssp. *dichotoma*

Ceropegia

a *Ceropegia ballyana*
b *Ceropegia bulbosa*
c *Ceropegia claviloba*
d *Ceropegia conrathii*
e *Ceropegia crassifolia* var. *crassifolia*
f *Ceropegia cufodontii*
g *Ceropegia linearis* ssp. *woodii*
h *Ceropegia humbertii*

a *Ceropegia denticulata*
b *Ceropegia distincta*
c *Ceropegia fortuita*
d *Ceropegia fusca*
e *Ceropegia hirsuta*
f *Ceropegia imbricata*
g *Ceropegia linearis* ssp. *woodii*
h *Ceropegia meleagris*

a *Ceropegia linearis* ssp. *linearis*
b *Ceropegia lugardiae*
c *Ceropegia madagascariensis*
d *Ceropegia meyeri-johannis*
e *Ceropegia nilotica*
f *Ceropegia occidentalis*
g *Ceropegia multiflora* ssp. *multiflora*
h *Ceropegia papillata*

a *Ceropegia pachystelma*

b *Ceropegia petignatii*

c *Ceropegia purpurascens*

d *Ceropegia somalensis*

e *Ceropegia sootepensis*

Ceropegia

a *Ceropegia rendallii*
b *Ceropegia rupicola*
c *Ceropegia saxatilis*
d *Ceropegia simoneae*
e *Ceropegia sobolifera* var. *sobolifera*
f *Ceropegia striata*
g *Ceropegia stapeliiformis* ssp. *stapeliiformis*
h *Ceropegia stentiae*

a *Ceropegia subaphylla*
b *Ceropegia yemenensis*
c *Ceropegia variegata*
d *Cibirhiza dhofarensis*
e *Cynanchum ampanihense*

Cynanchum

a *Cynanchum crassipedicellatum*
b *Cynanchum angavokeliense*
c *Cynanchum folotsioides*
d *Cynanchum gerrardii* var. *gerrardii*
e *Cynanchum luteifluens* var. *luteifluens*
f *Cynanchum messeri*

Further/Weitere *Cynanchum* spp.: XXII.f; XXXI.e,f,g; XXXVIII.a.

a *Cynanchum madagascariense*
b *Cynanchum sessiliflorum*
c *Cynanchum rossii*
d *Cynanchum mevei*
e *Cynanchum papillatum*

Further/Weitere *Cynanchum* spp.: XXII.f; XXXI.e,f,g; XXXVIII.a.

a *Dischidia bengalensis*
b *Cynanchum sigridae*
c *Dischidia astephana*
d *Dischidia imbricata*
e *Dischidia dolichantha*
f *Dischidia fruticulosa*

a *Dischidia lanceolata*

b *Dischidia hirsuta*

c *Dischidia ovata*

d *Dischidia major*

e *Dischidia major*

f *Dischidia punctata*

Dischidia

a *Dischidia livida*

b *Dischidia vidalii*

c *Dischidia truncata*

d *Dischidia sagittata*

e *Dischidia singularis*

a *Duvaliandra dioscoridis*
b *Dischidiopsis parasitica*
c *Duvalia angustiloba*
d *Duvalia gracilis*
e *Duvalia parviflora*
f *Duvalia polita*
g *Duvalia pubescens*

a *Echidnopsis cereiformis*
b *Duvalia sulcata* ssp. *sulcata*
c *Duvalia velutina*
d *Echidnopsis scutellata* ssp. *dhofarensis*
e *Echidnopsis angustiloba*
f *Echidnopsis virchowii* var. *virchowii*

a *Glossostelma brevilobum*

b *Fockea edulis*

c *Glossostelma brevilobum*

e *Glossostelma brevilobum*

d *Hoodia alstonii*

f *Cynanchum grandidieri (Folotsia grandiflora)*

Hoodia

a *Hoodia gordonii*
b *Hoodia gordonii*
c *Hoodia flava*
d *Hoodia pilifera* ssp. *pilifera*
e *Hoodia pedicellata*
f *Hoodia pilifera* ssp. *pillansii*

a *Hoodia juttae*

b *Hoya anulata*

c *Hoya benguetensis*

d *Hoya australis* ssp. *australis*

e *Hoya bordenii*

f *Hoya caudata*

a *Hoya fuscomarginata*

b *Hoya cinnamomifolia*

c *Hoya diversifolia*

d *Hoya imbricata*

e *Hoya globulosa*

f *Hoya heuschkeliana*

a *Hoya hypolasia*

b *Hoya inconspicua*

c *Hoya kerrii*

d *Hoya linearis*

e *Hoya limoniaca*

f *Hoya longifolia*

a *Hoya macgillivrayi*
b *Hoya meliflua*
c *Hoya meredithii*
d *Hoya magnifica*
e *Hoya nicholsoniae*
f *Hoya nummularioides*

a *Hoya purpureofusca*
b *Hoya obovata*
c *Hoya pachyclada*
d *Hoya pauciflora*
e *Hoya serpens*
f *Hoya thomsonii*
g *Hoya vitellina*

Huernia

a *Huernia humilis*
b *Huernia erinacea*
c *Huernia hallii*
d *Huernia hislopii* ssp. *hislopii*
e *Huernia hystrix* var. *hystrix*
f *Huernia kennedyana*
g *Huernia lavrani*

a *Huernia keniensis* var. *keniensis*
b *Huernia leachii*
c *Huernia lenewtonii*
d *Huernia macrocarpa*
e *Huernia nouhuysii*
f *Huernia saudi-arabica*
g *Huernia tanganyikensis*

Cynanchum, Huernia, Huerniopsis, Larryleachia

a *Huernia zebrina* ssp. *magniflora*

b *Huerniopsis atrosanguinea*

c *Huernia pillansii*

d *Huernia verekeri* var. *verekeri*

e *Cynanchum* (*Karimbolea*) *verrucosum*

f *Cynanchum* (*Karimbolea*) *macranthum*

g *Cynanchum* (*Karimbolea*) *macranthum*

h *Larryleachia perlata*

a *Notechidnopsis columnaris*
b *Larryleachia cactiformis*
c *Larryleachia marlothii*
d *Larryleachia picta*
e *Lavrania haagnerae*
f *Notechidnopsis tessellata*
g *Notechidnopsis columnaris* (l), *N. tessellata* (r)

Marsdenia, Matelea, Micholitzia, Ophionella

a *Marsdenia megalantha*

b *Micholitzia obcordata*

c *Ophionella arcuata* ssp. *arcuata*

d *Matelea cyclophylla*

e *Matelea cyclophylla*

f *Matelea cyclophylla*

a *Orbea abayensis*
b *Orbea carnosa*
c *Orbea laikipiensis*
d *Orbea lugardii*
e *Orbea miscella*
f *Orbea caudata* ssp. *rhodesiaca*
g *Orbea ciliata*

Orbea

a *Orbea deflersiana*
b *Orbea cooperi*
c *Orbea gemugofana*
d *Orbea gerstneri* ssp. *elongata*
e *Orbea laticorona*
f *Orbea lutea* ssp. *lutea*
g *Orbea maculata* ssp. *rangeana*

a *Orbea speciosa*
b *Orbea paradoxa*
c *Orbea schweinfurthii*
d *Orbea semitubiflora*
e *Orbea semota*
f *Orbea tubiformis*
g *Orbea variegata*

a *Petopentia natalensis*
b *Orbeanthus hardyi*
c *Pectinaria articulata* ssp. *articulata*
d *Pectinaria articulata* ssp. *asperiflora*
e *Pectinaria articulata* ssp. *borealis*
f *Piaranthus decorus* ssp. *cornutus*
g *Piaranthus decorus* ssp. *cornutus*

a *Cynanchum (Platykeleba) insigne*

b *Piaranthus comptus*

d *Piaranthus punctatus*

c *Piaranthus geminatus* var. *geminatus*

(below)

e *Piaranthus geminatus* var. *foetidus*

f *Pseudolithos caput-viperae*

g *Pseudolithos caput-viperae*

a *Quaqua acutiloba*
b *Pseudolithos dodsonianus*
c *Pseudolithos migiurtinus*
d *Quaqua dependens*
e *Pseudolithos migiurtinus*
f *Quaqua inversa*

a *Quaqua incarnata* b *Quaqua incarnata* c *Quaqua pulchra*
d *Quaqua pruinosa* e *Quaqua swanepoelii* f *Raphionacme globosa*
g *Quaqua linearis* h *Raphionacme globosa*

a *Raphionacme madiensis*
b *Raphionacme splendens*
c *Raphionacme grandiflora*
d *Raphionacme madiensis*
e *Raphionacme procumbens*
f *Rhytidocaulon fulleri*
g *Rhytidocaulon piliferum*
h *Rhytidocaulon sheilae*

a *Rhytidocaulon sheilae*
b *Sarcostemma antsiranense*
c *Sarcostemma arabicum*
d *Sarcostemma mulanjense*
e *Sarcostemma resiliens*
f *Sarcostemma socotrana*
g *Sarcostemma stoloniferum*
h *Sarcostemma vanlessenii*

Sarcostemma, Schizoglossum, Stapelia

a *Sarcostemma viminale* ssp. *australe*

b *Sarcostemma viminale* ssp. *orangeanum*

c *Sarcostemma brevipedicellatum*

d *Sarcostemma viminale* ssp. *odontolepis*

e *Sarcostemma viminale* ssp. *stipitaceum*

f *Sarcostemma viminale* ssp. *suberosum*

g *Schizoglossum atropurpureum* ssp. *atropurpureum*

h *Stapelia grandiflora*

a *Schlechterella abyssinica*
b *Stapelia acuminata*
c *Stapelia cedrimontana*
d *Stapelia engleriana*
e *Stapelia erectiflora* var. *erectiflora*
f *Stapelia flavopurpurea*
g *Stapelia gariepensis*
h *Stapelia gigantea*

Stapelia, Stapelianthus

a *Stapelia glanduliflora*
b *Stapelia obducta*
c *Stapelia similis*
d *Stapelia vetula*
e *Stapelianthus decaryi*
f *Stapelianthus hardyi*
g *Stapelianthus keraudreniae*
h *Stapelianthus madagascariensis*

a *Stapeliopsis exasperata*
b *Stapeliopsis saxatilis* ssp. *saxatilis*
c *Stapelia kwebensis*
d *Stapelianthus pilosus*
e *Stapeliopsis breviloba*
f *Stapeliopsis neronis*
g *Stenostelma capense*
h *Trachycalymma cristatum*

Stomatostemma, Tavaresia, Tridentea, Tromotriche

a *Stomatostemma monteiroae*
b *Tavaresia angolensis*
c *Tavaresia barklyi*
d *Tridentea dwequensis*
e *Tridentea jucunda*
f *Tridentea marientalensis* ssp. *albipilosa*
g *Tridentea virescens*
h *Tromotriche aperta*

a *Tromotriche baylissii*
b *Tromotriche choanantha*
c *Tromotriche longii*
d *Tromotriche longipes*
e *Tromotriche pedunculata*
f *Tromotriche revoluta*
g *Tromotriche thudichumii*
h *White-sloanea crassa*